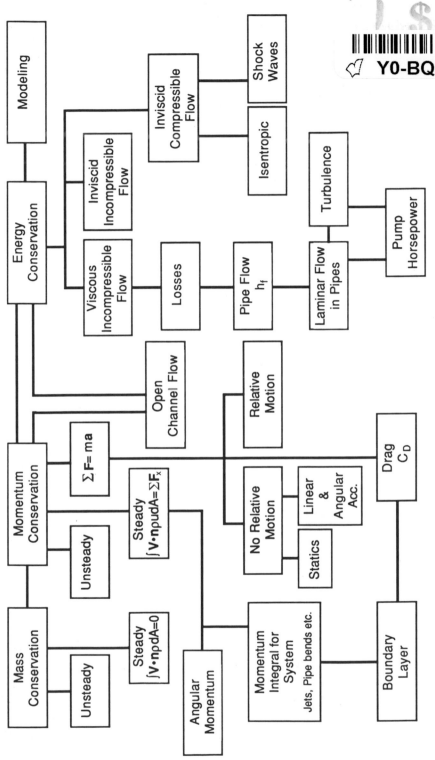

FLUID MECHANICS FUNDAMENTALS

FLUID MECHANICS FUNDAMENTALS

Walter R. Debler

The University of Michigan

PRENTICE HALL
Englewood Cliffs, New Jersey 07632

```
Debler, Walter R.
    Fluid mechanics fundamentals / Walter R. Debler.
        p.    cm.
    Includes index.
    ISBN 0-13-322371-X
    1. Fluid mechanics.    I. Title.
TA357.D425  1990
620.1'06--dc19                                          88-33056
                                                            CIP
```

Editorial/production supervision and
   interior design: Cheryl Adelmann Hadley
Manufacturing buyer: Mary Noonan
Page layout: Diane Koromhas
Cover art: Photo courtesy of Dr. J. Jacobs, from Ph.D. dissertation,
    J. Fluid Mechanics, UCLA 1986. Photo design by Rodney
    Hill.
Cover design: Wanda Lubelska
Cover information: When a fluid-fluid interface is accelerated in the
direction of the denser medium, the interface becomes unstable. The
acceleration, the density difference, and surface tension affect the
phenomenon. Here, an axisymmetric disturbance develops.

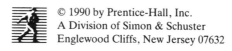

© 1990 by Prentice-Hall, Inc.
A Division of Simon & Schuster
Englewood Cliffs, New Jersey 07632

Printed in the United States of America
10 9 8 7 6 5 4 3 2 1

ISBN   0-13-322371-X

Prentice-Hall International (UK) Limited, *London*
Prentice-Hall of Australia Pty. Limited, *Sydney*
Prentice-Hall Canada Inc., *Toronto*
Prentice-Hall Hispanoamericana, S.A., *Mexico*
Prentice-Hall of India Private Limited, *New Delhi*
Prentice-Hall of Japan, Inc., *Tokyo*
Simon & Schuster Asia Pte. Ltd., *Singapore*
Editora Prentice-Hall do Brasil, Ltda, *Rio de Janeiro*

Contents

PREFACE   *xiii*

ACKNOWLEDGMENTS   *xvii*

1   FLUID PROPERTIES AND FLOW PHENOMENA   **1**

    1.1   Introduction   1

    1.2   Temperature and Pressure   6

    1.3   Density   8

    1.4   An Equation of State—The Ideal Gas Equation   9

    1.5   Intensive Properties of Fluids   10

    1.6   General Physical Properties   13

    1.7   Viscosity   18

    1.8   A Preview of Some Flows That Will Be Studied   23

2   PRESSURE DISTRIBUTIONS IN FLUIDS WITHOUT RELATIVE MOTIONS   **38**

    2.1   Introduction   38

    2.2   Hydrostatic Pressure Variation   39

    2.3   Manometers   47

2.4 Introduction to Forces on Planar and Curved Surfaces 51

2.5 Hydrostatic Forces and Moments On Planar Surfaces 53

2.6 Forces and Moments Due to Pressure Distributions on Surfaces, in General 63

2.7 Linear, Rigid-Body Acceleration 74

2.8 Accelerations and Pressures in Rotating Fluids 77

2.9 Summary 82

3 MASS CONSERVATION **97**

3.1 Introduction 97

3.2 Fluid Velocity 98

3.3 Volume Rate of Flow 100

3.4 Some Control Surfaces 105

3.5 Streamlines, Streaklines, and Pathlines 108

3.6 Conservation of Mass—General Ideas 114

3.7 Continuity Equation for Steady State 117

3.8 Unsteady Flow 124

3.9 Reynolds' Transport Theorem 127

3.10 Summary 128

4 KINEMATICS OF FLOW **136**

4.1 Introduction 136

4.2 The Concept of Rotation in Fluid Mechanics 137

4.3 Angular and Volume Deformation with Associated Consequences 139

4.4 Deformation and Rotation 140

4.5 Introduction to Acceleration 144

4.6 Acceleration and the Substantial Derivative 146

4.7 Stream Functions and Their Use 149

4.8 Differential Form of Mass Conservation 153

4.9 Summary 155

5 THE FLOW OF AN INVISCID FLUID   158

    5.1   Introduction   158

    5.2   The Bernoulli Equation for Steady Flow   159

    5.3   The Bernoulli Equation and Irrotational Flow   171

    5.4   Some Irrotational Flows   175

★5.5   Flow Nets   183

★5.6   The Euler Equations   186

★5.7   Some Applications of the Euler Equations   187

    5.8   Summary   193

6 THE CONSERVATION OF ENERGY   203

    6.1   Introduction   203

    6.2   The Flux of Kinetic Energy   204

    6.3   The Energy Equation   207

★6.4   Unsteady Energy Equation Applications   225

    6.5   Summary   230

7 MOMENTUM CONSERVATION   238

    7.1   Introduction   238

    7.2   The Momentum Equation   241

    7.3   Steady-State Momentum Applications   244

    7.4   Momentum Principle Applied to the Flow over a Flat Plate   256

    7.5   Conservation of Angular Momentum   259

    7.6   Application to Pumps and Turbines   262

    7.7   A Comparison of the Results of the Momentum and Energy Principles   274

★7.8   Examples of Unsteady Momentum Conservation   283

    7.9   Summary   287

* Optional section (see Preface).

8 SOME THEORETICAL PREDICTIONS OF THE EFFECTS OF VISCOSITY ON FLUID FLOW   **298**

8.1   Introduction   298

8.2   Flow between Plane Parallel Surfaces   300

8.3   Flow in Circular Conduits   306

8.4   Swirling Flow   315

8.5   An Unsteady Flow   320

8.6   Laminar Boundary Layers   323

★8.7   Laminar Boundary Layers over Two-Dimensional Bodies   331

★8.8   Flow between Almost Parallel Planes   340

8.9   Summary   347

9 THE USE OF DIMENSIONLESS PARAMETERS IN FLUID MECHANICS   **354**

9.1   Introduction   354

9.2   The π Theorem: A Proof   357

9.3   Choosing the Experimental Parameters   360

9.4   Calculation of the π Parameters   367

9.5   Interpretation of the Dimensionless Parameters   381

9.6   Modeling   385

9.7   Self-Similar Fluid Motions   392

9.8   Summary   396

10 FLOW IN CONDUITS (Internal Flows)   **404**

10.1   Introduction   404

10.2   Pipes in Series and Parallel   405

10.3   Minor Losses   408

10.4   Transition to Fully Developed Pipe Flow   412

10.5   Noncircular Conduits   413

10.6   Separation in Passages   415

10.7   Flow through Orifices   418

11 FLOW AROUND BODIES (External Flows)   426

11.1   Introduction 426

11.2   Drag on Bodies 431

11.3   Lift 436

11.4   Separation and Boundary-Layer Control 441

11.5   Summary 444

12 SECONDARY FLOWS AND FLOW STABILITY   451

12.1   Introduction 451

12.2   Basic Ideas 451

12.3   Summary 462

13 TURBULENT FLOW   464

13.1   Introduction 464

13.2   Turbulent Flow in Pipes 470

13.3   Flow over Flat Plates 486

13.4   Summary 495

14 COMPRESSIBLE FLOWS   499

14.1   Introduction 499

14.2   Sound Waves 500

14.3   The Speed of a Sound Wave 502

14.4   An Energy Equation for Gases 505

14.5   The Effect of Mach Number on a Flow 511

14.6   Nozzle Geometry and Mach Number Variation 512

14.7   Supersonic Flow 513

14.8   Shock Waves in Nozzles 522

14.9   Viscosity and Compressible Flow (Flow in Pipes) 533

14.10   Summary 538

15 OPEN CHANNEL FLOW   545

15.1   Introduction 545

15.2   Critical Conditions 550

15.3   Effect of Channel Cross Section on a Constant Flow   552

15.4   Nonrectangular Channel Section   554

15.5   Momentum Considerations in Open Channel Flows   555

15.6   Upstream and Downstream Effects   557

15.7   Flow Resistance in Open Channels   562

15.8   Free-Surface Profiles   567

15.9   Summary   574

16   THE NAVIER-STOKES EQUATIONS   580

16.1   Introduction   580

16.2   Equations of Motion   580

16.3   Boundary-Layer Equations   594

16.4   Some Similarity Solutions   602

16.5   Summary   609

17   FLOW MEASUREMENT   613

17.1   Introduction   613

17.2   Flow Visualization   614

17.3   Pressure Measurement   625

17.4   Velocity Measurement   627

17.5   Volume Flow Rate Measurement   634

17.6   Surface Elevation   640

17.7   Temperature Measurement   641

17.8   Viscosity Measurement   643

18   HYDRAULIC MACHINERY   645

18.1   Introduction   645

18.2   Similarity (or Affinity) Laws for Hydrodynamic Machines   646

18.3   Hydraulic Losses and Efficiency   652

18.4   Net Positive Suction Head (NPSH)   661

18.5   Performance of Pumps Connected in Series and in Parallel   666

18.6   Miscellaneous Considerations for Turbomachines   670

18.7   Positive-Displacement Pumps and Motors   674

18.8   Summary   677

APPENDICES

A.   Physical Properties   683

B.   Total Differentials   690

C.   Moving Control Volumes   694

D.   Pipe Networks   706

E.   Newton-Raphson Method of Solution for Pipe
       Networks   718

F.   Specific Heats, Enthalpy, and Entropy   727

G.   Integration Methods in Computer Programs   735

H.   Solutions to Falkner-Skan Equation by Runge-Kutta
       Method and Taylor Series Expansion   739

BIBLIOGRAPHY   743

INDEX   748

Preface

In writing this book I intended to present the material so that the basic principles of the subject are clearly stated, related to the reader's previous experiences, and repeatedly recalled. By this means the basic fabric of the subject should become apparent to the reader. The concepts and overall point of view are reinforced through the examples that have been provided. These, and the specifically related exercises that are located in special sections at the ends of the chapters, are designed to achieve two goals. First, the student should become involved in the development of the material so that the ideas are not superficially read. The significance and ease of the steps in the development should be realized by the reader through practice. Second, the exercises that are keyed to the text examples should cause the student to pause and reflect upon some of the aspects and assumptions associated with the work. The serious reader should not read past an example problem without having attempted the related exercises. Consequently, it is hoped that when the students are asked to solve some of the regular problems at the end of the chapters, they will not wonder, "Where do I start?"

In addition to the usual end-of-chapter problems intended to sharpen the reader's skills, exercises have been provided to reinforce the points made in most of the numbered examples in the text. Example 1.7.1, for example, will have associated with it Exercises 1.7.1a through c. These exercises will usually be in an order of increasing difficulty. The student should check the Exercises section at the end of the chapter after studying each of the examples and should work on any related exercises immediately. (The solutions to these exercises—not to be confused with the usual problems—are available to the reader in a separate booklet that can be an additional aid to learning the material.)*

---

* Figures referred to in Problems and Exercises are numbered according to the problems or exercises in which they occur (for example, Figure P.2.97 is discussed in Problem 97 of Chapter 2).

Solving engineering problems provides an effective way for students to increase their familiarity with the reading material, to explore topics, and to discover relationships that can be useful. This means that the problems should not emphasize formula manipulation; rather, they should lead to an understanding of the mechanics of the situation under examination, as well as the associated physical processes. Quite obviously, engineering entails a great deal of problem solving, but solving routine problems does not constitute the challenge of engineering. It is to be expected that in the future, readers who are now learning the material in this book will be facing problems they have not considered before. Consequently, it will be necessary for them to extend and adapt their existing information to cope with new situations. This requires more than a superficial understanding of the material. A comprehension of basic principles is needed. My efforts in preparing the structure, content, and tone of this book are directed at facilitating the acquisition of this understanding.

The reader of this book is assumed to have already learned a great deal from courses in mathematics, engineering, and the physical sciences. The concepts of Newton's calculus and his laws of motion should be familiar to the beginner in fluid mechanics. A course in the mechanics of materials that treats stress distributions and an introduction to thermodynamics will be useful to the tyro. This level of background will provide the foundation on which to build the concepts of fluid mechanics.

Each of the chapters of this book is intended to treat a particular aspect of the subject. Nevertheless, all of the topics need not be covered in order for the student to know a significant amount about the agents that govern the motion of fluids. The most basic or important material has been placed at the beginning of each chapter. In some chapters, certain topics that appear toward the end may, according to the needs of the reader, be left for a second reading. Such sections are marked with a large asterisk (★).

The book can well serve the student who is enrolled in a course of study that allows between 40 and 50 classroom hours for the material to be discussed. The needs of students differ, and some will be in programs that emphasize particular aspects of fluid mechanics. Thus, the subjects in the last nine chapters (internal flows, external flows, secondary flows, turbulence, gas dynamics, open channel flow, the Navier-Stokes equations, flow measurements, and hydraulic machinery) will undoubtedly be studied to different extents by students with different orientations. The sequence in which these chapters are read is not critical. They are mostly special applications of the material in the first half of the book. Consequently, the material in these chapters does not rely on the contents of the others in the second half of the book to any significant extent.

The ordering of the first nine chapters is a result of the author's experience in teaching civil and mechanical engineers. The chapter on dimensional analysis (Chap. 9) has been placed after the one in which viscous effects are discussed in order to give the reader additional background to appreciate the concepts.

Chapters 12 and 13, which discuss secondary flows and turbulence, might be reserved for another course if time and curriculum constraints prevent their inclusion in a first course. A course that has a strong mathematical orientation could require Chapter 16, on the Navier-Stokes equations, to follow Chap. 9. The chapter on flow measurements, Chap. 17, can be recommended to anyone who has completed Chap. 6.

Some of the material in the last nine chapters of the book could be included with the optional sections—those marked by an asterisk—in the first nine chapters to form the basis for a second course in fluid mechanics. Appendices D and E could be a part of such a course.

The contents of this book are presented in an inductive way. This process does not lend itself to a succinct manuscript, but I believe that it is the easier form for the beginner to learn the rudiments of a subject. In subsequent courses in fluid mechanics, the accumulated wisdom of the past will be presented in a general theory and the logical deductions will be efficiently made. The reader will then have had the requisite experience to appreciate the steps in the deductive process. This book is intended to provide the necessary background.

In writing this book I wished to convey some of the excitement the material holds for me. It is a subject that covers varied phenomena that manifest themselves in three-dimensional space. The results are sometimes difficult to visualize. It is for this reason that I have included many photographs.* To those engineers and scientists who have made them available to me, I am greatly indebted. If the written material measures up to these valued contributions to the book, it is due in large measure to the suggestions of Professor Daniel Jankowski, who read and criticized a large part of the manuscript. While he viewed this task as one of his professional responsibilities, I think of it as a generous token of our friendship. Other reviewers have aided me with their insights and specific suggestions, and I am grateful to them. Two people, Julan Jau and Dong Bai, have given me invaluable help by communicating the student's point of view. The first version of this book was edited by Ms. Joannie Reisman. Her comments and verve made the beginnings of this undertaking an easier matter.

Despite the help that I have received, any omissions and errors that may still be present in the work before the reader are my responsibility. Because such things are impediments to the flow of ideas, I shall be happy to acknowledge any suggestions for improvement that the reader gives me.

---

* The book *An Album of Fluid Motion,* by Milton Van Dyke, published by Parabolic Press, (P.O. Box 3032, Stanford CA 94305-0030) 1982, should be seen by all who wonder at the manifold ways in which natural phenomena occur.

Acknowledgments

It has been my good fortune to have experienced many good teachers during my student days. However, David Aptekhar, Robert Parry, Homer Newell, and Chia-Shun Yih deserve special comment. They, particularly, shared with me their apparent pleasure in finding explanations for the mysteries in the world around us. For this reason this book is dedicated to them.

1

Fluid Properties and Flow Phenomena

1.1 INTRODUCTION

The material in this book will prove to be challenging to the average student, but the very features of the material that present the challenges also provide the rewards. The return on the reader's investment of time and concentration will be an insight into a variety of physical processes that occur continually.

The reader may already be familiar with some interesting flow phenomena, and surely many more examples will be observed in the course of one's professional development as an engineer. Why does a stop sign flutter in a high wind? Why does a crumpled sheet of paper fall faster than an unfolded one of the same original size and weight? Why do steel ships float or airplanes fly? Can gasoline be vaporized within a fuel injection pump whose piston accelerates too rapidly? What is the effect of stream-lining on an automobile's performance? One attempts to provide answers to questions such as these by trying to understand the cause-and-effect relationships that are thought to occur in a flowing fluid. This process involves the application and understanding of fluid mechanics. It will include an examination of various principles and experimental results from physics and other sciences and the use of mathematics to interrelate many of the ideas.

Some of the flow cases just mentioned are associated with the physical properties of the fluid. In the remainder of this section some additional examples of fluid flows will be provided to give the reader some insight into the material that will follow in the book. (Occasionally a term that has a widely accepted meaning will be used in a special, technical sense. Later such terms will be given precise definitions so that no ambiguities can occur when they are used in a technical discussion.) A *liquid* is commonly defined as a substance that assumes the shape of its surrounding container. A *gas* is said to be a substance that completely fills a closed container. The intermolecular forces in a gas are assumed to be much weaker than those in a liquid, so that its

1

molecules are free to migrate throughout the container. Yet, both liquids and gases can be considered fluids if they meet a requirement about their resistance to distortion that will be discussed in a later section: A *fluid* is a substance that distorts continuously upon the application of a shear stress, no matter how small.

Leonardo da Vinci recognized that there is a great similarity between the flows of liquids and the flows of gases. Although this is true in many cases, there are also many differences. When a gas is flowing in a converging conduit, it must increase its speed as it moves, in accordance with the principles in Chapter 3. At the same time the speed is increasing, the gas pressure is decreasing, a conclusion that comes from Chapter 14. The changes of speed, density, and conduit shape are interrelated, so that the gas will reach a maximum speed that is determined by the temperature and pressure in the reservoir from which it came.

Some similar things happen when a liquid flows in a converging conduit. But it is possible that the pressure in the liquid will decrease to its temperature-dependent vapor pressure. The result might be that the liquid vaporizes in the conduit. This is usually undesirable and causes not only noise but also damage to the internal surfaces of the conduit. Moreover, the delivery of the liquid through the conduit will be reduced. Osborne Reynolds' observations of boiling in conduits were discussed in an 1881 article that included Fig. 1.1.1. Thus, low pressures in liquids can have consequences for the flow different from those of low pressures in gases.

The geometry of the passages or objects that form the boundaries of the fluid can also determine the nature of the flow. We shall see that the flow of a fluid in a tube can be orderly and predictable. If the same amount of fluid is allowed to flow, per unit of time, through a pipe of smaller diameter, the average speed of the fluid along the pipe's axis will increase. However, it is possible that the fluid motion will become disordered and random. The pressure changes along the axis of the pipe would then be significantly different from those of the larger pipe. This is because the flow in a round pipe is not always a stable flow. Under some conditions (determined by speed, size, density, and other fluid properties), the flow can change from an ordered one, in which

Figure 1.1.1 Drawing from Osborne Reynolds' paper explaining occurrence of water vapor at room temperature. Water flowing through the tube must increase speed as it passes through contraction near end. This increased speed, with associated higher kinetic energy of the water, is accompanied by lowering of pressure. If speed is high enough, pressure can reach vapor pressure at the ambient temperature. Vapor appears as many finely divided bubbles and occurs downstream of point of minimum pressure because of the finite time interval for these bubbles to grow to observable size. In expanding portion of tube, pressure is above the vapor pressure and vapor begins to go back into solution. Process is accompanied by a high-pitched sound. (Osborne Reynolds, *Papers on Mechanical and Physical Subjects,* Cambridge University Press, 1901)

the fluid particles move only parallel to the pipe's axis, to one that has random velocity fluctuations superimposed on the main flow. The term *laminar flow* is used to describe the former case, in which the fluid particles appear to move upon lamina. The latter situation is denoted as a *turbulent flow,* because its characteristics match well the meaning of that adjective. The features of these two types of flows were also discussed by Reynolds in the 1880s.

The occurrence of ordered and random flow can be observed in a wind-still room from the smoke pattern of a cigarette or of a candle with a poorly trimmed wick. The smoke rises from its source as a meandering filament and then changes abruptly into a disorganized plume of greater volume. Figure 1.1.2 presents a laminar-turbulent transition that clearly shows the change from order to chaos.

It is also necessary to mention that not only can the properties of the fluid determine some of the characteristics of the flow, but they can also determine whether or not flow occurs. If the surface tension of a liquid is very large, as is that of mercury, the liquid can emerge from the end of a vertical capillary tube as distinct drops. However, if the surface tension is low, like that of methyl alcohol, for example, the liquid will discharge as a cylindrical jet in most cases.

From the previous examples, the reader can infer that when one is faced with a problem in fluid mechanics, it is necessary to consider the particular nature of the situation. Such deliberations and the general principles of mechanics can help one to deduce the correct conclusions in a novel situation. Some examples follow that underscore this point.

Experience has taught us that it takes more power to pump a certain quantity of fluid through a pipe with a rough internal surface than through a comparable one with a smooth surface. Surface roughness is considered undesirable in most flow situations. However, when the speed of the fluid is low enough, the surface finish of the conduit no longer influences the flow.

The drag force on an airplane's wing serves as an introduction to an anomalous case concerning the effects of surface roughness. A significant reduction in this drag, with a resulting increase in speed or range of the aircraft, was achieved when it first

Figure 1.1.2 Growth of surface disturbances on liquid jet. As jet emerges from orifice, surface is smooth. Small waves, a disturbance, begin to grow. At first, growth has appearance of being orderly. There is one wave length in direction of jet's axis and another around its circumference. Waves become larger and less organized. Ultimately, large-amplitude undulations occur at random locations on liquid's surface. (Courtesy of Prof. J. W. Hoyt, San Diego State University)

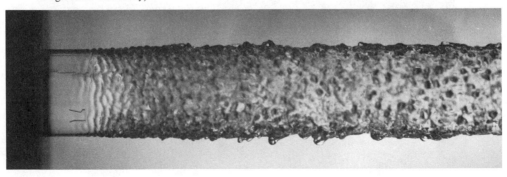

became the practice to make the rivet heads flush with the surface of the airplane. Then they no longer disturbed the flow, and the friction between the air and the aircraft was reduced. From this experience one might always strive for a smooth surface on a body to minimize the drag.

However, the experimental result for spheres that are moving at moderately high speeds does not support the use of smooth surfaces. The flow pattern past such a sphere is shown in Fig. 1.1.3. If the most forward part of the sphere were roughened, the resulting flow pattern could look like that in Fig. 1.1.4. The change of the flow pattern due to the presence of the roughness, and the resulting different pressure

Figure 1.1.3 Water originating from top flows past submerged sphere. As water moves over uppermost portion of sphere, it entrains a dye that seeps through carefully chosen points. Photograph shows that for the conditions of the experiment (free-stream speed = 25 cm/s, diameter = 20 cm) the water cannot flow over downstream half of sphere in same way as on upstream half. Sheet of dye entrained upstream flows smoothly upon sphere's surface at first. But upon reaching a location near sphere's equator, dye ceases to follow contour of sphere. Smooth sheet of dye is disrupted, seemingly torn apart. Surface flow is said to "separate" from boundary. (Fluid that moves away from boundary on downstream side of sphere is replaced by slow-moving fluid that swirls around in "separation region.") (Courtesy of H. Werlé, O.N.E.R.A., Paris, France)

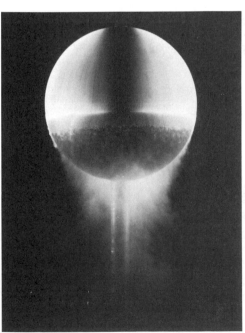

Figure 1.1.4 Under particular conditions, a free-stream speed of 150 cm/s in this case, sheet of dye on upstream part of sphere will contain small disturbances that interact sufficiently with the higher-velocity fluid slightly more remote from surface. This small amount of mixing is sufficient to add enough energy to liquid next to sphere's surface to permit it to flow into downstream portion of surface. Result is a reduced region of separation. The two different flow patterns in this figure and Fig. 1.1.3 are accompanied by different pressure distributions which yield different forces on sphere. (Courtesy of H. Werlé, O.N.E.R.A., Paris, France)

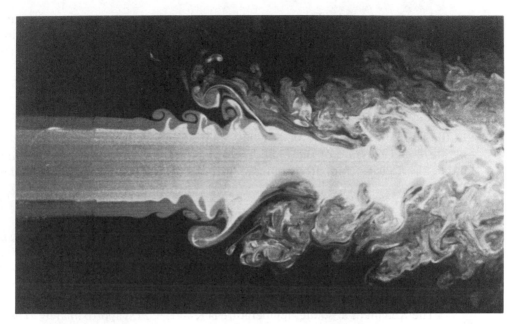

Figure 1.1.5 High-speed water jet is discharged within a tank of water and mixing occurs. Sheet of light passes through a center plane of jet and causes two different dyes to fluoresce to show mixing process. As jet leaves a transparent tube at top of photograph, disturbances grow on its periphery. Waves grow in amplitude and appear to curl over and "break." This entraps some of surrounding water. Disturbances at interface of jet and ambient liquid can be seen to be transmitted to interior of jet because of the two dyes used. The regular waves become irregular and continue to grow in amplitude, with concomitant increase in mixing. (From Dahm, Dimotakis and Frieter, from "Vortex Structure and Dynamics in the Near Field of a Coaxial Jet.", 1989)

distributions on the sphere, alter the drag on the sphere and explain the roughness paradox:

> The force on thin bodies due to a flowing fluid is largely due to fluid friction, and surface roughness increases this friction. For bodies that are wide in the direction normal to the oncoming flow, the fluid pressure distribution is the most important factor in determining the drag on the body. This distribution is significantly affected by the roughness on the body, with the result that a rough body can have a lower drag than a smooth one.

This anomalous case will be discussed further in Sec. 1.8; it is interesting to note here that one ascribes the pattern of indentations on golf balls to an attempt to reduce the drag on the balls and increase the probability of par.*

In the course of studying fluid mechanics, one learns that the flow of fluids is usually not a linear process. When the direction of flow is reversed, the resulting phenomena are not necessarily the same as before except for a change of direction. This is apparent if one recalls the difference between trying to blow out a candle by exhalation and by inhalation. The former process produces a jet, while the latter does not. Figure 1.1.5 shows a liquid jet. The breakdown of simple, regular disturbances at

---

* Professor Ascher Shapiro's book *Shape and Flow* (Anchor/Doubleday, New York, 1961) describes this phenomenon clearly. It is recommended reading.

the edge of, and within the jet, to complex and random ones is quite apparent. The motions of the fluid in the jet are large in comparison with the jet's diameter and are associated with nonlinear processes.

1.2 TEMPERATURE AND PRESSURE

In the next few sections we shall discuss fluid properties that depend upon the temperature and pressure of the fluid. The temperature T will usually be specified by either the Celsius (formerly centigrade) or the Fahrenheit scale. These scales are related by

$$°C = \tfrac{5}{9}(°F - 32)$$

Thus the average normal temperature of the human body is 37°C, or 98.6°F, while a common temperature for developing photographic film is 68°F, or 20°C. In the Celsius system the freezing point of water is taken as 0°C, and it is obvious that lower, or negative, temperatures can occur. The lowest temperature possible is about −273°C. This temperature is taken as zero on the Kelvin scale.

The Kelvin, or absolute, temperature is $K = °C + 273$, approximately. The absolute temperature that is associated with the Fahrenheit scale is the Rankine temperature; $R = °F + 460$, approximately.

Most of the physical properties of materials needed in fluid mechanics will be contained in tables in which the temperature will be listed in either the Celsius or the Fahrenheit scale. When one uses temperature in thermodynamic relationships, however, the absolute temperature (Kelvin or Rankine) must be used.

The ideas associated with pressure are familiar to most people. One encounters them when a pneumatic tire is inflated or when the barometer is read. There are a variety of systems for specifying the magnitude of a pressure. The pressure can be given in terms of atmospheres, pounds per square inch (psi), newtons (N) per square meter (pascals or Pa), or millimeters of mercury in the case of the barometer. Because one pascal is equivalent to a minuscule pressure unit, larger units of pressure such as kilopascals (kPa) or megapascals (MPa) are used. For example, standard, sea-level atmospheric air pressure is about 101.3 kPa.

The pressure p of a gas can also be discussed within the framework of kinetic theory of gases. In a perfect vacuum (zero pressure) there are no gas molecules to rebound from the container's walls. When gas molecules are introduced, the integrated intensity of the force on the wall to produce these rebounds is taken as a measure of the gas's pressure. As more and more gas molecules are added to a container, the number of rebounds per unit time increases and the pressure continually increases.

While the concept of pressure is a familiar one, it is necessary to be careful in specifying its value. The previous examples of the pressure in a pneumatic tire and the atmospheric pressure are not quite similar. The difference is the zero point for specifying the magnitude. The barometric pressure is an absolute pressure. One also measures pressures according to how far they exceed atmospheric pressure. This is a common situation. The air pressure specified for a present-day bicycle tire can be 50 to 90 psi, or 3 to 6 atmospheres. These pressures are implicitly the values that are in excess of local atmospheric pressure, for when the tire is completely deflated, there are molecules of air present on the inside that have all of the characteristics of the molecules in the atmosphere. The pressure in the uninflated tire is atmospheric pressure. As more air is added by the tire pump, the pressure in the tire increases above the atmospheric pressure, as indicated by the increased firmness of the tire.

The elastic stretching or distortion of a container of gas when different internal pressures are applied is often used as a means of specifying the amount of the pressure. It is quite usual to have a fluid inside a curved tube, one end of which is sealed and unsupported. As the pressure in the fluid is increased, the tube deforms in such a way as to diminish the curvature. A linkage converts this movement into the rotary motion of a pointer. This is the principle of a Bourdon tube pressure gauge. When the pressure inside the bent tube is the same as the local atmospheric pressure on the outside of the tube, there is no unbalanced force on it. This neutral position of the tube is used to calibrate the zero point on the circular scale of the gauge coincident with the pointer. Because of the way that a pressure gauge usually works, it is common to denote the pressure in excess of atmospheric pressure as *gauge pressure*. One should be quite explicit in specifying the value as absolute or gauge pressure by writing, for example, psia or psig.

The choice of atmospheric pressure as the equivalent of zero gauge pressure implies that negative gauge pressures will be encountered. If the local atmospheric pressure is 101.330 kPa and the gauge pressure is -2000 N/m^2, the absolute pressure would be 99.330 kPa. When one is dealing with negative gauge pressures, it is common to refer to them as *vacuum pressure*. For example, -2000 Pa is written as 2000 Pa vacuum.

Example 1.2.1:

The unit that is called the *bar* was defined as 10^6 dynes/cm^2. One may still find meteorological data given in this unit. At present it is defined as 0.986923 standard atmospheres, or a pressure equivalent to a barometric reading of 750.00 mm of mercury at 0°C. The equivalent of a bar in pascals can be found by converting dynes to newtons and centimeters to meters as follows:

$$1 \text{ dyne} = 1 \text{ (g-cm)/s}^2 \quad \text{and} \quad 1 \text{ g} = 0.001 \text{ kg}$$

Hence
$$1 \text{ dyne} = 10^{-5} \text{ kg-m/s}^2 = 10^{-5} \text{ N}$$

This result implies that

$$1 \text{ dyne/cm}^2 = 10^{-5} \text{ N/cm}^2$$
$$= 10^{-1} \text{ N/m}^2$$

The old definition of the bar can now be rewritten as

$$1 \text{ bar} = 10^6 \text{ dynes/cm}^2 = 10^5 \text{ N/m}^2$$
$$= 10^5 \text{ Pa}$$

For an answer in pounds per square inch (written as psi), one can proceed as follows:

$$1 \text{ lb} = 4.44822 \text{ N} \quad \text{and} \quad 1 \text{ ft} = 0.305 \text{ m}$$

so that
$$1 \text{ N/m}^2 = \frac{1}{4.44822} \frac{\text{lb}}{\text{ft}^2} (0.305)^2$$

$$= 0.02088 \frac{\text{lb}}{\text{ft}^2} = 0.02088 \text{ psf} = 1.45(10^{-4}) \text{ psi}$$

Hence

$$1 \text{ bar} = 10^5 (1.45)(10^{-4}) = 14.5 \text{ psi}$$

Since 1 bar = 0.986923 atmospheres,

$$1 \text{ atmosphere} = 14.69 \text{ psi}$$
$$= 2115.4 \text{ lb/ft}^2 \text{ (abbreviated psf)}$$

If, for some very special reason, the desired conversion is to "kilograms-force" per square meter, the calculation is as outlined below:*

$$1 \text{ lb} = \frac{1}{2.2} \text{ kgf (kgf = "kilograms" of force)}$$

Hence, using the calculation in the last conversion,

$$1 \text{ bar} = 2088 \text{ psf} = \frac{2088 \text{ kgf}}{2.2 \text{ ft}^2} = \frac{2088}{2.2} \frac{\text{kgf}}{(0.305)^2 \text{ m}^2}$$
$$= 10.203 \text{ kgf/m}^2$$

1.3 DENSITY

Density denotes the mass per unit volume (in kilograms per cubic meter, or slugs per cubic foot) of a substance. Quite commonly, the symbol for density is the Greek letter rho, ρ. The *specific gravity* (abbreviated SG) of a substance is ρ/ρ_{H_2O}. Tables A.4 to A.6 in Appendix A list the densities of a variety of gases and liquids. The values in Table A.6 could be used to find the mass of the hydrogen and the oxygen molecule in English or SI units. This can be done if one recalls that for gases at 0°C and 101.33 kPa,† 22.4 liters (1 liter = 0.001 m³) contain $6.023(10^{23})$ molecules. One could use Avogadro's number to find the number of molecules in one liter of hydrogen and one liter of oxygen at 0°C and 101.33 kPa.

The mass of a particle of gas is thought to be an invariant property of the particle. It is, in effect, the amount of matter that the particle possesses. Hence, if under

---

* The "kilogram-force" will not be used again in this book, just as the term *pound-mass* will be avoided whenever possible. In the International System of Units—the SI system—the basic units are mass, length, and time, whose magnitudes are the kilogram, meter, and second, respectively. Force is defined in terms of the basic units, so that 1 newton of force is equivalent to 1 kg-m/s². In a parallel way, we shall adopt as basic units the slug, foot, and second when working with the popular engineering (British) system; therefore, the unit of force, the pound, will be treated as a defined unit that is equivalent to 1 slug-ft/s². In conformance with this point of view, the pound-mass will be treated as 1/32.2 slug.

The kilogram is the unit of mass which is equal to the mass of an internationally recognized prototype preserved in a location outside Paris. This piece contains a definite number of atoms of a platinum alloy. Thus, a merchant who places a copy of this standard on one end of a beam balance and has an equal mass of some vegetable on the other end will be offering the customer a kilogram of the foodstuff.

In lifting the kilogram standard from its place of storage to its position on the balance, the merchant notes that certain muscles must exert a force on the object. Gravitational attraction must be overcome. Even though we may recognize that the force to lift the object would be less on the moon, we often carelessly relate the kilogram standard of the shopkeeper to a particular force, the force to support the object in a familiar gravitational field. This relationship is so frequent that we ultimately equate, sometimes unknowingly, weight with mass. For this reason the term *kilogram* to denote units of mass and force can lead to unnecessary confusion. Despite this fact, we must reconcile ourselves to seeing packages labeled to read "Net weight: 1 pound = 453.6 grams" for a considerable time to come. The interested student is urged to read *Gravity,* by George Gamow (Anchor/Doubleday, New York, 1962), for new insights into the matter.

† These values of T and p prescribe *standard conditions* for a gas. However, other "standards" exist. For example, at sea level a *standard atmosphere* specifies 288.2°K and 101.33 kPa.

Fluid Properties and Flow Phenomena   Chap. 1

conditions of higher pressure one finds that a liter of gas does not have 2.69(10²²)
molecules—the number under standard conditions—but rather 3 times as many, one
would be led to the conclusion that the mass per liter of the gas under these circumstances must be 3 times that of the value at standard conditions.

The symbol γ will be used to denote the weight per unit volume of a substance.
From the previous discussion, it follows that $\gamma = \rho g$, in which g is the local gravitational
constant.

1.4 AN EQUATION OF STATE—THE IDEAL GAS EQUATION

If one knows the absolute pressure of a gas, in some suitable system of units, and the
absolute temperature T, then most of the thermodynamic properties of the gas (e.g.,
specific heat, internal energy, density) can be determined. An equation that relates the
pressure, temperature, and density of a gas is termed an *equation of state*. One simple
and useful form of such an equation is the *ideal gas equation*. It is a simple and useful
way of specifying the density of a gas if the pressure and temperature are known. The
mathematical expression for this is

$$\frac{p}{\rho} = RT \tag{1.4.1}$$

where R is a constant that depends on the molecular weight of the gas. This is but one
of many such equations of state, some of which appear to be more suitable than others
in particular ranges of temperature and pressure. The ideal gas equation is useful for
pressures that are near atmospheric pressure. In this regime it encompasses the experimental results that are known as Charles' law and Boyle's law. Equation (1.4.1) must
be dimensionally consistent, and so it is common to express p in pascals, ρ in kg/m³,
T in degrees Kelvin, and R in (m)(N)/[(kg)(K)] or, equivalently, joules/[(kg)(K)].

Example 1.4.1:

In the United States of America it is common to specify the pressure in psf after converting from psi, the density in slugs/ft³, and the temperature in °R.* Sometimes the gas
constant in the ideal gas equation is given as R in ft-lb/slug-°R. Show that if p and T are
prescribed in this way but the weight per unit volume, $\gamma = \rho g$, is used, Eq. (1.4.1) would be
written as

$$\frac{p}{\gamma} = \hat{R}T \tag{1.4.2}$$

in which $\hat{R}$ has units of ft-lb/(lb of gas)(°R) and $R = g\hat{R}$ would relate the gas constants R
and $\hat{R}$. (The use of $\hat{R}$, with its associated units, persists because of its long history.) Since
Eq. (1.4.1) can be multiplied by $1/g$, one has $p/\rho g = RT/g$, or $p/\gamma = RT/g$. Hence
$\hat{R} = R/g$, as implied by Eq. (1.4.2).

The curves of constant temperature that are given by Eq. (1.4.1) relate p and $1/\rho$
by hyperbolas. While this equation is a good model of the interrelationship between
these variables at high temperatures and low pressures, at higher pressures the inter-

---

*A great deal of difficulty can be circumvented if all problem data are converted, at the outset, to
standard units for mass, length, time, and force in the SI and English systems. These, once again, are
kilograms, meters, seconds, and newtons and slugs, feet, seconds, and pounds, respectively.

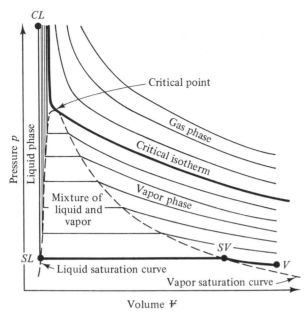

Figure 1.4.1 Isotherms shown as parameter of pressure-specific volume diagram for a pure substance. At *critical point,* pure liquid and pure vapor have the same properties and so are indistinguishable from one another. Diagram does not show state points for which substance is a solid. One can anticipate temperature lines that demarcate solid-liquid and solid-vapor transitions. State point where solid, liquid, and vapor phases are all in equilibrium is the *triple point.* (By permission, from M. K. Zemansky, "Heat and Thermodynamics," 2nd edition, McGraw-Hill Book Company, 1943)

molecular distances are reduced sufficiently that one molecule affects another. Van der Waals forces of molecular repulsion are encountered. At some *critical temperature,* different for each gas, a pressure is reached at which the associated intermolecular distances between the atoms become small enough that some atoms are attracted to one another. (This is the effect of *London forces.*) The medium then takes on the characteristics of a liquid. This is the critical point shown in Fig. 1.4.1. The ideal gas equation is useful at temperatures that are greatly in excess of the critical temperature. Below the critical temperature there are regions of the p versus $1/\rho$ diagram for a single chemical substance in which a pure liquid, a pure gas, or mixtures of gas and liquid can exist.

The reader who has had a course in thermodynamics should be able to recall the general features of the "steam tables." Such numerical data are, in effect, derived from a complicated equation of state, one that includes the possibility of a solid phase. Of course, regions in which the liquid and gas phases exist simultaneously must also be included.

1.5 INTENSIVE PROPERTIES OF FLUIDS

The discussion of the density of a fluid is a convenient introduction to some other physical properties of a fluid that are important in our work. These characteristics are similar to the reciprocal of the density, the volume per unit mass, or *specific volume,* because one specifies them on the basis of the quantity of the property per unit of mass of the substance. We could calculate the density of 7.3 kg of mercury that occupied a volume of $0.537(10^{-3})$ m^3. This would make $\rho = 7.3/[0.537(10^{-3})]$ kg/m^3. The reciprocal of this density, the specific volume, would be $0.735(10^{-4})$ m^3/kg. This latter figure is distinctive for mercury and should be the result regardless of the amount of the material that is being examined. Such a material property that is defined on a unit mass basis is called an *intensive property*. The characteristic that makes a property intensive is not its "per unit mass" basis. It is only required that the property be independent of the total mass of the substance that is being considered. This means that pH can also be considered to be an intensive property. However, weight, electrical conductivity, and magnetic attraction, which depend upon the total mass of the material, are examples of *extensive properties*.

When the calculation of the density of mercury was made in the last paragraph, it was assumed that the mercury was homogeneous. That is, we assumed that the mercury was exactly identical at every place within the $0.537(10^{-3})$ m³. If there were slight variations from place to place, say, from the bottom to the top, the density that was calculated would have been the average value for the entire volume. Now, one might pour off the top half of the mercury from its container and determine the density of the remainder. The mass of the mercury at hand would be divided by the volume that it occupied. The density found might differ slightly from the previous value that was obtained with the larger volume. If one wanted a better determination of the density of the mercury at the bottom of the vessel, a still smaller volume of the material near the bottom could be examined for its mass and the appropriate volume measured. This process of subdividing—say, reducing by half—could be continued, in principle, until one obtained volumes that had dimensions comparable to the mean free distance between the atoms. At that stage the process of halving the volume might not result in exactly halving the number of atoms that would contribute to the resulting measurement of the mass. One would expect this to occur because of the random positions that the atoms assume in the region in which they are enclosed. At any one instant of time the number of atoms in a given volume of space need not be exactly the same as in another volume of the same size. For this reason one would expect the mass per unit volume to fluctuate a bit as one approached volumes that have length scales of the order of the mean free distance between the molecules. The situation is illustrated in Fig. 1.5.1. There one sees that as the volume of the sample is reduced, the density approaches a specific value. The volume is small enough that any variations in the density (i.e., homogeneity) do not influence the local measurement. One has then a density which is appropriate for the particular location of interest. Smaller volumes do not improve the numerical value, because the number of molecules that contribute in a

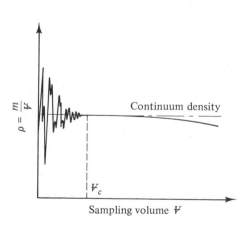

Sampling volume V

Figure 1.5.1 A property of a material should vary continuously with all its coordinates and, thus, with size of specimen. (This continuous variation is zero in a homogeneous substance.) Hence, change in property should be negligible near a particular location, and one speaks of the property at a *material point*. If, as figure shows, volume around a particular point in space becomes increasingly small, variation in a property—density, in this case—ceases to be continuous with size of specimen. When this size regime is reached, one no longer considers material to be a *continuum,* for which the material properties, deformations, stresses, etc., are continuous functions of space and time.

random or probabilistic way is too small to give a significant and consistent sample size. For volumes below a certain critical size, the concept of density—mass per unit volume—loses some of its significance. What might this critical volume be? The answer is not difficult to estimate.

Example 1.5.1:

If one were to assume that 100 atoms would give a statistically significant sample of the density, one could use Avogadro's number to determine the critical minimum volume of the sample. The atomic weight of mercury is 200.61. Its specific volume is $0.735(10^{-4})$ m^3/kg. Consequently, 0.20061 kg of this material would constitute a mole and contain $6.023(10^{23})$ atoms. The volume of this amount of mercury would be $0.20061(0.735)(10^{-4})$ m^3. One hundred atoms of the material would occupy proportionately less volume. The result is $[0.20061(0.735)(10^{-4})](100)/6.023(10^{23})$. The cube root of this number is the length of the cube that would contain the 100 atoms; it is $1.34(10^{-9})$ m, or $1.34(10^{-3})$ microns. (To compare the size of that region recall that the diameter of a human hair is about 60 microns).

It is useful to be conscious of the size of the region that is necessary for a representative value of the density to be specified at a location. We will then understand the meaning of the word *point* in connection with our work. Clearly we shall not use the geometric definition of the word. We should recognize that in the study of mechanics we are implicitly talking about a *"material" point,* a minute region surrounding a location that may interest us. The region is large enough that significant averages of the quantities to be measured can be obtained. For volumes that are larger than a "material" point, the volume per unit mass may change because of some spatial inhomogeneity. Thus, density may vary from point to point in the region of study. This region could be the atmosphere, a river, a pipeline, or a blood vessel, among other things.

Whenever one is dealing with a material in regions that are larger than a material point, for which there are sufficient molecules to form a statistical mean at such a point, it is common to say that one is using a region that is a *continuum*. This means that one can determine the physical properties and thermodynamic variables, among other things, of every point of the material with the aid of a suitable model. (In outer space, or at very low pressures in the laboratory, very few molecules of gas can be present in a cubic meter, so that a continuum would not exist in these cases.) We shall encounter some situations in which the fluid, by virtue of its motion, changes from a liquid into a vapor. One would then have a continuum for the liquid phase, and another for the vapor phase.

There are other intensive properties of fluids that will be of interest to us. The specific heat of a material at constant pressure, c_p, and that at constant volume, c_v, are examples. These quantities have units of joules/[(kg)(K)] or sometimes Btu/[(slug)(°R)]. The definitions of a joule, a Btu, and other units are given in the front of this book.

We shall also utilize the internal energy per unit mass in our study. The kinetic energy per unit mass is one-half the square of the speed. Momentum is defined as mass times velocity. Thus the momentum per unit mass, an intensive characteristic of the fluid for specified flow conditions, would be mass times velocity divided by mass, or simply the velocity. Because velocity is a vector quantity, the momentum per unit mass has three components, one for each of the coordinate directions.

1.6 GENERAL PHYSICAL PROPERTIES

Tables A.1 to A.7 give most of the pertinent characteristics of the materials mentioned in this book. The properties listed in these tables are quite common ones. Only one, surface tension (Table A.5), may require some detailed comments.

The concept of surface tension is one way to model the result of intermolecular attractive forces that act upon the molecules at an interface between one substance and another. The molecules at the interface experience attraction to the adjacent molecules of the same species and different attraction to the molecules of the second species that are located immediately on the other side of the interface. The net effect of these different attractions is that the molecules are somewhat constrained and their mobility is reduced. One can think of the constraining mechanism as an elastic "skin" or membrane that separates the fluids at the interface. It is this membrane that encapsulates a liquid drop in a gas, a gas bubble in a liquid, or a liquid bubble in another liquid. We imagine that as the interface is distorted, stresses are induced in this membrane in such a way that all of the forces on the membrane are related through Newton's laws of motion. In succeeding chapters we shall often use these laws for predicting the motion of fluids. For the present we shall apply Newton's laws to a static interface to determine the effect of its geometry on the stresses induced in the membrane model of the interface. *Surface tension* is the name given to these stresses.

Figure 1.6.1 shows a section of the surface of a pendant drop, such as could appear at a water faucet. We note that such a drop is not spherical but has two different radii of curvature. The segment of the interface is treated as a free body, and forces are summed in the direction of the local normal to the surface. In Fig. 1.6.1 the surface tension σ is applied to the edges of the membrane, because it is an interfacial effect; σ has units of force per unit length. (Normally, stresses are given in units of force per unit

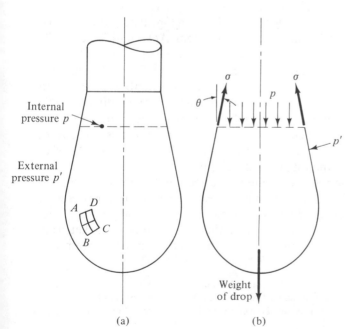

Internal
pressure p

External
pressure p'

(a)

(b)

Weight
of drop

Figure 1.6.1 A pendant drop in a gravity field can be at rest because of the equilibrium of the forces associated with gravitational attraction, the pressure exerted upon it by the surrounding liquid, and a tensile force that is assumed to exist in a surface layer having the dimensions of a molecule. We call this effect of the molecular forces near the interface between the liquid and the gas that surrounds it (or a solid with which it might be in contact) *surface tension.* Equilibrium of these forces interrelates geometry of drop and its specific weight with pressure at section of interest and local value of surface tension. (Angle θ is determined by segment of drop under study.)

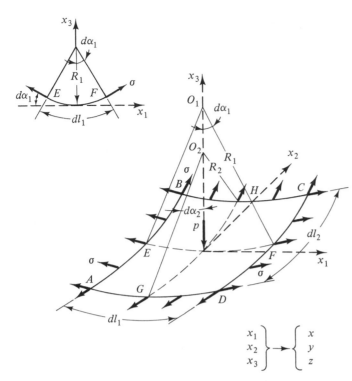

Figure 1.6.2 Examination of an element of surface of drop leads to another relationship between surface tension, local difference between internal and external pressure, and radii of curvature of the surface element, R_1 and R_2. Because thickness of this element is assumed to become vanishingly small, its weight does not enter into local analysis. (Adapted from J. W. Daily and D. R. F. Harteman, *Fluid Dynamics,* Fig. Prob. 1-6, by permission of Addison-Wesley Publishing Company, Inc., 1973)

area. However, when considering surface tension, the membrane is considered to be infinitesimally thin and a unit thickness of the membrane is employed, in effect.) It is assumed that σ is independent of direction, and this is in concert with other experimental evidence.* The length along edges AD and BC in Fig. 1.6.2 is the product of the radius and the included angle, which is $R_1(2d\alpha_1)$. Thus, in Fig. 1.6.2, which shows a section cut through the element of the surface, the force along AD is $\sigma(R_1)(2d\alpha_1)$. This gives a force component in the n direction of $\sigma(R_1)(2d\alpha_1)\sin(d\alpha_2)$. For small angles, the sine of an angle is equal to the angle itself. Hence AD and BC combine to give a force component of $2\sigma R_1(2d\alpha_1)\,d\alpha_2$. In the same way, it is possible to show that along AB and CD a total force component of $2\sigma R_2(2d\alpha_2)\,d\alpha_1$ occurs in the n direction. The weight of the element of fluid depends upon the volume of the element and, as we shall see in Chap. 2, such forces become negligibly small in comparison with surface forces as the element is reduced to a point. In view of this we shall treat the weight of the surface section as unimportant in the following development.

Also acting in the n direction is a component of force due to a pressure differential across the film. In Fig. 1.6.2 this difference is $(p-p')$ if p is the internal pressure and p' the external one. This pressure differential results in a force component in the n direction of $-(p-p')[R_1(2d\alpha_1)R_2(2d\alpha_2)]$, in which the term in square brackets is the area of the element of the surface. The minus sign for the coefficient of this component

---

*See the film "Surface Tension in Fluid Mechanics," available from Encyclopaedia Britannica Education Corporation, 425 North Michigan Avenue, Chicago, IL 60611. Other excellent films are also available from this and other sources, such as the University of Iowa and the American Society of Mechanical Engineers (ASME).

is due to the fact that if p is greater than p' the component acts in the negative n direction. By summing the components of the surface tension and pressure differential in the positive n direction, we get

$$2[\sigma R_1(2d\alpha_1)]\,d\alpha_2 + 2[\sigma R_2(2d\alpha_2)]\,d\alpha_1 - (p - p')[R_1(2d\alpha_1)R_2(2d\alpha_2)] = 0$$

or
$$p - p' = \sigma\!\left(\frac{1}{R_1} + \frac{1}{R_2}\right) \tag{1.6.1}$$

For a spherical drop, $R_1 = R_2$. (For a cylindrical stream of liquid, one of the radii, the one which corresponds to a generator of the cylindrical surface, is infinite.) As the radius decreases, the pressure differential must increase according to Eq. (1.6.1). Figure 1.6.3 depicts the surface tension stress for a spherical drop or bubble.

The rise of a liquid in a tube of small diameter, a capillary tube, will depend upon σ and the angle of contact, θ, also shown in Fig. 1.6.3. The length of the line of contact of the liquid with the tube is $2\pi R$. This results in a force component in the z direction of $(2\pi R)\sigma \cos\theta$. A column of liquid of height h is supported by this force component. The column weighs $\gamma \forall$, where γ is the weight per unit volume and $\forall$ is the volume. Thus the equilibrium of the surface tension and gravity (i.e., weight) forces requires

$$2\pi R\sigma \cos\theta = \gamma(\pi R^2 h)$$

so that
$$\sigma \cos\theta = \frac{\gamma R h}{2} \tag{1.6.2}$$

Consequently, when one does not wish a meniscus to rise appreciably in a tube, a large value of R is chosen. It is believed that trees, even very tall ones, send water to their highest branches by means of surface tension effects. Hence, the capillary passages must be extremely fine. In Chapter 2 we shall see how the liquid pressure within these capillaries, or at point A in Fig. 1.6.3, is affected by the height of the liquid column.

Example 1.6.1:

If a tube with an internal diameter of 1 mm has water inside it in the manner shown in Fig. 1.6.3, what is the maximum capillary rise that can be expected? (Immediately, one should write $D = 0.001$ m to convert the length to meters.)

Solution Equation 1.6.2 indicates that for a fixed γ, R, and σ, the value of h is a maximum for $\theta = 0$. For $\sigma = 0.072$ N/m from Table A.7,

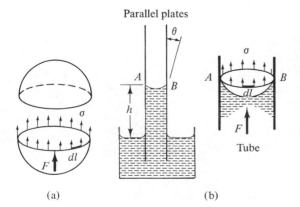

Parallel plates

Tube

(a)

(b)

Figure 1.6.3 Some shapes of interfaces between liquids, gases, and solids will depend upon situation at hand. For a spherical drop (a), the two radii of curvature are equal. Liquid in a tube (b) will have contact angle determined by the characteristics of the liquid, the gas (usually the unsaturated vapor of the liquid), and the solid. If *contact angle* θ of a drop on a solid is greater than 90°, one says that liquid *wets* solid because its molecules will adhere to surface and not roll off when surface is inclined to the horizontal.

$$h = \frac{(0.072)(2)}{(0.001/2)(9.8)(1000)} = 0.00294 \text{ m,}$$
$$= 2.94 \text{ mm}$$

The value of σ is sensitive to minute quantities of different chemicals and variations of temperature and electric charge. A detergent decreases the surface tension of water so that it flows evenly from glassware or developed photographic film instead of remaining dispersed over the surface in drops, which, when they dry, will form noticeable spots. Figures 1.6.4 and 1.6.5 are other examples of the effects of surface tension.

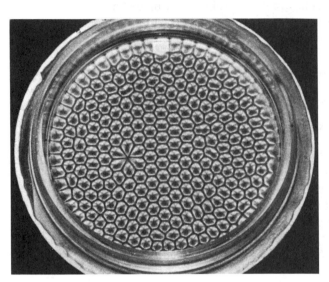

Figure 1.6.4 Magnitude of surface tension σ, and consequently its effect upon shape of interface, will depend upon chemical composition, electric charge, and temperature. Hence, temperature variations along a horizontal surface would result in a net force being applied to a liquid layer. Such a force could be brought into equilibrium if a slight motion were to occur within liquid layer, because then shear stresses would be induced. Photograph shows a particular form of such a motion in a very thin layer of liquid. Small metallic flakes, a few microns in thickness, reflect the light differently depending upon orientation within slowly moving layer. Hexagonal shapes are affected by circular boundary and blemish in surface of bottom heating plate just to left of center. (Courtesy of Prof. E. L. Koschmieder, The University of Texas at Austin)

Figure 1.6.5 When a liquid jet is formed by a circular orifice and enters a gas, often air, one speaks of a *free jet,* in contrast to a *submerged jet* as in Fig. 1.1.5. Effect of surrounding medium on a free jet is less than on a submerged jet because of smaller density and viscosity of ambient gas. Analysis of dynamics of jet alone indicates that once surface is distorted, due to omnipresent disturbances, perhaps, distortion will grow along jet's axis. Surface develops axisymmetric waves of ever-increasing amplitude. Ultimately, drops are formed. It has been shown that there is a wavelength for which disturbance grows fastest. (By permission, from W. Debler and D. Yu, Proc. Roy. Soc. (London) A415, pp. 107–119, 1988)

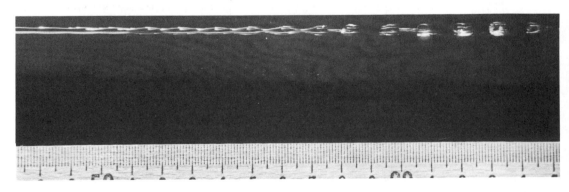

We now turn to the second liquid property that may require some additional discussion. Most liquids are considered, as a first approximation, not to become denser with increasing pressure when the temperature is kept constant. If this is so, one calls the liquid *incompressible*. As one increases the pressure on a volume of the liquid, the volume remains almost unchanged. (This does not preclude the possibility of a change of volume with a change of temperature while the pressure is held constant.) However, when extremely high pressures are applied to liquids, minute changes of volume can be detected. For each increment of pressure Δp, a volume of fluid V will undergo a change of volume ΔV. This process of volume change is elastic, and for a positive change of pressure, ΔV will be negative. One relates Δp and the fraction of volume change (called volume strain) by

$$\Delta p = -K \left(\frac{\Delta V}{V} \right) \quad (1.6.3)$$

where K is the *bulk modulus*. Usually, K is given for an isothermal compression and is called the *isothermal bulk modulus*. The value of K varies with pressure, as one can infer from Table A.7.

Example 1.6.2:

What is the percent change of volume of water at a pressure of 1000 atmospheres for a change of pressure of 2 atmospheres?

Solution   From Table A.7 and Eq. (1.6.3),

$$|\Delta V/V| = \Delta p/K = \frac{2(14.7 \text{ psi})}{420{,}000 \text{ psi}}$$

Therefore, $\Delta V/V = 0.00007,$

which gives a percent volume change of 0.007 for these *two* atmospheres of change of pressure.* (Note that Δp was not expressed in psf because K was listed in psi.)

The values given in Table A.7 are taken from experiments that are performed at constant temperature. This is only one way that a compression can occur; however, it is the manner in which laboratory tests can be conducted in a feasible manner.

On the basis of tests, sound has been found to be propagated through gases and liquids by means of a compression and expansion process that is not isothermal. The experimental data are consistent with a theory that assumes an absence of heat transfer, an *adiabatic process*. If one knows the speed of sound in a liquid, one can infer from the theory the value of the adiabatic compressibility.

The speed of sound is thought to be related to changes of pressure and the associated changes in density. (This is discussed in Chapter 14.) The mathematical statement of this relationship is

$$c = \sqrt{\frac{\Delta p}{\Delta \rho}}$$

for a reversible and adiabatic process. If K_a is the adiabatic bulk modulus and we note that $\Delta \rho / \rho = -\Delta V/V$, we can use a definition comparable to Eq. (1.6.3) to give

---

* "Compressibility" values listed in some tables give the percent change in volume for a change of *one* atmosphere of pressure.

$$c = \sqrt{\frac{K_a}{\rho}} \qquad (1.6.4)$$

This relationship is used to determine K_a from c and ρ. K_a is slightly lower than the isothermal compressibility values tabulated in this book.

1.7 VISCOSITY

One of the most important physical properties of a fluid is its viscosity. We shall see that this property does, in fact, serve to define a fluid. For an elastic material, it is known that when a shear stress is applied, such as those arising when a torque is applied to a shaft, a deformation, or strain, occurs. But for a particular stress, there is a definite strain. If the stress is not increased, the strain does not increase. Hence, for a given value of shear stress, the change of strain per unit of time is zero.

The material under test in Fig. 1.7.1 is not an elastic material. It is a material that we shall call a *fluid*. In this figure there are two concentric cylinders with an oil in the annulus. The falling weight exerts, in effect, a tangential force on the outer cylinder which transfers a shear stress—(tangential force)/area—to the fluid. As long as this stress is applied, the cylinder rotates at a specific angular velocity. There is a unique strain rate (strain per unit of time), which is determined from the angular velocity and the fluid velocity at the boundaries of the container. This existence of a nonzero strain rate is one of the things that differentiate a fluid from an elastic solid. If a cooking oil is tested in the annulus, it gives curve I in Fig. 1.7.2, while a heavy syrup yields data resulting in curve II. The second curve indicates that for a given shear stress, or torque applied to the cylinder, the syrup has a different strain rate, or angular velocity, than the oil. However, both curves share two important characteristics: the curves go

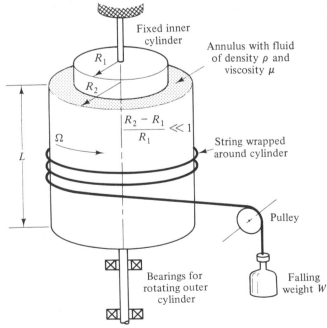

Fixed inner cylinder

R_1

R_2

Annulus with fluid of density ρ and viscosity μ

$\dfrac{|R_2 - R_1|}{R_1} \ll 1$

Ω

L

String wrapped around cylinder

Pulley

Bearings for rotating outer cylinder

Falling weight W

Figure 1.7.1 Viscometer propelled by falling weight that has reached its (constant) terminal speed.

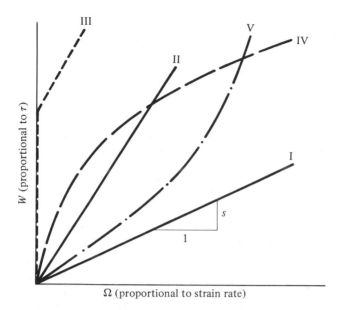

Figure 1.7.2 Rheological characteristics of various fluids that could be observed in a viscometer like that in Fig. 1.7.1. I, II: Newtonian fluids. III: Bingham fluid (plastic). IV, V: Non-Newtonian fluids. s = slope proportional to μ.

through the origin, and they are linear. We shall use the former characteristic to define a *fluid*; it is considered to be a substance that deforms continuously when subjected to shear stresses, no matter how small they are. It is not difficult to show that sand does not meet the definition of a fluid, while honey does, by comparing what happens when an amount of each is poured onto a horizontal surface.

This definition of a fluid includes both gases and liquids. Some common liquids, however, have a characteristic that excludes them from being strictly classified as a fluid. An acrylic paint is a case in point. When it is applied to a vertical wall, one must brush vigorously to spread it out. The force of gravity manifests itself as a finite shear stress distribution that is established within the paint film. The intensity of these gravity-induced stresses is insufficient to cause the paint to flow. Thus the material requires a nonzero stress level before it begins to deform continuously. Such a material is often referred to as a *Bingham fluid* (curve III). A thick suspension of clay and water also exhibits the characteristics of a Bingham fluid.

Some materials behave intermediately to the two types just discussed. Fluid "behavior" is characterized by a *constitutive equation* that relates the stresses induced in a fluid and its motions. Two constitutive relationships are given as curves IV and V. In the former case the slope of the stress versus rate of strain curve decreases with increasing values of strain rate. Grease is a complex composite of soap and oil. It has viscous characteristics similar to that illustrated by curve IV. Under the action of gravity, there is little tendency for grease to flow out of its container. If this relationship between stress and strain rate were to be extended to high strain rates, one would predict enormous shear stresses in high-speed bearings (e.g., the front wheel bearings of cars). But these shear stresses are not realized, and the frictional restraining torque that causes excessive power losses are fortunately absent. This means that the slope of the stress–strain rate curve must be less at these high strain rates than it was when the grease was almost imperceptibly oozing from an open, inverted container. A fluid with this characteristic of a decreasing slope $d\tau/d(du/dy)$ with increasing strain rate du/dy is

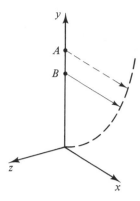

Figure 1.7.3 Flow conditions for applying simplest form of Newton's law of viscosity. Broken curve represents variation of fluid speed with position within fluid, such that $\mathbf{V} = u(y, t)\mathbf{i}$, where t = time.

called a *strain-thinning* material. This material characteristic is illustrated by curve IV in Fig. 1.7.2, whereas curve V depicts a *strain-thickening* fluid.

Materials in which the stress versus strain rate relationship is linear, such as curves I and II, are called *Newtonian fluids*. They form the bulk of the materials one deals with as an engineer. With few exceptions, we shall limit our study in this book to Newtonian fluids. Water, most oils, and air are examples of Newtonian fluids. Blood is not Newtonian in its behavior. Its constituents, the red and white corpuscles (platelike bodies that are transported by the plasma), are sometimes as large as the vessels in which they flow. This leads to different flow characteristics in capillaries and veins, something that does not occur if a fluid is Newtonian.

Newton considered the effects of fluid motion on the stress field, and one of his constitutive relationships was stated as the following:

"The resistance arising from the want of lubricity of a fluid is, other things being equal, proportional to the velocity with which the parts of a fluid are separated from one another."[*]

If one has a fluid that is moving only in one direction, say, the x direction shown in Fig. 1.7.3, so that at fixed positions in the fluid the velocity varies only with y, then the particle in Fig. 1.7.3 that moves through point A will be moving away from a slower one that passes through point B. The type of velocity distribution shown in Fig. 1.7.3 makes it apparent that one can express Newton's "velocity with which the parts of a

---

[*] Isaac Newton focused a great deal of his attention on the motion of the planets. A question that was being discussed in his time was the existence of an atmosphere in outer space. Newton thought that any atmosphere in the regions around distant planets must exert a restraining force, or torque, on the planets. A representation of these actions on the planets could be introduced into the equations of motion for their orbits. Then, if the results of the predicted motions agreed with observations, one might conclude that there existed a material between earth and the particular planet that exerted an effect on the planet due to the constitutive relationship of the medium, similar to that which had been assumed. Purposely, Newton assumed a host of relationships for stress and strain rate and then predicted the motions of certain planets. None of his predictions was borne out by observations. But his results did not make him despair. Rather, he surmised that since he had tried a reasonable variety of relationships between stress and strain rate and none of them seemed suitable for predicting the motion of bodies, there was probably no material in the outer atmosphere. The actual orbits seemed to agree with the predictions that were based upon a vacuous outer space.

Fluid Properties and Flow Phenomena   Chap. 1

fluid are separated from one another" as a derivative. This would lead to an expression for the shear stress τ that is induced in the fluid to be written as

$$\tau = \mu \frac{du}{dy} \tag{1.7.1}$$

in which μ is a constant that is determined by the fluid under consideration at the prevailing temperature and pressure.* The fluid's velocity can be written as $V = u(y, t)\mathbf{i}$, where t = time. This relationship between stress and strain rate is quite appealing. The stress is a linear function of the strain rate. Hence, if the stress in a fluid is constant, the rate of change of velocity with the y coordinate would be a constant, provided that the coefficient of viscosity, μ, was also a constant. The proportionality constant, μ, is a property of the particular fluid, but depends upon the temperature and pressure. It will be shown in Chap. 8 that Eq. (1.7.1) is in agreement with the experimental findings for many common liquids and gases. The experiments are performed in such a way that u is a function only of the coordinate that is perpendicular to its direction. For more complex flow patterns, the shear stress, or stresses, would include additional terms. (See Chap. 16.) The following example will serve to illustrate the use of Eq. (1.7.1) in a specific case.

Example 1.7.1:

Let us return to the situation that is illustrated in Fig. 1.7.1. We shall determine the torque that must be exerted on the outer cylinder, which rotates at a prescribed rate.

Solution   The geometry of the particular apparatus is $L = 200$ mm, $R_1 = 50$ mm, and $R_2 = 55$ mm. (Immediately one should write $L = 0.2$ m, $R_1 = 0.05$ m, and $R_2 = 0.055$ m for preferred length units.) Because the gap, $R_2 - R_1$, is small with respect to R_1, one can approximate the circular geometry of the annulus by a simpler, planar geometry. This is shown in Fig. 1.7.4. (A similar simplification occurs in surveying. One can determine the size and location of a piece of land by using plane trigonometry, but aerial navigation over long distances requires the use of spherical trigonometry.) In this simplification the annulus has been "cut" and "rolled out" into a strip. The velocity of the fluid in this model of actual flow is given in the figure. The length of the rotating cylinder is also very long with respect to $R_2 - R_1$ in order to avoid a major influence of the end regions on the general flow pattern.

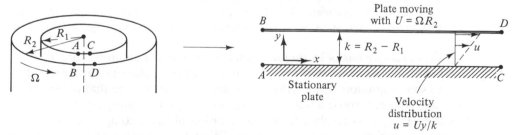

Figure 1.7.4 Transformation of narrow annular region to infinite strip. The cylinders (a) may be thought of as cut on the dashed line and rolled flat to form the moving and stationary plates (b). Note the correspondence of points.

---

* The units of μ are kg/m-s, and τ will be in pascals for u and y in m/s and m, respectively. However, it is still common to have μ given in gm/cm-s (a *poise*). Hence, 10 poise = 1 kg/m-s.

One thing that must be apparent if one examines Fig. 1.7.4 is that the fluid velocity is represented as being equal to the velocity of the boundaries at each of the solid boundaries that contain the fluid. Is this representation realistic? This question is an important one. Between 1830 and 1900, investigators tried to determine the relationship between the speed of the solid boundaries that contain the fluid and the velocity of the fluid that is immediately adjacent to the boundary. A great number of diverse experiments led to the conclusion that the fluid immediately adjacent to a boundary has the velocity of the boundary. Thus, if the boundary is stationary, the velocity of the fluid there is zero. If the boundary is moving with a tangential speed of U, the fluid has that velocity there at that boundary. This experimental result is called the *"no slip" condition.* It is now quite easy to determine the shear stress in accordance with Eq. (1.7.1). For the model of the flow shown in Fig. 1.7.4, the shear stress must be constant at every position within the gap. Otherwise, the force that is applied to successive layers of fluid would vary from the moving surface to the stationary one. We reject this possibility because the net force applied to the nonaccelerating fluid must be zero. This will be discussed in further detail in Chap. 8. If the shear stress does not change in the gap, then Eq. (1.7.1) would predict that the velocity gradient, du/dy, does not vary. This implies that the velocity profile (shown in Fig. 1.7.4) is a linear one, because such a profile has a constant slope. The slope is found to be

$$\frac{du}{dy} = \frac{u(y_2) - u(y_1)}{y_2 - y_1} = \frac{0.055\Omega - 0}{0.055 - 0.050}$$

The viscosity μ for the material is given here as 2.3×10^{-3} N-s/m^2. (Information for a variety of substances is contained in Figure A.1.) Equation (1.7.1) also predicts the shear stress on the solid boundaries. This relationship will yield a stress in SI units if this system of units is used to prescribe the other variables. Therefore, we write

$$\tau = (2.3)(10^{-3})\left(\frac{0.055\Omega}{0.005}\right) = 2.53(10^{-2})\Omega \quad \text{Pa}$$

The torque on the stationary boundary is the product of shear stress, area, and lever arm. The area is the surface area of the cylinder on which the shear stresses act. Consequently,

$$\text{Torque} = [2.53(10^{-2})\Omega][2\pi R_2 L]R_2$$

$$= 2.53(10^{-2})\Omega 2\pi (0.055)^2 (0.2) \quad \text{N-m}$$

Thus the torque varies linearly with Ω (in rad/s). Later on we shall make a better determination of the velocity profile in the annulus on the basis of the constancy of the moment between the two cylinders. (It will not be linear for large differences of radii, but it will vary approximately so for very small differences of the radii, say $(R_2 - R_1)/R_1 < 0.01$.)

This example has been useful because we examined an application of Newton's law of viscosity, which in turn caused us to contemplate the situation associated with the "no slip" condition. If we reflect upon the example, we see that the shear stress τ was a special shear stress; it was a shear stress acting on a plane perpendicular to the y direction. This stress resulted from the variation of the x component of velocity, u, with respect to distances in the y direction. The shear stresses in a fluid are, in general, the result of the variations of all three velocity components with position. Thus, Eq. (1.7.1) can be thought of as a limiting case. A more complete point of view will be presented by Eq. (4.3.2), after the necessary preliminary concepts are introduced. The complete constitutive equation for a Newtonian fluid will be presented in Chapter 16.

Example 1.7.2:

A rod of diameter 1.500 in. slides within a bearing with internal diameter 1.518 in. and 6 inches long at a speed of 2 in/s. A lubricant with the viscosity of castor oil at 80°F can be assumed to fill the clearance space between the journal and its bearing. What force is needed to translate the rod?

Solution Our first step is to convert the data to the pound-foot-second-slug system $(R_1 = (1.5/2)/12$ ft; $R_2 = (1.518/2)/12$ ft; $L = 0.5$ ft; $u = 2/12$ ft/s). The force will be an axial one. Because the radial clearance is 0.00075 ft and the shaft's radius is 0.0625 ft, a ratio of about 1/80, we shall treat the problem as the flow between two parallel plates separated by a distance of 0.00075 ft. The plate is $\frac{6}{12}$ ft long—this is in the direction of motion. The width of the plate is the circumference of the "rolled-out" cylinder, $\pi(1.5/12)$ ft. Equation (1.7.1) gives the shear stress on the shaft, or the moving plate that we are using to simulate the shaft's surface:

$$\tau = \mu \frac{du}{dy} = \mu \frac{0.167 - 0}{0.00075 - 0} = 222\mu \qquad \text{lb/ft}^2$$

(The stress is in lb/ft² in our formulas even though it is often specified in lb/in² elsewhere.) Figure A.1 gives a value of μ at about 80°F as 10^{-2} lb-s/ft² or, equivalently in other units, 10^{-2} slugs/ft-s. Thus

$$\tau = (10^{-2})(222) = 2.22 \text{ lb/ft}^2$$

and
$$F = \tau(A) = 2.22[0.5(0.125\pi)] = 0.436 \text{ lb}$$

This is the force due to the induced shear stresses acting on the lateral surface of the cylinder. It is also the force that must be applied at the end of the shaft: for a constant speed of translation the net force (action and reaction) on the shaft must be zero if there is to be no acceleration of it.

The value of μ is affected only slightly by variations of pressure, as long as the pressures are not extremely high or low. We shall have a need in our work for the ratio of the viscosity, μ (also called the *absolute viscosity*), to the mass density of the material, ρ. This quantity is called the *kinematic viscosity, ν,*[*] and Fig. A.2 gives values of this fluid property for various temperatures. Its value will depend on p through the associated effect on ρ.

1.8 A PREVIEW OF SOME FLOWS THAT WILL BE STUDIED

In this section several fluid flows will be discussed so that the reader can gain an additional inkling of the variety of phenomena that are possible. The examples are also intended to show some of the current views of the physical processes that occur.

One of the most common flow situations is that which occurs inside a pipe. The importance of such flows in engineering is not difficult to appreciate. There were several investigations during the nineteenth century of the influence of pipe size and volume discharge (in m³/s, ft³/s, or gallons per minute—abbreviated *gpm*) on the decrease in pressure that occurs along the length of the conduit. Various facts were

---

[*] ν will be in m²/s in the SI system. The cgs system has ν in units of cm²/s, which is called a *stoke*. Hence, 1 m²/s = 10^4 stokes.

established, but some of them appeared to be contradictory. George Stokes produced a theoretical analysis for pipe flow that related the change of pressure per unit length of pipe in a linear way with the volume per unit time of fluid flowing through the conduit. However, the experimental observations that he had available to him indicated that this linear relationship might not be the case. Stokes examined the critical features of his analysis and could discover no defects in it. Some experiments by Poiseuille would have supported Stokes' conclusion, but Stokes had compared his results with some other experiments, which we now believe were not pertinent to his results. The work of Hagen on pipe flow showed that different correlations must be made for small and large rates of volume discharge.

Toward the latter part of the nineteenth century, Osborne Reynolds provided additional experiments and an explanation for the apparently contradictory outcomes of pipe flow tests. He observed that the speed of the fluid through the conduit was not the only parameter that influenced the magnitude of the pressure gradient along the length of the conduit. The pipe's size, specified by its diameter, and the kinematic viscosity of the fluid were also important in a very special way. The general effects of pipe size had been investigated before, but Reynolds was able to provide a unity to the situation that had not existed previously. Briefly, he said that the product of the pipe diameter and the average fluid speed divided by the fluid's kinematic viscosity is a dimensionless ratio, and would serve as an indicator of the physical processes that were occurring in the flow.

If this ratio, now called the *Reynolds number,* were below a certain critical value, say 2000, the flow in the pipe would be quite orderly; it would appear to be moving along surfaces that are concentric shells. The term *laminar flow* was used by Reynolds to denote this flow regime. If the Reynolds number for the pipe flow is above the critical value, one could expect the fluid motions to be more complicated. There would be motions along the axis of the conduit, just as one would expect, but fluctuating motions in all directions would also be present. These motions would increase the rate of mixing within the conduit so that fluid flowing along the centerline at one instant of time might be found near the wall at a later instant. While we call such motions *turbulent* today, Reynolds termed them *sinuous,* to portray the movement of fluid back and forth across the glass pipe's centerline.

Accompanying the appearance of such random, or turbulent, motions, Reynolds found an increase in the axial pressure gradient. When the Reynolds number was below the critical value, the pressure drop was directly proportional to the average fluid speed (volume rate of flow divided by the conduit's cross-sectional area). For flows that had a Reynolds number above the critical value, the pressure drop per unit length of pipe was nearly proportional to the square of the average velocity. Figure 1.8.1 illustrates this situation. If the average velocity of water in a pipe with an internal diameter of 2.5 cm is 1.2 m/s, the Reynolds number would be greater than 2000, and one could expect the flow to be turbulent at 20°C.

As the pressure in the fluid decreases in the direction of the flow, a location along the pipe could be reached where the pressure in a flowing liquid has been reduced to its vapor pressure. This vapor pressure is a function of the ambient temperature. Many people are familiar with the fact that in mountainous areas, where the air pressure is low, the boiling temperature of water is lower than at sea level. It follows, then, that for sufficiently low pressures, water can boil (i.e., vaporize) around room tempera-

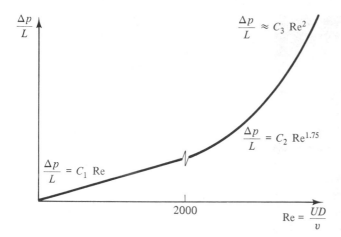

$$\frac{\Delta p}{L} \approx C_3 \, \text{Re}^2$$

$$\frac{\Delta p}{L} = C_2 \, \text{Re}^{1.75}$$

$$\frac{\Delta p}{L} = C_1 \, \text{Re}$$

2000

$$\text{Re} = \frac{UD}{v}$$

Figure 1.8.1 Pressure drop per unit of pipe length versus Reynolds number based upon pipe's diameter.

ture, say 20°C. The process by which a liquid flashes into vapor without large additions of heat is called *cavitation*. (Refer to Fig. 1.1.1.) The large increase in specific volume of the gaseous phase will serve to affect the flow process itself.

There are many places in a system of flowing fluid where the local pressure can be extremely low. This is the case when there are abrupt changes in the path of the fluid particles such as could be encountered in the bend of a pipe or at the inlet of a pump. The change of flow direction is associated with an acceleration of the fluid. The force to achieve this acceleration can come from a lowering of the pressure in the vicinity of the accelerating fluid. If cavitation occurs to a large extent at the inlet of a pump, the pump cannot deliver the required amount of liquid in a given time. The reappearance of the liquid phase after cavitation occurs is accompanied by the emission of light and sound, the latter of which is easily detected. In addition, the material surfaces that are adjacent to the region where the vapor is going back into solution suffer noticeable pitting and erosion. This cavitation damage can significantly shorten the useful life of pumps, hydraulic turbines that are used for power generation, and ships' propellers. Figure 1.8.2 shows the formation of a vapor trail at a propeller's

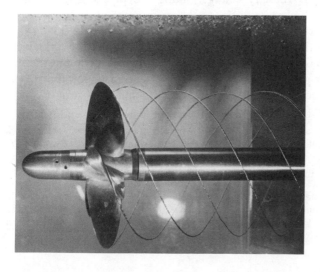

Figure 1.8.2 Five-blade high-speed propeller (operating below its design speed) that shows the common tip vortex cavitation. A three-blade high-speed propeller (operating at its design speed) is also shown. Propeller is almost *fully cavitating* (i.e., vapor bubbles do not collapse on blades). The extensive region of vapor is called *cloud cavitation*. (Courtesy of Prof. Claus Kruppa, Technical University of Berlin, F.R.G.)

Figure 1.8.3 Cavitation damage can occur on low-pressure sides of vanes of pump. At low levels of cavitation, operation of pump may not be seriously impaired, but constant erosion process that can occur on some materials can lead to significant deterioration of efficiency, structural weakness, and rotational unbalance. (Courtesy of Prof. Helmut Siekmann, Technical University of Berlin, F.R.G.)

tips. Often the surfaces of such machinery components are clad with materials that are resistant to attack by the processes of cavitation. This can be an expensive procedure, but it is justified by the high cost of subsequent shutdowns for maintenance.* Figure 1.8.3 shows the effects of prolonged operation of a turbomachine under cavitation conditions.

When gases are flowing along a pipeline, the associated pressure drop can be coupled with a decrease in the density of the gas. We shall see in Chap. 3 that for the flow of a gas in a conduit of constant diameter, if the density decreases, the velocity of the gas must increase. If the pipe is long enough, one finds that the velocity of the moving gas will increase to the speed of sound of the particular gas. This sonic speed, *c,* will be given in Chap. 14 as $c^2 = kRT$, in which $k = c_p/c_v$. (The ratio of the actual speed of the gas to the sound speed in the medium is called the *Mach number.*) In that chapter the reasons why the gas velocity changes along the pipe's length will be discussed also.

The interrelationships among velocity, density, and pressures in a flowing gas give rise to a number of interesting phenomena. In contrast to a liquid, in which the density is constant and in which the speed is reduced when it flows in a conduit that has an ever-increasing diameter, a gas may enter a diverging section under the proper conditions and increase in velocity. Speeds in excess of the speed of sound can be

---

* The release of dissolved gases as the liquid's pressure is reduced—something that precedes the occurrence of cavitation—can also be initiated by creating a low pressure as the result of an acceleration. If a small can of vacuum-packed fruit or vegetable juice, say, 6 fluid ounces, is held in one hand and given a sudden blow with the other on its end, an unexpected sound occurs. This is not due to the actual contact of the hand with the container. Because the liquid is vacuum-packed, it is at a pressure less than atmospheric. The sudden blow from the hand accelerates the can and the liquid. In order for the necessary force to be induced on one end of the contained liquid to accelerate it, the pressure is reduced at the end of the can that is opposite to the one that was struck. This lowering of the pressure causes some evolution of gas, which dissolves again immediately after the experiment. The sound that is heard is the result of the release and subsequent implosion of this gas.

achieved. Such a *supersonic* flow may proceed into regions where a sudden change of pressure occurs. The surface across which this jump in pressure occurs is called a *shock wave*. The density and temperature also change markedly. After the shock wave, the speed of the gas is less than the local speed of sound; it is *subsonic*.

Another type of coupling among temperature, pressure, and velocity occurs when a carbon dioxide fire extinguisher is used. Initially, the gas is at rest inside the bottle. The gas is under high pressure at the ambient, or room, temperature. When the gas is released, its temperature is reduced considerably, so that solid particles of carbon dioxide form at the mouth of the spray nozzle. We shall discuss a similar situation further in Chap. 14. Figure 1.8.4 shows a complex case of supersonic nozzle flow. The strong variations in density are made evident by utilizing the changes of refractive index that are associated therewith.

The speed of sound has been mentioned previously. Our ability to calculate its value in gases is a consequence of applying some of the observations made with water. The transmission of sound is modeled by sound waves moving with associated high- and low-pressure regions that mark the passage of the wave. Such a process is analogous to the way in which surface waves propagate in shallow water channels. The increase in pressure that occurs in a gas is similar to the increase in water depth as a

Figure 1.8.4 Reservoir at left (shown with a circular border because of form of lens used to record picture) is connected to a converging nozzle that merges smoothly with a diverging passage. Septum that contained high-pressure gas in reservoir has been ruptured, and this gas moves toward exit of convergent-divergent nozzle at right. Photograph shows last stages of transient flow. Transparent test section has been illuminated from behind with parallel beam of light. The different densities of the gas refract the light differently, causing some interference between rays that causes the different shades of gray in this *shadowgraph*. Darkest regions correspond to locations where changes in density gradient are large. Near end of nozzle, expanding shock wave (i.e., a zone across which there is a sharp change of pressure and density) is sweeping ahead of it the gas that originally was in nozzle. Several shock waves have already passed through this gas, and its state, judging by the mottled appearance of the region, appears to be quite agitated. (Courtesy of Dr. H.-O. Amann, Fraunhofer Institute for High-Speed Dynamics, Ernst Mach Institute, Freiburg, F.R.G.)

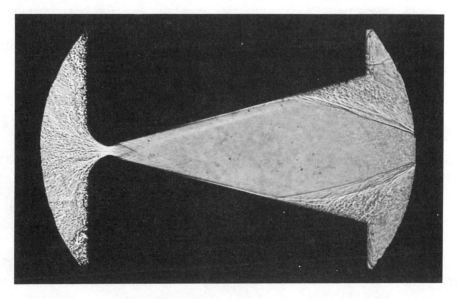

wave arrives at a place. When the crest of the wave passes, it is followed by the wave trough. This is equivalent to the low pressure (rarefaction) portion of a sound wave.

The generation of water waves requires the input of energy into the system by a wave generator. Near such a device the potential energy of the liquid is increased by the raising and lowering of the normally horizontal interface between the liquid and the gas, or other liquid. Since these waves usually disappear at large distances from the point of generation, the power used to create them is dissipated by viscosity. A ship can serve as the wave generator, as is illustrated by Fig. 1.8.5. The power needed to produce the waves is the product of the ship's speed and a quantity called *wave drag*. It then becomes reasonable to consider reducing the power that the ship must provide to move by reducing the waves that form at the bow and stern. This could yield a greater fuel economy or a higher ship speed from the power plant. Such economic considerations led to the development of the bulbous bows shown in Fig. 1.8.6 that markedly reduced the wave drag on large ships.

To be sure, some of the power that is developed by the ship's motors must be expended to overcome the frictional resistance of the hull as it glides through the water. In some respects, the sides of a ship can be thought of as a portion of a very large plate which has fluid flowing over it. The viscous drag on such a plate will vary with the speed of the flow in different ways, depending on whether the speed is slow or fast. Again, the Reynolds number will be the important parameter, and its magnitude will serve as the gauge to tell us if we are in the "low-speed" or "high-speed" regime.

In the case of the flow over a flat plate, the length of the plate is used in the Reynolds number instead of the diameter, which is used when one is dealing with pipes. If the Reynolds number for the flow over a plate becomes quite large, say, greater than 10^4, one observes that the portion of the fluid that is influenced by the motion of the plate becomes quite small. It appears that there is a variation of velocity,

Figure 1.8.5 Two sister ships demonstrate effect on wave generation of altering form of bow. Ship that produces fewer waves should waste less energy. (Courtesy of Prof. Takao Inui, University of Tokyo, SNAME Transactions, Vol. 70, p. 282 (1962), by permission)

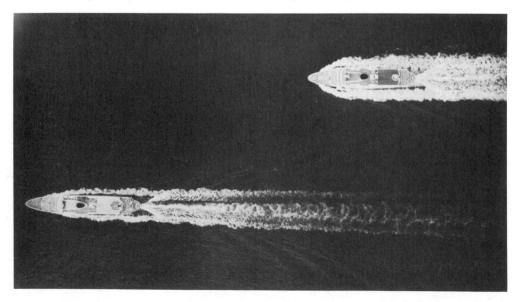

| (a) | (b) |

Figure 1.8.6 Form of the ship's bows used in demonstration pictured in Fig. 1.8.5. Bulbous bow, the shape of which was developed in laboratory tests, reduces formation of bow waves, a principal source for inefficient operation. (Courtesy of Prof. Takao Inui, University of Tokyo)

with resultant viscous shear stresses, in only a thin layer near the surface, or boundary, of the plate. This *boundary layer,* where the effects of the no-slip condition greatly affect the velocity profile, grows in thickness along the plate in the flow direction, yet remains surprisingly thin. Figure 1.8.7 shows a portion of a flow that is retarded by a stationary plane surface. This is the boundary layer. It is a small portion of the total flow region.

One significant way of reducing the frictional resistance of a hull is to allow it to rise out of the water while the vessel is moving at high speed. A winglike hydrofoil remains submerged to support the hull's weight and cargo through the generation of the necessary vertical component of force on the foil. Because of the size and shape of the foil, the drag force is much less than that which would be on the submerged hull. Such high-speed performance will create the possibility that the water flowing over the foil will cause the pressure on the foil to be reduced to the ambient vapor pressure of the flowing water. Cavitation would ensue. The hydrofoil sections are designed so that they produce the necessary lifting force while they are completely within a cloud of water vapor. The vapor phase reverts to the liquid phase aft of the foil. The cavitation damage that occurs as the vapor bubbles collapse is avoided by this mode of *supercavitating* operation. (The nature of bubble collapse is shown in Figure 1.8.8.) The viscous drag on hydrofoils can be made quite low, and raising the hull out of the water significantly reduces the wave drag.

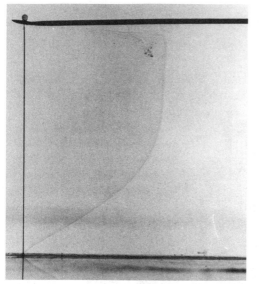

Figure 1.8.7 Darkening of liquid that has been swept past vertical (tellurium) wire is due to presence of metallic ions that went into solution when an electric pulse was applied to the wire. Coloring of water produces a display of velocity distribution near rigid boundary along bottom. From tests using many other techniques and criteria—often, pressure drop in a round tube or torque needed to rotate a cylinder containing a liquid in an annulus—it was concluded around 1900 that the best theoretical assumption for a continuum was that a fluid did not move relative to a rigid boundary. The distance from the boundary within which the fluid motion is influenced by the presence of the boundary was called the *boundary-layer thickness* by Ludwig Prandtl in 1904. (Courtesy of Dr.-Ing. M. Strunz, Stuttgart University, F.R.G.)

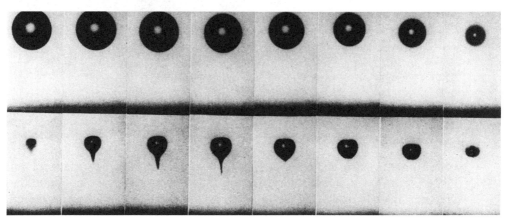

Figure 1.8.8 When vapor bubbles collapse in the presence of a rigid wall, final stages are accompanied by formation of jetlike, swirling flow on side of bubble nearest wall. Calculations based upon the implosion of a spherical cavity indicate that extremely high pressures are generated. While such high pressures may not be reached with an actual bubble, which has an evolution of the surface nearest the wall and a corresponding involution of the opposite surface, the jets that are produced by this change of bubble shape are estimated to have speeds in the range of 100–200 m/s. The momentum of these jets that are directed toward the wall could be responsible for the deterioration of the surface of the wall under cavitation conditions. (Courtesy of Prof. W. Lauterborn, Göttingen University)

The shape of a wing for low-speed airplanes is essentially that of a two-dimensional body if its span is sufficiently long. The amount of drag force that is exerted on it is produced in two ways. One way will be the magnitude of the shear stresses. These stresses act upon a surface that has a prescribed size; altering this surface may change the flow and thereby reduce the shear stresses. If, in reshaping the surface, we also increase its area, we may have decreased the stresses—force per unit area—but actually increased the force components, which are the sum of the products of the local shear stresses and the appropriate elements of area. One finds that the normal stresses (pressures) on the surface are also changed with each change of body shape. These pressures act on surface elements that have various orientations with the direction of flow, with the result that different resistance force components (in the direction of flow) are induced. Hence the process of reducing the drag force includes not only the shear stresses and the surface area, but also the normal stresses and the orientations of the surface elements with the flow direction.

The situation is partially characterized by Fig. 1.8.9. Two plates are illustrated. The plate in Fig. 1.8.9a is aligned with the flow, so that the only stresses that act on the plate in the flow direction are the shear stresses. Such shear stresses are most intense near the leading edge of the plate. The plate in Fig. 1.8.9b is situated perpendicular to the flow. The stresses that act in the flow direction and contribute to the drag are the normal stresses. The pressure on the upstream side of the plate is highest at the plane of symmetry. Here the vertical velocity must be zero, since it is neither up nor down on the plane of symmetry. Because the flow cannot penetrate the plate, the horizontal component of velocity is also zero. Places where the velocity is zero are called the *stagnation points,* and the associated high pressure is called the *stagnation pressure.* As one moves away from the stagnation point toward the edge of the plate, the pressure decreases. If the fluid is moving at a sufficiently high speed at the edge of the plate, it cannot suddenly change its direction and flow along the back of the plate. The location where the fluid first fails to flow along the body in the same direction as the upstream flow is called the *separation point* or *line.* In Figure 1.8.9b the separation line is the top and bottom edges of the vertical plate. On bluff bodies such as the sphere shown in Figs. 1.1.3 and 1.1.4, the position of the separation line usually depends on the flow Reynolds number. But on bodies with sharp edges, e.g., a disk, the separation line remains fixed at the edge, because of the severe directional changes of the surface that exist there. The separated region aft of the body has a pressure within it that is very nearly equal to the low pressure at the edge of the plate where separation occurred. Thus there is a distribution of relatively low pressure on the back side. The associated resultant force on the plate is due to the normal stresses alone.

Figure 1.8.9 Sources of drag on flat plates. (a) Plate aligned with flow: drag due to shear stresses. Streamlines are slightly divergent because plate is present. (b) Plate normal to flow: drag due to normal stresses. Streamlines curve to accommodate plate.

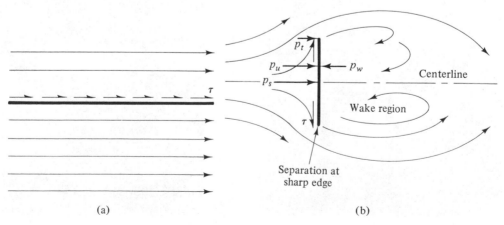

(a)   (b)

Most flows past submerged bodies are combinations of the two extreme cases of flat plates just mentioned. Reconsider now the flow past a sphere that is illustrated in Fig. 1.1.3. There is a region of separation behind the sphere where the pressure is low. One could calculate the drag on the sphere if the shear stresses and pressures were prescribed at each place on the surface. A similar situation is depicted in Fig. 1.1.4. Here the flow conditions have been changed slightly so that the extent of the separation region has been reduced. This can be accomplished by having the velocity of the fluid in Fig. 1.1.4 greater than that in Fig. 1.1.3 or by running both at the same speed and roughening the surface of the second sphere.

While the laminar or turbulent state of the boundary layer can significantly alter the drag on bodies subjected to conditions for which the Reynolds number is high, entirely different effects occur in low–Reynolds number flows. Then altering the body's shape by streamlining, for example, serves to increase the drag in an unexpected way. Figure 1.8.10a shows the flow pattern for a flow at a low Reynolds number. The absence of a significant amount of separation of the flow from the body is obvious in this figure.

In the course of the next chapters, the explanations of the mechanics of many flow cases are presented. The process will include citing the various assumptions, observations, and laws of physics that are inherent in the analyses. One of the purposes of these presentations is to give the reader the background necessary to examine new situations successfully. A routine or perfunctory analysis may not include an important

Figure 1.8.10(a) (left)   Here, streaks from reflecting particles that flow past a body appear to form parts of streamlines. Body is actually moving, and camera is moving at same speed. (In effect, body is at rest with respect to camera, and nearly stationary particles at far right and far left now appear to be moving from left to right.) As body moves along tube, some particles are accelerated to move past it and others decelerate after moving past constriction. With body and camera moving as a unit, an unsteady flow is made to appear steady.

Figure 1.8.10(b) (right)   Camera was stationary and had exposure setting that caused body to be slightly blurred, but one can visualize how fluid particles are forced out of the way by displacement of advancing body at one end, and how they are drawn toward receding body at other end. Path lines shown here are reminiscent of patterns of iron filings that record lines of magnetic field of a bar magnet. (Courtesy of Madeline Coutanceau, University of Poitiers, France, from J. Fluid Mech., Vol. 107, pp. 339 –373 (1981) with Patrick Thizon, by permission)

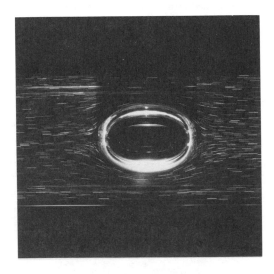 

parameter or phenomenon. The results of such an incomplete analysis could be an incorrect numerical result or an erroneous description of the flow that in reality occurs.

The variety in fluid mechanics is what may give the beginner some difficulty or anxiety. It is also the source of challenge and satisfaction for those who understand the basic principles, which in fact are not abstruse. For example, one notes that two dissimilar photographs of the same flow are given in Fig. 1.8.10a and b. A recognition of the fact that one picture was taken while moving with the object, while the other was taken with a stationary camera, allows the viewer to account for their differences in appearance.

EXERCISES

1.2.1(a) How many *pascals* correspond to 1 *psi*?

(b) If 1 bar corresponds to 0.986923 atmospheres (750 mm of mercury), how many millimeters of mercury are there in 1 atmosphere? How many millibars are in 1 atmosphere?

(c) What is the equivalent of 1 bar in kPa?

1.4.1(a) What is $\hat{R}$ for air in English units? ($g = 32.2$ ft/s^2.)

(b) Write in Table A.6 the value for R for air in SI units (m^2/s^2-K). What is the value of $\hat{R}$?

(c) What is the value of $\hat{R}$ in English units on the moon, where the gravitational constant is one-sixth that of the earth's? What is the weight per cubic foot of air on the moon? (Use 100°F and a pressure of 50 psia.)

1.5.1(a) What volume do 200 mercury atoms fill? Use the specific volume of mercury to deduce its density.

(b) What has a density of 1000 kg/m^3? One mole of this substance in the gaseous state has a mass of 0.018 kg. How many moles of this substance are contained in 1 m^3? How many molecules are in 1 m^3? What volume contains 100 molecules? What is the length of a side of a cube containing this volume?

(c) Air has a density of 1.225 kg/m^3 at 1 atmosphere and 15°C. What is the volume of the region that contains 100 molecules of air under these conditions? What would be the length of a side of a cube of this volume?

1.6.1(a) What would have been the *maximum capillary rise* to be expected if methyl alcohol were in the tube?

(b) What must be the radius of a tube to *ensure* that the rise of water in it will be 1 mm or less? What is the radius for a maximum of 0.1 mm or less?

(c) Two vertical glass plates are separated by a dis-

tance of 0.5 mm. They are placed vertically into a dish of water. What is the maximum height to which the water will rise between the plates? (This problem will require a derivation. Let z be the vertical axis and y the horizontal axis *in the plane of your paper*. Use a unit distance of 1 m in the x direction.)

1.6.2(a) What would be the change in volume of a 4-liter container of water that underwent a 2-atmosphere change of pressure? What side length, in mm, would correspond to a cube for this volume change? (Assume that the container did not deform during this pressure change because the container was a thick-walled cylinder and the pressure was applied by a massive piston.)

(b) The isothermal compressibility of methanol at 5000 atmospheres, in percent $\Delta V/V$ for 1 atmosphere of change, is given as 0.002. What is the corresponding isothermal bulk modulus?

(c) If the container was a thick-walled cylinder with an initial volume of 200 liters that was filled with water at atmospheric pressure, what would be the change of volume of the water if it had been compressed by the admission of methane gas at a constant pressure of 100 atmospheres? If the temperature of the methane is 20°C during the compression, what was the mass of the gas that entered the cylinder? What would be the volume of the gas that entered if it were at 20°C and 1 atmosphere?

(d) The speed of sound in water and that in mercury are measured as 1498 and 1450 m/s, respectively. Compare the isothermal and adiabatic bulk moduli for these liquids. Is the latter greater than the former?

1.7.1(a) Determine the torque needed to drive the viscometer at 100 revolutions in 11.5 s.

(b) A different internal cylinder is to be employed in the apparatus. Its outside diameter is 90 mm. How should the torque on it vary with the angular velocity?

(c) The machine is fitted with a different rotating cylinder having an inside diameter of 105 mm. What is the expected torque on the inside cylinder ($D_{inside} = 100$ mm.)? What is the torque on the outside cylinder? Should these two torques be the same? Explain any anomalies that have developed.

(d) A different liquid is put into the original viscometer and a torque of 0.12 N-m is needed to turn the outer cylinder at 100 revolutions in 86.2 s. What is the viscosity of the liquid in the test chamber? What torque would you expect for 100 revolutions in 43.1 s?

1.7.2(a) What is the force per square inch (i.e., per unit area) that must be applied to the circular cross section of the end of the rod?

(b) What would be the approximate temperature of the lubricant (assume castor oil) if the force to move the rod were one-third of that in the example?

(c) If the system has a vertical axis and the rod descends at the rate of 0.25 in/s due to its own weight, what would be the radial clearance between the rod and the cylindrical sleeve that guides it? Assume the rod is 2 ft long and made of aluminum and has a constant diameter of 1.500 in. Assume the viscosity of the example.

(d) Refer to the previous exercise and prepare a graph that could be used to measure the diameter of bearings. Plot the rate of rod descent versus the inside diameter of the bearing. Assume an operating temperature of 80°F.

PROBLEMS

1.1 A container for a rocket propellent weighs 950 N at the launching site, where $g = 9.8$ m/s². What would it weigh on Venus, where $g = 8.9$ m/s²? What would be its mass on Venus?

1.2 An experimental package weighing 150 N is supported by springs that deflect 2 mm under this weight at the rocket launching site, where $g = 9.8$ m/s². If this package and its support springs were put into operation on the earth's moon ($g = 1.61$ m/s²), what would be the expected spring deflection?

1.3 Calculate the mass of the experimental package referred to in Prob. 1.2 at the time when it is launched from the earth and when it arrives on the moon.

1.4 A particular seismic experiment is being planned for execution on the lunar surface ($g = 1.61$ m/s²). During the testing program on the earth ($g = 9.8$ m/s²), it was found that a weight of 78 N was required to balance the apparatus so that it remained horizontal. What volume of steel (SG = 7.8) must be added in order that the apparatus remain horizontal on the moon?

1.5 Calculate the density of air at 100 kPa and 20°C. What fraction of the density of water, at the same conditions, is this?

1.6 What are the volume and density of 1 mole of helium at 100°F and 20 psia? Determine the density of nitrogen under the same conditions. Is the ratio of the densities of these two gases proportional to the ratio of their molecular weights?

1.7 A mole of CO_2 is in a container at 20°C and 50 kPa. What is the density and weight of the gas at a location on the equator where $g = 9.78$ m/s²? What is its density and weight at the north pole where $g = 9.83$ m/s²?

1.8 A mole of CO_2 is in a container at 20°C and 50 kPa. What is the volume of the container? How many molecules are in the container? What is the mass of the CO_2 under these conditions? What is the density?

1.9 What is the volume of 1 mole of nitrogen at 100 kPa and 20°C? How many molecules of nitrogen exist in a volume of (0.1 mm)³? Is this number sufficient to give a meaningful statistical sample? Justify your response by using your knowledge of statistics or data about the size of recent polls or surveys.

1.10 If a droplet of oil rises in water, one can argue that the pressure in the drop decreases in such a way that it is nearly always the same as that of the water in which it is rising. Give substance to this argument.

1.11 A gas bubble rises in a vertical column used in a chemical process. Assume that the process occurs at constant temperature, and employ the ideal gas law for the contents of the bubble and the fact that $\Delta p = \sigma(2/R)$ to show that the bubble's radius cannot remain constant. Does the radius increase or decrease with elevation? Explain.

1.12 A needle with diameter d is supported on water by surface tension. Determine the maximum size that can be supported in this fashion by assuming that the limiting configuration has the needle wetted over its entire lower half and the water makes contact with the needle in the manner shown in Fig. P.1.12. (The contact angle is such that the air-water interface intersects the needle tangentially.)

1.13 Two gas bubbles, one with a diameter of 1 mm and the other of 2 mm, exist at the same elevation in a tank of water. In which bubble is the gas pressure greater? How much greater?

1.14 Using Fig. 1.6.3 show that the relationship between the capillary rise in a tube and the diameter is

$$h = \frac{4\sigma}{\gamma d} \cos \alpha$$

(The contact angle will depend upon the conditions at the interface, including the effects of contaminating substances, temperature, and similar factors, as well as the tube diameter. $\cos \alpha$ varies between $\frac{1}{6}$ and $\frac{3}{4}$ as the diameter decreases from 20 mm to 5 mm.)

1.15 The expression given in Prob. 1.14 can be used to estimate the error that surface tension can contribute to readings of certain instruments. For example, pressure can be measured by observing the height of the column of liquid that the pressure can support. The instrument that measures pressure this way is called a manometer. Find the allowable vari-

ation of diameter in a glass tube with a nominal diameter of 10 mm so that the variation of the capillary effect is no greater than 0.1 mm along the length of the tube when it contains water in contact with air. Assume 20°C and $\cos \alpha = \frac{2}{3}$.

1.16 Water at 68°F flows in a horizontal pipe with an internal diameter of 6 in. at such a rate (i.e., volume rate of flow) that the pressure decreases 1 pound per square inch (psi) in every 100 ft of pipe. At what distance from a point where the pressure is 75 psi absolute (psia) will the pressure in the pipe equal the vapor pressure? What do you expect occurs at this location?

1.17 What is the percent change of volume of the water at the end of 100 ft for the situation described in Prob. 1.17?

1.18 If air is flowing in a horizontal pipe with an internal diameter of 15 cm, it is possible that the pressure drops at the rate of 0.25 kPa per meter of length. As we shall learn when compressible flows are studied, the ratio between pressure changes and density changes for air can be expressed as $1.4(R)(T \text{ in K})$ under certain conditions that include the absence of heat transfer to the gas. If the pressure in the conduit is 200 kPa at the point where the temperature is 25°C, calculate the density there. What is the change of density per meter in that region for the pressure gradient given? What is the percent change in the density? Give a reason why the air could be considered to be effectively incompressible in this case when considering a conduit 100 m long.

1.19 If the flow of air is isothermal, then the ratio p/ρ is a constant along the conduit. (Note that the equation of state requires that $p/\rho = RT$ at each point for an ideal gas. Since T is assumed to be constant along the pipeline, so must p/ρ.) Starting with $p/\rho = $ constant, show that $dp/d\rho$ is equal to RT. Is the compressibility of air greater or less when the flow

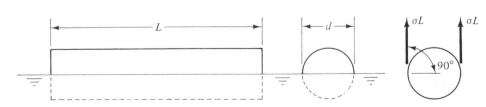

Figure P.1.12

is isothermal in comparison with the situation treated in Prob. 1.18?

1.20 The speed of sound in air is closely given by the square root of the expression $dp/d\rho$ given in Prob. 1.18, whereas the speed of sound in liquids or solids is approximated by $(K/\rho)^{0.5}$. Compare the speeds of sound in air, water, steel, and brass.

1.21 When one descends 1 m in water, the pressure increases 9.8 kPa.* At what depth would the density of the water have changed 2 percent?

1.22 It is intended to slide a plate past a stationary plate so that a gap of 1 mm is maintained. The moving plate is 500 mm wide and 800 mm long. Castor oil at 60°C fills the gap between the plates. At a sliding speed of 0.1 m/s, what is the force that must be exerted to pull the moving plate? Is there a force on the stationary plate? Explain.

1.23 A plate that is 400 mm wide and 1100 mm long is supported by an oil film of constant thickness that is on an inclined plane that has an angle of 10° with the horizontal. What is the normal stress on the film due to the presence of the plate if the plate is made of steel (SG = 7.9) and has a uniform thickness of 5 mm? What is the shear stress that is induced in the fluid when the plate moves at constant speed?

1.24 If a plate similar to the one described in Prob. 1.23 but weighing 1800 N were located on the 10° incline above an oil film having a viscosity of 0.04 kg/m-s, what would be the speed of its descent if the film were 0.2 mm thick?

1.25 A plate 1 mm in thickness, 1000 mm in length, and 400 mm in width is falling vertically in a slot that is 10 mm wide in such a way that one surface of the plate is maintained at a distance of 4 mm from one of the walls of the slot. The plate's speed is constant at 0.1 m/s, and it is made of steel. Calculate the viscosity of the oil that fills the space on both sides of the plate.

1.26 The apparatus described in Prob. 1.25 is

*This rate of pressure rise will depend on the density of the liquid, in accordance with the principles treated in Chap. 2.

altered so that the plate moves horizontally, with the 4-mm gap below the plate. In the lower gap there is a fluid with a viscosity of 10^{-3} N-s/m², whereas in the upper gap the fluid has a viscosity of $1.1(10^{-1})$ kg/m-s. What is the shear force on the plate if it translates at 0.1 m/s?

1.27 Two horizontal plates are separated by a distance of 0.09 in. The upper plate moves toward the right at 15 ft/s, while the lower one moves in the same direction at 10 ft/s. What is the shear stress of the upper plate if a fluid with a viscosity of $2(10^{-4})$ slug/ft-s (or lb-s/ft²) is between the plates? What would be the stress on the lower plate if it were moving to the left at 10 ft/s while the upper plate moved as before?

1.28 A film of water 0.25 in. deep is covered by a plate that moves at 5 ft/s. Above this plate there is a horizontal space that has a height of 1 in. and is filled with air. What is the total force on the plate? What percent of this force is due to the action of the air? How would this percent change if the height of the air layer were increased to 12 in., and then to 10 ft? What can you conclude about the contribution of the air to the total shear force in the plate?

1.29 A horizontal layer of water has a depth of 10 mm. Above it there is a layer of air which is bounded above by a horizontally moving plate that is 10 mm from the water's surface. This plate is moving at 2 m/s. Determine the velocity profile in the air and in the water. Assume 20°C. (Note that at the interface, the velocity and the shear stress in the two fluids must be continuous. What would be the result if one permitted the shear stress in the liquid, at the interface, to be greater than that in the gas? Disregard surface tension.)

1.30 For the dimensions given in Prob. 1.29, let the stationary surface be located above the air, and let the water rest on a plate that is moving at 0.3 m/s. What is the velocity profile in the air and in the water? Has the air layer served to appreciably affect the velocity profile in the water? Do the results of this problem support the assumption that is used at many liquid-gas interfaces where surface tension effects are absent that $\tau \approx 0$?

1.31 A steel shaft with a diameter of 60 mm is 200 mm long. It is supported vertically by two sleeve bearings that are 180 mm long. (The bearing separation distance is unimportant.) The clearance between shaft and bearing is 0.05 mm. The bearings are lubricated with castor oil at 40°C. Determine the speed at which the shaft moves due to its own weight.

1.32 One way to measure the viscosity of a fluid is to place a cylindrical bob concentrically inside a cylindrical cup with the liquid specimen. The cup is rotated at a known speed, and the torque on the bob, which is generated through the shear stress that results from the flow in the annulus, is measured. Determine the viscosity of the liquid in such a viscometer if the length of the bob is 4 in. and its diameter is 2 in., the gap is 0.125 in., the cup rotates at 30 rpm, and the ambient temperature is 68°F. The measured torque is 0.5 lb-ft. Neglect any effects at the bottom of the cup, which is vertical.

1.33 A horizontal circular shaft moves axially inside a sleeve bearing so that a clearance of 0.003 in. is maintained. An air supply, and appropriate bearing design, keep the shaft in this position. Calculate the force on the shaft to move it at the rate of 2 ft/s. (The shaft has a diameter of 4 in., and the bearing is 12 in. long. It operates at 80°F on air within the annulus between the journal and the bearing.)

1.34 If the shaft in Prob. 1.33 is rotating at 1200 rpm, what is the horsepower necessary to overcome bearing friction alone? If water had been used as the bearing lubricant instead of air, what would be the horsepower required to simply rotate the shaft at 1200 rpm?

1.35 Grease is a complex substance which has a latticelike structure of a soap containing an oil within its interstices. At low rates of shear, the structure collapses and the oil serves as a lubricant. The fact that grease has a resistance to deformation that depends upon the rate at which the deformation is applied can be appreciated by noting that in a sleeve bearing with an internal diameter of 1.5 in. and a gap between it and the shaft of 0.006 in., grease could be used as the lubricant. At room temperature the absolute viscosity is approximately that of castor oil when the rate of deformation is extremely small (i.e., when trying to pour it out of a container). If this viscosity were to remain constant when the shaft is turning at 10 revolutions per second, what would be the horsepower necessary per inch of bearing length just to keep the shaft rotating? (This rate of rotation would correspond roughly to that necessary to move an automobile at highway speeds. Accordingly, the horsepower calculated would be the friction horsepower for one wheel bearing if the grease did not change its viscosity with the rate of deformation.)

1.36 A four-wheel vehicle weighing 2000 lb is designed to have its weight equally distributed on its four tires. During a stopping maneuver, one of the tires "hydroplanes." This results when a thin film of water—0.004 in., in this case—comes between the tire and the road surface. Determine the force on the tire in the direction of motion. Assume that the tire is not rotating—the brake is "locked"—and the surface in contact with the water is a plane that is 11 in. long and 6 in. wide. At the time of braking, the vehicle speed was 30 mph (44 ft/s) and the viscosity of the water was $2(10^{-5})$ slug/ft-s. Use a coefficient of static friction of 0.8 to estimate the shearing force at the tire-road interface for a "locked" wheel under dry conditions.

2

Pressure Distributions in Fluids without Relative Motions

2.1 INTRODUCTION

The topics in this chapter are principally concerned with an analysis of the conditions within a fluid that is at rest. When a body of fluid is motionless, there are no relative motions within it, so that Newton's law of viscosity, Eq. (1.7.1), would require an absence of shear stresses within the fluid body. Consequently, the only stresses that could occur within the fluid are normal stresses, which must be equal to the thermodynamic pressure if no relative motions exist. Thus, the absence of any motions within a fluid simplifies an examination of it. It becomes the study of fluid statics, for which pressure is the only important stress within the fluid. (When one is concerned with liquids, one also has to account for surface tension effects at liquid-liquid and liquid-gas interfaces.)

Not only is the balance of forces within a fluid that is without relative motions relatively simple, but the other laws of nature that need to be considered in a more general case are implicitly or easily satisfied. Items like mass conservation and energy relationships are unnecessary to consider at length. It is for these reasons that the topic of this chapter usually serves to introduce the fluid mechanics tyro into the beginning aspects of the subject.

Rigid-body motions of a fluid will also be considered at the end of this chapter. Once again, no relative motions are included, so that no shear stresses will be present in the fluid. In these cases, however, an analysis of the pressure distribution includes the acceleration that the fluid is undergoing. Such cases will be relatively simple extensions of what has been done in the previous sections of the chapter.

Once the pressure distribution within a fluid is known, it is possible to determine the forces that are exerted on the boundary of a fluid whose shape is known. This is an important application of the material in this chapter. The work in the next few sections lays the basis for such calculations.

2.2 HYDROSTATIC PRESSURE VARIATION

Our first goal will be to predict the differences in pressure that can occur between two locations in a static fluid. However, we must first inquire into what is meant by the phrase *pressure at a point* and how this pressure itself might depend upon the coordinate directions. Figure 2.2.1 helps us to focus our thoughts on this question. If one wants to know the pressure at the corner of a container filled with a liquid, how are the pressures measured at O, O', and O'' related? Would these measurements make it possible to infer the pressure reading that was recorded on a nonorthogonal plane at a location near the corner?

The situation that we wish to consider is shown in Fig. 2.2.2. The coordinate axes are located at an arbitrary point in the fluid, and the plane ABC has some arbitrary orientation with respect to the coordinate axes. These have no prescribed direction except that the z axis has been chosen to be positive upward for convenience. Thus gravity acts in the negative z direction. We assume that the value of the pressure at point P on the infinitesimal plane ABC is known. Because of the minute size of ABC,

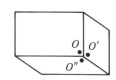

Figure 2.2.1 Pressure at corner of a box in a gravity field.

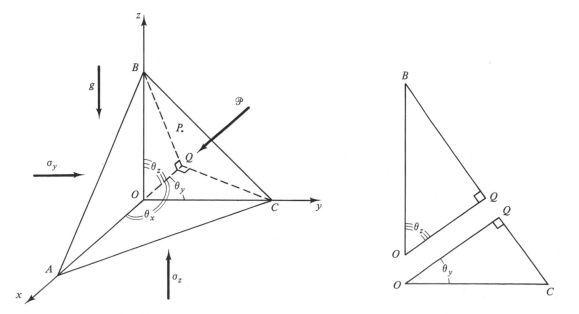

Figure 2.2.2 Pressure on tetrahedron of fluid in a gravity field. Normal stress at Q equals pressure in a static fluid.

this pressure can be assumed to be constant over that surface. Pressure is a stress, and since pressure always acts normal to a surface, it is referred to as a *normal* stress. (If the fluid had relative motions between the particles, the total normal stress would be influenced by those motions.) We now inquire about the value of the normal stress on the three coordinate planes, *BOC*, *AOB*, and *AOC*, that are near the point *P* but have different orientations in space than plane *ABC*. We shall show that the normal stress on each of these planes is the same. This common value is the negative of the pressure for a fluid at rest. (The negative sign arises from the convention of having a positive normal stress if it acts in the direction of the normal to the surface. The pressure on the tetrahedron acts against the surfaces of the fluid within the tetrahedron, so that it acts in the $-\mathbf{n}$ direction.)

To demonstrate the independence of the static pressure from direction, let us apply Newton's laws of motion to the mass of the body within the tetrahedron. This small body has infinitesimal forces acting on each of its surfaces. Because the infinitesimal body of fluid is not accelerating, the sum of all the appropriate infinitesimal force components, including a body force due to the gravitational field, is zero for each of the coordinate directions. We begin by considering the force on the plane *ABC* and its component in the positive *z* direction. The force is due to a stress $\mathcal{P}$ that acts toward the material within the tetrahedron. We write this infinitesimal force as $(dF)_{ABC}$, so that

$$(dF)_{ABC} = \mathcal{P}\mathcal{A}_{ABC} \qquad (2.2.1)$$

in which the script letter $\mathcal{A}$ denotes the area of a plane surface. For the configuration of Fig. 2.2.2, the positive *z* component of this force is

$$(dF_z)_{ABC} = -\mathcal{P}\mathcal{A}_{ABC} \cos \theta_z \qquad (2.2.2)$$

in which θ_z is the angle between the line *OQ*, which was constructed, and *OB*. The former line was drawn perpendicular to plane *ABC* through point *O*, so that it is parallel to the line of action of $\mathcal{P}$. The latter line has a length of *dz*. If σ_z is used to designate the as yet unknown normal stress that is acting on plane *AOC*, we recognize that it acts on the material within the tetrahedron in the positive *z* direction if σ_z also presses against the surface *AOC* from outside the tetrahedron. Hence the positive *z* component of the resulting infinitesimal force on this surface is $\sigma_z \mathcal{A}_{AOC}$. Because this last expression and $(dF_z)_{ABC}$ are the only surface forces that act in the *z* direction for this discussion, we inquire next about the body forces that are acting in the *z* direction. This is the weight of the tetrahedron and is written as

$$\text{Weight} = \frac{\gamma \mathcal{A}_{ABC}(OQ)}{3} \qquad (2.2.3)$$

in which γ is the specific weight of the material within the tetrahedron. The summation of the force components for the positive *z* direction is

$$\sigma_z \mathcal{A}_{AOC} - \mathcal{P}\mathcal{A}_{ABC} \cos \theta_z - \frac{\gamma \mathcal{A}_{ABC}(OQ)}{3} = 0 \qquad (2.2.4)$$

We can relate $\mathcal{A}_{AOC}$ to $\mathcal{A}_{ABC}$ as follows. If the base is *ABC* and the perpendicular height is *OQ*, Eq. (2.2.3) implies that

$$\text{Volume of tetrahedron} = \frac{\mathcal{A}_{ABC}(OQ)}{3} \tag{2.2.5}$$

But we can also write volume $= \mathcal{A}_{AOC}(OB)/3$. These two expressions for the volume of a tetrahedron give us, if we equate them,

$$\mathcal{A}_{ABC}(OQ) = \mathcal{A}_{AOC}(OB)$$

or
$$\mathcal{A}_{AOC} = \mathcal{A}_{ABC}\frac{OQ}{OB} \tag{2.2.6}$$

The insert to Fig. 2.2.2 aids one to see that

$$\frac{OQ}{OB} = \cos\theta_z \tag{2.2.7}$$

since OB is the hypotenuse of the triangle QOB and the angle at Q is 90° because of the perpendicular construction of the line OQ. If we simplify Eq. (2.2.6) with Eq. (2.2.7), we have

$$\mathcal{A}_{AOC} = \mathcal{A}_{ABC} \cos\theta_z \tag{2.2.8}$$

If this relationship is substituted in the second term of Eq. (2.2.4) and Eq. (2.2.6) is used for $\mathcal{A}_{ABC}(OQ)$ in the third term, one has

$$\sigma_z - \mathcal{P} - \frac{\gamma(OB)}{3} = 0 \tag{2.2.9}$$

as the result. (The common factor $\mathcal{A}_{AOC}$ has been canceled.) If the tetrahedron is now allowed to shrink to a material point, the distance OB will vanish, so that

$$\sigma_z = \mathcal{P} \tag{2.2.10}$$

in this limit. This means that at a point the stress on an arbitrary plane through the point is the same as the stress on a plane that is perpendicular to the z direction.

Moreover, since the z direction was the only special one and gravity played no role as we shrank the tetrahedron to a point, it can be concluded that $\sigma_x = \mathcal{P} = \sigma_y$ also in this same limit. (One could reach this same conclusion by using $\Sigma F_x = 0$ and $\Sigma F_y = 0$ in a fashion similar to Eq. (2.2.4).) Because all the σ's are equal to $\mathcal{P}$ for an arbitrary orientation of plane ABC, we say that the stress $\mathcal{P}$ at a point in a static fluid is independent of direction. In fact, this stress $\mathcal{P}$ is in equilibrium with the pressure that must exist within the fluid at the point in question. Moreover $\mathcal{P} = p$, the pressure, if the fluid is at rest. This leads us to say that the pressure in a static fluid is a scalar because it is independent of direction.

We now turn from a consideration of the conditions at a point to the pressure changes that could occur when one moves away from a point in any of the three coordinate directions. We seek to predict the pressure at an arbitrary point P in a fluid if we know the pressure at a reference point R. We call this latter pressure p_R. The situation is described in Fig. 2.2.3. The pressure at the midpoint of plane $BCGF$ and plane $ADHE$ will be estimated from the value at R. Once we have the value for the pressures at the centers of these planes, we can multiply them by the appropriate areas and know thereby the forces that are acting on faces $BCGF$ and $ADHE$. These forces act parallel to the z direction, and we can use Newton's laws to relate these forces. The result which we desire follows from such a relationship.

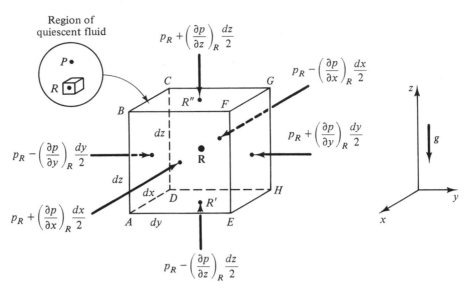

Figure 2.2.3 Pressure variation on surfaces of an infinitesimal parallelepiped of fluid.

The pressure at point R'' can be estimated from that at point R by using the ideas associated with the partial derivative. (A short discussion of total differentials and partial derivatives is given in Appendix B.) It is seen from Fig. 2.2.3 that R'' has the same x and y coordinates as point R. Only the z coordinate is different. The difference is $dz/2$. Because R'' is on the positive z side of R, we write

$$(\text{Pressure})_{R''} = (\text{pressure})_R + \begin{pmatrix} \text{variation of pressure} \\ \text{per unit distance} \end{pmatrix}_R \begin{pmatrix} \text{distance} \\ \text{in question} \end{pmatrix}$$

(2.2.11)

or

$$p_{R''} = p_R + \left(\frac{\partial p}{\partial z}\right)_R \frac{dz}{2}$$

When the same logic is applied to point R', it must be remembered that R' is on the $-z$ side of point R, so that

$$p_{R'} = p_R - \left(\frac{\partial p}{\partial z}\right)_R \frac{dz}{2}$$

(2.2.12)

Consequently, the sum of the forces in the z direction for the parallelepiped *ABCDEFGH* is

$$\left[p_R - \left(\frac{\partial p}{\partial z}\right)_R \frac{dz}{2}\right] dx\, dy - \left[p_R + \left(\frac{\partial p}{\partial z}\right)_R \frac{dz}{2}\right] dx\, dy - \gamma\, dx\, dy\, dz = 0$$

After the appropriate terms are added and subtracted, one has

$$-\left(\frac{\partial p}{\partial z}\right)_R dx\, dy\, dz = \gamma\, dx\, dy\, dz$$

If we divide by the volume of the parallelepiped, $dx\, dy\, dz$, which is assumed to be nonzero, the conclusion is

Pressure Distributions in Fluids without Relative Motions   Chap. 2

$$\left(\frac{\partial p}{\partial z}\right)_R = -\gamma \qquad (2.2.13)$$

This result shows that when gravity acts in the negative z direction, the variation of pressure with respect to distances in the positive z direction is equal to the negative of the specific weight of the fluid. (This means that as z increases, p decreases.)

This analysis used $-\gamma$ as the body force per unit volume in the positive z direction. For the coordinate system shown in Fig. 2.2.3, the body force components in the x and y directions are necessarily zero. This fact allows us to infer from the development that led to Eq. (2.2.13) that

$$\left(\frac{\partial p}{\partial x}\right)_R = 0 = \left(\frac{\partial p}{\partial y}\right)_R \qquad (2.2.14)$$

This statement implies directly that the changes in pressure are zero for changes in the x direction in a static fluid. The same is true for the other horizontal direction, y. Furthermore, the point R is quite general, so that the subscript can be suppressed. However, in Eqs. (2.2.13) and (2.2.14) the subscript does remind one that when using the relationships one must prescribe the location for evaluating the terms in the derivative.

Example 2.2.1:

In a saline solution, the specific weight $\gamma = 64 \text{ lb/ft}^3 - (0.1 \text{ lb/ft}^4)(z)$ is found to vary with z, where z is in feet. Determine $\partial p/\partial z$ for this static fluid at $z = 0$, $z = 10$ ft, and $z = 10.01$ ft. (The last calculation is intended to show that a representative value of $\partial p/\partial z$ exists near a point.)

Solution Since $\partial p/\partial z = -\gamma = -(64 - 0.1z) \text{ lb/ft}^3$,

$$\text{At } z = 0: \frac{\partial p}{\partial z} = -64 \text{ lb/ft}^3$$

$$\text{At } z = 10: \frac{\partial p}{\partial z} = -63 \text{ lb/ft}^3$$

$$\text{At } z = 10.01: \frac{\partial p}{\partial z} = -(64 - 1.001) = -62.999 \text{ lb/ft}^3$$

Thus, near $z = 10$:

$$\frac{\partial p}{\partial z} \approx -63 \text{ lb/ft}^3$$

We have seen through Eqs. (2.2.13) and (2.2.14) that the pressure p varies with x, y, and z in a particular way for a static fluid. With this information, we can construct the surfaces of constant pressure. Because the pressure does not change in the x and y directions, we know that the pressure does not change in a horizontal plane. Equation (2.2.13) tells us that the pressure decreases with increasing elevation. This information leads us to conclude that the surfaces of constant pressure are horizontal and the value of the constant pressure increases as one moves toward the center of the earth. This will be the case for a static fluid. (We recognize that this is in agreement with some of our experiences while swimming.)

Now is an ideal time to consider more generally the task of determining surfaces of constant pressure, even though in the hydrostatic case we have already determined this useful result. As a consequence of the foregoing discussions in this chapter, we have concluded that we expect p to be a function of x, y, and z usually, but that for the hydrostatic case, p is a function of the z coordinate only. These verbal statements are expressed mathematically as $p = p(x, y, z)$ and (for our static case) $p = p(z)$. The associated mathematical expressions remind us that we are operating in a continuum, because for any set of coordinates, the continuous function $p(x, y, z)$—whatever it might be—will produce for us the pressure at the point given by those coordinates. If we keep in mind the idea of a point function and a total differential, we can write

$$p - p_R \approx dp = \left(\frac{\partial p}{\partial x}\right)_R dx + \left(\frac{\partial p}{\partial y}\right)_R dy + \left(\frac{\partial p}{\partial x}\right)_R dz \qquad (2.2.15)$$

in which p is the pressure of a point dx, dy, and dz away from R. (dp will, in general, include $(\partial p / \partial t)\, dt$. However, we *assume* $\partial p / \partial t = 0$, something that would not be true in a pressure cooker that was heating up.)

Naturally, the approximation improves as dx, dy, and dz approach zero. Because dx, dy, and dz result from taking differences, it must be recognized that these quantities may be either positive or negative, depending upon the two points in question.

As one reads Eq. (2.2.15), it can be recognized that the partial derivatives $\partial p / \partial x$, $\partial p / \partial y$, and $\partial p / \partial z$ have been specified by the physical laws for a static fluid through Eqs. (2.2.13) and (2.2.14). Thus, these partial derivatives are numbers, numbers which may have different values at different points of space. If we have inserted the numbers for the derivatives that are appropriate to the region around points P and R into Eq. (2.2.15), we can determine dp just as soon as the values of dx, dy, and dz are specified. Equally well, one could specify a small value of dp and a set of values for dx and dy. Then one could use Eq. (2.2.15) to specify the required value of dz. Furthermore, one might set the value of dp to zero. In effect, a search would be initiated for all of the points R which had the same pressure as the point P. The locus of all these points would be a surface of constant pressure.

If the derivatives $\partial p / \partial x$, $\partial p / \partial y$, and $\partial p / \partial z$ are constants, say C_1, C_2, and C_3, Eq. (2.2.15) becomes particularly easy to integrate. One would integrate from point R as follows.

$$\int dp = C_1 \int dx + C_2 \int dy + C_3 \int dz \qquad (2.2.16)$$

or $\qquad\qquad p - p_R = C_1(x - x_R) + C_2(y - y_R) + C_3(z - z_R) \qquad (2.2.17)$

Equation (2.2.17) will find several applications in this chapter, but at this time we shall restrict our view to looking for points with coordinates (x, y, z) such that the pressure p is the same as that at point R. We want $p = p_R$. From Eq. (2.2.17), we have then

$$0 = C_1 x + C_2 y + C_3 z - (C_1 x_R + C_2 y_R + C_3 z_R) \qquad (2.2.18)$$

Because x_R, y_R, and z_R are specified and consequently known quantities, Eq. (2.2.18) represents an interrelationship between the coordinates of all the points in the fluid that have a pressure of p_R. Note that the term in parentheses is a constant. The mathematical expression represented by Eq. (2.2.18) defines a *plane*. Finally, if C_1 and C_2 are zero, this plane is one of $z = $ constant. This is the case for a hydrostatic fluid, and while we have reached the same conclusion by less formal means previously, we shall find that the ideas that led to Eq. (2.2.18) will prove useful to us again in the future.

The expression for the pressure at any point in a static fluid, Eq. (2.2.17), is useful in finding the pressure difference between any point and point R. If the pressure at point R is known, one can find the pressure at any other point directly.

Example 2.2.2:

Find the pressure at the bottom of a vertical sealed tube that is 6 m long (see Fig. 2.2.4). The top one-third of the tube is filled with oil (SG = 0.86), and the remainder is filled with water. The top of the tube is open to the atmosphere.

Solution This is a case for which C_1 and C_2 are zero in Eq. (2.2.17), since this situation corresponds to the one yielding Eqs. (2.2.13) and (2.2.14). We begin by locating the origin of the coordinate system. Any point will suffice, and in this example a point at the interface between the two liquids is convenient. Equation (2.2.17) reduces to

$$p - p_R = -\gamma_{oil}(z - z_R) \qquad \text{in the region for } z \geq 0.$$

The reference pressure is the atmospheric pressure that exists 2 m above the interface. We shall use gauge pressure in this problem. Since atmospheric pressure is zero gauge pressure,

$$p_R = 0 \qquad \text{at } z_R = 2$$

Then $p = -0.86(1000)(9.802)(z - 2)$. The pressure at the interface, $z = 0$, consequently is $+(0.86)(1000)(9.802)(2)$ Pa.* Even with our location of the coordinate system, Eq. (2.2.18) can be written for $z \leq 0$ as

$$p - p_{R'} = -\gamma_{H_2O}(z - z_{R'})$$

in which the prime on R serves as a reminder that the reference point might not always be the same as that just used. This is because our current equation demands a reference point in the local medium. Such a point is at the interface, where $z = 0$ and the pressure is the one just found. In choosing such a reference point assume the pressure to be continuous across the interface, where no effects of surface tension exist, because the interface is flat. Accordingly,

$$p - 1.72(1000)(9.8) = -(1000)(9.8)(z)$$

At the bottom of the tank, $z = -4$ m; we then have the pressure at this point,

$$p = 4(1000)(9.8) + 1.72(1000)(9.8)$$
$$= 56.06(1000) \text{ Pa} \qquad \text{gauge pressure}$$
$$= 56.06 \text{ kPa} \qquad \text{gauge pressure}$$

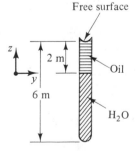

Free surface

Figure 2.2.4 Vertical cylinder filled with layers of oil and water (Example 2.2.2).

2 m

Oil

6 m

H_2O

* Often, the gravitational constant for the SI system will be approximated by the number 9.8 in the work that follows.

(It would be instructive for the reader to rework this example by placing the coordinate system at the top of the tube.)

Some liquids have densities, or specific weights, that are not constant. (Gravity is considered to be constant.) In Example 2.2.1, the specific weight γ varied with the vertical coordinate z. Here the density could be expressed as $\rho = \rho_0 - \beta z$, in which β is a constant gradient. In this case, the integration of Eq. (2.2.16) yields

$$p - p_R = -\int \gamma \, dz = -g \int (\rho_0 - \beta z) \, dz$$
$$= -g\,\rho_0(z - z_R) + g\left(\frac{\beta}{2}\right)(z^2 - z_R^2) \qquad (2.2.19)$$

since $\partial p/\partial x$ and $\partial p/\partial y$ were both zero. An incorrect equation would have been the result had the expression for ρ been inserted directly into Eq. (2.2.18).

Example 2.2.3:

A tank of liquid sodium is heated in such a way from the top that the material remains effectively at rest. Find the pressure difference between a point at the top of the tank and the bottom of the tank if the tank is 3 m deep with sodium. If the pressure at the free surface is equivalent to 0.3 atmospheres above the ambient atmospheric pressure outside of the closed tank, what is the pressure at the bottom of the tank? Refer to the accompanying figure for the variation of density with elevation.

Solution For a coordinate system that is located at the bottom of the tank,

$$\rho = 930 - (930 - 880)\left(\frac{z}{3}\right) \qquad \text{kg/m}^3$$

(Check to see if this linear relationship gives the results of Fig. 2.2.5 at $z = 0$ and 3.) Starting from Eq. (2.2.13), we have

$$\frac{dp}{dz} = -g\rho = -g\left(930 - \frac{50z}{3}\right)$$

Then Eq. (2.2.16) gives, upon integration (refer also to Eq. (2.2.19)),

$$p_B - p_T = -930g(z_B - z_T) + \left(\frac{25g}{3}\right)(z_B^2 - z_T^2)$$
$$= 9.8[(930)(3) + (\tfrac{25}{3})(-3^2)]$$
$$= 9.8[930(3) - 25(3)]$$
$$= 26.661(10^3 \text{ N/m}^2)$$
$$= 26.66 \text{ kPa}$$

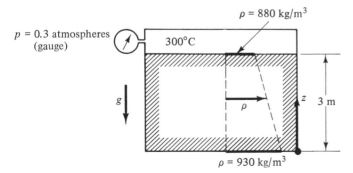

$p = 0.3$ atmospheres (gauge)

$\rho = 880$ kg/m³

300°C

g

z 3 m

ρ

$\rho = 930$ kg/m³

Figure 2.2.5 Container partially filled with a liquid whose density varies linearly with elevation due, perhaps, to differences of solute concentration (Example 2.2.3).

We are given that $p_T = 0.3$ atmospheres, and in Sec. 1.2 we developed the relationships 1 bar $= 10^5$ Pa and 1 atmosphere $= 1/0.997$ bars. This gives

$$p_T = 0.3\left(\frac{1}{0.997}\right)(10^5) \text{ Pa} = 30.09 \text{ kPa}$$

From the previous calculation, we now find

$$p_B = (26.66 + 30.09) \text{ kPa}$$

This is a gauge pressure. The absolute pressure must include the local atmospheric pressure, which is assumed in this instance to be 1 standard atmosphere, 101.3 kPa.

Gases, too, have variable density. This density depends upon the pressure and temperature according to some equation of state, and these quantities may be dependent upon the type of compression or expansion process that a gas undergoes in a particular situation. One possible case to consider is that of an isothermal atmosphere. At some location, say point 1, the ideal gas law would be written as $p_1/\rho_1 = RT_1$. At another point, one would have $p_2/\rho_2 = RT_2$. But if the gas is isothermal, $T_1 = T_2$, and then

$$\frac{p_1}{\rho_1} = \frac{p_2}{\rho_2} = \text{constant} = \frac{p}{\rho} \qquad \text{in general} \tag{2.2.20}$$

This last relationship between p and ρ can be incorporated into Eq. (2.2.13) to give the variation of pressure with elevation. We start with this equation and note that $\partial p/\partial x$ and $\partial p/\partial y$ are zero in accordance with the physics of a static fluid, so that

$$dp = -(\rho)g(dz)$$

and using $\rho = p/C$ from Eq. (2.2.20), where C is a constant, we find

$$\frac{dp}{p} = \left(-\frac{g}{C}\right)(dz) \tag{2.2.21}$$

This statement can easily be integrated to give a logarithmic variation between the pressure and the elevation for an isothermal gas, or

$$p_2 = p_1 \exp\left[-\frac{g(z_2 - z_1)}{p_1/\rho_1}\right] \tag{2.2.22}$$

2.3 MANOMETERS

A simple device that uses the principle that has just been treated can be used to measure the pressure in a fluid. The instrument in question is called a *manometer,* and it is used to determine many things ranging from blood pressure in human beings to the pressure at various locations on an aircraft's wing. An understanding of manometric measurements should now be easy, because the fluid that is within the manometer will generally be at rest when the instrument is being used.

The container shown in Fig. 2.3.1a has some undefined boundaries which surround the fluid at rest. As we have seen, the difference in pressure between points P

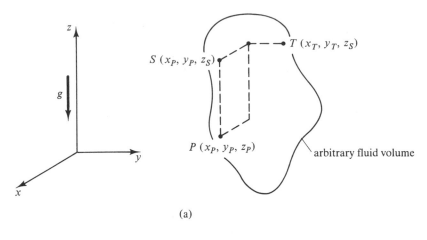

(a)

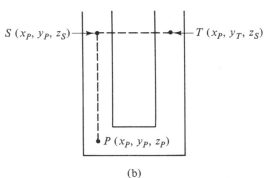

(b)

Figure 2.3.1 (a) Fluid with specific weight γ in container of arbitrary shape. (b) Plane-tube container bent into U shape. (Center-line of bend is usually a plane surface.)

and S is the same as that between points P and T, because it is specified in the figure that points S and T are at the same elevation. This pressure difference is given by

$$p_P - p_S = \gamma(z_S - z_P)$$

or

$$p_P = p_S + \gamma(z_S - z_P) \tag{2.3.1}$$

The same relationship would hold for points P and S in Figs. 2.3.1b and 2.3.2. In this latter situation, we shall apply the hydrostatic law for pressure variation in a vessel having two or more fluids of different densities. Point P is just above the interface

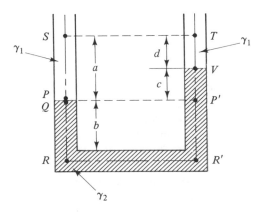

Figure 2.3.2 U tube containing two different fluids to facilitate calculation of pressure at different points. γ_1, γ_2 = specific weight.

between the two fluids. If Q is located immediately below the interface, the pressure there will approach that of P as these two points themselves approach the interface. This is, strictly speaking, not true, because a jump in the pressure across the interface could occur if the effects of surface tension are present due to significant curvature of the interface. We shall *assume,* however, that the tube that confines the fluid is sufficiently large that the interface is without appreciable curvature. The pressure at point R can be found in terms of that at Q (or, from what has just been said, at P), so that

$$p_R = p_P + \gamma_2 b = p_S + \gamma_1 a + \gamma_2 b$$

Now, finally, we would like to have an expression for the pressure at point T. It should be possible to continue the same process which just yielded p_R in terms of p_S as well as a, b, γ_1, and γ_2 to find p_T. If we follow along the circuit toward point T, we will find that at R' the pressure is the same as that at R because they are at the same elevation. Also, the pressure at P' is less than that at R', or R, by the same amount that the pressure at R is greater than that at point P. From this, we see that it was unnecessary to proceed from point P to point P' via points R and R'. One could move horizontally across in a fluid of a manometer without changing the pressure, provided that if we had taken the complete circuit, we would *not* have passed into a fluid of another density. The pressure at the interface, where again the effects of surface tension are assumed to be negligible, is

$$p_V = p_{P'} - \gamma_2 c = p_P - \gamma_2 c$$

The pressure at point T is but another step in the procedure, so that

$$
\begin{aligned}
p_T &= p_V - \gamma_1 d \\
&= (p_P - \gamma_2 c) - \gamma_1 d \\
&= (p_S + \gamma_1 a) - \gamma_2 c - \gamma_1 d \\
&= p_S + c(\gamma_1 - \gamma_2)
\end{aligned}
\tag{2.3.2}
$$

This gives the pressure difference between points T and S in terms of the lengths of the fluid columns and their respective specific weights. The algebraic signs are positive or negative in accordance with our understanding that pressure increases as one descends in a fluid, and vice versa.

Many forms of manometers (other than the U tube, which has just been discussed) are used, depending on the particular application. All, however, are based on the concepts that we have developed. The logic and method of analysis for them are equally simple. For example, the manometer of Fig. 2.3.2 looks like that shown in Fig. 2.3.3a. The pressure at points S and T is equal in this particular illustration. If we increase the pressure at S above that at T, the manometer might appear as in Fig. 2.3.3b. Now instead of trying to read the length of the column c in Fig. 2.3.3b, one might record distances e' and e, which correspond to the distances that the interfaces have gone down and up, respectively, from the position at zero pressure differential. If the bore of the tube were the same in both sides, or "legs," of the manometer, one would conclude—from volume conservation—that distances e and e' are equal. (This is so because the same volume of fluid with specific weight γ_2 has moved downward as upward.) This would mean that it would not be necessary to record both e and e' when making a reading. Instead, one could proceed by writing

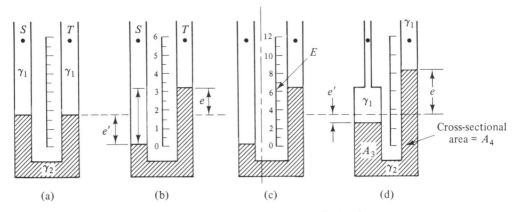

Figure 2.3.3 Conversion of U-tube manometer to single-tube manometer.

$$p_T = p_S - (e + e')(\gamma_2 - \gamma_1) = p_S - (2e)(\gamma_2 - \gamma_1) \tag{2.3.3}$$

Equally well, one could alter the markings on the manometer. One might obliterate the markings on the left and replace the markings on the right with numbers exactly twice those which were there originally. The manometer would be used by reading the markings shown in Fig. 2.3.3c as E. One would then write

$$p_T = p_S - E(\gamma_2 - \gamma_1) \qquad E = 2e \tag{2.3.4}$$

(The manometer scale might be marked 6 cm but a ruler would show a distance of 3 cm.) This system has advantages. The principal one is that only one liquid level must be recorded. Such a manometer is called a *single-tube manometer,* behind which there is, of course, a second tube that is not visible. If the tubes in the two legs of the manometer are not the same diameter or area, the factor 2 in Eq. (2.3.3) has to be changed. For legs of unequal area, the relationship

$$p_T = p_S - e\left(1 + \frac{A_4}{A_3}\right)(\gamma_2 - \gamma_1) \tag{2.3.5}$$

can be derived for the manometer shown in Fig. 2.3.3d. Even though the distance e could be measured, the top of the manometer fluid would be adjacent to a number corresponding to $e(1 + A_4/A_3)$ on a commercial single-tube manometer, since the distance e' is now included in this result.

Example 2.3.1:

A U-tube manometer with a constant bore is filled with mercury and connected, through suitable traps for environmental safety, to a system in which water is at different pressures. What is this pressure difference if the levels of the mercury columns are vertically 19 in. apart (see Fig. 2.3.4)? (Reading $= \frac{19}{12} = 1.5833$ ft.)

Solution   Equation (2.3.2) would indicate

$$p_S - p_T = \text{pressure difference} = (1.5833)(62.4)(1 - 13.6).$$

This gives a Δp of 1245 lb/ft², or 8.65 psi. The higher pressure would be associated with the part of the system that was connected to the depressed column of mercury. (What percent error would have been incurred by not including the specific gravity of the system's fluid, water?)

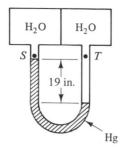

Figure 2.3.4 U-tube manometer being used to measure a pressure difference (Example 2.3.1).

2.4 INTRODUCTION TO FORCES ON PLANAR AND CURVED SURFACES

The forces that the fluid exerts on a surface are due to the action of stresses that are distributed over the surface, rather than one or more loads applied at distinct locations. Often, this distribution is not known initially and we must solve for it. We have found that for a static fluid, the pressure changes only with z, the vertical coordinate. This would tell us that the variation of the pressure on the vertical plane surface of Fig. 2.4.1 is $p = p_0 - \gamma z$, in which p_0 is the pressure at the free surface and the origin of z is in the free surface. It will prove convenient to introduce the coordinate Z, the distance from the free surface *down* to the point in question. Then $z = -Z$ and $p = p_0 + \gamma Z$. Here we shall be concerned only with the gauge pressure, γZ, because our interest is in the induced forces that are in excess (positive or negative) of the forces associated with atmospheric pressure.* (One should also bear in mind that most of the ideas concerning the forces induced by a normal stress, e.g., pressure, could be transferred without major difficulty to the forces from shear stresses that are associated with a flowing fluid. In such a case, however, the shear stresses would have to be

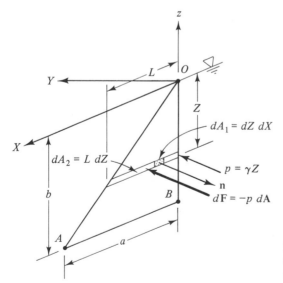

Figure 2.4.1 Infinitesimal hydrostatic forces on one side of a vertical, triangular surface. XY plane is horizontal.

* If both sides of the surface are wetted by the fluid, the same pressures will exist on the opposite sides of the surface. Hence the net force on both sides will be zero.

known by some means, and the pressure variation would likely be different from that of a fluid at rest. Thus, it behooves us to concentrate on the physical principles of the material that we are developing and avoid the tendency to focus on any associated geometric analogues.)

Returning to Fig. 2.4.1, we see that the element of force on the infinitesimal surface dA is

$$d\mathbf{F} = -p\,d\mathbf{A} \qquad (2.4.1)$$

This formula is written in vector notation because $d\mathbf{F}$ is a vector that is due to the result of a positive pressure acting in the direction opposite to the normal to the element that has an area of dA. This area can also be designated as $|d\mathbf{A}|$.* The treatment of $d\mathbf{A}$ as a vector conveys the idea of the importance of the surface's orientation. The normal to the surface, $\mathbf{n}$, is directed from the surface to the fluid whose pressure acts on the surface. If the sign of p is positive, then $d\mathbf{F}$ acts toward the side of the surface of interest in the direction opposite to $\mathbf{n}$. This is the reason for the *minus* sign of Eq. (2.4.1).

The area dA that is considered in Eq. (2.4.1) is infinitesimal in size. It is the area on which p is, in effect, constant. This equation provides a means of relating p with $d\mathbf{F}$. To apply Eq. (2.4.1), we must specify p and the magnitude of $d\mathbf{A}$ in order to obtain the magnitude of $d\mathbf{F}$. One could employ $dA = dA_1 = dX\,dZ$. (If the pressure is constant over a finite section of a plane surface, the element of force on that section is found by multiplying the pressure by the area on which p is constant.)

The hydrostatic pressure variation is now applied to the situation depicted in Fig. 2.4.1. The pressure is constant on horizontal planes, and so it is constant on the horizontal strip $L\,dZ = dA_2$. Hence the element of force on this strip due to p has a magnitude of $p\,dA_2$ and is designated dF. This element of force acts against the plate and normal to it. If we write $dF = pL\,dZ$, we are requiring that a positive dF act against the plate in the manner indicated in Fig. 2.4.1. All the dF's on the various area elements, regardless of the position Z on the plate, are parallel in this figure. In this case, the resultant, normally a vector sum, is simply a scalar sum of all the horizontal components of $d\mathbf{F}$. This is why we can simply write that

$$F_{\text{horiz}} = \int dF_{\text{horiz}} = \int pL\,dZ$$

All that one need do now is substitute the appropriate expressions for p and L. The pressure p poses no difficulty. We write it as $p = p_0 + \gamma Z$ so that the gauge pressure is $p = \gamma Z$ if p_0 is taken as the atmospheric pressure at $Z = 0$. Next, we examine the geometry. The length of the strip, L, varies with Z, since $L = (a/b)(Z)$. (Not all situations will be conceptually this simple.) The limits of integration are the limits on Z, from 0 to b. The force that is sought acts normal to the triangular surface. If the sign of the result from the integration is positive, the force will act against the plate in accordance with the conventions given previously.

On a curved surface, one must take special care to use Eq. 2.4.1 correctly. It is a vector equation; components must be used to sum the elements of force.

Example 2.4.1:

Find the vertical force on the semicircular surface shown in Fig. 2.4.2.

---

* The magnitude of a vector is often denoted in this way. $d\mathbf{A} = \mathbf{n}\,dA$, in which the normal to dA is directed *toward* the fluid particles that produce the pressure.

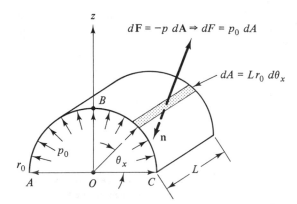

$$dF = -p\,dA \Rightarrow dF = p_0\,dA$$

$$dA = L r_0\,d\theta_x$$

Figure 2.4.2 Uniform pressure on interior of a cylindrical surface and infinitesimal force that is induced.

Solution We shall assume that the pressure is constant everywhere inside the vessel due to the high pressure of the contained gas. (Then the small variations in pressure due to elevation changes, i.e., gravitational effects, will not change the local pressure appreciably.) The elemental force on the shaded area is $dF = -p_0\,dA$. The area element is L long and ds wide. (Is $ds = r_0\,d\theta_x$?) Thus the magnitude of dF is $p_0 L(r_0\,d\theta_x)$ and its direction is radially outward at an angle of θ_x with the x axis. (If this area element had been located at point B, dF would be acting in the z direction, and if the area element were at point C, dF would be acting horizontally. The need for using components, rather than simply summing the magnitudes of the various dF's, is apparent.) The vertical component of any dF is

$$dF_z = |dF| \sin\theta_x = L(p_0 r_0\,d\theta_x)\sin\theta_x$$

Now we can sum the dF_z, a scalar component of the vector dF. This gives the total F_z due to the pressure distribution as

$$F_z = \int_0^\pi L p_0 r_0 \sin\theta_x\,d\theta_x = L p_0 r_0 \int_0^\pi \sin\theta_x\,d\theta_x$$

Show that $F_z = 2 L p_0 r_0$.

2.5 HYDROSTATIC FORCES AND MOMENTS ON PLANAR SURFACES

We have been able to deduce the way pressure changes in a static fluid of constant density as

$$p = p_R - \gamma z \tag{2.5.1}$$

or

$$p = p_R + \gamma Z \tag{2.5.2}$$

in which p_R is the pressure at a reference point and z is measured from there in the direction opposite to that of gravitational attraction, whereas Z is positive downward. Now we shall use this "hydrostatic law" to determine the forces on a variety of submerged surfaces. Again, it should be mentioned that the principles of physics for the problem are contained completely in Eq. (2.5.1) and the statements implicitly in Eq. (2.4.1). Any difficult aspects of the problems in this section are solely due to the geometric features of the particular problem.

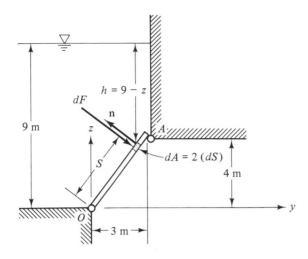

Figure 2.5.1 Geometry and infinitesimal force component needed to determine hydrostatic load on inclined rectangular gate 2 m wide, with hinge at O and line contact at A.

We already encountered one example of the use of Eq. (2.5.1) when the vertical surface of Fig. 2.4.1 was contemplated. In Fig. 2.5.1 we treat an inclined surface and determine the force that is induced on the surface by the fluid above it. In solving for this force we commence by solving for dF and write $d\mathbf{F} = -p\,d\mathbf{A}$:

$$|d\mathbf{F}| = |-p\,d\mathbf{A}| = p(2dS)$$
$$= \gamma h(2dS) = \gamma(9 - z)(2dS) \quad \text{for } h = Z$$

Now we can relate the distance along the gate from point O, which is designated S, to the vertical distance above O that corresponds to S. This relationship is given by the geometry of similar triangles: $S/z = 5/4$. From this, dF can be written as a function of S only:

$$|d\mathbf{F}| = \gamma\left(9 - \frac{4S}{5}\right)(2dS)$$

The total force on the gate is the algebraic sum of all of these elemental forces, since all of the elements are parallel to the normal vector of the surface. This is then the direction with which this total force would be aligned.

$$|\mathbf{F}| = 2\gamma \int_0^5 \left(9 - \frac{4S}{5}\right) dS \tag{2.5.3}$$

If the gate had had a triangular shape, the problem would be no more difficult, but it would be one step longer. The situation is depicted in Fig. 2.5.2. The plane has an edge view similar to that of Fig. 2.5.1. Thus, the essential difference in this instance is the area element, $d\mathbf{A}$. The width, instead of being constant (2 m, in Fig. 2.5.1), now varies with the position S. When S is zero, the length of the elemental area is 11 m; where S is 5 m, the length is zero. This means that

$$L = (11)\left(1 - \frac{S}{5}\right) \quad |d\mathbf{F}| = |-p\,d\mathbf{A}| \quad \text{with} \quad |d\mathbf{A}| = L\,dS *$$

---

* The statement $d\mathbf{F} = -p\,d\mathbf{A}$ is fundamental to all the problems in this section. After writing it, one must examine the geometry in order to write useful statements about dA, the area on which p is locally constant. Then one can proceed toward a solution.

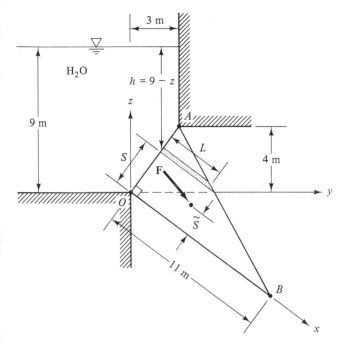

Figure 2.5.2 Diagram to aid in determining force and moment induced on a triangular plane surface by a hydrostatic pressure distribution. xy plane is horizontal.

Consequently,

$$|\mathbf{F}| = F = \gamma \int (h)L\, dS \quad \text{for } h = Z$$

$$= \gamma \int_0^5 (9-z)(11)\left(1-\frac{S}{5}\right) dS \qquad (2.5.4)$$

$$= \gamma \int_0^5 \left(9-\frac{4S}{5}\right)(11)\left(1-\frac{S}{5}\right) dS$$

In Fig. 2.5.1, the magnitude of the elemental force dF which is due to the pressure distribution acts at a distance S from the origin of the system at O. There is, consequently, an induced moment about the x axis that is designated $(dM_x)_0$. This is the x component of the infinitesimal moment vector $(d\mathbf{M})_0$. (We shall consider the other component at a later time.) $d\mathbf{M}_0$ equals $\mathbf{r} \times d\mathbf{F}$ so that

$$(dM_x)_0 = -S\, dF$$

in which the minus sign that ensues from $\mathbf{r} \times d\mathbf{F}$ is in agreement with the "right-hand rule" for such vector (cross) products. Since $d\mathbf{F}$ was given in the equation just prior to Eq. (2.5.3), it follows that

$$|dM_x|_0 = S[\gamma\left(9-\frac{4S}{5}\right)(2)(dS)]$$

and
$$|M_x|_0 = 2\gamma \int_0^5 S\left(9-\frac{4S}{5}\right) dS$$

In Fig. 2.5.4, $|dM_x|_0$ is again $S(dF)$, so that one has

$$|M_x|_0 = \gamma \int_0^5 S\left[\left(9-\frac{4S}{5}\right)(11)\left(1-\frac{S}{5}\right)\right] dS$$

by using the dF that is inherent in Eq. (2.5.4).

If one asked for the moment about an axis from O to A in Fig. 2.5.2, it would follow that

$$|dM_{OA}| = x\, dF$$

in which x takes on all values from 0 to L, with the latter value being determined by S. This requires that one must first sum along the strip, whose length is L, for a given distance S and then sum up all of the results from the strips as S varies from 0 to 5. This means that

$$|M_{OA}| = \int_{S=0}^{5} \int_{x=0}^{L} x\, dF \quad \text{with} \quad dF = p\,(dS)\, dx$$

$$= \int_{S=0}^{5} \int_{x=0}^{L} x\left[\gamma\left(9 - \frac{4S}{5}\right) dS\right] dx$$

$$= \gamma \int_{0}^{5} \left[\int_{0}^{L} x\, dx\right]\left(9 - \frac{4S}{5}\right) dS,$$

$$= \gamma \int_{0}^{5} \frac{L}{2}\left(9 - \frac{4S}{5}\right) L\, dS$$

One would have to insert the expression $(11)(1 - S/5)$ for L into this integral and proceed as before to obtain the moment. Note that the integrand and final result will differ from that in Eq. (2.5.4) by the equivalent of $L/2$. This outcome is not unexpected, since the dF's, expressed as $p\,(dS)\, dx$, are constant for $0 \le x \le L$ for each S. There is, then, a uniform loading along the length L, and the moment about one end of such a strip should be $(L/2)(\text{load})$.

We now apply what we have learned from Figs. 2.5.1 and 2.5.2 to more complex geometries so that we can deduce the forces that are induced on a variety of plane and curved surfaces. We begin by examining the surface shown in Fig. 2.5.3, which is

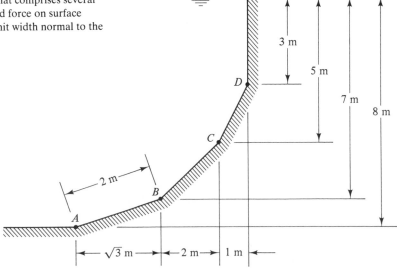

Figure 2.5.3 Calculation of force components on surface that comprises several planar elements. Find force on surface *ABCD*, which has unit width normal to the page.

composed of plane segments. From this figure we shall infer certain general features of all problems concerning hydrostatic forces on submerged surfaces of all shapes.

The force on this segmented surface $ABCD$ can be calculated by finding the force on AB, BC, and CD. The method is much the same as that which was just employed in Fig. 2.5.1. The force on each of these surfaces is perpendicular to that surface. Figure 2.5.4 shows the elemental forces on each of the three inclined segments. The results are

$$F_{AB} = \gamma \int_0^2 \left(8 - \frac{S}{2}\right) dS \qquad F_{BC} = \gamma \int_0^{2\sqrt{2}} \left(7 - \frac{S'}{\sqrt{2}}\right) dS'$$

$$F_{CD} = \gamma \int_0^{\sqrt{5}} \left(5 - \frac{2}{\sqrt{5}} S''\right) dS''$$

in which S is measured along AB from point A, S' is measured along BC from B, and S'' is measured along CD from C. The resultant force on surface $ABCD$ is obtained by the *vector* summation of $\mathbf{F}_{AB}$, $\mathbf{F}_{BC}$, and $\mathbf{F}_{CD}$. This means that the resultant of $\mathbf{F}_{AB}$, $\mathbf{F}_{BC}$, and $\mathbf{F}_{CD}$ is found by adding the x and z components of these forces. Therefore, using the angles indicated in Fig. 2.5.4,

$$F_x = F_{AB} \cos \alpha_x + F_{BC} \cos \alpha_x' + F_{CD} \cos \alpha_x''$$

and

$$F_z = -F_{AB} \cos \alpha_z - F_{BC} \cos \alpha_z' - F_{CD} \cos \alpha_z''$$

with

$$|\mathbf{F}| = \sqrt{F_x^2 + F_z^2}$$

We shall now determine F_x and F_z by a somewhat different procedure. Instead of calculating $\mathbf{F}_{AB}$ and then finding the x component of this force, we shall, at the outset,

Figure 2.5.4 Coordinates and reference normals for a surface of Fig. 2.5.3.

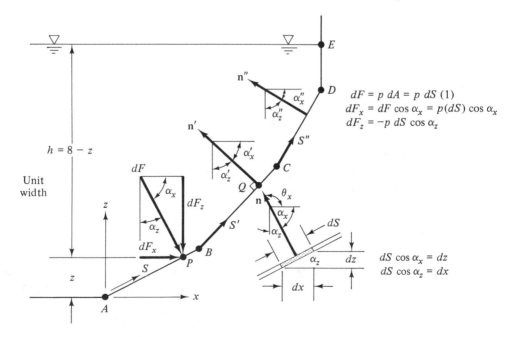

determine the x component of the elemental forces along AB and sum them to obtain the total component. What interests us here is not the final numerical value, but some observations about the numerical procedure and the general conclusions that can be drawn.

The situation is depicted in Fig. 2.5.4, where the pressure p is again hydrostatic. There the point P, along AB, has associated with it an infinitesimal force with components dF_x and dF_z. The former component is

$$dF_x = p\,(\text{width})\,dS\,\cos\alpha_x = p\,(\text{width})\,dz$$

because of the geometric relationships that are illustrated in the inset in Fig. 2.5.4. The use of the angles α_x, α_x', and so on, is a quite natural choice for a two-dimensional geometry. However, in order to establish a *sign convention* that will have broad application, we use the angle θ_x—see the inset in Fig. 2.5.4—instead of α_x. The latter angle is a natural one when the forces are summed. However, θ_x, *which is the angle between the local normal to the surface and the x axis*, is more appropriate when attention is focused upon the surface and its orientation. Because $\cos\alpha_x = -\cos\theta_x$, only the sign in the last result is changed. This leads us to write

$$dF_x = -p\,(dA)\,\cos\theta_x \qquad (2.5.5)$$

in which $dA = (\text{width})\,dS$. Note that the magnitude of $(dA)\cos\theta_x$ is the projection of the area dA onto a plane that is perpendicular to the x axis. This projection is $(dz)(\text{width})$ in this case. The algebraic sign of $(dA)\cos\theta_x$ is *determined by the sign of the x component of the normal* to the surface, **n**. In Fig. 2.5.4, the normals to surfaces AB, BC, and CD have negative x and positive z components.

The coordinate z is defined as the vertical distance from A to P, a point on the inclined surface. For the depth of 8 m from the free surface to point A

$$p = h\gamma = (8 - z)\gamma$$

and $(dA)\cos\theta_x = -(dz)(1)$ for a unit width, so that

$$(dF_x)_{AB} = \gamma(8 - z)\,dz$$

from Eq. (2.5.5). Finally,

$$(F_x)_{AB} = \gamma \int_0^1 (8 - z)\,dz \qquad (2.5.6)$$

Along BC, a calculation of dF_x at some point Q is dependent upon the pressure that is determined by the depth of submersion, h. This is $h = (8 - z)$ again, if z is still measured from point A. The reader should show that

$$(F_x)_{BC} = \gamma \int_1^3 (8 - z)\,dz \qquad (2.5.7)$$

and should be sure to justify the limits of integration.

This result for the x component of the force is interesting, because its form is nearly identical to that of $(F_x)_{AB}$. In a similar way, along CD, $(dF_x)_{CD} = p\,(dz)$, so that

$$(F_x)_{CD} = \gamma \int_3^5 (8 - z)\,dz \qquad (2.5.8)$$

Pressure Distributions in Fluids without Relative Motions   Chap. 2

in which the origin of z is, as before, at A. The three x components of the force on AB, BC, and CD are given by Eqs. (2.5.6), (2.5.7), and (2.5.8). These integrals can be combined into one so that

$$(F_x)_{AD} = \gamma \int_0^5 (8 - z)\, dz \qquad (2.5.9)$$

Note that the variable of integration is z, the vertical coordinate, and that the angles θ_x, θ_x', and θ_x'' did not enter into the calculation of $(F_x)_{AB}$, $(F_x)_{BC}$, $(F_x)_{CD}$, or $(F_x)_{AD}$. The last statement means that Eq. (2.5.9) would be true even if there were many more segments to the surface AD. This integral statement would be valid for an infinite number of segments, a smooth curve. One can also surmise that the magnitude of the horizontal force on AD is the same as on $A'D'$ for Fig. 2.5.5.

The conclusion reached for the situation depicted in Fig. 2.5.5 can be generalized as follows:

> The horizontal component of the force on any surface submerged in a static fluid is *equivalent* to that on a vertical projection of the surface.[*]

The forces which were distributed over surface AD in Fig. 2.5.4 also have vertical components that will now be considered. This figure shows that

$$dF_z = -p\, dS\, \cos\theta_z \quad \text{and} \quad dS\, \cos\theta_z = dx$$

since θ_z is the angle between the normal and the z axis. Because $p = \gamma h$, the elemental vertical component of the force of any section of AD is

$$dF_z = -\gamma h\, dx$$

The geometric significance of this formulation is shown in Fig. 2.5.6. This figure indicates that the vertical component of the elemental force at any point P, dF_z, has a magnitude that is equivalent to the product of γ and the element of volume between the position of P and the location of the liquid's free surface, dV. The total vertical component is obtained by summing these elemental forces over the entire surface.

The foregoing discussion leads to the following conclusion:

> The magnitude of the vertical force component on a submerged surface in a static fluid of constant density is *equivalent* to the weight of the fluid in the volume located in the region above that surface, up to the position of the liquid-air interface.

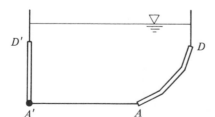

Figure 2.5.5 Planar and nonplanar surfaces with same vertical projection producing same horizontal force component.

[*] The word *equivalent* has been used in contradistinction to the term "same" because only magnitudes are under discussion.

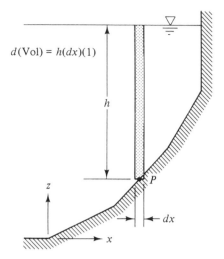

$d(\text{Vol}) = h(dx)(1)$

h

z

P

dx

x

Figure 2.5.6 Infinitesimal contributions to hydrostatically induced vertical force on nonplanar surface.

This method of calculating the vertical component can be applied to Fig. 2.5.7, in which the curved surface is the result of an infinite set of planar elements. The magnitude of the downward force on ABC must be equal to the upward force on $A'B'C'$, since the pressure is the same at comparable points on identical surfaces. But the downward force on ABC is

$$|F_z|_{AC} = \gamma \Psi_{ABCDEOA}$$

so that the upward force on $A'B'C'$ is

$$|F_z|_{A'C'} = \gamma \Psi_{A'B'C'D'E'O'A'}$$

This latter volume is the volume of an *equivalent* amount of fluid, since it is not geometrically located within the fluid container. This does not surprise us if we understand the concepts leading to the procedure that is given above.

One should keep in mind, however, that if the pressure distribution is not due to a static incompressible fluid, one can still determine the force on a surface. However, one must resort to first principles to solve for the force components. The next example illustrates this.

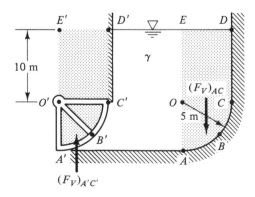

E' D' E D

γ

10 m

$(F_V)_{AC}$

O' C' O C

5 m

B'

A' A

B

$(F_V)_{A'C'}$

10 m $\perp$ page

Figure 2.5.7 Effect of surface's orientation on induced vertical force component due to a hydrostatic pressure distribution. Width is 10 m normal to page.

Example 2.5.1:

Pressures are measured on the sides of a structure as shown in Fig. 2.5.8. Consider these pressures to occur over the entire panel or segment, for convenience, even though there is a discontinuity at each junction. These pressures are due to a wind that is flowing from left to right. Determine the wind force on the structure per foot of width normal to the page.

Solution   This force is **F**, and it equals $F_x\mathbf{i} + F_z\mathbf{k}$. We shall calculate F_x on each of the segments by starting with Eq. (2.5.5). Because p and θ_x are constant on each segment, the integrals reduce to

$$F_x = -(p)(A)\cos\theta_x$$

Then

    I. $F_x = -(4.22 \text{ lb/ft}^2)(5 \text{ ft})(1 \text{ ft})\cos 180°$
         $= 21.1 \text{ lb}$
      $F_z = -(4.22 \text{ lb/ft}^2)(5 \text{ ft})(1 \text{ ft})\cos 90°$
         $= 0$

   II. $F_x = -(1.688 \text{ lb/ft}^2)(10 \text{ ft})(1 \text{ ft})\cos 144°$
         $= +13.66 \text{ lb}$
      $F_z = -(1.688 \text{ lb/ft}^2)(10 \text{ ft})(1 \text{ ft})\cos 54°$
         $= -9.92 \text{ lb}$

Since the normal vector makes an angle of 54° with the $+z$ axis, $\theta_z = 54°$

  III. $F_x = -(-2.954 \text{ lb/ft}^2)(10 \text{ ft})(1 \text{ ft})\cos 108°$
         $= -9.13 \text{ lb}$
      $F_z = -(-2.954 \text{ lb/ft}^2)(10 \text{ ft})(1 \text{ ft})\cos 18°$
         $= 28.09 \text{ lb}$

  VI. $F_x = -(0.422 \text{ lb/ft}^2)(5 \text{ ft})(1 \text{ ft})\cos 0°$
         $= -2.11 \text{ lb}$
      $F_z = -(0.422 \text{ lb/ft}^2)(5 \text{ ft})(1 \text{ ft})\cos 90°$
         $= 0$

(Show that on segments IV and V the results are $+9.13$ and -3.41 lb for the F_x's and 28.09 and -2.48 lb for the F_z's.)

Finally, the vector summation of all the forces on the six segments is found by the summation of all the individual vector components. (Show that this yields $\mathbf{F} = 29.2\mathbf{i} + 43.8\mathbf{j}$ lb.)

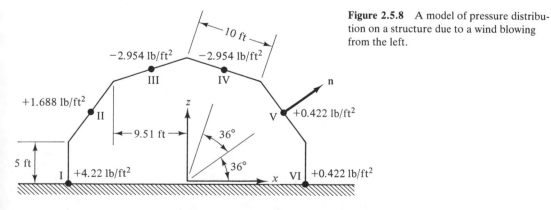

Figure 2.5.8   A model of pressure distribution on a structure due to a wind blowing from the left.

If the surface of Example 2.5.1 had been semicircular and the pressure had been given by $p = C(1 - 4\sin^2\theta_x)$, one could employ Eq. (2.5.5) to calculate F_x. Then $dA = r\,d\theta_x$, so that $dF_x = -p(r\,d\theta_x)\cos\theta_x$. This result is the same as if one had calculated the centripetal force as $dF = p(r\,d\theta_x)$ on a unit width and found the positive x component of this. These considerations lead to

$$F_x = -\int_0^\pi C(1 - 4\sin^2\theta_x)(r = \text{const})(L)\cos\theta_x\,d\theta_x$$

This x component will be found to be zero. One could have expected this result, since the pressure distribution is symmetric with respect to the z axis.

The determination of a concentrated force that is the static equivalent of the loading on a surface due to a pressure distribution is a first step in an evaluation of the forces needed to support the surface (i.e., the reactions). The following example illustrates this point.

Example 2.5.2:

The pressure distribution on the vertical surface shown in Fig. 2.5.9 is an approximation of what could occur when water flows past it in a special way. (The "nonhydrostatic" pressure distribution shown in Fig. 2.5.9 was chosen to underscore the generality of the ideas. They will be applied shortly to hydrostatic pressure distributions too.) Determine the reactions at locations A and B.

Solution The equilibrium of forces in the y and z directions is trivially satisfied. For the x direction, we can write

$$R_A + R_B = \text{load applied by pressure distribution}$$
$$= \int p\,dA = 9802\int_0^2 z(z^2 - 5z + 6)(3dz) \tag{2.5.10}$$
$$= 9802(8)\text{ N}$$

For the equilibrium of the moments, we can select any convenient point, say B, as a reference. Then

$$2R_A = \int z\,dF = 9802\int_0^2 z^2(z^2 - 5z + 6)(3dz) \tag{2.5.11}$$
$$R_A = 9802(3.6)\text{ N}$$

If, instead of the hydrodynamic loading that was prescribed in this example, we wished to employ a single force that would give the same reactions at A and B, what should be its magnitude? Where should be the point of application along AB? These

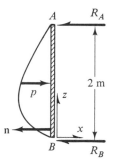

Figure 2.5.9 Loading of vertical plate by nonhydrostatic pressure distribution and associated forces needed to support the plate. Plate is 3 m wide normal to page.

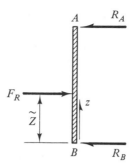

Figure 2.5.10 Statically *equivalent* loading of plate with a pressure distribution. (Note that supporting forces R_A and R_B are the same as in Fig. 2.5.9.)

questions refer to the values required for F_R and $\tilde{Z}$ in Fig. 2.5.10. Because R_A and R_B are to be the same as for the actual distributed load, one says that Fig. 2.5.10 represents the *statically equivalent loading* to that in Fig. 2.5.9. We recognize that the concentrated load, F_R, will bend and stress the plate differently from the distributed load. That, however, does not concern us. We demand only that the reactions at A and B be the same for the two loadings. The answer to our quest for F_R and $\tilde{Z}$ is straightforward. The equation $\Sigma F_z = 0$ is satisfied identically. The same is true for the y direction. The x direction provides

$$F_R = R_A + R_B \tag{2.5.12}$$

The equilibrium of the moments about point B requires that

$$\tilde{Z}(F_R) = 2(R_A) \tag{2.5.13}$$

Upon comparing Eqs. (2.5.10) and (2.5.12), we see that

$$F_R = \int (p)(dA)$$

Equations (2.5.11) and (2.5.13) require that

$$\tilde{Z}(F_R) = \int z\,(dF) = \int z\,(p)\,dA$$

Hence, F_R is equal to the magnitude of the total force applied to the plate by the pressure distribution. It is the *resultant force. The location of this resultant (force) is seen to be not arbitrary. It is located so that its moment about a reference point is equivalent to that of the actual distribution.* This requirement is inherent in all that follows in the next section.

2.6 FORCES AND MOMENTS DUE TO PRESSURE DISTRIBUTIONS ON SURFACES, IN GENERAL

For the case of the planar surface just considered, we have determined that the distributed load would have the same reactions at points A and B if the resultant load were applied as a concentrated force at a particular location $\tilde{Z}$, instead. We shall now treat nonplanar surfaces with pressure distributions to find the moments that result. If the surface is nonplanar, one *must,* at the outset, use infinitesimal force components. This is the significant feature of this section. The material that follows can be applied to the effects of pressure that occur on the wing of an airplane. Not only the lift and drag on it are of interest; one might want the optimal position of the main wing spar.

One criterion for such a "best" position could be a minimization of the torsional stresses in this spar.

We begin by finding the moment about point O in Fig. 2.6.1 due to the infinitesimal force $d\mathbf{F}$ that is applied at a point on the surface with coordinates x, y, z. This point has a radius vector $\mathbf{r} = x\mathbf{i} + y\mathbf{j} + z\mathbf{k}$. In addition, $d\mathbf{F} = dF_x\mathbf{i} + dF_y\mathbf{j} + dF_z\mathbf{k}$. For any pressure distributon, $dF_x = -p\,dA\,\cos\theta_x$, and we can see that $dA\,\cos\theta_x$ is the projection of the area onto the yz plane with an associated algebraic sign that is determined by the x component of the normal to the surface. We call this projected area dA_x because its normal is parallel to the x direction. One can think of dA_x as being the x component of $(dA)\mathbf{n}$, in which $\mathbf{n}$ is the normal to an infinitesimal area element of magnitude dA. The normal is pointed away from the surface to the particular location of interest and toward the fluid whose local pressure is p, just as in Eq. (2.4.1). Expressions for dF_y and dF_z are

$$dF_y = -p\,dA_y$$

and
$$dF_z = -p\,dA_z$$

in which
$$dA_y = dA\,\cos\theta_y$$
$$= \text{(projection of } dA \text{ onto a plane with a normal}$$
$$\text{in the } y \text{ direction)(algebraic sign from } \cos\theta_y)$$

The algebraic sign is associated with the sign of the y component of the normal to the surface. Similar things are true of dA_z.

The moment about a point due to $d\mathbf{F}$ and $\mathbf{r}$ is

$$d\mathbf{M_o} = dM_x\mathbf{i} + dM_y\mathbf{j} + dM_z\mathbf{k}$$

in which
$$dM_x = y\,dF_z - z\,dF_y$$
$$dM_y = z\,dF_x - x\,dF_z \qquad (2.6.1)$$
$$dM_z = x\,dF_y - y\,dF_x$$

(These elemental moments are the moments associated with the appropriate force components and perpendicular distances. The "right-hand" sign convention is implicit in the above equations.) These expressions for the components of $d\mathbf{M}$ are contained in

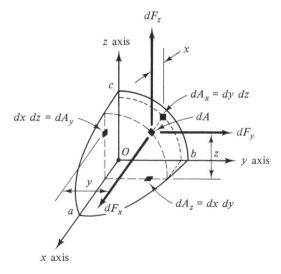

Figure 2.6.1 Generalized, three-dimensional surface (ellipsoid $x^2/a^2 + y^2/b^2 + z^2/c^2 = 1$) with infinitesimal surface tractions, dF_x, dF_y, dF_z. Center is 9 ft below free surface.

$$dM = r \times dF = \begin{vmatrix} i & j & k \\ x & y & z \\ dF_x & dF_y & dF_z \end{vmatrix} \qquad (2.6.2)$$

An integration of the moment components over the surface of a body—the incremental forces being due to a distribution of pressure on the surface—provides us with

$$M_x = \int_{\mathcal{A}} y\, dF_z - \int_{\mathcal{A}} z\, dF_y$$

$$M_y = \int_{\mathcal{A}} z\, dF_x - \int_{\mathcal{A}} x\, dF_z \qquad (2.6.3)$$

$$M_z = \int_{\mathcal{A}} x\, dF_y - \int_{\mathcal{A}} y\, dF_x$$

in which $\mathcal{A}$ reminds us that this integration is over the surface area of interest. Each of these integrals would have to be evaluated for a three-dimensional problem to obtain an answer for **M**. We did that in Example 2.5.2 and subsequently sought to "locate" the resultant through a defined distance $\tilde{Z}$. We shall now do something similar, but more general, and define certain distances $\tilde{Y}_x$, $\tilde{Z}_x$, and so on, by

$$M_x = \tilde{Y}_x F_z - \tilde{Z}_x F_y$$

$$M_y = \tilde{Z}_y F_{xy} F_z \qquad (2.6.4)$$

$$M_z = \tilde{X}_z F_y - \tilde{Y}_z F_x$$

These distances $\tilde{Y}_x$, $\tilde{Z}_y$, and so on,* are such that, when multiplied by the appropriate force component, they give the components of the moment that would be induced by the distributed pressure loading that actually occurs. Hence, by comparing Eqs. (2.6.3) and (2.6.4)

$$\tilde{Y}_x = \frac{1}{F_z} \int y\, dF_z \qquad \text{etc.} \qquad (2.6.5)$$

A two-dimensional example follows.

Example 2.6.1:

A preliminary design of a curved surface AB is shown in Fig. 2.6.2. Locate point O so that the pressure distribution on AB does not create any moment about an axis through O (perpendicular to the page).

Solution   We first locate a coordinate system at a convenient location. This place is not always an obvious one. We select point A because this gives the surface a simple equation. The x and z force components follow directly as

$$dF_x = -p\, dA_x = -p\,(\text{width})(-dz)$$

Then $$F_x = \int p\,(\text{width})(dz)$$

* The notation $\tilde{Y}_x$ should not intimidate the reader. It denotes the y position from the x axis of the force component in the z direction—the component corresponding to the letter that is not explicitly contained in $\tilde{Y}_x$. Similarly, $\tilde{Z}_y$ is the z position from the y axis of the x component of the force. These coordinates are often designated as those of the *"center of pressure."* Sometimes they are used to specify the *"lines of action"* of F_x, F_y, and F_z.

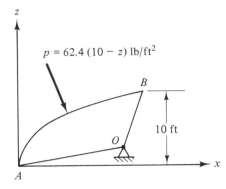

$p = 62.4\,(10 - z)\ \text{lb/ft}^2$

B

$10\ \text{ft}$

O

A

x

z

Figure 2.6.2 Surface AB with pressure distribution that must be supported at a single point, point O. Surface is 20 ft wide normal to page; and on AB, $x = z^2/4$.

and

$$F_z = -\int p\,(\text{width})(dx)$$

$$F_x = \int_0^{10} 62.4(10 - z)(20dz)$$

and

$$F_z = -\int_0^{25} 62.4(10 - 2x^{1/2})(20dx)$$

Now it is useful to calculate the M_y about point A and determine $\tilde{Z}_y$ and $\tilde{X}_y$. These are the distances where the resultant force's components F_x and F_z would, in effect, act. Consequently, at the points where these lines intersect there would be zero moment due to these components of $\mathbf{F}_R$. (Should the support, point O, be located at this junction, in view of the design objective?) A review of Eqs. (2.6.3) and (2.6.4) provides the following:

$$\tilde{Z}_y\,F_x = \int z\,dF_x = \int_0^{10} z[(62.4)(10 - z)(20dz)]$$

and

$$\tilde{X}_y\,F_z = \int x\,dF_z = \int_0^{25} x[62.4(10 - 2x^{1/2})(20dx)]$$

Since F_x and F_z were just previously found, the distances $\tilde{Z}_y$ and $\tilde{X}_y$ can be determined. These, under the conditions of this design problem, are the distances sought.

Example 2.6.2:

An octant of an ellipsoid (refer to Fig. 2.6.1) has a pressure distribution that depends, in this case, only on z, $p = 62.4(9 - z)$ lb/ft^2. Find $\tilde{X}_y$ and $\tilde{Y}_x$.

Solution Once again, Eqs. (2.6.3) and (2.6.4) give the definitions of these distances:

$$\tilde{X}_y = \frac{1}{F_z}\int x\,dF_z = \frac{1}{F_z}\int x(-p\,dA\ \cos\theta_z)$$

Now $dA \cos\theta_z$ for the surface ($\theta_z \le \pi/2$) is nonnegative. It is the magnitude of the projected area of dA onto the xy plane. This is illustrated in Fig. 2.6.1 as $dx\,dy$. (The sign of dA_z is positive, since the normal of the particular surface has a positive z component.) Consequently,

$$F_z\tilde{X}_y = \int x[-62.4(9 - z)\,dA_z]$$

We can see from the illustration that $dA_z = dx\,dy$, and the equation for an ellipsoid relates z with x and y on the surface. The result is

$$F_z\tilde{X}_y = \int_0^b \left(\int_0^{a(1 - y^2/b^2)^{1/2}} x\left\{ -62.4\left[9 - c\left(1 - \frac{x^2}{a^2} - \frac{y^2}{b^2}\right)^{1/2} \right] \right\} dx \right) dy$$

$\tilde{X}_y$ follows directly since F_z can be found from a similar integral—only the x coefficient would be missing. The actual computations may be lengthy, or some special integration techniques may be required, but the work can be done. The steps to the solution for $\tilde{Y}_x$ would be very little different from those above.

Figure 2.6.1 can aid us to find F_x and F_y on the curved surface that is shown there. (In what follows, one can assume that a general surface is displayed there.) The x component of force is $\int -p\,dA\,\cos\theta_x$, and $dA\,\cos\theta_x$ is, as before, dA_x. Hence, just as in the case of F_z, one would find the force components by integrating the pressure distribution over the appropriate projected area. For F_x, this projection is area cOb in Fig. 2.6.1. Thus, we can formally restate a previous result:

> The horizontal component of the force on any curved surface is the same as that on the appropriate plane vertical projection.

The distances $\tilde{Z}_y$ and $\tilde{Y}_z$ for the F_x on the curved surface will, from the definitions given by the associated integrals, be the same as for the F_x on the plane surface projection, $\mathcal{A}_x$, of the curved surface. The utility of these planar areas in hydrostatic cases lies with some of their geometric properties. Some of these properties for simple shapes are tabulated in handbooks; the use of such tables would obviate the evaluation of the integrals that arise in the work.

The fact that the vertical force on one side of a surface exposed to a quiescent fluid has a magnitude equivalent to the weight of the fluid above it agrees with our physical intuition about the matter. We came to that conclusion after studying the surface in Fig. 2.5.6 that has curvature in one plane only. An examination of the three-dimensional problem in Example 2.6.2 would lead to the same conclusion. Because we can associate the elemental vertical force component dF_z of a fluid at rest with an elemental weight, we can think of the distances $\tilde{X}_y$ and $\tilde{Y}_x$ as the coordinates of the center of mass of the appropriate fluid volume.

Example 2.6.3:

A metal tank is shown in perspective in Fig. 2.6.3. Three views are also given in Fig. 2.6.4, in which all of the dimensions are in feet. The tank is completely filled with water and

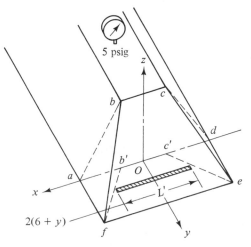

Figure 2.6.3 Geometry of three-dimensional tank completely full of water (Example 2.6.3).

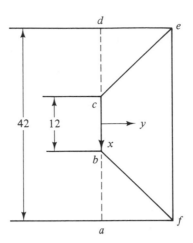

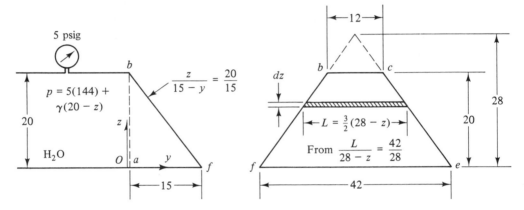

Figure 2.6.4 Projections of tank of Fig. 2.6.3 to facilitate calculation of resultant forces on an inclined surface and their lines of action.

pressurized at the top to 5 psig. Determine the force on the sloping surface *fbcef* that is due to the gauge pressure that acts on it. Find the "line of action" for each of the force components.

Solution

$$F_y = -\int p \, dA_y = -\int p(-dx\,dz) = \int_{A_{abcda}} [5(144) + \gamma(20 - z)] \, dx \, dz$$

$$= 5(144)\left(\frac{12 + 42}{2}\right)(20) + \gamma \int_0^{20} (20 - z)L \, dz$$

$$= 72(54)(10^2) + \gamma \int_0^{20} (20 - z)(\tfrac{3}{2})(28 - z) \, dz$$

$$= 3.888(10^5) + 62.4 \int_0^{20} \tfrac{2}{3}(560 - 48z + z^2) \, dz$$

$$= 3.888(10^5) + 41.6\left(560z - 24z^2 + \frac{z^3}{3}\right)_0^{20}$$

$$= 3.888(10^5) + 41.6(400)(28 - 24 + \tfrac{20}{3})$$

$$= 3.888(10^5) + 1.775(10^5) = 5.663(10^5) \qquad \text{lb}$$

$$F_z = -\int p \, dA_z = \int p \, dx \, dy = \int_{A_{gb'c'eg}} [5(144) + \gamma(20 - z)] \, dx \, dy$$

$$= 72(10)A_{gb'c'eg} + \gamma \int_{A_{gb'c'eg}} (20 - z) \, dx \, dy$$

$$= 72(10)\left(\frac{12 + 42}{2}\right)15 + 62.4 \int_0^{15} \left[20 - \frac{20}{15}(15 - y)\right]L' \, dy$$

$$= 2.916(10^5) + 62.4 \int_0^{15} [20 - \tfrac{20}{15}(15 - y)](2)(6 + y) \, dy$$

$$= 2.916(10^5) + 62.4\left[120 \int_0^{15} (6 + y) \, dy - \tfrac{40}{15}\int_0^5 (90 + 9y - y^2) \, dy\right]$$

$$= 2.916(10^5) + 62.4\left[120\left(6y + \frac{y^2}{2}\right) - \frac{40}{15}\left(90y + \frac{9y^2}{2} - \frac{y^3}{3}\right)\right]_0^{15}$$

$$= 2.916(10^5) + 13.104(10^5) = 16.02(10^5) \qquad \text{lb}$$

The contribution of dF_y to the moment about the x axis is $\int z \, dF_y$, which has been defined as $\tilde{Z}_x F_y$. Thus,

$$\tilde{Z}_x F_y = \int z \, dF_y = \int_{A_{abcda}} z[5(144) + \gamma(20 - z)] \, dx \, dz$$

in which the expression for dF_y is the same as the integrand for F_y. On surface $abcda$, the infinitesimal area element could be $L \, dz$ instead of $dx \, dz$. Therefore,

$$\tilde{Z}_x F_y = \int_0^{20} 5(144)zL \, dz + \gamma \int_0^{20} (20 - z)L \, dz$$

with
$$L = \tfrac{3}{2}(28 - z)$$

or
$$\tilde{Z}_x F_y = 720\left(\frac{3}{2}\right)\left[28\frac{z^2}{2} - \frac{z^3}{3}\right]_0^{20} + \gamma\left(\frac{3}{2}\right)\left[\frac{560z^2}{2} - \frac{48z^3}{3} + \frac{z^4}{4}\right]_0^{20}$$

$$= 3.177(10^6) + 2.246(10^6) = 5.423(10^6) \qquad \text{lb-ft}$$

Hence,
$$\tilde{Z}_x = \frac{\int z \, dF_y}{F_y} = \frac{54.23(10^5)}{5.663(10^5)} = 9.58 \qquad \text{ft}$$

In much the same way,

$$\tilde{Y}_x F_z = \int y \, dF_z = \int_{A_{gb'c'eg}} y[5(144) + \gamma(20 - z)] \, dx \, dy$$

On $gb'c'eg$, the area element could be $L' \, dy$ instead of $dx \, dy$, because it too is the projection of a constant pressure surface. Also, z and y are interrelated on the plane $gbceg$—where the pressure acts—through $z = 20(15 - y)/15$. Hence,

$$\tilde{Y}_x F_z = \int_0^{15} 5(144)[2(6 + y)]y \, dy + \gamma \int_0^{15} [20 - \tfrac{20}{15}(15 - y)][2(6 + y)]y \, dy$$

$$= 14.4(10^2)\left[3y^2 + \frac{y^3}{3}\right]_0^{15} + \frac{2\gamma}{15}(20)\int_0^{15} (6y^2 + y^3) \, dy$$

$$= 2.6(10^6) + 3.23(10^6) = 5.83(10^6) \qquad \text{lb-ft}$$

Consequently,
$$\tilde{Y}_x = \frac{\int y \, F_z}{F_z} = \frac{58.3(10^5)}{16.02(10^5)} = 3.64 \qquad \text{ft}$$

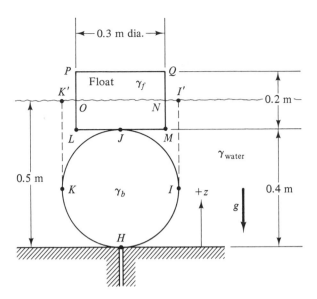

Figure 2.6.5 Ball-float system of Example 2.6.4 to illustrate hydrostatic forces on submerged and partially submerged bodies. Hole at H is at atmospheric pressure.

Example 2.6.4:

What must be the specific gravity of the float in Fig. 2.6.5 so that the ball valve to which it is attached begins to lift off when the water is 0.5 m deep?

Solution To solve the problem, we seek the condition in which the reaction at the hole begins to vanish. At this condition, the downward body force—the "weight" of the ball-float system—will be balanced by the forces which act on all the surfaces of the system. The weight of the system, due to gravitational attraction of the bodies, is

$$(\gamma_f \, \forall_{PQML} + \gamma_b \, \forall_{JIHK})$$

The remaining forces in the system are as follows:

Force (up) on surface $LJM = \gamma_{\text{water}} \, \forall_{ONML}$

$$= \gamma_{\text{water}} \left(\begin{array}{c} O \qquad\qquad N \\ \text{\rule{2cm}{0.8cm}} \\ L \qquad\qquad M \end{array} \right)$$

$$= \gamma_{\text{water}} \left[(0.3)^2 \frac{\pi}{4}(0.1) \right]$$

Force (down) on surface $KJI = \gamma_{\text{water}}(\forall \text{ "above" } KJI)$

$$= \gamma_{\text{water}} \left(\begin{array}{c} K' \qquad I' \\ \\ J \\ K \qquad\qquad I \end{array} \right)$$

$$= \gamma_{\text{water}} \left[\frac{\pi(0.4)^2(0.25)}{4} - \frac{4}{6}\pi(0.2)^3 \right]$$

Force (up) on surface $KHI = \gamma_{\text{water}}(\forall \text{ "above" } KHI)$

$$= \gamma_{\text{water}}$$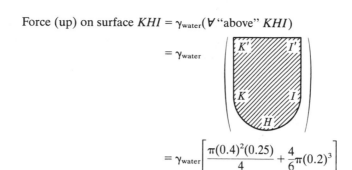

$$= \gamma_{\text{water}}\left[\frac{\pi(0.4)^2(0.25)}{4} + \frac{4}{6}\pi(0.2)^3\right]$$

(The vertical force on the ball is, consequently, the vector sum of the last two expressions.)

Force on ball's surface $= +\gamma_{\text{water}}$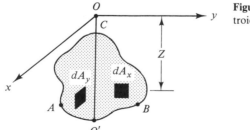

$$= \gamma_{\text{water}}\,\forall_{KJIH}$$

$$= \gamma_{\text{water}}\,\forall_{\text{sphere}}$$

$$= \gamma_{\text{water}}\left[\frac{4}{3}\pi(0.2)^3\right]$$

The forces on the surfaces of the ball and float must be balanced by the weight of the assembly at incipient "liftoff." A numerical expression for this statement is

$$-\left\{\gamma_f\left[(0.3)^2\frac{\pi}{4}(0.2)\right] + \gamma_b\left[\frac{4}{3}\pi(0.2)^3\right]\right\} + \gamma_{\text{water}}\left[\frac{4}{3}\pi(0.2)^3\right] + \gamma_{\text{water}}\left[(0.3)^2\frac{\pi}{4}(0.1)\right] = 0$$

Note that upward forces have been associated with the positive sign. If $\gamma_b/\gamma_{\text{water}}$ is known, $\gamma_f/\gamma_{\text{water}}$ can be determined; the latter is the specific gravity of the float.

The net vertical force on a submerged body that is due to the effects of hydrostatic pressure is called the *buoyant force* and is referred to as the body's *buoyancy* in the particular medium. If the body is completely submerged, the buoyant force is equivalent to the weight of the fluid displaced by the body.

In the very special case of a hydrostatic pressure distribution, $\partial p/\partial z = -\gamma$, the expression for F_x on a curved surface takes a form that suggests another means for evaluation. If Z is measured downward from the free surface, one has

$$|F_x| = |\int p\, dA_x| = \gamma|\int Z\, dA_x|$$

for constant γ. The terms in the integrand are shown in Fig. 2.6.6. Even though one is

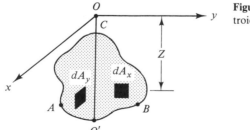

Figure 2.6.6 Definition diagram for centroids

dealing with the magnitude of the x component of the force on a curved surface, the integral requires an integration over the planar region $O'BC$. The integral $\int Z\,dA_x$ is used to define the centroidal distance $\bar{Z}$ of this plane area. This is in accord with the definition from integral calculus. These observations permit one to write

$$|F_x| = \gamma|\textstyle\int Z\,dA_x| = \gamma\bar{Z}A_x = (\gamma\bar{Z})A_x \qquad (2.6.6)$$

This relationship implies that if $\bar{Z}$ and A_x were known, then the magnitude of F_x could be obtained by multiplying the pressure at the centroidal depth $\bar{Z}$ by the appropriate projected area. Thus, in Fig. 2.6.7, in which the gate is not a curved surface,

$$F_y = [(H - 3)\gamma][6(1)]$$

since the centroid of the rectangular gate is 3 m from the bottom.

If one wishes to determine the moment induced by the pressure distribution associated with the plane surface in Fig. 2.6.7, a suitable starting point would be

$$M_x = \textstyle\int y\,(dF_z) - \int z\,(dF_y)$$

from equation group (2.6.3). In this particular case, $Z = -z$ and $dF_z = 0$, so that for axes located at E,

$$(M_x)_E = \textstyle\int Z(\gamma Z)\,dA_y = \gamma\int Z^2\,dA_y$$

After referring again to Fig. 2.6.6, it becomes clear that the last integral is the definition of the planar moment of inertia of the area, with respect to the x axis. This quantity is sometimes designated I_{xx}. If it is remembered that a plane surface—not necessarily rectangular—and the associated pressure distribution are being considered now, the x component of the moment through E can be written as

$$(M_x)_E = \gamma I_{xx}$$

For this special surface, the moment of the pressure distribution with respect to the same point is $\tilde{Z}_x F_y$. One can equate these expressions for the moment and obtain

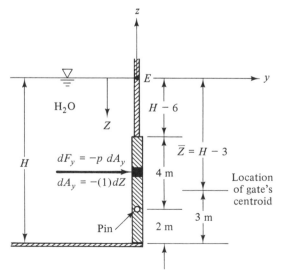

Figure 2.6.7 Hydrostatic forces on a vertical planar gate that can rotate about a horizontal axis. Gate is of unit width normal to page.

Pressure Distributions in Fluids without Relative Motions   Chap. 2

$$(\tilde{Z}_x)(\gamma \overline{Z})A_y = \gamma I_{xx}$$

so that

$$\tilde{Z}_x = \frac{I_{xx}}{\overline{Z}A_y}$$

This result is concise but potentially disconcerting to the reader who does not understand its origins and meaning. It gives, for the purposes of static equilibrium, the location of the resultant load on a submerged planar surface. It does so in terms of the geometric properties of this surface. If these geometric properties are readily available, one should use the results. If they are not, or if the pressure distribution is not hydrostatic, or if doubts arise about the applicability to a particular case, one should use the basic principles given by Eq. (2.5.5) and (2.6.3).

The last expression can be extended to give helpful qualitative and quantitative results. If I_{xx} is expressed about an axis through the centroid of the planar area (and parallel to the free surface), then

$$I_{xx} = I_{\overline{xx}} + (\overline{Z})^2 A_y$$

This means that

$$\tilde{Z}_x = \overline{Z} + \frac{I_{\overline{xx}}}{\overline{Z}A_y} \qquad (2.6.7)$$

This equation provides a more convenient means of finding $\tilde{Z}_x$, because $I_{\overline{xx}}$ is often tabulated. Equally important, however, is the fact that $\tilde{Z}_x \geq \overline{Z}$, with $\tilde{Z}_x = \overline{Z}$ the limiting case as the depth of the centroid of planar area, $\overline{Z}$, increases without limit. This qualitative relationship between $\tilde{Z}_x$ and $\overline{Z}$—two different conceptual ideas—is a useful rule of thumb, provided that the significance of $\tilde{Z}_x$ and $\overline{Z}$ is understood.

Example 2.6.5:

Find the depth H of the fluid at which the gate shown in Figure 2.6.7 is in equilibrium— rotating neither clockwise nor counterclockwise.

Solution   After a bit of reflection, it becomes obvious that the resultant force associated with the pressure distribution must go through O. O is located at $(H - 2)$ from the free surface, according to the figure. Thus,

$$\tilde{Z}_x = (H - 3) + \frac{(1)(6)^3/2}{(H-3)[6(1)]}$$

from Eq. (2.6.7), and $\tilde{Z} = (H - 2)$, the position of the pin.

Does $H = 6$ satisfy these two relationships?

Example 2.6.6:

The effective location of the vertical force component on surface AC in Figure 2.5.7, $\tilde{Y}_x$, can be found by finding the center of mass of the fluid in $ABCDEOA$. Since the liquid is assumed to be homogeneous in density, this position is equivalent to the centroid of the volume $ABCDEOA$. This means that, for y measured horizontally from O,

$$\tilde{Y}_x[(F_z)_{AC}] = -\gamma\left(\int_{ABCOA} y \, d\mathbf{V} + \int_{OCDEO} y \, d\mathbf{V} \right)$$

with

$$(F_z)_{AC} = -\gamma(\mathbf{V})_{ABCDEOA}$$

Consequently, the specific weight can be canceled to give

$$\tilde{Y}_x = \frac{\bar{y}_1 \forall_1 + \bar{y}_2 \forall_2}{\forall_1 + \forall_2}$$

in which $\bar{y}_1$ is the distance to the centroid of the volume $ABCOA$ from point O and $\bar{y}_2$ is the appropriate centroidal distance of the volume $OCDEO$. The geometry of Fig. 2.5.7 gives

$$\tilde{Y}_x = \frac{\left(\frac{4}{3}\frac{5}{\pi}\right)\left(\frac{\pi 5^2}{4}\right) + \frac{5}{2}(5)(10)}{\pi 5^2/4 + 50}$$
$$= 2.61 \text{ m}$$

2.7 LINEAR, RIGID-BODY ACCELERATION

Under conditions that can be easily achieved, it is possible to accelerate a container of fluid so that its contents move as a rigid body, i.e., so that there are no relative motions within the fluid. Consequently, we must conclude that there would be no shear stresses in the fluid; only normal stresses would occur. Thus, the situation is sufficiently similar to a quiescent fluid that theoretical analysis is quite easy. These stresses will now be calculated under the *assumption* that the acceleration has been constant for a sufficiently long period of time that any transient motions have ceased.

The situation is depicted in Fig. 2.7.1, in which a tank of fluid is accelerating upward at a rate of a_z. The actual configuration of the container is unimportant now, and we should concentrate our attention on the small element of fluid particles that is shown as an inset in Fig. 2.7.1. This element of fluid is in a gravity field acting in the $-z$ direction. Accordingly, the body force on the fluid in the $+z$ direction is $-\rho g (d\forall)$, in which $d\forall$ is the volume of the element $dx\, dy\, dz$. If we designate the pressure at the center of the element as p, the pressures at the upper and lower surfaces are as shown

Figure 2.7.1 Representation of container filled with water and being accelerated upward. Detail shows infinitesimal element of fluid and pressures on its horizontal surfaces.

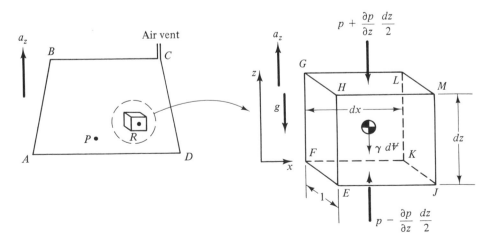

Pressure Distributions in Fluids without Relative Motions   Chap. 2

in Fig. 2.7.1 (in accordance with the assumptions of a continuum and the concept of differentials from the calculus). The net upward force due to the surface stresses is

$$\left(p - \frac{\partial p}{\partial z}\frac{dz}{2}\right) dx\,(1) - \left(p + \frac{\partial p}{\partial z}\frac{dz}{2}\right) dx\,(1) = -\frac{\partial p}{\partial z} dx\,dz \quad \text{for } dy = 1$$

since only pressure is to be considered. The acceleration in the $+z$ direction (upward) has been given as a_z. If a_z is positive, then the container is accelerating upward. But if in some situation a_z is negative, then the container will be accelerating in the $-z$ direction, which is downward. (If the container is moving upward during this time of negative a_z, we say that deceleration is occurring.) Thus, when we write a_z in subsequent expressions, we recognize the generality of what is being expressed, and we shall allow the sign of a_z to correspond with the particular situation at hand. The expressions for body and surface forces can now be related to the acceleration by Newton's laws:

$$-\left(\frac{\partial p}{\partial z}\right) dx\,dz - \rho g\,dx\,dz = a_z \rho\,dx\,dz$$

or

$$\frac{\partial p}{\partial z} = -\rho(a_z + g)^* \tag{2.7.1}$$

After examining the role of the term ρg in Eq. (2.7.1), one can infer that when the fluid is accelerating in a horizontal direction, taken as the x direction,† there is a variation of pressure in that direction. If we refer to Fig. 2.7.1 again but let $a_x \neq 0$, we deduce from Eq. (2.7.1) that

$$\frac{\partial p}{\partial x} = -\rho a_x \tag{2.7.2}$$

The pressure is assumed to vary continuously in the fluid, so that one can combine Eqs. (2.7.1) and (2.7.2) with the general statement for the pressure differential $dp = (\partial p/\partial x)\,dx + (\partial p/\partial z)\,dz$ to obtain

$$dp = -\rho a_x\,dx - \rho(a_z + g)\,dz \tag{2.7.3}$$

This result can be integrated between two points, say, R and P, in Fig. 2.7.1 to give

$$p_P - p_R = -\rho a_x(x_P - x_R) - \rho(a_z + g)(z_P - z_R) \tag{2.7.4}$$

Equation (2.7.3) could also be used to locate two neighboring points that were at the same pressure. Then the value of dp in Eq. (2.7.3) would be zero, so that for a prescribed set of values for a_x, a_z, and g one could determine the required interrelationship between dx and dz. The ratio of these values is

$$\frac{dz}{dx} = -\frac{a_x}{a_z + g} \tag{2.7.5}$$

---

* The reader is advised to determine the pressure gradient in the vertical direction, $\partial p/\partial z$, if the body is accelerating upward with a magnitude of 3 m/s². [Ans.: $(-12.82)(10^3)$ kg/m²-s²] What is the vertical pressure gradient when the body is accelerating downward with a magnitude of 3 m/s²? [Ans.: $(-6.82)(10^3)$ kg/m²-s²] What is the pressure gradient when the downward acceleration has a magnitude of 9.802 m/s²? [Ans.: Zero]

† In defining the x direction in this way, a_y will be identically zero.

But the ratio of the vertical displacement between two points to the horizontal distance between them is the slope of the line connecting them. This implies that Eq. (2.7.5) gives the slope of the surfaces of constant pressure in a fluid undergoing acceleration. (If p_P were set equal to p_R in Eq. (2.7.4), one would obtain the equation for planes of constant pressure.) If there is a gas-liquid interface in the container, it is often assumed that the gas is everywhere at the same pressure because of its small density with respect to the liquid. The interface is then taken as a surface of constant pressure.

Example 2.7.1:

Determine the pressure at location Q in Fig. 2.7.2 when the container is accelerating with $a_x = 4.9$ m/s^2.

Solution When the vessel accelerates, the free surface will be inclined at an angle given by Eq. (2.7.5); consequently,

$$\tan \theta = -\frac{4.9}{0+9.8} = -\frac{1}{2}$$

While the slope of the liquid's surface is known, the position of point R, where the pressure is the same as the air's (atmospheric pressure, because of the vent hole), remains unspecified. The location of R (or S) can be determined if one observes that the volume of the liquid has remained unchanged during the acceleration. It will be *assumed* initially that the air-water interface is below the corner C, as indicated in Fig. 2.7.2, rather than to the right of it, while the tank is accelerating. The geometric properties of this figure give us

$$\frac{PR}{PO} = \frac{1}{2} = \frac{PR}{0.75/2} \qquad PR = \frac{0.75}{4} = 0.188$$

The length of the wetted surface RQ is now known, so that the pressure p_Q can be determined from Eq. (2.7.4). (There is no need in this particular case to consider the possibility of point P being located to the right of point C, because the length PR was found to be less than PC.)

$$p_Q - p_{atm} = -(1000)(4.9)(0-0) - (1000)(0+9.8)[0-(0.3+0.188)] = 4.78 \text{ kPa}$$

(Determine the value of a_x so that the surface at the right just reaches the corner at A. *Ans.:* 7.84 m/s^2.)

The work in this section has been a prelude to much that will follow. The simplest case of Newton's laws—statics—was considered previously. Currently, we have con-

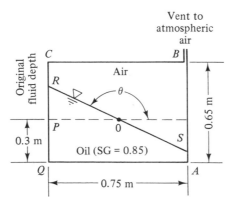

Vent to atmospheric air

Figure 2.7.2 Partially filled container whose liquid is being accelerated rectilinearly as a rigid body. a_x and a_z are constant.

sidered the linear acceleration of fluid particles, and this acceleration was constant throughout the fluid region. In Sec. 2.8, centripetal acceleration will be treated. In examining it, we shall once again consider the forces and accelerations to which an infinitesimal fluid particle is subjected.

2.8 ACCELERATION AND PRESSURE IN ROTATING FLUIDS

Rigid-body rotation is the name given to the motion when the fluid is prescribed to move without relative motion about an axis of rotation. This is just as it would be if the fluid were a rigid body. The tangential velocity V_θ is given by

$$V_\theta = \Omega r \tag{2.8.1}$$

and the radial velocity is zero.

Equation (2.8.1) implies no relative motion between fluid particles, so that only normal stresses occur.

Because point P in Fig. 2.8.1 is at the center of the region $ABCDA$, which circumscribes a set of particles that hold our interest for the moment, it is convenient to give P the coordinates of r, θ, and z. This means that the arc AB has a length of $(r - dr/2)\, d\theta$ and arc DC, $(r + dr/2)\, d\theta$. The faces BC and AD are radial rays with a length of dr. The pressure p is assumed to be known at point P, so that a representative value can be specified on the four faces whose traces are shown in Fig. 2.8.1. These values are obtained from those at P by the appropriate Taylor series. To determine the sum of the forces in the $+r$ direction, we must include the $+r$ components of the forces on AB, DC, BC, and AD. The last two faces are inclined in the θ direction, as shown, by the angle $\pm d\theta/2$. One can easily deduce that on BC the normal force must be

$$\left(p + \frac{\partial p}{\partial \theta}\frac{d\theta}{2}\right)(dr)\,dz$$

and the component of this force in the $+r$ direction is

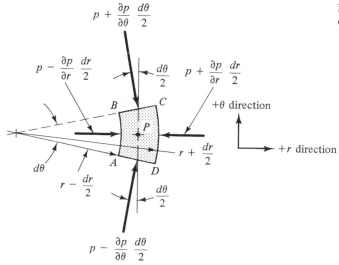

Figure 2.8.1 Free-body diagram of packet of fluid particles (cylindrical coordinates).

$$\left(p + \frac{\partial p}{\partial \theta}\frac{d\theta}{2}\right) dr\,dz\ \sin\frac{d\theta}{2} = \left(p + \frac{\partial p}{\partial \theta}\frac{d\theta}{2}\right) dr\,dz\frac{d\theta}{2} \qquad \text{if } \frac{d\theta}{2} \to 0$$

The $+r$ contribution of the forces on BC and AD is then

$$p\,dr\,dz\,d\theta$$

because of the minus sign in the expression for the pressure on AD. The contribution to the forces in the $+r$ direction on AB is

$$+\left(p - \frac{\partial p}{\partial r}\frac{dr}{2}\right)\left(r - \frac{dr}{2}\right) d\theta\,dz$$

and on DC, $$-\left(p + \frac{\partial p}{\partial r}\frac{dr}{2}\right)\left(r + \frac{dr}{2}\right) d\theta\,dz\,*$$

The total of the surface forces in the $+r$ direction on the four above-mentioned faces is $-(\partial p/\partial r)(r\,dr\,d\theta\,dz)$. Because gravity acts perpendicular to the $r\theta$ plane, there is no component of the "weight" force, the only body force considerd here, in either the $+\theta$ or the $+r$ direction. Also, we are considering a situation in which only pressure occurs. Hence, we can equate the total force found above with the product of the mass and the $+r$ acceleration of the packet of fluid. This means

$$-\frac{\partial p}{\partial r}(r\,dr\,d\theta\,dz) = \rho(r\,dr\,d\theta\,dz)a_r, \qquad (2.8.2)$$

in which a_r is the radial acceleration and $r\,dr\,d\theta\,dz$ is the volume of the fluid region.

We know that when a particle is executing a circular path, the radial acceleration has a magnitude of V_θ^2/r, in which V_θ is the speed in the tangential—or θ—direction; the direction of this acceleration is toward the center (in the $-r$ direction). In short,

$$a_r = -\frac{V_\theta^2}{r} \qquad (2.8.3)$$

The result of the last two equations is

$$\frac{\partial p}{\partial r} = \rho\left(\frac{V_\theta^2}{r}\right) \qquad (2.8.4)$$

We shall use this result often. Note that the right-hand side of Eq. (2.8.4) is never negative. Thus, this equation indicates that for the circular flows under consideration, the pressure increases with increasing r. For a fluid inside a container that has been rotating for a long time, the fluid will be rotating as a rigid body because of the effect of the walls and viscosity. In this case, $V_\theta = \Omega r$ and $\partial p/\partial r$ could be found without difficulty.

If one were now to apply Newton's laws in the z direction and allow for an acceleration in this direction, one would again conclude that

$$\frac{\partial p}{\partial z} = -\rho(g + a_z) \qquad (2.8.5)$$

---

* Explain the negative sign.

From this, one can find that

$$dp = \rho \frac{V_\theta^2}{r} dr - \rho(g + a_z) dz \qquad (2.8.6)$$

because the partial derivatives in $dp = (\partial p/\partial r)\, dr + (\partial p/\partial z)\, dz$ have just been determined in Eqs. (2.8.4) and (2.8.5). For rigid-body rotation,

$$dp = \rho \Omega^2 r\, dr - \rho(g + a_z)\, dz \qquad (2.8.7)$$

An integration of this equation produces

$$p_2 - p_1 = \rho \frac{\Omega^2}{2}(r_2^2 - r_1^2) - \rho(g + a_z)(z_2 - z_1) \qquad (2.8.8)$$

On surfaces of constant pressure, dp in Eq. (2.8.7) is zero. Hence, for $V_\theta = \Omega r$,

$$0 = \rho \Omega^2 r\, dr - \rho(g + a_z)\, dz$$

so that

$$(\text{constant})_1 = \rho \frac{\Omega^2 r^2}{2} - \rho(g + a_z)z$$

or

$$z - (\text{constant})_2 = \frac{\Omega^2 r^2}{2(g + a_z)} \qquad (2.8.9)$$

The surfaces of constant pressure for a fluid that is rotating about an axis parallel to the line of action of gravity has been shown in Eq. (2.8.9) to be a paraboloid of revolution. The $(\text{constant})_2$ provides for a whole family of surfaces. The interface between the air and the liquid for an open circular container should be a constant-pressure surface, a surface of atmospheric pressure. Thus, we have predicted its shape.

Example 2.8.1

A cylinder of water with a closed base and top is rotating as shown in Fig. 2.8.2. Determine the pressures at points A, B, and C.

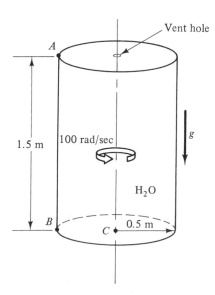

Figure 2.8.2 Closed container rotating about a vertical axis. (If the container is completely closed, it can be assumed that the pressure on the axis of rotation at the uppermost point in the container is the same as it was before the rotation began. This assumes that the liquid is incompressible and the temperature of the few molecules of vapor remains constant.)

Solution Equation (2.8.8) is appropriate here. Using the notation of the figure and remembering that along the axis there are no effects due to the rotation, we can find the pressure at point C immediately as

$$p_{vent} - p_C = \rho \frac{\Omega^2}{2}(0^2 - 0^2) - \rho g (1.5 - 0)$$

$$= p_{atm}$$

At point A, one has the same elevation as at the vent hole, so that

$$p_{vent} - p_A = \rho \frac{\Omega^2}{2}(0^2 - 0.5^2)$$

The pressure at point B is also obtainable directly:

$$p_{vent} - p_B = \rho \frac{\Omega^2}{2}(0^2 - 0.5^2) - \rho g (1.5 - 0)$$

Plot the variation in pressure from point A to point B. Is it similar to a hydrostatic variation? Explain.

The solutions of all such problems follow a similar scheme. To facilitate the calculations,

1. Provide yourself with a careful drawing that describes the situation.
2. Establish a reference pressure and locate its position, (x_R, r_R), in the fluid region.
3. Locate a convenient coordinate system, one which need not have x_R and r_R equal to zero.
4. Remember that the volume of the fluid is conserved until conditions dictate that some of it must spill out of the container.*
5. Employ Eq. (2.8.8), as appropriate.

Example 2.8.2

What is the value of Ω for p_A to be at atmospheric pressure for the U tube shown in Fig. 2.8.3?

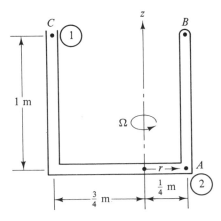

Figure 2.8.3   U tube rotating about a vertical axis. It is open at left side and closed at the right. Tube was initially full.

*If none has spilled out, the distance that the apex of the paraboloid recedes on the centerline is equal to the distance that the liquid rises along the wall. Can you prove this?

Solution Equation 2.8.8 again applies, because in its development the geometry of the container played no role.

If $p_A = 0$ gauge, then $p_A = p_{atm} = p_2$ in Eq. (2.8.8). Also, $p_1 = p_{atm}$ and $a_z = 0$. Hence,

$$0 = \frac{\rho\Omega^2}{2}\left[\left(\frac{1}{4}\right)^2 - \left(\frac{3}{4}\right)^2\right] - \rho(g)(-1)$$

$$\frac{2g}{(9-1)/16} = \Omega^2 = 2(9.8)\frac{16}{8} = 9.8(4)$$

and
$$\Omega = 6.26 \text{ rad/s}$$

What is p_B? We know that $p_A - p_B = \gamma(1 \text{ m})$. Then

$$p_B = p_A - (9802 \text{ N/m}^3)(1 \text{ m}) \qquad \text{Pa}$$

A sufficiently high Ω could give a value to p_A that would cause p_B to be p_{vapor}. Higher values of Ω would mean that the liquid in column AB would start to flow out and leave a vapor pocket at B. The extent of this pocket would depend on Ω. (Must p_A be in absolute pressure units when using p_{vapor}?)

Example 2.8.3

Find the surfaces of constant pressure for a cylinder of fluid that is rotating about a horizontal axis as shown in Fig. 2.8.4.

Solution Previously, when gravity was parallel to the axis of rotation, it was determined that the relationship between the radial pressure gradient and the radial acceleration was

$$\frac{\partial p}{\partial r}r\,dr\,d\theta = \rho r\,dr\,d\theta\,a_r$$

Because gravity now has a component in the radial direction, we can write

$$-\rho g\,\sin\theta\,(r\,dr\,d\theta) - \frac{\partial p}{\partial r}(r\,dr\,d\theta) = \rho r\,dr\,d\theta\,(a_r)$$

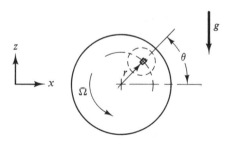

Figure 2.8.4 A cylinder of fluid that is rotating about a horizontal axis.

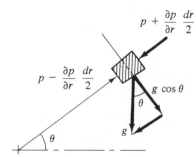

or

$$\rho g \, \sin \theta + \frac{\partial p}{\partial r} = \rho \Omega^2 r$$

since

$$a_r = \frac{-V_\theta^2}{r} = -r\Omega^2 \qquad \text{if } V_\theta = \Omega r$$

Because gravity has a component in the θ direction, it can be expected that $\partial p/\partial\theta \neq 0$. If the θ equation of $\mathbf{F} = m\mathbf{a}$ is written, one has

$$\left(p - \frac{\partial p}{\partial \theta} \frac{d\theta}{2} \right) dr - \left(p + \frac{\partial p}{\partial \theta} \frac{d\theta}{2} \right) dr - \rho g r \, dr \, d\theta \, \cos\theta = 0$$

if there is no acceleration in the θ direction. This results in

$$\frac{\partial p}{\partial \theta} = -(r \, \cos\theta)(\rho g)$$

Hence, from

$$dp = \frac{\partial p}{\partial r} dr + \frac{\partial p}{\partial \theta} d\theta,$$

$$dp = \rho \Omega^2 r \, dr - \gamma(\sin\theta \, dr + r \, \cos\theta \, d\theta)$$

so that

$$dp = \rho \Omega^2 d\left(\frac{r^2}{2}\right) - \gamma d(r \, \sin\theta)$$

For $dp = 0$ (i.e., a constant-pressure surface),

$$0 = \rho \Omega^2 d\left(\frac{r^2}{2}\right) - \gamma d(z)$$

An integration yields

$$\text{const}_1 = \rho \frac{\Omega^2}{2}(r^2) - \gamma z$$

or

$$\text{const}_2^2 = \frac{\rho \Omega^2}{2}(x^2) + \frac{\rho \Omega^2}{2}\left[z^2 - \underbrace{\frac{2g}{\Omega^2} z + \left(\frac{g}{\Omega^2}\right)^2}_{\text{I}} \right]$$

in which term I was selected to make the quantity in brackets in the equation a perfect square. Any terms needed in the last equation in order to preserve the previous equality were absorbed into the value of const_2^2. Does this last equation describe a circular cylindrical surface? Is its axis displaced from the axis of rotation? By how much?

2.9 SUMMARY

When there are no relative motions in a Newtonian fluid, it is hypothesized that shear stresses will be absent. Moreover, any normal stresses that occur will be free of any influence of such motions, so that the only normal stress will be the thermodynamic pressure. This pressure is isotropic.

Equations (2.7.1) and (2.7.2) give the relationship between spatial pressure variations and the acceleration and fluid density. They also include the effect of a body force due to gravitational attraction. These equations include the case for a static fluid. A manometer is a practical application of this case.

When the pressure variation is known, this function can be integrated to determine the pressure at various points within the fluid, provided that a (reference) pressure is stipulated at some point.

Figure 2.9.1 Parabolic mirror for astronomical observations that was constructed with a pool of mercury rotating at the correct speed. (Courtesy of Prof. E. Borra, Laval University)

Once the pressure distribution in a fluid is known, the forces that this pressure induces on surfaces can be found by integrating infinitesimal force components that are expressed in the manner of Eq. (2.5.5). If the surface is a plane surface, then one can orient the coordinate axes in such a way that two axes lie in the plane of the surface. This procedure can simplify some of the computations, because only one significant scalar equation results. For curved surfaces, however, one must use two or more force components. Moments are also induced upon surfaces submerged in a fluid. These too are in component form and must be integrated according to equation group (2.6.3). The calculation of such moments and the associated forces on a surface leads to the definition of the center of pressure, or the statically equivalent lines of action of the resultant forces.

If the fluid is at rest, then certain rules can be proved, e.g., that the vertical force on a submerged surface is equivalent to the weight of the fluid above it. However, the general aspects of the material presented in this chapter could be used whenever any normal stress, not necessarily a hydrostatic pressure distribution, is encountered.

The study of fluids rotating as a rigid body is interesting because the rotation introduces a centripetal acceleration. An analysis of the problem, with shear stresses necessarily absent, shows that the pressure increases as one moves away from the center of streamline curvature. Figure 2.9.1 shows a rotating tank of mercury. The parabolic surface of the mercury serves as a mirror for an astronomical telescope.

EXERCISES

2.2.1(a) What is $\partial p / \partial z$ at $z = 5$ ft and at 5 ft $\frac{1}{8}$ in.?
(b) Let $p = p_1 \exp[-gz/(p_1/\rho_1)] + C$, with C a constant. Determine C if the subscript 1 denotes conditions at sea level in a standard atmosphere ($p_1 = 101$ kPa and $T = 15°C$). What is $\partial p / \partial z$ at $z = 0$ and 1000 m?

2.2.2(a) Solve for the pressure at the interface and at the bottom of the tube if the coordinate system has its origin at the free surface.
(b) What is the pressure in the tube at a point 2 m from the bottom?
(c) If the tube is sealed at the top and the pressure

there is 300 kPa, what is the pressure at the bottom of the tube if the top pressure (300 kPa) is gauge pressure? If it is absolute pressure?

(d) Again, the tube is sealed at the top, but this time water fills the entire tube. The pressure at the top of the tube is the vapor pressure at 20°C. What is the pressure at the bottom of the tube? Is it gauge pressure?

2.2.3(a) What is the pressure in the middle of the tank?

(b) Determine the pressure at the bottom of the tank if the origin of the coordinate system is placed at the liquid-gas interface, where the gauge pressure is 0.3 atmospheres.

(c) If a liquid increases in density with increasing pressure (recall the bulk modulus), one can show that $z = K/g\rho + C_1$ and $p = -K \ln [C_2(z - C_1)]$. The latter relationship results from using the former with Eq. (2.2.13). What is the density of sea water at 1000 fathoms, assuming that its salinity is independent of depth? This salinity gives an SG = 1.03 at the free surface. What is the pressure there? What pressure would have been calculated if no compressibility had been used? (Note that K is the isothermal bulk modulus. Use an average value between 1 and 250 atmospheres. Must you use absolute pressure?)

(d) A research submarine records a water pressure of 5000 psi on a pressure gauge outside the hull. Assuming an average bulk modulus as in Exercise 2.2.3c, what is the submarine's depth? What depth would be inferred if one had assumed the density of water to be constant?

(e) Use Eq. (2.2.13) and the differential form of Eq. (1.6.3) to derive the expressions that are given in Exercise 2.2.3c.

2.3.1(a) Trichloroethylene—similar to CCl₄ but safer for humans—is used in the manometer instead of mercury. For the pressure difference of 8.65 psi, what would be the difference of elevation of the columns of manometer fluid?

(b) The system is to undergo maintenance, and air at 85°F and an average pressure of 40 psia is used in the system. What would be the difference of elevation of the trichloroethylene manometer for the 8.65 psi pressure difference in the system? What error would have been incurred if the system's fluid, air, had been neglected in the calculation?

(c) With water in the system, the manometer is inverted and an oil with an SG = 0.85 is used as the manometer fluid. What should be the pressure difference in the system if the levels of the manometer fluid differ by 39.37 in.?

2.4.1(a) For $p_0 = 0.05$ psig, $r_0 = 30$ ft, and $L = 180$ ft, what is F_z? What is the change of pressure from the bottom of this semicircular structure—perhaps a tennis practice area—to its highest point assuming a temperature of 70°F and assuming $p \approx$ constant? Would this hydrostatic pressure change affect the pressure on the inside surface of the curved structure appreciably? What would be the percent change?

(b) What are the magnitude and direction of the force due to p_0 on the plane whose trace is AOC and which has a length of L? How does this force relate to the F_z of the example? What is the net force on the structure—curved and plane surfaces—due to the internal pressure?

2.5.1(a) Explain the minus signs in the calculations for the force components in panel II. Explain also the two negative signs in the calculations for panel III.

(b) Affirm the values given for F_x and F_z on sections IV and V.

(c) Verify the force (both components) on sections I through VI.

2.5.2(a) Evaluate Eqs. (2.5.10) and (2.5.11) to obtain numerical results.

(b) If the plate were only 1 m high, the pressure distribution might be represented by $z(z^2 - 3z + 2)$. What would the reactions then be at A and B?

(c) What is the vertical distance below the free surface where the total force on the triangular plate in Fig. 2.4.1, in effect, acts?

(d) What is the moment about an x axis drawn through point O due to the pressure distribution in Fig. 2.5.1. (*Hint:* Are the elemental moments equal to $s(dF)$?) With this information, determine the distance from point O, as measured along OA, where the resultant force, in effect, acts.

(e) Refer to Exercise 2.5.2d but use Fig. 2.5.2 when answering the questions.

2.6.1(a) Determine F_x and $\tilde{Z}_y$.

(b) Determine the vertical force resultant and its "line of action."

(c) What would have been the vertical force resultant and its line of action if the surface whose trace is AB had been a plane?

2.6.2(a) What are F_z and $\tilde{X}_y$?

(b) Find $\tilde{Y}_x$.

(c) If the surface had been the octant of a sphere that had its center 10 ft below the surface of a lake, what would have been the effective line of action of the vertical component that was due to the pressure on the curved surface?

2.6.4(a) The ball is made of wood, SG = 0.9. What must the specific gravity of the float be for incipient motion of the ball-float system at the depth indicated?

(b) A new design of the ball-float system is proposed. The diameter of the ball is to be 0.3 m, the same as the diameter of the cylindrical float. What must be the density of the float if the ball has a SG = 0.9 for incipient "liftoff" at the water level of 0.5 m?

(c) The float in an alternative design is the same as in Fig. 2.6.5; however, the body that it supports is an inverted cone whose base is 0.3 m in diameter and 0.4 m in altitude. The tip of the cone is in the vent hole at the plane surface that is 0.5 m below the surface of the water. For a cone made of wood (SG = 0.9), what must be the specific gravity of the float?

2.6.5(a) Determine the moment about the x axis (perpendicular to the page) through point E due to the pressure distribution on the plane gate. What is the effective line of action, as measured from E, of the horizontal force on the gate?

(b) Where is the centroid of the plane area that constitutes the gate? What is the moment of inertia of the plane gate about an axis through E? Use this information to determine the "line of action" of the horizontal resultant.

(c) Compare the results and the effort associated with Exercises 2.6.5a and b. Which method do you prefer, and why?

2.7.1(a) What is the pressure at point Q when the free surface at the right just reaches point A?

(b) Determine the value of the horizontal acceleration so that the surface at the left reaches the corner at point C.

(c) What is the pressure at point Q for the conditions of Exercise 2.7.1b?

(d) What is the pressure at points C and Q for a horizontal acceleration of $1.2g$?

(e) Repeat Exercise 2.7.1a but keep $a_x = 4.9$ m/s². What is a_z? What is p_Q?

2.8.1(a) For an angular velocity of 12 rad/s, what is the pressure at point B?

(b) What must be the angular velocity if the pressure at point B is 2 atmospheres gauge pressure?

(c) If the tank were only half full of liquid, the free surface would become a paraboloid of revolution. Why? For an angular velocity of 2 rad/s, what is the equation of this surface? Where does this surface touch the sides of the tank? What is the pressure at point B? Where is the location of the free surface at $r = 0$? What is the pressure at point C?

(d) In the Exercise 2.8.1c, what is the ratio of the increase of elevation along the tank wall—the increase over the static condition—to the decrease in height along the axis? In view of this, what should be the depth of fluid along the axis when the rotational speed is increased so that the fluid reaches point A? What is the angular velocity for this situation?

2.8.2(a) What is the angular velocity if the pressure at A is 50 kPa *absolute pressure*?

(b) What must be the angular velocity if the pressure at B is 2 psi above the vapor pressure of water at 68°F?

(c) If the axis of rotation is moved $\frac{1}{2}$ m to the left, what is the pressure at A and B for the angular velocity in the example?

PROBLEMS

2.1 An underground water source is at atmospheric pressure. The inlet pipe of a pump that is located at ground level should not have a static pressure (i.e., a pressure when no water is flowing) that is less than 5 psi above the vapor pressure. In Fig. P.2.1, what can be the maximum value of H if the water temperature is 50°F? What would be the maximum height H if the water were coming from a reservoir of industrial hot water at 190°F?

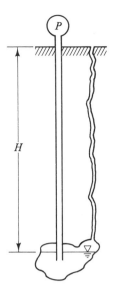

Figure P.2.1

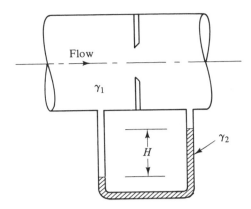

Figure P.2.3

2.2 A tube connecting two reservoirs is shown in Fig. P.2.2. What is the pressure difference between the reservoirs for $SG_1 = 2.3$, $SG_2 = 3.9$, $SG_3 = 1.0$, $SG_4 = 2.3$, and $H_1 = 52$ in., $H_2 = 16$ in., $H_3 = 9$ in., and $H_4 = 37$ in.?

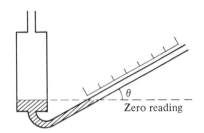

Figure P.2.4

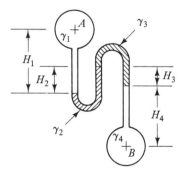

Figure P.2.2

2.5 The summer water temperature in the Arctic Ocean is given by $T = 5[1 - \tanh(Z - 10)]$, in which the depth, Z, is in meters and T is in degrees Celsius. In this range of temperatures, an approximation to the density in kg/m^3 is given by $\rho = 999.973 - (T - 4)^2/125$. What is the pressure gradient $\partial p / \partial Z$? Compare this gradient at 0, 5, 10, and 100 m. Compare the pressure at 10 m with that obtained by using the density at 5 m as an "average" (i.e., constant) value from the free surface to 10 m. [The integral of $\tanh x$ is ln $\cosh x$; $(\tanh x)^2$ has the integral $x - \tanh x$.]

2.3 A manometer filled with an oil ($SG = 3.0$) is used to measure the pressure differential across an orifice in a pipe. The reading, H, in Fig. P.2.3 is 43.2 cm. What is the difference in pressure in this case?

2.4 An inclined-tube manometer ($\theta = 30°$ with respect to the horizontal) is used to measure the pressure differential in a chimney. Water is in the manometer. If a scale along the inclined tube reads 7.4 in., what is the pressure differential being measured? The ratio of the diameters of the tubes in Fig. P.2.4 is 8 : 1.

2.6 A cubical tank, 1 ft on a side, is filled five-sixths full through an opening at the top which is subsequently sealed airtight. There is a small hole in the bottom of the tank through which the contents—water—slowly leak. Keeping the pressure of the air above the water in mind, determine the final water depth in the tank. (Assume that no air enters the tank through the hole in the bottom.)

2.7 The mean change of temperature with height (called the *lapse rate*) in the atmosphere near the earth is about 0.7°C per 100 m of elevation. For hydrostatic conditions, determine the change of atmospheric pressure that can be expected in 100 m by using the expression for ρ from the ideal gas law and $T = (T_{surface} + 273) + (-0.007)z$ for z in meters and T in kelvins. What must be the change of elevation to lower the boiling point of water 2°C? Let $T_{surface} = 18°C$ and the pressure there = 101 kPa.

2.8 In Prob. 2.7, what would be the results if there were a zero lapse rate (i.e., an isothermal atmosphere)?

2.9 With a negative lapse rate, how does the density of dry air change with elevation? What must the lapse rate be in order for the density to be independent of elevation?

2.10 Calculate the force on the surface due to the presence of water that is located on one side of the vertical plane shown in Fig. P.2.10. How would the calculation procedure be different if the upper edge, *CA*, were 5 m below the free surface?

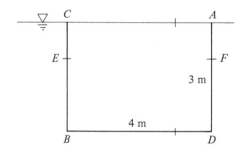

Figure P.2.10

2.11 Calculate the moment about an axis through line *CA* due to the hydrostatic pressure on one side of the plane shown in Fig. P.2.10.

2.12 Calculate the moment about an axis through line *CB* due to the hydrostatic pressure on one side of the plane shown in Fig. P.2.10.

2.13 Determine the position of a horizontal line, say, *EF*, in Fig. P.2.10 that divides the plate into two regions such that the force due to the hydrostatic pressure distribution is the same on each region.

2.14 Determine the position of a horizontal line, say, *EF*, in Fig. P.2.10 that divides the plate so that the moment about that line due to the hydrostatic pressure is the same for each region. Since the hydrostatic loading induces no moment about the line *EF* just located, where could one place a knife edge to resist the force on the plane associated with the loading?

2.15 Calculate the force on the plane due to the presence of water that is located on one face of the vertical surface shown in Fig. P.2.15.

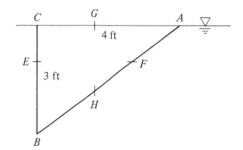

Figure P.2.15

2.16 Calculate the moment about an axis through line *CA* due to the hydrostatic pressure on one side of the plane shown in Fig. P.2.15.

2.17 Calculate the moment about an axis through the line *CB* due to the hydrostatic pressure on one side of the plane shown in Fig. P.2.15.

2.18 Determine the position of a horizontal line, say, *EF*, in Fig. P.2.15 that divides the plate into two regions such that the force due to the hydrostatic pressure is the same on each region.

2.19 Determine the position of a vertical line, say, *GH*, in Fig. P.2.15 that divides the plate into two regions such that the force due to the hydrostatic pressure is the same on each region.

2.20 Determine the position of a vertical line, say, *GH*, in Fig. P.2.15 that divides the plate so that the moment about that line due to the hydrostatic pressure is the same for each region. Since the hydrostatic loading induces no moment about the line *GH* just located, where could one place a knife edge to resist the force on the plane associated with the loading?

Problems

2.21 A submerged vertical plate ($H = 10$ ft, $r = 3$ ft) in Fig. P.2.21 has water on one side of it. Determine the force on the plate. If, in another installation, oil (SG = 0.84) wets the plate down to its centerline and water is below the centerline, what is the force on the plate?

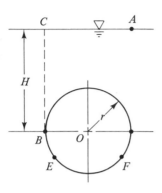

Figure P.2.21

2.22 What is the moment on the plate in Fig. P.2.21 about a horizontal axis through the centerline if water is on one side of the plate? What will be moment about the axis CB? ($H = 10$ ft, $r = 3$ ft.)

2.23 What is the moment induced on one side of the plate in Fig. P.2.23 about the axis BD by the hydrostatic pressure?

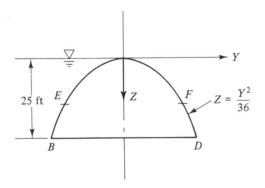

Figure P.2.23

2.24 Determine the location of a horizontal line in Fig. P.2.23, say, EF, such that the net moment due to the hydrostatic pressure distribution is zero. Would this position be suitable for placing an axis to pivot the surface? Why?

2.25 Let the specific gravity of a liquid increase linearly from 0.8 to 1.2 in 3 m. What is the force on the plate in Fig. P.2.10? How does this value for the force compare with the value calculated by using the average specific gravity over the surface?

2.26 What is the force F (per meter perpendicular to the page) that is necessary to hold the inclined gate closed if the tank is filled with water? Refer to Fig. P.2.26.

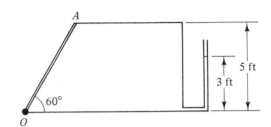

Figure P.2.26

2.27 When a fluid is in motion, the pressure distribution in the vertical direction might not be hydrostatic. For the flow under a gate, the pressure P and Q must be atmospheric (zero gauge). An approximation for the gauge pressure along PQ could be $p = (Z - h)(1 - Z/L)^{0.5}$. Determine the reactions at O to hold the gate in place. See Fig. P.2.27.

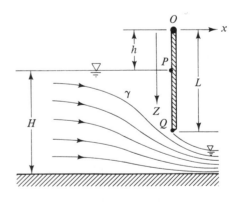

Figure P.2.27

2.28 Figure P.2.10 gives the vertical projection of a plane that makes an angle of 30° with the vertical. Calculate the total force on the plane if there is water on the upper side. What are the horizontal and vertical components of this force? Repeat the calculations for an angle of 45°. Compare the results for the resultant forces, as well as the horizontal and vertical components. Do the comparisons agree with your intuition? Explain.

2.29 Figure P.2.15 gives the vertical projection of a plane that makes an angle of 30° with the vertical. Calculate the total force on the plane if there is water on the upper side. What are the horizontal and vertical components of this force?

2.30 What are the x, y, and z components of the force on the surface ABC that is submerged in a pool as shown in Fig. P.2.30? [Volume of pyramid = (area of base)(height)/3.] Determine the moment about O.

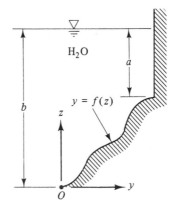

Figure P.2.31

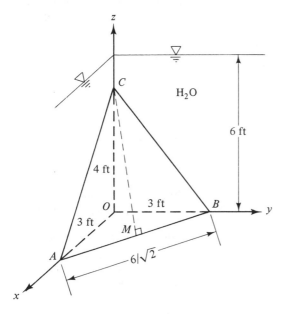

Figure P.2.30

2.31 The curve represented by $f(z)$ in Fig. P.2.31 is $(z/3)^{0.5}$. Determine the force components on this surface as well as their lines of action. The value of b is 12 ft; $a = 0$.

2.32 The curve represented by $f(z)$ in Fig. P.2.31 is $[9 - (z - 3)^2]^{0.5}$. ($b = 3$; $a = 0$). Determine the horizontal and vertical force components on the surface as well as their lines of action with polar coordinates, components, and $dF = p(dA)$, where dA is an element of the surface area, which is given by $r(d\theta)$(width).

Determine the moment about an axis through $(y, z) = (0, 3)$. What are the lines of action for the force components? Repeat the problem using Cartesian coordinates, and compare the procedural steps.

2.33 The curve represented by $f(z)$ in Fig. P.2.31 is $7 \sin(z/2k)$. ($a = 3$; $b = 12$; $k = 9$.) Determine F_z and F_y on this surface, as well as the moment induced about the x axis. Give the coordinates of the center of pressure for F_z and F_y.

2.34 If one horizontal plate is moving toward another below it, the pressure within the fluid layer between the plates is not hydrostatic. Its value is determined by the plate's motion and is given by $p = 0.5\rho\beta^2(b^2 - x^2)$, in which $\beta^2 = (3W/2\rho b^3)$ and the horizontal length of the plates is $2b$. (This is an approximation which results from letting the acceleration terms become small. Thus, the pressure distribution would support the moving plate at its terminal speed.) Find the force exerted on the upper plate by the pressure distribution and determine, thereby, the lift force on the plate. (W = plate's weight.)

2.35 If a plate of length b rests upon a horizontal surface as in Fig. P.2.35, then it will fall toward that surface, ultimately at constant angular velocity, in such a way as to create a pressure distribution within the fluid wedge of $p = 0.5\rho(\beta^2 + \dot{\beta} \cos 2\theta)(b^2 - r^2)$, with $\beta = -\dot{\theta}/2\theta$ and $\theta = C_2 \exp[-4t(Mg/\rho b^3)^{0.5}]$. What is the force normal to the plate? What is the moment on the plate due to this pressure dis-

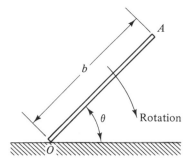

Figure P.2.35

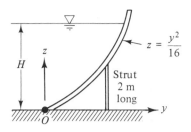

Figure P.2.37

tribution? What are the horizontal and vertical forces at point O? The horizontal reaction at O is due to friction and is dependent upon the vertical force. Is it possible that the horizontal force reaction at O will vanish? If this happens, what will prevent horizontal translation of the plate at point O? Can this conclusion be tested easily with a piece of stiff cardboard? M = plate's mass.

2.36 The retaining wall shown in Fig. P.2.36 has water on the left to an elevation of 3.5 m. Calculate the force in the struts if they are placed at 3 m intervals along the length of the retaining wall.

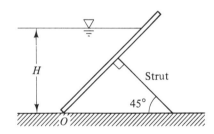

Figure P.2.36

2.37 The retaining wall shown in Fig. P.2.36 is designed so that the struts will fail if they must support a load of more than 500 N. The supports are placed 3 m apart. At what height of water will the supports fail?

2.38 The retaining wall shown in Fig. P.2.37 is designed so that the struts will fail if they must support a load of more than 500 N. The supports are placed 3 m apart. At what height of water will the supports fail?

2.39 The axis of a 10-sided cylinder is perpendicular to a slowly moving airstream. Each side of the polygon that forms the cross section is 1 cm long. One face is perpendicular to the airstream; it will be called panel I. The pressure is measured at the center of this panel as well as at the centers of the other nine. These faces are designated by roman numerals and proceed counterclockwise from I. The pressures are

$$p_I = 1.0 \text{ Pa}$$
$$p_{II} = p_X = 0.15 \text{ Pa}$$
$$p_{III} = p_{IX} = -0.6 \text{ Pa}$$
$$p_{IV} = p_V = p_{VI} = p_{VII} = p_{VIII} = -0.4 \text{ Pa}$$

Determine the force per unit length on this small body due to this low-speed pressure distribution. (Later, we shall put the phrases *low-speed* and *high-speed* into perspective.)

2.40 A house has a side view that makes it appear to be half of a hexagon. The vertical walls are 9 ft high. The roof is 18 ft long (measured along a line that is inclined 30° to the horizontal). The dimension of the house that is perpendicular to the page, its length, is 25 ft. The pressure is measured at the base of one wall, say, the left, and equals 4.22 lb/ft². At the center of the right and left roof segments the pressure is -3.38 lb/ft². At the base of the right wall the pressure is -0.42 lb/ft². What is the force on this house? Does the roof appear to be "pushed in" or "pushed out"?

2.41 The curve $g(z)$ that generates the surface in Fig. P.2.41 is $(z/2)$. Determine the force components on the surface if $\theta = 1.5\pi$ and $b = 10$ ft. What are their lines of action? The fluid is below the surface.

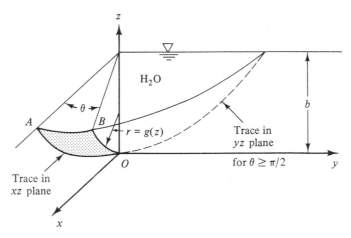

z

H_2O

A B

θ

$r = g(z)$

b

Trace in
yz plane

for $\theta \geq \pi/2$

y

O

Trace in
xz plane

x

Figure P.2.41

2.42 The curve $g(z)$ that generates the surface in Fig. P.2.41 is $[9 - (z - 3)^2]^{0.5}$, and $b = 3$. Determine the force components on the surface if θ is $\pi/2$. What are their lines of action? The fluid is above the surface.

2.43 The curve $g(z)$ that generates the surface in Fig. P.2.41 is $(0.7z)^{1/2}$, and $b = 4$ m. Determine the force components on the surface and the moment about the origin if θ is $\pi/2$. The fluid is above the surface.

2.44 The curve $g(z)$ that generates the surface in Fig. P.2.41 is $\sin(\pi z/2b)$, and $b = 5$ m. Determine the z reaction at O, the y reaction at A, and the x reaction at B needed to support the surface against the hydrostatic load if θ is $\pi/2$. The fluid is below the surface.

2.45 The pressure distribution on a circular cylinder has been measured, with the following results:

θ, degrees	0	20	40	60	80
p, in multiples of $\rho U^2/2$	1	0.6	−0.35	−1.55	−2.35
θ	100	120	140	160	180
p	−2.05	−0.18	−0.18	−0.18	−0.18

Determine the force on the cylinder in the flow direction. (At $x \to \infty$, the velocity is $\mathbf{V} = -U\mathbf{i}$.) Use $r = 1$ and a unit width. Assume the pressure distribution is symmetric about $y = 0$. In terms of the pressure units given, the drag coefficient for the cylinder will be

obtained by dividing the pressure force (in the flow direction) by the frontal area of the cylinder, which is the product of the diameter and the width.

2.46 The pressure distribution on a sphere is given by the following data, in which θ is the angle with respect to a radius drawn in the direction from which the flow originates. (The pressure readings are axisymmetric.)

θ, degrees	0	20	40	60	80
p, in multiples of $\rho U^2/2$	1.0	0.75	0.08	−0.68	−1.15
θ	100	120	140	160	180
p	−1.10	−0.65	0.15	0.10	0.10

Determine the force on the sphere in the flow direction. In terms of the units of pressure used here, the drag coefficient for the sphere is this drag force divided by the frontal area of the sphere, πr^2. Use $r = 1$.

2.47 A hemispherical surface forms the bottom of a vertical tank that is filled with water and opened to the atmosphere. The tank is 10 ft in diameter, and the top of the dome that forms the bottom is 16 ft below the free surface. What is the vertical force on the dome due to the water above it? If the upper part of the tank is closed off and pressurized to 20 psi, what is the total force on the submerged dome at the bottom of the tank?

Problems

91

2.48 A circular cylindrical tank is positioned so that its centerline is horizontal and at the level of the free surface of a liquid in which it is submerged. The tank is 2 m in diameter and 4 m long, with planar end plates. Calculate the vertical force on the cylinder. What is the vertical force on the tank just before it is fully submerged and still horizontal? What is the force on the circular end plates under this condition?

2.49 A spherical shell of steel that is 2 m in diameter and 1 cm thick is supported in water below the free surface at 10°C. Will the shell rise or sink when the supports are removed if it contains air at 10°C?

2.50 A spherical shell of steel 4 ft in diameter and 0.5 in. thick is held submerged in water at 40°F by an anchor chain. What force is in the chain? If a leak develops in the shell, how much water must leak inside before the sphere begins to sink?

2.51 The center of the float described in Prob. 2.50 (with no leak) must float 20 ft above the bottom of a stream that flows with $V = 7$ ft/s. This flow causes a horizontal force of magnitude $3.66D^2 V^2/8$, in which D is the diameter of the sphere. What length of anchor chain is needed to fasten the float to the bottom.

2.52 What diameter must a spherical balloon have to lift 500 lb in the air at 20°C if it is filled with hydrogen? Will the required size increase or decrease if helium is used? Assume that the cost per kilogram of helium is 10 times that of hydrogen. Which gas is cheaper to use in a balloon to lift a given weight? (Give the ratio of costs.)

2.53 A design for a research submarine to explore at great depths calls for a steel cylinder to be constructed that is 15 ft long and 6 ft in diameter. Ninety percent of the volume will be occupied by kerosene (SG = 0.85), so that the steel hull can withstand the hydrostatic pressure. Four hundred pounds of equipment is to be placed in the remaining space. The steel hull (SG = 7.9) is to be $\frac{1}{4}$ in. thick. Will the submarine float or sink in water at 68°F and atmospheric pressure? Assuming that the weight of the equipment is evenly distributed over its volume, what volume of kerosene would have to be added by removing equip-

ment if the vessel is too heavy to float. What volume of kerosene must be replaced by water to make the vessel neutrally buoyant at 68°F if it is too light as first designed?

2.54 A float with a triangular cross section, as shown in Fig. P.2.54, is in the position shown when the air-water interface is at the top edge, as shown. Determine all the reactions at O.

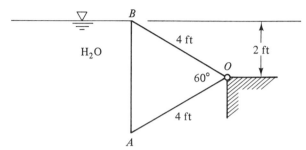

Figure P.2.54

2.55 If the float in Prob. 2.54 is replaced by one with a circular cross section and a 4-ft diameter, determine the density of the float if the free surface is to remain 2 ft above point O and to coincide with the upper horizontal tangent to the cylinder.

2.56 A triangular float such as the one shown in Fig. P.2.54 retains water and a 2-ft-deep layer of oil that has its free surface coincident with point B. What are the vertical and horizontal forces on the float due to the fluid? What is the weight per unit length of the float, assuming that it is homogeneous? What are the reactions at point O? Must a moment be applied at O to keep the float in the position shown?

2.57 A vertical plate 0.8 m high and 20 m long divides a deep pool into two parts as shown in Fig. P.2.57, in order to constrain a layer of oil. Determine the difference in elevation be-

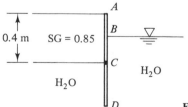

Figure P.2.57

tween the surface of the oil on one side of the plate and the surface of the water on the other. Is there any horizontal force on the plate due to the liquids? Is there any moment induced on the plate?

2.58 A circular cylindrical float with a diameter of 0.6 m serves to retain an oil layer 0.3 m deep. If the top of the float coincides with the free surface of the oil, find the vertical and horizontal forces on the cylinder. Refer to Fig. P.2.58.

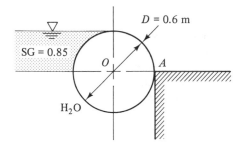

Figure P.2.58

2.59 A barge with a rectangular cross section has a beam, i.e., width, of 10 m. It is loaded in a sheltered cove on the Atlantic Ocean so that its keel is 2 m below the water's surface. It then proceeds up a river to discharge its cargo. What difference in freeboard will occur when the barge goes from the ocean (SG = 1.030) to the river (SG = 1.005)?

2.60 The chimney in Fig. P.2.60 allows the hot gases at B to rise to point A without mixing with the atmosphere. If the flue gas— essentially air—is at 200°F in the entire chimney, what is the difference in pressure be- tween point B', a location outside the chimney at its base, and B, a corresponding point within the chimney? The pressure at A is 14.7 psia, and the atmospheric air temper- ature is 68°F. Express the pressure difference in inches of water. (A manometer similar to that shown in Fig. P.2.4, called a *draft gauge*, would utilize water to measure the pressure difference.) What would be the effect on the chimney's performance if its height were doubled?

2.61 A drop of oil-solvent mixture has a specific gravity of 1.1. At what depth will it float to indicate wave motions if the water increases in density due to temperature changes with depth according to the relationship $\rho = 1000[1.0005 + d(10^{-5})]$? (The depth d is in meters, and ρ is in kg/m^3.)

2.62 Use the density gradient of Prob. 2.61 to esti- mate the buoyant force on a sphere of surface water, 1 m in diameter, that is released from the bottom of a lake 30 m deep. (This "sphere" could be a discharge of wastewater from a water-purifying system.)

2.63 A rigid insulated tank of methane is lowered into the Pacific Ocean to a depth of 2 km. Is the tank more or less buoyant at this depth than on the surface? Explain.

2.64 Water is in the container in Fig. P.2.64, and air, at atmospheric pressure, is above the in- terface. $AE = 20$ ft, $AC = 6$ ft, $AB = 4$ ft, and the tank is 8 ft wide. What value of a_x is necessary for the free surface to coincide with point C? (Let $a_z = 0$.) The vent is open to the atmosphere.

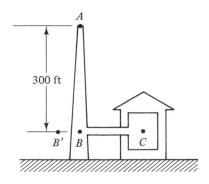

Figure P.2.60

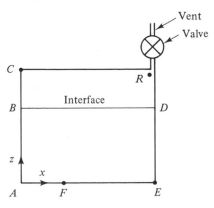

Figure P.2.64

Problems

2.65 For the tank described in Prob. 2.64, what must a_x become for the interface to coincide with point E? Sketch the interface when a_x is twice the value just calculated.

2.66 What is the pressure at points A, C, and E under the conditions of Prob. 2.64? What is the force on surfaces AC and ER? Why are they not equal in magnitude?

2.67 For the conditions described in Prob. 2.64 but with a layer of oil (SG = 0.86) above the water that is 1 ft deep when the tank is at rest, what acceleration a_x is necessary to have the oil-water interface reach point C? What are the slopes of the air-oil and oil-water interfaces?

2.68 For the configuration described in Prob. 2.64, determine the pressure at points A, C, and E when $a_x = 3g$ and $a_z = 9g$.

2.69 The oil-water-air system of Prob. 2.67 is subjected to $a_x = 2g$ and $a_z = 9g$. What pressure at locations A, C, and E can be expected?

2.70 In a vessel similar to the one shown in Fig. P.2.64, $AE = 5$ cm, $AC = 20$ cm, and $AB = 18$ cm. Water at 20°C is at the bottom, and the confined space above (the vent is closed) is at 10 kPa. With $a_x = 0$ and $a_z = 5g$, what is the slope of the interface? What is the pressure at points A and C?

2.71 If the container in Prob. 2.64 has a diameter of AE, what vertical acceleration can be allowed if the seam connecting the circular bottom to the vertical wall can withstand a force of 10 N per centimeter of length?

2.72 Are there any values for the acceleration of the vessel described in Prob. 2.68 for which the pressure is the same at all points in the liquid?

2.73 What is the pressure at the bottom of the flask described in Prob. 2.70 if it is accelerated downward at $3g$? What acceleration is necessary for cavitation to be possible at 20°C (vapor pressure = 2.34 kPa)?

2.74 The U tube shown in Fig. P.2.74 is accelerated with $a_x = \frac{1}{2}g$ and $a_z = 0$. Water is in the tube, and the vent is open to the atmosphere. What will be the position of the water in the tubes if $AC = ER = 25$ in., $AB = 12$ in., and $AE = 9$ in.? Will the positions be the same if, prior to the accelerations, the vent is closed? (*Note:* In this condition, the air in section DR

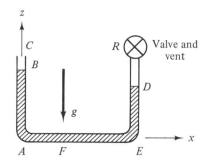

Figure P.2.74

will have a pressure that is dependent upon its volume.)

2.75 A U tube similar to the one shown in Fig. P.2.74 ($AC = ER = 24$ in. and $AE = 9$ in.) is completely filled with water, and the vent is closed. What is the pressure at C, A, E, and R when the tube is accelerated to the left at $\frac{1}{2}g$?

2.76 If in Prob. 2.75 the tube were accelerated to the right at $\frac{1}{2}g$, what would be the pressure at C, A, E, and R? (Can you assume that no liquid has spilled out of the tube?)

2.77 If mercury were to replace the water in Prob. 2.75, what value of acceleration with $a_z = 0$ would be necessary for the pressure at R to be nearly zero? If the mercury begins to recede from R, a space of nearly zero pressure—actually, the vapor pressure of mercury at the existing temperature—will exist above the liquid column in the tube ER. What value of acceleration is necessary, $a_z = 0$, for the tube RE to be emptied of mercury?

2.78 Tanker trailers for hauling gasoline (SG = 0.85) have interior compartments for structural rigidity. The trailer shown in Fig. P.2.78 has only its middle compartment completely full of gasoline (except for a small volume of air that will remain at atmospheric

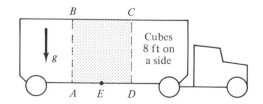

Figure P.2.78 Truck undergoes constant deceleration.

pressure). During an emergency stop, the trailer decelerates at $(1.1)(32.2 \text{ ft/s}^2)$. Where on the compartment walls is the maximum pressure induced? What is the resultant force on the bulkhead having the higher load during the stop?

2.79 A circular tank with a cross section similar to that sketched in Fig. P.2.64 is rotated about a vertical axis through F, which coincides with the axis of the tank. The dimensions are $AE = 2$ m, $AF = 1$ m, $AC = 1.2$ m, and $AB = 0.8$ m. At what speed of angular rotation will the free surface climb to points C and R? (The vent is open to the atmosphere.)

2.80 Under the conditions of Prob. 2.79, show that the lowest point of the air-water interface is at 0.4 m.

2.81 Under the conditions of Prob. 2.79, what is the pressure at locations, C, A, and F (which is on the axis of rotation)? (*Hint:* Use the result of Prob. 2.80 and the fact that there is no vertical acceleration to determine the pressures without having to find Ω.)

2.82 If the tank described in Prob. 2.79 is rotated at 12 rad/s and the vent is kept open to the atmosphere, how much liquid can be expected to be lost?

2.83 A circular tank with a diameter of 3 ft and a height of 6 ft is filled half with water and half with an oil (SG = 0.87). The tank has a tight cover with a vent hole at the center. When the tank is rotating at 5 rad/s, what is the pressure distribution along the vertical wall? What is the shape of the oil-water interface?

2.84 What angular speed is necessary for the free surface in Prob. 2.79 to reach the bottom? How much fluid would be spilled?

2.85 Plans are under consideration to cast concrete pipe in a vertical mold 20 ft high and 8 ft in diameter. At what speed must the mold be rotated if the difference in pipe wall thickness is to be no more than $\frac{1}{4}$ in. from top to bottom? The wall thickness at the top should be 5 in. What will be the weight of the finished section of pipe at this minimum speed? (Assume a specific gravity of 1.5.)

2.86 A manufacturer wishes to cast parabolic mirrors by rotating molten glass (SG = 2.65) while letting it cool sufficiently to form the mirror. If a mirror with a diameter of 30 cm is

to be cast in this way, what will be its focal length for a speed of rotation of 6 rad/s?

2.87 A U tube similar to that shown in Fig. P.2.74 is being rotated around a vertical axis through point F with a speed of 9 rad/s. The vent is open. If, prior to rotation, $AF = 0.2$ m, $AE = 1$ m, $AC = AR = 2$ m, and $AB = DE = 1$ m, what are the liquid levels during rotation?

2.88 A U tube (Fig. P.2.74) with dimensions $AC = AB = ER = ED = 18$ in. and $AE = 24$ in. is filled with mercury. It rotates around "leg" ER at 5.5 rad/s, and the vent is closed. What is the pressure at C, A, E, and R? At what rate of rotation will the U tube have zero pressure at R? At what angular speed will leg ER be empty of mercury?

2.89 A U tube exactly like the one described in Prob. 2.88 is rotated around "leg" ER. What is the pressure at C, A, E, and R?

2.90 The tank whose geometry is described in Prob. 2.79 is rotated at 3 rad/s at the same time it is being accelerated vertically at 10 m/s^2. What is the form of the water's free surface? What are the maximum and minimum pressures on the bottom of the tank?

2.91 A tank having a circular cross section with diameter 18 in. is being rotated along its axis, which is horizontal. Water completely fills the tank. (It is closed at both ends.) What are the surfaces of constant pressure?

2.92 A square box (0.3 m × 0.3 m, and 0.75 m high) is rotated around the center of its base, as shown in Fig. P.2.92, at 6 rad/s. Find the

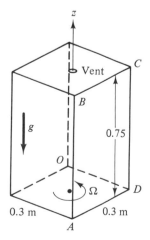

Figure P.2.92

pressure distribution along wall *ABCD*, assuming the container is completely full of water. (There is a lid, with a vent hole to the atmosphere in the middle.) Find the force on *ABCD*. Find the moment about *AD* due to the pressure distribution on *ABCD*. Find the moment about *AB* due to the existing pressure distribution. What is the "line of action" of the force on *ABCD*?

2.93 A capped oil well is 25,000 ft deep and has oil (SG = 0.8) in the lowermost 5000 ft. Assume that the natural gas that fills the remainder of the well has properties of methane at 100°F, the temperature at the top of the well. If the pressure there is 2000 psi, what is the pressure at the bottom of the well if the density of the gas is considered to be constant at the value at the top? What is the pressure at the bottom of the well if the gas is assumed to be isothermal and the density changes with pressure?

2.94 A circular cylinder (SG = 0.9) 4 in. long floats with its axis vertical at an interface between water and oil. The top of the cylinder is 1 in. above the interface. What is the density of the oil? What would be the density of the oil if the cylinder floated vertically at its midpoint? Do the previous answers indicate a linear scale if the float serves as a hydrometer?

2.95 A vertical cylinder, open at the top, with a diameter of 12 in. and a height of 18 in. is rotating about its axis. If it is initially two-thirds full of water, at what speed of rotation will the liquid begin spilling from the upper rim of the cylinder? What is the speed of rotation when the center of the cylinder's base first becomes free of water? How much water is spilled between these two speeds?

2.96 Test tubes 6 in. long are rotated in a centri-fuge so that they are perpendicular to the axis of rotation and their closed ends are 12 in. from the axis of rotation. If a tube is filled with a serum (SG = 0.7), what is the pressure at the closed end when the rotational speed is 2000 rpm? (Assume atmospheric pressure at the end nearest the rotation axis.)

2.97 The rectangular vertical gate in Fig. P.2.97 is 12 ft high and 8 ft wide and is hinged along the lower edge and held vertical and closed by a horizontal spring that bears against it at a point 9 ft above the hinge. The gate is part of a large, open tank that contains water whose depth should not be too great. Accordingly, the spring that pushes against the gate from the outside—where the pressure is atmospheric pressure—is given a preload that will just balance the hydrostatic force of the water when it is at the top of the gate. What is the force in the spring when the water is at the level of the hinge?

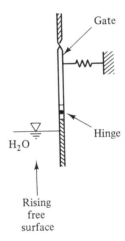

Figure P.2.97

3

Mass Conservation

3.1 INTRODUCTION

It is our goal to construct a model that describes the flow of a fluid in specific situations with reasonable accuracy. This will provide us with a method of predicting quantities such as velocity and temperature that are of interest to us. Many elements will have to be included in the model, including the appropriate physical characteristics of the medium. We shall start the development of a mathematical analogue by using, as best we can, the established laws of nature. Remember that these "laws" are usually the distilled result of many observations of natural phenomena over many years. In stating these laws, we are not as sure that they are correct as we are sure that the statements have not, so far, been found to be incorrect. Certainly, future generations may show a "law" to be a special case of another "law," or they may more restrictively define the conditions under which it is applicable. Thus, keeping in mind that the following laws are our current understanding of what is occurring, we shall attempt to learn what conclusions can be drawn from

1. The law of conservation of mass
2. The law of conservation of momentum
3. The law of conservation of energy

We now focus our discussion upon the concepts associated with mass conservation. It is the least abstruse of these laws and will permit us to focus on the logic of the development. Mass conservation will require the derivatives of the velocity components to be interrelated in a special way. This means that an arbitrary specification of the velocity field could be in violation of this law.

The word *conservation* in these laws has a special connotation in this context: The conversion of mass to energy, according to Einstein's relationship $E = mc^2$, is precluded by the laws of conservation for mass and energy that are used in this book. Also, the law of momentum conservation does not necessarily imply that the momentum of the fluid remains constant; it is an abbreviated version of Newton's law of motion. This, we know, states that the time rate of change of momentum of a particular group of fluid particles is equal to the force that has been applied to those particles. This collection of particles is called the *system*. (It is the mass of material upon which we shall concentrate our attention to study its behavior.) Equally well, we recognize that the energy of a system might not be conserved if energy is transferred to it in the form of heat. With these preliminary observations about the explicit and implicit meanings of the terms we shall be using, we now turn to a detailed examination of these laws and the ways that we can use them to give us meaningful results. This effort will occupy us in this chapter and the next four chapters. Much of the content of Chap. 3 is concerned with a fluid's velocity. It is appropriate that we begin our discussion with this flow variable.

3.2 FLUID VELOCITY

The velocity of a collection of particles may be difficult to measure directly. Often it is their position as a function of time that is determined. From this information, the velocity is then inferred. For example, the expressions

$$x = 5t + 7 \qquad y = 2t \qquad z = 0 \qquad (3.2.1)$$

could define the position of a collection of particles that were clustered about a point of space, at, say, $t = 0$. That point would then be

$$x = 7 \qquad y = 0 \qquad z = 0$$

These particles will have different positions at $t = \frac{1}{2}$, 1, 2, 3, and 5, which would also be determined by Eq. (3.2.1) in this instance. The relationships given by Eq. (3.2.1) are neither special nor unique. The expressions

$$x = 5e^{+kt} \qquad y = 2e^{-kt} \qquad z = 0 \qquad (3.2.2)$$

could also have been used to describe the relationships, provided that one wished to study that particular problem. Equations (3.2.1) and (3.2.2) illustrate the role of time t as a parameter in specifying the position of the particle in question. Tabular data that would be comparable to equations such as (3.2.1) or (3.2.2) could be collected by the proper observation of a weather balloon. At each time of observation, its position, (x, y, z), could be established. From these data one would be able to say something about the direction and variation of winds during the period of observation.

Because Eq. (3.2.1) or Eq. (3.2.2) gives the coordinates of a very small, moving volume of fluid particles as a function of time, we can determine how this position changes with time. Indeed, we can calculate and determine the time rate of change of this change of position. For example, we might determine the change of the x position of the particle between $t = \frac{1}{2}$ and $t = 1$. Then we could calculate the time rate of change of position in this time interval (i.e., $\Delta x / \Delta t$). This value should approximate the x component of velocity. One need not use finite differences to give these time rates of change of the x, y, or z coordinates for expressions like Eq. (3.2.1). One could differentiate these equations with respect to time. Also, it is apparent that these time rates of the position change are the velocity components of the associated fluid particles. We can call this velocity $\mathbf{V} = V_x\mathbf{i} + V_y\mathbf{j} + V_z\mathbf{k}$. For the expressions given by Eq. (3.2.1), we would have $dx/dt = V_x = 5$, as well as $V_y = 2$ and $V_z = 0$. Following a similar train of thought, one can inquire into the meaning of the second time derivative of the equations that specify the position of the fluid particles. These derivatives are d^2x/dt^2, d^2y/dt^2, and d^2z/dt^2, which are all zero in the case of Eq. (3.2.1), but nonzero for Eq. (3.2.2). Since such second derivatives are time rates of change of the first derivatives, the velocity components, we can conclude that they are accelerations. Thus:

$$\frac{d^2x}{dt^2} = a_x \qquad \frac{d^2y}{dt^2} = a_y \qquad \frac{d^2z}{dt^2} = a_z$$

These are the components of the acceleration of the fluid particles whose positions are specified as functions of time similar to Eqs. (3.2.1) and (3.2.2). This acceleration is represented by the vector $\mathbf{a} = a_x\mathbf{i} + a_y\mathbf{j} + a_z\mathbf{k}$. It would be a straightforward task to use Eq. (3.2.2) to determine the associated acceleration components. (Do you find $dx/dt = 5ke^{kt}$ and $d^2x/dt^2 = 5k^2e^{kt}$?) The specification of the position of fluid particles as a function of time, and the subsequent determination of velocity and acceleration, is called the *material,* or *Lagrangian,* description of the flow. In order to give the position of a great many fluid particles (minute packets containing many fluid molecules), it would be necessary to have at hand an equally great number of equations, or a method of generating them. This requirement is not the only one that makes the Lagrangian system of marginal use in the study of fluid mechanics. In general, one is not interested in following the history of specific fluid particles. This method of describing a flow has been presented primarily because it is the one commonly used in a course in dynamics and in the development of the equations in Chap. 2. Because it is assumed that the reader is familiar with this material, it has served as a useful introduction to the concepts that follow.

Usually, we inquire about the conditions that are present at particular locations in the fluid. For example, we would like to know the pressure distribution on the surfaces of an airfoil. It should be possible to know this without knowing which specific fluid particles are passing over the foil. Similarly, the flow through a conduit with branches and bends can be described effectively without following the progress of individual fluid particles. This can be accomplished by adopting the *spatial,* or *Eulerian,* description of fluid motion. In this case, the positions of certain particles are not specified. One uses expressions for the velocity components (in the x, y, and z directions, for example) that could be experienced at fixed locations, or measuring stations, in the region occupied by the fluid. These velocity components (u, v, and w, with $\mathbf{V} = u\mathbf{i} + v\mathbf{j} + w\mathbf{k}$) can be thought of as the velocity components that could be recorded at specified locations in space and which are due to the passage of unspecified fluid particles through those locations. We can say that some fluid particle must have had these velocity components, but we do not know which particle. Some examples of Eulerian specifications of velocity are:

$$\begin{aligned} u &= (-2y + 7x) \\ v &= (-7y + 8x) \\ w &= 5 \end{aligned} \qquad (3.2.3)$$

or
$$u = (-2y^2 + 5x)$$
$$v = (-7y + 3x^3) \qquad (3.2.4)$$
$$w = 2z$$

or
$$u_r = u_{radial} = \frac{5(1 - e^{-0.1t})}{r}$$

$$u_\theta = u_{tangential} = \frac{3(e^{-0.2t})}{r} \qquad (3.2.5)$$

$$u_z = 0$$

The last case is for a flow in which a cylindrical polar coordinate system (r, θ, z) is used to specify the locations even though, in this instance, there is no variation of the velocity in the θ direction (i.e., u_r and u_θ depend only on r and t). Equation (3.2.3) is called a *two-dimensional velocity field,* because the velocity depends on only two of the three coordinate directions. Using the number of space dimensions needed to describe the velocity as the criterion, Eq. (3.2.4) is an example of a three-dimensional velocity; and $u = (-2y^2 + 5x - 3z)$, $v = (-5y + 2x^3 + e^z)$, and $w = 0$ would also be a three-dimensional flow even though there are only two velocity components that are nonzero. There is a time dependency in the flow given by Eq. (3.2.5). The velocity will vary at points with a fixed radius as the time changes. Such a dependency of the velocity on time at a fixed point of space is the sole criterion for saying that the velocity is *unsteady.* Even though the velocities in a flow are steady, one might encounter a situation wherein the temperature changes with time at points within the flow. This would be a case of an unsteady temperature distribution. One would have an unsteady flow due to the unsteady temperature. (Undoubtedly the density of the fluid would also change with time.) The values of temperature at the spatial points in the region of interest comprise the *temperature field.* Similarly, one speaks of the *velocity field,* to denote the values of the velocity within the region under study. If a flow can be considered to be without variation over a significant region of space, it is termed a *uniform* flow in that region. (One strives to create a uniform flow at the entrance of a wind tunnel's test section. Why should this be so?)

It is worthwhile to add that other quantities, such as temperature and pressure, can be specified, or measured, as functions of time and space. If all the observed flow variables are independent of time, one has a *steady flow* and not just a steady velocity field.

3.3 VOLUME RATE OF FLOW

We shall observe for a moment the water flowing out of the water tap shown in Fig. 3.3.1, in order to determine the volume per unit of time (e.g., in m³/s or ft³/s, abbreviated cfs) that issues from the faucet. This *quantity per unit time,* called Q, could be determined with a pail and a watch. Imagine now that the water is flowing through a grid of wires, like a sieve, which at t_0 start to color the water.* After a short interval of time, $t_1 - t_0$, the dyeing of the water ceases. Thus at time t_1, there is a short cylinder of colored water in the flowing stream. This situation is shown in Fig. 3.3.1, where, in

---

* The water is assumed to have a small amount of thymol blue added to it. The wires are connected to a battery, and the electric current causes a change of pH in the neutrally titrated solution. A color change occurs that is similar to that of a phenolphthalein indication for alkalies.

Mass Conservation   Chap. 3

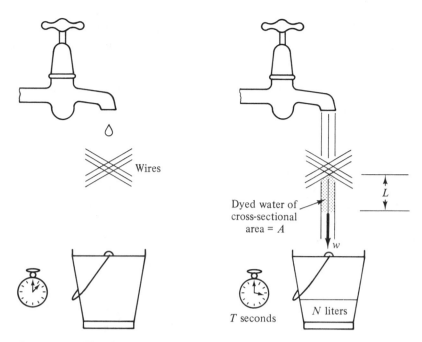

Figure 3.3.1 Flow from faucet into pail illustrating that volume rate of flow of fluid jet is equal to product of cross-sectional area of jet and its (uniform) speed. $w =$ vertical velocity of the water.

addition, some geometric features of the stream are given. The length of the colored cylinder is L, so that the fluid particles that were first dyed have traveled a distance L in the time interval $t_1 - t_0$. We conclude that the speed of the water must have been $w = L/(t_1 - t_0)$ if the time interval is short enough that one can neglect gravitational acceleration. Now, with a pail and a watch, we could determine the amount of water flowing during each second of time. This quantity has been called Q. But Q and the volume of colored water are related. The volume of this colored water is LA, in which A is the cross-sectional area of the water jet. Thus a volume of LA has passed through the mesh of wires and has been colored during the interval $t_1 - t_0$. This means that the volume per unit time that has flowed through the wire gauze is $LA/(t_1 - t_0)$. This is, of course, Q; it and our previous deduction that $w = L/(t_1 - t_0)$ lead us to the relationship

$$Q = wA \qquad (3.3.1)$$

In arriving at this conclusion about the rate of volume flow through a cross section with area A, we had the condition that the speed was constant over the entire cross-sectional area. This is manifested by the horizontal, upper, and lower surfaces of the dyed cylinder of fluid in Fig. 3.3.1. Often the speed of the fluid is not constant over the particular surface that is of interest to us. For example, the air coming through an opening in a ventilating duct may have a higher speed in one corner of the opening than in another if there have been many turns and bends in the duct previous to the opening in question. When such nonuniform velocities are present, Eq. (3.3.1) would have meaning only in the sense that the value of w in this equation would be an *average value*. The equation would define this average value if Q and A were known. We could obtain Q with the aid of our pail and a watch; however, we shall attempt to assess Q by returning to Eq. (3.3.1) and apply it to a very small element of area dA through which the speed of the fluid can be considered to be a constant value, even though w varies over the entire section of area A.

A situation for which the fluid velocity is not uniform involves a "no-slip" condition at the walls of a stationary conduit. (Recall that this condition requires that the fluid particles next to a solid boundary have the velocity of the boundary.) A way of proceeding is to consider the case shown in Fig. 3.3.2. In Fig. 3.3.2b, the nonuniform distribution in the conduit, which is assumed to have a cross section independent of the horizontal coordinate, has been modeled not as a continuous distribution (as in Fig. 3.3.2a) but by a number of small regions, each of which has a constant velocity. With the same logic as is used in the calculus, we assume that in the limit of increasing the number of subdivisions with decreasing cross-sectional area, one can calculate the flow rate for the nonconstant velocity distribution. Thus we write

$$dQ = w \, dA \tag{3.3.2}$$

instead of Eq. (3.3.1), so that

$$\frac{\text{Volume}}{\text{Time}} = Q = \sum_{i=1}^{N} w_i \,(dA_i) = \int_A w \,(dA) \tag{3.3.3}$$

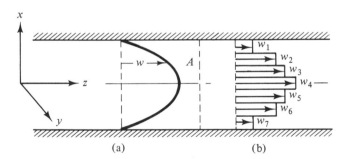

(a)

(b)

Figure 3.3.2 Approximations of continuous velocity distributions by stepwise varying distributions. (a) Continuous distribution in xz plane and (b) "stepped" distribution model; A = area across which flow rate is to be determined. (c) "Stepped" distribution for unit width. (d) Axisymmetric "stepped" distribution.

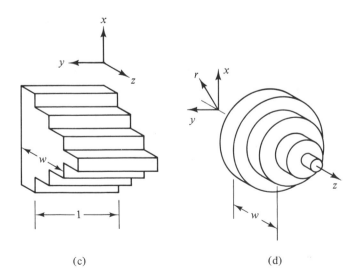

(c)

(d)

Mass Conservation   Chap. 3

The symbol w is used for the velocity at the appropriate station. The element of area dA will be different in different situations, depending on the particular geometry. For example, if the conduit is two-dimensional and there is no variation in the z direction, which is normal to the page (as in Fig. 3.3.2c), one can take a unit distance as the dimension of the conduit in the y direction. For this two-dimensional case, $dA = dx(1)$. (On this element of area, the fluid's velocity is prescribed and the flow across the area can then be found.)

Figure 3.3.3 illustrates a particular axisymmetric flow in a circular conduit. The element of area over which the fluid velocity can be considered constant is shown in Fig. 3.3.3c, and the "stepped" model of the continuous velocity distribution is shown in Fig. 3.3.2d.

Example 3.3.1

Applying Eq. (3.3.3) to the situation depicted in Fig. 3.3.3 yields

$$Q = \int w \, dA = \int_0^{r_0} w_{max} \left[1 - \left(\frac{r}{r_0} \right)^2 \right] (2\pi r \, dr)$$

$$= 2\pi w_{max} \int_0^{r_0} \left[r \, dr - \left(\frac{r}{r_0} \right)^2 r \, dr \right] = \frac{w_{max} \pi r_0^2}{2}$$

The average fluid speed is

$$\frac{Q}{A} = \frac{\dfrac{w_{max}}{2} \pi r_0^2}{\pi r_0^2} = \frac{w_{max}}{2}$$

in this case.

Figure 3.3.3 Axisymmetric velocity distribution. (a) Conduit with axial velocity $w = w_{max}[1 - (r/r_0)^2]$. (b) Velocity profile, w versus r. (c) $dA = 2\pi r \, dr$. For each radius, $w = $ const.

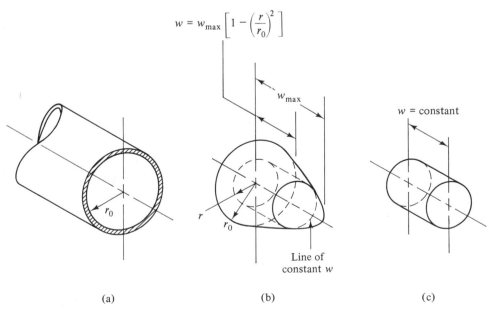

(a)   (b)   (c)

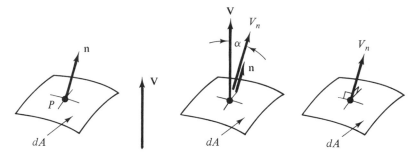

Figure 3.3.4 Relationship beween velocity vector **V** at point P on a control surface and unit normal vector **n** to area dA at that point. $V_n = |V| \cos \alpha$.

It is necessary to add the important fact that in Eqs. (3.3.1) and (3.3.2), the magnitude of the speed, w, must be the one associated with the velocity component that is perpendicular to the area element across which the flow rate is desired. Figure 3.3.4 shows such an area element at a point in space where the velocity is not normal to the area element. In this case the *normal component* of the velocity, V_n, must be used in a relationship like Eq. (3.3.2). The reason for this is clear after a moment of reflection. The *tangential component* of the velocity, **V**, which is designated V_t, serves to move the fluid particles parallel to the mouth of the "opening," dA, but it is V_n that transports the particles through the "opening."

Example 3.3.2

If a liquid flows from a tank through an opening, called an *orifice*, it will do so initially as a cylinder of liquid. This is a liquid jet, and the fluid velocity can be calculated in a manner similar to that for an object moving due to gravitation attraction. For the case shown in Fig. 3.3.5, the fluid velocity in the jet is given as $V = \sqrt{2g(H - z)}$. (A two-dimensional slot is *assumed*, so that it is convenient to *assume* a unit width perpendicular to the page.) Determine Q per unit width for this flow.

Solution We shall employ $Q = \int u\, dA$, because the velocity is not constant. Near the orifice, where the jet width is $2b$, the velocity is nearly horizontal. Then $u = V$, so that

$$Q = \int_{-b}^{+b} \sqrt{2g}\, \sqrt{H - z}\,(dz)(1)$$

or

$$Q = -\tfrac{2}{3} \sqrt{2g}\, [(H - z)^{3/2}]\big|_{-b}^{+b}$$

$$= \tfrac{2}{3} \sqrt{2g}\, [(H + b)^{3/2} - (H - b)^{3/2}]$$

$$= \frac{2H}{3} \sqrt{2gH}\left[\left(1 + \frac{b}{H}\right)^{3/2} - \left(1 - \frac{b}{H}\right)^{3/2}\right]$$

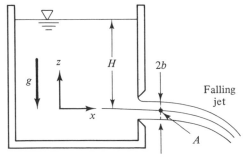

Figure 3.3.5 Liquid jet issuing from horizontal slot in tank. A = area across which flow rate is desired.

The average velocity, $Q/2b$, would be

$$u_{avg} = \frac{H}{3b} \sqrt{2gH} \left[\left(1 + \frac{b}{H}\right)^{3/2} - \left(1 - \frac{b}{H}\right)^{3/2} \right]$$

If one expanded the terms with the exponent $\frac{3}{2}$ by using the binomial theorem, one could obtain convenient estimates for u_{avg} for small values of b/H.

3.4 SOME CONTROL SURFACES

It should be noted in Fig. 3.3.1 that while A was designated as the cross-sectional area of the jet, it is *also* the area of that part of the wire grid through which the fluid is moving. It is in this latter sense—a surface that is specified in space at our choosing, across which fluid is moving—that we shall usually consider the area A. In various problems, we shall consider different geometrical surfaces, depending upon which flow rate is of interest to us. Thus the choice of the surface, called a *control surface,* may depend on our specific interests for a particular problem.*

A case that illustrates the choices available when one considers a control surface is given in Fig. 3.4.1. Here a candle with a chimney to protect the flame from the wind is shown. The chimney rests upon protrusions on the base to allow some air to enter. The oxygen in this air supports the combustion at the top of the candle. If an investigator were interested in determining the quantity of air flowing through these lower openings, the person would imagine a surface such as A_1 and conduct a survey of the air velocity at the points that make up A_1. One might, on some occasion, choose A_2 as the surface over which to make some measurements. This surface differs from A_1 in that it includes the outside of the glass chimney. Because the velocity component normal to the surface of the chimney is zero, the amount of air flowing across A_1 and A_2 would be the same. If the surface on which the measurements might be made were extended to include the base of the candleholder, one would have surface A_3. Again, the amount of air flowing across A_3 would be the same as that of A_1, only because the base is assumed to be nonporous. An extension of the measuring surface to A_4, however, requires special considerations. When one selects the geometry of the surface over the top of the chimney, the choice should be guided by the method that gives the most information in an efficient way. In Fig. 3.4.1, surface A_5, which accounts for part of A_4, appears superior to A_{5a}. This is so because it is relatively easy to measure the velocity component normal to the plane of the chimney's mouth.

If we assume that the velocity is axisymmetrical about the z axis, the velocity distributions across A_5 and A_1 (both of which are parts of the surface that completely surrounds the flame) are as shown in Fig. 3.4.2. The axial component of the velocity across A_5 is largest at the center, where it is directly above the heat source, and at the same time farthest from the stationary walls of the chimney. There we believe that the velocity must be zero because of the no-penetration and no-slip conditions. Once the buoyant jet, or plume, is out of the chimney, it will expand in width by ingesting some of the surrounding air. (This process is similar to what happens to the gases

---

*The word *control* in English has conveyed the meaning of "check" or "verify" in addition to the more usual current meanings of "regulate" or "constrain." The concept of checking should be kept in mind when using the term *control surface.*

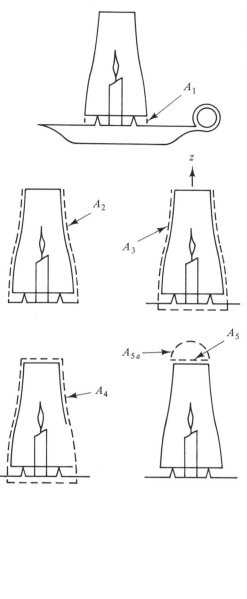

Figure 3.4.1 Variety of surfaces that could be part of a control surface for analysis of flow into and from a lamp chimney. A_1: cylindrical surface; A_2: surface of revolution; A_3: A_2 plus base of candleholder; A_4: A_3 plus top opening of chimney; A_5: circular plane surface; A_{5a}: hemispherical cap.

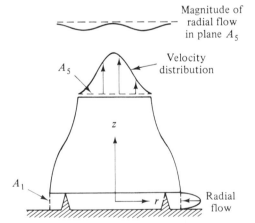

Figure 3.4.2 Surfaces across which air flows into and from a lamp chimney. z = direction of axial flow.

coming from a smokestack. The entrainment of the surrounding air serves to dilute the concentration of these gases.) Consequently, there is some radial inflow toward the axis at positions above the chimney. Hence the radial flow is negative. The velocity of the air leaving the chimney across A_5 will make a smooth transition to that existing above the chimney, where there is flow into the rising smoke plume. Thus some small radial component of velocity is depicted as being present on A_5. (We can anticipate that this radial component, which is in the plane of the control surface that we have selected, does not contribute to the movement of material across surface A_5.) Figure 3.4.2 also shows the velocity in a vertical section of the lower surface A_1. Here only the radial component is shown, because the horizontal surface of the candle-holder will inhibit axial (in this case, vertical) motions. We should remark upon one more thing in Fig. 3.4.2: the velocity profile shown for surface A_1 is understandably thought of as an *inflow* to the region bounded by the (total) control surface A_4, and the flow across A_5 can be thought of as the *outflow*.

If a plate with a small hole in it is now laid over the top of the chimney to coincide with the surface A_5, then a new axial velocity distribution, shown in Fig. 3.4.3, will result. If a nonporous band is placed around the base of the chimney, then the radial velocity across A_1 will vanish. The candle will also start to smoke, because its supply of oxygen has been cut off. It will again glow brightly only if the plate at the top of the chimney is removed. Then the axial velocity distribution might be as shown in Fig. 3.4.4. We see that this velocity is positive in the central section and negative along the chimney walls. The flow of positive velocity is the outflow, and the flow of negative velocity, the inflow. Hence, across any section of the control surface, both the magnitude and direction of the flow will be important to record for our work. If mass conservation is applicable to the present case, we would find it reasonable that the positive and negative portions of the velocity profile shown in Fig. 3.4.4 are not completely unrelated. This interrelationship will be developed in section 3.7.

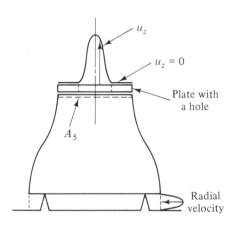

Figure 3.4.3 Effect of a particular flow requirement—a plate occludes a portion of the opening in a lamp chimney—on velocity distribution over the segment of the control surface A_5. Axial velocity $u_z = 0$ over a portion of A_5.

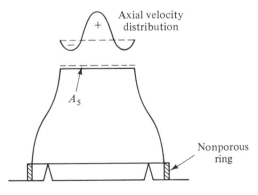

Figure 3.4.4 Velocity distribution on a segment of control surface A_5 due to blocking inlet to lamp chimney.

3.5 STREAMLINES, STREAKLINES, AND PATHLINES

The candleholder and chimney that served as an illustration in the discussion of control surfaces can be useful to explain another set of terms used in fluid mechanics. In Fig. 3.5.1, the candle is illustrated with air entering beneath the chimney and rising through its top. At each location within the chimney, one could, in principle, determine the velocity of the fluid (i.e., air, or air and combustion products) at a particular instant of time. Then if a diagram were constructed with the direction of the velocity shown at each station, the resulting picture would be many times more complicated than Fig. 3.5.1a. The picture would be more like that shown in Fig. 3.5.2, where iron filings have been strewn upon a sheet of paper and magnets below the paper have given the elongated metal particles distinct orientations. The pattern of the metallic particles is akin to the orientation of the myriad velocity vectors within the chimney of Fig. 3.5.1. If we were to sketch a line in Fig. 3.5.2 that was tangent to the iron particles, we would be mapping the magnetic field.

In a similar way, we are interested in the lines drawn tangent to the velocity vectors in a particular flow. Such lines serve to describe the velocity field in the fluid. We call these lines *streamlines*. This is shown in Fig. 3.5.1b and c. While the word *streamline* that we use in our everyday language connotes a smoothly shaped or graceful body, here we shall be using the word *streamline* in this more restricted sense.

This definition of *streamline* provides us with a definite way to construct it. At a point P, the velocity of the fluid that is at P is written as

$$\mathbf{V} = u\mathbf{i} + v\mathbf{j} + w\mathbf{k} \tag{3.5.1}$$

(Remember that u, v, and w are functions of x, y, and z and possibly t.) The tangent to the streamline will be considered to be a directed line segment, $\mathbf{ds}$, such that

$$\mathbf{ds} = (dx)\mathbf{i} + (dy)\mathbf{j} + (dz)\mathbf{k} \tag{3.5.2}$$

Figure 3.5.1 Various flow situations illustrating the difference between streamlines, pathlines, and streaklines. (a) Velocity vectors at several points, and line drawn tangent to them. (b) Pattern of streamlines for candleholder and chimney. (c) Possible streamlines when base of chimney is closed off. (d) Positions of smoke at two different times in an unsteady flow. P, P' = positions of same soot particle.

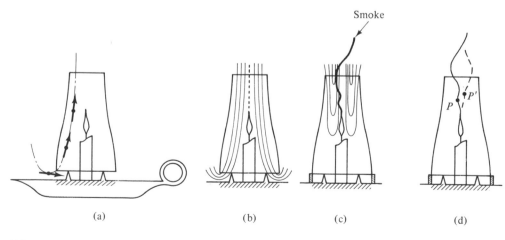

(a) (b) (c) (d)

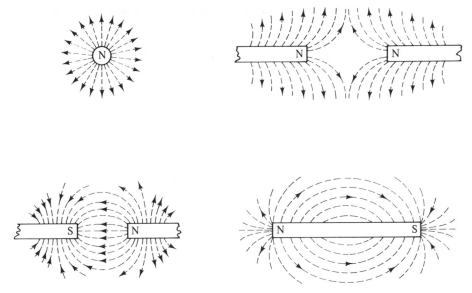

Figure 3.5.2 Inferring magnetic fields from orientation of magnetic fillings.

This is a convenient representation because we know that two vectors are parallel if their components are proportional. This would be the case for the vectors **V** and **ds** if

$$\frac{dx}{u} = \frac{dy}{v} = \frac{dz}{w} \qquad (3.5.3)$$

(If any velocity component is zero, then the corresponding component of **ds** must be zero also.) The method of using this equation, as well as its consequences, will become obvious from an example.

Example 3.5.1

Construct the streamlines for the situation where the velocity is given as $\mathbf{V} = (Ay)\mathbf{i} + (Bx)\mathbf{j}$.

Solution This distribution implies that the z component of velocity, w, must be zero. Thus, in order for Eq. (3.5.3) not to become useless (with a zero in the denominator), dz must be set to zero so as to have an indeterminate form (i.e., 0/0) with a value that is prescribed by the terms at the left of Eq. (3.5.3). Setting dz equal to zero means that the streamlines do not have a z variation; the streamlines are embedded in planes parallel to the xy plane. With this preliminary detail now resolved, we write for Eq. (3.5.3)

$$\frac{dx}{Ay} = \frac{dy}{Bx}$$

The solution of this differential equation is not difficult, and one obtains

$$\frac{Bx^2}{2} - \frac{Ay^2}{2} = \text{constant}$$

The coefficients A and B are usually specified. Hence, for any value of the constant that we choose to assign, we have the equation for a particular streamline. For another value of the constant, we obtain another streamline. This process continues as long as we can construct streamlines that are of interest to us. While A and B are often constants, as in this particular example, they could also be functions of time, such as $5t$ or $2t^2$. If this were so, we would still proceed as we have just done; however, we would bear in mind that the equation would give us different curves for each value of the time t, since the coefficients A and B would be changing. Streamlines are an instantaneous picture of an aspect of the geometry of the flow. For an unsteady velocity, the picture changes from instant to instant because the velocity at a point depends on time.

A noticeably unsteady flow can occur when we close off the lower opening of the chimney in Fig. 3.5.1. Even though air now tries to flow in through the top of the chimney near its rim, there may be a lack of oxygen near the flame so that the candle starts to smoke and flicker slightly. The rising column of smoke forms a line which is called a *streakline* because all of the particles that compose it have passed through a common point of space, the wick. Figure 3.5.1d shows two positions of the filament of smoke that rises from the candle under conditions of incomplete combustion. Assume that a particular particle of soot can be identified and it is located at point P. A brief time interval later this particle is in a new position, corresponding to P'. In this time interval, the particle in question has moved along some path from P to P'. This path of a particular particle is appropriately called a *pathline.* It gives the locus of points of space through which a particular particle has passed. It is a location-time description of a particle.

We shall learn to appreciate the fact that *streamlines, pathlines, and streaklines are coincident only for steady velocities.* Each type of line gives certain information, and each is made visible by particular experimental techniques. It is our task to recognize in a particular situation which kind of line we have and what information can be gleaned from it. The following material should make the differences between the various lines, and their utility, clear to the reader. A reexamination of Fig. 1.8.10a and b may prove to be useful also.

Example 3.5.2

In order to map winds, balloons are sent aloft and their positions recorded from several observation points. From these data, the flight path of a balloon is known. Does this path correspond to a streakline, a streamline, or a pathline?

Solution This would be a *pathline* associated with the particles of air that are attached to the outside of the balloon. Because the winds are generally unsteady, we would not expect this curve to conform to a streamline.

Example 3.5.3

A tanker is anchored for some time and a long, thin oil slick that leads back to the vessel is observed some distance from the ship. Is this slick similar to a pathline?

Solution No. Since all of the oil particles emanate from the ship's position, the locus of points making up the oil slick forms a streakline. (This would be similar to the smoke plume from a candle's wick.) If the current of water flowing past the ship—and carrying the oil away—is unsteady, this streakline will correspond to neither a streamline nor a pathline.

Example 3.5.4

A ship model is being maneuvered in a basin. Floating on the surface of the water are many particles that reflect light. While the ship is moving, a bright light is shone onto the surface of the water and a photograph is made with an exposure time of $\frac{1}{10}$ s. Even though this is not a long time interval, it is long enough that, as the reflecting particles move due to the ship's motions, short and bright streaks are recorded on the film. Do these streaks give an indication of pathlines and/or streamlines? If the shutter were left open for 10 s while the ship model was executing a sharp turn, would the photographic record of the streaks associated with the movement of the surface particles give the experimenter streaklines, streamlines, or pathlines?

Solution   The short exposure of the film to the light reflecting from the particles would produce many short streaks on the film. The length and direction of these streaks would indicate the magnitude and direction of the velocity of the water at the surface. The pattern of lines that were tangent to these short lines would be an indication of the streamlines at the instant the photograph was made. Because the ship model is being maneuvered, the flow will be unsteady, so that the streamlines and pathlines are not identical. However, the lines that would be produced on the film by some few peculiarly bright, or otherwise identifiable, particles during an interval of 10 s would be path lines. They would be the record of the location, in a horizontal plane, of those particles during that time.

Example 3.5.5

If a velocity distribution were specified as

$$V = Ax\mathbf{i} - Ay\mathbf{j}$$

we would infer that the z component of velocity, w, is to be taken as zero. We have

$$u = Ax \qquad v = -Ay$$

so it is easy to solve the differential equation for the streamlines, Eq. (3.5.3), or

$$\frac{dx}{Ax} = -\frac{dy}{Ay}$$

to obtain the curves $xy = $ constant. It will be of interest to solve for the components of velocity along the points in space associated with some particular streamline, say, $xy = 4$. This choice does not restrict our selection of the value for A. Let it be equal to 2. Since the velocity field that is specified by V is time-independent, we expect that the streamlines just found will also be pathlines. The following discussion should underscore this fact.

The u and v components have been prescribed at each point of space and will be associated with whatever particle passes through these points. Because the path of the particle is tangent to its current velocity, the slope of the pathline is the same as the streamline at a point at the moment the particle passes through the point. An instant later, the particle has moved to a new position. It has moved along its path that has the same slope as the streamline, since the velocity field is, in this case, time-independent. At this new point, the situation repeats itself, because of the steady velocity. Thus, in such a flow the streamline through a point and the pathline of a particle that has gone through that point always have the same slope. The two curves coincide.

The table below gives some velocities that a particle passing through (1, 4) would experience.

x	y
1.00	4.00
1.01	3.96
1.02	3.92
1.03	3.88

so that

u	v	V	Avg. speed	Distance between adjacent points
2.0	−8.00	8.246⎫		$[(0.01)^2 + (0.04)^2]^{1/2}$
2.02	−7.92	8.174⎰⎱	8.210	
2.04	−7.84	8.101⎰⎱	8.138	0.04123
2.06	−7.76	8.029⎰	8.065	0.04123

If a particle of fluid were at the location in space corresponding to (1, 4) at time $t = 0$, we would expect to find it at (1.01, 3.96) after a duration of 0.04123/8.21 units of time. After an additional 0.04123/8.138 units of time, the particle would be found at a location of (1.02, 3.92). One could use these facts to determine the acceleration of the particle.

At a later time, some other particle would pass through (1, 4). Because of the steady velocity field, it would pass through the same points as its predecessor. This means that the streakline generated by all of the particles that pass through (1, 4) will have a shape exactly like that of the streamline.

If one generalizes the line of argument that was used in Example 3.5.5, one could conclude that streamlines, pathlines, and streaklines are coincident in steady flows. We shall now see that, when the flow is unsteady, the geometric equivalence of streamlines, pathlines, and streaklines no longer generally exists.

Example 3.5.6

For a velocity distribution given by $\mathbf{V} = A\mathbf{i} + B\mathbf{j}$, the streamlines will be straight lines if A and B are constants. This would also be true if $A = C_1t$ and $B = C_2$. In this case, the streamlines would be

$$y = \frac{C_2 x}{C_1 t} + \frac{C_3}{C_1 t} = \frac{K_1 x}{t} + \frac{K_2}{t} \tag{3.5.4}$$

The slope of the streamlines would decrease with time. The value of u (C_1t, in this case) will be the x component of the velocity of the particle that is passing through the appropriate point at a particular instant. Since u and v in this example do not depend on x and y, the particle will have an associated x-directed velocity that depends only on t. At each point of space, it will be C_1t. Using X and Y to denote the coordinates of a particle's position, we can write

$$\frac{dX}{dt} = C_1 t \quad \text{and} \quad \frac{dY}{dt} = C_2$$

with the result that

$$X - X_0 = \frac{C_1 t^2}{2} \quad \text{and} \quad Y - Y_0 = C_2 t \tag{3.5.5}$$

in which X_0 and Y_0 give the location of the particle at $t = 0$. These equations can be rewritten as

$$X - X_0 = \frac{C_1(Y - Y_0)^2}{2C_2^2} \tag{3.5.6}$$

Since (X_0, Y_0) is different for each particle, we have an infinite set of parabolas, for our pathlines. They are pathlines because they give the time history of the position of a particle.

If we examine three such paths, we shall learn something about the associated streakline. Let the first particle be located at the origin when $t = 0$. This means that $X_0 = 0 = Y_0$ and the path of the particle that was at the origin at $t = 0$ is

$$X = \frac{C_1 t^2}{2} \qquad Y = C_2 t \tag{3.5.7}$$

Assume that we can identify the particle that went through the origin at $t = 1$. Then its coordinates are, using the expressions for X and Y in terms of t,

$$0 - X_0 = \frac{C_1(1)^2}{2} \quad \text{and} \quad 0 - Y_0 = C_2(1)$$

These equations then become

$$\left(X + \frac{C_1}{2}\right) = \frac{C_1 t^2}{2} \quad \text{and} \quad Y + C_2 = C_2 t \tag{3.5.8}$$

and are applicable to the particle that was at the origin at $t = 1$. An inquiry into the initial position of the particle that passes through the origin at $t = 2$ will give

$$X + 2C_1 = \frac{C_1 t^2}{2} \tag{3.5.9}$$

and

$$Y + 2C_2 = C_2 t$$

as the parametric equations for this particle. At $t = 2$, the three particles will be located from these equations as follows. Equation (3.5.7) applies to the particle that passed through the origin at $t = 0$:

$$X = \frac{C_1(2)^2}{2} \qquad Y = C_2(2) \qquad \text{at } t = 2$$

Equation (3.5.8) pertains to the particle that passed through the origin at $t = 1$:

$$X = C_1 \frac{(3)}{2} \qquad Y = C_2 \qquad \text{at } t = 2$$

Equation (3.5.9) is appropriate for the particle that passed through the origin at $t = 2$:

$$X = 0 \qquad Y = 0$$

These positions are plotted in Fig. 3.5.3. The line drawn through them connects the positions of particles that passed through the origin. This is the streakline through the

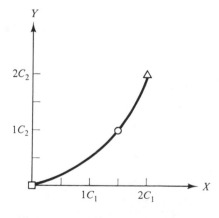

Figure 3.5.3 Plot of streakline at $t = 2$, from which one can infer that streaklines are neither streamlines nor pathlines for the particular flow.

origin. The reader may choose to show that these three points are not on a straight line. Hence, they could not coincide with a streamline. Could these points coincide with a parabola of the form $X - a = K(Y - b)^2$, which is the equation for the pathlines? The reader should establish that this is not so. Hence, the streakline can be neither a streamline nor a pathline for this unsteady flow.

3.6 CONSERVATION OF MASS—GENERAL IDEAS

One often hears the statement, "Mass cannot be created or destroyed." This idea was behind the experiments of Lavoisier when he explained the apparent disappearance of material during a combustion process. In this case, the original mass of material—with its associated number of atoms—was altered in form by the chemical reaction; but, if one was careful to account for all of the products of reaction, the original number of elemental atoms, and consequently the original mass, would still be present at the end.

This paraphrasing of the law of mass conservation is restricted to cases where the mass is not transformed into energy, a process that we now recognize as possible, and which we excluded in the introductory remarks. Let us now see how we can use the concept of mass conservation in our present course of study. We shall begin by considering the flow of a fluid in a conduit of varying cross-sectional area, as is shown in Fig. 3.6.1.

The conduit in Fig. 3.6.1 is assumed to be full of fluid flowing from the left to the right. For the moment, we shall focus our attention on a short segment of the pipe, imagining that we have instantaneously dyed a section of the fluid a distinctive color by which we can identify it at a later time. This mass of fluid is shown by the shaded region in Fig. 3.6.2 at t_0, which is the *material volume* for the particular mass of fluid that we shall now study. It has moved to the position shown in Fig. 3.6.3 during a brief interval of time, $\Delta t = t_1 - t_0$. The material volume occupies a different region of space at t_1. Now the fact that the mass of colored fluid must be the same in Figs. 3.6.2 and 3.6.3 will be important in helping us to formulate a mathematical statement that will represent the law of mass conservation. Initially, we can say that the mass of fluid represented by the material volume $ABCDA$ in Fig. 3.6.4, which combines parts of the previous figures, must be equal to the mass within material volume $A'B'C'D'A'$. Letting m = mass, this fact can be stated as

$$m_{ABCDA} = m_{A'B'C'D'A'} \tag{3.6.1}$$

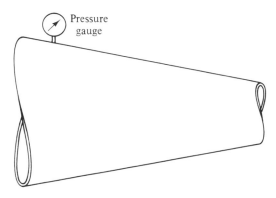
Pressure gauge

Figure 3.6.1 Conduit with varying diameter.

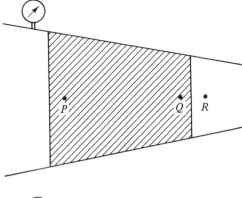

Figure 3.6.2 Representation of the fluid within the conduit showing fixed points of space P, Q, and R, and a material volume of dyed fluid at $t = t_0$.

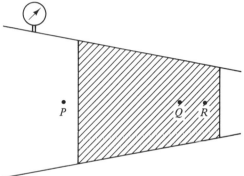

Figure 3.6.3 Representation of same location within the conduit showing the material volume of dyed fluid at t_1 with respect to the fixed points P, Q, and R.

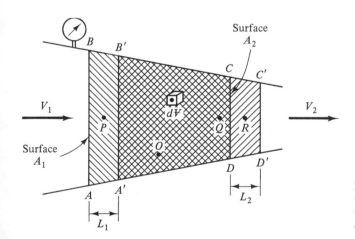

Figure 3.6.4 Superposition of representations of material volume at t_0 and t_1 to develop logic for deriving an equation of mass conservation.

but $$m_{ABCDA} = (\text{mass in region of space } ABB'A'A \text{ at } t = t_0)$$

$$+ (\text{mass in region of space } A'B'CDA' \text{ at } t = t_0) \qquad (3.6.2a)$$

Similarly,

$$m_{A'B'C'D'A'} = (\text{mass in region of space } A'B'CDA' \text{ at } t = t_1)$$

$$+ (\text{mass in region of space } DCC'D'D \text{ at } t = t_1) \qquad (3.6.2b)$$

These last two equations can be subtracted from one another to give

(mass in regions $A'B'CDA'$ and $DCC'D'D)_{t_1}$

$$- \text{(mass in regions } ABB'A'A \text{ and } A'B'CDA')_{t_0} = \Delta m \qquad (3.6.3)$$

This is the change of an extensive property of the system, its total mass, in the time interval Δt. The principle of mass conservation tells us that this change must be zero. But in order to prepare for later developments, we have written the right-hand side of Eq. (3.6.3) as Δm and not zero.

Equation (3.6.3) requires that we determine the mass in several regions of space at two different times. However, one of these regions, $A'B'CDA'$, must be used at both the initial and the final time. We begin by examining this region first. The mass in region $A'B'CDA'$ can be evaluated at t_1 if one can ascertain the average density within small subdivisions of the region, $d\mathcal{V}$ in Fig. 3.6.4. The density ρ is an intensive property; one obtains the local contribution to the total mass by forming the product $\rho \, d\mathcal{V}$. This observation lets us write two of the calculations implicit in Eq. (3.6.3) as

$$(m_{A'B'CDA'})_{t=t_1} - (m_{A'B'CDA'})_{t=t_0} = \int (\rho \, d\mathcal{V})_{t_1} - \int (\rho \, d\mathcal{V})_{t_0}$$

in which the integrals are evaluated at the points within $A'B'CDA'$. This expression is now written in more compact (and meaningful) form as

$$\int_{A'B'CDA'} \rho(x, y, z, t_1) \, d\mathcal{V} - \int_{A'B'CDA'} \rho(x, y, z, t_0) \, d\mathcal{V} = \Delta m_{A'B'CDA'}$$

Since these two integrals would be evaluated at the same set of points—i.e., those within a common volume—one could write this difference of integrals as the integral of the difference:

$$\int_{A'B'CDA'} [(\rho)_{t_1} - (\rho)_{t_0}] \, d\mathcal{V} = \Delta m_{A'B'CDA'}$$

If we recall that $t_1 - t_0 = \Delta t$, a small interval of time, we can introduce this Δt into the integral if it is convenient. (It will prove to be so.) Note that the integration is over space, so that a constant multiplier of Δt would not affect the integration process if one introduced it into the numerator and denominator as

$$\Delta t \int_{A'B'CDA'} \left[\frac{(\rho)_{t_0 + \Delta t} - (\rho)_{t_0}}{\Delta t} \right] d\mathcal{V} = \Delta m_{A'B'CDA'}$$

As $\Delta t \to 0$, the term within the square brackets becomes $\partial \rho / \partial t$. Moreover, as $\Delta t \to 0$, the region $A'B'CDA'$ becomes larger and larger and approaches region $ABCDA$, the initial volume of our system. With this notation, we can write Eq. (3.6.4) as

$$\Delta t \left[\int_{CV} \left(\frac{\partial \rho}{\partial t} \right)_{t_0} d\mathcal{V} \right] = \Delta m_{CV, \, \Delta t \to 0} \qquad (3.6.5)$$

(The material volume at the initial instant of the analysis coincides with the *control volume,* or CV.)

Equation (3.6.5) facilitates a rearrangement of Eq. (3.6.3) to the form

$$\lim_{\Delta t \to 0} \frac{\Delta m}{\Delta t} = \left[\int_{CV} \left(\frac{\partial \rho}{\partial t} \right)_{t_0} d\mathcal{V} \right] + \frac{1}{\Delta t} [(m_{C'D'DCC'})_{t_1} - (m_{ABB'A'A})_{t_0}] \qquad (3.6.6)$$

Mass Conservation   Chap. 3

The term on the left-hand side is the time rate of change of the mass of our system (i.e., within the material volume). Thus, the time rate of change of the total mass of a system, an extensive property of the system, is obtained by summing two terms. The first term is an integral that records the changes within the control volume. For the present, we shall concentrate on the second term. It will be shown to be equivalent to the rate at which mass flows out of the control volume across the control surfaces at t_0.

3.7 CONTINUITY EQUATION FOR STEADY STATE

We shall *assume* that the flow is steady; this is a special case of Eq. (3.6.6) that occurs often. Moreover, this case allows us to concentrate upon the last two terms of Eq. (3.6.6), the mass terms for regions $C'D'DCC$ and $ABB'A'A$. These terms require that the variables of the problem be evaluated at different points of space. It is recognized that for the assumed steady flow these variables can have different values from point to point in a region, but the variables of the problem do not change at a *particular* point as the time changes. While we shall allow the density, for example, not to be the same at points P, Q, and R in Fig. 3.6.4, we stipulate that the density at any point within the flow field does not change with time. This means that $\partial\rho/\partial t$ is zero at all points in the fluid. (This would be true for the time derivatives of all other flow variables for a truly steady flow and not just one with a steady density.)

For the present, we shall also *assume* that the flow is uniform at each of the cross sections of the conduit. Thus, the fluid properties and velocity can differ along the axis of the conduit, but they are all the same value along a cross section such as AB in Fig. 3.6.4. The uniformity of the velocity at each section makes it easy to determine the volume rate of flow across a section because no formal integration is required to evaluate some of the remaining terms in Eq. (3.6.6). The volume of fluid in region $ABB'A'A$ in Fig. 3.6.4 is the volume rate of flow multiplied by the time interval. This means that the mass in region $ABB'A'A$ in Fig. 3.6.4 must be the product of the average value of ρ in L_1, designated as $\bar\rho_{L_1}$, and the volume of $ABB'A'A$, or

$$m_{ABB'A'A} = \bar\rho_{L_1} V_1 A_1 (t_1 - t_0) \tag{3.7.1}$$

Similarly, the mass in region $C'D'DCC'$ can be written as $\bar\rho_{L_2} V_2 A_2 (t_1 - t_0)$ if it is remembered that the fluid's density may be changing as it flows across the surface labeled AB to the surface designated by CD that has an area A_2, and if V_2 is the velocity component normal to that surface. (Again, the normal velocity component has been *assumed* to be constant across the applicable segment of the control surface.) Now that we have evaluated the two nonzero terms in the right-hand side of Eq. (3.6.6), the results can be substituted in that equation to yield

$$\lim_{\Delta t \to 0} \frac{\Delta m}{\Delta t} = 0 + \frac{1}{\Delta t}(\bar\rho_{L_2} V_2 A_2 \Delta t - \bar\rho_{L_1} V_1 \Delta t) \tag{3.7.2}$$

This equation expresses the idea that the mass of fluid flowing into the control volume equals the amount leaving it for each instant of time. (This is true only for steady flow, which was one of the *assumptions* in the present case.)

The brief time interval Δt can be canceled on the right-hand side of Eq. (3.7.2) before the limit is taken. In the limit $\Delta t \to 0$, L_1 and $L_2 \to 0$ and the average values that are required are the values on the segments of the control surface. Consequently, for the case of steady, uniform flow,

$$0 = \rho_2 V_2 A_2 - \rho_1 V_1 A_1 \qquad (3.7.3a)$$

This equation is called the *continuity equation,* for steady, uniform flow. (Volume conservation at a point will provide a differential equation that is also commonly referred to as the continuity equation.) *If* the density is constant along the length of the conduit, $\rho_1 = \rho_2$, so that

$$V_1 A_1 = V_2 A_2 \qquad (3.7.3b)$$

for a fluid of constant density in steady flow. A fluid in which the particles do not change in density as they move from place to place is called an *incompressible fluid.*

Air that is subject to high pressure changes is usually considered to be compressible. Water and oil are so slightly compressible under the influence of pressure that they are considered to be almost incompressible. However, water does change its volume slightly with changes in temperature, and this results in buoyancy effects that are the cause of some fluid flows in nature. Still, we can consider water to be incompressible because its density is almost pressure-independent. However, the density of a liquid can also change due to diffusion of some solutes, depending on the concentration gradients. If we were to add NaCl to a flowing stream, we would want to conserve the mass of all the constituents. Because there is no apparent change of volume of water when adding NaCl to make a saline solution, volume conservation can also be invoked. In the following derivations and problems, an attempt will be made to specify clearly the situation with regard to the density. In this way the student should become more familiar (and secure) with this aspect of the subject.

One additional and important observation can be made concerning Eq. (3.7.2). In this case we require that mass conservation be observed without following the fluid particles, as, for example, in Eq. (3.6.3). Instead, we restrict our attention to a fixed surface in space. This is the surface that coincides with the surface of the material volume for the mass that is of interest to us at the initial time t_0. Subsequently, we continue to examine only this surface. This is the *control surface,* where we do our "accounting" to assure ourselves that mass conservation is occurring. In this process we keep account of the velocity and density of new particles that are continuously entering (and then leaving) the *control volume* (i.e., the region in space that was occupied by the fluid at the initial time, t_0—region *ABCD* in Fig. 3.6.4, for example).

Because now we are no longer concentrating on a fixed number of atoms of fluid, Eq. (3.7.2) does not apply to a closed system as did Eq. (3.6.3). We say that Eq. (3.7.2) is applied to the control surfaces of an *open system.* An important feature of much of our subsequent work will be the utilization of open systems.

The conduit shown in Fig. 3.6.1 was a common, but simple, situation. A more complex situation is shown in Fig. 3.7.1. For this case we apply the same reasoning that led us to Eq. (3.7.3a). This gives the result that the rates of mass flowing into and out of the region must be equal for *steady flow,* or

$$\rho_1 V_1 A_1 + \rho_3 V_3 A_3 = \rho_2 V_2 A_2 + \rho_4 V_4 A_4 \qquad (3.7.4)$$

for densities and velocities that are uniform (i.e., constant) *over their respective areas.*

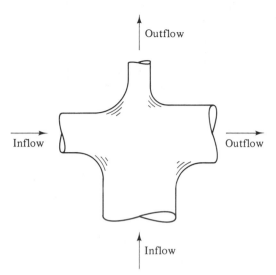

Figure 3.7.1 Portion of a conduit through which flow enters and leaves at various locations.

This statement is rewritten for increased lucidity with all of the terms on one side of the equality sign as

$$-\rho_1 V_1 A_1 - \rho_3 V_3 A_3 + \rho_2 V_2 A_2 + \rho_4 V_4 A_4 = 0 \tag{3.7.5}$$

If we consider our control volume and on each section of the control surface construct an *outward* normal, **n,** to the control volume, we have the situation depicted in Fig. 3.7.2. Then the flow rate *leaving* the control volume is, using vector notation, $(\mathbf{V \cdot n})\,dA$. No subscripts are used, because we shall understand that the scalar operation is performed at each and every appropriate section. The surfaces A_2 and A_4 in Fig. 3.7.2 are such sections. But what about sections A_1 and A_3? The operation $\mathbf{V \cdot n}$ can be applied here too, because the result will still be the mass rate of flow leaving the control volume.* (This may seem strange, but at section A_1, the term $\mathbf{V \cdot n}$ is equal to $(|\mathbf{V}_1|)(|\mathbf{n}_1|)(\cos \alpha)$, if we recall the definition of $\cos \alpha$ in Fig. 3.3.4). In view of this, we can see that the mass rate of flow leaving section A_1 is $[-(\text{magnitude of normal component of } \mathbf{V}_1)(A_1 \rho_1)]$. This minus coefficient is all-important. It denotes a negative "leaving" of mass from the region, per unit time. It is not too difficult to consider this negative "leaving" as the rate of mass entering the region. One can view Eq. (3.7.5) as a statement of the requirement that the *total* rate of mass *leaving* a control volume must be zero if the flow is steady. This is stated mathematically as

$$\Sigma \rho_i (V_i \cos \alpha_i)(\Delta A_i) = \sum_A \rho(\mathbf{V \cdot n})(\Delta A) = 0 \tag{3.7.6a}$$

for finite elements of area ΔA_i. This expression does not preclude the possibility that the velocity varies over the cross sections of our control surface. Then the ΔA_i are the area elements in which ρ_i and $V_i \cos \alpha_i$ are constant. This result is similar to that of Eq. (3.3.3) for cases similar to those in Fig. 3.3.2.

---

* Our use of vector operations will be limited. Although the use of Newton's laws of motion would make it convenient to use vector methods, we shall be dealing mostly with problems in one or possibly two dimensions, so that the advantages of vector methods are not fully realized. Nevertheless, the geometric concepts associated with vector operations are often helpful in understanding, or generalizing, a statement of a problem or a derivation. In view of this, several results will be stated in vector notation.

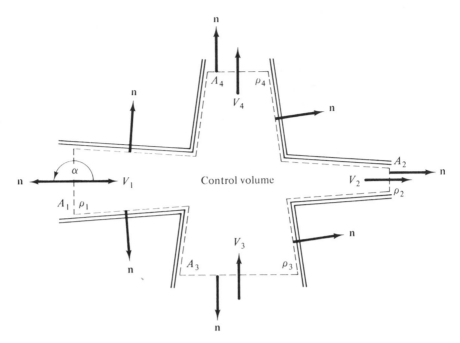

Figure 3.7.2 Representation of portion of conduit with multiple inflows and outflows to assist in analysis of mass flow rates.

For a continuously varying $\mathbf{V}$ and/or ρ, an infinitesimal dA would be required. Then, for a *steady flow*,

$$\int_{CS} \rho(\mathbf{V \cdot n})\, dA = 0 \qquad (3.7.6b)$$

If this equation is not satisfied, either the flow is unsteady or the values used in the integral are incomplete or incorrect.

Example 3.7.1

Apply Eq. (3.7.6a) to the flow situation shown in Fig. 3.7.3.

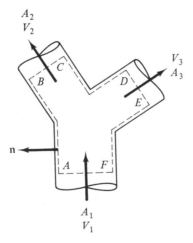

Figure 3.7.3 Analysis of flow in a pipe wye begins with introduction of a control volume. $\mathbf{n}$ = outer normal to control volume; $A_1 = 0.01$ m²; $A_2 = 0.004$ m²; $A_3 = 0.007$ m²; $V_1 = 3$ m/s; $V_3 = 2$ m/s. Find V_2.

Solution

Across surface:	$\mathbf{V} \cdot \mathbf{n} =$
AB	0 (because flow is tangent to wall)
BC	$+V_2$ (because of the given direction of $\mathbf{V}_2$)
CD	0
DE	$+2$ m/s
EF	0
FA	-3 m/s

so Eq. (3.7.6a) becomes

$$\rho(0)A_{AB} + \rho_2 V_2(0.004) + \rho(0)A_{CD} + \rho_3(2)(0.007) + \rho(0)A_{EF} + \rho_1(-3)(0.01) = 0$$

Example 3.7.2

Determine the values of $\mathbf{V} \cdot \mathbf{n}$ for the situation in Fig. 3.7.4.

Solution

On surface:	$\mathbf{V} \cdot \mathbf{n} =$
1	$-V_1$
2	$+V_2$
3	0 }
4	0 } (because velocity component normal to a wall is 0)
5	0 }
6	0 } (because velocity is tangent to a streamline)

If ρ *is constant* over all the surfaces that are appropriate to an application of Eq. (3.7.6b), it can be factored out of the integral so that for a *steady flow*

$$\int_{\text{CS}} \mathbf{V} \cdot \mathbf{n}\, dA = 0 \tag{3.7.7}$$

This is a statement of volume flow rate conservation.

Example 3.7.3

For the velocity distribution $\mathbf{V} = (4y - 7x)\mathbf{i} + (7y - 8x)\mathbf{j}$,

(a) Draw the velocity components perpendicular to planes having traces AB, BC, CD, and DA shown in Fig. 3.7.5a.

Solution The velocity component perpendicular to AB is u. This component is $(4y - 7x)$.

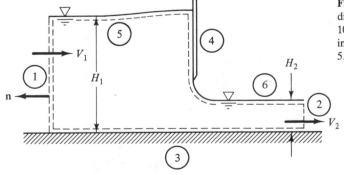

Figure 3.7.4 Control volume for two-dimensional flow past a sluice gate. Width is 100 m normal to the paper. Surface 1: entering flow; surface 4: the sluice gate; surfaces 5, 6: free surfaces.

Along AB, $x = 0$. Thus, $u = 4y$ on AB. These velocity distributions are illustrated in Fig. 3.7.5b.

The velocity component perpendicular to BC is v. Along BC, $v = 7(1) - 8x$.

The component perpendicular to CD is $u = 4y - 7(1)$, and the component perpendicular to AD is $v = 7(0) - 8x$.

(b) Determine the value of $\int \mathbf{V} \cdot \mathbf{n} \, dA$ over surfaces AB, BC, CD, and AD.

Solution 1. On AB, the outer normal $\mathbf{n} = -\mathbf{i}$, and so

$$\int \mathbf{V} \cdot \mathbf{n} \, dA = \int [(4y - 7x)\mathbf{i} + (7y - 8x)\mathbf{j}] \cdot (-\mathbf{i})[dy \,(1)]$$

for a unit distance perpendicular to the xy plane. The result, for $x = 0$, is

$$-\int_0^1 (4y - 7x) \, dy = -\frac{4y^2}{2}\bigg|_0^1 = \underline{-2}$$

2. On BC, $\mathbf{n} = \mathbf{j}$, and so

$$\int \mathbf{V} \cdot \mathbf{n} \, dA = \int_0^1 [7(1) - 8x] \, dx$$

or

$$\left(7x - \frac{8x^2}{2}\right)\bigg|_0^1 = 7 - 4 = \underline{3}$$

3. On DC, $\mathbf{n} = +\mathbf{i}$, and so

$$\int \mathbf{V} \cdot \mathbf{n} \, dA = \int_0^1 (4y - 7)(1) \, dy$$

or

$$\left(\frac{4y^2}{2} - 7y\right)\bigg|_0^1 = 2 - 7 = \underline{-5}$$

Figure 3.7.5 (a) Region in a particular flow field in which a control volume has been drawn to assess the net rate of mass efflux. (b) Normal velocity components on the various portions of the control surface.

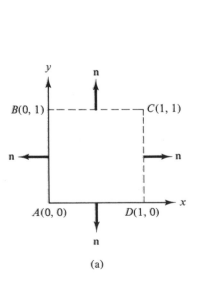

(a)

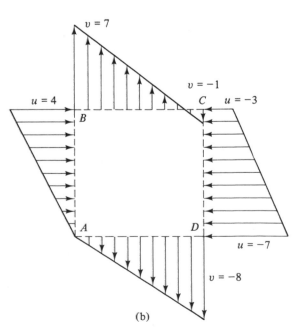

(b)

Mass Conservation   Chap. 3

4. On AD, $\mathbf{n} = -\mathbf{j}$, and so

$$\int \mathbf{V} \cdot \mathbf{n} \, dA = -\int_0^1 [7(0) - 8x] \, dx$$

or

$$\frac{8x^2}{2}\Big|_0^1 = \underline{\underline{4}}$$

(c) For a steady, incompressible flow,

$$\int_{CS} \mathbf{V} \cdot \mathbf{n} \, dA = 0$$

in which the area of the control surface includes surfaces AB, BC, DC, and CA. This means that the sum of the four numerical results in part (b) must be zero. Is it?

Example 3.7.4

When a parallel flow with constant speed moves past a flat plate aligned with the flow, one has occasion to calculate the flow through a series of surfaces in order to verify that the principle of mass conservation is being satisfied. There are some surfaces across which it is obvious how to determine the flow rate, and other surfaces can be represented by combinations of these for convenience of calculation.

Figure 3.7.6 illustrates the situation at hand. In this case the plate forms one surface—perhaps a trivial one, because we sense that the flow across an impermeable surface should be zero. However, in some interesting situations the plate can be considered to be porous, and there exists a velocity component normal to the plate that accounts for the fluid being expelled or absorbed across the surface of the plate.

We shall now apply Eq. (3.7.6b), but ρ will be considered to be constant. Our equation will have terms corresponding to each of the nontrivial portions of the control surface. Across surface OA, which is *assumed* to have a unit width, we can write

$$\rho Q_{OA} = \rho \int dQ = -\rho \int_0^\delta u(1) \, dy = -\rho \int_0^\delta U \, dy = -\rho U \int_0^\delta dy = -\rho U \delta$$

This result is a simple application of Eq. (3.7.6b), because the flow can be assumed to be parallel to the x axis as it comes onto the plate, and $\mathbf{V} \cdot \mathbf{n} = -u$. The element of area, dy (1), is perpendicular to this velocity component.

The surface BC is drawn at some arbitrary station along the plate. The position of point B is chosen to coincide with the place where the velocity, in the direction of the

Figure 3.7.6 Flow past an impermeable flat plate, with no flow in the z direction. U = upstream velocity to which plate is aligned.

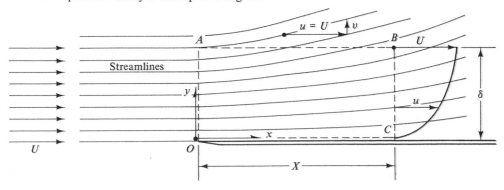

plate, has increased from zero (the value at the plate) to the value U, which is the speed that one finds in regions where the flow is, for all practical purposes, unaffected by the presence of the plate. The distance BC is a measure of the distance over which the rigid boundary influences the flow. This distance is called the *boundary layer,* and we shall refer to this example again when boundary layers are presented in detail. The velocity profile for the x component u is given in Fig. 3.7.6. There may also be a y component of velocity at station BC; however, it is of no immediate concern here, since the area element of interest is dy (1). This elemental area is perpendicular to the x direction, and the velocity component normal to it is u. Hence we need only specify the u component at station BC if we wish to write $\int \mathbf{V} \cdot \mathbf{n} \, dA$ and calculate the flow rate in accordance with Eq. (3.7.6b). This is

$$\rho Q_{BC} = \rho \int_0^\delta u \, dy \qquad u = u(y) \text{ on } BC$$

Across surface OC, the calculation of the flow rate is simple. Because the surface is impermeable, there is no velocity component normal to the surface. Thus $\mathbf{V} \cdot \mathbf{n}$ is zero at each point on the surface, and the integral in Eq. (3.7.6b) is a trivial summation of zeros.

The situation across surface AB is more complicated. Here also the velocity is not perpendicular to the elements of surface dx, which subdivide the total surface AB. There is a y component of velocity, v, which generally is not zero. (The manner in which v varies with x is a matter that must be reserved for Chapter 16.) Because the area element dx has a normal that is parallel to the y direction, one must use the velocity component that is parallel to this direction in order to satisfy the logic that led to Eq. (3.7.6b). This component is v. Consequently, the flow rate across surface AB is

$$\rho Q_{AB} = \rho \int_0^X v \, dx \qquad v = v(x, \delta) \text{ on } AB$$

We shall interrelate the various flow rates that have just been calculated. The nature of this interconnection will be determined by our demand that mass conservation be observed for the flow. If the flow is steady, the total efflux across the control surface $OABCO$ in Fig. 3.7.6 must be zero. Therefore

$$-\rho U \delta + \rho \int_0^\delta u \, dy + \rho \int_0^X v \, dx = 0$$

This can be rewritten as

$$-\rho \int_0^\delta (U - u) \, dy + \rho \int_0^X v \, dx = 0 \qquad (3.7.8)$$

3.8 UNSTEADY FLOW

Until now, we have emphasized steady flows. This gave us the opportunity to become acquainted with the concept of mass conservation in its simplest form. With the derivation of an integral expression for the rate of change of mass in a steady flow, Eq. (3.7.6b), we can include this expression in the general statement made by Eq. (3.6.6). In that expression the last two terms are equivalent to the integral of Eq. (3.7.6b). Because of this result, we can write the equation that represents $\Delta m / \Delta t$ as the sum of two integrals which, because of the physics that we have experienced, are equal to zero. This general statement of mass conservation—and note that it excludes mass diffusion—is

$$\int_{CV} \left(\frac{\partial \rho}{\partial t} \right) d\mathcal{V} + \int_{CS} \rho \mathbf{V} \cdot \mathbf{n} \, dA = 0 \qquad (3.8.1)$$

This result relates the total time rate of change of the mass of fluid that was within the control volume at t_0. Thus,

1. If there is a rate of increase within the control volume, the first integral in Eq. (3.8.1) will be positive.
2. This equation will require the second integral to be negative in order for the sum to be zero.
3. If the second integral, the rate of mass efflux from the control volume, is negative, we recognize that there must be a net rate of mass influx into the control volume.
4. This seems to be a logical conclusion to the assumption in statement 1.

Equation (3.8.1) could also have been written as

$$\frac{\partial}{\partial t}\left(\int_{CV} \rho \, d\forall\right) + \int_{CS} \rho \mathbf{V} \cdot \mathbf{n} \, dA = 0 \qquad (3.8.2)$$

because the differentiation and integration required in the first integral relate to different variables—i.e., time and space variables. The term inside the parentheses is the instantaneous mass within the control volume.

Example 3.8.1

The case of an emptying vessel is shown in Fig. 3.8.1. Liquid enters the tank at the rate of Q_1 (m^3/s) through a pipe of area A_1. It leaves the container, which has a cross section of area A_0, through a pipe of area A_2. Apply Eq. (3.8.2) to interrelate the variables for this situation.

Solution If the control surface is the one indicated by the dashed line *EFIJGHE*, then conditions are changing within the control volume for $Q_1 \neq Q_2$. Equation (3.8.2) states that

$$\text{Rate of mass increase in CV} = -\int \rho \mathbf{V} \cdot \mathbf{n} \, dA$$

This integral, for the chosen control surface, leads to $-(V_2 A_2 - V_1 A_1)\rho$ or $+(Q_1 - Q_2)\rho$. If $Q_1 > Q_2$, there will be a positive rate of mass increase within the control volume. Since it has been *assumed* that a liquid of constant density is flowing,

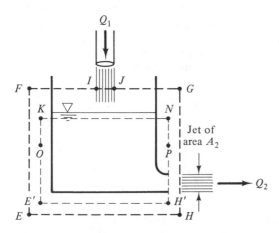

Figure 3.8.1 Flow out of a tank that is being filled from above so that liquid level in tank could be unsteady. *KN*: free surface of area A_0; *EF*: control surface segment; $E'H'$: another possible control surface segment.

$$\frac{\partial}{\partial t} (\int \rho \, dV) = \rho \frac{d}{dt} \text{(volume of liquid in CV)}$$

Because ρ is constant, we can summarize our previous remarks as

$$\frac{dV}{dt} = Q_1 - Q_2$$

For constant flow rates, the implication of this step is

$$V = V_{t=0} + (Q_1 - Q_2)(t)$$

At the particular instant when the liquid surface was at KN, one might have chosen the control surface to coincide with the dashed line $EKNH$. The free surface will be rising above KN for $Q_1 > Q_2$. But within this control volume there are no changes with time if Q_1 and Q_2 are constant. Then

$$\int \rho \mathbf{V} \cdot \mathbf{n} \, dA = -V_1 A_1 \rho + V_2 A_2 \rho + (A_0 - A_1) V_s \rho$$

in which V_s is the rate at which the liquid surface is rising past the surface KN at that instant. The three terms on the right-hand side of this equation are equal to $\partial(\int \rho \, dV)/\partial t$, which is zero for the control volume $E'KNH'E'$.

A control volume could have been selected with a horizontal surface through OP. Again, if Q_2 were steady, one would have constant conditions within the control volume, and the result of applying the equation of mass conservation would be

$$-V_0 A_0 \rho + \rho Q_2 = 0$$

in which V_0 is the downward flow, assumed uniform, within the vessel toward the outlet.

It can be concluded from this discussion that in some problems there is no unique choice for the control surface. Each one may require a small variation of approach to use a physical principle. Some may be easier than others, but the best choice can be determined only from experience.

Example 3.8.2

If we decided to fill a container with a liquid of varying density, e.g., water with a changing sodium chloride content, we could proceed in the manner shown in Fig. 3.8.2. Two vats are filled initially with incompressible liquids which when mixed have a volume equal to the sum of the two original volumes.* In vat II, there is a mixing device that keeps the contents homogeneous. Liquid is withdrawn from this vat, possibly with the aid of a pump, at the rate Q_2. Vat I is a source of liquid of constant density $\rho_1 = \bar{\rho}_1$, which flows into vat II either by gravity or by means of a pump. If initially $\bar{\rho}_2 > \bar{\rho}_1$, the contents of vat II will be diluted with increasing time. The density of the liquid that is issuing out of this two-tank system, ρ_2, will be a monotonically decreasing function of t. The opposite will be true if $\bar{\rho}_2 < \bar{\rho}_1$. Check the validity of the differential equations that follow.

For vat II:

$$\begin{aligned}
\bar{\rho}_1 Q_1 - \rho_2 Q_2 &= \frac{d}{dt}(\rho_2 V_2) \\
&= \frac{\rho_2 \, dV_2}{dt} + V_2 \frac{d\rho_2}{dt}
\end{aligned} \qquad (3.8.3)$$

The overbar on ρ_1 should serve as a reminder that this term is not time-dependent according to Fig. 3.8.2.

---

* This is not exactly so for all liquids; e.g., alcohol and water.

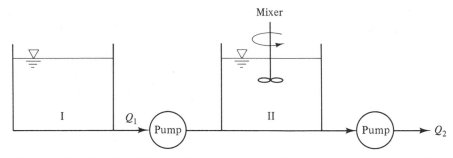

Figure 3.8.2 Method of filling a tank at rate Q_2 with a liquid whose density varies continuously. Vat I: $\rho = \bar{\rho}_1$. Vat II: $\rho_{2, t=0} = \bar{\rho}_2$.

Because the fluid is incompressible, volume conservation requires that

$$Q_1 - Q_2 = \frac{d\mathcal{V}_2}{dt} \tag{3.8.4}$$

which implies for constant Q's

$$\mathcal{V}_2 = (\mathcal{V}_2)_0 + (Q_1 - Q_2)(t) \tag{3.8.5}$$

in which the subscript 0 denotes the condition at $t = 0$. Equations (3.8.4) and (3.8.5), combined with Eq. (3.8.3), result in

$$Q_1(\bar{\rho}_1 - \rho_2)\, dt = [(\mathcal{V}_2)_0 + (Q_1 - Q_2)t]\, d\rho_2 \tag{3.8.6}$$

This equation can be integrated to yield ρ_2 as a function of t.

3.9 REYNOLDS' TRANSPORT THEOREM

Equations (3.8.1) and (3.8.2) are the result of logical steps that can be applied to other situations. In the aforementioned equations, ρ was a particular scalar property that was subject to change within the control volume and/or was transported out of the control volume. This change of, and transport of, ρ in the two different integrals formed the total change rate of the mass of the system under consideration, an extensive property. We shall now deal with a particular intensive property—and the time rate of change of its associated extensive property. This analysis will point out the generality of the conclusion that is reached.

The time rate of change of the internal energy of a system will be designated by $\Delta(\text{IE})/\Delta t$. This rate of change of the extensive property will be associated with the intensive property of the internal energy per unit mass, to be designated by the symbol e. The regions shown in Fig. 3.6.4 are used here to facilitate our development of an expression for $\Delta(\text{IE})/\Delta t$:

$$\frac{\Delta(\text{IE})}{\Delta t} = \frac{1}{\Delta t}\left[\begin{pmatrix}\text{change of internal energy}\\ \text{within } A'B'CDA' \text{ in } \Delta t\end{pmatrix}\right.$$

$$\left. + \begin{pmatrix}\text{internal energy within}\\ C'D'DCC' \text{ at } t_1\end{pmatrix} - \begin{pmatrix}\text{internal energy within}\\ ABB'A'A \text{ at } t_0\end{pmatrix}\right]$$

In the limit as $\Delta t \to 0$, the first term within the square brackets is $\Delta t\,[\partial(\int_{\text{CV}} (\rho e)\, d\mathcal{V})/\partial t]$, in which ρe is, in effect, the internal energy per unit volume in the region $A'B'CDA'$, which approaches *ABCDA,* the control volume, as $\Delta t \to 0$. The two last terms within the square brackets for our expression for $\Delta(\text{IE})/\Delta t$ can be written as

$$\Delta t \left[(\overline{\rho e})_{L_2}(V_2 A_2) - (\overline{\rho e})_{L_1}(V_1 A_1) \right]$$

if one is considering, at first, a uniform flow at each cross section. The extension to nonuniform flows follows in the same way as the one that led to Eq. (3.7.6b). Because this is true, we can write

$$\frac{\Delta(\text{IE})}{\Delta t} = \frac{\partial}{\partial t} \int_{\text{CV}} (e\rho) \, d\mathbb{V} + \int_{\text{CS}} (e\rho) \mathbf{V} \cdot \mathbf{n} \, dA \tag{3.9.1}$$

This means that the rate of change of the extensive property internal energy, $\Delta(\text{IE})/\Delta t$, is obtained by using the intensive property e that is multiplied by ρ in each of the integrals. From this result one can infer that to obtain the rate of change of any *extensive property*—call it E, for example—one first determines the associated *intensive property*—call it i—and writes

$$\frac{\Delta E}{\Delta t} = \frac{\partial}{\partial t} \int_{\text{CV}} (i\rho) \, d\mathbb{V} + \int_{\text{CS}} (i\rho) \mathbf{V} \cdot \mathbf{n} \, dA \tag{3.9.2}$$

This expression gives the content of the first axiom of Reynolds' "general equations of motion of any entity." If one were concerned about the rate of change of the kinetic energy of the fluid within a region, then E would be the total kinetic energy and i would equal $V^2/2$. Similarly, if E were the potential energy of a system, i would be gz, if z is measured positively upward.

3.10 SUMMARY

The requirement that mass must be conserved is expressed mathematically by Eq. (3.8.1). This equation requires an integration over a control volume and its control surface. Besides having information about the density of the fluid at the appropriate points, one must know the velocity there too. The velocity expressions that one uses are usually not those of a particle, but those existing at a point within the region of interest.

Moreover, one is usually not interested in the path lines that tracer particles in the flow make evident. Streak lines are produced by several methods, including the stream of smoke coming from a nozzle and flowing over a body, but usually a different type of line, the streamline, is desired. Only in a steady flow are these three lines coincident.

Once the velocity distribution in a region of flow is known, it is possible to determine the volume rate of flow by integrating the normal velocity component to an appropriate segment of a control surface. This flow rate and the area of the related surface through which the flow issues determine the average velocity of flow through that area.

Finally, we have concluded that the average rate of change of an extensive property of a material volume can be mathematically expressed in terms of two integrals just as the change of mass of that material volume is expressed. The result is Reynolds' transport equation, Eq. (3.9.2).

EXERCISES

3.3.1(a) If the centerline velocity of a tube with a diameter of 1 cm is 6 cm/s, what is the volume rate of flow if the velocity profile is similar to that given in the example? What is the average velocity?

(b) A measurement shows that the flow rate in a plastic pipe with an inside diameter of $\frac{3}{4}$ in. is 0.7 gpm. (About 231 in.^{3} = 1 gallon.) If the axial velocity is a paraboloid, what is the centerline velocity? What is the average velocity?

(c) A parabolic velocity distribution will exist in a round conduit if the product of the average velocity, the diameter, and the reciprocal of the kinematic viscosity is less than 2000. (Recall from Chap. 1 that this product is termed the *Reynolds number.*) Using a value of 10^{-6} m^2/s as the kinematic viscosity of water, calculate the maximum velocity in a tube (internal diameter = 1 cm) for a parabolic velocity distribution. What is the corresponding volume rate of flow?

(d) Are the data in Exercise 3.3.1b consistent with the parabolic velocity distribution that is also prescribed? Use the Reynolds number criterion in Exercise 3.3.1c and a water temperature of 75°F.

(e) If the Reynolds number for a round pipe exceeds 2000, the flow will not be laminar with a parabolic velocity profile. A turbulent pipe flow has a velocity distribution that can be approximated by $w = w_{max}(1 - r/r_0)^n$, in which n can be taken as $\frac{1}{7}$ and r_0 is the radius of the pipe. What volume rate of flow for water can be expected if the Reynolds number is 50,000 in a tube with a diameter of 2.5 cm? What is the ratio w_{max}/w_{avg} for this flow? What is this ratio for a parabolic flow?

3.3.2(a) What is Q if $H = 2$ m, $2b = 4$ cm, and the width of the slot is 60 cm? What is u_{avg}?

(b) What must be the height of the reservoir above the centerline of the jet if Q is to be 2.4 m^3/s for a slot 2 m long when $2b = 6$ cm?

(c) Find Q and u_{avg} for the case in which b/H becomes quite small (i.e., retain only terms linear in b/H in any expansions). Are you satisfied with the approximate results?

3.5.1(a) Let $A = 2$ and $B = 1$. Find the equation of the streamline passing through the point $(3, 0)$.

What is the velocity there? What is the slope of the streamline through this point?

(b) Let $A = (4 - t)$ and $B = 2$. Draw the streamlines through the point $(2, 0)$ for $t = 0, 2, 4, 6$. Where would you expect to find a particle at $t = 0.1$ that had been at $(2, 0)$ at $t = 0$?

(c) Refer to Exercise 3.5.1b, and in time steps of 0.05, locate the positions of the particle that was at $(2, 0)$ at $t = 0$. Use between 5 and 10 intervals. If a line were to connect these points, would it be a streamline, a streakline, a path line, or some combination of the foregoing?

(d) Use $A = 5t + 1$ and $B = 2t^2$. At $t = 1$, what is the equation of the streamline through $(3, 0)$? What is the x coordinate on this streamline (at $t = 1$) for $y = 0$? What is the streamline at $t = 2$ through the point that was just determined? At $t = 2$, what is the equation of the streamline through $(3, 0)$? Is this a steady flow? Explain.

3.5.4(a) Filaments of smoke are introduced into a wind tunnel in which a new automobile is being tested. The lines of smoke flow around the body to aid the desinger in understanding the related airflow. If steady conditions are possible during the test, are streamlines, streaklines, or pathlines being madc visible by the smoke? What would be your answer if the motor that powered the tunnel were suddenly turned off?

(b) A ship model is being towed in a long tank in which small plastic beads float upon the surface. What should be the orientation of a streak on a photographic negative that was caused by the reflected light from a particle near the bow of the model while it was being towed at steady speed? (The camera was stationary and a shutter speed of about 0.1 s was used. Reference the direction with respect to the bow-stern axis.) What is the orientation of the streaks made by the particles adjacent to the midship sides of the model? Do these streaks on the negative correspond to segments of path lines? Do all of the short streaks on a particular negative constitute portions of streamlines? Explain.

(c) If the towing carriage that is used to pull the model in Exercise 3.5.4b is also equipped to inject dye into the water just ahead of the model's bow, what would the image of this dye show on a photograph that was taken from the moving carriage? (Are streamlines, streaklines, or pathlines being made visible? Qualify your answers as appropriate.) Assume a constant towing speed.

(d) If a platinum wire is inserted into the flow of a slightly acetic water solution, hydrogen bubbles will form if the battery is properly connected. These bubbles will be swept off as a sheet from an uninsulated wire but as a fine line if the wire is coated with a dielectric paint, save for one point where it has been removed. Assume that streaks of bubbles are generated by having the electric potential on and off for 1-s intervals. If the water flow is steady, what do the collection of these bubble streaks constitute? If the flow in question is past an oscillating cylinder, would a high-speed $(\frac{1}{1000}$ s) photograph show streamlines? Explain. In this case, what would be the locus of the individual bubbles at the end of a particular streak? These bubbles were recorded by high-speed motion pictures.

3.5.5(a) Determine the y coordinate for $x = 1.04$ on the streamline $xy = 4$. What are u, v, and $|\mathbf{V}|$ for this point? What was the time for a particle to travel along this streamline from $x = 1.03$ to 1.04?

(b) What was the acceleration of the particle that moved from $x = 1.00$ to $x = 1.01$ along the streamline $xy = 4$?

3.5.6(a) Let K (constant of integration) $= -1.0$, 1.2. For $C_1 = 1$ and $C_2 = 2$, show that the points corresponding to Fig. 3.5.3 do not lie on a straight line by drawing the streamlines. Draw the streamlines for $t = \frac{1}{2}, 1, \frac{3}{2}, 2$. Are the streamlines straight lines? Show that they do not correspond to points on the appropriate equation for the pathline.

(b) Let $C_1 = 1$ and $C_2 = 2$. Draw the streamlines for $t = \frac{1}{2}, 1, \frac{3}{2}, 2$. What is the equation of a particle that was at $(3, 2)$ at $t = 0$? Make a graph—not necessarily exact—of this equation to show its general shape.

(c) What are X_0 and Y_0 for the particle that goes through the origin at $t = \frac{3}{2}$? What is the equation of the path of this particle? Where will it be at $t = 2$? If it is fitting that this point be plotted in Fig. 3.5.3, do so.

(d) Find the streamline through $(0, \frac{1}{2})$ for $C_1 = 1$ and $C_2 = 2$.

3.7.1(a) What is V_2 if $\rho_1 = \rho_2 = \rho_3$? What is V_2 if $\rho_2 = 2\rho_1$ and $\rho_3 = 3\rho_1$?

(b) If in Fig. 3.7.3 one were to change V_3 so that it had the same magnitude but opposite direction, what would V_2 be for $\rho = $ constant?

(c) If, in Fig. 3.7.3, $V_2 = 0$, would there be any constraint on the density at sections 1 and 3? If the density at section 1 is 0.001 kg/m³, what is ρ_3?

3.7.2(a) Evaluate Eq. (3.7.6b) for $\rho = $ constant, assuming a unit width (perpendicular to the page) and elevations at sections 1 and 2 of H_1 and H_2, respectively.

(b) What is V_2 if $H_1 = 10$ ft, $H_2 = 3$ ft, and there is a constraint that $0.5V_1^2 + 10(32.2) = 0.5V_2^2 + 3(32.2)$? What should be the dimension of V_2 in order that the dimension of each term in this constraint be the same? (The dimension of 32.2 is ft/s².) What is Q at sections 1 and 2?

3.7.3(a) Sketch the distribution of the velocity component normal to line AB in Figure 3.7.5a for $\mathbf{V} = (7x - 2y)\mathbf{i} + (3x - 7y)\mathbf{j}$. What is $\mathbf{V} \cdot \mathbf{n}$ along this line? Determine $\int_{AB} (\mathbf{V} \cdot \mathbf{n}) \, dA$. Is this a steady flow?

(b) For the $\mathbf{V}$ given in Exercise 3.7.3a, is $\int (\mathbf{V} \cdot \mathbf{n}) \, dA = 0$ for the surface $AB + BC + CD + DA$? (Refer to Fig. 3.7.5a.) Should this integral vanish if the density is constant? Does the expression for $\mathbf{V}$ describe an incompressible flow?

3.7.4(a) What is $\int_0^x v \, dx$ if $u = Uy/\delta$ for $0 \le y \le \delta$? What is the value of u at $y = 0$ and $y = \delta$? Do these values appear to be satisfactory in view of the discussion of the no-slip condition and definition of δ?

PROBLEMS

3.1 A rectangular duct 24 in. wide and 8 in. high has an average velocity in its cross section of 8 ft/s. What is the volume rate of flow? If the temperature of the flowing air is 70°F and the pressure is 14 psia, what is the mass rate of flow?

3.2 An open channel flow (consider it to be two-dimensional, with a width of 8 m perpendicular to the page) must pass below a gate as shown in Fig. 3.7.4. Upstream, the depth is 6 m. The discharge is 4.8 m³/s. Past the gate, the depth is 0.6 m. What is the upstream velocity. What is the downstream velocity?

3.3 The velocity distribution in the cross section of a river (at a bend) is measured as shown in Fig. P.3.3. The figures in the boxes are the measured stream velocities. (Assume these values apply to the entire area of the appropriate element.) What is the total volume rate of flow? What is the average velocity?

3.4 Draw the streamlines for the following velocity fields with $w = 0$:

$u = 5,$ $\quad v = 0$ $\qquad u = 0.2x, v = 0$
$u = 0.3y, v = 0$ $\qquad u = 0,$ $\quad v = 5$
$u = 0,$ $\quad v = 5x$ $\qquad u = 5x,$ $\quad v = 2x$
$u = x,$ $\quad v = -y$ $\qquad u = y,$ $\quad v = x$

3.5 What is the form of the streamlines given by $u_r = 0, u_\theta = 5/r$; and by $u_r = 16/r, u_\theta = 0$? What is the shape of the streamlines for $u_r = 16/r, u_\theta = 5/r$?

3.6 The rectangular cross section of a wind tunnel changes from 8 ft × 16 ft at the inlet to 2 ft × 4 ft in the test section. Because the wind speeds are low, we can assume that the air acts as though it is incompressible. If the wind speed is 30 ft/s in the test section, what is the air speed at the inlet of the wind tunnel?

3.7 Water is flowing through a round pipe with a 6-in. diameter at a speed of 9.5 ft/s. This pipe is reduced subsequently in diameter to 3 in. What is the average velocity in the reduced section?

3.8 A horizontal tube that is square in cross sec-

tion narrows from 1 ft² to $\frac{1}{4}$ ft². This decrease in cross section occurs over a distance of 8 ft. If the average velocity of water is 3 ft/s at the larger section, what should the average water speed be at the smaller section? What should the average speed be at the mid-distance of the reducing section?

3.9 A reducing section in a pipe changes the diameter from 8 in. to 4 in. This occurs smoothly over a distance, along the centerline, of 18 in. What is the velocity at the inlet and exit sections for an oil flow of 1.75 ft³/s? What is the expected average oil velocity at the midpoint of the reducing section?

3.10 Refer to Prob. 3.9 and write an expression for the average velocity as a function of the distance along the reducing section.

3.11 A gas enters a heat exchanger through a circular duct of area 0.196 m² with a density of 0.95 kg/m³ at an average speed of 6 m/s. Its density is decreased to 0.89 kg/m³ at the outlet of the device, where the exit area is 0.283 m². What is its average velocity at the outlet?

3.12 If the heat exchanger in Prob. 3.11 incurs a malfunction so that the gas does not change in density, what will be the exit velocity for the inlet conditions that were given?

3.13 Pure water enters a mixing tank steadily at the rate of 1.1 ft³/s. Sodium chloride is being added at the rate of 0.01 lb/s, and the two substances are thoroughly mixed before being discharged. What is the density of the solution that is being discharged at a constant rate?

3.14 Fresh water enters a mixing tank at the rate of 1.6 ft³/s, and a saline solution with a specific gravity of 1.025 is added at the rate of 0.5 ft³/s. The process is one of steady state. What must be the volume rate of discharge, and

Figure P.3.3

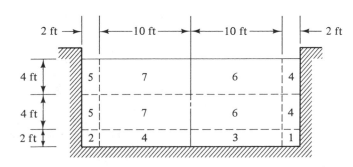

what would be the specific gravity of the discharged solution?

3.15 A gas mixture with a density of 0.67 kg/m³ flows steadily into a vessel with an inlet port of 0.196 m² at a velocity of 15 m/s. A flow meter at the outlet of the tank indicates a volume rate of flow of 4 m³/s. What is the discharge density?

3.16 A gas is moving in a circular passage whose cross-sectional area is increasing in the flow direction. At one section, the area, velocity, and density are 0.196 cm², 3.7 m/s, and 3.7 kg/m³, respectively. Downstream, the velocity is 468.2 m/s and the area is 0.257 cm². What is the density there?

3.17 Because of a change in downstream conditions in the flow nozzle described in Prob. 3.17, the velocity becomes 205 m/s while the area has remained the same. What is the density under these conditions?

3.18 A shock wave occurs in a nozzle in which a gas is flowing. This disturbance has a very minute thickness, so that the phenomenon occurs without change of cross-sectional area. Upstream of the shock, the velocity is 448 m/s and the density is 2.11 kg/m³; downstream, the density is 4.22 kg/m³. What is the velocity downstream of the shock wave?

3.19 A conduit, in which an incompressible oil flows at 2 m³/s, divides into a tube that is 800 mm in diameter and one that is 600 mm in diameter. If one can assume that the velocity in each of these sections is constant, determine what velocity is in the smaller tube, if in it the velocity is one-half that in the larger tube.

3.20 A tube that is 1000 mm in diameter is blocked by relatively thick plates in which there are 100 holes with a diameter of 20 mm. If the velocity in the unobstructed tube is 2 m/s, what is an estimate of the velocity of the water in these small holes?

3.21 Two streamlines are 5 mm apart at one section of a two-dimensional flow where the velocity is 0.6 m/s. Subsequently, the same streamlines are 7 mm apart. Estimate the velocity at a location midway between these streamlines.

3.22 A two-dimensional flow has radial streamlines directed toward the origin, where a sink

is located. At a distance of 11 cm, the radial velocity is 0.2 m/s and the streamline spacing is 1.8 cm. What should be the spacing of these same streamlines at a radius of 5 cm? What should be the velocity at a radius of 5 cm?

3.23 Draw a streamline for the velocity distribution given in Example 3.5.1. Let the value of the streamline constant be unity, and let A and B equal 2. Draw also the streamlines associated with $A = 2 - t$ and $B = 2$ for $t = 0, 1, 2, 4$; let the streamline constant equal 1.

3.24 Are streamlines, pathlines, or streaklines being indicated in the following experiments? (Note that sometimes they are all coincident; state where this is applicable.)
  (a) Smoke lines around a model of an automobile in a wind tunnel operating at constant speed
  (b) Smoke lines around the same model as in part (a) but during the time when the wind tunnel is changing speed.

3.25 In the cases below, what is being indicated by the flow visualization that is described? Does either represent a streamline? Explain.
  (a) The smoke from a smokestack of a ship at rest but with a steady breeze
  (b) The same situation as in part (a), but with the ship moving at constant speed

3.26 Refer to Example 3.5.5 and let $u = y$ and $v = x$. Determine the general equation for the associated streamlines. What is the equation for the streamline through $(2, 0)$? How much time will it take a particle going through $(2, 0)$ to reach $x = 2.02$? Is this a steady flow? Is the flow one-, two-, or three-dimensional?

3.27 If the slope of a streamline is $2x/y$, what is the equation of the streamlines? What are u and v at $x = 5$?

3.28 If, in Example 3.5.6, the values of A and B are $C_1 t^2$ and C_2, respectively, determine the streamlines, pathlines, and streaklines. Do they coincide?

3.29 Determine the streamlines for $u = xt$, $v = -yt$. What is the specific equation for the streamline passing through $(1, 2)$ at $t = 1$? Plot the path of the particle that was at this point at that time. (Note that a closed-form solution may not be apparent. Use graphic means, if necessary, employing small time steps.)

3.30 Refer to Fig. 3.7.2, and let V_3 have the same magnitude but opposite direction. Find V_2.

3.31 In Example 3.7.2, what relationship does mass conservation give between V_1, V_2, H, and H_2?

3.32 Refer to example 3.7.3, but instead of using Fig. 3.7.5a, let the coordinates of the intersections be $(0,0)$, $(0,1)$, $(2,1)$, and $(2,0)$. Find the flow across each of the four faces (unit width perpendicular to page). Is $\int_A \mathbf{V} \cdot \mathbf{n}\, dA = 0$?

3.33 Refer to Fig. 3.7.5a, but let $u = x^2 - 3$ and $v = -2xy$. Show that $\int_A \mathbf{V} \cdot \mathbf{n}\, dA = 0$.

3.34 Refer to Eq. (3.7.8), and evaluate $\int_0^X v\, dx$ for $u = U \sin(\pi y/2\delta)$. As $X \to 0$, $\delta \to 0$; does this give a satisfactory result for $\int_0^X v\, dx$?

3.35 Rewrite Eq. (3.7.8) to include a uniform suction velocity on the plate V_s. What must be the value of V_s if it is required that there be no flow across surface AB in Fig. 3.7.6?

3.36 The axial velocity w in a narrow annulus between two concentric cylinders of radii r_i and r_o can approximate $w/W = 1 - Y/(r_o - r_i)$ if the inner cylinder is moving axially with a speed W and the gap is $(r_o - r_i)$. Y is measured from the inner cylinder. Determine the volume rate of flow and the average velocity.

3.37 If the inner cylinder in Prob. 3.36 is rotated with speed Ω while the outer one is at rest, the tangential velocity V_θ in the annulus is $\Omega(r_i^2 r_o^2/r - r_i^2 r)/(r_o^2 - r_i^2)$ for sufficiently small Ω. What is the volume per unit time being transported around the annulus? (Use a unit length of cylinder.) Check to see that the given velocity profile does, indeed, give the desired fluid speeds at $r = r_i$ and r_o.

3.38 The tangential velocity distribution approximated in Fig. 1.7.4 is actually $v_\theta = Ar + B/r$. Determine the constants A and B using the fact that at $r = R_2$, $v_\theta = \Omega R_2$, and that at $r = R_1$, $v_\theta = 0$. Let $R_2 - R_1 = h$. Determine the volume rate of flow across a radial section $R_1 \le r \le R_2$. (Use a unit height in the axial direction.)

3.39 If, in Prob. 3.38, $h/R_1 \ll 1$, then the effects of the boundary's curvature might be negligible. Should this be the case, then we would expect that v_θ could be approximated by $\Omega R_2 y/h$, in which y is measured from the inner cylinder $(0 \le y \le h)$. What should be

the ratio h/R_1 for the curvilinear flow to be approximated adequately by the expression $v_\theta = \Omega R_2 y/h$? (*Hint:* Use a series expansion for v_θ in terms of h/R_1.)

3.40 The velocity in a two-dimensional channel of height b can be described by $u = Uy/b$, in which u is the component of velocity that is parallel to the stationary lower surface from which y is measured. The upper surface is moving in the $+x$ direction with U. (If U is negative, the upper plate is moving in the $-x$ direction.) Determine the total volume rate of flow and the average velocity. (Use a unit width perpendicular to the page.)

3.41 A horizontal, two-dimensional channel that is 10 mm high has fine reflecting particles suspended in it which give an indication of the velocity distribution. At distances of 1, 3, 5, 7, and 9 mm from the lower surface, the velocity component parallel to the surface is 2.11, 4.23, 6.29, 8.41, and 10.50 mm/s, respectively. What is the associated volume rate of flow? What is the average velocity?

3.42 If a pressure gradient exists between two horizontal, parallel plates separated by a distance $2b$, the velocity component parallel to the pressure gradient could be described by $u = (u_{max})[1 - (y/b)^2]$ for $-b \le y \le b$. Determine the related volume rate of flow and the average velocity. (Assume a unit width in the other horizontal direction.)

3.43 If a fine platinum wire is stretched vertically across a channel formed by two horizontal plates in which water flows, hydrogen bubbles could be formed through electrolysis. The movement of these bubbles could give the following values of velocity:

y, cm =	-0.9	-0.7	-0.5	-0.3	-0.1
u,cm/s =	0.181	0.523	0.742	0.824	0.987

y, cm =	$+0.1$	$+0.3$	$+0.5$	$+0.7$	$+0.9$
u,cm/s =	1.006	0.801	0.761	0.505	0.208

What is the associated volume rate of flow in a channel having a 2-cm distance between the plates so that $-1 \le y \le 1$? What is the average velocity?

3.44 A round pipe with a radius of 1.75 in. has an axial velocity distribution of $u = u_{max}[1 - (r/1.75)^2]$, in which u is in feet per second and r in inches. What is the volume rate of flow along the pipe? What is the average velocity?

3.45 In a tube with a diameter of 14 mm, the following axial velocities were measured at the corresponding radii: $u = 1.0$ cm/s, $r = 0$ mm; 0.9184, 2; 0.6735, 4; and 0.2653, 6. With these data, estimate the volume rate of flow. What is the average velocity in the tube?

3.46 If, instead of the velocity profile given in Example 3.3.1, one observed $w/w_{max} = (1 - r/r_0)^{1/7}$, what would be the volume rate of flow? Recall that through "integration by parts" one can write $\int x(1 + x)^n \, dx$ as $(1 + x)^{n+2}/(n + 2) - (1 + x)^{n+1}/(n + 1)$. What is the average value of the velocity for this velocity profile?

3.47 Measurements in an air duct with a diameter of 1.4 m show velocities of 5.00, 4.77, 4.43, and 3.79 m/s at radii of 0, 0.2, 0.4, and 0.6 m. What volume rate of flow do these readings imply? What is the average velocity?

3.48 The flow along a flat plate that is aligned with the wind will be such that the velocity component parallel to the plate will vary from zero at the plate to the speed of the fluid upstream of the plate in a relatively small distance that shall be called δ. (The magnitude of this distance perpendicular from the wall, δ, will depend on the distance from the leading edge, x.) One good approximation of the velocity distribution across BC in Fig. 3.7.6 is $u = U_\infty[2(y/\delta) - (y/\delta)^2]$. What is the flow across the surface BC? If the flow across OA is $U_\infty(\delta)(1)$, what must be the value of the flow across AB? Is this flow into or out of the control volume? Does this imply that the streamlines that cross AB have a positive or a negative slope?

3.49 Instead of a quadratic (or any polynomial) function of y/δ for the velocity profile over a flat plate, as in Prob. 3.48, one could use $u = U_\infty \sin(\pi y/2\delta)$ to good advantage. Use this function to determine the flow across BC and compare it with the value $\frac{2}{3}U_\infty \delta$, which is the result for the quadratic distribution, above. What is the average velocity across BC for the sine and quadratic functions?

3.50 If a fluid flows over a porous plate, the distribution of the velocity component parallel to the plate will depend upon the amount of fluid that is drawn through the porous surface. This effect will be evident in the value of δ^* in the expression $u = U_\infty[1 - \exp(-y/\delta^*)]$, in which U_∞ is the velocity parallel to the plate at large values of y, the distance perpendicular to the plate. (u will depend on x through δ^*.) Determine the volume rate of flow across BC for $\delta = 4\delta^*$ in Fig. 3.7.6. If the flow across OA is $(U_\infty)(4\delta^*)(1)$, what must be the flow across AB?

3.51 The velocities and distances in a direction perpendicular to a flat plate along BC in Fig. 3.7.6 are measured as

y, cm	0.2	0.4	0.6	0.8	1.0	1.2	1.4
u/U_∞	0.181	0.330	0.451	0.550	0.632	0.699	0.753

y, cm	1.6	1.8	2.0	2.4	2.8	3.2	4
u/U_∞	0.798	0.835	0.865	0.909	0.939	0.959	0.982

Determine the flow rate across BC for $\delta = 4$ cm and $U_\infty = 20$ m/s.

3.52 After a fluid passes over a flat plate, there is a "defect" in the velocity profile taken at downstream locations. This is due to the no-slip condition on the plate, an effect that disappears slowly downstream. If one measures y perpendicular to the plane of the plate, the following values could result from measurements taken approximately one plate length downstream of the plate's trailing edge:

y, cm	0	0.2	0.4	0.6	0.8	1.0
U, m/s	5.00	5.20	5.74	6.01	7.38	8.16

y, cm	1.2	1.4	1.6	1.8	2.0	2.2
U, m/s	8.81	9.24	9.61	9.80	9.91	9.96

What is the volume rate of flow across the height $2.2 \geq y \geq 0$? (Assume a unit width.)

3.53 The unsteady velocity field given by $u = 5 \exp(-at)$ and $v = 2 \exp(-at)$ has streamlines with a geometry that is independent of time. Describe them and the pathlines.

3.54 For the flow $u = 5 \exp(-2t)$ and $v = 2 \exp(-t)$, the streamlines change their slope with time. Describe the pathlines. Does the streakline through the origin coincide with a streamline or a pathline?

3.55 A flow with $u = 2 \exp(-at)$ and $v = 5[1 - \exp(-at)]$ has streamlines that are straight lines. Describe the flow at small and very large times. What are the pathlines? What

Mass Conservation   Chap. 3

can you say about the streakline through the origin?

3.56 A flow is specified by $u = \omega \sin \omega t$ and $v = \omega \cos \omega t$. What are the streamlines? What is the form of the pathlines? Define the pathlines for a particle that is at the origin at $t = 0$, a second that is at the origin at $\omega t = \pi/6$, a third that is there at $\omega t = \pi/4$, and a fourth at $\omega t = \pi/3$. Where are these particles at $\omega t = \pi/3$? Show that all these points are on a circle of unit radius whose center moves about the origin with a frequency of ω. Are the streamlines, pathlines, and streaklines coincident for this flow?

3.57 The flow in a pipe bend will not exhibit a velocity profile that is axisymmetric about the axis of the bend. Secondary flows induced in the bend will destroy the symmetry. To test a computer program that would take measured values of the velocity profile and integrate them to obtain the flow rate Q, the following velocity profile was devised (it does not satisfy the no-slip condition at the wall, but that is a minor blemish in testing the computer program):

$$u_{\text{axial}} = 1 + r(\cos\theta) \qquad 0 \le r \le 1$$

(Standard polar coordinates are used.)
(a) Plot u_{axial} against r for $\theta = 0, \pi/2, \pi, 3\pi/2$.
(b) Using $Q = \iint u_{\text{axial}} r (dr\, d\theta)$, determine Q. What is V_{avg}?
(c) Does the value of V_{avg} surprise you? Could you deduce its value by noting the symmetry, or lack thereof, of the velocity around $r = 0$?

3.58 (a) Show that, for the particles that have a path through the origin in Example 3.5.6, the pathlines are associated with the parametric equations

$$x = \frac{C_1}{2}(t^2 - \tau^2) \quad \text{and} \quad y = C_2(t - \tau)$$

in which τ is the time when the particle passes through the origin.
(b) What are the parametric equations for the particle that will pass through the ori-

gin at $t = 3$? Where is the particle at $t = 2$? Where would this particle be in Fig. 3.5.3? Where would it and the other three particles shown in Fig. 3.5.3 be located at $t = 3$? Draw the streakline through the origin at $t = 3$.

3.59 For the expanding chamber shown in Fig. P.3.59, an application of the equation for mass conservation to the indicated control volume yields

$$\frac{\partial \bar\rho}{\partial t} = \frac{-\bar\rho(12)}{L_0 + Vt}$$

Verify this result. If $L_0 = 0.15$ at $t = 0$ when $\bar\rho = 2.8$, solve for $\partial\bar\rho/\partial t$. Integrate the equation given to obtain $\bar\rho = \bar\rho(t)$.

Note also that the mass in the cylinder is constant, so that $L_0 A(2.8) = \text{mass} = (L_0 + 12t)A\bar\rho$. Does your integrated equation agree with this result?

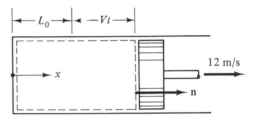

Figure P.3.59 $u = 12x$, with $x \le (L_0 + Vt)$; $\mathbf{V \cdot n} = 12$; $\rho(t)$ is uniform at $\bar\rho(t)$.

3.60 Water enters a horizontal two-dimensional slot, whose height is $2h$, with a uniform speed of U. At some distance downstream, the velocity is measured in the lower half of the slot as $u = U \sin(\pi y/2h)$, in which y is measured from the lower wall. (The velocity profile can be assumed to be symmetric about the centerline of the slot.) Has any water leaked into (or out of) the slot between the point where the water enters and the point where the velocity measurements are made? (A proper answer will include a magnitude and a sign.)

Problems

4

___

Kinematics of Flow

4.1 INTRODUCTION

Chapter 3 dealt with velocity and velocity variations without any concern about the forces that accompany the flows. Thus, aspects of the flow kinematics, but not the dynamics, have been discussed. Although we shall be concerned with the latter for most of the remainder of this book, the former will still be the focus of Chap. 4.

The determination of the values of the flow quantities (e.g., the velocity, density, temperature, or pressure) may be the end result desired; they might be found from a set of separate measurements made at particular locations within the flow field. We may also be interested in using these values to describe other quantities, such as acceleration or rates of diffusion. In this chapter we shall examine the ways of presenting information about measured quantities in the flow of a fluid. The material that we shall treat can be likened to some special tools that must be made in order that some larger task can be completed well and efficiently at a later date. Even though the concepts that will be discussed are not in themselves part of the subject of fluid mechanics, they are helpful, and often crucial, to understanding it.

When one attempts to model a particular flow, it is often convenient to construct a mathematical expression that produces the values of the appropriate physical quantity that is expected at the points of the flow field. These values must be in accordance with the physical principles that govern the instance under consideration. For example, we shall see that such mathematical results provide us with a means of estimating the stresses that are induced at various points in a flow. Consequently, one would be able to predict the acceleration of fluid particles at various positions because of the dependency of the acceleration on the local stress through Newton's laws of motion.

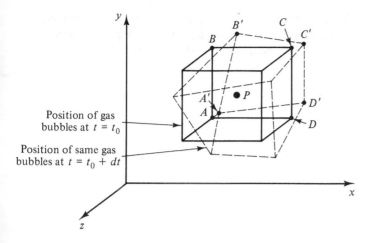

Position of gas bubbles at $t = t_0$

Position of same gas bubbles at $t = t_0 + dt$

Figure 4.2.1 Material volume in shape of a parallelepiped can be distorted as it moves in space. $ABCDA$ = position of gas bubbles at $t = t_0$. $A'B'C'D'A'$ = position of same bubbles at $t = t_0 + dt$.

4.2 THE CONCEPT OF ROTATION IN FLUID MECHANICS

If were able to delineate a small cube, as in Fig. 4.2.1, with fine platinum wires, and as some water flowed past point P an electric current were passed briefly through the wires, small gas bubbles would form on the wires. These would be swept off the wires by the flowing water, and a rhombohedron-like shape with edges made up of gas bubbles would be seen to be moving away from the location of point P. The reason the figure, which is originally a cube, would be distorted is that there is a velocity difference between various points in the original cube. This is the general case, and consequently, some parts of the cube's surface move farther than others in a brief interval of time. The result of this is that the cube is distorted in the flow.

This situation is schematically presented in Fig. 4.2.2, where only face $ABCD$ of Fig. 4.2.1 is pictured. If the fluid particles on line BC are moving relative to point P, then particle E in Fig. 4.2.2 has moved a distance in the x direction in time dt which is

Figure 4.2.2 Projections onto xy plane of relative displacements of a material volume. These displacements are termed "relative" because they exclude the displacements of the center of mass of the material volume. Consequently, they are relative to that point. P' = location of fluid particle at $t_0 + dt$ that was at point P at time t_0.

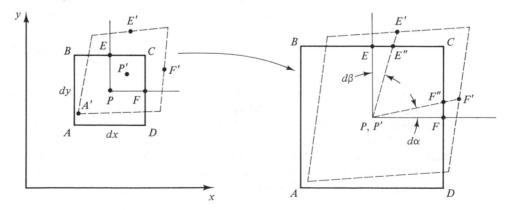

equal to $[u + (\partial u/\partial y)(dy/2)]\, dt$. Since the fluid that was originally at point P has moved a distance $u\,(dt)$, the relative x-directed displacement of the particle which was at point E, with respect to the particle which was originally at P, is the product of the relative velocity and the time,

$$\frac{\partial u}{\partial y}\frac{dy}{2}\, dt \approx EE''$$

(The error induced in using point E'' instead of E' vanishes as $dt \rightarrow 0$.) This relative displacement is shown in Fig. 4.2.2. We see that line $P'E''$ has rotated clockwise from PE. The small angle $d\beta$ is given by

$$\tan(d\beta) = \frac{EE''}{P'E} \approx d\beta \qquad \text{for small angles}$$

or

$$d\beta \approx \frac{\dfrac{\partial u}{\partial y}\dfrac{dy}{2}\, dt}{dy/2}$$

so that

$$\frac{(d\beta)_{\text{clockwise}}}{dt} = \frac{\partial u}{\partial y} \tag{4.2.1}$$

In the same way, because v varies from point P to point F, line PF in Fig. 4.2.2 rotates counterclockwise in relation to $P'F'$ at a rate

$$\frac{(d\alpha)_{\text{counterclockwise}}}{dt} = \frac{\partial v}{\partial x} \tag{4.2.2}$$

The average rate at which lines PE and PF are rotating in a counterclockwise direction is designated ω_z. The subscript z denotes an axis which is perpendicular to the xy plane. This definition of ω_z can be interpreted as the rate of rotation of the diagonal AC about point P in Fig. 4.2.2. Hence

$$\omega_z = \frac{1}{2}\left(\frac{\partial v}{\partial x} - \frac{\partial u}{\partial y}\right) \tag{4.2.3}$$

in which the minus sign on the second of the partial derivatives is necessary because positive values for $\partial u/\partial y$ would yield a clockwise rotation. A completely similar development would show that the rotation rates of fluid particles about the x and y directions are

$$\omega_x = \frac{1}{2}\left(\frac{\partial w}{\partial y} - \frac{\partial v}{\partial z}\right) \qquad \omega_y = \frac{1}{2}\left(\frac{\partial u}{\partial z} - \frac{\partial w}{\partial x}\right) \tag{4.2.4}$$

The rate of rotation is evaluated at a point in the fluid and could be nonzero for streamlines that are straight as well as for streamlines that are circles. In general, the rates of rotation of a fluid particle will not be zero. This is especially true of a viscous fluid, which is the usual case with which one is confronted. Nevertheless, in Chap. 5 we shall inquire into the consequence that follows if the rate of rotation of all of the particles in the region of flow is zero. This means that then the entire flow field would have the property that is called *irrotationality*.

4.3 ANGULAR AND VOLUME DEFORMATION WITH ASSOCIATED CONSEQUENCES

Equations (4.2.1) and (4.2.2) showed, in conjunction with Fig. 4.2.2, the manner in which one could calculate the angular rotation of a particle about its own axis. We shall now examine briefly some additional consequences of having a velocity field with nonzero spatial velocity derivates such as $\partial u/\partial x$ and $\partial u/\partial y$.

We can assess the amount of *angular deformation,* designated as *d,* by referring to Fig. 4.2.2 and Eqs. (4.2.1) and (4.2.2) again. The sum of the clockwise rotation rate of $d\beta$ and the counterclockwise rotation rate of $d\alpha$ is the rate at which the right angle *EPF* in Fig. 4.2.2 is decreasing. Hence

$$\text{Rate of angular strain} = (\text{constant})\left(\frac{\partial v}{\partial x} + \frac{\partial u}{\partial y}\right) \tag{4.3.1}$$

This angular deformation can be useful to us. If we *assume* a linear coupling between this rate of angular deformation and the magnitude of an induced shear stress, we would write

$$\tau = \text{shear stress} = \mu\left(\frac{\partial v}{\partial x} + \frac{\partial u}{\partial y}\right) \tag{4.3.2}$$

in which μ is a constant of proportionality, called the *coefficient of viscosity* (or *absolute viscosity*). If the flow does not have a v component, or $\partial v/\partial x = 0$, Eq. (4.3.2) reduces to the expression for Newton's law of viscosity that was introduced in Chap. 1. Equation (4.3.2) is for a two-dimensional flow, and additional terms would result for three-dimensional cases. Equation (4.3.2) and related formulas will be treated in detail in Chap. 16.

We now turn to the part of the relative velocity that is related to the *rate of volume expansion.* Figure 4.3.1 will be useful in this discussion. The x-directed change in the length of a packet of fluid, a material volume, that is moving with $\mathbf{V}$ at its center will be $(\partial u/\partial x)(dx/2)(dt)(2)$. This is caused by the relative velocities at the extremes of the packet; the last factor of 2 is associated with the fact that the extension over the distance $(+dx/2)$ is matched by that over the distance $(-dx/2)$. This implies that as the movement of side *CD* in Fig. 4.3.1 occurred in the x direction, a volume was also being displaced by the surface whose trace is *CD* in Fig. 4.3.1. The altitude of the displaced region is $(\partial u/\partial x)(dx\, dt)$, and the area is the side of the parallelepiped that is perpendicular to the x axis. This is $dy\, dz$, as Fig. 4.3.1 illustrates. Hence, during the interval dt, the volume displaced by surface *CD* and *AB* is $(\partial u/\partial x)(dx)(dy)(dz)(dt)$. This volume change is due to the u variation with x. Similarly, the amount of volume change of the infinitesimal volume, $d\forall = dx\, dy\, dz$, due to the v variation in the y direction is given by the quantity $(\partial v/\partial y)(dy)\, dt\, dx\, dz$.

(Note that $\partial v/\partial y$ changes in value over the surface of the element under consideration, and the value chosen to be used in the calculation is that which occurs at the center of the region of interest. This is in agreement with our notions of linear changes for variations from the reference point if the system is linear. Furthermore, we expect that there will be first-order contributions to the changes in the value of a function, even though it is not linear, if the variation of the independent variables is kept infinitesimal. We would expect this by considering a Taylor series expansion.)

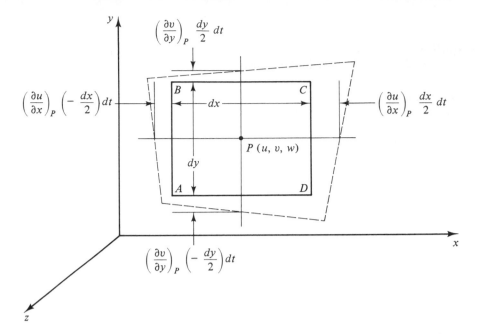

Figure 4.3.1 Change of size of material volume from which the rate of volume change can be deduced. $ABCD$: Size of packet cross section at $t = 0$. Size at $t_0 + dt$ indicated by dashed lines. Positions of packet superimposed to show relative motions only.

Thus, for the three-dimensional case associated with Fig. 4.3.1, we see that the change in the material volume $d\Psi$ in the time interval dt will be

$$(\Delta\Psi \text{ in } dt) = \left(\frac{\partial u}{\partial x} dx\, dy\, dz + \frac{\partial v}{\partial y} dy\, dx\, dz + \frac{\partial w}{\partial z} dz\, dx\, dy \right) dt$$

if the volume change due to the variation of w in the z direction is added to the other two terms just found. This result leads us to the determination of the time rate of change of the volume, $d\Psi/dt$; this would be

$$\frac{d\Psi}{dt} = \left(\frac{\partial u}{\partial x} + \frac{\partial v}{\partial y} + \frac{\partial w}{\partial z} \right) dx\, dy\, dz \qquad (4.3.3)$$

If we are concerned with fluids that are *incompressible,* we would be logical in saying that a collection of fluid molecules, a material volume, should not change its volume. We would demand that $d\Psi/dt = 0$. Equation (4.3.3) would then require that

$$\frac{\partial u}{\partial x} + \frac{\partial v}{\partial y} + \frac{\partial w}{\partial z} = 0 \qquad (4.3.4)$$

for such an incompressible fluid, since $dx\, dy\, dz = d\Psi$ is arbitrarily small, but not zero. Equation (4.3.4) is called the *continuity equation* in fluid mechanics. It requires that the components of velocity be interdependent if volume changes are to be avoided.

4.4 DEFORMATION AND ROTATION

We shall now formalize the descriptive discussions of Secs. 4.2 and 4.3 by writing some definitions and related equations. They will aid us in finding the local angular velocity and angular deformation of the fluid. The relative velocity between two neighboring points has components in the x, y, and z directions that are written as du, dv, and dw. If one is considering a particular instant of time, $dt = 0$, there are only spatial variations of the velocity components. The differences in the velocity components at two points separated by dx, dy, and dz are

$$du = \frac{\partial u}{\partial x} dx + \frac{\partial u}{\partial y} dy + \frac{\partial u}{\partial z} dz$$

$$dv = \frac{\partial v}{\partial x} dx + \frac{\partial v}{\partial y} dy + \frac{\partial v}{\partial z} dz \qquad (4.4.1)$$

and

$$dw = \frac{\partial w}{\partial x} dx + \frac{\partial w}{\partial y} dy + \frac{\partial w}{\partial z} dz$$

The terms on the right-hand side of these equations will now be rearranged, without changing their value. After doing this we shall discover that we have expressed the relative velocity components in terms of angular velocity and deformation, as well as volume distortion.

In each of these equations, one can add and then subtract a specific pair of terms—the terms designated Ia and Ib—in a formal way to obtain

$$du = \frac{1}{2}\underbrace{\left(\frac{\partial u}{\partial x} + \frac{\partial u}{\partial x}\right)}_{\text{II}} dx + \frac{1}{2}\underbrace{\left(\frac{\partial u}{\partial y} + \frac{\partial v}{\partial x}\right)}_{\text{III}} dy \overset{\text{Ia}}{-} \frac{1}{2}\left(\frac{\partial v}{\partial x} - \frac{\partial u}{\partial y}\right) dy$$

$$+ \frac{1}{2}\underbrace{\left(\frac{\partial u}{\partial z} + \frac{\partial w}{\partial x}\right)}_{\text{III}} dz \overset{\text{Ib}}{-} \frac{1}{2}\left(\frac{\partial w}{\partial x} - \frac{\partial u}{\partial z}\right) dz$$

$$dv = \frac{1}{2}\underbrace{\left(\frac{\partial v}{\partial x} + \frac{\partial u}{\partial y}\right)}_{\text{III}} dx \overset{\text{Ia}}{+} \frac{1}{2}\left(\frac{\partial v}{\partial x} - \frac{\partial u}{\partial y}\right) dx + \frac{1}{2}\underbrace{\left(\frac{\partial v}{\partial y} + \frac{\partial v}{\partial y}\right)}_{\text{II}} dy$$

$$+ \frac{1}{2}\underbrace{\left(\frac{\partial v}{\partial z} + \frac{\partial w}{\partial y}\right)}_{\text{III}} dz \overset{\text{Ib}}{-} \frac{1}{2}\left(\frac{\partial w}{\partial y} - \frac{\partial v}{\partial z}\right) dz \qquad (4.4.2)$$

and

$$dw = \frac{1}{2}\underbrace{\left(\frac{\partial w}{\partial x} + \frac{\partial u}{\partial z}\right)}_{\text{III}} dx \overset{\text{Ia}}{-} \frac{1}{2}\left(\frac{\partial u}{\partial z} - \frac{\partial w}{\partial x}\right) dx + \frac{1}{2}\underbrace{\left(\frac{\partial w}{\partial y} + \frac{\partial v}{\partial z}\right)}_{\text{III}} dy$$

$$+ \frac{1}{2}\left(\frac{\partial w}{\partial y} - \frac{\partial v}{\partial z}\right) dy + \frac{1}{2}\underbrace{\left(\frac{\partial w}{\partial z} + \frac{\partial w}{\partial z}\right)}_{\text{II}} dz$$

This subdivision of du, dv, and dw in this fashion could be purely arbitrary, but there is a good reason for it. The partitioning has yielded various terms that have important physical meanings. (The terms designated II have been interpreted in Sec. 4.3 to mean the partial time rates of volume change of a material volume that is located at a specified point. The terms designated III are the time rates of angular distortion—in planes perpendicular to the x, y, and z axes—of the material volume. We saw that, too, in Sec. 4.3. The terms not labeled either II or III are the time rates of angular rotation of the fluid particles about their own axes. This interpretation was introduced in Sec. 4.2.) We shall shortly give an example showing how to calculate these terms and how to visualize them. However, it is worth saying in passing that the three parts of Eq. (4.4.2) are a statement of the kinematic rule that any relative velocity field can be decomposed so as to represent the rates of volume change, angular deformation, and rotation.

This correspondence between the elements of the relative velocity, $d\mathbf{V} = (du)\mathbf{i} + (dv)\mathbf{j} + (dw)\mathbf{k}$, can be established in general; however, it will be done here for the limited case of two-dimensional flow. This will simplify the presentation. We prescribe that there are no variations in the z direction and that w is constant. Consequently, the partial derivative of any variable with respect to z is zero. Also, the derivative of w with respect to any coordinate is zero. Then the equations of set (4.4.2) reduce to

$$du = \frac{1}{2}\underbrace{\left(\frac{\partial u}{\partial x} + \frac{\partial u}{\partial x}\right)}_{\text{II}} dx + \frac{1}{2}\underbrace{\left(\frac{\partial u}{\partial y} + \frac{\partial v}{\partial x}\right)}_{\text{III}} dy - \frac{1}{2}\left(\frac{\partial v}{\partial x} - \frac{\partial u}{\partial y}\right) dy$$

(4.4.3)

$$dv = \frac{1}{2}\underbrace{\left(\frac{\partial v}{\partial x} + \frac{\partial u}{\partial y}\right)}_{\text{III}} dx + \frac{1}{2}\underbrace{\left(\frac{\partial v}{\partial y} + \frac{\partial v}{\partial y}\right)}_{\text{II}} dy + \frac{1}{2}\left(\frac{\partial v}{\partial x} - \frac{\partial u}{\partial y}\right) dx$$

and, of course,

$$dw = 0$$

in which the terms labeled II and III have the same meaning as before. This result has four unique terms that are denoted as

$$\bar{e} = \frac{1}{2}\left(\frac{\partial u}{\partial x} + \frac{\partial u}{\partial x}\right) \qquad \tilde{e} = \frac{1}{2}\left(\frac{\partial v}{\partial y} + \frac{\partial v}{\partial y}\right)$$

$$d = \frac{1}{2}\left(\frac{\partial v}{\partial x} + \frac{\partial u}{\partial y}\right)$$

(4.4.4)

and

$$\omega = \frac{1}{2}\left(\frac{\partial v}{\partial x} - \frac{\partial u}{\partial y}\right)$$

(4.4.5)

Finally, one lets

$$e = \bar{e} + \tilde{e} = \frac{\partial u}{\partial x} + \frac{\partial v}{\partial y}$$

(4.4.6)

With these definitions, the expressions labeled as Eq. (4.4.3) can be written as

$$du = (\bar{e})\, dx + (d)\, dy - (\omega)\, dy$$

$$dv = (\tilde{e})\, dy + (d)\, dx + (\omega)\, dx$$

(4.4.7)

The terms $\bar{e}$ and $\tilde{e}$ are the time rates of *volume strain* in the x and y directions, respectively, for our two-dimensional model, just as in Sec. 4.3. The terms d and ω correspond to the time rate of *angular deformation* of a fluid particle and the time rate of *rotation* of a particle as it "tumbles" while moving along its path (not necessarily circular). The following example is intended to illuminate the concepts of volume strain, angular deformation, and rotation.

Example 4.4.1

The decomposition of relative velocity components into rotation, angular deformation, and volume deformation terms will be done and interpreted pictorially for the velocity distribution

$$\mathbf{V} = (-2y + 7x)\mathbf{i} + (-7y + 8x)\mathbf{j}$$

Hence, $u = -2y + 7x$; $v = -7y + 8x$.

	A	B	C	D
u	$\dfrac{-7}{2}$	-1	$\dfrac{7}{2}$	1
v	-4	$\dfrac{-7}{2}$	4	$\dfrac{7}{2}$

The values of these components are given in the accompanying table for the points $A = (-\frac{1}{2}, 0)$, $B = (0, \frac{1}{2})$, $C = (\frac{1}{2}, 0)$, and $D = (0, -\frac{1}{2})$. These values are obtained from the expressions for u and v, which also yield

$$\omega = \frac{1}{2}\left(\frac{\partial v}{\partial x} - \frac{\partial u}{\partial y}\right) = \frac{1}{2}(8 + 2) = 5$$

$$d = \frac{1}{2}\left(\frac{\partial v}{\partial x} + \frac{\partial u}{\partial y}\right) = \frac{1}{2}(8 - 2) = 3$$

ω can be used to determine the portion of the relative velocity associated with particle rotation, and d is the average rate at which an angle drawn within the particle is changing. Because in this example the velocity components vary linearly with position, we shall have values for $\bar{e}$, $\tilde{e}$, d, and ω that are constants for the entire xy plane. This will mean that the interpretation for the terms in Eqs. (4.4.4) to (4.4.7) can be conveniently explored over finite distances even though we *usually* think of $\bar{e}$, $\tilde{e}$, d, and ω as being associated with a point and the infinitesimal region around it. (In such a minute region, the velocity components will vary approximately linearly with position, regardless of the mathematical form of these components. This is in concert with the concepts inherent in the calculus.) Around the origin, we have chosen several points and calculated the velocity there. Because $\mathbf{V}(0, 0) = 0$, we can view the velocities at, say, A, B, C, and D as relative velocities with respect to the origin. We shall subdivide du into u_{ed}, u_{ad}, and u_{ro}, in which the subscripts denote the rates of *elongation deformation, angular deformation, and rotation.* * Hence, from Eq. (4.4.7),

$$[u_{\mathrm{ed}}, u_{\mathrm{ad}}, u_{\mathrm{ro}}] = [(\bar{e})\,dx, (d)\,dy, -(\omega)\,dy]$$

A similar notation and correspondence will be applied to v. Figure 4.4.1a gives the location of the points. At B, for example, $u_B = -1 = 0 + \frac{3}{2} - \frac{5}{2}$. The reason for the subdivision of u_B in these proportions to give $u_{\mathrm{ed}} = 0$, and so on, should become apparent after a thorough examination of all of the drawings in Fig. 4.4.1 in conjunction with the text below.

The velocity component depicted in Fig. 4.4.1b is one associated with rigid-body rotation. Here one has $u_{\mathrm{ro}} = -\omega y$ and $v_{\mathrm{ro}} = \omega x$. In Fig. 4.4.1c, the axis AC would be tending to rotate in a counterclockwise direction about O, the center of the particle, while BD would be rotating clockwise. The rate at which the angle would be decreasing is $\dot{\theta}$, which has been defined as $2d$ in Fig. 4.4.1c.

From this we can see that the value of d will tell us how rapidly a square, in this two-dimensional example, of fluid particles will be evolving into a rhombus, or a parallelogram in general. (It is this distortion that is associated with the generation of shear stresses.)

* The subscripts are abbreviations and do not denote components in a coordinate system.

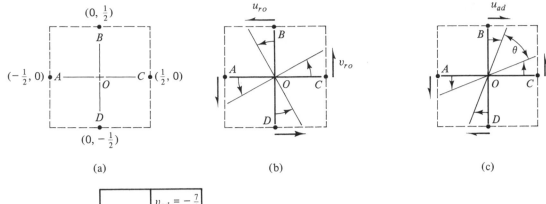

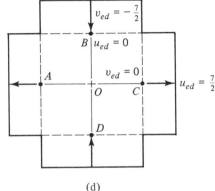

Figure 4.4.1 (a) Location of reference points for a material volume. (b) Rotational components of relative velocity $\mathbf{V}_{ro}$ for $\mathbf{V} = (-2y + 7x)\mathbf{i} + (-7y + 8x)\mathbf{j}$. $u_{ro} = -\frac{5}{2}$; at B, $u_{ro} = -\omega y$. At C, $v_{ro} = \frac{5}{2}$, $v_{ro} = \omega x$. (c) Components of relative velocity yielding angular deformation $\mathbf{V}_{ad}$ for this example. At A, B, C, or D, $u_{ad} = u - u_{ro}$ and $v_{ad} = v - v_{ro}$. $u_{ad} = \frac{3}{2}$ at B from $d = \dot{\theta}/2 = \frac{1}{2}(v_{ad}/x + u_{ad}/y)$ and $x = y = \frac{1}{2}$ with $v_{ad} = \frac{3}{2}$ at C. (d) Components of relative velocity associated with elongational deformation $\mathbf{V}_{ed}$ for this particular case. Note that u_{ed} and v_{ed} are the "average" values.

Figure 4.4.1d shows that at point B there is a v_{ed} that is negative; the line BD is contracting. On the other hand, AC is expanding. Both rates have the same magnitude. This is as it should be for an incompressible fluid so that volume is conserved.

It was mentioned earlier that our velocity field is linear, and this causes the associated derivatives to be constants. This means, for example, that while generally $du = (\bar{e}) dx + (d) dy - (\omega) dy$, from Eq. (4.4.7), we can simplify matters in our case because the coefficients of dx and dy are constants. We can use finite x's and y's and write, after recalling that $\mathbf{V}(0, 0) = 0$,

$$u_{\text{relative to origin}} = u$$

$$= (\bar{e})x + (d)y - (\omega)y$$

$$= u_{ed} + u_{ad} + u_{ro}$$

The reader should check that the numbers in Fig. 4.4.1 agree with this result.

4.5 INTRODUCTION TO ACCELERATION

A description of flow that concentrates on the motion of a particle (i.e., a Lagrangian description) permits a direct determination of the acceleration of that particle by differentiating expressions, as was shown in the text following Eq. (3.2.2). Differentiation with respect to time does not have the same significance (i.e., acceleration of a particle) if the description of the flow is Eulerian [spatial; see Eqs. (3.2.3) to (3.2.5)]. Then something else must be done, as we now shall see.

Using a particular example, we shall examine the concept of the acceleration of a fluid particle in a region where the flow is considered to be steady. Our example is the steady flow of compressible gas in a converging-diverging nozzle, as shown in Fig. 4.5.1. The equations that result from Newton's laws and assumed relationships between pressure and density can be used to predict the velocity, pressure, and density at various points of the nozzle, once its geometry is specified. These equations will be developed in Chap. 14. However, for now we shall present some results to provide the motivation and data for the development at hand. These data are indicated in Fig. 4.5.1. They are the velocity component in the x direction and the density.

From Fig. 4.5.1, it can be seen that the density changes as one observes different points along the nozzle's axis. The axial speed u changes also, but the flow is steady. If one were to measure the density, or any other flow parameter, at an arbitrary point in the nozzle, it has been assumed here that the value would not vary with time.

However, let us now concentrate on the observer who cannot look at various, fixed points in the nozzle. This observer is required to move with the flow. (We are adopting a Lagrangian point of view for a brief instant.) Consider the fluid particle that is at point A in the nozzle at some time t_0. After a short interval of time, the particle is at point B.

Example 4.5.1

Estimate the amount of time necessary for a particle to move from point A to point B in the nozzle.

Solution

$$\Delta t = \frac{\Delta x}{u_{avg}}$$

$$= \frac{1.5 - 1.0}{(447 + 334)/2} = \frac{1}{781} \text{ s}$$

Figure 4.5.1 Velocities and densities of a gas at various positions in a nozzle from which the time rate of change of a material volume's density can be deduced. x = axial position along nozzle, ft; ρ = density, slugs/ft^3; u = velocity, ft/s.

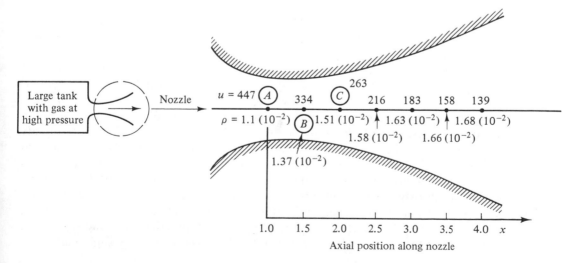

As $\Delta x \rightarrow 0$, the time interval becomes the distance traveled by the particle, divided by the associated local speed, $\Delta x/u$.

In going from point A to point B, the density of a small material volume has also changed. At point A it was ρ_A, and at point B the density of the material volume—air, for example—has become ρ_B. This means that in the time necessary for the particles to go from point A to B, the density has changed from ρ_A to ρ_B, or $\Delta\rho$ in Δt. This implies a $\Delta\rho$ in the time interval $\Delta x/u$. Hence, we can show that the time rate of change of the density of the material volume that has gone from A to B is $[u(d\rho/dx)]_A$ at point A. (The total differential is used here because ρ is assumed to vary only in the axial direction, x. The subscript A defines the point at which u and $d\rho/dx$ are evaluated. A derivative has been written instead of a finite difference because the associated functional relationships are assumed to exist.) Equally well we could show that the time rate of change of the axial speed of the particle that has gone from A to B is $u(du/dx)$. This, of course, implies that the acceleration in the axial direction of the nozzle, at point A, is $[u(du/dx)]_A$.

One should pause and reflect upon what has just been developed. In the simple one-dimensional situation (i.e., where changes occur only with respect to the axial direction, the x coordinate), we have used information that can be measured at geometric points in the region of flow (e.g., u and du/dx) to make statements about the acceleration of a particle of fluid that is passing through the geometric location in question. Usually one uses information at fixed points in space in fluid mechanics for use with the appropriate laws of nature. This is the reason why the development of an expression for acceleration when using a Eulerian description of the flow is important.

4.6 ACCELERATION AND THE SUBSTANTIAL DERIVATIVE

A natural extension of the previous discussion gives the general expression for the differences in the x component of velocity by using the total differential. This is

$$du = \frac{\partial u}{\partial x}dx + \frac{\partial u}{\partial y}dy + \frac{\partial u}{\partial z}dz + \frac{\partial u}{\partial t}dt \qquad (4.6.1)$$

Note that the last term has been included to allow the possibility of an unsteady flow. This term was excluded in Eq. (4.4.1), in which a specific instant of time, $dt = 0$, was considered.

This equation states that the x component of velocity can vary as dx, dy, dz, and dt change, and the rates of change—$\partial u/\partial y$, $\partial u/\partial t$, and so on—will have to be specified. Then the difference in u between any two points that are separated by dx, dy, and dz for two closely spaced instants of time dt can be found from Eq. (4.6.1) by using the appropriate distances and time interval. Although such a calculation can be readily completed, it would be less useful than the one we shall now initiate. To do this, we will not choose dx, dy, and dz in an arbitrary way. Instead, we choose two points (x_o, y_o, z_o) and (x_f, y_f, z_f) that are separated by the appropriate distances dx, dy, and dz but have a special characteristic: these points are the original and final locations of a particle that was at (x_o, y_o, z_o) at the beginning of the time interval dt and at (x_f, y_f, z_f) at the end of that time interval. The point (x_f, y_f, z_f) is one of a multitude that surround (x_o, y_o, z_o), but in this case the distance $ds = \sqrt{dx^2 + dy^2 + dz^2}$ was traversed by some particle in

dt. If u, v, and w are the velocity components of the particle, specified at some place intermediate to the two points in question, the x displacement of the particle in dt is

$$dx = u\,dt \tag{4.6.2}$$

Similarly, the other displacements are

$$dy = v\,dt \tag{4.6.3}$$

and

$$dz = w\,dt \tag{4.6.4}$$

These special distances—displacements associated with a particle from the reference point (x_o, y_o, z_o)—give the differential velocity du associated with these displacements a unique meaning. Thus we can write

$$du = \frac{\partial u}{\partial x}(u\,dt) + \frac{\partial u}{\partial y}(v\,dt) + \frac{\partial u}{\partial z}(w\,dt) + \frac{\partial u}{\partial t}dt \tag{4.6.5}$$

This du is now, by virtue of choice of dx, dy, and dz, the difference of velocity that some particle experienced during dt when it traversed the distance between (x_o, y_o, z_o) and (x_f, y_f, z_f). Thus, dividing through by dt in Eq. (4.6.5) gives the *time rate of change of u for this particle*; this is

$$\frac{du}{dt} = u\frac{\partial u}{\partial x} + v\frac{\partial u}{\partial y} + w\frac{\partial u}{\partial z} + \frac{\partial u}{\partial t} \tag{4.6.6a}$$

This defines du/dt as the x component of acceleration of the particle that passed through (x_o, y_o, z_o), the place where u, v, w, and the appropriate derivatives are evaluated. This means that Eq. (4.6.6a) defines a_x at x_o, y_o, z_o. This equation is a generalization of the specific result appropriate to Fig. 4.5.1. Because of the special values of du, dv, and dw given by Eqs. (4.6.2) to (4.6.4) that are associated with the displacement of a particle, the collection of terms on the right-hand side of Eq. (4.6.6a) is called the *substantial derivative of u* and is abbreviated by Du/Dt. Hence,

$$\frac{Du}{Dt} = u\frac{\partial u}{\partial x} + v\frac{\partial u}{\partial y} + w\frac{\partial u}{\partial z} + \frac{\partial u}{\partial t} \tag{4.6.6b}$$

We can apply the same reasoning that led to Eq. (4.6.6a) to verify that the a_y and a_z that exist at x_o, y_o, z_o can be written as

$$\frac{Dv}{Dt} = u\frac{\partial v}{\partial x} + v\frac{\partial v}{\partial y} + w\frac{\partial v}{\partial z} + \frac{\partial v}{\partial t} \tag{4.6.7}$$

and

$$\frac{Dw}{Dt} = u\frac{\partial w}{\partial x} + v\frac{\partial w}{\partial y} + w\frac{\partial w}{\partial z} + \frac{\partial w}{\partial t} \tag{4.6.8}$$

(What would be the expression for the substantial derivative of ρ? What would be its physical interpretation?)

Example 4.6.1

For the velocity distribution $u = 5x$, $v = -5y$, and $w = 0$, determine the associated accelerations.

Solution One has, from Eqs. (4.6.6b), (4.6.7), and (4.6.8),

$$a_x = \frac{Du}{Dt} = (5x)(5) + (-5y)(0) + 0(0) + 0$$

$$a_y = \frac{Dv}{Dt} = 5x(0) + (-5y)(-5) + 0(0) + 0$$

$$a_z = 0$$

Hence
$$|\mathbf{a}| = \sqrt{(-25x)^2 + (25y)^2} = 25\sqrt{x^2 + y^2}$$

The direction of this acceleration is given by the angle θ_{acc},

$$\theta_{acc} = \arctan\frac{a_y}{a_x}$$

The angle of the velocity, with respect to the x axis, is

$$\theta_{vel} = \arctan\frac{v}{u}$$

The use of polar coordinates in a particular instance would require one additional consideration in the determination of velocity components: the effect on the acceleration due to the change of direction of the particle. This leads to an additional contribution to the radial acceleration (i.e., either centrifugal or centripetal). However, if the acceleration, a vector, is known, one can readily transform it to other coordinate systems in a manner similar to the method that follows.

Example 4.6.2

The determination of the radial component of the acceleration of Example 4.6.1 at $(1, 2)$ is facilitated by Fig. 4.6.1. The r and θ components of acceleration can be written as

$$a_r = a_x \cos\theta + a_y \sin\theta \tag{4.6.9}$$

and
$$a_\theta = -a_x \sin\theta + a_y \cos\theta \tag{4.6.10}$$

We could now use the a_x and a_y from Example 4.6.1 and solve for a_r and a_θ at $(x, y) = (1, 2)$ if we wished.

A development of radial and angular acceleration from the basic principles of dynamics would give

$$a_r = \frac{\partial u_r}{\partial t} + u_r\frac{\partial u_r}{\partial r} + \frac{u_\theta}{r}\frac{\partial u_r}{\partial \theta} - \frac{u_\theta^2}{r} \tag{4.6.11}$$

and

$$a_\theta = \frac{\partial u_\theta}{\partial t} + u_r\frac{\partial u_\theta}{\partial r} + \frac{u_\theta}{r}\frac{\partial u_\theta}{\partial \theta} + \frac{u_r u_\theta}{r} \tag{4.6.12}$$

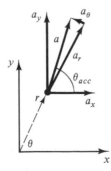

Figure 4.6.1 Decomposition of acceleration into components.

Kinematics of Flow   Chap. 4

for a flow in cylindrical polar coordinates that has $w = 0$ and no z dependence. [Do the last terms in Eqs. (4.6.11) and (4.6.12) represent centrifugal and Coriolis contributions, respectively?]

We shall use the concepts and the relationship that were developed in this section extensively in the chapters that follow. Now, however, we turn to another application of the ideas of a total differential and a point function.

4.7 STREAM FUNCTIONS AND THEIR USE

A streamline was defined as one of the lines which can be drawn tangent to a set of velocity vectors in a region where a fluid is flowing. A number of examples have been given for which the streamlines could be found from the velocity field. We shall build upon this information now. First of all, we shall *assume* that we know the form of a number of streamlines in a particular situation to provide a point of departure for the discussion. A number of representative streamlines are shown in Fig. 4.7.1 for a two-dimensional flow under a barrier that is called a *sluice gate*. The streamlines have been numbered from the bottom to the top at the left of the figure. The velocity at the upstream location is assumed to be independent of height for this example. Because the streamlines that are shown have been drawn equispaced in this constant-velocity region, the volume flow rate is the same between any two adjacent streamlines. We shall call the total flow Q, and we could then designate the eleven streamlines as $(0.0)Q$, $(0.1)Q$, $(0.2)Q$, and so on, in this particular situation. Such a designation would be applicable along the entire length of the streamline because there is no flow across the streamlines, by their definition. Each of these numbers is a constant that identifies a streamline, and it is usual to call it ψ. The streamlines for the flow under a sluice gate have been redesignated with ψ_i in Fig. 4.7.2, to be more general; the subscripts on ψ are used as a reminder that the value varies from streamline to streamline. In Fig. 4.7.2, a portion of the flow field where the velocity is not uniform has been enlarged. In this enlargement two neighboring streamlines have been de-

Figure 4.7.1 Two-dimensional flow of a liquid under a sluice gate showing streamlines.

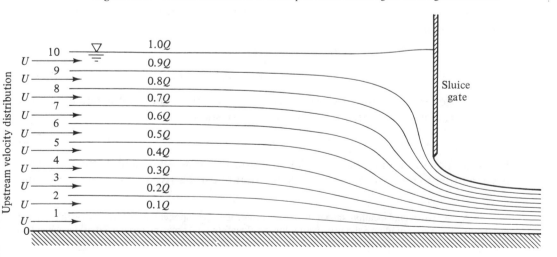

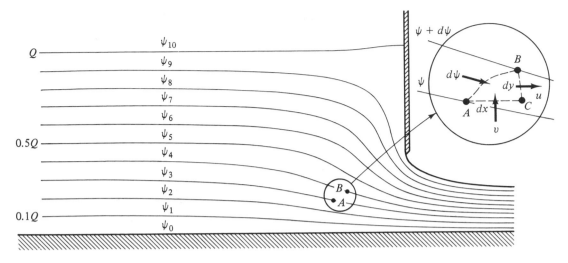

Figure 4.7.2 Flow between two streamlines, ψ and $\psi + d\psi$, with the path between two points A and B on these streamlines that is prescribed by Cartesian coordinates.

picted. One has been marked ψ, and the other, $\psi + d\psi$, so that the flow between the two streamlines is $(\psi + d\psi) - \psi$, or $d\psi$. This is the flow rate passing across the boundary AB of the infinitesimal control volume ABC. For the particular figure that has been drawn, this flow rate is entering the control volume; hence, the flow rate leaving the control volume across AB is $-d\psi$. Next, we shall examine the flow rates across the other two surfaces of our control volume to obtain an important equality. If at the point of our interest the velocity is $\mathbf{V} = u\mathbf{i} + v\mathbf{j}$, with u and v not restricted concerning their algebraic sign, the flow out of the control volume would be obtained by first taking $\mathbf{V} \cdot \mathbf{n}$ on the particular surfaces and then multiplying by the appropriate areas. In the present case, this yields

$$\text{Flow rate out* of control volume across } AC = -v \, dx \, (1)$$
$$\text{Flow rate out of control volume across } CB = +u \, dy \, (1)$$

The sum of all the flow rates—out of all the parts of the control surface—must be zero for this incompressible flow, so that

$$-d\psi - v(dx) + u(dy) = 0$$

or
$$d\psi = -v(dx) + u(dy) \tag{4.7.1}$$

This equation is important, as we shall shortly realize. It is a differential equation for ψ and tells us that along a streamline, where ψ is constant,

$$0 = -v \, dx + u \, dy \tag{4.7.2}$$

From this equation, one can deduce the relationship between dx and dy that one must have in order to move from one point to another but still arrive on the same streamline

---

* At the particular location shown in our example, the y component of velocity, v, would have a negative value. This would give the proper sign for the flow rate out, according to our intuition, when this negative number would be inserted in the right-hand side of the equation.

after the change of position. We can show that Eq. (4.7.2) gives the slope of a streamline, the value v/u. This is also the slope of the velocity $\mathbf{V} = u\mathbf{i} + v\mathbf{j}$. This is as it should be, according to the definition of a streamline. This is the reason why the curves for a constant ψ are called *streamlines*. These curves of ψ coincide with the streamlines. The function ψ is usually called the *stream function*. Equation (4.7.1) is even more useful once we remember that we are prescribing that the streamlines are determined by some, as yet unspecified, function of the coordinates of space. Because $\psi = \psi(x, y)$, we can write the total differential* as

$$d\psi = \frac{\partial \psi}{\partial x} dx + \frac{\partial \psi}{\partial y} dy \tag{4.7.3}$$

If the coefficients of dx and dy are compared in Eqs. (4.7.1) and (4.7.3), we find that

$$u = \frac{\partial \psi}{\partial y} \tag{4.7.4}$$

and

$$v = -\frac{\partial \psi}{\partial x} \tag{4.7.5}$$

These relationships between the velocity components and the partial derivatives of ψ were our goal. Now we shall utilize these results.

Example 4.7.1

The stream function

$$\psi = Uy \left(1 - \frac{a^2}{x^2 + y^2} \right) \tag{4.7.6}$$

will be examined to find its properties.

Solution The streamline $\psi = 0$ is easy to describe. According to Eq. (4.7.6), ψ will be zero for $y = 0$ and $x^2 + y^2 = a^2$. The former condition gives the x axis† as the locus of one set of points for this streamline, while the latter condition prescribes a circle of radius a. Other streamlines could be plotted by assigning values to ψ and solving for the required values of x and y. Note that as $x^2 + y^2$ becomes very large, ψ tends toward the value of Uy. According to Eqs. (4.7.4) and (4.7.5), one has

$$u = \frac{\partial \psi}{\partial y} = U - Ua^2 \left[\frac{y(-2y)}{(x^2 + y^2)^2} + \frac{1}{x^2 + y^2} \right] \tag{4.7.7}$$

and

$$v = -\frac{\partial \psi}{\partial x} = +Ua^2 \left[\frac{y(-2x)}{(x^2 + y^2)^2} \right] \tag{4.7.8}$$

These equations indicate that the flow tends to become uniform and horizontal (i.e., u = constant and $v = 0$) at large distances from the origin.

Example 4.7.2

If one were given $\psi = 3xy$, one could plot the curves $\psi = 0$, $\psi = 1$, $\psi = 2$, $\psi = 5$, and so on. They would be hyperbolas. (Recall that the flow rate between adjacent streamlines is the difference in the value of ψ between the streamlines.) For this function of ψ,

---

* Total differentials are reviewed in Appendix B.
† Omitting the origin, because there the term in parentheses in Eq. (4.7.6) is singular.

$$u = \frac{\partial}{\partial y}(3xy) = 3x$$

and
$$v = -\frac{\partial}{\partial x}(3xy) = -3y$$

Example 4.7.3

If $u = 5y$ and $v = -2x$, ψ can be found because

$$u = 5y = \frac{\partial \psi}{\partial y}$$

and upon integration, $\psi = \frac{5}{2}y^2 + f(x)$. Also, $v = -2x = -\partial\psi/\partial x$, so that $\psi = +x^2 + g(y)$. [Why $f(x)$ in one case and $g(y)$ in the other?] If we compare the flow expressions for ψ, we conclude that $f(x) = x^2 + $ constant and $g(y) = \frac{5}{2}y^2 + $ constant. Thus, $\psi = x^2 + \frac{5}{2}y^2 + $ constant. It is a simple exercise to show that u and v necessarily satisfy volume conservation at a point. We shall do this in Sec. 4.8 in greater generality than when Eq. (4.3.4) was derived.

In some situations, plane polar coordinates (r, θ) may be more convenient than Cartesian coordinates. With the aid of Fig. 4.7.3, along with the line of reasoning which yielded Eqs. (4.7.3) and (4.7.5), one can show that

$$u_r = \frac{1}{r}\frac{\partial \psi}{\partial \theta} \tag{4.7.9}$$

and
$$u_\theta = -\frac{\partial \psi}{\partial r} \tag{4.7.10}$$

[Note that in Fig. 4.7.3, $AC = r\,d\theta$ and $CB = dr$. Equation (4.7.3) would give $d\psi = (\partial\psi/\partial r)\,dr + (\partial\psi/\partial\theta)\,d\theta$ for polar coordinates, and an examination of Fig. 4.7.3 would lead to $d\psi - u_r(r)\,d\theta + u_\theta\,dr = 0$ for volume conservation of a steady flow through region ABC. An identification of the coefficients of dr and $d\theta$ in these two results yields Eqs. (4.7.9) and (4.7.10).]

Figure 4.7.3 Path between two points prescribed by polar coordinates on streamlines ψ and $\psi + d\psi$.

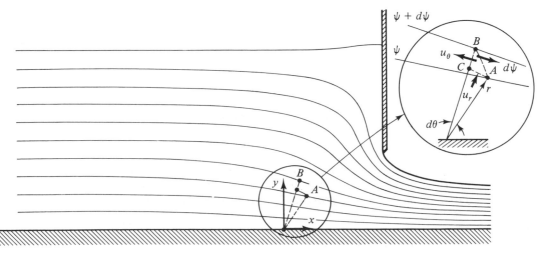

4.8 DIFFERENTIAL FORM OF MASS CONSERVATION

The concept of mass conservation was discussed in Chap. 3. There, starting with the ideas behind Reynolds' transport theorem, we derived the requirement that

$$\frac{\partial}{\partial t} \int(\rho \, d\forall) + \int\rho\mathbf{V}\cdot\mathbf{n} \, dA = 0 \qquad (4.8.1a)$$

and for the special case of steady flow,

$$\int\rho\mathbf{V}\cdot\mathbf{n} \, dA = 0 \qquad (4.8.1b)$$

The integral in Eq. (4.8.1b) must be evaluated at all points of the control surface. We applied this equation to a variety of situations in order to gain familiarity with it. In Example 3.7.3, an incompressible fluid was considered and the velocity was prescribed by the vector function

$$\mathbf{V} = (4y - 7x)\mathbf{i} + (7y - 8x)\mathbf{j} \qquad (4.8.2)$$

An integration around a finite control surface in that example showed that Eq. (4.8.1b) was satisfied by this velocity distribution. The consequence of this is the assurance that Eq. (4.8.2) could, indeed, be a description of an incompressible flow. The requirement for mass conservation could also be found to be satisfied by this velocity field for a much smaller control volume. It is the purpose of this section to derive a condition that the vector function for a velocity field must obey to satisfy mass conservation at a material point. This condition is a differential equation that can be derived by starting with Eq. (4.8.1a).

Near a particular point of space, the variations of density and velocity are determined by the various appropriate derivatives of the functions. The control volume in Fig. 4.8.1 is infinitesimal in extent, so that derivatives will be used to determine the differences in density and velocity that could exist at different parts of the control surface. These values, with their derivatives, will be used to evaluate the surface integral in Eq. (4.8.1a) for the infinitesimal region depicted in Fig. 4.8.1. On surface AB of this figure, $\rho\mathbf{V}\cdot\mathbf{n}$ has a value equal to

$$-\left[\rho u - \frac{\partial(\rho u)}{\partial x}\frac{dx}{2}\right]$$

while on segment BC the value is

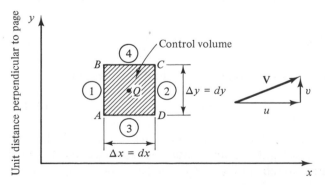

Unit distance perpendicular to page

Figure 4.8.1 Application of mass conservation to an infinitesimal material volume $\Delta x \, \Delta y$ (1).

$$+ \left[\rho v + \frac{\partial(\rho v)}{\partial y} \frac{dy}{2} \right]$$

If expressions for $\rho \mathbf{V} \cdot \mathbf{n}$ are included for the remaining segments of the control surface, CD and AD, one can evaluate the surface integral in Eq. (4.8.1a) as

$$\int \rho \mathbf{V} \cdot \mathbf{n} \, dA = - \left[\rho u - \frac{\partial(\rho u)}{\partial x} \frac{dx}{2} \right] dy \, (1) + \left[\rho v + \frac{\partial(\rho v)}{\partial y} \frac{dy}{2} \right] dx \, (1)$$

$$+ \left[\rho u + \frac{\partial(\rho u)}{\partial x} \frac{dx}{2} \right] dy \, (1) - \left[\rho v - \frac{\partial(\rho v)}{\partial y} \frac{dy}{2} \right] dx \, (1)$$

The time derivative of the volume integral in the equation for mass conservation, Eq. (4.8.1a), can be represented as

$$\frac{\partial(\rho_{\text{avg}} \, \forall)}{\partial t} = \left(\frac{\partial \rho_Q}{\partial t} \right)(dx \, dy)(1)$$

in which ρ_Q is the density at the center of the infinitesimal control volume.

These steps that are legitimate for an infinitesimal region permit Eq. (4.8.1a) to be expressed in two dimensions as

$$\frac{\partial \rho}{\partial t} dx \, dy + \left[\frac{\partial(\rho u)}{\partial x} + \frac{\partial(\rho v)}{\partial y} \right] dx \, dy = 0$$

in which all the variables are evaluated at the point in question, point Q. This expression can be rewritten on a unit volume basis by dividing through by the infinitesimal volume, $d\forall [= dx \, dy \, (1)]$. The result is

$$\frac{\partial \rho}{\partial t} + \frac{\partial}{\partial x}(\rho u) + \frac{\partial}{\partial y}(\rho v) = 0 \qquad (4.8.3)$$

in which u, v, and ρ are the velocity components and density, respectively, at the center of the element. This is the *equation of mass continuity* in differential form. It applies at each point of a region where a two-dimensional flow occurs. For a three-dimensional flow, one can write the equation for mass conservation at a point as

$$\frac{\partial \rho}{\partial t} + \frac{\partial(\rho u)}{\partial x} + \frac{\partial(\rho v)}{\partial y} + \frac{\partial(\rho w)}{\partial z} = 0 \qquad (4.8.4)$$

After performing the indicated differentiation of product terms and regrouping the results, this equation has the form

$$\frac{\partial \rho}{\partial t} + \mathbf{V} \cdot \text{grad} \, \rho + \rho \, \text{div} \, \mathbf{V} = 0 \qquad (4.8.5)$$

In Cartesian coordinates, div $\mathbf{V} = 0$ for an incompressible fluid is expressed as

$$\frac{\partial u}{\partial x} + \frac{\partial v}{\partial y} + \frac{\partial w}{\partial z} = 0 \qquad (4.8.6)$$

which is called the *continuity equation*. While this equation is in Cartesian coordinates, a similar set of steps can be performed if the control volume is expressed in cylindrical polar coordinates, as is the control volume in Fig. 2.8.1. (Just as in Sec. 2.8, one must take care to account for the varying surface area in the r direction when adapting the Reynolds transport theorem to this situation.) The result of such an analysis is

$$\frac{\partial \rho}{\partial t} + \underbrace{\frac{(\rho u_r)}{r} + \frac{\partial(\rho u_r)}{\partial r}}_{\dfrac{1}{r}\dfrac{\partial(r\rho u_r)}{\partial r}} + \frac{1}{r}\frac{\partial(\rho u_\theta)}{\partial \theta} + \frac{\partial(\rho u_z)}{\partial z} = 0 \tag{4.8.7}$$

or $\qquad \left(\dfrac{\partial \rho}{\partial t} + u_r \dfrac{\partial \rho}{\partial r} + \dfrac{u_\theta}{r}\dfrac{\partial \rho}{\partial \theta} + u_z \dfrac{\partial \rho}{\partial z}\right) + \rho\left[\dfrac{1}{r}\dfrac{\partial(ru_r)}{\partial r} + \dfrac{1}{r}\dfrac{\partial u_\theta}{\partial \theta} + \dfrac{\partial u_z}{\partial z}\right] = 0$

so that for an incompressible fluid, for which $D\rho/Dt = 0$,

$$\left[\frac{1}{r}\frac{\partial}{\partial r}(ru_r) + \frac{1}{r}\frac{\partial u_\theta}{\partial \theta} + \frac{\partial u_z}{\partial z}\right] = 0 \tag{4.8.8}$$

in cylindrical polar coordinates. This last result for an incompressible fluid could also be obtained directly if the expression for div $\mathbf{V} = 0$ had been written in cylindrical polar coordinates.

4.9 SUMMARY

The differential velocity between two adjacent points, $d\mathbf{V} = du\,\mathbf{i} + dv\,\mathbf{j} + dw\,\mathbf{k}$, can be expressed as $d\mathbf{V} = d\mathbf{V}_{\text{rotation}} + d\mathbf{V}_{\text{distortion}}$. The rate of rotation is given by $\boldsymbol{\omega}$ whose components for a two-dimensional case are given by Eq. (4.2.3). The relative velocity associated with distortion is made up of terms that can be identified with shape deformation [Eq. (4.3.1)] and volume distortion [Eq. (4.3.3)]. All of these quantities are discussed in Example 4.4.1.

The Eulerian description of fluid motion includes an expression for the acceleration of the fluid at a spatial point that is not a simple time derivative. The expressions for the three components of acceleration are given by Eqs. (4.6.6b) to (4.6.8).

Mass conservation for a two-dimensional incompressible flow provides the basis for the existence of a stream function for each such velocity field. The value of the stream function, ψ, will depend upon x and y only for the steady flows treated in this chapter. The lines given by $\psi = \text{constant}$ are tangent to streamlines. The velocity components are given by $u = +\partial\psi/\partial y$ and $v = -\partial\psi/\partial x$.

The law of mass conservation must also be obeyed at each location within the fluid. In Sec. 4.8, Reynolds' transport theorem for a finite material volume is applied to an infinitesimal control volume. The result for Cartesian coordinates is $\partial\rho/\partial t + \partial(\rho u)/\partial x + \partial(\rho v)/\partial y + \partial(\rho w)/\partial z = 0$, which is the differential form of the integral expression developed in Chap. 3.

EXERCISES

4.4.1(a) Use Eqs. (4.4.3) through (4.4.7) to show that $V_{\text{rel}} = (d)\,dx + (e)\,dy + (\omega)\,dx$. Next, show that since $v(0,0) = 0$ and e, d, and ω are constant, this equation can be interpreted so that $v_{\text{ad}} = (d)x$, $v_{\text{ed}} = (e)y$, $v_{\text{ro}} = (\omega)x$. Evaluate $\mathbf{V}_{\text{ed}}$ at points C and D.

(b) Calculate v_{ad}, v_{ed}, and v_{ro} at points C and D. Do these values sum to v_C and v_D, respectively?

(c) For the velocity $\mathbf{V} = (3x - 2y)\mathbf{i} + (9x - 3y)\mathbf{j}$, determine u, v, w, and d at $(0.5, 0)$. What is u_{rel} at $(0, 0.5)$? Decompose this relative velocity component into u_{ed}, u_{ad}, and u_{ro}—since $\mathbf{V}(0, 0) = 0$—and give values for u_{ed}, u_{ad}, and u_{ro}.

(d) For the situation described in this example, plot the velocity at points A, B, C, and D to a suitable scale on a drawing of the square $ABCD$. (Start by finding the velocity at the corners of the square. This will help in the drawing of the distorted figure.) At $t = 1$, where are the particles that were at A, B, C, and D when $t = 0$? At $t = 1$, draw the distorted particle $A'B'C'D'$ that was $ABCD$.

4.5.1(a) Estimate the amount of time that is needed for a particle to move from point B to point C.

4.6.1(a) What is the acceleration of the fluid particles at $(0, 0, 0)$ and $(1, 2, 1)$?

(b) For $(u, v, w) = (5x, -2y, -3z)$, give expressions for the three components of acceleration of fluid particles.

(c) For $(u, v, w) = (5xt, -2yt, -3xt)$, give expressions for the components of acceleration. What is the acceleration of the fluid particles at $(0, 0, 0)$ and $(1, 2, 1)$ at $t = 0$ and $t = 2$?

4.6.2(a) Give the values for a_r and a_θ at $(1, 2)$.

(b) For the velocity distribution used in Example 4.6.2, is $\theta_{acc} = y/x$? Does your answer to this question give you some insight into the values for a_r and a_θ that were obtained in the last equation?

(c) For $u = 3y$ and $v = -3x$, is $\mathbf{V} \cdot \mathbf{r} = 0$? What does this tell you about the orientation of $\mathbf{V}$ with respect to $\mathbf{r}$?

4.7.1(a) What are u and v at $(x, y) = (-a, 0)$, $(0, a)$, and $(a, 0)$?

(b) Determine u and v on the circle $r = a$. What are u_r and u_θ on $r = a$? (*Hint:* Use coordinate transformations similar to Eqs. (4.6.9) and (4.6.10). Comment on the value of u_r. What is u_θ?

(c) On $y = 0$, Eq. (4.7.8) gives v as identically zero. Plot u/U—hence $|\mathbf{V}|/U$—for $x \leq -a$ on $y = 0$. At what value of a/x does $u/U = 0.99$? This result should indicate the range of upstream influence that a body will have on a flow.

PROBLEMS

4.1 For $u = 3x$, $v = -3y$, solve for u_{ed}, u_{ad}, u_{ro}, v_{ed}, v_{ad}, v_{ro} at the points $(1, 1)$, $(2, 1)$, $(0, 1)$, $(1, 2)$, and $(1, 0)$. What are the values of ω and d at $(1, 1)$?

4.2 For

$$u = U\left[1 - \frac{a^2(x^2 - y^2)}{(x^2 + y^2)^2}\right]$$

and $\quad v = Ua^2\left[-\frac{2xy}{(x^2 + y^2)^2}\right]$

solve for u_{ed}, u_{ad}, u_{ro}, v_{ed}, v_{ad}, and v_{ro} at the points $(5, 0)$, $(5.1, 0)$, $(4.9, 0)$, $(5, 0.1)$, $(5, -0.1)$. What are the values of ω and d at $(5, 0)$? Use $U = 2$ and $a = 5$.

4.3 In Fig. 4.5.1, the axial velocity data that are presented could be approximated by $u = 447/x^{0.8}$. What is du? What is the expected value of the change of u at $x = 2.5$?

4.4 The density values given in Fig. 4.5.1 are coarsely approximated by $\rho = 1.1(10^{-2})(x^{0.35})$. What is $d\rho$ at $x = 2.5$?

4.5 Use the approximate functional relationships for the velocity and density given in Probs. 4.3 and 4.4 to determine $d\rho$ and $d\rho/dt$. What is the latter value at $x = 2.5$?

4.6 For each of the velocity distributions given in Prob. 3.4, determine a_x and a_y.

4.7 For $u = Cx/(x^2 + y^2)$ and $v = Cy/(x^2 + y^2)$, find a_x and a_y.

4.8 For the velocity distribution given by $u = 3x$ and $v = -3y$, find a_x and a_y.

4.9 The flow $u = 5x$, $v = -5y$ that was introduced in Example 4.6.1 can be written in polar coordinates as

$$u = 5r \cos\theta \quad \text{and} \quad v = -5r \sin\theta$$

These Cartesian velocity components yield u_r and u_θ through the same transformation of vector components that yielded Eqs. (4.6.9) and (4.6.10). Solve for u_r and u_θ and use these expressions in Eqs. (4.6.11) and (4.6.12) to determine a_r and a_θ. Do these values agree with those in Example 4.6.2?

4.10 The stream function given by Eq. (4.7.6) can be rewritten in polar coordinates as

$$\psi = Ur \sin \theta \left(1 - \frac{a^2}{r^2}\right)$$

Use Eqs. (4.7.9) and (4.7.10) to find u_r and u_θ. What are the values of these components on $r = 1$?

4.11 Carry out the derivation of Eqs. (4.7.9) and (4.7.10).

4.12 For $\psi = 0.5 \ln (x^2 + y^2)$, what are u, v, a_x, and a_y? For $\psi = \ln r$, what are the polar velocity components? How are these results related to the Cartesian velocity components just found?

4.13 For $\psi = \arctan (y/x)$, what are u and v? Do these components satisfy the continuity equation everywhere?

4.14 Determine ψ for the velocity distributions given in Examples 3.5.1 and 3.5.5.

4.15 What is the associated expression for ψ if $u = 2Ay$ and $v = 2Ax$?

4.16 Which of the velocity profiles given in Prob. 3.4 satisfy the volume constraint on an incompressible fluid?

4.17 Given that the components listed below are associated with an incompressible flow, determine the missing component as completely as you can.

$$u = -5x, \quad v = 5y, \quad w = ?$$

$$u = -2y, \quad v = 2x, \quad w = ?$$

$$u = -\frac{3y}{(x^2 + y^2)}, \quad v = \frac{3x}{(x^2 + y^2)}, \quad w = ?$$

$$u = x + y + z, \quad v = x + y + z, \quad w = ?$$

$$u = 2xy, \quad v = y + z^2, \quad w = ?$$

4.18 Equation (4.7.6) gives a case where the origin must be excluded from the $\psi = 0$ line. What are u and v at the origin?

4.19 Determine if the u and v which are given in Eqs. (4.7.7) and (4.7.8) satisfy the equation of continuity, $\partial u/\partial x + \partial v/\partial y = 0$. Is this result also true at the origin?

4.20 Show that the result of Prob. 4.19 was not fortuitous. Substitute the velocity components given by Eqs. (4.7.4) and (4.7.5) in terms of the derivatives of ψ directly into the continuity equation to obtain

$$\frac{\partial^2 \psi}{\partial x\, \partial y} - \frac{\partial^2 \psi}{\partial y\, \partial x} = 0$$

Under what conditions is this equation always satisfied?

4.21 A stream function may be associated with an unsteady flow, since the analysis that led to Eq. (4.8.4) is not restricted to any particular instant of time. Describe the streamlines for:

$$\psi = (1)(\ln r)(e^{-0.1t}) + (2)(\theta)(1 - e^{-0.05t})$$

in polar coordinates (r, θ). Are they circles for $t \to 0$ and radial lines for $t \to \infty$? Are they spirals in general?

5

The Flow of an Inviscid Fluid

5.1 INTRODUCTION

Chapter 2 was concerned with the application of Newton's laws of motion when there was no acceleration or when rigid-body motions were occurring with accelerations. Now, we are ready to extend our perspectives by considering flows of a special fluid. It is a hypothetical fluid that is described by the adjective *inviscid,* which denotes the absence of viscosity. The existence of viscosity (i.e., fluid friction) was proposed, and its ramifications were explored, by Newton. About 50 years later (1738), Daniel Bernoulli* employed Newton's laws of motion to establish a theory governing fluid flow. Because of the complexity of the problem, he did not include viscosity in the development. His many useful results seemed to justify his assumptions. For almost 150 years, scientists and mathematicians were able to use his theory to solve problems concerning waves, tides, jets that issued from pipes, and many other topics. We shall see in Chap. 14 that the basis of Bernoulli's work can also be applied to aspects of the flow of gases that are moving at very high speeds. This appears to indicate that in a great number of cases the principal effects can be described by first neglecting the influence of viscosity. In some situations, however, the exclusion of viscosity in an analysis leads to results that cannot be reconciled with observations. For instance, one consequence of using Bernoulli's equation is the absence of drag on a two-dimensional body, such as a long, circular cylinder. The existence of such a drag becomes evident to anyone who extends an arm out of a window of a moving vehicle.

Some important ideas about the effects of viscosity were provided by the experiments of Poiseuille and the analysis of Stokes, both around 1850. (These accomplish-

---

* Daniel's father, Johann, and his uncle, Jakob, were prominent mathematicians of their time.

ments will be discussed in Chap. 8.) Finally, in 1904, Ludwig Prandtl, an elastician by training, applied his great insight to the physics of fluid flow and provided a rationale for one's ability to neglect viscosity in some cases and its crucial inclusion in other cases. (Prandtl's work will also be discussed in Chap. 8.) However, now we shall examine the importance of Bernoulli's contribution, one which greatly enhances our ability to analyze and explain some of the phenomena that are observed.

5.2 THE BERNOULLI EQUATION FOR STEADY FLOW

We have already used Newton's laws of motion to relate the pressure to the motion of a fluid when it is undergoing a constant, rectilinear acceleration. The motion associated with rigid-body rotation was also discussed in a previous chapter (Chap. 2). The pressure distributions in the fluid that resulted from these motions were found there with the use of the relation $\mathbf{F} = m\mathbf{a}$. We shall now apply Newton's laws to a somewhat more general situation. The result, even with its limiting assumptions, will be of great utility. In arriving at this result we shall use the same approach that was used previously with success.

As an aid to the following discussion, it will be assumed that we have drawn the streamlines of a *steady* flow by means of a sequence of many photographs and a suitable flow visualization technique. For convenience, we shall consider the flow to be two-dimensional (i.e., all of the streamlines can be expressed in terms of x and y and there is no variation in the z direction). One such xy plane is shown in Fig. 5.2.1. It is possible to identify a segment of fluid, such as material volume $ABCDA$, that appears in one of several photographs taken, say, at microsecond intervals. In the next image of the flow, this material volume would be at another position along the streamline and might be recognized as $A'B'C'D'A'$. (Henceforth, we shall consider the material volume to be a fluid particle.) Because we shall be dealing with streamlines in a steady flow, we recognize that the path of the particle $ABCDA$ will be coincident with a streamline. Thus it will arrive at O' along a streamline at an appropriate time $t_0 + dt$. During the time interval between the two locations of the particle, it may have changed its speed as well as its direction of motion. It also may have deformed and rotated.

Such changes of speed and direction are associated with an acceleration of the particle $ABCDA$. We shall soon compute this acceleration, as well as the forces that appear to be acting on the particle. Finally, we shall interrelate these two calculations with the aid of Newton's laws of motion.

Figure 5.2.1 has a coordinate system to facilitate the computations associated with Newton's laws of motion. At some point O along a streamline, where the particle $ABCDA$ is located at time t_0, we draw perpendicular axes that are aligned in the directions of the tangent to the streamline and the normal to the streamline. The decision to place the coordinate axes in this way may appear to be arbitrary, but it will later prove to be useful. (Figure 5.2.2 gives an example of an experimentally obtained representation of streamlines for the slow flow of a fluid past a rectangular plate along with a coordinate system. The vertical line is a strut which supports the horizontal plate shown in the center of the figure. The photograph illustrates the variation of streamline spacing—i.e., fluid speed—and direction. An optical technique called *interferometry* was used to make the streamlines visible.)

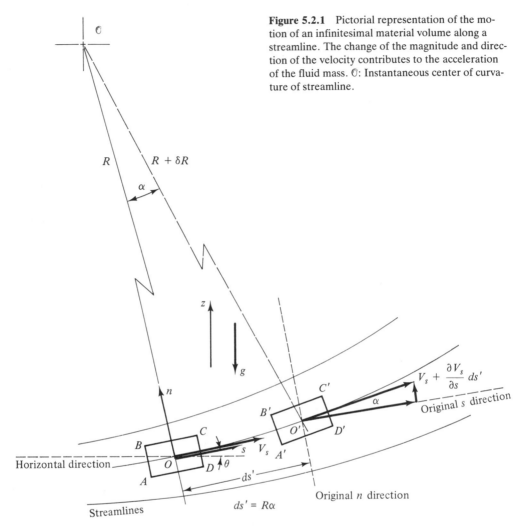

Figure 5.2.1 Pictorial representation of the motion of an infinitesimal material volume along a streamline. The change of the magnitude and direction of the velocity contributes to the acceleration of the fluid mass. ⓪: Instantaneous center of curvature of streamline.

Figure 5.2.2 Streamlines of the flow past a rectangular bar with a superimposed *sn* coordinate system.

The Flow of an Inviscid Fluid   Chap. 5

By referring to Fig. 5.2.1, one can determine the associated acceleration of the fluid particle that must have occurred while it changed location along the streamline. The particle changed speed in accordance with the rate of speed change that existed at the particular flow position, O. If this rate of speed change is given in terms of the distance traversed by the particle along the streamline, the change of speed from position O to O' is $(\partial V_s/\partial s)(ds')$, in which ds' (not to be confused with ds, the length of the element in the s direction) is the distance that the particle moved along the streamline in the time between t_0 and $t_0 + dt$. This change of speed is shown in Fig. 5.2.1. However, the acceleration of the particle is not obtained by knowing only the change of speed during the time interval dt. The direction is also important. When the particle is at position O', its velocity must be tangent to the streamline at that position. But the velocity component at O' that is parallel to the coordinate direction *originally prescribed* is

$$\left(V_s + \frac{\partial V_s}{\partial s} ds'\right) \cos \alpha$$

The angle α, the change in the direction of the streamline, is small, so that its cosine is nearly unity. Moreover, if the distance ds' is taken smaller and smaller as the time interval between the observed positions O and O' is reduced, α approaches zero and the cosine of the angle approaches 1.0.

This time interval, dt, will be the quotient of the distance traveled by the particle and the average speed during the elapsed time,

$$\frac{ds'}{\frac{1}{2}\left\{V_s + \left[V_s + \left(\frac{\partial V_s}{\partial s}\right)(ds')\right]\right\}}$$

The acceleration of the particle in the s direction is the change of speed in this direction, divided by the time during which the change has occurred. This is

$$\lim_{ds' \to 0} \frac{(\partial V_s/\partial s)(ds')}{ds'/\frac{1}{2}\left\{V_s + \left[V_s + \left(\frac{\partial V_s}{\partial s}\right)(ds')\right]\right\}} = V_s\left(\frac{\partial V_s}{\partial s}\right)$$

The procedure for determining the acceleration of a particle that was just employed is similar to that which was discussed in Sec. 4.6. Ultimately, one does not follow a particular particle and note the change of velocity. Rather, the observed values of the velocity and its derivatives at a point in space, V_s and $\partial V_s/\partial s$, are used in the calculation.

The mass of the particle is $\rho \, d\forall$, or $\rho \, dn \, ds$ (1). Thus we can write the product of the mass m and the component of the acceleration in the s direction as

$$ma_s = V_s \frac{\partial V_s}{\partial s} \rho \, ds \, dn \tag{5.2.1}$$

This quantity must be equal to the sum of all of the forces that act in the positive s direction, according to Newton's laws. Figure 5.2.3 shows the normal forces, due to pressure, that act on the surfaces AB and CD. (We have assumed that when the normal and tangential coordinate system was introduced at point O, the particle was delineated with sides that were parallel to these directions.) The component of the body

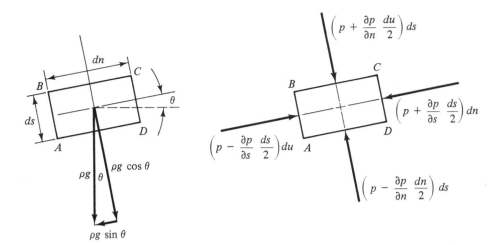

Figure 5.2.3 Body forces (a) and surface forces (b) on a material volume (fluid particle) flowing along a streamline.

force in the $+s$ direction is shown in Fig. 5.2.3 as $-\rho g (\sin \theta)\, ds\, dn$. Consequently, the forces in the $+s$ direction that result from these surface stresses and the appropriate body force are

$$\left(p - \frac{\partial p}{\partial s}\frac{ds}{2}\right) dn - \left(p + \frac{\partial p}{\partial s}\frac{ds}{2}\right) dn - \rho g (\sin \theta)\, ds\, dn \qquad (5.2.2)$$

But $\sin \theta$ is related to the angle of inclination of the streamline with the horizontal (this is shown in Fig. 5.2.4), so that

$\sin \theta =$ change of elevation per unit distance traveled along the streamline

$$= \frac{\partial z}{\partial s}$$

This implies that the body force in the $+s$ direction is

$$-\rho g \frac{\partial z}{\partial s}\, ds\, dn \qquad (5.2.3)$$

The surface and body forces, as represented by expressions (5.2.2) and (5.2.3), can be equated to the product of mass and acceleration, Eq. (5.2.1), in accordance with Newton's laws of motion:

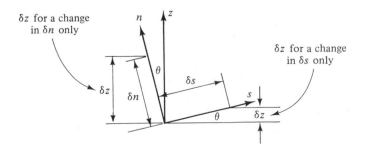

Figure 5.2.4 Relationship between the (vertical) z axis and the sn coordinate system. Streamline makes an angle θ with horizontal at point O. Thus, $\sin \theta = \partial z / \partial s$; $\cos \theta = \partial z / \partial n$.

The Flow of an Inviscid Fluid   Chap. 5

$$\left(p - \frac{\partial p}{\partial s}\frac{ds}{2}\right)dn - \left(p + \frac{\partial p}{\partial s}\frac{ds}{2}\right)dn - \rho g\frac{\partial z}{\partial s}\,ds\,dn = ds\,dn\,\rho\left(V_s\frac{\partial V_s}{\partial s}\right)$$

or, per unit volume,

$$-\frac{\partial p}{\partial s} - \rho g\frac{\partial z}{\partial s} = \rho V_s\frac{\partial V_s}{\partial s}$$

$$= \rho\frac{\partial(V_s^2/2)}{\partial s}$$

This can be rewritten as

$$\frac{\partial}{\partial s}\left(\frac{V_s^2}{2} + \frac{p}{\rho} + gz\right) = 0 \tag{5.2.4}$$

if one divides through by ρ, which is now *assumed* to be a constant. This implies that the term in parentheses, a scalar, does not change in value along the streamline. This collection of terms is often called the *Bernoulli constant*. Thus,

$$\frac{V_s^2}{2} + \frac{p}{\rho} + gz = \text{Constant along a streamline} \tag{5.2.5}$$

As we shall soon see, the constant in this equation can be different for each streamline in the flow.

In the derivation of Eq. (5.2.4), Newton's laws of motion were used at the outset. Finally, we integrated Eq. (5.2.4) and obtained the Bernoulli equation, (Eq. (5.2.5). It may be recalled from a course in dynamics that the equations for Newton's laws of motion can be integrated to give an equation that is termed the *work-energy equation*. Thus, it is not surprising that Eq. (5.2.5) has terms in it that remind us of energy quantities. The first term, $V_s^2/2$, is the kinetic energy per unit of mass of a fluid particle. (This is easily understood when one recalls that kinetic energy is usually written as $mV^2/2$ for a particle of mass m.) In a similar fashion, the term gz is the potential energy per unit mass of a particle. The term p/ρ represents the work done by the pressure field on the fluid particles. This quantity will be discussed further in Chap. 6. It was convenient, however, to introduce at this time the idea that the Bernoulli equation is a particular form of a more general result, the energy equation, that is deduced from the principle of the conservation of energy. The Bernoulli equation is an energy equation, in which it is assumed that the work done on the fluid by the pressure field is entirely transformed into changes of the kinetic and potential energies of the fluid. One could say that there is no dissipation for these two forms of energy in the system. In Chap. 6, we shall expand our views to allow for the transfer of energy to the fluid particles in the form of heat. Energy will also be permitted to be transferred to the fluid by the application of mechanical work. Finally, the concept of internal energy will be employed. This will mean that the work done by the pressure field within the fluid may result in changes of kinetic, potential, and internal energy of the fluid, as well as transferring energy to an external system (as in a generator) and/or rejecting heat to the surroundings (as in a steam condenser).

The constant in Eq. (5.2.5) can be evaluated at a particular point on a streamline where V_s, p, and z are known. If this point is designated point 1, it follows that

$$\frac{V^2}{2} + \frac{p}{\rho} + gz = \frac{V_1^2}{2} + \frac{p_1}{\rho} + gz_1 \tag{5.2.6}$$

Note that the subscript s on the velocity term, V, has been suppressed. It is unnecessary if we understand the meaning of V. The terms on the left-hand side of Eq. (5.2.6) could be used to evaluate the conditions at some second point, say, point 2, on the same streamline.

It is also common to write Eq. (5.2.6) in an alternative form by dividing through by the gravitational constant g. This gives

$$\frac{V_1^2}{2g} + \frac{p_1}{\gamma} + z_1 = \frac{V_2^2}{2g} + \frac{p_2}{\gamma} + z_2 \qquad (5.2.7)$$

In view of the discussion about energy that appeared above, one can think of Eq. (5.2.7) as containing terms that are equivalent to the energy per unit weight flowing per second.

Example 5.2.1

The water flowing through a nozzle changes its kinetic energy considerably (see Fig. 5.2.5). Experience with such a device has shown that viscosity plays a minor role in this accelerating flow and the Bernoulli equation gives useful results. Apply Eq. (5.2.7) to this nozzle flow.

Solution If this equation is utilized along the streamline between points 1 and 2, it is written as

$$\frac{V_1^2}{2g} + \frac{p_1}{\gamma} + z_1 = \frac{V_2^2}{2g} + \frac{p_2}{\gamma} + z_2$$

Now we shall set p_2 equal to atmospheric pressure, because the pressure everywhere on the surface of the jet is atmospheric pressure. (Because the jet is neither expanding nor contracting, the pressure on the centerline should also be atmospheric pressure if surface tension is neglected. This would imply that there is no change of liquid pressure in the z direction. Such a pressure variation would not be hydrostatic, but it is satisfactory because the jet is accelerating downward due to "free fall." Hence, we are justified in writing $\partial p/\partial z = 0$ in the jet.) We write $p_2 = 0$ gauge pressure. Inside the nozzle, where the flow is horizontal, there is no vertical acceleration, and the pressure changes hydrostatically even though the fluid is not at rest. (In Chap. 8, we shall see that this important conclusion is also valid when Newtonian viscosity is considered.) Thus the pressure on the centerline of the nozzle's inlet can be expressed as

$$p_1 = p_g - \frac{D_1 \gamma}{2}$$

in which p_g is the gauge pressure at the bottom of the pipe. Because the nozzle axis has been taken as horizontal, $z_1 = z_2$, so that

$$\frac{p_g - D_1 \gamma/2}{\gamma} + \frac{V_1^2}{2g} = \frac{0}{\gamma} + \frac{V_2^2}{2g}$$

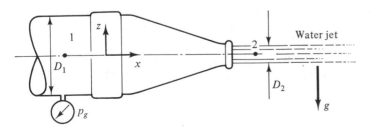

Water jet

Figure 5.2.5 Discharge of water from a horizontal nozzle, with gauge at bottom of conduit.

The Flow of an Inviscid Fluid   Chap. 5

if gauge pressures are used throughout. Volume conservation for an incompressible fluid gives us, if constant fluid velocities are assumed over sections 1 and 2,

$$V_1 A_1 = V_2 A_2$$

or

$$V_1 = \frac{V_2 A_2}{A_1} = V_2\left(\frac{D_2}{D_1}\right)^2$$

This result can be incorporated into the expression for the Bernoulli equation, and V_1 can be eliminated. Hence,

$$\frac{V_2^2}{2g}\left[1 - \left(\frac{D_2}{D_1}\right)^4\right] = \frac{p_g}{\gamma} - \frac{D_1}{2} = \frac{p_1}{\gamma}$$

Often, the pressure at location 1, on the centerline, will be given directly, or in the case of small pipes and large pressures, the term $D_1/2$ will be neglected with respect to p_g/γ. The reader may wish to solve for V_2 if $D_1 = 12$ cm, $D_2 = 4$ cm, and $p_g = 4.0(10^4)$ Pa. Be sure to use meters, kilograms, and seconds as the units for calculations. (*Ans.:* 8.93 m/s)

We shall now discuss the flow in a converging-diverging section by means of the Bernoulli equation (see Fig. 5.2.6). The flow contracts over a short length of conduit and then expands slowly to its original diameter. The flow follows along the tube's walls as they reduce in diameter. There is a relatively small amount of viscous effect, because this contraction section is short. The downstream section has an expansion that is sufficiently gradual that the fluid can expand laterally in order to keep moving along the walls without large eddies forming near the wall that would give rise to dissipative effects with associated pressure changes. This device is called a *Venturi tube* and is useful for metering rates of flow, because there is very little change of the sum $V^2/2 + p/\rho + gz$ through the device.

Example 5.2.2

Determine the volume rate of flow for the manometer reading given in Fig. 5.2.6.

Solution The Bernoulli equation can be written between points 1 and 2:

$$\frac{V_1^2}{2g} + \frac{p_1}{\gamma} = \frac{V_2^2}{2g} + \frac{p_2}{\gamma}$$

and volume conservation gives

$$V_1(D_1^2) = V_2(D_2^2)$$

Pressures p_1 and p_2 are related by the hydrostatic law, so that

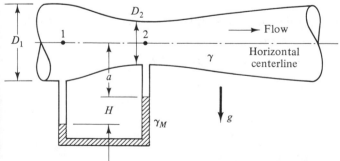

Figure 5.2.6 Flow of a liquid through a horizontal Venturi tube with a manometer.

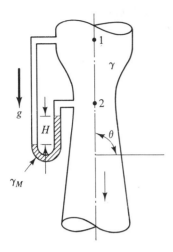

Figure 5.2.7 Flow through a vertical Venturi tube. θ is 90° to the horizontal.

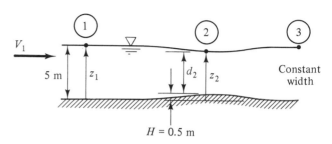

Figure 5.2.8 Liquid flowing in a wide, horizontal channel with a change in bottom profile. $Q = 10$ m³/s per meter of width.

$$p_1 + a\gamma + H\gamma - H\gamma_M - a\gamma = p_2$$

This result allows $p_1 - p_2$ to be replaced by $H(\gamma_M - \gamma)$ in the Bernoulli equation. Also, V_2 in that equation is equal to $V_1(D_1^2/D_2^2)$ from the law of continuity. This gives the equation for V_1:

$$\frac{V_1^2}{2g}\left[\left(\frac{D_1}{D_2}\right)^4 - 1\right] = \frac{H(\gamma_M - \gamma)}{\gamma}$$

so that

$$Q = \frac{\pi D_1^2}{4}\left[\frac{2gH(\gamma_M - \gamma)/\gamma}{(D_1/D_2)^4 - 1}\right]^{1/2} \tag{5.2.8}$$

(Now apply the Bernoulli equation to the vertical Venturi tube shown in Fig. 5.2.7 and deduce the same conclusion reached in this example. Does this result suggest that the same equation for Q could result if $\theta \neq 90°$?)

In the next example, we shall apply the Bernoulli equation to a situation where one would expect that there is only a small change in the elevation of the free surface between the two locations of interest.

Example 5.2.3

Determine the height of the fluid as it passes over a gradual change of bottom topography in a channel of unit width (see Fig. 5.2.8).

Solution   The Bernoulli equation gives, between points 1 and 2,

$$\frac{V_1^2}{2g} + \frac{p_{\text{atm}}}{\gamma} + z_1 = \frac{V_2^2}{2g} + \frac{p_{\text{atm}}}{\gamma} + z_2$$

$$V_1 = \tfrac{10}{5} = 2 \qquad \text{m/s}$$

$$\frac{2^2}{2(9.8)} + 5 = \frac{V_2^2}{2g} + z_2$$

This equation must be solved for z_2. It is required that the sum $V^2/2g + p/\gamma + z$ at point 2 be the same as at point 1. This sum is often termed the *total head* for an incompressible fluid. (The term *Bernoulli constant* is also used.) V_2 and z_2 are related by the requirement for volume conservation, for an incompressible fluid:

$$V_1 z_1 = V_2(z_2 - H)$$

$$V_2 = \frac{5(2)}{z_2 - 0.5}$$

However, it is useful not to use z as the vertical coordinate. Rather, we shall use the elevation of the free surface with respect to the channel's bottom. Therefore, the fluid's depth d will be introduced into the problem; then $z_2 = d_2 + H$. The Bernoulli equation can then be written as

$$\frac{2^2}{2(9.8)} + 5 = \frac{100}{2(9.8)(d_2^2)} + (d_2 + 0.5) \tag{5.2.9}$$

This equation can be solved by assuming values for d_2 and plotting the results for the right-hand side as a function of d_2. One then plots on the same axes the known value of the total head given by the left-hand side of the equation. The result of such an effort is shown by the solid line in Fig. 5.2.9, along with the dashed vertical line. The intersections of these two curves are solutions, because the left side of Eq. (5.2.9) equals the right at these points.

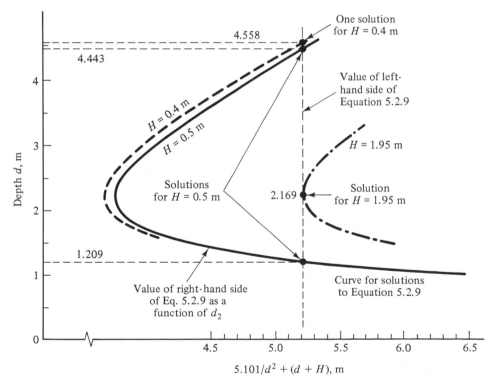

Figure 5.2.9 Representation of depth d of liquid flowing over hump of height H in a horizontal channel. Solid curve: curve for right-hand side of Eq. (5.2.9) for $H = 0.5$ m. Dashed vertical line: value of left-hand side of Eq. (5.2.9). This is the initial total head.

The fact that two solutions are seemingly possible in Example 5.2.3 appears curious. One would expect a unique solution. This would be the case if an additional criterion were introduced. This and related topics will be discussed in Chap. 15, where the flow in open channels is discussed; for now, we can gain some insight into the matter by finding the solution to the problem at the point where the hump's height is 0.4 m. This would correspond to some location between points 1 and 2. The answers would be given by the dashed curve in Fig. 5.2.9 for which one value of depth d is 4.558 m. Thus, as the hump increases in height from 0 to 0.5 m, the depth of the water over the hump decreases continuously from 5 m to 4.443 m. The other solution, 1.209 m, for $H = 0.5$ would be encountered only—if at all—if the hump's apex were to be very much higher, say, 1.95 m; then, as the hump receded to an elevation of 0.5 m, the solution for the depth could be 1.209 m. If the hump were this 1.95 m, the solution would be that indicated by the dot-dash curve in Fig. 5.2.9. For such a hump the solution curve is tangent to the vertical line whose value represents the upstream total head. (A depth of 2.169 m would be indicated.) Once the hump is high enough that this critical condition is achieved, then it would be possible for the high-velocity, low-depth flow—the lower portion of the solid curve—to occur downstream of this apex. Since this was not the case in Example 5.2.3, the depth of 4.443 m, and not 1.209 m, is the result that we sought for the hump of 0.5 m under the conditions that were given.

The next example deals with the rise of pressure at the tip of a body that is associated with the impact of fluid against this point. Such a situation arises when still air is struck by the moving airspeed indicator of an airplane. A comparable situation would occur when a particle of air in a blowing wind strikes the tip of a weather instrument. This air can be assumed to be moving at a uniform speed over considerable distances in the neighborhood of the meteorological instrument. (With respect to the pilot and the moving airplane, the air is moving toward the aircraft at a uniform speed.) The device that is discussed in the example that follows is appropriately called an *impact tube*. The term *Pitot tube* is also used. A design modification to this device allows two pressure measurements to be made with the same device, which is then called a *Pitot-static tube*. These devices can be analyzed effectively with the Bernoulli equation, because experience shows that the fluid changes its speed from that of the free stream to zero at the probe's tip in a short distance, about 10 probe radii. Hence, there is not much distance over which shear stresses can act in order to alter markedly the predictions of the (inviscid) Bernoulli equation.

Example 5.2.4

Apply the Bernoulli equation to the case of the flow in a conduit such as the one illustrated in Fig. 5.2.10. The fluid is water, and there is a nonconstant axial velocity distribution. A small tube is inserted into the stream to probe the cross section for the pressure at point 2. The impact tube is connected to a manometer, which, in turn, is connected to a "wall tap" slightly upstream of the tip of the probe.

Solution   Such an instrument can be used to determine the velocity profile in the conduit. For example, if the tip of the tube is at some location within the pipe, say, point 2, then if point 1 (where the fluid speed is to be determined) is not far distant from the tip of the probe, there will be insignificant viscous effects. We *assume* this to be the case. The application of Eq. (5.2.6) between points 1 and 2 on the same streamline in Fig. 5.2.10 gives

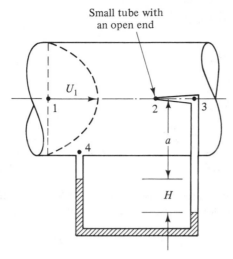

Small tube with an open end

Figure 5.2.10 Using a Pitot, or impact, tube in a duct to determine the velocity of the flowing fluid. $\gamma_{\text{manometer}} = (\text{SG})\gamma_{H_2O}$.

$$p_2 - p_1 = \rho\frac{(U_1^2 - 0^2)}{2}$$

in which p_2 is the pressure at the tip of the probe where the fluid is at rest. If p_2 and p_1 were known, or their difference, then U_1 would be known. The use of Newton's laws, in the vertical direction, provides this pressure difference. Because the flow is specified to be horizontal, the pressure varies from point 1 to 4 hydrostatically. Thus, we can verify that

$$p_1 + a\gamma_{H_2O} + H(\text{SG})(\gamma_{H_2O}) - H\gamma_{H_2O} - a\gamma_{H_2O} = p_3 = p_2$$

Note that the location of the probe (distance a) is unimportant. Solving for U_1 in terms of H, γ_{H_2O}, and $(\text{SG})\gamma_{H_2O}$, one obtains

$$U_1 = \sqrt{2g[(\text{SG}) - 1]H} \tag{5.2.10}$$

In doing this problem, we wrote the Bernoulli equation for two points on the *same* streamline. Had we carelessly used points 4 and 2 in the equation, an incorrect solution would have been the result.

The following example is one in which the free-surface elevation changes considerably as the liquid flows under a gate. There is some decrease in the total head in the actual process, but it will be slight when compared with the large changes of potential and kinetic energy. For this reason, we can apply the Bernoulli equation to the situation. In doing so we shall recognize that the small dissipative mechanisms will give a slightly different experimental result from the one that we are predicting with an equation that neglects viscous effects. We shall *assume,* when using the Bernoulli equation, that the fluid particles immediately adjacent to a solid boundary can slip relative to the boundary. Later, we shall modify this assumption and invoke the no-slip condition so that our present results do not apply in a minute layer next to the boundary.

Example 5.2.5

Apply Eq. (5.2.7) to analyze the flow under the sluice gate shown in Fig. 5.2.11. *Assume* that the flow is horizontal and independent of depth at points far upstream and downstream (e.g., 1 and 2 as well as 5 and 6).

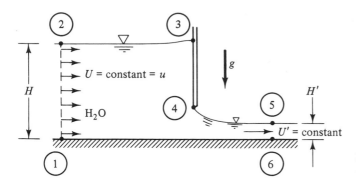

Figure 5.2.11 Flow of a liquid under a sluice gate in a horizontal channel.

Solution Points 2, 3, 4, and 5 are on the same streamline. On this streamline, $V^2/2g + p/\gamma + z$ will be constant for an inviscid fluid. At point 2,

$$\frac{V^2}{2g} + \frac{p}{\gamma} + z = \frac{U_2^2}{2g} + \frac{(p = 0 \text{ gauge*})}{62.4} + H$$

and at point 3, $V = 0$; hence,

$$\frac{V^2}{2g} + \frac{p}{\gamma} + z = \frac{0^2}{2g} + \frac{0}{62.4} + \text{elevation}_3$$

At point 4, the speed is unknown. The pressure is 1 atm. Thus

$$\frac{V^2}{2g} + \frac{p}{\gamma} + z = \frac{V_4^2}{2g} + \frac{0}{\gamma} + (\text{height of opening})$$

The right-hand side of this expression has only one known quantity: the pressure, which is zero gauge pressure. At point 5, V = desired speed of fluid, U', so that the terms in the Bernoulli equation are

$$\frac{V^2}{2g} + \frac{p}{\gamma} + z = \frac{U'^2}{2g} + \frac{0}{\gamma} + (\text{water depth})$$
$$= \frac{U'^2}{2g} + 0 + H'$$

Now the conditions at point 2 and 5 can be related, because they are on the same streamline. Consequently,

$$\frac{V_2^2}{2g} + H_2 = \frac{V_5^2}{2g} + H_5$$

But V_2, H_2, V_5, and H_5 are also related by the continuity equation. For a channel of constant width between 2 and 5, we have

$$V_2(H_2) = (V_5)H_5$$

or, in terms of the values at 2 and 5,

$$UH = U'H'$$

This allows us to write the Bernoulli equation as

$$\frac{V_2^2}{2g} + H_2 = \frac{V_2^2}{2g}\left(\frac{H_2}{H_5}\right)^2 + H_5$$

* The pressure must be in units consistent with γ so that the terms $V^2/2g$, p/γ, and z all have the same units.

or, using the symbolic values at 2 and 5,

$$\frac{U^2}{2g} + H = \frac{U^2}{2g}\left(\frac{H}{H'}\right)^2 + H' \tag{5.2.11}$$

from which one can solve for H' for prescribed values of U and H.

We shall now focus our attention on the streamline between points 1 and 6 in Fig. 5.2.11. Because the flow is horizontal at the upstream station, there is no vertical acceleration there. Thus, despite the fact that the fluid is in motion in Example 5.2.5, the pressure variation in the vertical direction is the same as a hydrostatic variation. At point 1, we have $z = 0$, by choice of a datum for elevation. This means that $V^2/2g + p/\gamma + z$ can be written as

$$\frac{U^2}{2g} + \frac{p_2 + H\gamma}{\gamma} + 0$$

and at point 6, $V^2/2g + p/\gamma + z$ becomes

$$\frac{U'^2}{2g} + \frac{p_5 + H'\gamma}{\gamma} + 0$$

The pressures p_2 and p_5 are zero gauge. Then the Bernoulli equation between points 1 and 6 on the same streamline is

$$\frac{U^2}{2g} + H = \frac{U'^2}{2g} + H'$$

This result is the same as that given in Example 5.2.4 but obtained for a *different* streamline. This means that in this *particular* case, $V^2/2g + p/\gamma + z$ is the *same* constant along two streamlines. If this is so for all streamlines, the flow field is termed *irrotational*. This very special characteristic of the velocity field was mentioned in Sec. 4.2; its consequence will be discussed in Sec. 5.3.

The reader is encouraged to extend this work by determining the fluid's elevation at point 3. Justify that $\mathbf{V} = 0$ is here. (*Ans.:* $H_3 = H + U^2/2g$) The locations where $\mathbf{V} = 0$ are called *stagnation points*.

5.3 THE BERNOULLI EQUATION AND IRROTATIONAL FLOW

The application of Newton's laws of motion in the streamline direction yielded a very useful result. If the fluid is *inviscid* and the flow *steady,* the sum

$$\frac{V^2}{2} + \frac{p}{\rho} + gz = \text{constant for a given streamline} \tag{5.3.1}$$

We can learn something about how the numerical value of the constant varies from streamline to streamline if we refer to Fig. 5.2.1. We shall now concentrate on the direction that is perpendicular to the streamline, the n direction. We shall need to employ

$$\Sigma F_n = ma_n \tag{5.3.2}$$

The acceleration a_n is determined by considering the fundamental definition of acceleration, and referring to Fig. 5.2.1. Thus

$$a_n = \frac{(V_n)_{O'} - (V_n)_O}{\Delta t}$$

$$= \frac{\left(V_s + \frac{\partial V_s}{\partial s} ds'\right) \sin\alpha - 0}{ds'/V_s} = +\frac{V_s^2}{ds'} \sin\alpha + \left(\begin{array}{c} \text{higher-order} \\ \text{terms in } ds' \end{array}\right)$$

$$= +\frac{V_s^2}{ds'}(\alpha) \qquad \text{for small } \alpha$$

But in Fig. 5.2.1, angle α is ds'/R, so that

$$a_n = \frac{V_s^2}{R}$$

[Compare this result with Eq. (2.8.3).] In a manner akin to that which led to Eqs. (5.2.2) and (5.2.3), we can show that

1. The net surface force in the positive n direction is $-(\partial p/\partial n)\, dn\, ds$, and
2. The body force in the positive n direction is $-\rho g\, dn\, ds\, \cos\theta$ or $-\rho g\, dn\, ds\, (\partial z/\partial n)$.

These expressions for a_n and the forces in the n direction permit us to write Eq. (5.3.2) as

$$-\frac{\partial p}{\partial n} - \rho g \frac{\partial z}{\partial n} = \rho \frac{V_s^2}{R} \tag{5.3.3}$$

This equation differs from Eq. (5.2.4). However, it can be rewritten as

$$-\frac{\partial}{\partial n}\left(\frac{p}{\rho} + gz + \frac{V_s^2}{2}\right) = 0 \tag{5.3.4}$$

for ρ = constant *if* a justifiable reason can be found to write

$$\frac{V_s}{R} = \frac{\partial V_s}{\partial n} \tag{5.3.5}$$

Equation (5.3.4) means that

$$\left(\frac{p}{\rho} + gz + \frac{V_s^2}{2}\right) = \text{constant} \tag{5.3.6}$$

as one moves *perpendicular* to a streamline. We have seen previously that Eq. (5.3.6) is true *along* a streamline for certain assumed conditions. This means that *if* Eq. (5.3.5) applies to an entire flow field, the equation

$$\frac{p}{\rho} + gz + \frac{V_s^2}{2} = \text{constant}$$

applies not only along a streamline but also perpendicular to a streamline. Consequently, the Bernoulli equation would have the same constant at all points of the field for such a special situation.

We shall now see that Eq. (5.3.5)—and consequently Eq. (5.3.4)—is satisfied when the rate of rotation of each fluid particle about its center of mass as it translates

The Flow of an Inviscid Fluid   Chap. 5

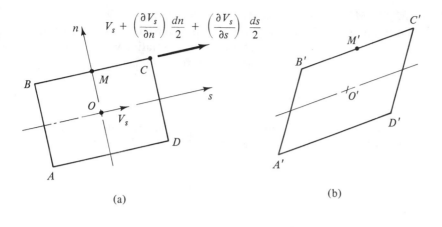

(a)

(b)

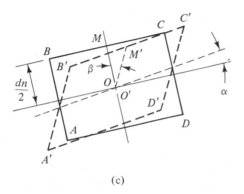

(c)

Figure 5.3.1 (a) Infinitesimal material volume (fluid particle) flowing in the streamline direction, s. (b) Distorted material volume that has moved from O to O' along a streamline. (c) Superposition of initial and final shapes of the fluid particle to show the rotation of the particle about its own axis as it translates along the streamline.

along a path* is zero. In Fig. 5.3.1, we can observe that the fluid particle $ABCDA$ may have distorted slightly in moving to point O'. Its configuration is delineated by the shape $A'B'C'D'A'$ in this figure. It can be seen that the line which was parallel to the streamline at point O has rotated, counterclockwise, by the amount α. Because $\alpha = ds'/R$ and $ds' = V_s(\Delta t)$, we have that $\alpha/\Delta t = V_s/R$. Furthermore, we shall designate $\alpha/\Delta t$ as $\dot{\alpha}$, so that

$$\dot{\alpha} = \left\{ \begin{array}{c} \text{rate of } \textit{counterclockwise} \text{ rotation of the line parallel} \\ \text{with the original streamline direction} \end{array} \right\} = \frac{V_s}{R}$$

A line that was originally perpendicular to the streamline direction at point O in Fig. 5.2.1 can rotate also. This rate of rotation need not be the same as that just found. When the particle arrives at point O', this line, OM in Fig. 5.3.1a, need not have remained perpendicular to the streamline direction. Its rate of rotation will now be examined.

Figure 5.3.1c shows the shape of a particle that was originally at point O in Fig. 5.3.1a, superimposed on the shape of the particle when it arrives at point O', shown in

---

* Such a rotation is inherent in any motion. We have shown in Chap. 4 that any displacement of a deformable body can be decomposed into a rigid-body rotation, a distortion, and an implicit translation. The distortion term can consist of a volume change as well as angular distortions. The present treatment follows the spirit of Sec. 4.2 but uses s and n coordinates instead of x and y.

Fig. 5.3.1b. Points M and M' need not coincide, even though points O and O' were superimposed. This is because the distensible fluid body may have been moving so that point M would be moving relative to point O. The amount of relative velocity of point M, in the direction of the streamline, and with respect to point O, is

$$(V_s)_{M \text{ relative to } O} = \frac{\partial V_s}{\partial n} \frac{dn}{2}$$

This means that in an interval of time dt, point M has increased its s distance from point O by the amount MM', which can be written, except for higher-order terms, as

$$MM' = \frac{\partial V_s}{\partial n} \frac{dn}{2} dt$$

In view of this, the clockwise angle β is

$$\beta = \frac{MM'}{dn/2} = \frac{\partial V_s}{\partial n} dt$$

so that

$$-\dot{\beta} = \begin{Bmatrix} \text{rate of } \textit{counterclockwise} \text{ rotation of the} \\ \text{line perpendicular to a streamline} \end{Bmatrix} = -\frac{\partial V_s}{\partial n}$$

The minus sign used in this last expression is necessary because $\dot{\beta}$ is a clockwise rate of rotation. Consequently, the average rate of rotation of the s and n axes in the counterclockwise direction is $(\dot{\alpha} - \dot{\beta})/2$, or $(V_s/R - \partial V_s/\partial n)/2$. For an irrotational flow, this quantity is zero, by definition. Indeed, had we not used the coordinates along, and normal to, the streamline, we would have set ω, defined by Eqs. (4.2.3) and (4.2.4), equal to zero—as it must be, for an irrotational flow. Hence, the condition to make Eq. (5.3.5) valid is that the flow field be *irrotational*.

The following examples are given to emphasize the need to utilize the Bernoulli equation along a streamline if the flow is not irrotational. For those flows that can be considered to be irrotational, the ability to employ the added scope of the Bernoulli equation can be useful.

Example 5.3.1

The sluice gate treated in Example 5.2.5 had a uniform flow upstream and downstream. The former was given as $u = U = $ constant. If this sole component of the upstream velocity is used in the equations for fluid rotation, the only component of ω which is not trivially zero would be $\omega_y = \frac{1}{2}(\partial u/\partial z - \partial w/\partial x)$, because $v = 0 = w$. Since u is a constant at the upstream station, ω_y is also zero. This leads to the conclusion that the flow is irrotational upstream of the gate. It is this irrotationality that was responsible for the result in the discussion following Example 5.2.5. There, $V^2/2g + p/\gamma + z$ was shown to be equal to the same constant, $U^2/2g + H$, along two different streamlines. In the next example, we shall deal with a situation where this is not so.

Example 5.3.2

Assume that instead of a velocity profile that is uniform upstream of a sluice gate, the velocity is as illustrated in Fig. 5.3.2. While ω_x and ω_z are identically zero from their definitions in Eqs. (4.2.3) and (4.2.4), ω_y is not zero. It is, according to Eq. (4.2.4),

$$\omega_y = \frac{1}{2}\left(\frac{\partial u}{\partial z} - \frac{\partial w}{\partial x}\right)$$

The Flow of an Inviscid Fluid   Chap. 5

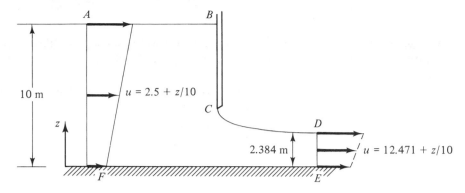

Figure 5.3.2 Flow under a sluice gate of a liquid having nonuniform velocity distribution.

which becomes

$$\omega_y = \frac{1}{2}\left[\frac{\partial(2.5 + z/10)}{\partial z} - \frac{\partial(0)}{\partial x}\right] = \frac{1}{20}$$

For this instance, a similar calculation at the downstream station would show that $\omega_y = \frac{1}{20}$ there also. The velocity given there also satisfies mass conservation at a point, and it yields the same total flow rate as that which exists upstream of the gate. Since ω_y is not zero, even though the other two components of rotation are, we now realize that we can apply the Bernoulli equation only between points on the same streamline, such as A and D or F and E. (Do so and compare the results.)

Example 5.3.3

It will be pointed out in Sec. 5.4 that the circular flow $u_\theta = A/r$ (with $u_r = 0$) is irrotational. This can be verified by determining ω_z from the equivalent Cartesian form of this velocity,

$$u = -\frac{Ax}{x^2 + y^2} \quad \text{and} \quad v = \frac{Ay}{x^2 + y^2}$$

This implies that the Bernoulli equation could be used between points on two different streamlines, in this case, one point at $r = r_1$ and the other at $r = r_2$. The Bernoulli equation would then read

$$\frac{1}{2g}\frac{A^2}{r_1^2} + z_1 + \frac{p_1}{\gamma} = \frac{1}{2g}\frac{A^2}{r_2^2} + z_2 + \frac{p_2}{\gamma}$$

If one chose to model the swirling flow into a drain, such as that of a bathtub, the pressure on the free surface would be everywhere the same, so that $p_1 = p_2 = p_{atm}$. If point 2 were far from the drain (i.e., if $r_2 \rightarrow \infty$), the result would be

$$z_\infty - z_1 = \frac{1}{2g}\frac{A^2}{r_1^2} \tag{5.3.7}$$

This shows that as r_1 decreases toward the axis of the swirl, the surface elevation z_1 must also decrease. (The origin would be excluded, however.)

5.4 SOME IRROTATIONAL FLOWS

We have seen that the Bernoulli equation can provide valuable information when it is applied between two points along the same streamline. Furthermore, it has been pointed out that, under the condition of irrotational flow, the Bernoulli equation applies between points on different streamlines. Because such a possibility might lead to additional and useful results, it is natural to inquire about the nature of irrotational flows.

Rotationality in a flow has been defined in Chap. 4 as a vector quantity that has three components; in Cartesian coordinates, they are

$$\omega_x = \frac{1}{2}\left(\frac{\partial w}{\partial y} - \frac{\partial v}{\partial z}\right)$$

$$\omega_y = \frac{1}{2}\left(\frac{\partial u}{\partial z} - \frac{\partial w}{\partial x}\right) \tag{5.4.1}$$

and

$$\omega_z = \frac{1}{2}\left(\frac{\partial v}{\partial x} - \frac{\partial u}{\partial y}\right)$$

These three equations are also related to the components which result from "operating" on the velocity vector, $\mathbf{V} = u\mathbf{i} + v\mathbf{j} + w\mathbf{k}$, with the "curl" operator. This operator produces another vector that is defined in the following way:

$$\text{curl } \mathbf{V} = \left(\frac{\partial w}{\partial y} - \frac{\partial v}{\partial z}\right)\mathbf{i} + \left(\frac{\partial u}{\partial z} - \frac{\partial w}{\partial x}\right)\mathbf{j} + \left(\frac{\partial v}{\partial x} - \frac{\partial u}{\partial y}\right)\mathbf{k} \tag{5.4.2}$$

We note that the components of the rotation vector $\boldsymbol{\omega}$ are $\frac{1}{2}$ times the components of curl $\mathbf{V}$. (The vector curl $\mathbf{V}$ itself is called the *vorticity vector*.) It also can be verified that the form of curl $\mathbf{V}$ can be rewritten, for convenience, as

$$\text{curl } \mathbf{V} = \begin{vmatrix} \mathbf{i} & \mathbf{j} & \mathbf{k} \\ \dfrac{\partial}{\partial x} & \dfrac{\partial}{\partial y} & \dfrac{\partial}{\partial z} \\ u & v & w \end{vmatrix} \tag{5.4.3}$$

If one desires to consider irrotational flows, one requires each of the components in Eq. (5.4.2) to be zero. This is the same as writing

$$\text{curl } \mathbf{V} = 0 \tag{5.4.4}$$

There are many flows for which Eq. (5.4.4) is satisfied.

The case of $\mathbf{V} = U\mathbf{i}$ is one case, and $\mathbf{V} = [-Ax/(x^2 + y^2)]\mathbf{i} + [Ay/(x^2 + y^2)]\mathbf{j}$ from Example 5.3.3 is another. The latter example results from writing the x and y velocity components for a flow that has circular streamlines about the origin such that $u_\theta = A/r$, when written in plane polar coordinates. [We note in passing that another flow pattern with circular streamlines, $u_\theta = \Omega r$, does not have the property curl $\mathbf{V} = 0$. However, in this section we shall concentrate on flows for which Eq. (5.4.4) is guaranteed to apply.] How can $\mathbf{V}$ be specified so that the applicability of Eq. (5.4.4) is satisfied?

One needs some way to specify $\mathbf{V}$ such that the operation of curl $\mathbf{V}$ will always yield zero. A result from vector differentiation gives us the needed clue. For this reason we shall now examine the mathematical concept called the *gradient operator*. The gradient of a scalar function, such as temperature, $T(x, y, z, t)$, is a vector which is defined as

$$\text{grad } T = \frac{\partial T}{\partial x}\mathbf{i} + \frac{\partial T}{\partial y}\mathbf{j} + \frac{\partial T}{\partial z}\mathbf{k}, \quad \text{in Cartesian coordinates}$$

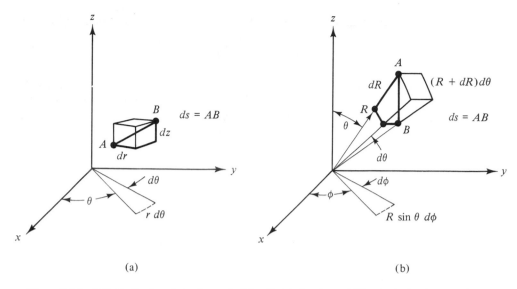

Figure 5.4.1 Infinitestinal spatial volumes in (a) cylindrical polar and (b) spherical polar coordinates.

The differentiation is performed with respect to the arguments that are the units of length in the chosen coordinate system. For example, the arc length ds is given in Cartesian coordinates as

$$ds^2 = dx^2 + dy^2 + dz^2$$

In cylindrical coordinates (r, θ, z), the arc length is, according to Fig. 5.4.1a,

$$ds^2 = (dr)^2 + (r\,d\theta)^2 + (dz)^2$$

and in spherical polar coordinates (R, θ, ϕ), one has from Fig. 5.4.1b,

$$ds^2 = (dR)^2 + (R\,d\theta)^2 + (R\,\sin\theta\,d\phi)^2$$

Hence, in a spherical polar coordinate system, the gradient of the scalar Φ is*

$$\text{grad}\,\Phi = \frac{\partial\Phi}{\partial R}\mathbf{i}_R + \frac{1}{R}\frac{\partial\Phi}{\partial\theta}\mathbf{i}_\theta + \frac{1}{R\,\sin\theta}\frac{\partial\Phi}{\partial\phi}\mathbf{i}_\phi$$

in which $\mathbf{i}_R$, $\mathbf{i}_\theta$, and $\mathbf{i}_\phi$ are the unit base vectors in the R, θ, and ϕ directions, respectively.

The reason for introducing the gradient of a scalar is the result that always occurs when one performs the operation

$$\text{curl}\,(\text{grad}\,\Phi)$$

in which Φ is a scalar function of (x, y, z, t). The result is zero, as we shall immediately demonstrate. If we define $\mathbf{V} = \text{grad}\,\Phi$, or

$$\mathbf{V} = \frac{\partial\Phi}{\partial x}\mathbf{i} + \frac{\partial\Phi}{\partial y}\mathbf{j} + \frac{\partial\Phi}{\partial z}\mathbf{k} = u\mathbf{i} + v\mathbf{j} + w\mathbf{k} \tag{5.4.5}$$

---

*Note that the uppercase symbol Φ is a scalar function and the lowercase symbol ϕ is the azimuthal angle (i.e., longitude).

and insert the resulting $\partial\Phi/\partial x$, $\partial\Phi/\partial y$, and $\partial\Phi/\partial z$ into Eq. (5.4.3), we find that

$$\text{curl}\,(\text{grad}\,\Phi) = \text{curl}\left(\frac{\partial\Phi}{\partial x}\mathbf{i} + \frac{\partial\Phi}{\partial y}\mathbf{j} + \frac{\partial\Phi}{\partial z}\mathbf{k}\right) = 0 \qquad (5.4.6)$$

The zero as the result of curl (grad Φ) will be apparent if $\partial\Phi/\partial x$, $\partial\Phi/\partial y$, and $\partial\Phi/\partial z$ are substituted for u, v, and w in Eq. (5.4.3). (Try it and see!) The result of Eq. (5.4.6) means that if we define a scalar function Φ by

$$u = \frac{\partial\Phi}{\partial x}$$

$$v = \frac{\partial\Phi}{\partial y} \qquad (5.4.7)$$

$$w = \frac{\partial\Phi}{\partial z}$$

we shall obtain a velocity field that is irrotational. This will be true except, perhaps, at points where the functions are singular (i.e., become infinite in magnitude). This is the procedure for which we searched.

We are now inclined to ask, "Which functions $\Phi(x, y, z, t)$ are useful to us?" The word *useful* is subject to interpretation, but we know that a given velocity field, such as $\mathbf{V} = \text{grad}\,\Phi$, can be associated with an incompressible fluid only if div $\mathbf{V} = 0$, which is written in Cartesian coordinates as

$$\frac{\partial u}{\partial x} + \frac{\partial v}{\partial y} + \frac{\partial w}{\partial z} = 0 \qquad (5.4.8)$$

With u, v, and w defined by Eq. (5.4.7) for an irrotational flow, we see that incompressibility as expressed in Eq. (5.4.8) requires that

$$\frac{\partial}{\partial x}\left(\frac{\partial\Phi}{\partial x}\right) + \frac{\partial}{\partial y}\left(\frac{\partial\Phi}{\partial y}\right) + \frac{\partial}{\partial z}\left(\frac{\partial\Phi}{\partial z}\right) = 0$$

This last result can be written as

$$\frac{\partial^2(\Phi)}{\partial x^2} + \frac{\partial^2(\Phi)}{\partial y^2} + \frac{\partial^2(\Phi)}{\partial z^2} = 0$$

or

$$\left(\frac{\partial^2}{\partial x^2} + \frac{\partial^2}{\partial y^2} + \frac{\partial^2}{\partial z^2}\right)(\Phi) = 0$$

Sometimes one defines $\nabla^2 = \partial^2/\partial x^2 + \partial^2/\partial y^2 + \partial^2/\partial z^2$ and calls it the *Laplace operator*. For Φ to be an acceptable function, in the sense that an irrotational and incompressible flow field will ensue, we require that Φ satisfy

$$\nabla^2\Phi = 0 \qquad (5.4.9)$$

Equation (5.4.9), the *Laplace equation*, has interesting consequences, because it is a linear equation. If one has two functions Φ_1 and Φ_2, both of which satisfy Eq. (5.4.9), then one can show that the function $\Phi = \Phi_1 + \Phi_2$ also satisfies Eq. (5.4.9). This is so because we have stipulated that $\nabla^2\Phi_1 = 0$ and $\nabla^2\Phi_2 = 0$. Because the Laplace operator is linear, we can easily show that

$$\nabla^2\Phi = \nabla^2(\Phi_1 + \Phi_2) = \nabla^2\Phi_1 + \nabla^2\Phi_2 = 0 + 0 = 0$$

The result need not be limited to only Φ_1 and Φ_2. Any number of Φ's can be used. Hopefully, the functions Φ_i $(i = 1, 2, \ldots)$ that are used and combined will give a worthwhile result. We shall now examine a few solutions of the Laplace equation that are relevant to fluid mechanics.

1. The function

$$\Phi_1 = Ux \tag{5.4.10}$$

is the velocity potential for a *uniform* flow in the $+x$ direction, because

$$u = \frac{\partial \Phi_1}{\partial x} = U \quad \text{and} \quad v = \frac{\partial \Phi_1}{\partial y} = 0 \quad w = 0$$

2. The function

$$\Phi_2 = A (\ln r) \quad \text{for} \quad r = \sqrt{x^2 + y^2} \tag{5.4.11}$$

is the potential for a *two-dimensional source,* because

$$u_r = \frac{\partial \Phi_2}{\partial r} = \frac{A}{r} \quad \text{and} \quad u_\theta = \frac{\partial \Phi_2}{r \, \partial \theta} = 0$$

This radial velocity—in the xy plane—is constant on a circle of radius r. The flow across a control surface that is a circular cylinder having a unit length in the z direction is $\int u_r r \, d\theta$. Hence,

$$Q = (\text{volume per unit length in } z \text{ direction}) = \int_0^{2\pi} \frac{A}{r} r \, d\theta$$

This yields $A = Q/2\pi$, and A is called the *strength* of the source. Of course, $A > 0$ for a source. [The term *line source* is also used for the flow given by Eq. (5.4.11). The point from which the flow appears to emanate is on the z axis. This axis is the line referred to in the term *line source.* This two-dimensional flow is in contrast to a three-dimensional—or point—source, to be given by Eq. (5.4.14).]

Example 5.4.1

For $\Phi = 7[\ln (x^2 + y^2)^{1/2}] = 3.5[\ln (x^2 + y^2)]$,

$$u = \frac{\partial \Phi}{\partial x} = \frac{3.5(2x)}{x^2 + y^2} \quad \text{and} \quad v = \frac{3.5(2y)}{x^2 + y^2}$$

The slope of the streamline, v/u, becomes y/x. This is the slope of a line through the origin. (Note the singularity at the origin.)

3. The function

$$\Phi_3 = -B (\ln r), \quad B > 0 \tag{5.4.12}$$

is a *two-dimensional, or line, sink,* because only u_r is nonzero. It is negative, so that a sink is, in effect, a negative source.

4. The function

$$\Phi_4 = C\theta \tag{5.4.13}$$

is a *two-dimensional vortex,* because

$$u_\theta = \frac{\partial \Phi_4}{r \, \partial \theta} = \frac{C}{r} \quad \text{and} \quad u_r = 0$$

A source and a vortex can be combined to form a spiral flow.

5. The function

$$\Phi_5 = -\frac{D}{R} \quad D > 0 \quad \text{and} \quad R = (x^2 + y^2 + z^2)^{1/2} \tag{5.4.14}$$

is a *three-dimensional,* or *point,* source because

$$u_R = \frac{\partial \Phi_5}{\partial R} = \frac{D}{R^2}$$

This radial velocity, with "point symmetry" about the origin, is constant on a spherical control surface of radius R. The flow across this surface would be the product of this velocity and the surface area of a sphere, $4\pi R^2$. Thus, $Q = (D/R^2)(4\pi R^2)$, and $D = Q/4\pi$, the *strength* of the point source.

It is to be expected that the list of functions which satisfy $\nabla^2 \Phi = 0$, and which can give interesting flow patterns, could be very long. The few that have been listed can give us a hint of their importance.

If Φ_1 and Φ_2 are combined to form

$$\Phi = Ux + A[\ln (x^2 + y^2)^{1/2}] \tag{5.4.15}$$

one has a source in a uniform flow. This gives

$$\mathbf{V} = \frac{\partial \Phi}{\partial x}\mathbf{i} + \frac{\partial \Phi}{\partial y}\mathbf{j} = u\mathbf{i} + v\mathbf{j}$$

so that

$$u = U + \frac{Ax}{x^2 + y^2}$$

and

$$v = \frac{Ay}{x^2 + y^2}$$

One notes that $u = u_1 + u_2 = \partial\Phi_1/\partial x + \partial\Phi_2/\partial x, \ldots$ so that the velocity components are a linear combination of the components from the various potential functions, Φ_1, Φ_2, and so on. This is illustrated in Fig. 5.4.2.

Figure 5.4.2 also shows some of the velocity components at different points. For large values of $r = (x^2 + y^2)^{1/2}$, the effect of the source is small and the flow is nearly that which is due to the uniform flow U alone. However, as r decreases, the velocity of the source becomes increasingly strong, so that at some point on the negative x axis there is a flow to the left from the source that has the same magnitude as the uniform flow U, which is moving to the right. The result is $u = 0$. Because the x axis is a line of symmetry for this case, $v = 0$. This means that at some point on the x axis, $\mathbf{V} = 0$, and this will be a stagnation point. One could, as we expect, find the streamlines from the velocity field. These streamlines have been sketched for the case under examination in Fig. 5.4.2. The streamline through the stagnation point, for this case, divides the flow into two regimes. One part contains fluid from the uniform field, and the other has the

The Flow of an Inviscid Fluid   Chap. 5

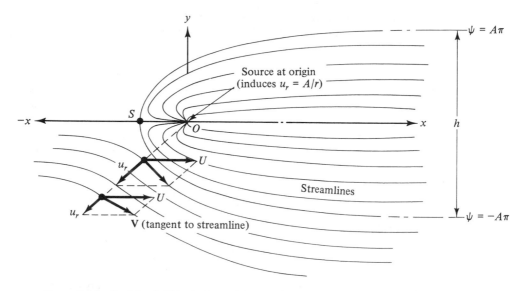

Figure 5.4.2 Rankine half body formed by a uniform flow and a line (two-dimensional) source. Generally: $\mathbf{V}$ = vector sum of U and u_r. Flow at $x \to -\infty$: $\mathbf{V} \to U\mathbf{i}$. $h = 2\pi A/U =$ asymptotic width.

fluid from the source which is being swept downstream. Since no fluid passes across a streamline, the two streamlines $\psi = A\pi$ and $\psi = -A\pi$ appear to delineate a two-dimensional body with stagnation point at the upstream end. The downstream portion goes off to infinity at a constant width, $h = A(2\pi)/U$. (This relationship for the width can be verified in problem 5.68 at the end of this section.) It is often interesting to generate other "body shapes" by various combinations of "sources" and "sinks." However, it is not our present purpose to do so. Rather, we choose to recognize that the list of all possible Φ's such that $\nabla^2 \Phi = 0$ gives what are called *indirect results*. This is to say, one obtains a flow pattern and then must examine it to see if it has some properties, other than incompressibility and irrotationality, that are useful. For example, "What boundary conditions does Φ satisfy?" The following example is a case in point.

Example 5.4.2

We have seen in Chap. 4 (refer to Example 4.7.1) that a stream function in the form

$$\psi = 5r(\sin \theta)\left(1 - \frac{9}{r^2}\right) \tag{5.4.16}$$

gives a streamline array that appears to simulate the flow over a cylinder of radius 3 that is coincident with $\psi = 0$. We shall now discuss the flow generated by this stream function.

    Equations (4.7.4) to (4.7.5) and (4.7.9) to (4.7.10) give the u and v, or u_r and u_θ, components of velocity. If one takes the time to determine the rotation in the flow, $\frac{1}{2}(\partial v/\partial x - \partial u/\partial y)$, the result will be found to be zero. Hence, Eq. (5.4.16) represents an irrotational flow* with the consequence that $V^2/2 + p/\rho + gz$ is a constant throughout the entire flow field. We can determine this constant at any position, including one that is far from the body ($r \to \infty$) where V can be determined to be 5. The pressure there, the

---
* Consequently, a velocity potential also exists. It is $\Phi = Ux[1 + 9/(x^2 + y^2)]$. Is $\nabla^2 \Phi = 0$?

ambient pressure, will be designated p_∞. We shall *assume* the z axis to be perpendicular to the page. This permits the Bernoulli equation to be written for this problem as

$$\frac{5^2}{2} + \frac{p_\infty}{\rho} = \frac{V^2}{2} + \frac{p}{\rho}$$

in which V and p are the velocity and pressure at any point in the flow. The points on the circle $r = 3$ are of particular interest, because this closed streamline can be thought of as a simulation of a circular cylinder.

We can find on $r = 3$ that $u_r = 0$, as anticipated, and that $u_\theta = -2(5) \sin \theta$.

The pressure on the simulated cylinder of radius 3 follows from the Bernoulli equation,

$$\frac{5^2}{2} + \frac{p_\infty}{\rho} = \frac{[2(5) \sin \theta]^2}{2} + \frac{p}{\rho}$$

which leads to

$$\frac{p - p_\infty}{\rho(5^2)/2} = 1 - 4 \sin^2 \theta \qquad (5.4.17)$$

This equation is plotted in Fig. 5.4.3, along with the experimentally determined values of $(p - p_\infty)/(\rho U^2/2)$, which is called the *pressure coefficient*. One observes from this figure that the actual pressure distribution on a cylinder differs from the distribution that results from Eq. (5.4.17). Still, the experimental and analytical results are in reasonable agreement up to $\theta = 60°$. The reason why the pressure curves associated with Eq. (5.4.17) and experiments differ toward the rear of the cylinder is that the simple theoretical model, Eq. (5.4.16), requires that the flow be symmetrical fore and aft of the body. This symmetry is not completely realized in most flows of practical interest. The fluid does not flow along the aft part of the body in the same way it does in the forward part. Hence, Eq. (5.4.16) is only a crude model for the flow.

Until now, we have found functions Φ that satisfied the Laplace equation and examined the flow fields that they represented. If the Φ's produced realistic flow patterns, they would be examined further. This is called the *"indirect"* method because the flow field follows from the function Φ. A "direct" method would start with a given

Figure 5.4.3 Theoretical and measured pressure distributions around a circular cylinder, $(p - p_\infty)/(\rho U_\infty^2/2)$. Dot-dash curve: theoretical distribution from Eq. (5.4.17). Dashed curve: approximate distribution in air for $U_\infty = 1.63$ m/s and $D = 1$ m. Solid curve: approximate distribution in air for $U_\infty = 4.6$ m/s and $D = 1$ m. (Courtesy of Dover Publications, Inc.)

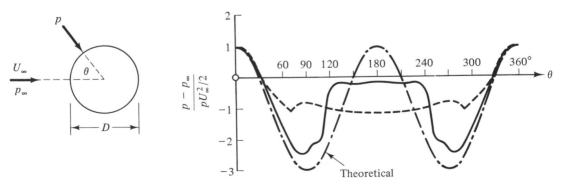

The Flow of an Inviscid Fluid   Chap. 5

flow field and determine the function Φ. In solving the "direct" problem, the equation $\nabla^2 \Phi = 0$ will have to be satisfied within the geometric region of interest. The associated boundary conditions for that region will also affect the function Φ that must be determined. At a solid boundary with inward normal (outward for the fluid which is in contact with the boundary), we require that the fluid not penetrate the boundary. This requires $\mathbf{V} \cdot \mathbf{n} = 0$ at the boundary.*

★5.5 FLOW NETS

A solution to the problem of finding the proper function $\Phi(x, y, z, t)$ to model a particular flow may be easier to obtain if some of the general characteristics of Φ are more evident. This function will be a constant for a certain set of values x and y in a steady, two-dimensional flow. If we restrict ourselves to a two-dimensional case, the assumed continuity of Φ allows us to write

$$d\Phi = \frac{\partial \Phi}{\partial x} dx + \frac{\partial \Phi}{\partial y} dy \qquad (5.5.1)$$

On lines where $\Phi = $ constant, $d\Phi = 0$, so that

$$0 = \frac{\partial \Phi}{\partial x} dx + \frac{\partial \Phi}{\partial y} dy \qquad (5.5.2)$$

is the differential equation for the lines of $\Phi = $ constant. Now $u = \partial \Phi / \partial x$ and $v = \partial \Phi / \partial y$, and so Eq. (5.5.2) becomes

$$\frac{dy}{dx} = \frac{-\partial \Phi / \partial x}{\partial \Phi / \partial y} = \frac{-u}{v} \qquad (5.5.3)$$

The slope of the $\Phi = $ constant lines, which is given in Eq. (5.5.3), is the negative reciprocal of the slope of a streamline v/u. This means that the lines $\Phi = $ constant are orthogonal to the streamlines, as shown in Fig. 5.5.1.

Figure 5.5.1 also shows something else about lines of $\Phi = $ constant and $\psi = $ constant. Two Φ lines which differ by a value $\Delta \Phi = K$ appear to form two opposite sides of a squarelike shape whose other two sides are ψ lines which differ in value by the same value K. It is for this reason that we say that the Φ and ψ lines form *curvilinear squares*. This collection of "squares" is called a *flow net*. The fact that the figures have equal median lines can be easily demonstrated.

At point P in the inset to Fig. 5.5.1,

$$\operatorname{grad} \psi = \left(\frac{\partial \psi}{\partial x}\right)_P \mathbf{i} + \left(\frac{\partial \psi}{\partial y}\right)_P \mathbf{j}$$

and

$$\operatorname{grad} \Phi = \left(\frac{\partial \Phi}{\partial x}\right)_P \mathbf{i} + \left(\frac{\partial \Phi}{\partial y}\right)_P \mathbf{j}$$

---

* The Laplace equation, $\nabla^2 \Phi = 0$, occurs often in the study of physical processes. For steady-state heat conduction in a solid material with a thermal conductivity which is independent of direction, the Fourier law of heat conduction yields $\nabla^2 T = 0$ as the governing equation. Boundary conditions could be the temperatures which are imposed on the surface, and/or amount of heat transferred at the exterior surfaces. The latter condition involves the normal derivative of the temperature at the surface.

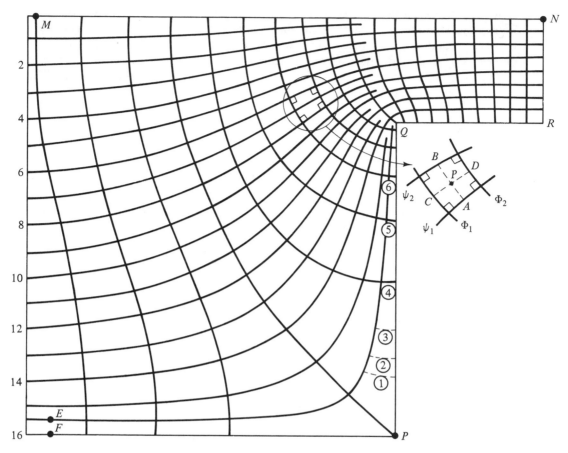

Figure 5.5.1 Irrotational flow pattern for the two-dimensional flow in a 4 to 1 contraction. The lines of constant Ψ and Φ nearly form the desired curvilinear squares in this preliminary construction of the flow net. This means that some additional modifications to these lines could be made to improve the drawing and, therefore, the solution to the flow problem.

The gradient of a scalar has a magnitude which is the rate of change of the scalar with respect to the distance that is perpendicular to the lines on which the scalar is constant. At point P, one has for the magnitudes

$$|\text{grad } \psi| = \frac{\psi_2 - \psi_1}{\overline{AB}} = \frac{K}{\overline{AB}}$$

and

$$|\text{grad } \Phi| = \frac{\Phi_2 - \Phi_1}{\overline{CD}} = \frac{K}{\overline{CD}}$$

Thus, if $|\text{grad } \psi| = |\text{grad } \Phi|$, the distances $\overline{AB}$ and $\overline{CD}$ are equal. This must be the case, since

$$|\text{grad } \psi| = \sqrt{\left(\frac{\partial \psi}{\partial x}\right)^2 + \left(\frac{\partial \psi}{\partial y}\right)^2} = \sqrt{(-v)^2 + (u)^2}$$

and

$$|\text{grad } \Phi| = \sqrt{\left(\frac{\partial \Phi}{\partial x}\right)^2 + \left(\frac{\partial \Phi}{\partial y}\right)^2} = \sqrt{u^2 + v^2}$$

Hence, if we have an irrotational flow within certain boundaries, we could attempt to draw a flow net of "curvilinear squares" (i.e., when figures having, approximately, equal median lines and 90° angles at the corners result) with lines of $\Phi = $ constant and $\psi = $ constant. When such a flow net is completed, one has the required distribution of Φ and ψ lines. Either set of lines can be used to give the velocity field from*

$$\frac{\partial \Phi}{\partial x} = u = \frac{\partial \psi}{\partial y} \qquad (5.5.4)$$

$$\frac{\partial \Phi}{\partial y} = v = -\frac{\partial \psi}{\partial x} \qquad (5.5.5)$$

After u and v are specified, the Bernoulli equation can be introduced and the associated pressures in the fluid can be determined. The following example shows this.

Example 5.5.1

Figure 5.5.1 shows the streamline pattern in a two-dimensional contraction. Also shown in the figure are six locations where the distances between two adjacent streamlines were measured. These distances, according to an arbitrary scale, are the following:

$$\text{width}_1 = 42 \qquad \text{width}_2 = 33 \qquad \text{width}_3 = 25$$
$$\text{width}_4 = 19 \qquad \text{width}_5 = 12 \qquad \text{width}_6 = 7$$

The units are immaterial. These numbers will be compared with the widths between the same pairs of streamlines in the wide section at the left where the flow is uniform. Here the width between E and F is 16. The velocity between E and F, where it is constant, is denoted as V_0. Since no flow crosses the streamlines, the velocity will be inversely proportional to the width between them. Hence $V_1 = V_0(16/42)$, and so on. This will correspond to the average velocity between the associated streamlines. The collection of these average velocities might occur on a streamline that was between the two that were used to determine the velocities. On such a streamline, the value of $V^2/2 + p/\rho$ would be constant— gravity acts perpendicular to the page—and this would be $V_0^2/2 + p_0/\rho$. Accordingly,

$$\frac{p - p_0}{\rho V_0^2/2} = 1 - \left(\frac{V}{V_0}\right)^2$$

in which p and V are to be evaluated at any point in the flow. Again, the term on the left is the pressure coefficient. With the foregoing in mind, the reader should verify the following results.

Section	Velocity	Pressure coefficient
1	$0.380V_0$	0.854
2	$0.485V_0$	0.765
3	$0.640V_0$	0.590
4	$0.842V_0$	0.290
5	$1.333V_0$	−0.777
6	$2.286V_0$	−4.224

*It is also true that the rate of flow between streamlines ψ_2 and ψ_1 in Fig. 5.5.1 is $\psi_2 - \psi_1$. Hence, the speed at point P in Fig. 5.5.1 is $(\psi_2 - \psi_1)/AB$. The direction of flow is normal to AB, in the limit as AB goes to zero. This velocity could also be used in the same way as the results of Eqs. (5.5.4) and (5.5.5) are used.

A plot of the dimensionless pressure coefficient along the step would give an approximation to the pressure distribution there.* If a realistic pressure distribution were available along the entire step, it could be integrated to give the force that was being exerted on the step by the flowing fluid. Plotting this pressure distribution would also show where the pressure reached certain values of interest, say, the vapor pressure, or perhaps p_0.

★5.6 THE EULER EQUATIONS

A great deal of information resulted when Newton's laws of motion were applied to a material volume. The equations which were derived for the streamline direction and the direction normal to the streamline, Eqs. (5.2.4) and (5.3.4), prescribe the manner in which the Bernoulli constant varies in these directions. While the streamline direction is not an unnatural one, it might not have occurred to the student who is new to the subject to consider it when developing the equations of motion. It may have seemed more obvious to select an xyz coordinate system which was oriented in some convenient way with respect to the flow geometry. Indeed, such a coordinate system is useful. The equations which result when we apply Newton's laws to an xyz system are worth developing. This will now be done.

In Fig. 5.6.1, an infinitesimal surface surrounds point P, an arbitrary point in an incompressible fluid which, for the present time, is considered to be without viscosity. We shall assume that we possess a complete description of the flow at point P. With this knowledge of the velocity components and all their derivatives, an estimate of the fluid acceleration is available if we use the methods given in Chap. 4. Also, the pressure can be estimated at the faces of the material volume by means of a Taylor series expansion. Such an expansion enables us to show that the pressure on the two faces of the surface of the material volume in Fig. 5.6.1, with normals parallel to the x direction, can be written as

$$p - \frac{\partial p}{\partial x}\frac{dx}{2} \quad \text{and} \quad p + \frac{\partial p}{\partial x}\frac{dx}{2}$$

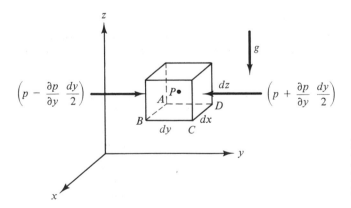

Figure 5.6.1   Fluid flow past point P through an infinitesimal control volume. Fluid density is ρ. Parallelepiped around point P corresponds to a minute material volume.

*Near the corner of the step, the fluid speed will tend toward infinity and the pressure coefficient to minus infinity. This will cause some difficulties in the numerical integrations of pressure in the vicinity of this corner. Elsewhere the results will be quite satisfactory.

if only the linear terms in the series are retained. These expressions lead us to conclude that the net force in the $+x$ direction that is due to the pressure can be written as $-(\partial p/\partial x)\,dx\,dy\,dz$. Again, $-\partial p/\partial x$ is the net force per unit volume, due to pressure, acting on the material volume in the x direction.

With this expression for the force in the $+x$ direction on the fluid that is within the material volume, we can equate the net force on the material volume to the product of the mass and acceleration to obtain

$$-\left(\frac{\partial p}{\partial x}\right) dx\,dy\,dz = a_x[dx\,dy\,dz\,(\rho)] \tag{5.6.1}$$

With a similar series of arguments, we can drive the equations in the y and z directions that are appropriate to an application of Newton's laws to Fig. 5.6.1. The results are

$$\rho(a_y)\,dx\,dy\,dz = -\frac{\partial p}{\partial y}\,dx\,dy\,dz \tag{5.6.2}$$

and

$$\rho(a_z)\,dx\,dy\,dz = -\frac{\partial p}{\partial z}\,dx\,dy\,dz - \rho g\,dx\,dy\,dz \tag{5.6.3}$$

One can divide the above equations by the volume of the infinitesimal material volume, $dx\,dy\,dz$, and use the fact that

$$a_x = u\frac{\partial u}{\partial x} + v\frac{\partial u}{\partial y} + w\frac{\partial u}{\partial z} + \frac{\partial u}{\partial t} \tag{5.6.4}$$

from Eq. (4.6.6a) to obtain

$$\rho\left(\frac{\partial u}{\partial t} + u\frac{\partial u}{\partial x} + v\frac{\partial u}{\partial y} + w\frac{\partial u}{\partial z}\right) = -\frac{\partial p}{\partial x} \tag{5.6.5}$$

In a similar fashion, we can use the expressions for a_y and a_z from Eqs. (4.6.7) and (4.6.8) to obtain

$$\rho\left(\frac{\partial v}{\partial t} + u\frac{\partial v}{\partial x} + v\frac{\partial v}{\partial y} + w\frac{\partial v}{\partial z}\right) = -\frac{\partial p}{\partial y} \tag{5.6.6}$$

and

$$\rho\left(\frac{\partial w}{\partial t} + u\frac{\partial w}{\partial x} + v\frac{\partial w}{\partial y} + w\frac{\partial w}{\partial z}\right) = -\frac{\partial p}{\partial z} - \rho g \tag{5.6.7}$$

These equations, called the *Euler equations,* are the desired results. We shall next manipulate and examine these last equations to produce some conclusions that have been found for more specialized cases. If this were the only result of the development that led to Eqs. (5.6.4) to (5.6.7), the effort would be questionable. However, this development is fundamental to constructing a more elaborate model of fluid motion that includes viscosity. Moreover, our efforts will produce a form of the Bernoulli equation that is applicable to unsteady flows.

★ 5.7 SOME APPLICATIONS OF THE EULER EQUATIONS

We next inquire whether the Euler equations can be used to specify the conditions under which the Bernoulli constant is the same constant for all streamlines. We shall use the Euler equations and the concept of the "rotation" of a particle about its center of gravity to specify a condition which gives the same Bernoulli constant at every point in the region of flow of an inviscid fluid in a gravitational field. This result is equivalent to that which was expressed by Eq. (5.3.5), when Newton's laws were applied to the direction normal to a streamline. The development which we shall undertake begins by recalling the concept of rotation from Chap. 4. In Eq. (5.4.1), the three components of the rotation vector are presented as:

$$\omega_x = \frac{1}{2}\left(\frac{\partial w}{\partial y} - \frac{\partial v}{\partial z}\right)$$

$$\omega_y = \frac{1}{2}\left(\frac{\partial u}{\partial z} - \frac{\partial w}{\partial x}\right) \tag{5.7.1}$$

$$\omega_z = \frac{1}{2}\left(\frac{\partial v}{\partial x} - \frac{\partial u}{\partial y}\right)$$

If the flow is prescribed to be irrotational, all of the components of $\boldsymbol{\omega}$ are zero, so that

$$\frac{\partial v}{\partial x} = +\frac{\partial u}{\partial y}$$

$$\frac{\partial w}{\partial y} = +\frac{\partial v}{\partial z} \tag{5.7.2}$$

$$\frac{\partial u}{\partial z} = +\frac{\partial w}{\partial x}$$

We shall now inquire what consequences this requirement for *irrotationality* produces through the use of the Euler equations, Eqs. (5.6.5) to (5.6.7). Let us first concentrate on Eq. (5.6.5) for a *steady flow*. This is

$$\rho\left(u\frac{\partial u}{\partial x} + v\frac{\partial u}{\partial y} + w\frac{\partial u}{\partial z}\right) = -\frac{\partial p}{\partial x}$$

By substituting for $\partial u/\partial y$ and $\partial u/\partial z$ the expressions that were shown to be equivalent to them in the first and third parts of Eq. (5.7.2), we have

$$\rho\left(u\frac{\partial u}{\partial x} + v\frac{\partial v}{\partial x} + w\frac{\partial w}{\partial x}\right)dx = -\frac{\partial p}{\partial x}dx$$

or, equivalently

$$\rho\frac{\partial}{\partial x}\left(\frac{u^2 + v^2 + w^2}{2}\right)dx = -\frac{\partial p}{\partial x}dx$$

We could also make the appropriate substitutions for $\partial v/\partial x$ and $\partial v/\partial z$ in the *steady-state* Euler equation for the y direction, keeping in mind that the flow is to be considered irrotational, to produce

$$\rho\frac{\partial}{\partial y}\left(\frac{u^2 + v^2 + w^2}{2}\right)dy = -\frac{\partial p}{\partial y}dy$$

The approach that has been used for the x and y directions could be repeated again for the z direction to rewrite the third Euler equation as

$$\rho\frac{\partial}{\partial z}\left(\frac{u^2 + v^2 + w^2}{2}\right)dz = -\frac{\partial p}{\partial z}dz - \rho g\,dz$$

(Note that no restrictions were placed on the dx, dy, and dz which were employed. We require only *irrotational* flow.)

These last three equations contain the key elements of our derivation. If we add them up, we obtain

$$\rho\left[\frac{\partial(V^2/2)}{\partial x}dx + \frac{\partial(V^2/2)}{\partial y}dy + \frac{\partial(V^2/2)}{\partial z}dz\right] = -\left(\frac{\partial p}{\partial x}dx + \frac{\partial p}{\partial y}dy + \frac{\partial p}{\partial z}dz\right) - \rho g\, dz$$

This statement implies that

$$\rho\, d\left(\frac{V^2}{2}\right) = -dp - \rho g\, dz$$

or

$$d\left(\frac{V^2}{2} + \frac{p}{\rho} + gz\right) = 0$$

because of the total differentials that are involved. This final equation states that the Bernoulli constant is, indeed, one constant throughout the entire (steady-state) flow field if the flow is *irrotational,* because the quantity within the parentheses does not change as we execute an *arbitrary change of position* given by dx, dy, and dz.

Our next effort will be to show that Eqs. (5.6.5) to (5.6.7) can be rewritten in a particular way that will give us a fact with which we are already familiar: the sum $V^2/2 + p/\rho + gz$ is constant along a streamline in *steady flow.* Now, along a *streamline,* it is known from Eq. (3.5.3) that

$$\frac{dx}{u} = \frac{dy}{v} = \frac{dz}{w} \tag{5.7.3}$$

Then $v = u\, dy/dx$ and $w = u\, dz/dx$. If these expressions for v and w are inserted into the steady-state version of Eq. (5.6.5), the result is

$$\rho\left(u\frac{\partial u}{\partial x} + u\frac{dy}{dx}\frac{\partial u}{\partial y} + u\frac{dz}{dx}\frac{\partial u}{\partial z}\right) = -\frac{\partial p}{\partial x}$$

If one formally multiplies both sides of the above equations by dx, the following result is obtained:

$$\rho u\left(\frac{\partial u}{\partial x}dx + \frac{\partial u}{\partial y}dy + \frac{\partial u}{\partial z}dz\right) = -\frac{\partial p}{\partial x}dx$$

By using our knowledge of a total differential to rewrite the term which appears in the parentheses, the following version results:

$$\rho u\,(du) = -\frac{\partial p}{\partial x}dx$$

or

$$\rho\, d\left(\frac{u^2}{2}\right) = -\left(\frac{\partial p}{\partial x}\right)dx$$

In a similar fashion, Eqs. (5.6.6) and (5.6.7) can be rewritten for the special case of a (steady-state) streamline by substituting $u = v\,(dx/dy)$ and $w = v\,(dz/dy)$ in the y equation, and $u = w\,(dx/dz)$ and $v = w\,(dy/dz)$ in the z equation. Such a procedure gives

$$\rho\, d\left(\frac{v^2}{2}\right) = -\left(\frac{\partial p}{\partial y}\right)dy$$

and
$$\rho d \left(\frac{w^2}{2}\right) = -\left(\frac{\partial p}{\partial z}\right) dz - \rho g \, dz$$

Our desired conclusion follows once we have added together the last three equations to obtain

$$\rho d \left(\frac{u^2 + v^2 + w^2}{2}\right) = -\left(\frac{\partial p}{\partial x} dx + \frac{\partial p}{\partial y} dy + \frac{\partial p}{\partial z} dz\right) - \rho g \, dz$$

Since we are treating a fluid with *constant density* (i.e., since ρ does not vary with the coordinate directions), we can divide through by ρ; and with a minor rearrangement of the terms, one obtains, because $dp = \partial p/\partial x \, dx + \partial p/\partial y \, dy + \partial p/\partial z \, dz$,

$$d \left(\frac{u^2 + v^2 + w^2}{2}\right) = -d \left(\frac{p}{\rho}\right) - g \, dz \tag{5.7.4}$$

This equation is true for very specific changes of coordinates dx, dy, and dz. It must be recalled that these coordinate steps are associated with those necessary to stay on a streamline. This is true because we have employed Eq. (5.7.3).

By bringing all of the terms in Eq. (5.7.4) onto one side of the equality sign, we can write the result as

$$d \left(\frac{\mathbf{V} \cdot \mathbf{V}}{2} + \frac{p}{\rho} + gz\right) = 0 \tag{5.7.5}$$

or

$$\frac{|\mathbf{V}|^2}{2} + \frac{p}{\rho} + gz = \text{constant} \tag{5.7.6}$$

along a *streamline*. We have already encountered this result.

If the pressure and density of the fluid are interrelated, as they can be for a gas, the Euler equations can be integrated to yield a more general form of the Bernoulli equation. The derivation of this result, under the *assumption* that the velocity is steady, will follow many of the same steps that led to Eq. (5.7.4).

Such a line of argument yields

$$\int d \left(\frac{u^2 + v^2 + w^2}{2}\right) = -\int \frac{1}{\rho} \left(\frac{\partial p}{\partial x} dx + \frac{\partial p}{\partial y} dy + \frac{\partial p}{\partial z} dz\right) - \int g \, dz \tag{5.7.7}$$

The first and last integrals are the same as those found before. The second one now contains the factor $1/\rho$, since it, as well as p, is assumed to vary with position. (If there is zero variation in ρ, nothing is lost, and the incompressible Bernoulli equation is recovered.)

*The integration of the second integral is possible if $\rho = \rho(p)$.** The term in parentheses on the right-hand side of Eq. (5.7.7) is simply an expression for the total differential dp. This means that Eq. (5.7.7) can be integrated along a streamline from point 1 to point 2:

$$\left(\frac{V_2^2}{2} + gz_2\right) - \left(\frac{V_1^2}{2} + gz_1\right) + \int_1^2 \frac{dp}{\rho} = 0$$

---

* A fluid of which this is true is called *barotropic*.

or
$$\frac{V_1^2}{2} + \int_0^1 \frac{dp}{\rho} + gz_1 = \frac{V_2^2}{2} + \int_0^2 \frac{dp}{\rho} + gz_2 \tag{5.7.8}$$

in which 0 is an arbitrary reference location. This expression states that $(V^2/2 + \int dp/\rho + gz)$ is a constant along a streamline. [We observe that if ρ is a constant, Eq. (5.7.6) is the result, in accordance with our expectations.]

We shall now include the effect of the unsteady terms, $\partial u/\partial t$, $\partial v/\partial t$, and $\partial w/\partial t$ in this expression and obtain the Bernoulli equation for unsteady flows.

With these unsteady terms included in the equations, the procedure that gave Eq. (5.7.5) would now give

$$\frac{\partial u}{\partial t} dx + \frac{\partial v}{\partial t} dy + \frac{\partial w}{\partial t} dz = -d \left(\frac{\mathbf{V} \cdot \mathbf{V}}{2} + \frac{p}{\rho} + gz \right) \tag{5.7.9}$$

as the applicable differential equation along the streamline. The left-hand side of this equation can be rewritten in vector form to give additional insight. If $d\mathbf{s} = dx\, \mathbf{i} + dy\, \mathbf{j} + dz\, \mathbf{k}$ is a tangent to the streamline (with an assumed direction along the streamline), we can write

$$\frac{\partial u}{\partial t} dx + \frac{\partial v}{\partial t} dy + \frac{\partial w}{\partial t} dz = \frac{\partial \mathbf{V}}{\partial t} \cdot d\mathbf{s}$$

Equation (5.7.9) can then be integrated between two points along the streamline and has the form for constant density

$$\int_1^2 \frac{\partial \mathbf{V}}{\partial t} \cdot d\mathbf{s} = - \left[\left(\frac{V_2^2}{2} + \frac{p_2}{\rho} + gz_2 \right) - \left(\frac{V_1^2}{2} + \frac{p_1}{\rho} + gz_1 \right) \right] \tag{5.7.10}$$

The right-hand side of this result appears to be the same as that which was extensively used previously. The left-hand side is nonzero for an unsteady flow. We shall now examine this expression. If V_{+s} is the speed of the fluid in the direction of the streamline that was assumed positive (i.e., from point 1 to point 2), the integral can be rewritten as

$$\int_1^2 \frac{\partial V_{+s}}{\partial t} ds = \left(\frac{V_1^2}{2} + \frac{p_1}{\rho} + gz_1 \right) - \left(\frac{V_2^2}{2} + \frac{p_2}{\rho} + gz_2 \right) \tag{5.7.11}$$

The only difference between this result and that for the steady case is the term on the left-hand side of Eq. (5.7.11). A minor difficulty with this expression can be experienced in evaluating the time derivative of the speed, along the streamline direction, over the change of location from point 1 to point 2. Even though the magnitude of the speed may be increasing with time, the direction of that speed must be kept in mind in determining the sign of the derivative. (V_{+s} is positive if it is in the direction $1 \rightarrow 2$ in this derivation.) The following example should make this point clear.

Example 5.7.1

The two tanks shown in Fig. 5.7.1 are connected by a pipe of area A_3 and length L_3. What should be the frequency of oscillation of the fluid system if viscosity is neglected?

Solution We can apply Eq. (5.7.11) directly to the case illustrated in this figure. The coordinate system shown in Fig. 5.7.1 has $z_2 = 0$. Volume conservation requires that $V_1 A_1 = V_2 A_2 = V_3 A_3$. If V_1 is dz_1/dt, it will be positive when the left-hand fluid column is

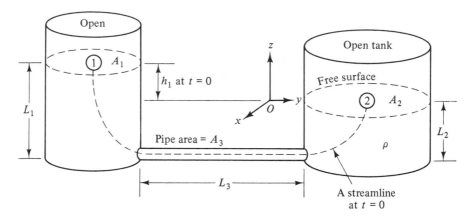

Figure 5.7.1 Unsteady flow in a conduit between two reservoirs. Dotted line: a streamline at instant of observation.

rising. Thus, its component in the direction from point 1 to point 2 is negative. This fact should be kept in mind as the integral is evaluated below:

$$\int_1^2 \frac{\partial V_{+s}}{\partial t} ds = \frac{\partial V_1}{\partial t} L_1 + \frac{\partial V_3}{\partial t} L_3 + \frac{\partial V_2}{\partial t} L_2$$

$$= \frac{\partial V_1}{\partial t} L_1 + \frac{\partial (V_1 A_1/A_3)}{\partial t} L_3 + \frac{\partial (V_1 A_1/A_2)}{\partial t} L_2$$

$$= \frac{\partial V_1}{\partial t}\left[L_1 + \frac{A_1 L_2}{A_2} + \frac{A_1 L_3}{A_3} \right]$$

$$= \frac{\partial}{\partial t}\left(-\frac{dz_1}{dt} \right)\left[L_1 + \frac{A_1 L_2}{A_2} + \frac{A_1 L_3}{A_3} \right]$$

If the constant in the square brackets is defined as K, one has

$$\int_1^2 \frac{\partial V_{+s}}{\partial t} ds = -K\frac{d^2 z_1}{dt^2}$$

Consequently, the unsteady Bernoulli equation for this case has the form

$$-K\frac{d^2 z_1}{dt^2} = \left(\frac{dz_1}{dt}\right)^2\left(\frac{1}{2}\right)\left[1 - \left(\frac{A_1}{A_2}\right)^2 \right] + gz_1$$

For the case of a manomenter tube of length L ($= L_1 + L_2 + L_3$) and $A_1 = A_2 = A_3$, the equation of motion reduces to

$$\frac{d^2 z_1}{dt^2} + \frac{g}{L}z_1 = 0$$

The solution is

$$z_1 = h_1 \cos\left(\sqrt{\frac{g}{L}} t \right)$$

for $z_1 = h_1$ at $t = 0$.

The Flow of an Inviscid Fluid   Chap. 5

5.8 SUMMARY

The Bernoulli equation was developed by integrating Newton's equations of motion along a streamline. If the direction of integration is normal to a streamline, the Bernoulli equation again results if the flow is irrotational (i.e., curl $\mathbf{V} = 0$). The forces that were included in these analyses are due to fluid pressure and gravity only. Viscous effects were neglected. In a number of applications where the viscous effects are small with respect to others, the Bernoulli equation can be very useful.

Irrotational flows are associated with the existence of a potential function Φ. If the flow is incompressible, then Φ must satisfy the Laplace equation $\nabla^2 \Phi = 0$. In two-dimensional incompressible flows, the lines of $\Phi = $ constant and $\psi = $ constant form the "flow net."

The Euler equations are also an expression of Newton's laws of motion for an inviscid fluid. They yield the Bernoulli equation if one integrates them between any two points in the fluid region if the flow is irrotational. The Bernoulli equation also follows from the Euler equations if they are integrated along a streamline, in accordance with what was learned earlier in the chapter.

The Euler equations contain temporal derivatives of the velocity components. If one includes them in the integration along a streamline, one can obtain a Bernoulli equation for such an unsteady flow. (One can also develop a Bernoulli equation for an unsteady irrotational flow, even though that was not demonstrated in this chapter. Then the potential function Φ will depend upon time as well as space. Consequently, the term $\partial\Phi/\partial t$ is included in such an unsteady irrotational result. Moreover, the Bernoulli constant will be a function of time.)

EXERCISES

5.2.1(a) What is V_2 if D_1 is 12 cm and $p_g = 200$ kPa? Use $D_2 = 6$ cm. Repeat the problem for a 2 cm nozzle diameter.

(b) Based on the results of the previous calculations, what *diameter ratio* would be required for one to neglect the upstream velocity contributions to the Bernoulli equation in such a nozzle problem?

(c) If $Q = 0.07$ m³/s and the pipe and nozzle diameters are 12 and 4 cm, respectively, what is the centerline pressure upstream of the nozzle, p_1? What would the gauge pressure p_g be at the bottom of a horizontal pipe that connected to the nozzle?

5.2.2(a) A particular Venturi tube design has the diameters in the ratio of 2 to 1 with the larger one 6 in. The flow of water is being measured, and a mercury manometer has a reading H of 37 in. What is the volume rate of flow?

(b) The same type of flow meter is used in another water system, but a manometer with a liquid having a specific gravity of 4 is used. If this manometer has a reading H of 17 in., what is the volume rate of flow?

(c) A third Venturi tube is used to measure the flow of an oil, SG = 0.85, and water is used as the manometer fluid. What should be the manometer reading when 1.25 cfs is flowing through the flow meter of Ex. 5.2.2a?

5.2.3(a) What is the speed of the water as it passes over the hump at the point of maximum elevation, $H = 0.5$ m?

(b) What would be the value of the depth at a point where the hump was 0.3 m above the floor channel?

(c) Assume that the channel had a flow of 10 m³/s but that its upstream depth was 4 m. What would be the depth of fluid as it passed over a hump of height 0.5 m?

(d) If, in a channel, the depth is measured at 4 m at

a hump when the upstream depth is 5 m, what is the volume rate of flow? The hump is 0.4 m high at the point of measurement.

5.2.4(a) Water is flowing in a horizontal pipe. A manometer connected to an impact tube and a wall tap contains mercury. What is the speed of the water at a location where the measurement system gives a reading of 11.5 in.?

(b) An impact tube is used to record the speed of air in a wind tunnel. A liquid with a specific gravity of 0.84 is used in the manometer. What should be the reading at a point where the air speed is 5 m/s?

(c) It is intended to use this impact tube to measure speeds lower than 15 m/s in the wind tunnel. What would be the lowest speed that should be considered if the manometer can be read to 0.01 mm?

5.3.2(a) Calculate the speed of the fluid at point D. What is the fluid speed at point E? Use the Bernoulli equation between points F and E to obtain the last result.

5.3.3(a) If $A = 7$ ft^2/s, what is the elevation, relative to the height at infinite radius, at radii of 1 and 2 ft?

(b) The depression of a free surface at a water drain is 0.375 in. at a radius of 1 in. Determine the value of A for this flow. What is the velocity along the free surface at $r = 1$ in.?

(c) Similar velocity distributions (i.e., irrotational) are observed in cyclonic storms if one does not include the region near the center, the "eye." If an 80 mph wind is observed at a point that is 1 mile from the center of the storm, what would be the pressure there, relative to the pressure at a point far from the center (*note:* as $r \to \infty$, $V \to 0$)?

5.4.1(a) Find $|\mathbf{V}|$ as a function of position for the given velocity potential.

(b) What are u and v for $\Phi = 6(\ln r)$? What are the radial and angular components of the velocity?

(c) Use Eqs. (4.7.4) and (4.7.5) and appropriate

integration to determine the stream function for the velocity potential given in Example 5.4.1. [*Note:* Equation (4.7.4) will also yield, upon integration, an arbitrary function of x, $f(x)$. Equation (4.7.5) produces an arbitrary function of y, $g(y)$, in addition to a specific relationship between x and y. By comparing the results of these two equations, one can specify $f(x)$ and $g(y)$. Often they are equal to one, and the same, constant.]

5.4.2(a) At what location(s) on the cylinder is the pressure coefficient zero? What is the pressure at this point? What is the fluid velocity there?

(b) You can show that at a point along the x axis that is about 10 radii from the origin, the fluid velocity is about 99 percent of the velocity at infinity. How far must one move away from the cylinder along the y axis for this to be true also?

(c) The dashed curve in Fig. 5.4.3 corresponds to the pressure distribution when there is a large wake behind the cylinder. Approximate this distribution by using Eq. (5.4.17) for $\theta < 50°$ and $\theta > 310°$; use a pressure coefficient of -1 in the region between. Estimate the drag on a cylinder of unit length if the fluid is air. What would be the drag if the cylinder were in water?

5.5.1(a) Where along the step between stations 1 and 6 does the pressure equal that at the upstream region OM?

(b) Plot the variation of pressure along centerline MN.

(c) Use the streamline labeled 14 as the contour of a wall for a wind tunnel's entrance section that should have a 4:1 contraction. Plot the variation of pressure along this curved wall as a function of distance along the wall. What are the maximum and minimum pressure gradients— change of pressure per unit length—encountered along the wall? (If the pressure gradients are positive and large, the flow of a viscous fluid will not tend to follow the wall. Rather, it separates and leaves as an eddy.)

PROBLEMS

5.1 In a steady flow, a particle passing a particular location has a speed of 3 m/s. In moving a distance of 1 cm along its path, its speed increases by 0.02 m/s. What is the acceleration?

Had the speed decreased by 0.03 m/s in the same distance, what would have been the acceleration? For a fluid with a density of 860 kg/m^3, what would be the change of pres-

The Flow of an Inviscid Fluid   Chap. 5

sure per meter for the two situations just described? (Neglect gravity and viscous effects.)

5.2 In a converging two-dimensional passage, the streamline spacing goes from $\frac{3}{4}$ in. to $\frac{1}{4}$ in. What should be the pressure difference between these two stations in terms of $\rho V_1^2/2$, if gravity effects are neglected?

5.3 A particular streamline in a water flow is horizontal between two points where the pressure differs by 50 psi. If the fluid's speed at the point of higher pressure is 5 ft/s, what is the speed at the other point? Assume an inviscid flow.

5.4 The pressure in a 1-in.-diameter hose is 50 psig on the centerline. A nozzle with a $\frac{1}{4}$-inch-diameter opening is at its end. What velocity of the water jet can be expected? Use the Bernoulli equation to predict the maximum height of the jet. Repeat, using Newton's laws of motion for a particle to predict the trajectory of a particle. Do the two answers correspond? Does this surprise you? Explain.

5.5 If the water jet from Prob. 5.4 must strike an object 100 ft horizontally away from the nozzle tip—at the same elevation—what must be the minimum hose pressure? At what angle must the nozzle be inclined to the horizontal?

5.6 The gauge pressure on the bottom of a horizontal pipe (5 cm inside diameter) is 300 kPa. What velocity can be expected at the exit of a nozzle that is 1 cm in diameter? Sea water (SG = 1.03) is flowing. What would be the expected jet velocity if the 300 kPa had been measured at the center of the conduit?

5.7 A fire hose with a diameter of 3 in. discharges through a 1-in.-diameter nozzle. The hydrant pressure—assume the velocity to be nearly zero there because of the size of the hydrant—is 56 psig. If the nozzle is directed at 60° to the horizontal, what is the maximum height and distance that the water will reach? At what angle should the hose be set with respect to the horizontal to achieve the maximum range of the jet?

5.8 In order to increase the amount of water delivered through a horizontal 3-in.-diameter hose from a pressurized tank at the same elevation, two changes appear feasible. Both would require the same financial investment.

One plan would install a new pump to double the reservoir pressure that supplied the hose. The alternative would consist of a second hose operating at the same reservoir pressure. Justify your choice of solution.

5.9 An open channel 3 m wide and 2 m deep has water flowing in it at 1 m/s. There is a hump in the bottom of the channel with a height of 0.4 m. (This is downstream of the location where the depth is 2 m.) Determine the free-surface elevation—perhaps more than one is possible—at the hump's apex.

5.10 In Prob. 5.9, the flow speed is described as subcritical if it is less than the speed of propagation of a small disturbance, $[g(\text{depth})]^{0.5}$. In this case one would expect the speed to remain subcritical at the hump; the depth at the hump is used to determine the small disturbance velocity there. Use the results from Prob. 5.9 to determine if $V/[g(\text{depth})]^{0.5} < 1$ upstream. Repeat for either of the two depths at the hump that appear to be possible from the Bernoulli equation.

5.11 If, during flow conditions, the water in a channel 20 ft wide and 3 ft deep is flowing at 10 ft/s and passes over a gradual excavation with a maximum depression of 8 in., what depth(s) can be expected to occur there? Refer to Prob. 5.10 in developing your answer.

5.12 What should be the depth of water in a section of a normally horizontal channel that has a hump of 1 m? Upstream of this change of bottom profile, the water flows at 10 m³/s per meter of width and the depth is 5 m.

5.13 Past the hump mentioned in Prob. 5.12, the channel's bottom begins a gradual rise. What depth can be expected when the change of elevation peaks at 1.95 m? What is the water's speed? What is the value of $V/\sqrt{g(\text{depth})}$ here?

5.14 The course of a river is being altered to improve barge navigation. At one section, it will be widened from 15 m to 20 m without changing its depth of 4 m. If the river normally flows at 60 m³/s, what will be the expected free-surface elevations at the wider portion?

5.15 A section of a river is to be deepened to provide a more constant depth along its route. The river is 200 m wide and discharges 1000 m³/s. If work begins to change the depth from 3 m to 4 m, what should be the effect on the elevation of the free surface, and consequently on shore facilities?

5.16 A large open storm drain (assume a rectangular cross section) flows down a hillside and changes its elevation 200 ft along 1 mile of its length. If the water depth is 3 ft upstream and there is a flow of 15 cfs. for each foot of width, what will be the expected depth 1 mile downstream? Comment on the result.

5.17 A siphon is to be used to regulate the level of water in a reservoir as shown in Fig. P.5.17. The siphon has a cross-sectional area of 1 m², and the reservoir has a surface area of 4000 m². Point D is located 6 m below B, and C is 2 m above B. What will be the rate of discharge when the reservoir level is 2.5 m above B? Estimate the time needed to decrease the elevation to B. (Assume no inflow to the reservoir, and neglect the deceleration effects associated with the unsteady position of point A. $A_C = 0.9A_D$.)

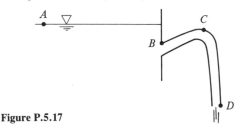

Figure P.5.17

5.18 For the reservoir described in Prob. 5.17, what elevation of the free surface must not be exceeded to avoid cavitation within the siphon?

5.19 A siphon system similar in geometry to that described in Prob. 5.17 must regulate the water level so that once the level of the free surface, point A, rises to that of point C, the siphon begins to flow and returns the level to point B in 4 min. If there is no additional flow into the reservoir, how many siphons must be provided to meet this requirement? For this number of siphons, what will be the time if 15 m³/s of water is flowing into the reservoir? (Neglect any acceleration effects on the problem due to the unsteady conditions of the reservoir.)

5.20 Water in a channel is controlled by a sluice gate so that upstream of the gate the eleva-

tion h_1 is 20 ft and the water speed is 3 ft/s. What depth h_2 and speed can be expected downstream of the gate?

5.21 In the same channel described in Prob. 5.20, the water depth upstream, h_1, is now 18 ft, and h_2 is measured at 3 ft at the measurement station downstream of the gate. What is the volume rate of flow per foot of channel width? What is the volume rate of flow if the channel is 42 ft wide? If the ratio h_1/h_2 is sufficiently large, can one neglect the upstream kinetic energy in the calculation for the flow rate?

5.22 As the opening of a sluice gate, distance b, is changed, the depth downstream of it, h_2, changes as $h_2 = Cb$, in which $C < 1$. For a channel with a depth of 4 m where the water speed is 1.5 m/s, find the value of b for $C = 0.61$.

5.23 If the volume rate of flow in Prob. 5.22 is to be maintained but both the upstream and downstream elevations could change, plot $\tilde{Q}/b\sqrt{2gh_1}$ as a function of b/h_1. From your curve, determine the value of h_1/b above which $\tilde{Q}$, flow/unit width, varies as $\sqrt{h_1}$. (b is fixed.)

5.24 As indicated in Fig. P.5.24, the downstream depth is less than the opening of the gate, b. The contraction coefficient h_2/b remains about 0.61 over a range of values of b/h_1. What must be the gate opening in order to have a discharge of 24 m³/s per meter of width? $h_1 = 8$ m.

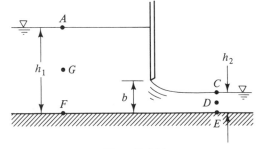

Figure P.5.24

5.25 The large reservoir in Fig. P.5.25 has a sluice gate at one end that is 20 m wide. The gate at P is open, and the water that discharges through it reaches a constant depth of 2 m when the reservoir depth is 11 m. How does

The Flow of an Inviscid Fluid   Chap. 5

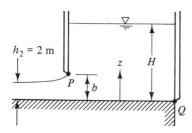

Figure P.5.25

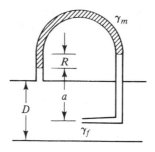

Figure P.5.29

the velocity vary with z and across this 2-m depth? What is the discharge? If the gate at the opposite end, Q, is opened so that a jet issues forth with a width of 20 m and a vertical extent of 2 m, will the velocity be the same at the upper and lower surfaces of the jet? What will be the pressure in the middle of the jet? Is $\partial p/\partial z = 0$ in the jet? Is $\partial p/\partial z = 0$ downstream from point P? Explain any differences in the two pressure gradients. Compare the discharges from P and Q.

5.26 The system shown in Fig. 5.2.10 is incorporated into a water pipe that is 18 in. in diameter. The manometer fluid is mercury, and the reading, H, is measured at 30.5 cm. What water speed is being sensed by the Pitot tube?

5.27 The speed of air in a ventilation shaft is being measured with a Pitot tube and manometer similar to that shown in Fig. 5.2.10. There is water in the manometer. What should be the change of the level of the manometer fluid for a wind speed of 30 ft/s?

5.28 Assume that one can detect a change in the difference of elevation of the liquid in a manometer corresponding to 0.002 cm. What would be the speed of air that would move the water in a manometer connected to a Pitot tube this distance from an initial zero reading? Can you suggest a method for measuring even lower air speeds? (This last question is "open-ended," so let your imagination have "free reign.") What fluctuations in speed could be measured if the air were moving at 30 m/s?

5.29 The velocity in a conduit is being sensed with an impact tube as shown in Fig. P.5.29. Determine the velocity V as a function of R. Does this result also depend on the distance a, the location in the conduit? If $\gamma_m > \gamma_f$, would it be desirable to locate the manometer below the conduit? Why? If this is the case, determine the dependence of V on R.

5.30 It is desired to measure the velocity of air with a Pitot-static tube. What is the minimum speed that can be sensed if the manometer that is used will give a reliable reading of pressure difference of $(0.0005/12)(0.80)(62.4)$ lb/ft^2? (This is the hydrostatic pressure due to a column of alcohol 0.0005 in. high.) Assume 68°F and standard barometric pressure. Repeat the calculation for a measurement of a water flow.

5.31 If a Pitot tube (also called an *impact* or a *total pressure tube*) is mounted on an airplane to measure the airspeed, what pressure difference—the total pressure minus the atmospheric pressure—would be recorded at a flight speed of 360 km/h? (Use an air temperature of $-20°C$ and an atmospheric pressure of 97 kPa.)

5.32 The velocity of water (or most any other fluid) flowing turbulently in a round pipe can be approximated by $V/V_{max} = (y/r_0)^{1/7}$, in which V_{max} is the velocity on the centerline and y is the distance from the wall of a tube of radius r_0. Use the criterion in Prob. 5.30 to determine the minimum distance from the wall where a meaningful measurement can be made if $V_{max} = 20$ ft/s and $r_0 = 6$ in. If the diameter of the Pitot-static tube is $\frac{1}{8}$ in., can one come close enough to the wall to obtain the finest measurement which the manometer is capable of sensing?

5.33 For the information in Prob. 5.32, determine the pressure readings from a Pitot-static tube that one should expect in a velocity survey at $r = 0$, 1 in., 3 in., and 5 in. Now draw circles of radii $r = \frac{1}{2}$ in., 2 in., and 4 in., and using the velocities obtained for the radii in Prob. 5.32 as representative values for the appropriate rings—with $V_{max}[\pi(0.5/12)^2]$ as an exception—determine the volume rate of flow for these discrete measurements. Compare this volume

Problems

flow-rate with the value of Q using an integration of the velocity distribution given in Prob. 5.32.

5.34 A horizontal Venturi meter with diameters of 6 in. and 3 in. is connected to a mercury U-tube manometer. A reading of 27 in. is recorded for a water flow. What is the volume rate of flow?

5.35 The same Venturi tube as in Prob. 5.34 passes 1.5 cfs. What reading can be expected on a mercury manometer? If carbon tetrachloride (SG = 1.6) were used in the manometer, what reading should be expected? What reading should occur if an oil (SG = 0.9) is used? Does your answer indicate that the U-tube manometer with oil cannot be installed below the Venturi tube? Explain.

5.36 Air can be metered in a Venturi tube if the velocity does not exceed a value that would result in significant effects due to compressibility. This limit is approximately 100 m/s. For a Venturi tube with a diameter ratio of 3/1, specify a manometer liquid to obtain maximum sensitivity of the reading to flow rate, and give the expected scale length of the manometer for this limiting speed.

5.37 A Venturi tube is to be designed to record a small flow, 600 cm³/s. It is to be connected to instrumentation that can tolerate a maximum pressure differential that is equivalent to a column of water 0.5 m high. If the area ratio between the largest and smallest sections of the tube is to be 4, what should be the design values of the diameters at these locations for a water flow?

5.38 The air required for an internal-combustion engine passes through a fuel metering system to produce the desired fuel/air ratio. In a carburetor, the air flows through a Venturi tube and the resulting low pressure there draws the fuel out of the float bowl (see Fig. P.5.38). For the engine conditions such that 0.1 pounds of air per second enter the carburetor, determine the size of the fuel orifice opening to give a fuel/air ratio, by weight, of 0.071. Assume $A_{jet}/A_{orifice} = 0.72$ and an air pressure at the orifice of 0.8 atm. (Let the SG of gasoline = 0.86.) What difference in fuel/air ratio can be expected if the float level is mistakenly adjusted so that the effective depth of fuel is 2.1 in.?

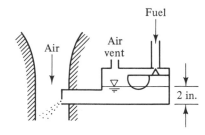

Figure P.5.38

5.39 A Venturi tube is to be used in a spraying device that should lift a liquid into an airstream flowing horizontally through the tube. A pipe connects the minimum cross section of the Venturi tube with the liquid, whose free surface is 200 mm below the axis of the tube and vented to the atmosphere. For a liquid with a specific gravity of 2 and air with a density of 1.2 kg/m³, what volume rate of airflow will be needed to bring the liquid into the airflow for mixing? The Venturi meter has an upstream diameter of 20 mm, where the pressure can be assumed to be 1 atm, and a diameter of 4 mm where the liquid enters.

5.40 A well-designed orifice has only very minor losses. The one shown in Fig. P.5.40 is to be used to meter air for a test stand of an internal-combustion engine with a displacement of 1550 cm³ (i.e., requiring this volume of air every 2 revolutions). If the engine is operating at 4000 rpm, what should be the reading of the water manometer?

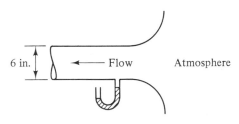

Figure P.5.40

5.41 A nozzle of the type shown in Fig. P.5.41 is being calibrated by connection to a tank 10 ft in diameter that is filled with water. The free surface in the tank is falling at the rate of 0.1 inch per second when the manometer reading is recorded. If the nozzle is well designed so that there is no energy loss, what should be the reading R?

The Flow of an Inviscid Fluid   Chap. 5

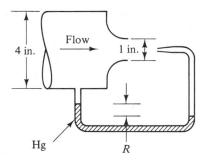

Figure P.5.41

5.42 A fountain is to be built which should have a jet of water to reach a height of 20 ft. The pressure in the water main (almost zero velocity there) is 65 psig. The jet, after it issues from a nozzle at the end of a 3-in.-diameter pipe, is 1 in. in diameter. What must be the pressure inside the 3-in. pipe just ahead of the nozzle? Is the water system capable of supplying this pressure?

5.43 A gasoline fuel-injector nozzle has an opening of A_0. It is closed by a spring, but it is open when the gauge pressure in the line, 4 mm in diameter, reaches a specified level—400 kPa, in this case. If $d_0 = 1$ mm and the device which closes off the nozzle opening does not affect the injection process, determine the volume rate of flow of gasoline (SG = 0.86) after injection starts. If the cylinder that is supplied by the nozzle has 400 cm³ of air, based on 20°C and 101.8 kPa, how long should the nozzle remain open to achieve a fuel/air ratio, by weight, of 0.074?

5.44 Water at 10 psig flows from a pipe 6 in. in diameter through a well-formed transition section that is connected to a 2-in.-diameter pipe. There is a flange and gasket at each end of the transition piece. If the gaskets become faulty—with age, perhaps—would either of the joints have a tendency to leak water when the flow rate is 1 cfs? Could air be sucked into the system with faulty gaskets for certain flow rates if upstream pressure remains at 10 psig?

5.45 A horizontal gutter 1 ft deep and 2 ft wide is completely full of water. There is an opening in the bottom with an effective area of 1 ft² every 110 ft. The water discharges through it as a jet into a large sewer pipe that is below ground level. What is the maximum rate of

rainfall that can occur so that the gutter does not overflow?

5.46 As a ball rotates, it induces a flow about it because fluid adheres to the surface and must perforce rotate. (This is the no-slip condition.) If a Ping-Pong ball is "cut" so that it spins as shown in Fig. P.5.46, predict whether the ball should reach the table earlier or later if its rotation rate is high. Support your point of view. (*Hint:* Consider the pressure at U and L when the ball's center is stationary and the air flows past the ball. For a ball rotating as shown, where is the air speed the highest and lowest?)

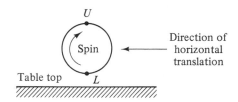

Figure P.5.46.

5.47 Water is being drawn upward by a pump from a pond. The pump should deliver 2 cfs, and the inlet pipe is 6 in. in diameter. What is the maximum elevation above the pond that the pump can be placed so that the inlet pressure does not go below 2 psia? The water temperature is 70°F. What would be the maximum elevation if the inlet pipe were 12 in. in diameter? Does the Bernoulli equation pose any restrictions or qualifications on the previous answers?

5.48 A water jet issues vertically at a speed of 14 m/s from a nozzle that is 30 mm in diameter. At a height of 4 m above the nozzle, an impact tube is inserted into the jet to record its velocity. What will be the impact pressure that will be sensed at the tip of the tube? Estimate the diameter of the jet at this elevation.

5.49 A pipe 1 ft in diameter discharges water against a round plate as shown in Fig. P.5.49. Determine the pressures at points O, M, and I when 8 ft³/s is flowing. Is the connection at I liable to be in tension or compression due to the pressure distribution only? Explain. (*Note:* $A_O = 2\pi(4)(\frac{1}{2}) \neq A_M$; plate gap = 6 in.)

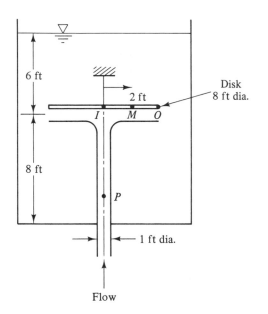

6 ft

2 ft

Disk
8 ft dia.

I M O

8 ft

P

1 ft dia.

Flow

Figure P.5.49

5.50 If the system shown in Fig. P.5.49 is drawing water radially inward toward I and subsequently toward P at 8 ft³/s, what are the pressures at I, M, and O? If the plate weighs 450 lb, would you recommend a tension or a compression member to keep it in place? Will cavitation occur at any of the points P, I, M, and O for this flow rate? At what points is cavitation first likely to occur? (Include point O in your consideration.)

5.51 Assume that the velocity distribution given in Example 5.3.2 exists downstream of the sluice gate, and determine ω_y. Is this velocity distribution one that gives the same volume rate of flow as occurs upstream of the gate?

5.52 The flow in a channel upstream of a sluice gate need not always have a constant velocity. Refer to Fig. 5.3.2, where the velocity at A is now 6 ft/s and decreases linearly to 4 ft/s at F. Can the Bernoulli equation be written between points A and D? Solve for V_D. What is Q? Use the new velocity distribution downstream of the gate, where it would be more easily measured.

5.53 The velocity distribution given in Example 5.3.3 can be used to model the winds in a cyclonic storm, except near the very center. (Actually, the velocity falls rapidly to zero near the "eye," or center of such a storm.)

Determine how the pressure varies with the distance from the center of a storm. Define the radius at which the air pressure would be three-quarters of the ambient pressure $p(r \to \infty) = p_\infty$. Assume constant density.

5.54 If one skims a spoon across the surface of a cup of coffee, two swirls are generated which move across the surface. These are generally called *vortices*. The velocity distribution approximates $V_\theta = V/r$ for $r \geq r_c$ and Ωr for $r \leq r_c$. With $V_\theta = 1$ ft/s at $r_c = 1.5$ in., determine the form of the free surface for such a vortex. (Note that for $r \geq r_c$, the velocity field is irrotational, so that the Bernoulli equation can be applied from streamline to streamline; for $r \leq r_c$, the Euler equations will give $\partial p/\partial z$, $\partial p/\partial r$, and $\partial p/\partial \theta$ in the total differential dp.)

5.55 A tornado has a swirl velocity distribution similar to that described in Prob. 5.54. If viscous effects and an updraft near the eye of the storm produce a value of $r_c = 50$ m—as defined in Prob. 5.54—where V_θ has a value of 150 km/h, what pressure can be expected at the eye of the storm? What will be the pressure 500 m from the eye? (Assume an air density of 1.2 kg/m³ and an atmospheric pressure 100 kPa far from the storm's center.) At the center of the storm, the pressure inside a building will remain close to the local ambient pressure. Compare the magnitude of the loading exerted on a flat roof by the storm's maximum pressure decrease with that of a snow load by determining the equivalent depth of snow. (Assume that 7 in. of snow will exert the same pressure as 1 in. of water.)

5.56 The reader can assume that the velocity distribution given in Example 5.3.3 exists in a vortex separator. In such a conical device the fluid is introduced tangentially, say, at the bottom of the cone. This causes all of the fluid therein to rotate. If a lighter fluid, oil, for example, is present in a mixture with water, it will move toward the axis of rotation and rise toward the top of the cone, where it can be removed. (A second exhaust port at the outer edge of the case is provided for the "purified" water.) What value of the constant in the equation for the velocity would give vapor pressure ($T = 20°C$) at $r = 1$ cm if atmospheric pressure exists at $r = 60$ cm?

The Flow of an Inviscid Fluid   Chap. 5

5.57 An elbow meter is used to measure the volume rate of flow by measuring the pressure difference between the inside and outside radii of a pipe bend. This flow case could be modeled two-dimensionally as shown in Fig. P.5.57. Let $r_2 = 4$ ft and $r_1 = 3$ ft. For a uniform inlet velocity of 9 ft/s at point O and a velocity distribution in the bend after point O given as $V_\theta = K/r$, determine K from mass conservation considerations. What is $p_2 - p_1$ for a water flow rate of Q?

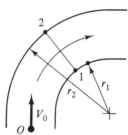

Figure P.5.57 Horizontal view of elbow meter.

5.58 If the pressure at point O in Fig. P.5.57 is 10 psi vacuum, what volume rate of flow will cause cavitation in the bend, which could simulate the inlet section of a pump? (Where do you expect this phenomenon to occur in the bend?)

5.59 The temperature in a room is given as $T = 20 + 0.2x - 0.4z$ (x and z are in meters). What is the temperature gradient in the room? What is the differential equation for surfaces of constant temperature? What geometric relationship does grad (T) have to this surface?

5.60 On the assumption that the temperature in a region can be written as $(T_0 - C/R^2)$, in which R is the radial distance, based on spherical geometry, from the origin, what is the temperature gradient? What are the surfaces of constant temperature? Is their normal parallel to the local temperature gradient?

5.61 If a stationary container is filled with a saline solution so that its density varies linearly from $\rho = 1020$ kg/m³ at the bottom to 1005 kg/m³ at the top, what is the density gradient if the liquid depth is 0.6 m? If this container is accelerated horizontally, the surfaces of constant density will tilt so that they coincide with the surfaces of constant pressure. For

$a_x = 0.3g$, what are these constant-density surfaces? What is the density gradient with this acceleration?

5.62 For the velocity potential given in Example 5.4.1, determine the value of u, v, and the speed at $(2, 1)$. What is the slope of the streamline? Determine if div $\mathbf{V} = 0$ everywhere. Comment on the result.

5.63 For the velocity potential given in Example 5.4.1, determine the acceleration of a particle at $(2, 1)$.

5.64 Determine u, v, and w for a three-dimensional source with $D = 10$. What is the speed at $(2, 1, 2)$? What is the direction of the streamline? Is div $\mathbf{V} = 0$ everywhere? What is the consequence on the flow field if $D = -10$?

5.65 Refer to Prob. 5.64 and determine u_R, u_θ, and u_ϕ, directly using spherical coordinates. What is $|\mathbf{V}|$?

5.66 Determine u, v, u_r, and u_θ for a two-dimensional vortex with $C = 5$. Evaluate your components at $(3, 4)$.

5.67 Describe the flow, by discussing the streamlines, that results from the combination of a two-dimensional source and a vortex.

5.68 If a source and a uniform flow are combined with $U = 6$ m/s and $A = 3.5$ m²/s, what are u and v at $(4, 3)$, $(-4, 3)$, $(-4, 0)$? Where are both u and v equal to zero—i.e., where is the stagnation point? Does $\mathbf{V} \to U\mathbf{i}$ at $x = 100$? What is the discharge from the source, and where does this fluid go? From $Q = \int \mathbf{V} \cdot \mathbf{n} \, dA$ determine the width at $x = 100$ of the stream that emanates from the source.

5.69 Where is the stagnation point for a three-dimensional source and a uniform flow in the x direction? [Let U and D in Eqs. (5.4.10) and (5.4.14) be arbitrary.] What is the velocity (speed and direction) as $x \to +\infty$?

5.70 Use the stream function given in Example 5.4.2 to determine u_r and u_θ on $r = 3$. Where on this circle is $p = p_\infty$? Where are the stagnation points on $r = 3$? Is $\omega_z = 0$?

5.71 Combine the stream function given by Eq. (5.4.16) with a two-dimensional vortex with $C = -2$. What are u_r and u_θ on $r = 3$? Where are the stagnation points? (Do they occur on $r = 3$ for all values of C?)

5.72 Use the addition of the vortex to the flow in

Prob. 5.71 to find the force on a circle at $r = 3$. How does this force vary with the value of C? (The constant C is proportional to a term called the *circulation,* which determines the lift. In a manner similar to a "slicing" golf ball, the greater swirl of the vortex, the greater the lift.)

5.73 If a source is placed at $(-1, 0)$, called point P, the stream function would be $A \theta_1$ and the velocity potential $A \ln r_1$. Let $A = 3$. The angle θ_1 is formed by a ray from P and the x axis. In this case, the radial velocity is $3/r_1$, in which r_1 is measured from P. Place another source with the same constant A at $(+1, 0)$, called point Q, and graphically determine the velocity due to (the sum of) sources at P and Q along $x = 0$. Is $x = 0$ a streamline? Can you justify calling the sources at P and Q images of one another? [*Hint:* Does the line $x = 0$ form a plane of symmetry (i.e., a reflection)?] What is the pressure at $(0, 5)$ with respect to the pressure at large x and y where $p = p_\infty$?

5.74 If a source is placed at $(-1, 0)$, called point P, and a sink with the same strength at $(+1, 0)$, called point Q, the streamlines will be circles passing through P and Q with the center along the y axis. Choose a value $A = -B = 4$ in Eqs. (5.4.11) and (5.4.12), but use r_1 as in Prob. 5.73 and let r_2 be measured from the sink. What is the velocity along the y axis? Could this velocity be consistent with circular streamlines whose centers are along the y axis?

5.75 Refer to Prob. 5.74 and relate r_1 and r_2 to r and θ by means of the cosine law. Use equation group (5.4.7) to determine u and v. What is the slope of the velocity? What is the slope of the streamlines? Is this slope the same as that of a circle with an origin on the y axis?

5.76 If a source is placed at $(-1, 0)$, called point P, and a sink of equal strength at $(+1, 0)$, called point Q, the streamlines will be circles. If a uniform flow is now added with a potential of Ux, a closed streamline is formed. This closed streamline could model a body that is submerged in a flow that has a uniform velocity U far from the body. This is called a *Rankine body.* Let $A = 8$ in Eq. (5.4.11) and $U = 4$. Show that the stagnation points are $(-\sqrt{5}, 0)$ and $(+\sqrt{5}, 0)$. Note that the streamlines for the source at P are given by $A \theta_1$, and deter-

mine the value of the streamline (due to the source, sink, and uniform flow) at the stagnation points. Find the points along the x and y axes where $\psi = 0$. These and all other points in the plane for which $\psi = 0$ constitute the curve of the Rankine body.

5.77 With the aid of a digital computer, let the distance between the source and sink, PQ, go from 2 to 1 and then to 0.5 while the value of A goes from 8 to 16 and then to 32, respectively. (Refer to Prob. 5.76 for the definition of the symbols.) Let U remain at 4. Find the stagnation points and the maximum magnitude of y on the streamline forming the Rankine body, $\psi = 0$. For this case, in which the product of the distance between the source-sink combination and the constant A remains a constant while the distance goes to zero, the Rankine body becomes more and more circular. Can you infer from the determined values for x_{stag} and y_{max} that the body is approaching a circle in the limit of $(A)(PQ) = $ constant as $PQ \to 0$? In this limit, the streamlines will be given by the form of Eq. (4.7.6).

5.78 If the two sources in Prob. 5.73 are combined with a vertical uniform flow (from top to bottom) given by $-Vy$ for its potential function, two stagnation points should ensue, as well as two half bodies similar to the one shown in Fig. 5.4.2. For the values of A given in Prob. 5.73 and $V = 32$, locate the stagnation points. (A digital computer may be helpful.) What is the velocity at $(0, 0)$ and $(0, -0.9)$? What is the pressure at these points with respect to the pressure in the uniform flow far from the sources, $(p - p_\infty)/\rho$?

5.79 Use a digital computer to determine the minimum spacing between the sources in Prob. 5.78 that would give the pressure change $(p - p_\infty)/\rho$ of -60 somewhere along the y axis. (This problem is intended to indicate that when a fluid passes between a row of bodies—pump vanes, for example—the pressure can decrease locally to such a value that local cavitation and other phenomena could occur.)

5.80 Use the flow net of Fig. 5.5.1 to determine the pressure variation along the straight line MN. Determine the force on this surface.

6

The Conservation of Energy

6.1 INTRODUCTION

The concepts introduced in this chapter are a natural extension of the results of Chap. 5. There, we presented the Bernoulli equation, which is a special case of the relationship between work and energy. This equation was the result of integrating Newton's laws of motion under restrictive conditions (e.g., absence of viscosity, incompressible or barotropic fluids, body forces restricted to gravitational effects, and steady flow). The result was that the sum of the terms $V^2/2 + p/\rho + gz$ in the steady-flow Bernoulli equation is constant along a streamline.

However, prior to this result we had written, in effect,

$$\rho \frac{\partial}{\partial s}\left(\frac{V^2}{2} + gz\right) = -\frac{\partial p}{\partial s} \tag{6.1.1}$$

This equation states that if the sum of the kinetic and potential energies of a unit mass of fluid is increasing, then $\partial p/\partial s$ must be negative. Hence, the pressure in the fluid must be decreasing along that portion of the streamline. It will be assumed that the speed of the fluid is increasing along the streamline, and that this is the reason why the left-hand side of Eq. (6.1.1) is positive. With these prescriptions of the pressure distribution and speed changes in the flow, we can conclude that the fluid surrounding a particular material volume exerts a force on it in the direction of motion. (If the pressure decreases in the flow direction, there is a net force on the material volume due to this pressure difference.) Thus, the surroundings would be doing work on the material volume, $\mathbf{F}\cdot d\mathbf{s}$, as it moves along its streamline. This work that is applied to a material volume manifests itself as an increase in the kinetic (and potential) energy of the minute system of fluid particles.

This interpretation of the Bernoulli equation can be summarized as follows. The sum of the kinetic and potential energies—terms that can be called *mechanical energy* because of their relationship with Newton's laws—and the rate of work done on a material volume during its motion are always completely in balance. Because we restricted ourselves to an inviscid fluid in arriving at this conclusion, the effect of a dissipative mechanism—fluid viscosity—was excluded from the analysis. If there is some way in which energy can be extracted from a packet of fluid in a dissipative way, we would expect that the conservative relationship between the change of kinetic and potential energy and the work expended on the packet of fluid would not remain in balance. A similar situation would exist if the material volume were to enter a region where it would receive a sudden influx of energy from some external power source, such as a pump. Then there would be a sharp jump in the total energy of the material volume, so that the sum $(V^2/2 + p/\rho + gz)_{\text{upstream}}$ that existed ahead of the pump would be less than the sum $(V^2/2 + p/\rho + gz)_{\text{downstream}}$ that occurs after the pump.

An understanding of the situation can be gained by studying the consequences of the first law of thermodynamics and deriving an equation that is sometimes called the *general energy equation*. In the pursuit of this result, we shall follow the path of reasoning that led to the equation for mass conservation, Eq. (3.8.1). In the development that follows, we shall need to determine the net flux of energy out of a control volume. Therefore, it is natural that we turn now to consider the flux of kinetic energy flowing across a section of a control surface.

6.2 THE FLUX OF KINETIC ENERGY

Sometimes a portion of the control surface is associated with the plane of a conduit's cross section, as in Fig. 6.2.1. A velocity distribution is also given in this figure for purposes of discussion. This particular velocity profile has fluid moving both to the left and to the right. In view of this, it can be expected that there is a flux of kinetic energy—the time rate of flow of kinetic energy—in both directions. It might be of interest to know the net flux of kinetic energy over a cross section in a prescribed direction. In order to determine such a flux, it is necessary to assign a positive and a negative direction to the flow so that the sign of the final result will tell us the direction of the total flux of kinetic energy that passes through the cross section in question. This sense of direction is prescribed in the present example on the surface *CD*, which is a

Figure 6.2.1 Nonuniform velocity profile for stationary lower boundary and moving upper boundary: $u = U_0[1 - (y/b)^2] - U_0(y + b)/4b$. Solid curve at right: Velocity profile across boundary *CD*.

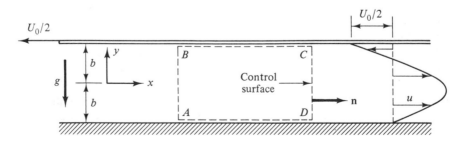

segment of the entire control surface, *ABCDA,* by an outer normal, **n**. (The unit vector **n** points outward from the control volume.) If the velocity is in this direction on the surface, one will have a positive efflux of mass, as well as kinetic energy, from the control volume.

The quantity of fluid that is flowing in the direction of positive **n** is $(\mathbf{V \cdot n})(dA)(dt)$. The mass rate of flow is the product of the mass per unit volume and the volume per unit of time, $\rho(\mathbf{V \cdot n})(dA)$. The dimensions of this product are those of mass per unit time, which the reader can easily verify. Each unit of mass that flows has some kinetic energy (KE), and this could be quantified as the kinetic energy per unit mass. The product of this kinetic energy per unit mass and the mass per unit time flowing across an element of area dA would give the amount of kinetic energy flowing across dA per unit time. In terms of the dimensions of the quantities, we have (KE/mass)(mass/time) = KE/time. Thus, it is apparent that we need to know the kinetic energy per unit mass of the flowing fluid. This is not difficult to determine. Since the kinetic energy for a particle with mass m is expressed as

$$\text{KE} = \tfrac{1}{2}(m)(\text{speed of particle})^2$$

the KE/mass of the flowing fluid is $(\mathbf{V \cdot V})/2$. The foregoing discussion gives us the reason to write the flux of kinetic energy flowing across CD from the control volume as

$$\frac{\text{flow of KE}}{\text{second}} = \Sigma \left(\frac{\text{KE}}{\text{mass}} \right) \left(\frac{\text{flow of mass}}{\text{second}} \right)$$

or

$$\text{KE flux} = \dot{\text{KE}} = \int \frac{\mathbf{V \cdot V}}{2} \rho \mathbf{V \cdot n} \, dA \tag{6.2.1}$$

Example 6.2.1

Find the flux of kinetic energy across CD in Fig. 6.2.1 in the direction of positive **n** for the velocity distribution

$$u = U_0 \left[1 - \left(\frac{y}{b} \right)^2 \right]$$

if W is the width of the plates. The density will be assumed to be constant over surface CD.

Solution On CD, we have $\mathbf{V \cdot n} = u$ and $\mathbf{V \cdot V} = u^2$. Equation (6.2.1) provides us with

$$\dot{\text{KE}} = W \rho U_0^3 \frac{1}{2} \int_{-b}^{+b} \left[1 - \left(\frac{y}{b} \right)^2 \right]^3 dy$$

$$= \tfrac{16}{35} b W \rho U_0^3 \tag{6.2.2}$$

if one expands the cubic expression in the integral.

In this example, we have found the flux of kinetic energy across the surface CD by using the specified velocity distribution. It is reasonable to ask how this actual flux of kinetic energy would compare with the value that would be calculated using the average velocity. This would be

$$\dot{\overline{\text{KE}}} = \int_{-b}^{+b} \rho \overline{V} \, dA \left(\frac{\overline{V}^2}{2} \right) = \frac{1}{2} \rho \overline{V}^3 \int_{-b}^{+b} dA = \frac{1}{2} \rho \overline{V}^3 (2b) W$$

$$= \rho \overline{V}^3 b W \tag{6.2.3}$$

The overbar in $\overline{\dot{KE}}$ is to emphasize that it is based on the average velocity. A comparison between the actual kinetic energy flux and this value requires that one introduce the value of the average velocity into this expression. Q, and therefore the average velocity, can be found for the velocity distribution used in the example as

$$Q = \int_{-b}^{b} U_0\left[1 - \left(\frac{y}{b}\right)^2\right] W \, dy = WU_0\left(y - \frac{y^3}{3b^2}\right)\Big|_{-b}^{b}$$

$$= \tfrac{4}{3} U_0 W b$$

and
$$\text{Area} = 2bW$$

Thus,
$$V_{avg} = \overline{V} = \frac{Q}{A} = \frac{2}{3}U_0 \tag{6.2.4}$$

This value of $\overline{V}$ can now be inserted into Eq. (6.2.3) to give

$$\overline{\dot{KE}} = \rho bW (\tfrac{2}{3}U_0)^3 = 8\rho bW U_0^3/27 \tag{6.2.5}$$

Upon comparing the values for the kinetic energy flux calculated in Eqs. (6.2.2) and (6.2.5), it can be seen that when the actual, nonconstant velocity profile is used, the value of the flux is greater than that which is calculated using the average, or constant, value. This is always the case. Often, it is not difficult to determine the average velocity, and consequently a computation of the kinetic energy that utilizes the average velocity would be easy to make. This value would be less than the one that would result from using the existing velocity profile, but it could, nevertheless, be useful. Thus it seems natural to seek or calculate a set of "correction factors" to obtain equivalent results when using the two methods for calculating the flux of kinetic energy. Such a correction factor would be a constant, called α, for a particular velocity profile such that

$$\text{KE flux, actual} = \alpha(\text{KE flux using average velocity})$$

or
$$\alpha = \frac{\int \frac{\rho}{2} u^3 \, dA}{\int \frac{\rho}{2}(\overline{V})^3 \, dA} = \frac{1}{A}\int\left(\frac{u}{\overline{V}}\right)^3 dA \tag{6.2.6}$$

for ρ = constant. This ratio α is called the *kinetic energy correction factor*. Values for this coefficient can be tabulated for various velocity profiles. Then a detailed velocity survey might not be necessary if a good estimate of the actual flux of kinetic energy were desired and the general description of the flow fitted one of those whose kinetic energy had been cataloged.

Example 6.2.2

Under certain conditions, the flow in a round pipe can be described by

$$u = u_0\left(1 - \frac{r^2}{a^2}\right)$$

in which u_0 is the centerline speed and a is the pipe's radius. Show that the average velocity is $u_0/2$, and find the kinetic energy correction factor.

Solution The actual kinetic energy per unit time is

$$\dot{KE} = \int \frac{u^2}{2}\rho u\,(dA) = \frac{\rho}{2}u_0^3 \int_0^a \left(1 - \frac{r^2}{a^2}\right)^3 (2\pi r)(dr)$$

$$= \frac{\rho u_0^3 \pi a^2}{2} \int_0^a \left(1 - \frac{r^2}{a^2}\right)^3 \left(\frac{2r\,dr}{a^2}\right)$$

in which a^2 is introduced because the derivative of r^2/a^2 is $(2r\,dr)/a^2$. The integration is straightforward and yields

$$\dot{KE} = \frac{-u_0^3 \pi a^2}{8}\rho \left(1 - \frac{r^2}{a^2}\right)^4 \Big|_0^a = u_0^3 \frac{\pi a^2}{8}\rho$$

One can verify from $Q = \int u\,dA$ that $Q = \pi a^2 u_0/2$. Thus, $V_{avg} = u_0/2$ and

$$\overline{\dot{KE}} = \frac{V_{avg}^2}{2}[(\rho V_{avg})(\pi a^2)] = \frac{\rho}{2}(\pi a^2)\frac{u_0^3}{8}$$

and for $\alpha(\overline{\dot{KE}}) = \dot{KE}$,

$$\alpha = 2$$

6.3 THE ENERGY EQUATION

We begin by referring to the work of Sec. 6.2. We know that the expression

$$\int_{CS} \left(\frac{V^2}{2}\right)(\rho \mathbf{V} \cdot \mathbf{n})\,dA$$

represents the net kinetic energy efflux across the control surface. We extend this notion now and, upon reflection, conclude that the potential energy per unit mass is gz. It follows that

$$\int_{CS} (gz)(\rho \mathbf{V} \cdot \mathbf{n})\,dA$$

is the net efflux of potential energy across the control surface. Following this line of investigation one step further, we define e as the internal energy* of the fluid per unit mass, so that

$$\int_{CS} e(\rho \mathbf{V} \cdot \mathbf{n})\,dA$$

represents the net internal energy efflux across the control surface. We shall think of internal energy as being associated with the temperature of the fluid. This temperature might rise as a result of the transfer of energy as heat from the surroundings to the fluid in the control volume. An increase of temperature could also occur as a result of the fluid friction (i.e., viscosity) in a flowing system. However slight this elevation of temperature might be, it could be important in determining the ultimate conditions that occur in a flow.

With the recognition that we shall be concerned with three forms of energy—kinetic, potential, and internal—we can write the sum of the efflux of this combination of three energies as

---

* In some texts on thermodynamics, the internal energy is designated by u, but this could cause some confusion with our use of u as the x component of velocity.

$$\int_{CS} \left(\frac{V^2}{2} + gz + e \right) (\rho \mathbf{V} \cdot \mathbf{n}) \, dA$$

This integral is one part of the expression for the time rate of change of the total energy of a material volume, according to the Reynolds transport theorem. This was given by Eq. (3.9.2) and becomes, in this case

$$\frac{\Delta(\mathrm{KE} + \mathrm{PE} + \mathrm{IE})}{\Delta t} = \frac{\partial}{\partial t} \int_{CV} \left(\frac{V^2}{2} + gz + e \right) \rho \, d\forall$$

$$+ \int_{CS} \left(\frac{V^2}{2} + gz + e \right) \rho \mathbf{V} \cdot \mathbf{n} \, dA$$

We can relate the rate of change of the total energy to the material volume *ABCDEA* in Fig. 6.3.1 at t_0. This material volume is a general representation of any flow system to which energy is being added in the form of heat and work. The first law of thermodynamics states that for a closed system,

$$\left(\begin{matrix} \text{Work done on fluid per unit time} \\ \text{to change energy of fluid system} \end{matrix} \right) + \left(\begin{matrix} \text{heat added to fluid per unit time} \\ \text{to change energy of fluid system} \end{matrix} \right) = \left(\begin{matrix} \text{total change of energy of} \\ \text{fluid system per unit time} \end{matrix} \right)$$

The works of J. Joule and others have demonstrated that the result of doing work on a system (a material volume, for example) or adding heat to it, is an increase of the system's energy. One can say that energy was transferred to the system in the form of heat or work. The sources of this "energy in transition" are chemical and nuclear activities that release the energy that has been stored. Work could be realized because of a chemical process in a battery that produces the electricity to drive a motor. Equally well, combustion or nuclear fission could supply the energy to a steam engine. Subsequently, we shall not write the lengthy phrases in the previous statement of the first law. Instead, the phrases *work done on the fluid* and *heat added* or *heat transferred*

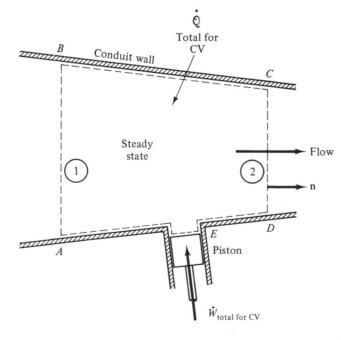

Figure 6.3.1 Control volume into which energy is added in the form of heat and mechanical work.

will be used. This phraseology includes implicitly the concept that work and heat are the agents for transferring energy to the system from some other source. We shall now examine each of the phrases within the parentheses and write an equivalent mathematical expression for it. In effect, we already have found an expression for the last phrase. We turn now to the first one.

Some of the boundaries in Fig. 6.3.1 have fluid flowing across them (e.g., *AB, CD*), and others have rigid boundaries (e.g., *BC*), while a portion of the boundary, such as location *E* in Fig. 6.3.1, may be moving. There is no net mass transfer across this moving boundary. If work is to be done at a boundary* on a fluid system, one must have

1. A stress, either normal or tangential, and
2. A motion of the boundary such that work, in fact, occurs per unit of time (work per unit time is designated *power,* defined as $\mathbf{F \cdot V}$)

It is useful to examine the work done at boundaries, such as location *E* in Fig. 6.3.1, which is associated with some external device such as a pump. If there is work on the fluid in the control volume, this work is greater than 0; for a turbine, or for any work on the surroundings of the control volume, this work is less than 0. For the piston shown in location *E* of Fig. 6.3.1, it can be concluded that the rate of work, or power, input to the fluid is

$$\dot{W} = (\text{speed of piston advance})(\text{pressure})(\text{area})$$

This is a particular case of a more general expression that can be written when normal stresses are considered for the time rate of work done on the fluid. Then the magnitude of the normal stress vector acting *against* the piston must be used in the following expression instead of the pressure:

$$\dot{W} = -(\text{pressure})(\mathbf{V \cdot n})(\text{area of action of stress})$$

The negative coefficient is necessary to obtain a positive value of $\dot{W}$ when $\mathbf{V \cdot n}$ is negative. (Note that $\mathbf{n}$ is an outer normal to the control volume.)

This last result can be applied to section *AB* also. Here the fluid to the left of the boundary (i.e., outside) is doing work on the fluid inside the control volume at the rate of

$$d\dot{W}'_{AB} = -[(p)_{AB}][-(|\mathbf{V}|)_{AB}](dA_{AB})$$

That is,
$$\dot{W}'_{AB} = \int_{AB} p(V)\, dA$$

is the rate of work being performed by the surroundings across *AB*, on the fluid in the control volume. The use of $\dot{W}'$ on *AB*, instead of simply $\dot{W}$, denotes work performed at a boundary across which fluid is passing. The work associated therewith is often referred to as *flow work*. Across *CD*, the situation is

$$\dot{W}'_{CD} = -\int_{CD} p(V)\, dA$$

---

*Work can also be done on volume elements of the fluid. If the fluid were magnetic, an alternating magnetic field could do work on it. Note that one can also view potential energy as work done against gravity forces.

since **n** and **V** are directed the same here. The minus sign shows that across CD the rate of work on the fluid in the control volume is negative. This implies that work is being done on the fluid outside the control volume, with the consequence that energy is being extracted from the control volume. The time rate of change of energy will be referred to as *power* hereafter.

Across BC, $\mathbf{V \cdot n} = 0$. This is also true over the major portion of the lower boundary in Fig. 6.3.1, except, of course, at location E, a fact that has been discussed already.

The result of this discussion of the work done on the fluid within the control volume allows us to summarize it as

$$\dot{W}' = -\int_{\text{cs}} p\,(\mathbf{V \cdot n})\,dA \qquad (6.3.1)$$

This is the rate of energy transferred to the mass within the control volume by the surrounding fluid due to the existing pressure field. The calculation of $\dot{W}'_{AB}$ and $\dot{W}'_{CD}$ can be done readily when the flow is steady. The rate of work being done by a reciprocating piston, such as in the illustration at point E, would be an unsteady quantity. One would use the integrated value over one cycle in that case to find the work being done on the fluid within the control volume for that cycle.

Certainly, it may be that work is done on the fluid within the control volume due to shear stresses. The moving boundary of Fig. 6.3.2 is a case in point. Some of the input power in this figure would have resulted in a movement of the plate and can be considered to be useful. But some portion may manifest itself as a temperature rise within the fluid because of fluid viscosity. A fraction of the energy transferred to the fluid as work could pass out of the system to the surroundings (i.e., could be carried away or transferred) in the form of heat. This thermal energy is considered to be "lost" to the system within the control volume because it is usually unfeasible to convert it to useful work.

Example 6.3.1

What is the power input to the fluid flowing between the plates in Fig. 6.3.2? Use a length of plate $= L$ and a unit width in the z direction.

Solution The power will be the product of the force that the plate exerts on the fluid in shear at $y = h$ and the speed of the surface in the direction of this shear. This means that

$$\dot{W} = (\text{shear stress})(\text{area})(U)$$

$$= \left(\frac{\mu U}{h}\right)(L)(1)(U) = \frac{\mu U^2 L}{h}$$

in dimensions of (force × length)/time.

$U \longrightarrow$

Figure 6.3.2 Mechanical power transfer to a fluid by means of a shear stress at a moving boundary. Moving plate is at top; stationary plate, at bottom. Dashed line: velocity distribution.

The moving plate should remind one that the blade of a rotating propeller can also do work on a fluid due to the tangential stresses.

At the outset of our discussion, the energy equation was written, in effect, as

$$\begin{pmatrix} \text{Power transferred to system} \\ \text{by mechanical means} \end{pmatrix} + \begin{pmatrix} \text{power transferred to system} \\ \text{by thermal means} \end{pmatrix} = \begin{pmatrix} \text{time rate of change of} \\ \text{total energy of fluid} \end{pmatrix}$$

or $\qquad\qquad A \qquad\qquad + \qquad\qquad B \qquad\qquad = \qquad\qquad C$

Items A and C have now been discussed. The rate of energy supplied to the control volume by heat will be designated simply as $\dot{\mathcal{Q}}_{\text{total}}$. This means that the energy equation could be written as

$$-\int_{\text{CS}} p\,(\mathbf{V}\cdot\mathbf{n})\,dA + \dot{\mathcal{Q}}_{\text{total across CS}}$$

$$= \int_{\text{CS}} \left(\frac{|\mathbf{V}|^2}{2} + gz + e\right)\rho(\mathbf{V}\cdot\mathbf{n})\,dA + \frac{\partial}{\partial t}\int_{\text{CV}} \left(\frac{|\mathbf{V}|^2}{2} + gz + e\right)\rho\,d\mathbf{V}$$

Finally, we separate the rate of work being done on the fluid in the control volume due to external devices—e.g., the piston in Fig. 6.3.1—from that occurring on segments of the control surface across which mass is flowing, the $\dot{W}'$ terms. Such segments are surfaces AB and CD in Fig. 6.3.1. Designating this power from an external device as $\dot{W}$ and using the pressure as the normal stress, we can write the previous equation as

$$-\int_{\text{CS}} p\,(\mathbf{V}\cdot\mathbf{n})\,dA + \dot{W} + \dot{\mathcal{Q}} = \int_{\text{CS}} \left(\frac{|\mathbf{V}|^2}{2} + gz + e\right)\rho(\mathbf{V}\cdot\mathbf{n})\,dA$$

$$+ \frac{\partial}{\partial t}\int_{\text{CV}} \left(\frac{|\mathbf{V}|^2}{2} + gz + e\right)\rho\,d\mathbf{V}$$

(In a viscous, incompressible fluid, the thermodynamic pressure is the average of the normal stresses. The normal stresses in the three orthogonal directions can differ from the pressure because of the effects of viscosity and velocity gradients. Chapter 16 treats this topic.) Now the surface integral on the left-hand side of the last equation, which represents the power associated with the flow across the control surface, can be combined with the one on the right merely by introducing p/ρ as the coefficient of p. The result is the energy equation for an open system:

$$\dot{W} + \dot{\mathcal{Q}} = \frac{\partial}{\partial t}\int_{\text{CV}} \left(\frac{|\mathbf{V}|^2}{2} + gz + e\right)\rho\,d\mathbf{V}$$

$$+ \int_{\text{CS}} \left(\frac{|\mathbf{V}|^2}{2} + gz + \frac{p}{\rho} + e\right)\rho(\mathbf{V}\cdot\mathbf{n})\,dA \qquad (6.3.2)$$

It is useful to replace the term gz by $g(\bar{z} + z')$, in which $\bar{z}$ corresponds to the location of the centroid of the conduit's cross section in Fig. 6.3.1 and z' is the vertical distance from this point. Thus, z' will have both positive and negative values. For a circular segment of the control surface, $\bar{z}$ will be at its center. If the velocity profile is symmetric about the axis of a circular duct, one would have, for $\mathbf{V}\cdot\mathbf{n} > 0$ and $\rho =$ constant at the cross section, for example, in Eq. (6.3.2),

$$\int \rho g z\, \mathbf{V}\cdot\mathbf{n}\, dA = \int |\mathbf{V}|\rho g z\, dA = \int |\mathbf{V}|\rho g \bar{z}\, dA + \int |\mathbf{V}|\rho g z'\, dA$$
$$= \rho g \bar{z}\, V_{\text{avg}}\, A + 0$$

because of the asymmetry of z' and the assumed symmetry of $|\mathbf{V}|$ with respect to $\bar{z}$.

For the situation shown in Fig. 6.3.1, some further, reasonable assumptions can be made to facilitate our work in specific problems. (It should be kept in mind, however, that the situation shown in Fig. 6.3.1 is a special, albeit quite common, case and Eq. (6.3.2) would be valid for problems in which the fluid flows in and out of more than two locations. We shall first employ Eq. (6.3.2) in the useful and simple case of *uniform flow at a single inlet and single outlet for a steady flow*—i.e.,

$$\frac{\partial}{\partial t} \int_{CV} \left(\frac{V^2}{2} + gz + e \right) \rho \, d\mathbf{V} = 0$$

At the *upstream* section,

1. Let V be the average velocity at the section. Call it V_1.
2. Let p be independent of location. Call it p_1. Note that we have accounted for the changes in pressure that can occur due to hydrostatic pressure by the introduction of z' with an implicit $p' = \gamma z'$.
3. Let ρ be independent of location. Call if ρ_1.
4. Let e be independent of location. Call it e_1.

We shall make similar assumptions at the downstream section, CD, and use the subscript 2 to designate the flow parameters. $\mathbf{V} \cdot \mathbf{n} = 0$ over a portion of the boundaries shown in Fig. 6.3.1. This means that our labor is less. Across AB,

$$\mathbf{V} \cdot \mathbf{n} = -V_1$$

and across CD,

$$\mathbf{V} \cdot \mathbf{n} = +V_2$$

With the *assumptions* that have been made above, Eq. (6.3.2) reduces to

$$\dot{W} + \dot{\mathcal{Q}} = V_2 \rho_2 A_2 \left(\alpha_2 \frac{V_2^2}{2} + g\bar{z}_2 + \frac{p_2}{\rho_2} + e_2 \right)$$
$$- V_1 \rho_1 A_1 \left(\alpha_1 \frac{V_1^2}{2} + g\bar{z}_1 + \frac{p_1}{\rho_1} + e_1 \right) \qquad (6.3.3)$$

in which $\bar{z}_1$ and $\bar{z}_2$ are the elevations of the centroids of the cross sections A_1 and A_2. (V_1 and V_2 are the average velocities at these sections even if the velocity is not strictly uniform. To allow this possibility, α_1 and α_2 have been included to properly account for the kinetic energy.) Each term on the right-hand side of Eq. (6.3.3) contains the mass rate of flow $\dot{m}$ as a coefficient. Equation (6.3.3) can be written as

$$\frac{\dot{W}}{\dot{m}} + \frac{\dot{\mathcal{Q}}}{\dot{m}} + \left(\alpha_1 \frac{V_1^2}{2} + \frac{p_1}{\rho_1} + g\bar{z}_1 \right) = \left(\alpha_2 \frac{V_2^2}{2} + \frac{p_2}{\rho_2} + g\bar{z}_2 \right) + (e_2 - e_1) \qquad (6.3.4)$$

This is an *intensive* equation in the sense that it is valid for each unit of mass flowing per unit time. Clearly, to obtain $\dot{W}$ one must multiply the quantity $\dot{W}/\dot{m}$ by $\dot{m}$.

The difference in the internal energies, $e_2 - e_1$, was purposely placed on the right-hand side of Eq. (6.3.4). One can think of Eq. (6.3.4) as a statement about the power associated with a flowing fluid in the following general way:

The sum of the fluid's original mechanical energy terms—the same as in the Bernoulli equation—plus the mechanical and thermal energy transferred from external sources yields the final mechanical energy of the fluid plus the energy difference due to internal energy changes.

An equivalent form of Eq. (6.3.4) is

$$\frac{\dot{W}}{g\dot{m}} + \frac{\dot{\mathcal{Q}}}{g\dot{m}} + \left(\alpha_1 \frac{V_1^2}{2g} + \frac{p_1}{\gamma_1} + \bar{z}_1\right)$$

$$= \left(\alpha_2 \frac{V_2^2}{2g} + \frac{p_2}{\gamma_2} + \bar{z}_2\right) + \frac{e_2 - e_1}{g} \tag{6.3.5}$$

The dimension of the terms in Eq. (6.3.5) appears to be a length. Actually, the terms have the dimensions of (power)/(weight of fluid flowing per unit time). (Note that $g\dot{m}$ is the weight rate of flow.)

If the fluid is a gas, the density ρ can vary. It is useful to identify the sum $p/\rho + e$ as the enthalpy h. The enthalpy is a tabulated function for most gases commonly used. If the specific heat is constant in the temperature range of interest, one can use $(h_2 - h_1) = c_p(T_2 - T_1)$. An incorporation of h into Eq. (6.3.3) results in

$$\dot{W} + \dot{\mathcal{Q}} + \dot{m}_1\left(\alpha_1 \frac{V_1^2}{2} + gz_1 + h_1\right) = \dot{m}_2\left(\alpha_2 \frac{V_2^2}{2} + gz_2 + h_2\right) \tag{6.3.6a}$$

In writing this equation for the flow in a conduit such that $\dot{m}_1 = \dot{m}_2$, one could conveniently divide by this mass rate of flow and obtain an equation similar to Eq. (6.3.4). The extension of Eq. (6.3.6a) to cases in which fluid is transported through more than two parts of the control surface follows the same logic. The result is

$$\dot{W} + \dot{\mathcal{Q}} + \sum_{i=1}^{N_{\text{in}}} \dot{m}_i\left(\alpha_i \frac{V_i^2}{2} + g\bar{z}_i + h_i\right) = \sum_{j=1}^{N_{\text{out}}} \dot{m}_j\left(\alpha_j \frac{V_j^2}{2} + g\bar{z}_j + h_j\right) \tag{6.3.6b}$$

N_{in} and N_{out} correspond to the number of locations on the control surface where $\mathbf{V}\cdot\mathbf{n} < 0$ and $\mathbf{V}\cdot\mathbf{n} > 0$, respectively.

Example 6.3.2

The flow into the large inlet of an air compressor is from the atmosphere, where the temperature is 20°C and the pressure is 101.4 kPa. What is the power input to the compressor if the outlet pressure is 506 kPa and the temperature is 191°C? Consider the machine to be well insulated. The outlet port of the compressor is 100 mm in diameter, and the air velocity is 15 m/s.

Solution The energy equation, Eq. (6.3.6a), and the mass rate of flow will be needed for the solution. The latter is $\dot{m} = \rho VA$, in which the terms are conveniently evaluated at the exit of the compressor for this problem. The density can be calculated from the given data by means of the equation of state, $p/\rho = RT$:

$$\dot{m} = \frac{(506)(10^3)}{(286.9)(464)}(15)(0.1)^2\left(\frac{\pi}{4}\right)$$
$$= 0.45 \text{ kg/s}$$

If we neglect any change in potential energy and note that, with the large inlet, the velocity there will be relatively small, the energy equation becomes

$$\dot{W} + 0 + 0.45\left[\frac{0^2}{2} + c_p(20 + 273)\right] = 0.45\left[\alpha_2\frac{15^2}{2} + c_p(191 + 273)\right]$$

$$\dot{W} = 77{,}308 \text{ W (watts)}$$

if one uses $c_p = 1004$ J/kg-K). This answer *assumes* a highly turbulent exit flow that has almost no variation in velocity across the exit duct, so that $\alpha_2 = 1$. The solution just obtained is the power (i.e., energy per unit time) transferred to the air. The power input into the compressor would be obtained by dividing $\dot{W}$ by the compressor's efficiency.

Example 6.3.3

The compressor in Example 6.3.2 delivers its output from a manifold equally to 16 pipes, each with a diameter of 25 mm. At the entrance to these pipes, the temperature and pressure are essentially the same as at the compressor's outlet. What is the air velocity near the entrance of these 25-mm pipes? The pipes decrease in diameter to a location where the pressure and temperature are 283.4 kPa and 120°C, respectively. What is the speed of the air and the diameter of the conduit at this location?

Solution If the flow in each of the 16 pipes is the same, then there will be $0.45/16 = 0.028$ kg/s flowing in each pipe. The density at the inlet to the pipes will be the same as at the outlet of the compressor,

$$\rho = \frac{p}{RT} = \frac{(506)(10^3)}{(286.9)(464)} = 3.80 \text{ kg/m}^3$$

Consequently, the velocity near the tube's inlet would be

$$V = \frac{\dot{m}}{\rho A} = \frac{0.028}{3.8(0.00049)} = 15 \text{ m/s}$$

(Clearly, the result should be the same if the energy equation were applied.) To find the velocity at the reduced section of this 25-mm pipe, we need the same equations as before. But while we know the density at the location in question,

$$\frac{(283.4)(10)^3}{(289.6)(393)} \text{ kg/m}^3$$

we know neither the area nor the velocity there. The latter variable is determined through the energy equation, Eq. (6.3.6a):

$$\frac{15^2}{2} + c_p(191 + 273) = \frac{\alpha V^2}{2} + c_p(120 + 273)$$

$$\alpha\frac{V^2}{2} = \frac{225}{2} + (1004)(191 - 120) \qquad V = 378 \text{ m/s} \quad \text{if} \quad \alpha = 1$$

Then
$$A = \frac{\dot{m}}{\rho V} = \frac{0.028}{2.49(378)} = 2.97(10^{-5}) \text{ m}^2$$

The diameter becomes

$$\left[(29.7)(10^{-6})\left(\frac{4}{\pi}\right)\right]^{1/2} = 6.15(10^{-3}) \text{ m} = 6.2 \text{ mm}$$

For an incompressible fluid, a liquid, $\rho_1 = \rho_2 = \rho = $ constant. Also, the change in internal energy manifested as a small rise in the liquid's temperature will not be available, practically, for conversion into any other form of useful energy. Thus the term

$e_2 - e_1$ in the energy equation, Eq. (6.3.4), is considered to be a *"loss"*—a short form for *loss of available energy.* * We can now reformulate the general energy equation for *liquids* as follows:

> The sum of the liquid's original mechanical energy terms plus energy transferred by mechanical and thermal means from external sources yields the final mechanical energy of the liquid plus the losses, the unavailable energy, associated with dissipative mechanisms.

The term $e_2 - e_1$ that represents this dissipation is not simply neglected. It can be a significant quantity, and an accounting must be made for it. The usual practice is to relate it to some easily measured quantity in a controlled experiment and then use the results in similar applications. The pressure drop in a horizontal section of an insulated pipe is a case in point. One could write for such a pipe

$$\left[\alpha_1 \frac{V_1^2}{2} + \frac{p_1}{\rho} + g(0)\right] = \left[\alpha_2 \frac{V_2^2}{2} + \frac{p_2}{\rho} + g(0)\right] + (e_2 - e_1)$$

For constant diameter with $V_1 = V_2$ and $\alpha_1 = \alpha_2$,

$$e_2 - e_1 = \frac{p_1 - p_2}{\rho} = \frac{\text{loss}}{\text{mass flow/second}}$$

Experience shows that these losses can be well represented, in general, by

$$-\dot{\mathcal{Q}}_{\text{viscous dissipation}} + (e_2 - e_1) = \frac{\text{loss}}{\text{mass flow/second}} = K_1\left(\frac{\text{length}}{\text{diameter}}\right)\left(\frac{V^2}{2}\right) \qquad (6.3.7a)$$

The constant K_1 in this equation is, in fact, not one value but can vary over a range depending on V, the viscosity and density of the fluid, and the pipe diameter and roughness. In terms of Eq. (6.3.5), one should write

$$-\frac{\dot{\mathcal{Q}}_{\text{viscous dissipation}}}{g} + \frac{e_2 - e_1}{g} = \frac{\text{loss}}{\text{weight flow/second}} = K_1\left(\frac{\text{length}}{\text{diameter}}\right)\left(\frac{V^2}{2g}\right) \qquad (6.3.7b)$$

(The constants in the last two equations are the same. The only difference between the two expressions is the inclusion of the gravitational constant g when dealing with the weight rate of flow.)

Regions where the liquid must flow through an intricate passage, say, in a valve, will also generate "losses." These are generally expressed as $e_2 - e_1 = K(V^2/2)$ or $(e_2 - e_1)/g = K(V^2/2g)$, depending on whether one is using Eq. (6.3.4) or Eq. (6.3.5). The constant K varies from configuration to configuration, and its values are tabulated in Table 10.3.1.

---

* If the control surface is not adiabatic, there could be a transfer of heat to the surroundings that would be associated with the mechanical energy that was being dissipated through the effects of viscosity. Thus, one could separate $\mathcal{Q}$ into a heat addition due to external sources and a heat addition—a negative value—due to a loss of heat from the control volume that is associated with viscous dissipation. The sum of the absolute value of this latter term and the increase of internal energy would properly give the energy "loss" for incompressible fluids. It is in this sense that the term *loss* should be understood. The consequence of this meaning is that a viscous fluid could remain at constant temperature—$\Delta e = 0$—if there were sufficient heat transfer, the agent for the energy change, to the surroundings, with the attendant "loss of available energy."

Example 6.3.4

Oil from a reservoir flows through a pipe that is 10 mm in diameter and 5 m long in order to lubricate and cool the gear teeth of a speed reducer (see Fig. 6.3.3). The oil passes through a well-designed orifice (i.e., one having almost no losses) before spraying into the atmosphere. Determine the volume rate of oil flow that would be available from the system *assuming* an adiabatic system.

Solution We shall write the energy equation between locations 1 and 4. Because there has been no opportunity for power to be introduced into the fluid and we are going to *assume* an insulated system, we can write

$$-\int_1 \left(\frac{V^2}{2} + gz + \frac{p}{\rho} + e\right)\rho \mathbf{V} \cdot \mathbf{n} \, dA = \int_4 \left(\frac{V^2}{2} + gz + \frac{p}{\rho} + e\right)\rho \mathbf{V} \cdot \mathbf{n} \, dA$$

At location 1, the velocity, pressure, and elevation at the control surface are constant. The same is, in effect, true at location 4, if $\bar{z}$ and α_4 are used:

$$\dot{m}_1\left(\frac{V_1^2}{2g} + \frac{p_1}{\gamma} + \bar{z}_1 + \frac{e_1}{g}\right) = \dot{m}_4\left(\alpha_4\frac{V_4^2}{2g} + \frac{p_4}{\gamma} + \bar{z}_4 + \frac{e_4}{g}\right)$$

or, since $\dot{m}_1 = \dot{m}_4$ and $(e_4 - e_1)/g = \text{losses}_{1\text{-}4}$,

$$\frac{0^2}{2g} + \frac{p_{\text{atm}}}{9802(0.85)} + 3 = \alpha_4\frac{V_4^2}{2g} + \frac{p_{\text{atm}}}{9802(0.85)} + 0 + \text{losses}_{1\text{-}4}$$

Hence,
$$3 = \alpha_4\frac{V_4^2}{2g} + \text{losses}_{1\text{-}4} \qquad (6.3.8)$$

These losses are due to the inability of the fluid to follow the contours of the boundaries at 2, where there is an abrupt change in geometry; to the viscous dissipation in the pipe between points 2 and 3; and to the inability of the flow to convert completely the pressure and velocity ahead of the nozzle to velocity at the nozzle's exit. The causes of these losses will be discussed in Chaps. 7 and 8. For the present, we shall accept the fact that these losses are some fraction, determined sometimes experimentally, of the kinetic energy at the appropriate section. Thus we write for the loss at section 2

$$\text{Loss}_2 = K_2\frac{V_{\text{pipe}}^2}{2g}$$

and $K_2 \approx 1$ for a poorly formed inlet. [The magnitude of the K terms will be discussed in

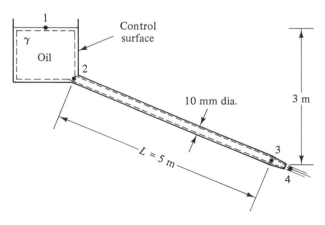

Figure 6.3.3 Discharge of a liquid from a reservoir through a conduit.

later chapters. These coefficients are introduced now only to give the reader an idea about the nature of the loss term in Eq. (6.3.4) or Eq. (6.3.5).] The losses in the conduit are

$$\text{Loss}_{2\text{-}3} = K_f \frac{V_{\text{pipe}}^2}{2g}$$

with K_f depending on the type of piping that is employed. If it is new and smooth, K_f would be*

$$\frac{0.316 L/D}{(V_{\text{pipe}} D/\nu)^{1/4}}$$

While K_f decreases with V_{pipe}, the conduit energy dissipation between locations 2 and 3 appears to vary nonlinearly with V_{pipe}, actually to the 1.75 power. The losses in a well-shaped nozzle will be small, and we shall assume that

$$\text{Loss}_{3\text{-}4} = 0.1 \frac{V_4^2}{2g}$$

If we recall that mass conservation must be observed,

$$V_4 A_{\text{jet}} = V_{\text{pipe}} A_{\text{pipe}} = Q$$

This specification of the losses leads one to rewrite the energy equation between locations 1 and 4 as

$$3 = \alpha_4 \frac{V_4^2}{2g} + \frac{V_4^2}{2g} \left(\frac{A_{\text{jet}}}{A_{\text{pipe}}}\right)^2 (K_2 + K_f) + 0.1 \frac{V_4^2}{2g}$$

This equation could be solved for V_4 as soon as K_f is expressed in terms of this variable. Then Q could be found as the product of V_4 and A_4. One observes that Eq. (6.3.8) expresses the change of energy, per unit weight of fluid, of a material volume moving from location 1 to location 4. This is *similar* to writing the Bernoulli equation for the no-loss condition between locations 1 and 4 on a streamline. This result will occur in many of the examples that follow.

Example 6.3.5

Water flows steadily from a mountain reservoir through a hydroelectric plant to a lake at a lower elevation. This is diagramed in Fig. 6.3.4. In this case, there is a transfer of energy from the fluid to the turbines in the power plant. (It is possible that there are also pumps in the plant which, during periods of the day when the electrical demand is so low that it is feasible and economical, pump water through the turbines up to the storage basin for utilization during subsequent peak-demand periods. Such installations are called *pump-storage plants*.) The energy equation can be used to determine the mechanical transfer of energy to the fluid. Thus, we write, using the lower lake as the elevation datum,

$$0 + \dot{W} - \int_A \left(\frac{V^2}{2} + \frac{p}{\rho} + gz + e\right) \rho \mathbf{V} \cdot \mathbf{n} \, dA = \int_E \left(\frac{V^2}{2} + \frac{p}{\rho} + gz + e\right) \rho \mathbf{V} \cdot \mathbf{n} \, dA$$

At the free surfaces of A and E, the velocity is assumed constant, albeit zero. The elevation is constant, the pressure is atmospheric pressure, and steady state is *assumed*.

---

* This relationship would be used if VD/ν is greater than 2000. Then one could be reasonably sure that the flow would be turbulent in the pipe. For $VD/\nu < 2000$, use $K_f = 64(L/D)/(VD/\nu)$. The basis for these constants is discussed in Chaps. 8 and 10.

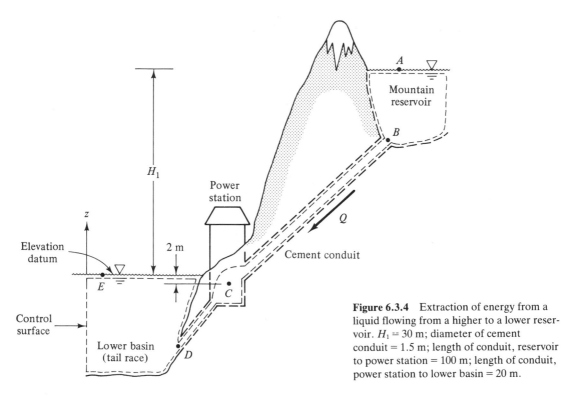

Figure 6.3.4 Extraction of energy from a liquid flowing from a higher to a lower reservoir. $H_1 = 30$ m; diameter of cement conduit = 1.5 m; length of conduit, reservoir to power station = 100 m; length of conduit, power station to lower basin = 20 m.

These facts lead one to evaluate the integrals in the energy equation, Eq. (6.3.5), as

$$0 + \frac{\dot{W}}{g\dot{m}} + \left(\frac{0^2}{2g} + \frac{0}{\gamma} + H_1 \right) = \left(\frac{0^2}{2g} + \frac{0}{\gamma} + 0 \right) + \text{losses}_{AE} \qquad (6.3.9)$$

The losses are the sum of those at the entrance to the conduit, the conduit losses, and the exit loss where the fluid enters the lower lake, or *tailwater*. There may be some losses in the machinery that produces the energy transfer $\dot{W}$ by mechanical means. But we shall not be concerned with them here; we shall consider $\dot{W}$ to be the net energy transferred to the fluid, a negative quantity in this case. The losses at the *tailrace*, point *D*, where the water from the draft tube *CD* enters the stilling basin, will be about

$$\frac{(1)(V_{\text{pipe}})^2}{2g}$$

since all of the kinetic energy in the pipe will be dissipated in the stilling basin, where the fluid velocity becomes nearly zero. Thus we can write for the energy equation

$$\frac{\dot{W}}{g\dot{m}} = -H_1 + \frac{V_{\text{pipe}}^2}{2g}(1 + K_{f_1} + K_{f_2} + 1)$$

If the flow rate Q is known, V_{pipe} can be found from a knowledge of the conduit's size. H_1 will normally be greater in magnitude than the losses, which are positive. If this is the case, $\dot{W}/g\dot{m}$ will be negative. This indicates that the energy per unit of weight of fluid flowing into the basin is negative and implies that there is energy being extracted from the fluid. If the pipe were insulated, the loss term would increase the water temperature somewhat. If the pipe transferred energy in the form of heat, some portion of the equivalent thermal energy would be transferred to the air and soil in the surroundings.

Point E is usually above the turbines at C—2 m, in this case—so that vapor pressure is not reached near the blades with the associated damaging effects of cavitation. Once again, uniform flows were *assumed* at the appropriate segments of the control volume. This gives Eq. (6.3.9) the form of energy change per unit weight for fluid particles flowing from station A to station E.

Example 6.3.6

Determine p_3 for the insulated mixing tee shown in Fig. 6.3.5. The water temperature at each inlet is 0.01°F less than the temperature at the outlet, section 3.

Solution   We shall solve for p_3 by first considering the mixing tee as a control volume and applying the energy equation in the form

$$\int \left(\frac{V^2}{2} + \frac{p}{\rho} + gz + e \right)(\rho \mathbf{V} \cdot \mathbf{n})\, dA = \dot{W} + \dot{\mathcal{Q}}$$

A steady-state situation is *assumed,* and one further notes that $\dot{W}$ and $\dot{\mathcal{Q}}$ are zero for this problem. (The insulation implies $\dot{\mathcal{Q}} = 0$, so that the change of temperature is not due to a heat addition but due to the change of mechanical energy into thermal energy that will, undoubtedly, be unavailable for further useful work. In this context one speaks of energy dissipation.) If one *assumes* also that the velocities and other properties are constant at segments 1, 2, and 3 of the control surface, the following is the value of the integral above:

$$-V_1 \rho_1 A_1 \left(\frac{V_1^2}{2} + \frac{p_1}{\rho_1} + g\bar{z}_1 + e_1 \right) - V_2 \rho_2 A_2 \left(\frac{V_2^2}{2} + \frac{p_2}{\rho_2} + g\bar{z}_2 + e_2 \right)$$

$$+ V_3 \rho_3 A_3 \left(\frac{V_3^2}{2} + \frac{p_3}{\rho_3} + g\bar{z}_3 + e_3 \right) = 0$$

Also, we have assumed $e_1 = e_2$ and $e_1 < e_3$, because we are given that (temperature)$_3$ > (temperature)$_1$ by 0.01°F. Since 778 ft-lb of energy is equivalent to the energy to raise one pound of water (at standard gravity) one degree Fahrenheit, we can see that as the water goes from 1 to 3 (or 2 to 3), the change of internal energy is 7.78 ft-lb/s for each pound flowing per second. This implies that

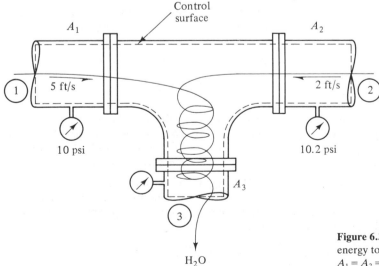

Control surface

A_1

A_2

① 5 ft/s

2 ft/s ②

10 psi

10.2 psi

A_3

③

H_2O

Figure 6.3.5 Conversion of mechanical energy to thermal energy in a mixing tee. $A_1 = A_2 = A_3 = \frac{1}{2}\,\text{ft}^2$.

$$\frac{e_3}{g} - \frac{e_1}{g} = 7.78$$

since this would be the change of internal energy (per pound per second flowing) due to the increased temperature and

$$\frac{e_3}{g} = \frac{e_2}{g} + 7.78$$

The quantities $V_i \rho_i A_i$ in the complete energy equation that was just written are related by mass conservation so that

$$g \int (\rho \mathbf{V} \cdot \mathbf{n}) \, dA = 0 = -\left[5\left(\frac{1}{2}\right)\gamma\right] - \left[2\left(\frac{1}{2}\right)\gamma\right] + \left[V_3\left(\frac{1}{2}\right)\gamma\right]$$

This result gives $V_3 = 7$ ft/s. These steps allow the evaluation of p_3 from the previous equation, if we prescribe $\bar{z}_1$, $\bar{z}_2$, and $\bar{z}_3$. We shall *assume* that the plane of the bend is horizontal, so that these centroidal locations can be taken to be zero. Therefore,

$$\left[5\left(\frac{1}{2}\right)\gamma\right]\left[\frac{5^2}{2g} + \frac{10(144)}{62.4} + \frac{e_1}{g}\right] + \left[2\left(\frac{1}{2}\right)\gamma\right]\left[\frac{2^2}{2g} + \frac{10.2(144)}{62.4} + \frac{e_2}{g}\right]$$

$$= \left[7\left(\frac{1}{2}\right)\gamma\right]\left[\frac{49}{2g} + \frac{p_3}{\gamma} + \left(\frac{e_2}{g} + 7.78\right)\right]$$

This result can now be simplified because of the previous relationship between the values of e. It reduces to

$$5(0.39 + 23.1) + 2(0.06 + 23.43) = 7\left(0.76 + \frac{p_3}{\gamma} + 7.78\right)$$

from which one can solve for p_3.

It has been mentioned previously that sometimes the losses are given in terms of $V^2/2g$ as $KV^2/2g$. In the previous example, K would have been found to be 10.23 because of the temperature data and the loss relationship, $K = (7)^2/2g = 7.78$. Recall that the sum $(V^2/2g + p/\gamma + z)$ is called the *total head*. Because of the losses in a flowing system, the total head decreases in the direction of flow. If we *assume*, however, an incompressible and inviscid fluid, and if $\dot{W}$ and $\dot{\mathfrak{Q}}$ are zero, then we have conditions for which the Bernoulli equation applies. It does not contain an internal energy term, so we can conclude that for the Bernoulli equation the total head, sometimes called H, remains constant.

Example 6.3.7

A pump takes water from a reservoir as part of a firefighting installation. There are two inlets in the pump's suction line, as shown in Fig. 6.3.6. It is specified that the pump should deliver 90 psig. Determine: (1) the maximum flow rate, and (2) the minimum power that must be supplied to the pump for the maximum flow rate condition.

Solution The energy equation, Eq. (6.3.2), is applied to a control surface that includes cross sections 3 and 4, where pressures and elevations are specified. Again, an adiabatic system is *assumed*. If the velocities along these sections are nearly constant and centroidal pressures are used, the equation can be written as

$$0 + 0 + \frac{V_3^2}{2g} + \frac{90(144)}{62.4} + 10 = \frac{V_4^2}{2g} + \frac{0}{\gamma} + 13 + \frac{e_4 - e_3}{g}$$

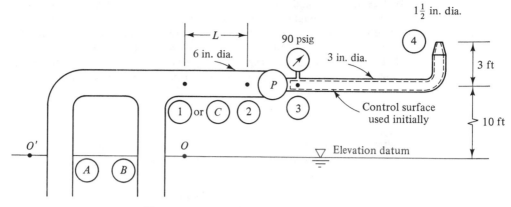

Figure 6.3.6 Pumping a liquid from a reservoir.

after dividing through by the weight rate of flow, γQ. "Continuity" relates the V's by the pipe diameters, so that $V_4 = V_3(3/1.5)^2$. This relationship can be incorporated into the energy equation to give

$$V_4^2 = 2g\left[\frac{90(144)}{62.4} - 3 - \text{losses}_{3\text{-}4}\right]$$

The maximum V_4 (and, hence, Q) will occur if the losses are minimized. An upper bound for V_4 entails the *assumption* that losses$_{3\text{-}4}$ = 0. With Q now known, one can proceed to the second task. However, before doing so we shall examine the flow at the inlet, as this will give a result with broad applicability.

The energy equation, with $\dot{W} = 0 = \dot{Q}$, for the control volume shown in Fig. 6.3.7 is

$$0 + 0 = \int\left(\frac{V^2}{2} + \frac{p}{\rho} + gz + e\right)\rho\mathbf{V\cdot n}\ dA = \int_A + \int_B + \int_C$$

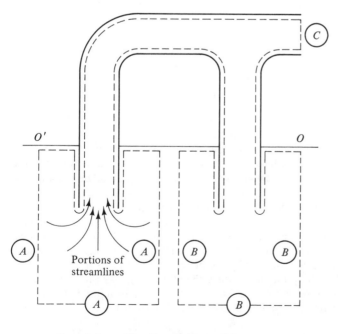

Figure 6.3.7 Control surface in an infinite reservoir.

On surface A, $V \approx 0$ and the sum $p/\rho + gz$ is always zero because of the assumption of $V \approx 0$, with the consequence that the pressure is "hydrostatic" far from the inlet to the pipe. Even though $V_A \approx 0$, $\dot{m}_A \neq 0$, because of the large area that makes up surface A.

Observe that the inlet to the pipe was *not* used as part of the control surface. Neither the fluid velocity nor the associated pressure is exactly known, because of the curved streamlines there. Thus, the evaluation of the integrals on control surfaces A, B, and C is

$$\dot{m}_A\, e_{O'} + \dot{m}_B\, e_O = \dot{m}_C\left(\frac{V_C^2}{2} + \frac{p_C}{\rho} + g\bar{z}_C + e_C\right)$$

Note that in writing the energy equation, it has been *assumed* that all particles of fluid at station C have the same internal energy per unit mass. It is natural for this control volume to have $e_{O'} = e_O$. Mass conservation provides us the relationship

$$\dot{m}_A + \dot{m}_B = \dot{m}_C$$

This statement allows the energy equation to be rewritten (and simplified) as

$$\frac{e_C - e_O}{g} = -\left(\frac{V_C^2}{2g} + \frac{p_C}{\gamma} + \bar{z}_C\right)$$

Because $e_C - e_O$ and V_C^2 are positive, as well as $\bar{z}_C$, in this case, one can expect that p_C must be a negative gauge pressure (i.e., a vacuum).

This last expression, an equation for each pound per second flowing at station C, serves as a basis for using the energy equation between two locations or points, rather than strictly integrating the results across segments of the control surface, when uniform flow conditions are assumed there. We have seen a similar result in the previous examples. Such an approach is employed next.

For those fluid particles going through pipe A, per pounds flowing per second,

$$\frac{V_{O'}^2}{2g} + \frac{p_{O'}}{\gamma} + z_{O'} = \frac{V_C^2}{2} + \frac{p_C}{\gamma} + z_C + \underbrace{\text{losses}_{O'\text{-}C}}_{(e_C - e_{O'})/g}$$

For the fluid that goes through pipe B,

$$\frac{V_O^2}{2g} + \frac{p_O}{\gamma} + z_O = \frac{V_C^2}{2g} + \frac{p_C}{\gamma} + z_C + \underbrace{\text{losses}_{O\text{-}C}}_{(e_C - e_O)/g}$$

$$V_O = V_{O'} \approx 0 \qquad p_O = p_{O'} = 0 \text{ gauge}$$

If the free surface is the elevation datum, $z_O = z_{O'} = 0$. This result implies that the losses in the path O'–C (pipe A) are equal to those in path O–C (pipe B). This includes the losses associated with the mixing of the streams at point C. Nothing has been said about the diameter, length, or internal obstructions in either pipe. All that we know is that in the parallel pipes, A and B, the losses—on a pounds per second basis—should be the same for each pipe.

Since we know Q from a previous step, p_C would follow once the losses are specified. If the losses are large, the pressure at point C could be very low. Indeed, the pressure there could approach the vapor pressure of the liquid, and it would "boil" at the local temperature. This is called *cavitation*. The damage shown in Fig. 1.8.3 is the result of such cavitation, which, it is thought, is produced by the bubbles appearing as in Figure 1.8.8.

The power required of the pump to produce the 90 psi of gauge pressure and the associated flow can be found now that Q and the losses have been examined. The energy equation is conveniently written from point O to 4, since V_4 is known:

$$\frac{\dot{W}}{Q\gamma} + \frac{0^2}{2g} + \frac{0}{\gamma} + 0 = \frac{V_4^2}{2g} + \frac{0}{\gamma} + 13 + \text{losses}_{0\text{-}4}$$

The *minimum* power would occur for zero losses. Then

$$\dot{W}_{\text{min}} = Q\gamma\left(\frac{V_4^2}{2g} + 13\right)$$

This is the power *into* the fluid. The pump would require, as a minimum, an input of $\dot{W}/$(pump efficiency).

Example 6.3.8

A 24-in.-diameter pipe divides at a Y section into two 12-in. pipes as shown in Fig. 6.3.8. The flow, in this case, divides evenly due to proper adjustment of the conditions farther downstream. Find the pressures at the exit of the Y section.

Solution We can begin by writing the energy equation for the control volume, which consists of the entire Y section (clearly, $\dot{W}$ is zero, since no energy is added to the fluid in the Y section):

$$(\dot{W}t)_1\left(\alpha_1\frac{V_1^2}{2g} + \frac{p_1}{\gamma} + \bar{z}_1\right) + 0 + 0 = (\dot{W}t)_2\left(\alpha_2\frac{V_2^2}{2g} + \frac{p_2}{\gamma} + \bar{z}_2\right)$$

$$+ (\dot{W}t)_3\left(\alpha_3\frac{V_3^2}{2g} + \frac{p_3}{\gamma} + \bar{z}_3\right)$$

$$+ \text{total losses}$$

in which $(\dot{W}t)_i$ is the weight rate of flow through the ith section. If this distribution were given—say, $(\dot{W}t)_2 = (\dot{W}t)_3 = 0.5(\dot{W}t)_1$—along with p_1, A_1, A_2, and A_3, one would have

$$\left(\alpha_1\frac{V_1^2}{2g} + \frac{p_1}{\gamma} + 0\right) = \frac{(\dot{W}t)_2}{(\dot{W}t)_1}\left(\alpha_2\frac{V_2^2}{2g} + \frac{p_2}{\gamma}\right) + \frac{(\dot{W}t)_3}{(\dot{W}t)_1}\left(\alpha_3\frac{V_3^2}{2g} + \frac{p_3}{\gamma}\right) + \frac{\text{total losses}}{(\dot{W}t)_1}$$

in which p_2/γ and p_3/γ are the only unknowns. (This assumes that the relationships for the losses are known.) While we shall assume that all the α's are unity, more information is needed in order to solve for p_2 and p_3. To do this, we envisage a control volume sur-

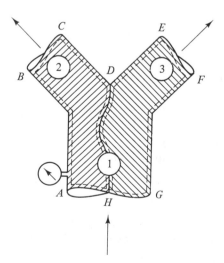

Figure 6.3.8 Control surfaces for the analysis of the flow in a pipe wye. Initially, surface DH is not utilized.

rounding, at a particular instant of time, the fluid particles that will leave via section 2. We do the same for the particles destined to leave via section 3. We have constructed two control volumes so that each pertains to a specific portion of the flow. (See Fig. 6.3.8) Along the interface DH, one control volume does work on the other, the net effect of which will manifest itself—if nonzero—in the loss terms. The energy equations for these two systems of particles follow from applying Eq. (6.3.5) to the appropriate control surface segments. These equations are

$$\left(\frac{V_1^2}{2g}+\frac{p_1}{\gamma}+z_1\right)=\left(\frac{V_2^2}{2g}+\frac{p_2}{\gamma}+z_2\right)+\frac{\text{losses}_{1\text{-}2}}{(\dot{W}t)_{1\text{-}2}}$$

and

$$\left(\frac{V_1^2}{2g}+\frac{p_1}{\gamma}+z_1\right)=\left(\frac{V_3^2}{2g}+\frac{p_3}{\gamma}+z_3\right)+\frac{\text{losses}_{1\text{-}3}}{(\dot{W}t)_{1\text{-}3}}$$

Once the loss terms are given—on the basis of previous experience—one can determine p_2 and p_3. While the sum of these equations yields the equation preceding, an important point to remember is that each of the last equations can be solved in turn. But in the first equation of this example there were two unknowns, p_2 and p_3, so that a solution was impossible at that stage.

Each of the two equations just given is intensive, because it refers to each pound of fluid flowing into regions 2 and 3. They describe, in effect, the energy history of particles that start at location 1 and end at location 2 or location 3. (Because a steady state is assumed, particles at 1 will assume the same values when they arrive at locations 2 and 3 as their predecessors.) Thus, the equations represent energy statements for each particle along a streamline in this steady-state case. Viewed in this light, the subdivision of the control volume into the two subregions is not such an arbitrary or capricious act.

If two streams that have the same physical properties *merge* and mix, one would have a similar problem to that just treated in this example, as well as that discussed in Example 6.3.7. If the two streams have different properties and thoroughly mix in a section before being discharged, one could proceed much as for the situation shown in Fig. 6.3.8:

(a) Mass conservation:

$$\rho_2\, Q_2 + \rho_3\, Q_3 = \rho_1\, Q_1$$

(b) Volume conservation for incompressible fluid:

$$Q_2 + Q_3 = Q_1$$

Given ρ_2, ρ_3, Q_2, and Q_3, one knows ρ_1 and Q_1.

(c) Energy conservation, assuming a horizontal pipe and α's $= 1$:

$$\rho_2 Q_2\left(\frac{V_2^2}{2g}+\frac{p_2}{\gamma_2}\right)+\rho_3 Q_3\left(\frac{V_3^2}{2g}+\frac{p_3}{\gamma_3}\right)=\rho_1 Q_1\left(\frac{V_1^2}{2g}+\frac{p_1}{\gamma_1}\right)+\text{total losses}$$

If p_2 and p_3 are given, p_1 can be found, since it is the only unknown. In prescribing these pressures, we shall demand that the "total head"

$$\frac{V_1^2}{2g}+\frac{p_1}{\gamma_1}+\bar{z}_1$$

be the same for each channel at the junction. ($\bar{z}_1$ would be excluded for a horizontal wye.) Clearly, if there were two inlets and two outlets, more information would be needed to solve the problem.

Equation (6.3.2) is repeated here because we shall now focus our attention on the first integral in it (in Sec. 6.3, this term was zero, because only steady flows were discussed):

$$\dot{Q} + \dot{W} = \frac{\partial}{\partial t}\int_{CV}\left(\frac{V^2}{2} + gz + e\right)d\forall + \int_{CS}\left(\frac{V^2}{2} + gz + e + \frac{p}{\rho}\right)\rho\mathbf{V}\cdot\mathbf{n}\,dA \qquad (6.4.1)$$

An analysis of the flow out of an emptying container is our first application of the energy equation for an unsteady case.

Example 6.4.1

Analyze the flow out of the container shown in Fig. 6.4.1, where h is the instantaneous distance between the free surface and the bottom.

Solution We shall use as the control volume region *BCDEFGB*, for which surface *GB* is located an arbitrary distance b $(b < h)$ above the bottom.* Hence there is a flow across surface *GB* that is downward and related to the rate at which the surface elevation changes. If we keep this in mind, we can write Eq. (6.4.1) as

$$\rho Q\left(\frac{p}{\rho} + gb + \frac{V^2}{2}\right)_{GB} = \rho Q\frac{V_{jet}^2}{2} + \frac{\partial}{\partial t}\int\rho\frac{V^2}{2}\,d\forall \qquad (6.4.2)$$

on the assumption that the effects of the internal energy e can be neglected. If the elevation h were constant, the pressure along surface *GB* would be constant at $(h - b)\gamma$. However, the column whose length is $h - b$ is decelerating downward, which is tantamount to accelerating upward. Consequently, the pressure on *GB* is $(h - b)\rho(g + a_z)$. The integral on the right-hand side of Eq. (6.4.2) can be differentiated and interpreted as follows:

$$\frac{\partial}{\partial t}\int\rho\frac{V^2}{2}\,d\forall = \int\left[\frac{\partial}{\partial t}\left(\frac{V^2}{2}\right)\right]\rho\,d\forall + \int\left(\frac{V^2}{2}\right)\frac{\partial}{\partial t}(\rho\,d\forall)$$

Because the mass within the control volume—the region below *GB*—is constant with respect to time, the last integral vanishes. In the second integral, the term in square brackets is constant within the control volume (i.e., almost independent of position) and can be treated as a coefficient of $\int\rho\,dV$. This integral is the mass in the control volume. This leads to

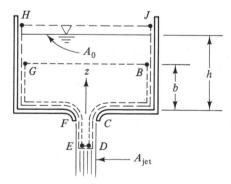

Figure 6.4.1 Unsteady discharge of a liquid from a tank. A_{jet} = jet area; A_0 = tank's cross-sectional area. (Assumed constant *unless* noted otherwise.)

*In Example 6.4.2, the problem will be reworked using region *HJBCDEFGH* to compare the solutions.

$$\rho Q \left[\left(\frac{h-b}{\rho} \right) \rho(g + a_z) + gb + \frac{V^2}{2} \right] = \rho Q \frac{V_{\text{jet}}^2}{2} + \rho A_0 b \frac{d}{dt} \left(\frac{V^2}{2} \right) \qquad (6.4.3)$$

In these equations, V, the velocity along GB as well as that within the control volume, is positive if it is upward. Because all of the liquid between surfaces GB and HJ is moving downward as if it were a solid body, $a_z = dV/dt$. Incorporating this notation into our work, the result is

$$\rho Q \left[hg + (h - b) \left(\frac{dV}{dt} \right) + \frac{V^2}{2} \right] = \rho Q \frac{V_{\text{jet}}^2}{2} + \rho A_0 b V \frac{dV}{dt} \qquad (6.4.4)$$

The flow rate Q is $(-dh/dt)A_0$, which is the rate of flow out of the control volume, $V_{\text{jet}} A_{\text{jet}}$. It is also the volume rate of flow into the control volume, $-VA$, for V positive upward. This observation allows the equation to be simplified to

$$hg + h \frac{dV}{dt} + \frac{V^2}{2} = \frac{V_{\text{jet}}^2}{2}$$

The relationship $V = dh/dt$ leads to $dV/dt = d^2h/dt^2$. This value can be employed and the differential equation written as

$$hg + h \frac{d^2h}{dt^2} + \frac{1}{2} \left(\frac{dh}{dt} \right)^2 = \frac{1}{2} \left(\frac{dh}{dt} \right)^2 \left(\frac{A_0}{A_{\text{jet}}} \right)^2$$

or

$$h \left(g + \frac{d^2h}{dt^2} \right) = \frac{1}{2} \left(\frac{dh}{dt} \right)^2 \left[\left(\frac{A_0}{A_{\text{jet}}} \right)^2 - 1 \right] \qquad (6.4.5)$$

If the tank is emptying slowly, one can expect $g \gg d^2h/dt^2$ and the latter term could be neglected with respect to the former. Such a simplification allows Eq. (6.4.5) to be integrated easily. (An even simpler result would be possible if $A_0/A_{\text{jet}} \gg 1$.)

If the vessel does not have a constant cross section A_0, then the proper functional relationship between the cross-sectional area and the liquid's elevation must be inserted for A_0 prior to the differentiation and integration. Such a shape would be necessary if one wished $dh/dt = $ constant. This would be desirable in a water clock.

Example 6.4.2

Perform an analysis of the system in Fig. 6.4.1, but use the control surface $HJBCDEFGH$ instead.

Solution The choice of the control surface GB in Example 6.4.1 resulted in a constant mass within the control volume even though the kinetic energy within it varied with time. An alternative and acceptable choice of control volume would have a control surface, HJ in Fig. 6.4.1, which is slightly above the free surface of the water. However, in this case there would be no flow across HJ. In addition to the change of kinetic energy within the control volume, there would be a change of potential energy.

The expression for the energy equation that is appropriate to the present control volume would be

$$0 = \rho Q \frac{V_{\text{jet}}^2}{2} + \frac{\partial}{\partial t} \int \rho \left(\frac{V^2}{2} + gz + e \right) d\forall$$

The change of internal energy within the control volume will be neglected once again. However, the change of potential energy is

$$\frac{\partial}{\partial t} \int \rho g z \, d\forall = \gamma \frac{\partial}{\partial t} \int z \, d\forall = \gamma \frac{\partial}{\partial t} (\bar{z}\forall)$$

$$= \gamma \frac{\partial}{\partial t} \left(\frac{h}{2} A_0 h \right) = \gamma A_0 h \frac{dh}{dt}$$

The change of kinetic energy within the control volume is not the same as in Example 6.4.1, because of the change of mass within the control volume. The expression is

$$\frac{\partial}{\partial t}\int \rho\frac{V^2}{2}\,dV = \int\left[\frac{\partial(V^2/2)}{\partial t}\right]\rho\,dV + \int\left(\frac{V^2}{2}\right)\frac{\partial}{\partial t}(\rho\,dV)$$

$$= \left(V\frac{dV}{dt}\right)\rho h A_0 + \left(\frac{V^2}{2}\right)\rho A_0\frac{dh}{dt}$$

$$= \rho A_0\left[h\left(\frac{dh}{dt}\frac{d^2h}{dt^2}\right) + \frac{1}{2}\left(\frac{dh}{dt}\right)^2\frac{dh}{dt}\right]$$

These quantities can now be combined according to the prescription given in the first equation of this example so that

$$0 = \rho\left(-\frac{dh}{dt}A_0\right)\frac{V_{jet}^2}{2} + \rho A_0\frac{dh}{dt}\left[h\frac{d^2h}{dt^2} + \frac{1}{2}\left(\frac{dh}{dt}\right)^2\right] + \rho g A_0 h\frac{dh}{dt}$$

The common factor $\rho A_0(dh/dt)$ can be excluded, since a nonzero transient solution is being sought. Also, a substitution for V_{jet}, in terms of $-dh/dt$, can be made, with the result that

$$0 = -\frac{1}{2}\left(\frac{dh}{dt}\right)^2\left[\left(\frac{A_0}{A_{jet}}\right)^2 - 1\right] + h\left(g + \frac{d^2h}{dt^2}\right)$$

This is equivalent to Eq. (6.4.5) in Example 6.4.1.

Example 6.4.3

A small, high-speed wind tunnel takes air from the atmosphere and passes it into a insulated tank having a volume of 12 m³. This has been evacuated to an absolute pressure of 50 kPa at 15°C before a valve is opened between it and the wind tunnel. Determine the rate of temperature change in the tank at the instant of valve opening. At this time the following conditions are observed at the partially opened valve, which then is the location of minimum cross-sectional area in the tunnel ($A = 3$ cm²): $V = 313.2$ m/s, $p = 53.5$ kPa, $T = -28.93$°C. (Assume that the contents of the tank are homogeneous.)

Solution A schematic sketch is shown in Fig. 6.4.2. An adiabatic process will be assumed. In addition, it is inferred that the tank is large enough that the kinetic energy within it is always nearly zero. Changes in potential energy will be neglected, too.

Upon substitution of the given data in the energy equation, and given the assumptions that were made, one has

$$0 = \frac{\partial}{\partial t}\int e\rho\,dV + \int\left(\frac{V^2}{2} + h\right)\rho\mathbf{V}\cdot\mathbf{n}\,dA$$

Figure 6.4.2 Unsteady flow of air through a wind tunnel.

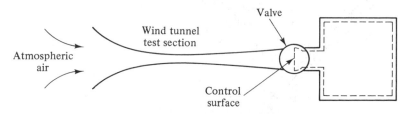

This can be expanded to read

$$0 = e\frac{\partial}{\partial t}\int \rho \, d\forall + \frac{\partial e}{\partial t}\int \rho \, d\forall + \int \left(\frac{V^2}{2} + e + \frac{p}{\rho}\right)\rho \mathbf{V} \cdot \mathbf{n}\, dA$$

because it is assumed that the properties in the tanks are homogeneous.

Mass conservation for the system requires that

$$e\left(\frac{\partial}{\partial t}\int \rho \, d\forall + \int \rho \mathbf{V} \cdot \mathbf{n}\, dA\right) = 0$$

or

$$e\frac{\partial}{\partial t}\int \rho \, d\forall + \int e\rho \mathbf{V} \cdot \mathbf{n}\, dA = 0$$

By combining this result from mass conservation with the previous energy relationship, one has

$$\frac{\partial e}{\partial t}(m_{CV}) = -\int \left(\frac{V^2}{2} + \frac{p}{\rho}\right)\rho \mathbf{V} \cdot \mathbf{n}\, dA$$

$$= -\left[(313.2)^2\left(\frac{\rho}{2}\right) + 53.5(10^3)\right](-313.2)(3)(10^{-4}) \quad\quad (6.4.6)$$

The density ρ in this expression can be found from the ideal gas law:

$$\rho = \frac{p}{RT} = \frac{53.5(10^3)}{(287)(273 - 28.9)} = 0.764 \text{ kg/m}^3$$

However, the initial density in the control volume must also be found. It is

$$\rho_0 = \frac{50(10^3)}{(287)(273 + 15)} = 0.605 \text{ kg/m}^3$$

This gives an initial mass of 7.26 kg $[= 0.605(12)]$. Because $\partial e/\partial t = \partial(c_v T + \text{const})/\partial t$, where c_v is assumed to be a constant $(= 717.4 \text{ J/kg-K})$, one finds that

$$\frac{\partial T}{\partial t} = \left(\frac{1}{717.4}\right)\left(\frac{1}{7.26}\right)(8548) = 1.53°\text{C/s}$$

Example 6.4.4

Find the equation of motion for the liquid in the manometer shown in Fig. 6.4.3. What is the period of oscillation if $\mu/\rho \approx 0$?

Solution If an adiabatic control volume is selected that envelops the manometer, the energy equation reduces to $0 = \partial/\partial t[(V^2/2 + gz + e)\rho \, d\forall]$. This expression will be evaluated after examining each of the three integrals that are part of it.

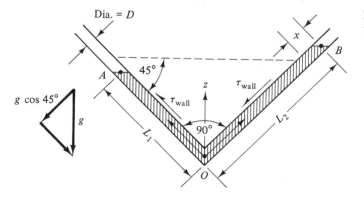

Figure 6.4.3 Oscillation of a liquid in a tube.

$$\frac{\partial}{\partial t}\int \frac{V^2}{2}\rho\,dV = \frac{d(V^2/2)}{dt}\int \rho\,dV$$

since it is assumed that the kinetic energy per unit mass is independent of the position inside the control volume. The result is $V(dV/dt)(\rho LA)$. To evaluate the time rate of change of potential energy, one may find it helpful to examine each part of the system. Thus

$$\frac{\partial}{\partial t}\int gz\rho\,dV = \gamma\frac{\partial}{\partial t}\left(\int_{L_1} z\,dV + \gamma\int_{L_2} z\,dV\right)$$

$$= \gamma\left[\frac{d}{dt}(\bar{z}_{L_1} V_{L_1}) + \frac{d}{dt}(\bar{z}_{L_2} V_{L_2})\right]$$

in which $\bar{z}$ is measured from the origin shown in Fig. 6.4.3 and use has been made of the definition of the centroid for the "legs" with length L_1 and L_2. But $\bar{z} = (L/2)(\cos 45°)$ and $V = LA$, so that

$$\frac{\partial}{\partial t}\int \gamma z\,dV = \frac{\gamma A}{2}\left[\frac{d}{dt}(L_1^2 + L_2^2)\right]\cos 45°$$

$$= \gamma A\,\cos 45°\left(L_1\frac{dL_1}{dt} + L_2\frac{dL_2}{dt}\right)$$

$$= \gamma A\,\cos 45°\frac{dL_2}{dt}(-L_1 + L_2)$$

since $dL_2/dt = -dL_1/dt$ in this case. We can reduce this expression to one in terms of the excursion of the right "leg" from its equilibrium position, x:

$$L_2 - L_1 = 2x \quad\text{and}\quad \frac{dL_2}{dt} = \frac{dx}{dt}$$

Therefore,
$$\frac{\partial}{\partial t}\int \gamma z\,dV = \gamma A\,(\cos 45°)(2x)\left(\frac{dx}{dt}\right)$$

The most difficult integral to evaluate, at this stage, is the last one. Again, spatial uniformity is assumed, so that

$$\frac{\partial}{\partial t}(\int e\rho\,dV) = \rho AL\left(\frac{de}{dt}\right) = \rho V L\dot{e}$$

This is the first time that $\dot{e}$ must be treated. To do this, consider for the moment a jet of fluid that comes from a reservoir through a pipe, as in Fig. 6.4.4. The increase in internal energy of each unit of mass of liquid is, from the definition associated with the energy equation, the loss. A parcel of fluid has gained this internal energy while moving down the

Figure 6.4.4 Net internal energy flux out of a control volume due to viscous effects. Shaded area in jet: 1 unit mass of liquid.

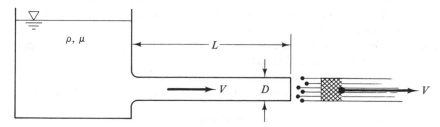

pipe of length L in the time L/V. Thus, the rate of energy increase would be $\dot{e} = (\text{loss})/(L/V)$. For conditions such that the term $VD\rho/\mu$ is less than 2000, the losses, e, can be expressed as $32\mu VL/\rho D^2$, so that $\dot{e}$ is $32\nu V^2/D^2$.* Thus, the energy equation can be written for this flow—the manometer under consideration could satisfy the constraint $VD\rho/\mu \ll 2000$—as

$$(\rho LA)V\frac{dV}{dt} + \left(\frac{32\nu}{D^2}\right)V^2(\rho LA) + g\rho A \, \cos 45°\left(2x\frac{dx}{dt}\right) = 0$$

This can be written as

$$V\frac{dV}{dt} + \left(\frac{32\nu}{D^2}\right)V^2 + \frac{2g}{L} \, \cos 45°\left(x\frac{dx}{dt}\right) = 0$$

One can observe that $V = dx/dt$. Then this equation is satisfied by $dx/dt = 0$, or

$$\frac{d^2x}{dt^2} + \left(\frac{32\nu}{D^2}\right)\frac{dx}{dt} + \left(\frac{2g}{L} \, \cos 45°\right)x = 0 \qquad (6.4.7)$$

If $\nu \approx 0$, the resulting equation describes simple harmonic motion,

$$x = A \, \sin \omega t + B \, \cos \omega t$$

wherein the period is

$$\mathcal{T} = \frac{2\pi}{\omega} = 2\pi \sqrt{\frac{L}{2g \, \cos 45°}}$$

The careful reader can see that the manometer in Fig. 6.4.3 need not have had symmetric angles. Comparable results could be obtained easily for an asymmetric case. Also, the oscillation of a liquid in two interconnected large reservoirs of different cross sections could be similarly determined if fluid viscosity is neglected. The role of the term $(32\nu/D^2)(dx/dt)$ will depend on the magnitude of the coefficient for the last term, $g(\text{const})/L$. If it is large enough, oscillation will not occur; only a slow return to the equilibrium position should ensue. Otherwise, the oscillations will be damped (i.e., the maximum amplitude will decrease with each swing of the liquid column.) This is the situation that is wanted for surge tanks. The reader is encouraged to review the theory of second-order ordinary differential equations, to appreciate the results of Eq. (6.4.7).

6.5 SUMMARY

The first law of thermodynamics relates the change of energy of a closed system with the amount of energy that is transferred to it from the surroundings in the form of heat and mechanical work, or their chemical and electrical equivalents. The time rate of such changes for a material volume can be expressed by the Reynolds transport theorem, and in this case the intensive properties are the kinetic, potential, and internal energies on a unit mass basis. If one subdivides the mechanical work added to the material volume into the work provided by mechanical devices associated with the control volume in question and the flow work occurring across its boundaries, one can include the flow work in the surface integral associated with the Reynolds transport theorem. Equation (6.3.2) is the result.

Certain obvious simplifications to this equation occur if the flow is one of steady state and the properties are uniform at the segments of the control surface across which

* Do the units yield (m-N/s)/kg in the SI system?

the fluid flows. Equation (6.3.6b) is the consequence of such simplifications, while Eq. (6.3.4) is appropriate to the special case when $N_{in} = 1 = N_{out}$.

If the flow is unsteady, both the surface and volume integrals in Eq. (6.3.2) have to be considered. The evaluation of these integrals may sometimes be facilitated through a favorable choice of the control volume. While such choices always affect the terms in the surface integral, in the unsteady case there are situations in which the surface integral can be made to vanish, or the derivative of the volume integral can be set to zero. The examples in Sec. 6.4 point this out.

EXERCISES

6.2.1(a) What is the time rate of flow of kinetic energy across CD if $U = 2$ ft/s, $W = 18$ in., $b = 1.5$ in., and the fluid is water?

(b) Let the velocity across CD be $u = +7(y + 6)/24$. What is the time rate of flow of kinetic energy across CD? What would be your answer if $u = -7(y + 6)/24$?

(c) Use the velocity distribution given in Fig. 6.2.1 to determine the time rate of flow of kinetic energy across CD.

6.2.2(a) What is the kinetic energy correction factor for the velocity distribution given in Example 6.2.1?

(b) For a flow between two parallel plates that are a distance k apart, the velocity is 5/2 ft/s. What is the kinetic energy correction factor α?

(c) What is α for the velocity distribution given in Fig. 6.2.1?

(d) For the flow given in Example 6.2.2, a laminar profile, the value of α was shown to be 2. This correction factor will approach unity as the velocity profile becomes more constant across the conduit. In a turbulent flow, this is the case and an approximate form of u is given by $U_0(1 - r/r_0)^N$, in which N is about $\frac{1}{7}$. What is the kinetic energy correction factor?

6.3.1(a) Determine the power necessary to rotate a bearing 50 mm in diameter with a clearance of 0.06 mm at 1000 rpm. The lubricant has a viscosity of 0.1 kg/m-s. The width of the bearing is 150 mm.

(b) What power is being expended along the *stationary* surface in Fig. 6.3.2?

6.3.2(a) What is the expected power input to the compressor if its efficiency is 81 percent?

(b) What is the power input to the air if the compressor operates at the pressure ratio in the example with the same atmospheric air, but the exit temperature is 180°C (duct size and air speed also constant)?

(c) For the pressures and temperatures *in the example,* what would be the power supplied to oxygen leaving the compressor at 12 m/s in an 80-mm duct?

6.3.3(a) Consider the flow in the pipes to be isentropic (i.e., the pressure and density at any point are related by $p/\rho^k = $ constant). What is the pressure at a location where the temperature is 150°C? What is the diameter there?

(b) What are the temperature and pressure at the location where the diameter is 10 mm? (Assume isentropic flow.)

6.3.4(a) Assume that the elevation difference between the reservoir and the nozzle is $H,$ and not 3 m. If K_f is as given in the example and oil with a kinematic viscosity of 10^{-4} m^2/s is used, what will be the value of H if the jet's velocity is 1.2 m/s? (Use diameter$_{jet} = 3$ mm).

(b) If the elevation difference is, in fact, 3 m, what is the speed of the oil at the jet when the same oil is used as in Exercise 6.3.4a? (It may be necessary to use an iterative solution.)

(c) For the case treated in Exercise 6.4.3b, what would be the speed of the jet if the connection at the reservoir were reworked so that K_2 became 0.3? What difference of elevation would achieve the same flow of lubricant? Does the cost of reworking the connection appear to be advisable?

6.3.5(a) What is the maximum power that can be generated—no losses—if H_1 is 100 ft?

(b) Consider $K_f = fL/D$, in which L and D are the length and diameter of the conduit respectively. The factor f is called the *friction factor* and varies with the Reynolds number and the roughness of the conduit when normalized with the pipe diameter. For this problem, let $f = 0.029$ and use the data given above to determine the power-generating capability of the installation under these conditions.

(c) Estimate the water pressure in the conduit immediately after the turbines in the generating station, point C. Use the 1.5-m conduit diameter.

6.3.6(a) What would be p_3 if the temperature rise across each flow path were 0.02°F? What would the corresponding loss coefficient, K, be for this situation?

(b) If, for a temperature rise of 0.01°F, there were a heat loss to the surroundings of 2.59 ft-lb/s for each pound of water flowing per second, what should be the value of p_3?

(c) If the same tee fitting were installed in a different circuit but there was no recorded temperature rise across the tee, perhaps because heat was transferred out, what would be the expected value of p_3?

PROBLEMS

6.1 A two-dimensional flow in a channel of height b could be $u = (0.5)(y/b)$ if the upper surface is moving at 0.5 m/s. What is the kinetic energy correction factor for this case?

6.2 Two horizontal plates that are separated by a distance b are filled with a flowing fluid having a velocity profile given by $Ky(y - b)$, in which y is measured normal to, and from, the lower plate. K in this expression is a constant determined by the maximum velocity in the layer. What is the associated kinetic energy correction factor? (Note: $K < 0$ for $u > 0$.)

6.3 If, for the conditions of Prob. 6.1, a pressure gradient existed along the channel, the velocity profile could be $u = (0.5)(y/b) - Ky(y - b)$, in which K is a constant associated with the pressure gradient. Determine the kinetic energy correction factor.

6.4 When there is a high value for the speed along the centerline of a smooth round pipe, U_0, the axial velocity profile can be closely approximated by $U_0(y/r_0)^{1/7}$. The variable y is the distance from the inside wall, $r_0 - r$. Find the average velocity in terms of the centerline value. How does it compare in magnitude with the profile given in Example 6.2.2? Find the kinetic energy correction factor using $r = r_0 - y$ in $dA = 2r\,dr$. How does this value compare with that found in Example 6.2.2?

6.5 For the flow in a river that is described in Fig. P.3.3, what is the kinetic energy correction factor?

6.6 Assume that the velocity distribution given in Prob. 6.2 can be applied to a channel that tapers slowly from a width b equal to 100 mm

to a width of 50 mm. The maximum velocity is 0.5 m/s where the width is 100 mm. (Assume that the breadth in the z direction—perpendicular to V and b—is 1 m.) Determine the pressure difference between these sections using the actual flux of kinetic energy. Repeat using the average velocity at each section and the appropriate kinetic energy correction factor.

6.7 Water flows steadily from one reservoir to another whose free surface is 50 ft lower. Give an expression for the losses in the conduit due to the flow.

6.8 A river is dammed so that a 4-m difference in elevation is created. A series of turbines that operate well at such conditions is installed to generate power. What output (maximum) can be expected in a river that flows at 80 m³/s? (Neglect the river's Δ(K.E.).)

6.9 Power is being generated in northern France by taking advantage of the variation in water depth associated with the tides. Assume that water is allowed to enter the Rance River when the English Channel is an average of 3 m higher than it for 4 hours per day. Water from the river flows into the Channel when it is depressed the same amount for the same period of time. The turbines in the facility which utilizes this flow are reversible. What is the maximum power that can be generated in this power plant if 80 m³/s is the influx and efflux of water? (Neglect river's K.E.)

6.10 A stream, in negotiating a bend, induces velocity components that are directed laterally across the stream, in opposite directions on the surface and bottom. This phenomenon induces flow losses. Shortly after a bend, the *total* speed has increased from 0.85 m/s to 0.90 m/s. Downstream of the bend, the depth has been reduced to 1.5 m. What are the losses in the bend for this channel of constant width?

6.11 Water in a pipe 3 in. in diameter (pipe sizes and threads have tended to remain in inches—or inch equivalents—even in countries which have long used the metric system) flows through a fully open globe valve at the rate of 0.5 ft³/s, with the result that there is a 3.5-psi drop in pressure. What is the loss of energy, expressed in feet of water? Express the loss as a constant K that multiplies the average kinetic energy term, $V^2/2g$.

6.12 Air at 20°C and 102 kPa flows through a duct 1 m in diameter that after turning 90° reduces to a diameter of 0.8 m. A water manometer registers a pressure difference between the larger and smaller portions of the duct equal to 16.3 cm when 30 m³/s of air is flowing. Does the upstream or downstream manometer tap sense the higher pressure? What is the loss in the duct expressed in meters of air, in meters of water, and as a constant K that multiplies the average kinetic energy term, $V^2/2g$, for the upstream section? (Let $\Delta z = 0$.)

6.13 When a fluid flows through a conduit, there are losses due to viscous effects. For a pipe with a relatively rough interior surface, the loss coefficient K (losses = $KV^2/2g$) is 0.017 per foot of length of a 6-in. diameter pipe, whereas the coefficient could be as low as 0.0075 for a much smoother surface. Determine the difference in power required to pump 4 ft³/s of oil (SG = 0.86) through a mile of the two kinds of pipes.

6.14 Refer to Prob. 6.13, and for old and new pipe, give the maximum horizontal distance between pumps which produce a pressure increase of 80 psig before the pressure is reduced to −13 psig. What is the percent increase in number of pumps required for the 4 ft³/s flow if the smooth pipe is allowed to become rough?

6.15 A 6-in. diameter smooth pipe is to be used to raise water from a pond. At what elevation must the pump be placed so that the

pressure at its inlet does not fall below −13.5 psig? The pipe is vertical and extends 30 ft below the water surface. A flow rate of 3.5 ft³/s is needed. (Loss = $f(L/D)V^2/2g$ and $f = 0.316/(VD/\nu)^{0.25}$.)

6.16 Steel conduit can be assumed to be relatively smooth if it is new. (With age, the quality of the surface will usually deteriorate.) Consider this to be the case, and assume that with time there is an increased pressure loss of 1 m of water when 1.2 m³/s flows 1 km through a tube that is 0.5 m in diameter and horizontally installed. If the losses are linearly proportional to length, in what length of new pipe will the pressure decrease 200 kPa? If in this range of fluid velocities the losses vary as the square of the average velocity, what length of "fouled" pipe will give a 200-kPa pressure drop for flow of 1.0 m³/s? (Refer to Prob. 6.13.)

6.17 A steel conduit 0.5 m in diameter is used in pumping water and potatoes with the aid of specially designed pumps. When water alone is flowing in the pipe, the pressure is measured along its length by observing the height to which a column of water rises in a glass tube attached along the top of the pipe. (The glass tubes are called *piezometer tubes,* and holes in the conduit through which pressure is sensed are *piezometer taps.*) In the first kilometer of the pipeline there is an increase in elevation 150 m and the elevation of the water in the piezometer tubes decreases the equivalent of 162 m of water due to the flow. Predict the difference in the elevation in the piezometer tubes for the same flow if in the next kilometer of pipeline its elevation decreases 150 m. For the entire 2 km distance, which piezometer tube has the highest water level? $Q = 1.2$ m³/s.

6.18 In Prob. 6.17, the pressure decreased the equivalent of 162 m of water as a pipeline went 1 km up an incline. What must the pressure at the upstream be in order that vapor does not begin to form at the point that is 150 m above the first piezometer tap? How much pump capacity, in kW, must be installed between this point and another that is 2 km downstream but at the same elevation and pressure?

6.19 The losses of 15 ft in a system of piping can be

expressed as $5.2V^2/2g$, in which V is the average speed in a pipe 8 in. in diameter. What is the rate of flow? If a pump is inserted into the piping system to compensate for the losses, and the flow rate is doubled, what is the horsepower of the pump?

6.20 If the pump in Prob. 6.19 is rated at 5 hp, what is the expected flow rate after it is installed?

6.21 Water flows from a reservoir through a hose connection that is 10 m below the free surface. The hose is horizontal and terminates in a nozzle which sprays a jet 20 mm in diameter into the atmosphere. The losses in the hose (50 mm in diameter) can be expressed as $2.1V^2/2g$, in which V is the average velocity in the hose. What is the rate of flow?

6.22 Water is being pumped from a large reservoir through a pipe into a second reservoir that has its free surface 50 ft above the other. When the pump is not running, the water flows in the opposite direction at the same velocity as when the pump was operating. Assume that the losses are independent of the flow direction. What is the power that must be supplied to the pump-motor system (assume an efficiency of 85 percent) when it is pumping 2 ft³/s?

6.23 A pump is located 2 m above an open pool of water. It discharges the water through a horizontal hose 50 mm in diameter that terminates in a nozzle that discharges a jet 20 mm in diameter. The gauge pressure just ahead of the nozzle is 300 kPa. The losses in the hose

system are $2.1V^2/2g$. What is the power rating of the pump?

6.24 A water main (6 in. in diameter) has a pressure of 60 psig at street level. At what elevation is the pressure at atmospheric pressure when there is no flow in the pipe? At what elevation does this occur when the flow is 1 and 4 ft³/s if the losses are $0.04V^2/2g$ per foot of pipe?

6.25 A piping system to supply water to a large city appears as in Fig. P.6.25. The conduit is 2 m in diameter, and the discharging jet (into a water-purification system) is 0.7 m in diameter. The losses in the system are $(0.01)LV^2/2g$ but $(0.5)V^2/2g$ at the entrance to the conduit from the reservoir. The pump increases the pressure 300 kPa. Plot as the ordinate the value of $(p/\gamma + z)$ at points A through E against the distance along the pipeline as abscissa. These points on the hydraulic grade line are useful for determining the operating characteristics of the system. Determine from your graph whether there are any points in the system where the pressure is zero gauge. Where is the minimum pressure in the system? (L is in meters.)

6.26 A mountain reservoir is connected to a lake in the valley, 110 m below, by a pipe that is 4 m in diameter. Water passes through an array of turbines to generate power. What is the flow rate when no power is being generated and the combined losses are $15V^2/2g$? (The discharge outlet is now just below the free surface of the lake.) Power is to be generated,

Figure P.6.25

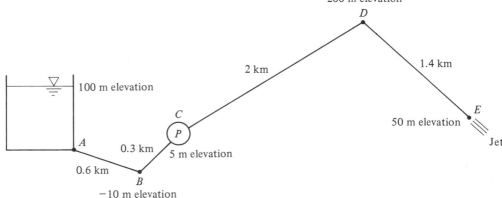

The Conservation of Energy   Chap. 6

without changing this expression for the losses; however, the flow rate will be one-half that when no power is extracted. What is the output of the system?

6.27 The system described in Prob. 6.26 operates for 8 hours a day. For 16 hours, when the power demand is low, water is pumped from the lake up to the reservoir so that in a daily cycle there is no net flow. How does this pumping power compare with the power generated on an hourly and daily basis?

6.28 Determine the energy saving over a period of 365 days if gate valves ($K = 0.19$) can be installed instead of the less expensive globe valves [$K = 5 = \text{losses}/(V^2/2g)$] for the flow in pipe 12 in. in diameter and a 12 ft^3/s flow of water. How many 100 W light bulbs could be illuminated with the savings?

6.29 Assume that it takes five complete turns to open a globe valve—a type used for sill cocks such as those to which one attaches a garden hose. When fully open, this valve causes an energy loss that is expressed as $10V^2/2g$. (Assume that the coefficient for $V^2/2g$ is (10)(5/(number of turns the valve is open)). For a pressure in the water mains (where zero velocity is assumed because of the size) of 60 psig, what rate of flow can be expected through a short horizontal hose 1 in. in diameter (whose losses are insignificant in comparison with the valve) when the valve is fully open, 1 turn open, and $\frac{1}{4}$ turn open?

6.30 Referring to Prob. 6.29 for specifications, determine the valve opening for a jet that reaches a height in a fountain that is one-half the maximum height possible with such a valve in the system.

6.31 If milk (SG = 0.95) is pumped through a cooler with fifty 180° bends in a 2-in. pipe, the resulting pressure drop associated with these bends must be added to the losses due to pipe friction. The loss in each bend is $2.2V^2/2g$. Determine the power needed because of these bends when 90 gallons per minute (gpm) is being pumped.

6.32 A 3-in.-diameter pipe is connected in series with a 6-in. pipe. At this sudden expansion, there is an energy loss of

$$\frac{[1 - (D_1/D_2)^2]^2 \, V_1^2}{2g}$$

in which the subscripts 1 and 2 refer to upstream and downstream conditions, respectively. Determine the pressure drop across such an expansion when $V_1 = 10$ ft/s for water.

6.33 If a 6-in. pipe is connected at its end to a 3-in. pipe, there will be a loss of energy as the flow adjusts to the new boundaries after a sudden contraction. If $A_2/A_1 = 0.25$, where the subscripts 1 and 2 refer to upstream and downstream conditions, there will be a loss of $0.33V_2^2/2g$. If a manometer between the two pipes gives a pressure difference of 5 in. of water, what is the rate of flow of air (68°F and 14.7 psia)?

6.34 A 3-in.-diameter pipe connects the outlet of a pump to another piece of equipment that also has a flange for 3-in. pipe. It is proposed to replace the long piece of pipe between the flanges with a 6-in. pipe in order to reduce the frictional losses, $h_L = f(L/D)V^2/2g$, in which $f = 0.316/(VD/\nu)^{0.25}$. "Minor losses" due to the sudden enlargement in diameter, and subsequent sudden contraction, will also occur. What is the length of 6-in. pipe for which the reduced frictional losses (in comparison with a 3-in. pipe) would be equal to the minor losses occurring at each end? (Refer to Probs. 6.32 and 6.33 for the values of the minor losses.) Let $Q = 0.5$ cfs.

6.35 High-speed open channel flows can be destructive to the channel's surface. Consequently, it is sometimes desired to slow the current, and consequently raise its elevation. This process, called a *hydraulic jump*, is associated with a loss, in meters or feet, of $(h_2 - h_1)^3/(4h_1 h_2)$, in which h_1 is the lower upstream depth. For a river flowing with a rate of 5 m/s at a depth of 2 m, what depth and speed can be expected downstream of such a "jump"? (Assume a unit width.)

6.36 The hydraulic jump referred to in Prob. 6.35 is one way to introduce a dissipative effect into an open channel flow. Another scheme is shown in Fig. P.6.36. The corrugated surfaces provide regions where the flow is accelerated and regions where the water is stilled and the dissipation process is carried to completion. For a series of five cascades, the loss in meters is 0.3. For a stream 2 m deep and a current of 4 m/s, what will be the depth and water speed downstream of the cascade?

0.5 m

Figure P.6.36

6.37 Assume that the dissipation per cascade mentioned in Prob. 6.36 can be linearly extended for design purposes. For the conditions described there, how many cascades must be provided to increase the depth to 3.6 m?

6.38 A particular design of exhaust system takes 4.00 kW from the output of a 1600-cm^3 (i.e., volume of air passing through the engine in two revolutions) internal-combustion engine running at 4800 rpm. Against what "back pressure" (i.e., pressure at the entrance to the exhaust system) is the engine operating under this condition? (Assume the air has a temperature of 250°C at 103 kPa to calculate the air's density at the entrance to the exhaust system, even though this pressure is an initial estimate.)

6.39 Air is circulating in a loop that is used as a wind tunnel. A 25-hp motor drives an axial-flow fan to produce an airstream of 100 ft/s in a test section that is 6 ft by 6 ft. If the tunnel is to operate continuously, how much heat must be removed from the air if the walls are well insulated? What rate of water must be circulated if the rise in coolant is to be 5°F? If the air is not cooled, what will be its temperature increase in 1 hour? It's volume is 39,000 ft^3.

6.40 The flow in a horizontal circular duct divides equally into two conduits as shown in Fig. 6.3.8. The large duct is 1 m in diameter, and the two smaller ones are 0.8 m. The upstream pressure is 10 kPa gauge where 1.5 m^3/s of air (20°C and 110 kPa) is flowing. What are p_2 and p_3 if the losses in each branch of the Y can be expressed as $1.7V_2^2/2g$?

6.41 Air is flowing in a square ventilating duct 3 ft wide, as shown in Fig. P.6.41. There are two square lateral ducts, each 3 ft wide, through which the air (68°F and 14.7 psia) flows equally to two rooms. If the air pressure at point 1 is 10 in. of water, determine the pressure that can be expected at points 2, 3, and 4. The losses in the first branch are $0.92V_1^2/2g$ for the flow from 1 to 3 and almost zero for

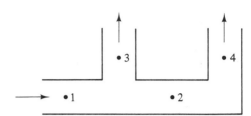

Figure P.6.41 Flow directions shown are for Prob. 6.41. The opposite directions apply for Prob. 6.42.

the flow from 1 to 2. At the elbow there is a loss of $1.13V_4^2/2g$. V_1 is 28 ft/s.

6.42 Refer to Fig. P.6.41. Air is now being drawn from points 3 and 4 by a fan that is located downstream from point 1. The air at points 3 and 4 is at 20°C and 101.1 kPa. Estimate the pressure at point 1 and the amount of air coming from points 3 and 4 if the flow at point 1 is 5400 cfm (ft^3/min). Use the following losses: flow from 4 to 2, $1.13V_2^2/2g$; flow from 2 to 1, $0.46V_1^2/2g$; flow from 3 to 1, $0.28V_1^2/2g$. The constants are average values and depend upon the ratio of the flow division at the branch, which will determine the type and size of dissipative eddies that are formed there.

6.43 Water flows in a Venturi tube that has diameters of 9 and 3 in. Tests show that 97 percent of the head available at the inlet is available at the exit and that the losses occur mainly in the diverging portion. A pressure gauge at the smallest section reads 1 psi vacuum when 4.5 cfs is flowing through the flow meter. What are the pressures upstream and downstream of it?

6.44 A vertical pipe has two pressure gauges that are 54 ft apart but read the same pressure. If the gauges have just been calibrated, what is the explanation of this occurrence?

6.45 A pump is located 5 ft above the surface of water in a reservoir where the pressure is atmospheric pressure. What should be the reading in a pipe at a point that is 30 ft above

The Conservation of Energy   Chap. 6

the pump if the losses in the pipe are equivalent to $1.8V^2/2g$? The average velocity in the pipe, V, is 8.7 ft/s, and $\dot{W}/g\dot{m} = 91.5$ ft.

6.46 A siphon is used to convey water from a reservoir to the inlet of an axial-flow turbine that is contained within the conduit (this is described as a *tube turbine*). The diameter of the tube is 42 in., and the elevation difference between the surface of the reservoir and the stilling basin downstream of the turbine is 18 ft. Neglecting all losses in the line except the $0.15V^2/2g$ "effective" pipe friction and $V^2/2g$ at the exit, what is the volume rate of flow if the turbine does not extract any energy from the water? For this same system, how does the developed power vary with the amount of water flowing through the siphon?

6.47 An artificial geyser for a park fountain is produced by a pump that propels a stream of water 275 ft in the air from a nozzle that is 2 in. in diameter and located 5 ft above the surface of the lake. What is the minimum power that must be supplied to the pump? If losses in the pump's piping system are $2.1/2g$ times the square of the velocity at the exit of the nozzle, what is the power that must be supplied to an 80 percent efficient pump?

6.48 A small opening—an orifice—in a tank is submerged below the free surface of a reservoir. A horizontal jet issuing from this opening enters a small cup that is at a distance Y below the orifice mentioned above. What is the horizontal distance X that the cup must be displaced from the orifice, whose K value = 0.1?

7

Momentum Conservation

7.1 INTRODUCTION

Our attention will now be directed to the third of the conservation laws, the law of conservation of momentum. Strictly speaking, momentum is not conserved in most situations. Rather, we shall relate changes of linear momentum of a given mass of fluid with the forces that act upon it. Some of the forces that act on the fluid are transmitted by walls and other boundaries of the fluid. The momentum equation that will be derived will permit us to calculate the forces that flowing fluids exert on walls and other boundaries of the fluid. This is so because such forces are equal in magnitude, but opposite in sign, to the forces which these boundaries exert on the fluid. The ability to calculate such forces is useful because it will allow us to estimate the drag forces on bodies such as buildings and parachutes, as well as the propulsive forces of machines similar to jet engines.

Before you become deeply involved with this topic, recall that particular forces on the walls of fluid containers have already been calculated in this text. This was done for static cases and for cases with rigid-body accelerations. Newton's laws were applied, and the normal stresses in the fluid were determined. From this information, the force on the surface in question was found by integrating the detailed pressure distribution that ensued. In principle, a similar approach to determining the force will be used when the fluid particles are accelerating in a general way.

As an example, consider the case of the sluice gate shown in Fig. 7.1.1. The energy equation can be used to determine the elevation h_2 from the h_1 and V_1 that are specified in Fig. 7.1.1. The equation of mass conservation is also used in this analysis. Now, if we knew that the streamlines for the flow under the sluice gate are similar to the lines that are included in Fig. 7.1.1, we could deduce the velocity at points near the

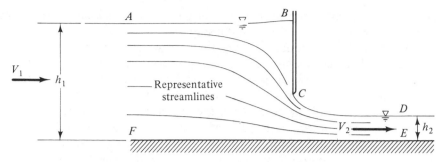

Figure 7.1.1 Flow under a sluice gate, with section AF far upstream of the gate.

gate. The fluid speed will vary inversely with streamline spacing. We can see from the pattern of Fig. 7.1.1 that near point B the fluid velocity is low (point B is in fact a stagnation point) and the speed increases as one approaches the edge of the sluice gate, point C. The information about the variation of velocity near the surface of the gate from B to C can be used to find the pressure variation very near the gate. The energy equation gives, assuming zero losses,

$$\frac{p}{\rho} = (z_A - z)g + \frac{V_A^2 - V^2}{2}$$

in which V and z are the velocity and elevation of various points near the surface of the gate. Let us assume that the calculation for p has been carried out for a particular streamline spacing and that the pressure distribution is found to be like that shown in Fig. 7.1.2. This pressure distribution is now assumed to be the same as that acting on the gate and can be integrated to obtain the resultant force of the fluid on the gate. This force is called $\mathcal{R}_x$. Sometimes a detailed analysis is not necessary in order to determine the integrated effect $\mathcal{R}_x$, for reasons that we shall soon discover. Moreover, sometimes such an analysis cannot be carried out, because of its complexity.

The material in this chapter is directed at making a determination of forces like $\mathcal{R}_x$ when only the upstream and downstream conditions are prescribed. The principle that we shall use deals with the relationship between the forces that are applied to a system of fluid particles and the time rate of change of momentum of these particles. This relationship is, of course, a statement of one of Newton's laws. We shall not apply the "laws of motion" to elements, or particles, of fluid as we did in Chap. 2 when the hydrostatic law was developed. We shall sum the momentum changes of a large col-

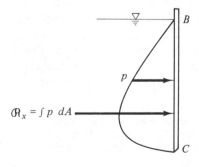

Figure 7.1.2 Pressure distribution on the face of the sluice gate due to the flowing liquid.

lection of material during a time interval and relate this change to the sum of the external forces that act on this material system.

Fortunately, the momentum equation that we shall derive and use does not rely upon a detailed knowledge of the flow. However, it does not give us detailed information about the flow, such as the particular pressure distribution. It does give us various overall, or integrated, results. While the actual shape of the pressure distribution may not be available to us from the momentum equation, the integral of the pressure distribution can be found in many cases. This is the resultant force. Hence, expecting to gain useful information from the momentum equation that may not be available from the energy equation directly, we now consider what is necessary for understanding and deriving a mathematical statement of this "law of nature."

It is useful to recall the definitions of momentum and the change of momentum for particles before engaging in a discussion of the momentum of the particles in a region of a continuum. *Momentum* is defined as $m\mathbf{V}$, in which m is the mass of the particle (or collection of particles) that has velocity $\mathbf{V}$. Since $\mathbf{V}$ is a vector with components u, v, and w, the momentum is also a vector, with components mu, mv, and mw for the x, y, and z directions. If a particle has a mass m in kilograms and an x component of velocity of 3 m/s, its x component of momentum is $3m$, in kg-m/s. If the v and w velocity components were prescribed, the y and z components of momentum could be similarly calculated. If the same particle subsequently had a different x component of velocity, say, 5 m/s, it would then have a momentum of $5m$, in kg-m/s. The change of the x-directed momentum between the final and the initial states would be $5m - 3m$ (kg-m/s), or $2m$ (kg-m/s). If the particle had had an initial x component of velocity of 3 m/s and subsequently $u = 2$ m/s, there would be a change of momentum between the final state and the initial state of $m(u_{final} - u_{initial})$, or $m(2 - 3) = (-1)m$, in kg-m/s. Even though momentum is a vector quantity, it is easy to write the momentum when the final x component of velocity is in the negative x direction. For example, if $u = -4$ m/s, then the associated momentum is $-4m$ (kg-m/s). If this is the final momentum of the particle and the initial momentum was $3m$ (kg-m/s), then the change of momentum would be $m(u_{final} - u_{initial})$, or $m[(-4) - (+3)] = -7m$ (kg-m/s).

Example 7.1.1

Determine the change in the y momentum for a particle of mass $2m$ if $v_{final} = 1.5$ m/s, $v_{initial} = -2.1$ m/s, and v is the y component of velocity.

Solution

$$\Delta(y \text{ momentum}) = 2m[1.5 - (-2.1)] = 7.2m \qquad \text{kg-m/s}$$

The rate of change of momentum of a system of particles is related to the external forces that act during the time that the momentum change occurs. Newton's second law of motion states, in effect, that the force that acts on the system is equivalent* to the change of momentum per unit time. Thus we shall be interested not only in the changes of momentum that occur, but also in the time interval over which these changes occur. In Example 7.1.1, the change in y momentum might have oc-

---

* The word *equivalent* was used instead of *equal* to emphasize that a force produces a change of momentum, but that force is not momentum. This is similar to the case of the force which a mass produces in a gravitational field. We commonly call this *weight,* but weight is not mass.

curred during 0.3 s. Then the time rate of change of momentum would have been $2m(3.6)/0.3 = 24m,$ in kg-m/s^2. (Remember that 1 newton (N), a unit of force, is equivalent to 1 kg-m/s^2.) If the momentum change in the example had occurred during 0.1 s, the reader could show that during this time an average force of $72m$ newtons was applied to the system. The two numerical values just determined are particular examples of a result that can be derived by starting with a statement of one of Newton's laws of motion:

$$\mathbf{F} = m\mathbf{a} = m\frac{d\mathbf{V}}{dt}$$

This expression can be integrated between two times to read

$$\frac{\int_{t_1}^{t_2} \mathbf{F}\, dt}{t_2 - t_1} = \frac{m(\mathbf{V}_{t_2} - \mathbf{V}_{t_1})}{t_2 - t_1}$$

if one introduces the time interval $t_2 - t_1$. We next define a temporal average of the force, $\mathbf{F}_{avg}$, as the term on the left-hand side of this equation. Then

$$\mathbf{F}_{avg} = \frac{m(\mathbf{V}_{t_2} - \mathbf{V}_{t_1})}{t_2 - t_1} \tag{7.1.1}$$

In much of the work that follows, we shall be concerned with a continuously acting force on a system of particles. This is true because we shall concentrate initially on steady-state situations in this chapter. Finally, it is worth noting that not only the magnitude of the force that produces the momentum change, but also its direction, can be found. If there has been a positive time rate of change of x momentum, the force on the system of particles is positive. It acts on the material volume in the positive x direction. It is the cause of the momentum change. The change is the result. The momentum equation expresses this cause-and-effect relationship.

7.2 THE MOMENTUM EQUATION

Before we start deriving a statement of Newton's momentum principle, his second law, let us recall the logical steps that were used in the discussions of mass conservation and energy conservation. We identify, by some means, a known quantity of fluid existing in a region $ABCDA$ at time t_0, as in Fig. 7.2.1. (This material volume obviously is not the converging conduit that we have used as an aid heretofore. While the associated control volume was specialized, it did not limit our results in any way, there is no reason now not to use the more general configuration of Fig. 7.2.1.) A small interval of time, perhaps 1 microsecond, is allowed to elapse, and the mass of fluid then occupies the region $A'B'C'D'A'$ at time t_1. In the interval $t_1 - t_0$, particle A, for example, will be displaced to point A'. (As $\Delta t \to 0$, region $A'B'C'D'A'$ approaches region $ABCDA$. At the initial and final times, the total quantity of a particular characteristic of the material is surveyed. The change of this quantity is related to something else by some "law" of nature. When we looked at mass conservation, we postulated that the change of the mass was zero. When energy conservation was studied, the total energy change—kinetic, potential, and internal—was related to the rates of heat and work

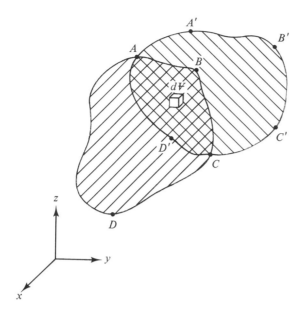

Figure 7.2.1 Material volumes associated with a particular collection of fluid particles within a large quantity of a flowing fluid. Material volume: specific region in space occupied by a definite mass of fluid at a given point in time. Control volume: a designated region in space through which fluid passes. Region $ABCD'A$ may contain different particles at t_0 and t_1; however, $\mathbf{V}$ is the same at each point if the flow is steady.

that were added to the material. Newton's second law states that the time rate of change of the momentum of a system of particles is due to the external forces which are exerted on the system. There is a direct relationship between the quantities, which can be expressed as an equation. Thus, it is natural that we turn to an evaluation of the change of the momentum for the system $ABCDA$ shown in Fig. 7.2.1.

We shall focus our attention initially on the rate of change of the linear momentum component in the x direction. It is to be recognized that momentum is a vector quantity and must be dealt with in terms of components. The procedure that we used in Chaps. 3 and 6 to produce equations that express some physical laws will be employed again.

Newton's laws postulate that the time rate of change of momentum for a system of particles is equivalent to the net force that is applied on that system of particles. With the aid of the Reynolds transport theorem, we can express one component of this relationship as

$$\Sigma F_x = \frac{\partial}{\partial t}\int_{CV}(\rho u)\,d\forall + \int_{CS}(\rho u)(\mathbf{V}\cdot\mathbf{n}\,dA) \tag{7.2.1}$$

We rewrite Eq. (7.2.1) to apply to all of the components of momentum as

$$\Sigma \mathbf{F} = \frac{\partial}{\partial t}\int_{CV}(\rho \mathbf{V})\,d\forall + \int_{CS}(\rho \mathbf{V})(\mathbf{V}\cdot\mathbf{n}\,dA) \tag{7.2.2}$$

Whenever possible, we shall attempt to calculate the forces which make up the summation $\Sigma\mathbf{F}$. Often there is at least one force which acts on the fluid that we cannot assess in detail. In this case we shall assign a resultant force, $\vec{\mathcal{F}} = \mathcal{F}_x\mathbf{i} + \mathcal{F}_y\mathbf{j} + \mathcal{F}_z\mathbf{k}$, to describe the net affect of the *unknown* force, or stress distribution. If all the other terms in Eq. (7.2.2) are known, one can solve for the components $\mathcal{F}_x$, $\mathcal{F}_y$, and $\mathcal{F}_z$. One will obtain a magnitude and a sign for these components. If this sign is negative, it simply means that the component acts in the opposite direction of the positive coordinate direction, in accordance with the normal usage of vectors in mechanics.

The successful use of Eq. (7.2.2) will depend upon the extent to which it is comprehended, as well as the way in which one uses it to analyze a given situation. A method of analysis employing momentum and force diagrams that will be discussed in Sec. 7.3 should prove helpful. In making our analysis of a problem and using Eq. (7.2.2), it will be necessary to think about the physical processes that are occurring so that reasonable decisions can be made about the forces that act on the fluid system.

These forces will be the result of stress distributions, some of which will be known quite well from the physics of the particular situation. Thus, when there is a region of horizontal flow, one can conclude that there is no vertical acceleration, with the consequence that the vertical pressure variation is hydrostatic. This fact was used in Chap. 6. Similarly, if one is dealing with a liquid jet, the pressure in the jet will be the ambient pressure throughout its cross section, if surface tension is neglected. Moreover, if one is concerned with flow in a closed horizontal conduit at a location where the pressure at the centerline is p, then one can conclude that the average pressure at that section is p also, because the pressure does not change from one side of the pipe to the other and it varies linearly from the top to the bottom.

There will also be unknown pressure distributions, such as those which occur on the curved portions of a pipe bend. Such a pressure distribution is the result of the type of flow in the passage. At the same time, this unknown pressure distribution must be such that it is in concert with the rate of momentum change that the fluid has experienced in altering its direction. One uses the momentum equation to calculate the integrated effect of this unknown pressure distribution. This is done in Sec. 7.3.

Prior to starting Sec. 7.3, where a number of problems that employ the momentum equation are treated, it is necessary to note that the actual velocity distribution at the control surface will influence the magnitude of the surface integral in Eq. (7.2.2), and consequently the force that is being applied to the fluid. This fact is recognized in some devices that increase the amount of thrust from jet engines. The following example illustrates this point.

Example 7.2.1

The steady flow in the circular duct shown in Fig. 7.2.2 is axisymmetrical but not of uniform speed across the duct. Determine the momentum per unit time passing across the surface represented by CD.

Solution The actual velocity u can be thought of as the sum of the average velocity $\bar{U}$ and a local variation u'. Thus

$$u = \bar{U} + u'$$

The momentum per unit time for an incompressible fluid crossing surface CD is

$$\int (\rho u)(\mathbf{V \cdot n}\, dA)$$

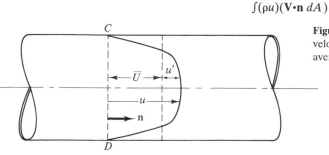

Figure 7.2.2 Decomposition of the axial velocity in a conduit, u, into the sum of the average velocity $\bar{U}$ and a local variation u'.

On surface CD, $\mathbf{V \cdot n} = u = \overline{U} + u'$, according to the notation of Fig. 7.2.2. For a fluid of constant density, this integral can be written as

$$\rho\int(\overline{U} + u')(\overline{U} + u')\,dA = \rho\int\overline{U}^2\,dA + \rho\int 2\overline{U}u'\,dA + \rho\int(u')^2\,dA \qquad (7.2.3)$$

The second integral on the right-hand side of the equation vanishes because

$$\overline{U}A = \int u\,dA = \int(\overline{U} + u')\,dA$$
$$= \int\overline{U}\,dA + \int u'\,dA$$
$$= \overline{U}A + \int u'\,dA \qquad (7.2.4)$$

Hence using the results of Eqs. (7.2.3) and (7.2.4), we may write

$$\frac{\text{Actual momentum}}{\text{Unit time}} = \rho\overline{U}^2 A + \rho\int(u')^2\,dA$$

The first term on the right-hand side is the rate of momentum flux, based on the average velocity. Because the integral has a positive integrand, it gives a positive value as long as $u' \neq 0$. We can now conclude that (actual momentum flux) $\geq$ (momentum flux based on an average velocity).

One defines a *momentum correction factor* β as the ratio of the actual momentum flux to the flux that is based on the average velocity. Example 7.2.1 leads one to the conclusion that $\beta \geq 1.0$. (The reader may wish to show a similar property of α, the kinetic energy correction factor.)

7.3 STEADY-STATE MOMENTUM APPLICATIONS

Our first applications of Eq. (7.2.2) involve the steady state only. This initial limited scope will permit an increased emphasis on the aspects of the momentum equation that some students find difficult, while still providing a wide variety of applications. The examples that follow have a standardized format. This is done in order to organize the thought processes associated with solving problems of this type. When one has gained sufficient experience and confidence, one may choose to employ some shortcuts or to consolidate the diagrams that are associated with the problems here into a smaller set. However, the reader is encouraged to follow closely the material during an initial perusal so that the subtleties and logic of the physics become evident.

In many of the applications that follow, the pressure distributions on the control volumes will be expressed in gauge pressure. The effect of atmospheric pressure on the external forces applied to the control volume will be zero. This is true because these contributions to the force components are self-canceling as one integrates over the *entire* surface. This is an important point.

Example 7.3.1

Find the force on the sluice gate that is the result of the flowing water for the conditions shown in Fig. 7.3.1a.

Solution (a) In doing problems of this kind it is wise to examine the situation and write an expression for the momenta without considering the force system at first. This is facilitated by drawing a special diagram similar to Fig. 7.3.1b. Such a diagram is dependent upon the choice of control volume.

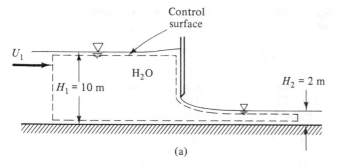

(a)

Figure 7.3.1(b)   Diagram of flow for Example 7.3.1 listing the quantities needed to evaluate the momentum integral.

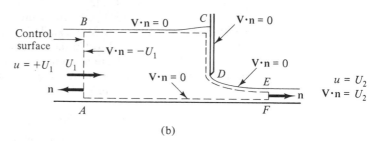

(b)

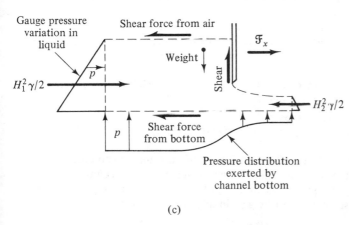

(c)

Figure 7.3.1(c)   Diagram of flow for Example 7.3.1 detailing the forces acting on the fluid within the control volume. $\mathcal{F}_x = +x$ component of force on fluid from gate. $eH_1^2\gamma/2$ = resultant pressure force, upstream. $H_2^2\gamma/2$ = hydrostatic force, downstream.

(b)   Now a separate diagram should be drawn, especially when beginning this topic, to delineate the force system acting on the fluid in the control volume. Note that $\mathcal{F}_x$ has been drawn in the $+x$ direction in Fig. 7.3.1c. This is done even if our intuition, which is sometimes deceiving, would tell us that the force of the gate on the fluid would be directed toward the left. With the direction that was chosen, the ultimate sign on the result for $\mathcal{F}_x$ will tell us whether it has a positive or a negative value. Accordingly, we shall know the direction in which this component acts.

(c)   With the information contained in Fig. 7.3.1b and c, the equality proposed by Newton, Eq. (7.2.2), can be employed. It is wise, at first, not to write Eq. (7.2.2) immediately and then make the appropriate substitutions. The probability would then be great that algebraic signs, as well as ideas about force and momentum, will be confused.

Since only the x direction is being considered in this problem, Eq. (7.2.1) can be employed. Also, Fig. 7.3.1b, which systematizes the correct expressions for u and $\mathbf{V} \cdot \mathbf{n}$, will prove to be valuable. Because the density is constant in this example, the integral $\int (\mathbf{V} \cdot \mathbf{n}\, dA)\rho u$ can be written as

$$\rho \int_{AB} (-U_1)(U_1)\, dA + \rho \int_{CD} (+U_2)(U_2)\, dA$$

Because the integrands are constants in this expression, one can factor them out and integrate only on dA across AB and CD. The results of these integrations would be A_1 and A_2, respectively. Thus the expression for the momentum flux is

$$\rho U_2^2 H_2(1) - \rho U_1^2 H_1(1)$$

in which $A_1 = H_1(1)$, the unity factor being associated with a unit width.

Now it is time to turn to Fig. 7.3.1c. The sum of the forces acting on the fluid within the control volume is denoted by $\Sigma \mathbf{F}$. This symbol should not be thought of as a numerical value. Rather, it should be viewed as a kind of operator that conveys the command, "Sum the forces acting *on* the fluid within the control volume." The segments of the control surfaces labeled AB and EF are taken far upstream and downstream, where the flow is horizontal. This fact permits us to treat the vertical pressure variation as linear. (Gauge pressure has been used.) Figure 7.3.1c helps us to write, without doubting,

$$\Sigma F_x = \frac{H_1^2 \gamma}{2} - \frac{H_2^2 \gamma}{2} + \left(\begin{array}{c} \text{shear forces neglected for now as} \\ \text{small in comparison with } \dfrac{H_1^2 \gamma}{2} \text{ and } \mathscr{F}_x \end{array} \right) + \mathscr{F}_x$$

One should observe that this expression contains only forces and the previous expression contains only momentum flux terms. In essence, there has been a conscious effort to separate the cause and effect parts of Newton's statement that relates forces and rates of momentum changes.

Finally we employ Newton's second law in the form of Eq. (7.2.1) and equate the previous two expressions with the result that

$$\mathscr{F}_x = \frac{H_2^2}{2}\gamma - \frac{H_1^2}{2}\gamma + \rho U_2^2 H_2 - \rho U_1^2 H_1$$

(d) $\mathscr{F}_x$ is the x component of the force* of the wall (i.e., the surroundings) on the system (i.e., the flowing fluid within the control volume). This means that the force that the fluid exerts on the wall, a *reaction,* is equal in magnitude and opposite in sign to $\mathscr{F}_x$. If we call the force of the fluid on the surroundings (the gate in this case) $\mathscr{R}_x$, we have

$$\mathscr{R}_x = -\mathscr{F}_x = -\left(\frac{H_2^2 \gamma}{2} - \frac{H_1^2 \gamma}{2} + \rho U_2^2 H_2 - \rho U_1^2 H_1 \right)$$

The use of $\mathscr{F}_x$ and $\mathscr{R}_x$ may seem to be redundant. Both are used in this and subsequent examples to stress the content of Eq. (7.2.2). Even though one often wants to know the forces on the *surroundings,* Eq. (7.2.2) applies to the *fluid* in the control volume. If in Eq. (7.2.1) one introduces $-\mathscr{R}_x$ for $\mathscr{F}_x$, one can easily make the sign errors or conceptual errors alluded to previously.

(e) The gate must be supported in such a way that the force $\vec{\mathscr{R}}$, $\mathscr{R}_x$ in this case, can be opposed. A free-body diagram of the gate should be drawn and the appropriate types of constraints (pins with no moment-carrying capacity, knife edges, and the like) included. If the gate that we have been discussing were cemented in place, the reactions could not be

---

* So that if $\mathscr{F}_x = +55$, we would know that the component on the fluid was in the $+x$ direction, and if $\mathscr{F}_x = -55$, the component on the fluid would be in the $-x$ direction.

found in a simple fashion. Nevertheless, one would require that the summation of all the forces acting on the gate be zero. The consequence of this is

$$(R_x)_{\text{due to supports}} + (\mathcal{R}_x)_{\text{due to water}} = 0$$

(The atmospheric pressure has not been explicitly included, since its effect cancels when considering both sides of the gate. Note that gauge pressures are illustrated in Fig. 7.3.1c.) It appears from this that R_x is directly related to $\mathcal{R}_x$ and, by action and reaction, to $\mathcal{F}_x$. It is understandable that some portion of R_x, a component of the supporting force, is needed to change the momentum of the fluid. Depending on the type and orientation of the problem, other loads on the structure such as wind forces or weight—if x were vertical, of course—would have to be included in the determination of R_x.

Example 7.3.2

The pipe tee shown in Fig. 7.3.2 will divert the flow in two directions, with the result that there will be a reaction, $\vec{\mathcal{R}} = \mathcal{R}_x\mathbf{i} + \mathcal{R}_y\mathbf{j}$, on the pipe due to the fluid. The procedure used

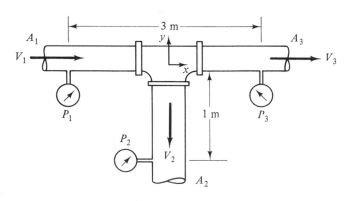

(a)

Figure 7.3.2(a) Flow of fluid in a pipe tee. $A_1 = A_2 = A_3 = 0.01 \text{ m}^2$, $p_1 = 5$ Pa, $V_1 = 5$ m/s, $V_3 = 1.5$ m/s.

Figure 7.3.2(b) Momentum diagram for Example 7.3.2.

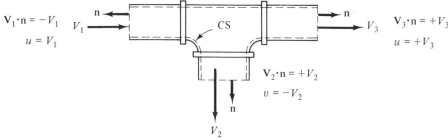

(b)

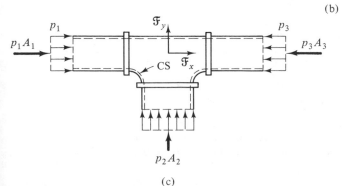

(c)

Figure 7.3.2(c) Force diagram for Example 7.3.2. $\mathcal{F}_y$ = force component of tee acting on fluid in control volume. Gravity acts in the $-z$ direction.

in Example 7.3.1 will be employed again to find this reaction. Figure 7.3.2a is a general schematic diagram.

Solution (a) Draw a momentum schematic diagram and evaluate the momentum integrals

$$\int \rho \mathbf{V} \cdot \mathbf{n}(u)\, dA \quad \text{and} \quad \int \rho \mathbf{V} \cdot \mathbf{n}(v)\, dA$$

Figure 7.3.2b is the schematic diagram of momenta. For the velocities in the x and y directions, the resulting integrals are

$$\int \rho \mathbf{V} \cdot \mathbf{n}(u)\, dA = -V_1 V_1 \rho A_1 + V_3 V_3 \rho A_2$$

and
$$\int \rho \mathbf{V} \cdot \mathbf{n}(v)\, dA = -\rho V_2^2 A_2$$

(b) Draw a force schematic diagram and evaluate

$$\Sigma F_x \quad \text{and} \quad \Sigma F_y$$

Figure 7.3.2c is the schematic diagram of forces. Here,

$$\mathscr{F}_x + p_1 A_1 - p_3 A_3 = \Sigma F_x$$
$$\mathscr{F}_y + p_2 A_2 = \Sigma F_y$$

(c) Write Newton's relationship for the *cause* of the momentum changes:

$$\Sigma F_x = \mathscr{F}_x + p_1 A_1 - p_3 A_3 = \rho V_3^2 A_3 - \rho V_1^2 A_1$$
$$\Sigma F_y = \mathscr{F}_y + p_2 A_2 = \rho V_2^2 A_2$$

(d) Compute the reactions by the fluid on the tee (once again, the pressures will be gauge pressure; the effect of atmospheric pressure on any force component will be self-canceling when one integrates around the complete control volume):

$$\mathscr{R}_x = -\mathscr{F}_x = p_1 A_1 - p_3 A_3 + \rho (V_1^2 A_1 - V_3^2 A_3)$$
$$\mathscr{R}_y = -\mathscr{F}_y = p_2 A_2 - \rho V_2^3 A_2$$

(e) Compute the force required to restrain the tee. The summations of forces on the tee in the x and y directions include $\mathscr{R}_x$ and $\mathscr{R}_y$. In the z direction, the weight of the water in the tee, as well as the weight of tee itself, must be included in the calculation of this support component.

Example 7.3.3

A stationary vane diverts a stream that we specify to be emanating from a slot of height h and width W. For initial simplicity, *consider* gravity to act perpendicular to the paper and *assume* that the fluid moves onto the vane without loss of energy. Find the force exerted on the vane by the fluid. The general schematic diagram is given in Fig. 7.3.3a.

Solution (a) Draw the momentum diagram and evaluate the integrals

$$\int_{\text{CS}} \rho \mathbf{V} \cdot \mathbf{n}(u)\, dA \quad \text{and} \quad \int_{\text{CS}} \rho \mathbf{V} \cdot \mathbf{n}(w)\, dA$$

Figure 7.3.3b is the schematic diagram of momenta. If one assumes no losses from surface AD to surface BC, atmospheric pressure at points A, B, C, and D, and uniform speed across surfaces AD and BC, then $|V_1| = |V_2|$. This means that

$$\int \rho \mathbf{V} \cdot \mathbf{n}(u)\, dA = -V_1 V_1 bW\rho + V_1 V_1 \cos \theta_x\, bW\rho$$

and
$$\int \rho \mathbf{V} \cdot \mathbf{n}(w)\, dA = V_2 V_2 \cos \theta_z\, bW\rho$$

(b) Draw the force diagram and evaluate ΣF_x and ΣF_z. Figure 7.3.3c is the schematic diagram of forces. Because $\mathscr{F}_x$ is the $+x$ component of $\vec{\mathscr{F}}$, the force of the vane on the fluid, one can anticipate $\mathscr{F}_x < 0$ in this case. In this figure, the effect of atmospheric

Momentum Conservation   Chap. 7

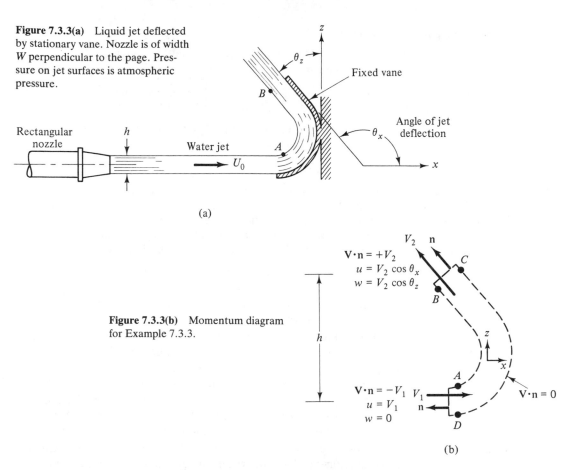

Figure 7.3.3(a) Liquid jet deflected by stationary vane. Nozzle is of width W perpendicular to the page. Pressure on jet surfaces is atmospheric pressure.

Rectangular nozzle

h

Water jet U_0

A

B

Fixed vane

θ_z

θ_x

Angle of jet deflection

z

x

(a)

Figure 7.3.3(b) Momentum diagram for Example 7.3.3.

V_2 n

$\mathbf{V} \cdot \mathbf{n} = +V_2$
$u = V_2 \cos \theta_x$
$w = V_2 \cos \theta_z$

C

B

h

A

z

x

$\mathbf{V} \cdot \mathbf{n} = -V_1$ V_1
$u = V_1$ n
$w = 0$

D

$\mathbf{V} \cdot \mathbf{n} = 0$

(b)

Absolute pressure on fluid in control volume exerted by vane

Gauge pressure distribution which yields $\mathcal{F}_x$

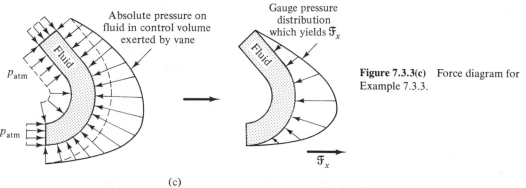

p_{atm}

Fluid

Fluid

p_{atm}

$\mathcal{F}_x$

$\mathcal{F}_x$

Figure 7.3.3(c) Force diagram for Example 7.3.3.

(c)

pressure on the net force on the fluid can be conveniently seen to be self-canceling with respect to each of the coordinate directions. This schematic indicates that $\Sigma F_x = \mathcal{F}_x$. (Note carefully once again that this equation states in *this* case that the summation of the x forces on the vane is equal to $\mathcal{F}_x$. Do not conclude that the symbol ΣF_x will always be replaced by $\mathcal{F}_x$!)

(c) Write Newton's relationship for the cause of momentum changes:

$$\Sigma F_x = \mathcal{F}_x = -V_1^2 \rho b W + V_2 (V_2 \cos \theta_x) \rho b W$$

(d) Compute the reactions:

$$\mathscr{R}_x = -\mathscr{F}_x = V_1^2 \rho b W - V_1^2 \rho b W \cos \theta_x = V_1^2 \rho b W (1 - \cos \theta_x)$$
$$= \rho Q V_1 (1 - \cos \theta_x)$$

Would $\mathscr{R}_z = -\mathscr{F}_z = -\rho Q V_2 \cos \theta_z - $ (weight of water), *if* gravity were in the $-z$ direction? In such a case, the use of U_0 and $\sqrt{U_0^2 - 2gh}$ instead of V_1 and V_2 would be necessary.

(e) Compute the forces required to restrain the vane. Equilibrium of the vane will require that

$$(R_x)_{\text{support}} + (\mathscr{R}_x)_{\text{water}} = 0$$

(The weight of the vane could naturally be included to find $(R_z)_{\text{support}}$ if gravity were acting in the $-z$ direction. The weight of the water would also have to be considered then.) Gravitational effects were not neglected in the expression for $\mathscr{R}_z$ that was presented in part (d). In that equation the water's weight was included. Since $\mathscr{R}_z$ is the total z component provided by the water against the vane, the vane's equilibrium requires that

$$(\mathscr{R}_z)_{\text{water}} - (\text{vane weight}) + (R_z)_{\text{support}} = 0$$

When a strong stream from a faucet impinges on a horizontal platter, the water tends to move radially away from the area of impact in a thin layer. A short distance away there is a circular discontinuity in the elevation, and at points farther from the region of impact the layer of water is thicker than before. Such a relatively sudden change in elevation occurs in open channel flow, too, and is important to the civil engineer.* In the transition region, where the depth may change five- to tenfold, there is a great deal of large-scale agitated motion of the water. The surface tends to become foamy, because a considerable amount of air is entrained. This region is one of strong mixing. Such a process would reasonably be one in which there is a great amount of energy dissipation. In fast-flowing rivers and channels, it may be desirable to cause such a sudden change in elevation—called a hydraulic jump—to minimize the scouring of the sides and bottom.

Example 7.3.4

A stream with depth y_1 and a current of velocity V_1 undergoes a hydraulic jump so that the downstream depth is y_2. Use the momentum equation to predict the value of y_2. The general schematic diagram is given in Fig. 7.3.4a.

Solution (a) Draw a momentum diagram (Fig. 7.3.4b) and evaluate the integral $\int_{CS} \rho \mathbf{V} \cdot \mathbf{n}(u) \, dA$:

$$\int_{CS} \rho \mathbf{V} \cdot \mathbf{n}(u) \, dA = \rho V_2^2 y_2 - \rho V_1^2 y_1$$

(b) Draw a force diagram and evaluate ΣF_x. Figure 7.4.3c is the schematic diagram for ΣF_x. The result is

$$\Sigma F_x = \frac{y_1^2 \gamma}{2} - \frac{y_2^2 \gamma}{2} - F_{\text{shear}}$$

* This jump in elevation is analogous to the jump in pressure that occurs across a shock wave in air, a matter of interest to the aeronautical engineer.

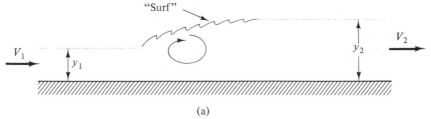

(a)

Figure 7.3.4(a) River channel showing stationary hydraulic jump. $V_1 =$ river speed upstream of discontinuity; $V_2 =$ speed downstream of discontinuity. Assume unit width perpendicular to paper.

Figure 7.3.4(b) Momentum diagram for Example 7.3.4.

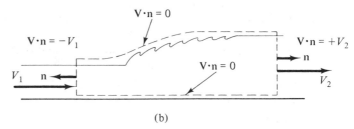

(b)

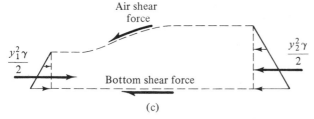

(c)

Figure 7.3.4(c) Force diagram for Example 7.3.4.

(c) Write Newton's relationship for the cause of momentum changes:

$$\Sigma F_x = -F_{shear} + \frac{y_1^2 \gamma}{2} - \frac{y_2^2 \gamma}{2} = \rho V_2^2 y_2 - \rho V_1^2 y_1 \qquad (7.3.1a)$$

This equation can be rearranged, with $V_2 y_2 = V_1 y_1$ from the law of continuity, to obtain

$$\frac{V_1^2}{gy_1} = \frac{1 - (y_2/y_1)^2}{2(y_1/y_2 - 1)} - \frac{F_{shear}}{\rho g y_1^2/2} \frac{1}{2(y_1/y_2 - 1)} \qquad (7.3.1b)$$

Neglect shear forces, for the present, because in comparison with other forces, they have only a modifying role in this case. Rivers will have a slight slope, so that for a coordinate system aligned with the bottom there will be a small weight component. It will be in a direction opposite to that of the bottom shear force. (Had the energy equation been used in the analysis, the loss term would have had to be estimated. If it were neglected, as being small, an incorrect final result would have ensued. The interrelationship between the energy equation and the momentum equation is discussed in Sec. 7.7.) Neglecting the shear effects gives

$$\frac{V_1^2}{gy_1} = \frac{1}{2}\left(\frac{y_2}{y_1}\right)\left(\frac{y_1 + y_2}{y_1}\right) \qquad (7.3.2)$$

We are able to predict y_2 in terms of y_1 from the value of $V_1/(gy_1)^{1/2}$ from this result:

$$\frac{y_2}{y_1} = \frac{-1 + \sqrt{1 + 8\,\mathrm{Fr}_1^2}}{2} \qquad (7.3.3)$$

in which Fr_1, called the *Froude number,* equals $V_1/(gy_1)^{1/2}$.[*] In this case, ΣF_x did not contain an unknown $\mathscr{F}_x$, whereas in Example 7.3.3, $\mathscr{F}_x$, the unknown force on the vane, was the only term in ΣF_x. This should caution the reader to determine *all* of the forces on the control surface and include any unknown ones in this summation. Remember that ΣF_x is not usually zero in a dynamic case—a mistake that is commonly made—but equal, when it includes all the forces acting on the system, to the change of x momentum.

In the next example we shall use a coordinate system that is translating at constant velocity. This Newtonian frame of reference will permit us to convert an unsteady problem into a steady one. Also, the system does not require the equations applicable to the problem to be altered.[†] Hence, Eq. (7.2.2) will be applied with all the velocities designated with respect to this moving frame. These ideas are used unconsciously by airline service personnel pouring coffee during a smooth flight. In an absolute frame of reference, the jet of coffee is unsteady because it occupies different points of space at different times. But with respect to a coordinate system located in the aircraft, the velocity in the stream of coffee could be considered constant, assuming, of course that the pourer has a steady hand.

Example 7.3.5

Analyze the force on a vane moving away from a nozzle at constant velocity. Use a coordinate system attached to the vane and refer all velocities to this system. This frame is translating with a speed of V_0. The general schematic is shown in Fig. 7.3.5a.

Solution (a) Draw a momentum diagram (all velocities are relative to the moving frame!) and evaluate the integral $\int_{CS} \mathbf{V} \cdot \mathbf{n} \rho(u) \, dA$. The schematic diagram of momenta is given in Fig. 7.3.5b.

At AB: $u = (U_0 - V_0)$ and $\mathbf{V} \cdot \mathbf{n} = -(U_0 - V_0)$

At CD: $u = (U_0 - V_0) \cos \theta_x$

and $\mathbf{V} \cdot \mathbf{n} = +(U_0 - V_0)$.

These values permit the momentum integral for the x direction to be written as

$$\int u \rho \mathbf{V} \cdot \mathbf{n} \, dA = \rho A_0 [-(U_0 - V_0)^2 + (U_0 - V_0)^2 \cos \theta_x].$$

(b) Draw a force diagram and evaluate ΣF_x and ΣF_y. The schematic diagram of forces is given in Fig. 7.3.5c. The solutions are

$$\Sigma F_x = \mathscr{F}_x \quad \text{and} \quad \Sigma F_y = \mathscr{F}_y$$

(Note again that in writing $\Sigma F_x = \mathscr{F}_x$ we are not simply interchanging symbols. We are stating that the sum of all of the x force components on the fluid is ΣF_x. This sum is equal only to the force that the vane exerts on the fluid, $\mathscr{F}_x$, for this particular case. The symbol ΣF_x should be viewed as an operator; it requires that one sum the forces!)

(c) Write Newton's relationship for the cause of momentum changes:

$$\Sigma F_x = \mathscr{F}_x = \rho A_0 (U_0 - V_0)[(U_0 - V_0)(\cos \theta_x - 1)] \tag{7.3.4}$$

---

[*] Later in this chapter we shall show that the speed of a small surface disturbance in a channel of depth y is $\sqrt{gy}$. Thus, the Froude number is the ratio between the current and the propagation speed of a small disturbance. It is nondimensional. [Note that the negative solution that is inherent in Eq. (7.3.2) has been rejected in Eq. (7.3.3).]

[†] In Appendix C, where moving coordinate frames are treated, this assertion will be verified.

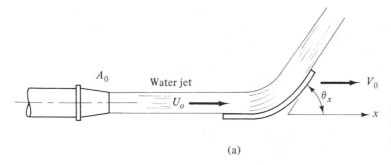

Water jet
A_0
U_o
θ_x
V_0
x

Figure 7.3.5(a) Liquid jet deflected by a moving vane. A_0 = nozzle area; U_0 = jet speed; V_0 = vane speed.

(a)

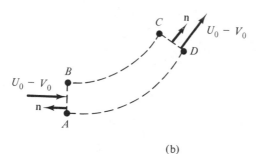

C n
$U_0 - V_0$
D
B
$U_0 - V_0$
n
A

Figure 7.3.5(b) Momentum diagram for Example 7.3.5.

(b)

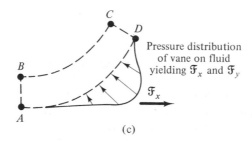

C
D
Pressure distribution of vane on fluid yielding $\mathcal{F}_x$ and $\mathcal{F}_y$
B
$\mathcal{F}_x$
A

Figure 7.3.5(c) Force diagram for Example 7.3.5.

(c)

in which $\rho A_0(U_0 - V_0)$ is the mass rate of flow of the fluid which is undergoing a change of momentum.

(d) Calculate the jet's reaction on the vane. There will be x and y reactions similar to those in Example 7.3.3.

(e) Calculate the power produced by the jet-vane system. Because the vane is not accelerating, the sum of all the forces acting on it, including $\vec{\mathcal{R}}$, must be zero. But $\vec{\mathcal{R}}$ produces work and power. This power is $\vec{\mathcal{R}} \cdot \mathbf{V}_{\text{vane}}$, in which $\mathbf{V}_{\text{vane}}$ is the velocity of the vane. This power can, in some other cases, be used in the energy equation to determine other quantities, such as pressure.

Equation (7.3.4) contains a mass rate of flow that is less than that emanating from the stationary nozzle. If there are a great many vanes, *all* of the nozzle's flow, $\rho U_0 A_0$, will change in momentum, so that the relation (force = momentum change rate) would be expressed as

$$\mathcal{F}_x = \rho U_0 A_0 [(U_0 - V_0)(\cos \theta_x - 1)]$$

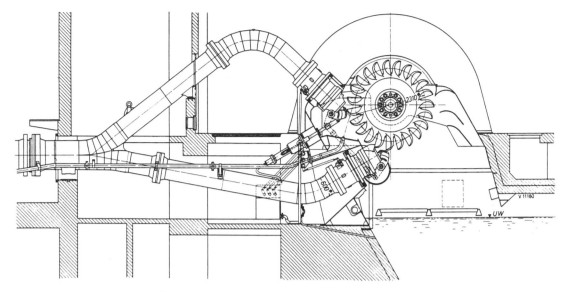

Figure 7.3.6 Pelton turbine with horizontal axis and two internally regulated jets with stream diverter, between which there is a braking jet. Operating conditions: $H = 890$ m, $Q = 10.1$ m³/s, $n = 500$ rpm, $P = 78,800$ kW. (*Courtesy of J.M. Voith GmbH, Heidenheim, F.R.G.*)

Figure 7.3.6, a drawing of a Pelton turbine installation, illustrates an application of Example 7.3.5.*

In Examples 7.3.1 through 7.3.5 the control volume was drawn around the fluid for emphasis. It was intended to reinforce the point that the momentum equations, as written, apply to the change of momentum of the fluid. The surface forces applied to this fluid are due to the fluid boundaries that usually consist of, in part, some solid device that is, in turn, connected to some larger unit. Depending upon the restraint of this unit, it may remain stationary or accelerate appropriately.

It was found that the force on the fluid was $\vec{\mathcal{F}}$, which caused a reaction $\vec{\mathcal{R}}$ on the containing surface. This reaction due to the fluid would have to be matched by a force from the surroundings, **R**, if the device were to remain fixed. Thus $\mathbf{R} = \vec{\mathcal{F}}$. This is understandable because ultimately the surroundings must be the source of the force that changes the fluid's momentum. It is for this reason, and the concept of action and reaction, that one is sometimes led to use an alternative control volume that encompasses the entire device within which the momentum of the fluid is changed. Thus, instead of Fig. 7.3.2c, one could choose Fig. 7.3.7. In this Figure, R_x and R_y are the forces induced on the control volume, the system of pipes in this case, to produce the momentum change for the fluid that is within it; and $\mathbf{R} = R_x \mathbf{i} + R_y \mathbf{j}$ is the force of the surroundings acting on the pipe to retain it. Similarly, Fig. 7.3.8 could be used as a starting point instead of Fig. 7.3.3c. Here, the magnitude and direction of the inlet and outlet variables (on the control surface) are specified; **R** is the force necessary to support the vane when the fluid is moving as shown. When dealing with a complex

---

* Strictly speaking, one should apply angular momentum, which will be discussed in Sec. 7.5, to find the torque induced on the rotor. However, the fact that the action of the vanes is restricted to almost a single, large radius allows one to model the problem successfully with linear momentum.

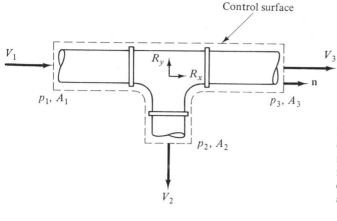

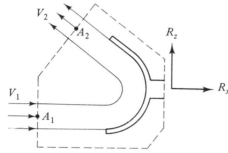

Figure 7.3.7 Control surface surrounding exterior of conduit, instead of its interior as in Fig. 7.3.2c, where the control volume surrounds only the fluid. Control surface here consists of lateral surfaces of pipe assembly and pipe cross sections.

Figure 7.3.8 Control surface surrounding support for vane to include vane within the control volume, instead of excluding the vane as in Fig. 7.3.3c. Flow paths inside control volume are not of concern.

device in which the fluid's momentum is altered, this procedure can be helpful to simplify the analysis. This approach is not limited to linear momentum problems only, as we shall see in Sec. 7.5.

The case discussed in Example 7.3.1 could also be considered using Fig. 7.3.9. Then R_x is the force component exerted on the fluid within the control volume by the surrounding media. If, after solving for R_x, it is concluded that its value is due to the forces contributed by the air through free-surface shear and by shear at the channel's bottom, and to the normal force exerted by the gate, one would have an estimate of the maximum contribution of the gate to R_x.

Figure 7.3.9 Control surface that surrounds the sluice gate, in contrast to the control surface in Fig. 7.3.1c.

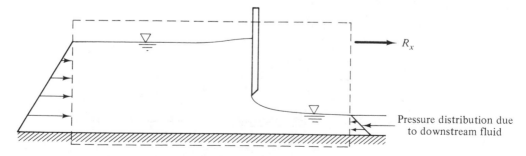

Pressure distribution due to downstream fluid

Example 7.3.6

An airplane (takeoff weight = 45,000 lb) is powered by two turbofan jet engines. It is in a vertical climb with a constant rate of ascent of 750 ft/s while the afterburner is operating. The engine's exhaust gases leave at a speed of 2500 ft/s relative to the aircraft while 450 lb/s of air passes through each engine. Determine the drag on the aircraft in this situation.

Solution   It is natural to have the control volume surround the entire aircraft. Because there is no acceleration of the control volume and a steady state relative to this control volume is *assumed,* we have

$$R_z - \text{weight} = \int (w) \rho \mathbf{V} \cdot \mathbf{n} \, dA$$

in which R_z is the force applied to the surface of the control volume and w is the z component of the relative velocity. R_z should be the drag on the aircraft exerted by the surrounding air. One would expect that R_z will be negative. Using the data provided, one has

$$R_z - 45,000 = 2 \left[(-750) \left(-\frac{450}{32.2} \right) + (-2500) \left(+\frac{450}{32.2} \right) \right]$$

(Justify the algebraic signs on the right-hand side of this equation. The z coordinate is positive upward.) Thus

$$R_z = -48,913 + 45,000 = -3913 \text{ lb}$$

7.4 MOMENTUM PRINCIPLE APPLIED TO THE FLOW OVER A FLAT PLATE

In several of the previous examples, the shear forces have been pictured in the diagrams, but they have been neglected in the calculations. We have said, "Neglect the shear forces as small with respect to the other known forces," or something similar, to indicate that they were being omitted from further consideration. It is now possible to examine a particular situation that will give some insight into the order of magnitude of the shear forces on a flat surface. With this information it may be possible to assess the validity of the phrase "small with respect to. . . ." Also, we shall deal with a velocity distribution that is not spatially uniform. Hence we shall be required to introduce this nonuniformity into Eq. (7.2.2).

A model for the situation that interests us is shown in Fig. 7.4.1. Here a stream of fluid is flowing over a flat plate. Upstream of the leading edge of the plate, the velocity

Figure 7.4.1   Flow past a plate that is aligned with the oncoming stream. The streamlines are deviated because of the no-slip condition at the plate, which retards the flow.

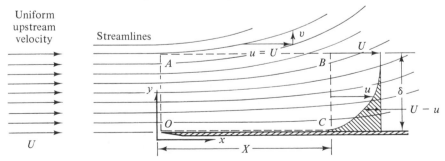

Momentum Conservation   Chap. 7

is considered to be uniform with a speed of U parallel to the plate. After the fluid reaches the plate, it is retarded in the vicinity of the plate. There is no motion relative to the plate where it and the fluid are in contact. In a segment X downstream from the leading edge, the velocity profile will be something like that shown in Fig. 7.4.1. Previously, in Chap. 3, the control volume $OABCO$ was used to find the mass rate of flow across the various segments of the control surface. It was found that a portion of the fluid flowing across vertical surface OA does not pass through the vertical surface BC. Consequently, it must pass out of the control volume via surface AB. In order for it to do this, there must be a vertical component of velocity along the points of surface AB.* The net effect of all of these vertical velocities, however small, is the mass transport needed to satisfy the principle of mass conservation for the region $OABCO$. (The volume flow rate passing across surface AB is $\int_0^\delta (U - u)\, dy$, which is equivalent to the shaded area in Fig. 7.4.1.) The horizontal, or u, component of velocity along surface AB will be very nearly equal to the free-stream speed, U. (The elevation of point B, δ, is specified so that $u = U$ at the associated plate position, X. For all other points along AB, the x component of velocity will also be equal to U. Section 8.6 discusses this point more fully.) Each particle of matter that passes across AB has in the positive x direction a momentum/mass equal to U, its x component of velocity. This fact is useful for calculating the flux of x momentum. It permits the integral $\int_{AB} \rho(\mathbf{V}\cdot\mathbf{n})(u)(dA)$ to be written as $U \int_{AB} \mathbf{V}\cdot\mathbf{n}\rho\, dA$. The integral in this expression is the one that was used to find the mass efflux across segment AB of the control surface. In the present case, we already know the value of this particular integral from the necessity of having mass conservation for the total control surface. The result is

$$\int_{AB} \mathbf{V}\cdot\mathbf{n}\rho\, dA = \int_{BC} \rho(U - u)\, dA$$

or, by multiplying both sides by U,

$$U\int_{AB} \mathbf{V}\cdot\mathbf{n}\rho\, dA = U\int_0^\delta \rho(U - u)\, dy \qquad (7.4.1)$$

in which dA on BC is dy for a unit width normal to the page. [On AB, $dA = dx(1)$.] The application of $\int\rho\mathbf{V}\cdot\mathbf{n}(u)\, dA$ to the remaining segments of the control surface (e.g., OA, BC, and OC) and the summing of the results with the right-hand side of Eq. (7.4.1) gives the total time rate of change of the x momentum. Subsequently, the force on the material in the control volume can be found. This sum is

$$\int_0^\delta -U(\rho U)\, dy + \int_0^\delta u\,(\rho u)\, dy + 0 + U\int_0^\delta \rho(U - u)\, dy$$

for the surfaces OA, BC, OC, and AB, respectively.

If we assume that there is no pressure gradient in the x direction, the forces due to pressure will be identical on surfaces OA and BC and will cancel one another when summed. Because we have *assumed* no velocity gradients at the distance δ from the

---

*We suspect that such a vertical velocity component may exist because with the slowing down of the flow near the boundary, the streamlines must diverge slightly there. This upward inclination of the streamlines near the plate continues for some extent into the remaining fluid above the plate. The slope of the streamlines give us the slope of the velocity vector, a fact which makes the existence of a v component reasonable to suspect.

plate, we can expect the shear stresses to be absent on surface AB. This means that, for the model that we are considering now, there is only one force on the fluid in the x direction. It is the shear force that the plate exerts on the fluid. We shall call this force $\mathcal{F}_x$, and using Newton's momentum relationship, we have, for $\rho = \text{constant}$,

$$\Sigma F_x = \mathcal{F}_x = -\rho U^2 \int_0^\delta \left(\frac{u}{U}\right)\left(1 - \frac{u}{U}\right) dy$$

The force on the plate is $-\mathcal{F}_x$, and we shall designate it as D, the *drag* on the plate. For a plate of unit width, the result is

$$D = \rho U^2 \int_0^\delta \left(\frac{u}{U}\right)\left(1 - \frac{u}{U}\right) dy \qquad (7.4.2)$$

To determine the drag, a knowledge of the velocity profile u/U is necessary. Extensive laboratory measurements have been made on many kinds of plates—smooth, rough, sharp, or blunt leading edges, and so on—to determine the kinds of velocity distributions that can be produced. In the special situation called *laminar flow*, which is sometimes associated with low-speed motion, the experimentally observed values are closely approximated by

$$\frac{u}{U} = \sin\left(\frac{\pi y}{2\delta}\right) \qquad (7.4.3)$$

(This function has the desired values of u at $y = 0$ and $y = \delta$ along with a zero derivative at the latter location.) If this relationship for u/U is used in Eq. (7.4.2), a value of the drag D would result that depends on δ. This is reasonable, because δ will in turn depend on the distance along the plate, X. Equation (7.4.3) is only one of the many viable expressions for the velocity distribution in which the argument, y/δ, could be thought of as the independent variable instead of the more obvious one, y. Because δ does not depend on y, but rather x, one could introduce it into Eq. (7.4.2) directly under the integral sign as

$$D = \rho U^2 \delta \int_0^\delta \frac{u}{U}\left(1 - \frac{u}{U}\right)\left(\frac{1}{\delta}\right)(dy) = \rho U^2 \delta \int_0^\delta \frac{u}{U}\left(1 - \frac{u}{U}\right) d\left(\frac{y}{\delta}\right)$$

Then, with the change of variable $\eta = y/\delta$, we can write

$$D = \rho U^2 \delta \int_0^1 \frac{u}{U}\left(1 - \frac{u}{U}\right) d\eta \qquad (7.4.4)$$

[Equation (7.4.3) would then be written as $u/U = \sin(\pi\eta/2)$.] If Eq. (7.4.4) is factored and rearranged, one has

$$\frac{D}{\frac{1}{2}\rho U^2 X} = \frac{2\delta}{X} \int_0^1 \frac{u}{U}\left(1 - \frac{u}{U}\right) d\eta \qquad (7.4.5)$$

The ratio Drag/$(\rho U^2(\text{area})/2)$ is called the *drag coefficient*, in general. In this case, however, it is called the *skin friction coefficient* C_f, since the shear stresses are the only agents yielding a drag force. The surface area used in Eq. (7.4.5) is $X(1)$, since the plate has a length of X and a unit width. This is the "wetted" area of one side of the plate and is not the frontal area that is usually used in a drag coefficient.

For the present, we shall anticipate the experimental results of measuring C_f for a variety of conditions. For the usual cases in engineering, one observes that

$$0.001 \le C_f \le 0.005 \tag{7.4.6}$$

With this information, we can make some estimate of the importance of the frictional drag that was neglected in a previous analysis of the hydraulic jump, even though the pressure changes in the flow direction are not exactly zero.

One of the known forces acting on the fluid within a hydraulic jump was the force due to gravity that yielded the hydrostatic pressure variations. The resultant of one distribution, at the upstream section, is $\gamma y_1^2/2$. (This would be nearly the case even if y were measured normal to a riverbed that has a slight slope with respect to the horizon; see Example 7.3.1.) It is convenient to compare the neglected shear term with it.

This ratio is, per unit width,

$$\frac{F_{\text{shear}}}{\gamma y_1^2/2} = \frac{C_f(\rho U_1^2 X)/2}{\rho g y_1^2/2} = \frac{C_f U_1^2}{g y_1} \frac{X}{y_2} \frac{y_2}{y_1}$$

(The downstream height y_2 has been artificially included for later convenience.) It is obvious that the flow on the bed of a stream under the roller of a hydraulic jump is not directly similar to the flow over a flat plate; however, with a properly chosen value of C_f, it might be possible to get a reasonable estimate of the ratio of forces which occur in our equation. If the value of $C_f = 0.005$ is used, the largest expected value for the shear friction coefficient for flat plate with a turbulent boundary layer will have been selected. This should serve to give us a useful "upper bound" estimate on the shear force that was neglected in Example 7.3.4. The ratio of the length of a hydraulic jump to the downstream depth, X/y_2, is about 5, in practice. If a rather large jump, $y_2/y_1 = 10$, is considered, our ratio becomes,

$$\frac{F_{\text{shear}}}{\text{Upstream gravity force}} = 0.005 \frac{U_1^2}{g y_1}(5)(10) = 0.25 \frac{U_1^2}{g y_1} \tag{7.4.7}$$

One can expect that $U_1^2/g y_1$ will be between 1 and 100. However, it is not necessary to use this fact, because we have already obtained a valuable result.

This rough estimate has shown that the term that was neglected in Eq. (7.3.1a) to obtain Eq. (7.3.2) has a maximum value of 14 percent of $V_1^2/g y_1$ when the geometric factor is included. (A closer examination of the skin friction data would show that a lower value of C_f should be used, so that an estimate of 3 percent is more realistic.)

7.5 CONSERVATION OF ANGULAR MOMENTUM

There is a result similar to Eq. (7.2.2) that is applied to rotating fluid systems. It is due to Euler and states:

> The time rate of the change of angular momentum of a mass of fluid is equal to the net torque that is applied to that fluid mass.

In order to apply this result to our work, it is necessary to recall the definition of *angular momentum*. The angular momentum per unit mass is $\mathbf{r} \times \mathbf{V}$. We need to bear

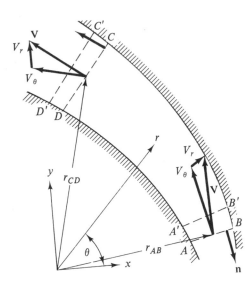

Figure 7.5.1 Flow in a curved channel. AB: Boundary of material volume at t_0. $A'B'$: boundary of the material volume at $t_0 + dt$.

this in mind even though we shall be considering motions in this section that are principally in one plane. This will often be the plane of the printed page. Even though this limits the range of applications, it is convenient initially because then only one scalar component—the one normal to the reference plane—needs to be considered. In this case, a control volume such as that shown in Fig. 7.5.1 is appropriate. The fluid is moving in the plane of the paper, and while an axial motion perpendicular to the paper is not excluded, it does not enter into our present considerations. The fluid is originally in region $ABCDA$ at t_0 and moves to $A'B'C'D'A'$ in the interval dt. Again, we shall *restrict* ourselves to steady motions. Thus, for the computations of the angular momentum in the region $A'B'CDA'$ that a part of the fluid occupies at both t_0 and $t_0 + dt$, the values are the same at the initial and final times and cancel when the difference of angular momentum is taken, just as similar cancellations took place when dealing with linear momentum, mass conservation, and energy conservation for steady flows.

The angular momentum expression that will reoccur in this section is rV_θ, since this is the z component of $\mathbf{r} \times \mathbf{V}$ in Fig. 7.5.1. The radius r is drawn perpendicular from the axis of rotation to the velocity vector. The V_θ component of velocity is perpendicular to r. Thus we write for the change of the z component of angular momentum per unit time:

$$\frac{\Sigma(rV_\theta m)_{CC'D'DC; t_0 + dt} - \Sigma(rV_\theta m)_{BB'A'AB; t_0}}{dt}$$

where m is the mass.

As the time interval goes to zero, we can rewrite this result as

$$\frac{dt \int \rho r V_\theta (V_{\text{out}})\, dA - dt \int \rho r V_\theta (V_{\text{in}})\, dA}{dt}$$

where V_{out} is the velocity out of the control volume and V_{in} is the velocity into the control volume; or*

---

*This is the z component from $\int_{CS} \rho(\mathbf{r} \times \mathbf{V})(\mathbf{V} \cdot \mathbf{n})\, dA$, the expression for the time rate of change of angular momentum from the Reynolds transport theorem, for the *assumed* steady flow.

$$\frac{dt \int_{\text{CS}} \rho r V_\theta (\mathbf{V \cdot n}) \, dA}{dt}$$

Therefore, following Euler, one can relate the integral in the numerator to the torque as follows:

$$\Sigma(\text{T})_z = \int_{\text{CS}} \rho r V_\theta (\mathbf{V \cdot n}) \, dA \qquad (7.5.1)$$

A positive torque must be applied to the fluid to obtain a positive time rate of change of angular momentum. If V_θ and r are nearly constant across AB and CD, respectively, and $(\mathbf{V \cdot n})$ on the surfaces can also be considered constant, this last result becomes

$$\Sigma(\text{torques})_z = \rho\{[(rV_\theta)_{CD}][(\mathbf{V \cdot n})A]_{CD} + [(rV_\theta)_{AB}][(\mathbf{V \cdot n})A]_{AB}\}$$

Mass conservation leads us to write, for an incompressible fluid,

$$[(\mathbf{V \cdot n})A]_{CD} = Q = -[(\mathbf{V \cdot n})A]_{AB}$$

Consequently the final result is

$$\Sigma(\text{torques})_z = \rho Q [(rV_\theta)_{\text{final}} - (rV_\theta)_{\text{initial}}] \qquad (7.5.2)$$

(Remember that V_θ is a signed quantity.)

Example 7.5.1

The conduit shown in Fig. 7.5.2a turns the flow through 135°. $Q = 50$ m³/s of water. Assume no losses. Determine whether $\vec{\mathscr{F}}$, the force of the conduit on the fluid in the control volume, passes through point O, the intersection of the axes of the inflow and outflow pipes.

Solution The momentum diagram is given in Fig. 7.5.2b. The law of continuity gives:

$$V_1 = 3.98 \text{ m/s} \quad \text{and} \quad V_2 = 7.07 \text{ m/s}$$

The energy equation for the control volume (between stations 1 and 2) yields, provided $\Delta z = 2$ m is included,

$$p_1 = 1.44 \text{ MPa} \quad \text{if} \quad p_2 = 1.4 \text{ MPa}$$
$$\int u \rho \mathbf{V \cdot n} \, dA = -449(10^3) \text{ N}$$
$$\int w \rho \mathbf{V \cdot n} \, dA = 250(10^3) \text{ N}$$

The force schematic shown in Figure 7.5.2c, along with the amounts just given, leads to

$$\mathscr{F}_x = -25.55(10^6) \text{ N}$$
$$\mathscr{F}_z = 7.64(10^6) \text{ N}$$

Hence
$$|\vec{\mathscr{F}}| = 26.67(10^6) \text{ N}$$

Point O is chosen as the origin of the system because then the $\mathbf{r \times V}$ terms [i.e., rV_θ in Eq. (7.5.2)] vanish. The torques, with respect to the y axis, exerted on the fluid are due to its weight and the resultant force applied by the conduit. The pressures exerted by the fluid in the inflow and outflow pipes induce forces on the fluid within the control volume that also intersect at O. Thus they produce no moment about this point. The result of taking moments about an axis through O is

$$[392.2(10^3)](2.2) - OP|\vec{\mathscr{F}}| = 0$$
$$OP = 0.029 \text{ m}$$

and
$$\arctan \alpha = 7.64/25.55$$

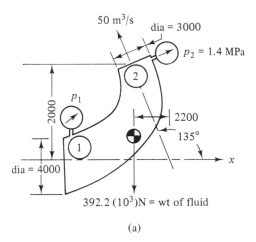

(a)

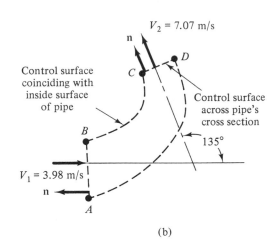

(b)

Figure 7.5.2(a) General schematic diagram of vertical pipe bend for Example 7.5.1. Rate of outflow is 50 m³/s. $p_2 = 1.4$ MPa. Weight of fluid $= 392.2(10^3)$ N. All distances in mm.

Figure 7.5.2(b) Control volume for analyzing the angular momentum changes of the fluid in a pipe bend. On surface AB: $\mathbf{V \cdot n} = -3.98$, $u = 3.98$, $w = 0$. On surface CD: $\mathbf{V \cdot n} = 7.07$, $u = 7.07 \cos 135°$, $w = 7.07 \cos 45°$.

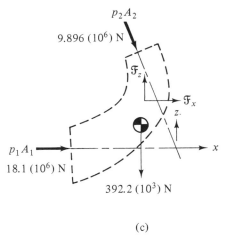

(c)

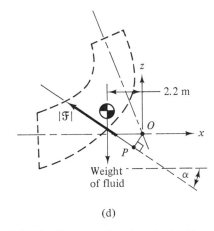

(d)

Figure 7.5.2(c) Forces induced on the fluid in a pipe bend.

Figure 7.5.2(d) Torques induced on the fluid in a pipe bend.

This use of the "moment of momentum" could lead to a desirable placement of the support for the curved conduit. (Should its weight be taken into account when treating the equilibrium of this piece?)

7.6 APPLICATION TO PUMPS AND TURBINES

The use of angular momentum was applied to the analysis of hydraulic machinery by Leonhard Euler. One can treat both *radial-* and *axial-flow machines* with Eq. (7.5.2). The former type is usually called either a *centrifugal pump*, a *centrifugal compressor*, or a *radial-flow turbine*. Pumps, compressors, and turbines can also be axial machines.

The fans that circulate air within a room and look like an airplane propeller are axial-flow types; those that distribute air to many rooms through duct systems are usually centrifugal blowers. Electric hair dryers have either a centrifugal or an axial blower.

We shall begin by discussing the centrifugal machine represented in Fig. 7.6.1. The fluid enters the "eye of the impeller, or runner, parallel to the axis of rotation and moves along the surface of the impeller to its periphery. It is collected there in a scroll case that conducts the fluid into the outlet pipe of the pump. Some inlet "swirl" may be imparted to the fluid by vanes, but this is usually not the case in a single-stage pump. The angle that the blades make with the tangent to the inlet circle, β_i, is specified for the designed flow rate and pump speed so that the fluid enters onto the impeller in a

Figure 7.6.1 General representation of a centrifugal pump. (The use of **W** to denote the relative velocity in a hydraulic machine is common in the U.S.A. It is not the z component of velocity as heretofore, just as **U** is not the x component.)

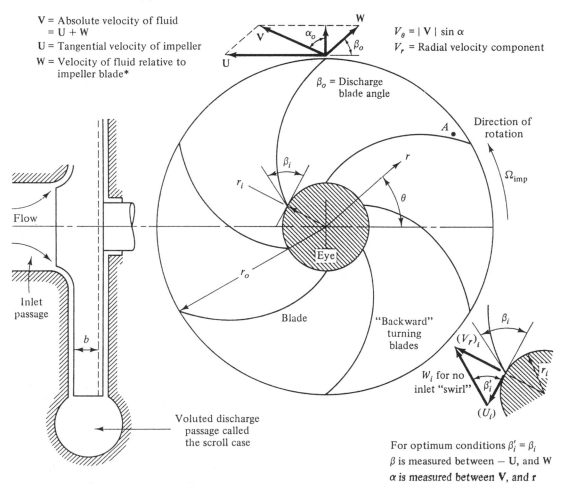

V = Absolute velocity of fluid
  = $U + W$
U = Tangential velocity of impeller
W = Velocity of fluid relative to impeller blade*

$V_\theta = |V| \sin \alpha$
V_r = Radial velocity component

β_o = Discharge blade angle

Direction of rotation

Ω_{imp}

Flow

Inlet passage

Blade

"Backward" turning blades

W_i for no inlet "swirl"

$(V_r)_i$

(U_i)

Voluted discharge passage called the scroll case

For optimum conditions $\beta_i' = \beta_i$
β is measured between $-U$, and W
α is measured between V, and r

*The use of **W** to denote the relative velocity in a hydraulic machine is common in the U.S.A. It is not the z component of velocity as heretofore, just as **U** is not the x component.

direction that is tangent to the impeller blades. This provides a smooth transition onto the impeller for the fluid and minimizes losses, called *shock*. The detail diagram in Fig. 7.6.1 shows the inlet geometry. If there is no inlet swirl, then the absolute velocity of the fluid entering the impeller is purely radial. The "backward" blading accommodates the flow for little "preswirl" and also keeps the absolute velocity of the fluid at the periphery of the impeller down to acceptable levels so that, once again, energy losses are minimized. Some centrifugal machines are designed with radial blades for ease of manufacture or strength, but they will then usually have curved portions at the inlet for good efficiency.

If the pump housing provides a constant distance b for the depth of the blade on the impeller, the volume rate of flow is

$$(V_r)_i(2\pi r_i) = (V_r)_o(2\pi r_o) = \frac{Q}{b} \tag{7.6.1}$$

in which V_r is the radial component of velocity. The torque that the impeller applies to the fluid is, according to Eq. (7.5.2),

$$T_z = \Sigma(\text{torque on fluid})_z = \rho Q[(rV_\theta)_o - (rV_\theta)_i] \tag{7.6.2}$$

(The frictional torques associated with seals and bearings must be accounted for properly to obtain the total torque applied to the pump's impeller shaft.) The torque and the pump speed, Ω_{imp}, give the rate of energy transfer. This rate of energy transfer will be called $\dot{W}$, as before. From $\dot{W} = \mathbf{T} \cdot \mathbf{\Omega}$, one has

$$\frac{\dot{W}}{\text{Weight flowing per second}} = \frac{T\Omega_{imp}}{Q\gamma} = \frac{\Omega_{imp}}{g}[(rV_\theta)_o - (rV_\theta)_i] \tag{7.6.3}$$

If there is no inlet swirl, the term $(V_\theta)_i$ is zero. (Note that a positive torque on the fluid with a positive value of Ω_{imp} in Fig. 7.6.1 would be associated. This gives a positive power input to the fluid for no inlet swirl.)

One can now apply the energy equation to a control volume from the inlet to the outlet of the pump and determine, for known values of Q and p_{inlet}, what the outlet pressure should be for a perfectly efficient impeller. To do this, one would set the losses to zero in a suitable expression of the energy equation,

$$\frac{\dot{W}}{Q\gamma} + \left(\frac{V_i^2}{2g} + \frac{p_i}{\gamma} + z_i\right) = \left(\frac{V_o^2}{2\gamma} + \frac{p_o}{\gamma} + z_o\right) + \text{losses}_{i\text{-}o} \tag{7.6.4}$$

The outlet pressure will be lowered when the pump is not operating at its intended speed or volume rate. The angle β_i will not be optimum, among other things, if a pump is running at the wrong speed. Viscous effects and the associated shear stresses will reduce the pump's efficiency. Unwanted, but unavoidable, flows within the pump can also be initiated. These are called *secondary flows*, and Fig. 7.6.2a shows a related secondary flow that is induced in front of a post attached to a flat plate or floor along which a fluid is flowing. The scouring effect of this secondary flow at the base of a model bridge pier can be seen in Fig. 7.6.2b.

The vector diagram at the periphery of the impeller in Fig. 7.6.1 shows a decomposition of the absolute velocity into **U** and **W**. The latter is the velocity relative to the rotating impeller, and it is customary in the United States to use the symbol **W** for it. Near the outlet of the pump, say, at point A in Fig. 7.6.1, one would expect the fluid to

(a)

(b)

Figure 7.6.2 (a) Secondary flow, called a *"horseshoe" vortex* in this case because of its shape, induced at junction of a circular cylinder and the floor on which it stands. Note how different the flow pattern becomes in regions away from the floor. (*Courtesy of H. Werlé, O.N.E.R.A., Paris, France.*) (b) Model study of bridge pier. Water comes from right to left. Sand is deposited in region behind cylinder where water speed is relatively small. Scouring effects of induced "horseshoe" vortex upstream of cylinder can be seen. Effects of such a vortex can also be seen in freshly driven snow at the base of a tree. (*Courtesy of Electricite de France, Chatou, France.*)

be moving tangentially to the blade of the impeller. Its speed would be W with respect to the impeller. But this particle at the arbitrary point A on the impeller is also moving at $\mathbf{U}$ with respect to a fixed frame of reference.* Keeping in mind that

> A fluid particle's absolute velocity is the vector sum of its velocity relative to a moving frame and the absolute velocity of the frame,

we write

$$\mathbf{V}_{abs} = \mathbf{W} + \mathbf{U} = \mathbf{V}$$

This is in addition to $\mathbf{V} = \mathbf{V}_r + \mathbf{V}_\theta$, in which the r and θ directions are associated with fixed coordinates. We shall use both vector decompositions to describe $\mathbf{V}$, because each will provide us with useful results.

As a fluid particle leaves the impeller, its absolute velocity, which is also its momentum per unit mass, remains unchanged if no forces have been exerted upon it in the time interval during which it is last on and just off the impeller. This is a good *assumption*. During the period when the fluid particle is off the impeller, it is natural and convenient to use r and θ components to describe $\mathbf{V}$. The radial component will be used when mass conservation is discussed. ($\mathbf{V} \cdot \mathbf{n}\, dA$ on a circle gives $V_r\, dA$.) The tangential component, V_θ, is needed for the torque and power calculations, as we have

---

* A person walking down the aisle of a moving bus has an absolute velocity that is the vector sum of the velocity relative to the vehicle and the bus's velocity relative to the ground.

seen previously. While the fluid particle is on the impeller, especially at its exit, its velocity, **W,** should be tangent to the surface of the blade. Therefore this W_o serves to prescribe β_o, the outlet blade angle.

If the pump's speed and discharge are suitably chosen, one would expect that the fluid should flow from the stationary inlet pipe onto the rotating impeller with a minimum of energy loss. (The flow should be "shockless.") To do this, it should pass onto the impeller with **W** tangent to the blade at its leading edge. One can see the interrelationship between the two ways of decomposing **V** by referring to Fig. 7.6.3a and b.

For a given flow rate Q and a geometry for the impeller that specifies r at the inlet and outlet, one can calculate V_r from Eq. (7.6.1). Once the preswirl in the inlet section is specified, V_i can be specified there in accordance with the vector diagram of Fig. 7.6.3a. The speed of rotation of the pump and r_i will prescribe U—its direction is circumferential—at the inlet. Using this value and the absolute velocity just found, one can find W_i, which is presented in Fig. 7.6.3a. This diagram shows the angle β_i for the blading that is equal to the angle made by the relative speed, W_i. This will produce a *"shockless" entry* for the given conditions. Similar interrelationships will exist between the velocity components and blade angles at points on the impeller such as point A in Fig. 7.6.3a. The following nomenclature is used in the analysis of such pumps:

$$\mathbf{U} = \text{tangential velocity of impeller}$$
$$\mathbf{V} = \text{absolute velocity} = \mathbf{U} + \mathbf{W}$$
$$V_r = \text{radial velocity component}$$

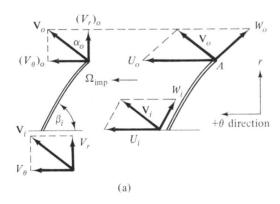

(a)

Figure 7.6.3 (a) Velocity diagrams for a centrifugal pump. (b) Velocity diagrams for a radial-inflow (Francis) turbine.

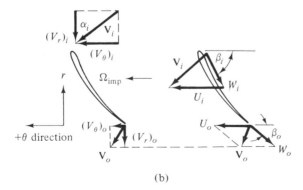

(b)

$$V_\theta = |\mathbf{V}| \sin \alpha$$
$\mathbf{W}$ = velocity of fluid relative to impeller blade*
α = angle measured between $\mathbf{V}$ and $\mathbf{r}$
β' = angle measured between $-\mathbf{U}$ and $\mathbf{W}$
β = angle between tangent to blade and $-\mathbf{U}$

For optimum conditions, $\beta_i' = \beta_i$.

Example 7.6.1

The impeller of a centrifugal pump is to be designed without inlet swirl and with

$$r_i = 20 \text{ mm} \qquad \beta_i = 55°$$
$$r_o = 170 \text{ mm} \qquad \beta_o = 45°$$

and a shaft rotation of 1750 rpm. The impeller is to be 15 mm deep (i.e., in the shaft direction). It will be assumed that the flow on the impeller does not vary in the direction parallel to the drive shaft, z. Determine the expected pressure rise, discharge, and power requirements for this design.

Solution Figure 7.6.4a is the velocity diagram at the inlet. From this figure we can find the following:

$$W_i = \frac{[1750(2\pi)/60](0.02)}{\cos 55°} = 6.39 \text{ m/s}$$
$$(V_r)_i = 3.67(\tan 55°) = 5.24 \text{ m/s}$$
$$\text{Discharge} = Q = 2\pi(0.02)(0.015)(5.24) = 0.01 \text{ m}^3/\text{s}$$

$(V_r)_o$ is obtained from Q; $V_r = Q/[2\pi r(\text{width})]$. This value, along with β_o and Ω_{imp}, prescribes the velocity diagram at the exit as shown in Fig. 7.6.4b. Here,

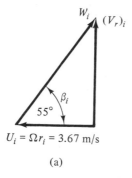

$U_i = \Omega r_i = 3.67$ m/s

(a)

Figure 7.6.4 (a) Velocity diagram at inlet of centrifugal pump impeller. $U_i = \Omega r_i = 3.67$ m/s. (b) Velocity diagram at outlet of centrifugal pump impeller.

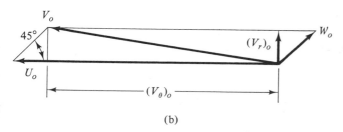

(b)

---

* The use of W to denote relative velocity in a hydraulic machine is common in the United States. It is not the z component of velocity as heretofore, just as U is not the x component.

$$U_o = \Omega r_o = 31.2 \text{ m/s}$$
$$(V_r)_o = 0.62 \text{ m/s}$$
$$W_o = 0.88 \text{ m/s}$$
$$V_o = 30.6 \text{ m/s}$$
$$(V_\theta)_o = 31.2 - (0.88)(\cos 45°) \text{ m/s}$$

Consequently, from $T = \rho Q(r_o V_o - r_i V_i)$,

$$T = \rho(0.01)[0.17(31.2 - 0.62) - 0.02(0)]$$
$$= 52.9 \text{ N-m} \quad \text{for} \quad \rho = 1000 \text{ kg/m}^3$$
$$\dot{W} = T\Omega = 9.68 \text{ kW} \quad \text{or} \quad \frac{9680}{Q\rho} = 968 \text{ N-m/kg}$$

This last value can be used in the energy equation. We shall find the expected Δp by assuming zero losses in the pump's passages and neglecting elevation changes:

$$\frac{p_2 - p_1}{\rho} = 968 + \frac{(5.24)^2}{2} - \frac{(30.6)^2}{2}$$

so that $(p_2 - p_1) = 454$ kPa. (Observe that the absolute fluid velocities were used to obtain this result.)

A lawn sprinkler is, in effect, a radial-outflow turbine.* The analysis of this machine is very similar to that of a centrifugal pump. One must, once again, express the absolute velocity of the fluid as it discharges into the air in terms of V_r and V_θ as well as W and U.

Example 7.6.2

An agricultural sprinkler has two jets, one at each end of an arm that is 8 ft long. This arm rotates horizontally about its midpoint. While the jet axes are at right angles to the arm, they are also oriented at 40° to the horizontal to increase the "reach" of the spray. Water leaves each nozzle (area = 0.1 ft^2) at a speed of 100 ft/s relative to the nozzle. The unit rotates at 120 rpm.

(a) What torque is applied to the water?
(b) What is the resistance torque (due to seals, bearings, wind, and so on) that acts on the sprinkler at the given speed of rotation?

Solution This is a steady-state problem, so that

$$T_z = \int (rV_\theta)\rho \mathbf{V} \cdot \mathbf{n} \, dA = \rho Q[(rV_\theta)_{\text{out}} - (rV_\theta)_{\text{in}}]$$

The inlet angular momentum can be taken as zero. Also,

$$Q = 2V_{\text{jet}} A_{\text{jet}} = 2(100)(0.1) \text{ cfs}$$

and

$$U = [2\pi(120/60)]4 = 50.27 \text{ ft/s}$$

It is useful to draw a representation of the velocity. This is given in the accompanying diagram. T_z is the torque exerted on the fluid. As a reaction, the fluid exerts a torque on the surroundings, M_z, that is opposite in direction and equal in magnitude to T_z. For the rotor, one has

$$\Sigma M_z \text{ on rotor} = (\text{polar moment of inertial})(\dot{\Omega})$$

---

*It is also called *Hero's turbine* or *Barker's mill*. The device has been described in writings dating from 250 B.C. Around 1740, Johann Segner in Germany and Robert Barker in England sought industrial applications for devices similar to this.

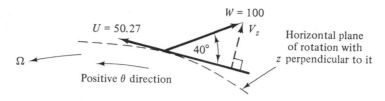

If $\dot{\Omega} = 0$, then

$$\text{(Torque applied by fluid to rotor)} + \text{(resistance torque)} = 0$$

Hence,

$$\text{Resistance torque} = -M_z = T_z = -4087 \text{ lb-ft}$$

The negative sign indicates that this torque is in the minus θ direction. It opposes the motion.

It is also appropriate to use Fig. 7.6.3a and b when discussing another type of turbine, the Francis turbine. This is a radial-inflow machine, and suitable guide vanes give the water the correct velocity as it enters the rotating set of blades. Figure 7.6.5 shows three sets of vanes and blades. The outermost blades are fixed. They provide the V_θ component, the inlet swirl, to the water. They are set at some nominal angle depending upon the usual (design) conditions associated with the source of water. The intermediate set of vanes are adjustable to meet the seasonal variations in the water source and are called *wicket gates*. The innermost set of vanes are the blades of the rotating member that is called the *runner*. After the water leaves this part of the

Figure 7.6.5 Section view of a Francis turbine. (*Courtesy of J.M. Voith GmbH, Heidenheim, F.R.G.*)

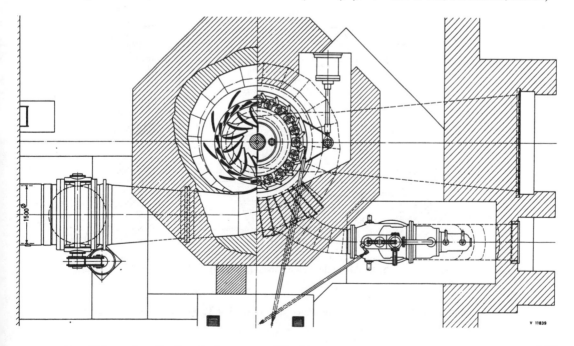

turbine assembly, it proceeds through the *draft tube* that takes it to the lower water level, called the *tailrace*. In the draft tube, the water will have as little swirl as possible. The torque applied to the fluid while it is on the runner is

$$T_z = \rho Q[(rV_\theta)_{\text{outlet}} - (rV_\theta)_{\text{inlet}}]$$

as before. (Again, V_θ is counterclockwise for a positive value.) We would have, for the case of no exit swirl,

$$T_z = \rho Q[-(rV_\theta)_{\text{inlet}}]$$

For a positive V_θ, this torque would be negative, and one recognizes that a negative torque applied to the fluid is a positive torque applied to the surroundings. If the product of T_z and Ω is negative, one would have a negative amount of energy transferred to the fluid per unit time.* This is a positive power output from the machine. We would expect that from a power-producing turbine. (Figure 7.6.3 may be helpful in the assignment of algebraic signs to V_θ and Ω.)

The principles that have been presented are also applicable to fans. A fan is an axial-flow machine, as is the Kaplan turbine shown in Figure 7.6.6; the following presentation will use such a propeller turbine to focus the discussion. It is very much like a windmill except that one has a set of wicket gates to provide an inlet swirl. This swirl will have been removed by the blades as the water leaves them and enters the

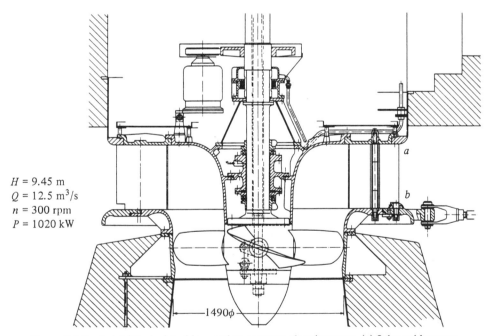

$H = 9.45$ m
$Q = 12.5$ m³/s
$n = 300$ rpm
$P = 1020$ kW

Figure 7.6.6 Axial (Kaplan) turbine used to generate electric power. (a) Inlet guide vanes. (b) Mechanism for setting angle of inlet guide vanes. (*Courtesy of J.M. Voith GmbH, Heidenheim, F.R.G.*)

* Remember that both T_z and Ω are signed quantities. For $V_\theta > 0$, we have $T_z < 0$. Ω is positive in Figs. 7.6.1 and 7.6.3.

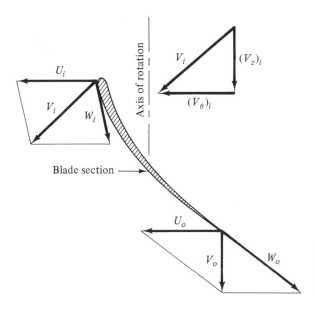

Figure 7.6.7 Velocity diagrams for an axial-flow machine for which there is no angular velocity for the fluid after it leaves the vane, i.e., no postswirl.

draft tube. In Fig. 7.6.6 the blade that is coincident with the centerline appears as a tip or end view. One needs to have other views of the blade, too, such as a root section view, because of the continuously changing blade shape from root to tip.

Figure 7.6.7 is an arbitrary section with the associated vector diagrams. The vectors are V_θ and V_z; the latter is directed along the axis of rotation. The radial velocity component—from blade root to tip—is *assumed* to be zero. This is largely true in most cases. Accordingly, mass conservation for the impeller will require that, at each radius, the axial flow component be the same at the inlet and outlet of the impeller. ($\mathbf{V \cdot n}\, dA$ on a disk that forms the plane of the propeller's rotation will yield $V_z\, dA$ with the result that $(V_z)_o = (V_z)_i$. At each radius for which vector diagrams are drawn, U will be the same at the entrance and exit segments of the blades so that $U_o = U_i$. Once again, the requirement of "shockless" entrance will be observed at the design conditions. Finally, Eqs. (7.6.2) to (7.6.4) can be applied. This is done in the following example.

Example 7.6.3

An axial-flow turbine (Kaplan turbine) like that shown in Fig. 7.6.6 has been designed to operate at 300 rpm and use 12.5 m³/s of water (a usable reservoir height of 9.45 m was available). The turbine's tip diameter is 1500 mm, and the hub diameter is 600 mm. The inlet vanes induce a swirl velocity upstream of the turbine that is approximated here by $V_\theta = 2.55/r$. There should be no swirl in the turbine's draft tube. What power can be expected from the installation, and what should be the blade angles for the turbine? (The blade at the right rotates toward you.)

Solution We shall *assume* that $\dot{W}/\dot{m}$ is a constant throughout the flow. Then we can use an equation for this average value,

$$\frac{\dot{W}}{\dot{m}} = \Omega\left(\frac{\text{torque}}{\dot{m}}\right) = \Omega[(rV_\theta)_{\text{outlet}} - (rV_\theta)_{\text{inlet}}]$$

It is prescribed that $(V_\theta)_{\text{outlet}} = 0$.

The axial velocity is

$$\frac{Q}{A} = \frac{12.5}{(\pi/4)(1.5^2 - 0.6^2)}$$

$$= \frac{12.5}{1.484} = 8.42 \text{ m/s}$$

The analysis that follows is for the flow at the blade's tip. A similar analysis would have to be done at all radii.

At the tip, the peripheral speed is

$$U = r\Omega = 0.75[5(2\pi)] = 23.56 \text{ m/s}$$

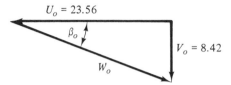

The associated velocity triangle is shown in the accompanying diagram. The diagram makes it evident that

$$\tan \beta_o = \frac{8.42}{23.56}$$

The prescribed inlet swirl condition leads to

$$V_\theta = \frac{2.55}{0.75} = 3.39 \text{ m/s at the blade tip,}$$

and
$$V_{\text{axial}} = 8.42 \text{ m/s}$$

follows in the space between the blade's root and tip if a uniform axial flow is assumed. These velocities can be portrayed in the diagram given here.

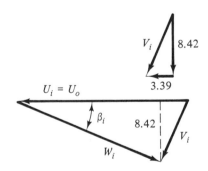

The direction of the relative velocity is gotten from

$$\tan(\beta_i) = \frac{8.42}{23.56 - 3.39}$$

and
$$\frac{\text{Power}}{\text{mass flow rate}} = 5(2\pi)\left[-\frac{3}{4(3.39)} + 0 \right] = -79.87 \text{ N-m/kg}$$

In concert with an *assumption* that each particle, whether it be at the root or the tip, has been subjected to the same change of angular momentum, we have power $= \dot{m}(\dot{W}/\dot{m})$, or

$$\text{Output Power} = 998 \text{ kW}$$

Figures 7.6.8 to 7.6.10 show the details of the turbine rotors that have been discussed in this section. The occurrence of cavitation is evident in Fig. 7.6.11. (See also Figs. 1.8.2, 1.8.3, and 1.8.8.)

Figure 7.6.8 Pelton turbine wheel for the "New Colgate" generating station in California. Notice the scalloped tips of the "buckets" that minimize interference with the water jet prior to a vane rotating into a near-optimum position with respect to the jet. (*Courtesy of J.M. Voith GmbH, Heidenheim, F.R.G.*)

Figure 7.6.9 Francis turbine "wheel" for the Paulo Afonso III dam in Brazil. The turbine was designed to produce 221 MW when operating at a head of 87.5 m with a flow of 284 m³/s while rotating at 138.3 rpm. (*Courtesy of J.M. Voith GmbH, Heidenheim, F.R.G.*)

Figure 7.6.10 A Kaplan turbine with adjustable blades. The shaft that connects it to the generator is visible at the left. (*Courtesy of J.M. Voith GmbH, Heidenheim, F.R.G.*)

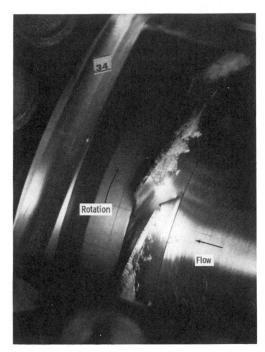

Rotation

Flow

Figure 7.6.11 Cavitation in an axial flow pump. NASA photograph.

7.7 A COMPARISON OF THE RESULTS OF THE MOMENTUM AND ENERGY PRINCIPLES

In solving for the forces that are exerted on a vane by a fluid jet (Example 7.3.3), the Bernoulli equation—or the energy equation—was used to determine the constancy of the fluid's speed as it traveled along the vane, assuming no energy losses or changes of potential energy. With this information, the momentum principle was employed to determine the unknown resultant force. Similarly, the energy equation, with loss terms, would be used to determine the pressure downstream of a pipe tee. Then one could find the fluid reactions on the tee with the momentum equation in a manner similar to that in Example 7.3.2. But when the change of elevation in a hydraulic jump was discussed, the momentum equation was used, and the energy equation was given minor attention because of the unknown losses of energy in the region where there is an abrupt change of elevation. It would have been unwise to neglect these undefined losses, because of their influence on the final result. Instead, the momentum equation played the major role, and a result was obtained by neglecting the bottom shear stresses (a factor that was subsequently shown to be less than some of the terms that were retained in reaching a solution). Subsequently, the elevation change found in this way could be used to determine the losses.

In this section, we shall focus on the interrelationship of momentum and energy equations to show some ways in which they complement one another. In order to do this, we shall discuss a moving disturbance that can be made steady by allowing the reference frame to translate with the speed of the disturbance. A surge wave will be the object of our study.* When surge waves occur in rivers whose elevation changes are

---

* For this work, Example 7.3.4 will serve as an introduction.

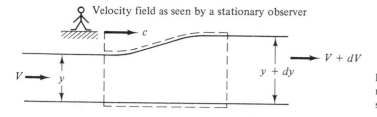

Velocity field as seen by a stationary observer

Figure 7.7.1 Schematic drawing of a moving control volume traveling with a surface discontinuity. Width = 1 unit.

caused by the tides, they are called *tidal bores.* Figure 7.7.1 shows the region of discontinuity in elevation of such a traveling wave, and it appears that there is some similarity in geometry with the hydraulic jump. In the figure, the region of immediate interest is enclosed in dashed lines. This region is moving with the advancing wave at a speed of c, relative to a stationary observer. The problem is an unsteady one, because conditions are changing with time at a fixed point in space. Because the speed of propagation, c, is assumed to be constant, we can transform this unsteady problem into a steady one by associating our coordinate system with the moving control volume so that all of the velocities are then expressed relative to the moving coordinate frame. These velocities are depicted in Fig. 7.7.2.

Using the notation of Fig. 7.7.2, we can apply the principle of mass conservation to the surfaces where mass is crossing. This gives for unit width

$$y(V - c) = (y + dy)(V - c + dV) = \text{constant} \tag{7.7.1}$$

or

$$0 = y\,dV + dy\,(V - c) + dy\,dV \tag{7.7.2}$$

For infinitesimal dy and dV, the last term is of "higher order" and is sometimes neglected. This situation will *not be assumed* to exist without comment in the development that follows.

Momentum conservation between the upstream and downstream locations gives

$$-F_{\text{shear}} + \frac{\gamma}{2}[y^2 - (y + dy)^2] = (V - c)y\rho[(V - c + dV) - (V - c)]$$

$$\frac{F_{\text{shear}}}{\gamma} + y\,dy + \frac{dy^2}{2} = \frac{-1}{g}(V - c)y\,dV \tag{7.7.3}$$

In Eq. (7.7.3), the expression for dV in Eq. (7.7.2) could be substituted. The result is

$$\frac{F_{\text{shear}}}{\gamma} + y\,dy + \frac{dy^2}{2} = \frac{1}{g}(V - c)^2 \frac{y\,dy}{y + dy}$$

which reduces to

$$(V - c)^2 = g\left(y + \frac{dy}{2}\right)\left(\frac{y + dy}{y}\right) + \frac{F_{\text{shear}}}{\rho\,dy}\frac{(y + dy)}{y} \tag{7.7.4a}$$

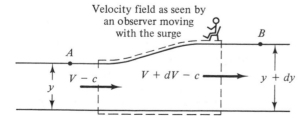

Velocity field as seen by an observer moving with the surge

Figure 7.7.2 Velocity field relative to a moving control volume.

Sec. 7.7   A Comparison of the Results of the Momentum and Energy Principles   **275**

Equation (7.7.4a) can be solved for the case of an infinitesimal dy, i.e., $dy \ll y$, to determine the relative speed of wave propagation.

However, the change in surface elevation need not be small. If $y = h_1$ and $dy = h_2 - h_1$, whereby h_2 is the downstream height of the liquid, one has for a *negligible shear* force

$$\frac{(V - c)^2}{gh_1} = \frac{1}{2}\left(\frac{h_2}{h_1}\right)\left(\frac{h_2}{h_1} + 1\right) \tag{7.7.4b}$$

This is similar to the result for a hydraulic jump that was obtained in Example 7.3.4. Equation (7.7.4b) is a consequence of using the principles associated with momentum, with the *assumption* that the bottom shear force is negligible. Let us now determine what energy conservation will give us.

Using conditions at point A in Fig. 7.7.2, the upstream location, and at B, downstream, one has

$$\frac{(V - c)^2}{2} + gy = \frac{[(V + dV) - c]^2}{2} + g(y + dy) + \text{losses}_{A\text{-}B}$$

and this reduces to

$$0 = dV\,(V - c) + \frac{dV^2}{2} + g\,dy + \text{losses}_{A\text{-}B} \tag{7.7.5}$$

Again, Eq. (7.7.2) can be used for eliminating dV from Eq. (7.7.5), with the result that

$$0 = (V - c)(-1)\frac{(dy)(V - c)}{y + dy} + \frac{dy^2\,(V - c)^2}{2(y + dy)^2} + g\,dy + \text{losses}_{A\text{-}B} \tag{7.7.6}$$

Equations (7.7.4a) and (7.7.6) are two expressions for $(V - c)^2$ that were obtained from different physical laws. If we equate the values of $(V - c)^2$ in these equations, we obtain

$$g\left(y + \frac{dy}{2}\right)\left(\frac{y + dy}{y}\right) + \frac{(F_{\text{shear}})}{\rho\,dy}\frac{(y + dy)}{y} = \frac{g(y + dy)^2}{y + dy/2} + \frac{(\text{losses})(y + dy)^2}{dy(y + dy/2)}$$

This reduces to

$$\frac{g\,(dy)^3}{4} + \frac{(F_{\text{shear}})}{\rho}\left(y + \frac{dy}{2}\right) = (\text{losses})(y + dy)y \tag{7.7.7}$$

If the losses were neglected, the shear force would have to be negative for $dy > 0$. For positive dy, the neglect of shear force would only minimize the losses. Since the magnitude of the shear force should be positive, the possibility of the losses being zero is rejected as an approximation for the equations that model the situation.

Example 7.7.1

An analysis of the flow associated with an injector is also interesting when one contemplates the roles of momentum and energy in a problem. Such a system is shown in Fig. 7.7.3a. The jet could come from a high-pressure steam source, and p_1 could be associated with the pressure in a condenser and the higher pressure, p_2, could be atmospheric pressure. The jet entrains fluid, and a mixing process ensues so that the large velocity

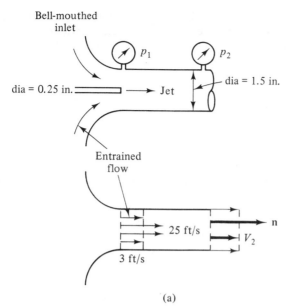

Bell-mouthed inlet

p_1 p_2

dia = 0.25 in. ═══ → Jet

dia = 1.5 in.

Entrained flow

25 ft/s

n

V_2

3 ft/s

(a)

Figure 7.7.3 (a) Schematic drawing (top) of an injector showing velocity distributions (bottom) at the inlet and at a downstream section. (b) Momentum and force diagrams for the injector.

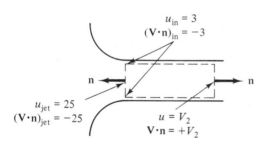

$u_{in} = 3$
$(\mathbf{V} \cdot \mathbf{n})_{in} = -3$

n   n

$u_{jet} = 25$
$(\mathbf{V} \cdot \mathbf{n})_{jet} = -25$

$u = V_2$
$\mathbf{V} \cdot \mathbf{n} = +V_2$

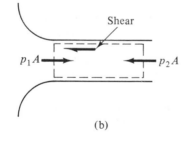

Shear

$p_1 A$ →   ← $p_2 A$

(b)

gradients shown in Fig. 7.7.3a are eliminated at a short distance downstream that will be called location 2. (The inlet velocity distribution may in practice not be as discontinuous as shown, but that would not vitiate the following discussion.)

Volume conservation requires that the nearly uniform velocity V_2 be related to the total incoming flow. Hence $V_2 A_2 = Q$, or

$$V_2 = \frac{25 A_{jet} + 3 A_{annulus}}{A_2} = 3.612 \text{ ft/s}$$

Momentum conservation would interrelate the momentum fluxes, the pressures, and the shear force acting along the conduit's surface. This distribution of the shear stresses and pressure along the duct's wall is unknown, so that only the resultant effect can be included. In this example, the units of the term denoted as "shear force" will be in pounds. The momentum equation for this case is

$$V_2 \rho V_2 \frac{(15)^2}{144} \frac{\pi}{4} - \left(3^2 \rho \frac{(1.5^2 - 0.25^2)}{144} \frac{\pi}{4} + 25^2 \rho \frac{0.25^2}{144} \frac{\pi}{4} \right) = (p_1 - p_2) \frac{\pi}{4} \frac{1.5^2}{144} - F_{shear}$$

If water ($\rho = 1.932$ slugs/ft^3) is flowing, this equation becomes

$$p_2 - p_1 = 25.2 - 81.5 F_{shear}$$

Energy conservation is applied next. The appropriate control surface must include the contributions of the jet and the entrained fluid in this steady-state problem. The application of Eq. (6.3.2) for a steady flow yields

$$3\rho \frac{\pi}{4} \frac{(1.5^2 - 0.25^2)}{144} \left[\frac{9}{2} + \frac{p_1}{\rho} + g(0) + e_1 \right] + 25\rho \frac{0.25^2}{144} \left(\frac{25^2}{2} + \frac{p_1}{\rho} + e_1 \right)$$

$$= V_2 \rho \frac{\pi}{4} \frac{1.5^2}{144} \left(\frac{V_2^2}{2} + \frac{p_2}{\rho} + e_2 \right)$$

This result can be simplified by using the equation for volume conservation. The conclusion is

$$57.2 = \frac{p_2 - p_1}{\rho} + (e_2 - e_1)$$

The term $e_2 - e_1$ in this equation represents the loss—per unit of $\dot{m}$—due to the mixing of the jet and the entrained fluid. One can see from the results of the momentum equation that the maximum pressure increase is 25.2 lb/ft^2 if water is being pumped by the injector. This would yield a minimum value of $(e_2 - e_1)$ of 44.15 ft^2/s^2 from the energy equation, above. Had one set $(e_2 - e_1)$ equal to zero in this energy equation to solve for the pressure and then used the momentum equation, the shear force would have been negative (i.e., aligned with the flow). This would be unsatisfactory.

Example 7.7.2

Perforated plates occur as baffles and supports in equipment through which fluids flow. With the aid of the momentum equation, we can estimate the effects of the size and number of holes in the plate on the pressure drop. (The plate, which could be in a heat exchanger, might have 50 tubes of diameter D and 100 holes of diameter d. One liquid would flow outside the tubes and through the holes, and another would flow within the tube. The entire bundle of tubes with their baffle plates would be contained in a duct of large diameter. We shall concentrate on the flow between this duct and the tubes.) The situation is schematically illustrated in Fig. 7.7.4a, which shows only one of the tubes and its adjacent holes.

Use the three conservation laws to determine $p_1 - p_2$, the pressure drop across a plate. Definitions associated with the diagram are as follows:

$$A_o = \text{area of individual hole} = \frac{\pi d^2}{4}$$

$$A_o' = 2A_o = \text{area of holes associated with } each \text{ tube} = 6 \left(\frac{\pi d^2 / 4}{3} \right)$$

$$A_1 = \text{flow area between hexagon and tube} = 6 \left(b \frac{\sqrt{3}}{2} b \right) - \frac{\pi D^2}{4}$$

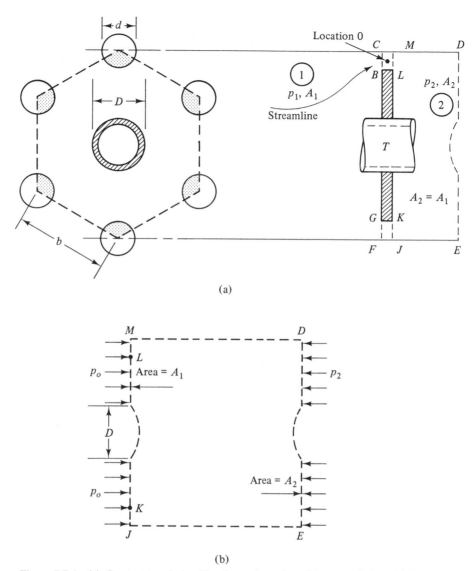

(a)

(b)

Figure 7.7.4 (a) Geometric relationships for a plate pierced by many holes. (b) Pressure distributions on control surfaces associated with the flow through a porous plate.

Note that the area of each hole that is assigned to one tube is $A_o/3$ but the total passage area associated with each tube is $A_o' = 2A_o$

Solution If the flow is considered to be without effective losses from location 1, which is in a region upstream of the hole, to location 0, which is in the hole, one writes the appropriate energy equation as

$$\frac{V_1^2}{2g} + \frac{p_1}{\gamma} = \frac{V_0^2}{2g} + \frac{p_0}{\gamma}$$

or

$$p_0 = p_1 + \rho \frac{V_1^2 - V_0^2}{2}$$

Losses will be *assumed* to occur in the downstream portion of the plate, and these losses will be the cause for the pressure drop $p_1 - p_2$.

The momentum equation, applied downstream of the plate, will be useful in relating p_2 to p_0 and also to p_1, in view of the last expression (see Fig. 7.7.4b). In attempting to write an expression for the forces on the fluid downstream of the plate, we *assume* the area of the perforation, A_o, is sufficiently large that the pressure in the perforation, p_o, exists in the eddy that occupies the entire rearward face of the plate. (We know that the pressure in the wake behind a sphere, for example, is effectively constant over the surface of the body that is covered by the wake.) Since p_0 also acts in the n partial holes ($n = 6$ hexagonal spacing), there is a positive force acting across the control surface $JKLM$ of $p_0 A_1$. This leads us to write for the control surface $JKLMDEJ$

$$p_0 A_1 - p_2 A_2 = \rho Q (V_2 - V_0)$$

If the previously determined expression for p_0 is substituted in this equation—and also A_1 for A_2 and V_1 for V_2—one finds that

$$A_1(p_1 - p_2) + A_1 \rho \frac{V_1^2 - V_0^2}{2} = \rho V_1 A_1 (V_1 - V_0)$$

or

$$\frac{p_1 - p_2}{\gamma} = \frac{V_1^2}{2g} \left(1 - \frac{A_1}{A_0'}\right)^2 \tag{7.7.8}$$

The common factor A_1 was canceled and V_1 was substituted for $V_o A_o'/A_1$ by using the continuity equation. V_1 is used in Eq. (7.7.8) because it is the velocity that would be specified to calculate the effectiveness of the heat transfer between the fluid passing through the perforations and the fluid flowing inside the tubes designated by T in Fig. 7.7.4a. (These tubes would also be installed in a hexagonal pattern.)

The loss for the sudden enlargement from area A_1 to A_2 in a pipeline can be derived by the method that led to Eq. (7.7.8). It is then common to use the velocity V_1 in the upstream passage—$V_1 = V_2 A_2/A_1$—in expressing this loss. The final expression is

$$\text{Losses} = \frac{V_1^2}{2g} \left(1 - \frac{A_1}{A_2}\right)^2 \tag{7.7.9}$$

When $A_2 \gg A_1$, such as for a pipe emptying into a reservoir, the result reduces to

$$\text{Losses} = \frac{1}{2g} V_1^2 (1) \tag{7.7.10}$$

The next case that will be examined combines the energy equation with a result coming from the angular momentum relationships that were introduced in Sec. 7.6. For a centrifugal pump, the velocity diagram at any point on the rotor's blades is given in Fig. 7.7.5. In this figure the magnitude of the absolute velocity, V, can be found by using either its absolute components in the r and θ directions, V_r and V_θ, or the relative components U and W. Just as in Sec. 7.6, $U = \Omega r$ is the tangential velocity component of the rotor and W is the speed of the fluid relative to the rotor.

The torque applied to the fluid between the inlet, defined here as point 1, and the outlet, point 2, is

$$T = \rho Q [(rV_\theta)_2 - (rV_\theta)_1]$$

so that the power supplied to the fluid by this process is $T\Omega$. Thus,

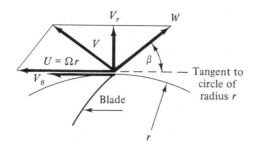

Figure 7.7.5 Diagram to aid in the development of the energy equation relative to a control volume affixed to a rotating impeller.

$$\dot{W} = \rho Q[(UV_\theta)_2 - (UV_\theta)_1] \qquad (7.7.11)$$

This statement will be incorporated into the energy equation after the product UV_θ is revised. To do so, we write V_r and V_θ in terms of U and W. These relationships are

$$V_\theta^2 = (U - W \cos\beta)^2 = U^2 - 2UW \cos\beta + W^2 \cos^2\beta$$

and

$$V_r^2 = W^2 \sin^2\beta$$

Because $V^2 = V_r^2 + V_\theta^2$,

$$V^2 = U^2 - 2UW \cos\beta + W^2 \qquad (7.7.12)$$

The angle β can be elminated by reintroducing $V_\theta (= U - W \cos\beta)$. The product UV_θ is $(U^2 - UW \cos\beta)$, and this gives Eq. (7.7.12) the form

$$UV_\theta = \frac{V^2 + U^2 - W^2}{2} \qquad (7.7.13)$$

With the use of the subscripts 1 and 2, we can use this equation to give the power supplied to the fluid expressed by Eq. (7.7.11):

$$\frac{\dot{W}}{\dot{m}} = \frac{V_2^2 - V_1^2}{2} + \frac{U_2^2 - U_1^2}{2} - \frac{W_2^2 - W_1^2}{2} \qquad (7.7.14)$$

The complete energy equation for an adiabatic machine will provide another expression for this power:

$$\frac{\dot{W}}{\dot{m}} = -\left(\frac{V_1^2}{2} + \frac{p_1}{\rho}\right) + \left(\frac{V_2^2}{2} + \frac{p_2}{\rho}\right) + (e_2 - e_1) \qquad (7.7.15)$$

In this equation, $e_2 - e_1$ is the "loss" per unit $\dot{m}$ as the liquid moves through the rotor. Gravity effects have been neglected. These last two equations can now be compared, with the result that

$$0 = \frac{W_2^2 - W_1^2}{2} + \frac{p_2 - p_1}{\rho} - \frac{U_2^2 - U_1^2}{2} + (e_2 - e_1) \qquad (7.7.16)$$

wherein $U = \Omega r$. The losses can include friction within the rotor's passages as well as any entrance and exit losses. Such losses occur because of the fluid's inability to follow the rotor's contours under many conditions. This can happen at the entrance to the rotor if its angular speed is not optimally matched with the flow rate. The finite thickness of the blades at the outlet of the rotor will usually result in a region directly behind the blades in which the velocity is considerably less than the average value of W

used in Eq. (7.7.16). (This region with a defect in the velocity profile is called the *wake* behind the blade. Usually, there are also many swirling fluid motions in the wake, and as these decay, their lessening kinetic energy usually leads to an increase in the unavailable energy.)

The pressure rise across the rotor that results from either Eq. (7.7.15) or Eq. (7.7.16) is the purpose for using a pump. Equation (7.7.15) makes it apparent that this pressure rise is only part of the energy supplied to the rotor. The ratio of the energy contained in the pressure change to the total energy supplied is called the *reaction*. Hydraulic machines can be classified according to the percent of reaction. A Pelton turbine is an example of a turbomachine for which $p_2 - p_1 = 0$. If a machine has zero reaction, it is called a pure *impulse machine*. Some axial-flow turbines are 50 percent reaction. Figure 7.7.6 is an example of the velocity triangles for this case.

If Eq. (7.7.14) is used to specify $\dot{W}/\dot{m}$ and $(p_2 - p_1)/\rho$ is obtained from Eq. (7.7.15), the ideal, or lossless, reaction is

$$(\% \text{ reaction})_{\text{ideal}} = \frac{[(U_2^2 - U_1^2) - (W_2^2 - W_1^2)](100)}{(V_2^2 - V_1^2) + (U_2^2 - U_1^2) - (W_2^2 - W_1^2)} \tag{7.7.17}$$

Example 7.7.3

Verify that the velocity triangle shown in Fig. 7.7.6 gives an ideal reaction of 50 percent.

Solution For the axial-flow machine, $U_1 = U_2$. Using the subscripts i for "in" and o for "out," Eq. (7.7.17) reduces to

$$\%R = \frac{(W_o^2 - W_i^2)(100)}{(W_o^2 - W_i^2) - (V_o^2 - V_i^2)}$$

In the present case,

$$V_i \sin \alpha_i = V_z = W_o \sin \beta_o$$

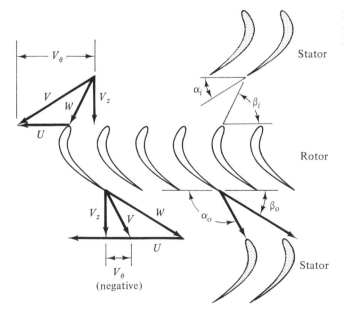

Figure 7.7.6 Diagram for evaluating the percent reaction for an axial-flow machine. $\alpha = \beta_o$; $\pi - \alpha_o = \beta_i$.

so that $V_i = W_o$ and $W_i = V_o$. Then

$$\%R = \frac{(W_o^2 - W_i^2)(100)}{(W_o^2 - W_i^2) - (W_i^2 - W_o^2)}$$

$$= 50$$

★ 7.8 EXAMPLES OF UNSTEADY MOMENTUM CONSERVATION

The importance of the examples in Secs. 7.3 and 7.4 lies not only in the procedures used and results obtained, but also in the experience and insights that were gained. We can now turn with confidence to the complete expression for momentum conservation, Eq. (7.2.2), which includes a time-dependent term. We do this by examining first one of the scalar equations associated with the forces that induce a change of the momentum of a material volume. The x component of Eq. (7.2.2) is

$$\Sigma F_x = \frac{\partial}{\partial t} \int_{CV} (u) \rho \, dV + \int_{CS} (u) \rho \mathbf{V} \cdot \mathbf{n} \, dA \tag{7.8.1}$$

We shall now see how this result is applied.

Example 7.8.1

Determine the reactions of the container on the fluid in Fig. 7.8.1 so as to determine the forces needed to hold the container in place.

Solution   Our first task is to decide on the appropriate control volume. We shall use *EFGHIE* in Fig. 7.8.1, which is similar to the control volume treated in Example 3.8.1, *EFHGE*. In this instance, the horizontal momentum of the fluid is being altered and the vertical momentum is also undergoing change. Thus, in addition to Eq. (7.8.1), one has

$$\Sigma F_z = \frac{\partial}{\partial t} \int (w) \rho \, dV + \int (w) \rho \mathbf{V} \cdot \mathbf{n} \, dA \tag{7.8.2}$$

The forces on the fluid include those of the container on the fluid plus gravitational attraction. If the force of the container on the fluid is

$$\vec{\mathcal{F}} = \mathcal{F}_x \mathbf{i} + \mathcal{F}_z \mathbf{k}$$

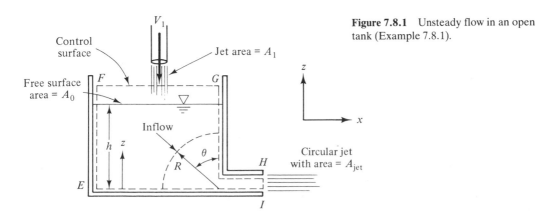

Figure 7.8.1   Unsteady flow in an open tank (Example 7.8.1).

Then the forces on the fluid in *excess* of atmospheric pressure are

$$\Sigma F_x = \mathscr{F}_x$$

and

$$\Sigma F_z = \mathscr{F}_z - (\gamma)(\text{instantaneous volume})$$

The integrals of Eq. (7.8.2) will now be considered. The more familiar integral will be treated first. Thus,

$$\int (w)\rho \mathbf{V} \cdot \mathbf{n}\, dA = -V_1(\rho)(-V_1)A_1$$

We now turn to the other integral in Eq. (7.8.2) and write it as

$$\frac{\partial}{\partial t}\int w\rho\, d\forall = \int \left(\frac{\partial w}{\partial t}\right) d(m) + \int w\frac{dm}{dt}$$

It will be assumed that $\partial w/\partial t$ and w are independent of location within the control volume and have values of $\partial W/\partial t$ and W, respectively. (The usual sign convention applies. W is positive if it is upward.) Using this simplificaton, the integral becomes

$$\frac{\partial}{\partial t}\int w\rho\, d\forall = \frac{\partial W}{\partial t}\rho A_0 h + W\rho A_0 \frac{dh}{dt}$$

(In using W, an average value, a momentum correction factor β could be introduced.) The momentum equation, Eq. (7.8.2), can now be written as

$$\mathscr{F}_z - \gamma h A_0 = \frac{dW}{dt}\rho A_0 h + W\rho A_0 \frac{dh}{dt} + V_1^2 \rho A_1 \tag{7.8.3}$$

In this example, W is not dh/dt, as it was in Example 6.4.1, because the surface elevation could be increasing at the same time there was discharge from the orifice at H. The value of V_1 will be important. Hence it would be useful to examine this point with the aid of the equation of continuity. A suitable form would be

$$-V_1 A_1 + V_{\text{jet}} A_{\text{jet}} + \frac{\partial}{\partial t}(\forall \text{ liquid in CV}) = 0$$

if V_1 is considered to have a positive value when the upper jet is pointed downward. This implies that $V_1 A_1$ is the flow into the control volume when it is positive. (This is in contrast to our usual convention for V's but agrees with that for $\mathbf{V} \cdot \mathbf{n}$ on the control surface.) $V_2 = V_{\text{jet}}$ is positive in this example, too. Since $\forall_{\text{CV}} = hA$, the continuity equation is

$$A_0 \frac{dh}{dt} = V_1 A_1 - V_{\text{jet}} A_{\text{jet}} = Q_{\text{in}} - Q_{\text{out}} \tag{7.8.4}$$

The value of V_{jet} can be expressed in terms of h and its derivatives by employing the methods of Example 6.4.2. Without undertaking such a complete analysis here, we shall approximate the result as $V_{\text{jet}} = \sqrt{2gh}$. This step leads to a differential equation for volume conservation,

$$\frac{dh}{dt} = \frac{V_1 A_1}{A_0} - \sqrt{2gh}\frac{A_{\text{jet}}}{A_0} \tag{7.8.5}$$

The integration of this equation will give h as a function of t. (The choice of the sign of V_1 to be positive if it is filling the tank facilitates the comprehension of this equation.)

In order to glean a usable result from Eq. (7.8.3), one needs an expression for W in terms of the given data. We shall let it equal the average velocity in the tank that is associated with the liquid flowing out of the orifice. This is

$$W = -(V_{\text{jet}})\left(\frac{A_{\text{jet}}}{A_0}\right) = -\sqrt{2gh}\left(\frac{A_{\text{jet}}}{A_0}\right) \tag{7.8.6}$$

This statement leads to the following revision of Eq. (7.8.3):

$$\mathscr{F}_z = -\left(\frac{A_{\text{jet}}}{A_0}\right)\rho A_0 h\frac{d}{dt}(\sqrt{2gh}) - \left(\frac{A_{\text{jet}}}{A_0}\right)\sqrt{2gh}\,\rho A_0\frac{dh}{dt} + V_1^2\rho A_1 + \gamma h A_0$$

or
$$\mathscr{F}_z = \rho\left(-\frac{3}{2}A_{\text{jet}}\sqrt{2gh}\frac{dh}{dt} + ghA_0 + V_1^2 A_1\right) \tag{7.8.7}$$

Equation (7.8.5) and its solution for h can now be coupled with this result to give $\mathscr{F}_z$ as a function of time. (Recall that $\mathscr{F}_z$ is the force component that the tank exerts on the fluid.)

For a calculation of ΣF_x in Eq. (7.8.1), it is necessary to decide on a value of u for the fluid particles within the container in order to evaluate $\partial(\int u\rho\,d\mathbf{V})/dt$. [It would not be unreasonable to use an expression like Eq. (7.8.6) in our work. The value would have to be "weighted," because near the top of the container u would be nearly zero. The component w would also go to zero near the bottom of the container. This implies that replacing w in Eq. (7.8.3) by W from Eq. (7.8.6) must surely require a momentum correction factor.]

At this time, the flow into the orifice will be approximated by a sink flow that is the negative of the source flow given by Eq. (5.4.14). Then Q enters the orifice through two octants and $u_R = -Q/\pi R^2$. The x momentum in these octants is $\int\rho u_R\sin\theta\cos\phi\,(d\mathbf{V})$, in which θ and ϕ are shown in Fig. 7.8.2. The volume element, $d\mathbf{V}$, is in spherical coordinates and equals $(R^2)(\sin\theta\,dR\,d\theta\,d\phi)$ in Fig. 7.8.2, with $0 \le \theta \le \pi/2$ and $\pi/2 \le \phi \le 3\pi/2$. Figure 7.8.2 defines the coordinate system. This integration is straightforward and gives the value $\rho QR/2$ for the x momentum. The value $R = h$ will be adopted as the effective extent of our sink for momentum purposes. The mass in the entire container is $\rho A_0 h$, and this should have the momentum just calculated for the sink flow. Consequently, an equivalent value of u, from the standpoint of momentum, is an average U based on momentum that is given by

$$U = \frac{\rho Qh/2}{\rho A_0 h} = \frac{V_{\text{jet}}}{2}\left(\frac{A_{\text{jet}}}{A_0}\right) = \frac{1}{2}\left(\frac{A_{\text{jet}}}{A_0}\right)\sqrt{2gh}$$

With this *assumption* formulated, $\partial(\int u\rho\,d\mathbf{V})/\partial t$ in Eq. (7.8.1) becomes, approximately,

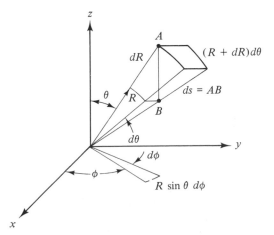

Figure 7.8.2 Spherical coordinate system.

$$\rho\left(\frac{dU}{dt}\,hA_0 + UA_0\frac{dh}{dt}\right)$$

From this point onward, the solution for $\mathcal{F}_x$ in Eq. (7.8.1) follows much as did $\mathcal{F}_z$ in Eq. (7.8.2). The reader is encouraged to carry out the steps.

Example 7.8.2

Can the oscillating manometer liquid that is shown in Fig. 7.8.3 be analyzed by means of the momentum equation?

Solution We shall use the momentum equation for the x and y directions. The only terms to consider for the x direction are

$$\Sigma F_x = \frac{\partial}{\partial t}\int u\rho\, d\forall$$

since there is no flow across the control surfaces. The pressures will be different around the circumference of the tube, and their values will have to reflect the effect of gravitational attraction on the liquid in the sloping arms. At their juncture, $\bar{p}_O$ is taken as the mean pressure across the tube's cross section. The pressure $\bar{p}_O$ is applied to the fluid by the walls of the tube near location O, which coincides with point O. Thus the summation of all of the x force components that act on the fluid within the control volume is

$$\bar{p}_O A - L_2 A\gamma\cos 45° - \tau_{\text{wall}}\,L_2(D\pi)$$

Note that the x-directed pressure forces on segment OA have been *assumed* to counterbalance the x component of weight for this branch. The only x momentum is in segment OB, and it is $u = dx/dt$. Therefore,

$$\frac{\partial}{\partial t}\int u\rho\, d\forall = \frac{d}{dt}\int\frac{dx}{dt}\rho L_2\, dA$$
$$= \frac{d^2 x}{dt^2}\rho L_2 A + \frac{dx}{dt}\,\rho\,\frac{dL_2}{dt}A$$

However, $dL_2/dt = dx/dt$, so that the x momentum equation becomes

$$\bar{p}_O A - L_2 A\gamma\cos 45° - \tau_{\text{wall}}\,L_2(D\pi) = A\rho\left[L_2\frac{d^2 x}{dt^2} + \left(\frac{dx}{dt}\right)^2\right]$$

It is observed that $v = -dx/dt$, so that the momentum equation for the y direction becomes

$$\bar{p}_O A - L_1 A\gamma\cos 45° + \tau_{\text{wall}}\,L_1(D\pi) = -\frac{d^2 x}{dt^2}\rho L_1 A - \frac{dx}{dt}\rho\frac{dL_1}{dt}A$$

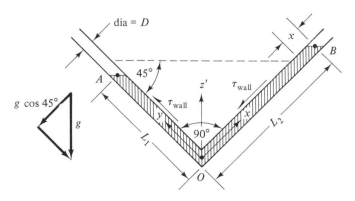

Figure 7.8.3 Fluid oscillations in a bent tube. z' in in the xy plane. Control volume completely encompasses tube.

Momentum Conservation   Chap. 7

Furthermore, $dL_1/dt = -dL_2/dt = -dx/dt$. The two momentum equations have $\bar{p}_O A$ in common. This can be eliminated between them so that

$$\frac{d^2x}{dt^2} + 4\frac{\tau_{\text{wall}}}{\rho D} + \left(\frac{2g\ \cos 45°}{L}\right)x = 0$$

if $\pi D^2/4$ is used for A. In Chap. 8 we shall determine that $4\tau_{\text{wall}}/\rho$ is $32\nu V/D$ when laminar flow occurs. By using dx/dt for V, the final result of the analysis is

$$\frac{d^2x}{dt^2} + \left(\frac{32\nu}{D^2}\right)\frac{dx}{dt} + \left(\frac{2g\ \cos 45°}{L}\right)x = 0$$

This is the same result as Eq. (6.4.7). The particular geometry shown in Fig. 7.8.3 was fortunate. Had the included angle between the manometer legs not been 90°, but, say, 60°, then there would have been x and y momentum in both legs. The decomposition of the pressure and gravity effects—all included in some unknown fashion in the force on the fluid, $\vec{\mathscr{F}}$—would not have been so convenient.

7.9 SUMMARY

The postulates of Newton that relate the forces applied to a material volume and its change of momentum can be expressed through the Reynolds transport theorem to produce Eq. (7.2.2). This vector equation is utilized by determining the associated components for the momentum and force terms. Both of these quantities must be completely assessed and then interrelated in the cause-and-effect relationship that is stated by Newton's second law. The equality sign in the mathematical statement of this law expresses the causality associated with this law of physics.

In this chapter, the forces acting on the control volume were examined. Often, an unknown force whose source one could anticipate could be assessed by knowing the change of momentum that it produced on the material volume at the particular instant of time that it occupied the control volume. By Newton's third law it is possible to deduce the corresponding forces that are being applied to the surroundings for a particular problem. A pipe elbow within which a fluid is flowing is an example of such surroundings. After the force that the elbow exerts on the fluid has been determined, the reaction of the fluid upon the elbow is known. Then one can do an analysis of the elbow to determine the forces that are necessary to constrain it in a particular position.

From such a series of steps, and a little thought, one can conclude that a control volume that includes the stationary pipe elbow, in this case, could also be used. Then the only material whose momentum was being changed would be the fluid, and this change could be determined. The forces applied to the system would be those external ones on the fluid and the pipe, including gravitational attraction and the force to keep the elbow in place. Equation (7.2.2) applied to such a control volume would then directly give an expression for the net force that the surroundings must apply to the pipe elbow to keep it stationary. However, for the dynamic analysis of compound systems that include moving fluids and containing vessels, some clarity in the procedure can occur if one first solves for the forces that the vessel exerts on the fluid to change its momentum, and then writes the equations of motion for the vessel with the inclusion of the reaction of the fluid on the vessel.

Euler's concept of relating torque with changes of angular momentum provided the means to study effectively both radial- and axial-flow hydrodynamic machines.

Centrifugal pumps and radial-inflow (Francis) turbines are of the former type, while axial-flow (Kaplan) turbines and pumps are typical of the latter. Only examples of machines employing liquids or low-speed air were cited in the chapter, but the same principles augmented with the inclusion of the compressibility effects associated with gases could be used to analyze the performance of gas turbines and compressors.

While most of the chapter was devoted to steady flows, Eq. (7.2.2) can be usefully applied to many unsteady cases. Then the time rate of change of momentum within the control volume must be assessed, and this usually requires some attention to detail.

EXERCISES

7.1.1(a) What would have been the change in y momentum if u_{final} was 7.1 m/s and $u_{initial}$ was 3.8 m/s? What would have been the change in x momentum? What would have been the total change of momentum? Put the total momentum answer in vector form.

(b) Initially, a group of particles with a mass of 0.027 kg has a velocity $V = 13.2i + 9.1j$. After 6 s, the velocity is $V = 17.7i + 4.6j$. What were the changes in the x and y momenta during this period?

(c) In Exercise 7.1.1b, what was the magnitude of the total momentum change? What was its direction?

7.2.1(a) For the flow in a round tube in which $dA = 2\pi dr$ the axial velocity can sometimes be described as $u = u_{max}[1 - (r/r_0)^2]$. In this case V_{avg} is $u_{max}/2$. What is the associated local value of u'?

(b) Remember that the derivative of $(r/r_0)^2$ is $2r(dr)/(r_0)^2$. What is the value of β for the velocity distribution given in Exercise 7.2.1a?

(c) The flow in a round pipe contains a sudden change in the velocity profile due to the confluence of two streams. The pipe has a diameter of 6 in. A high-speed (102 ft/s) airjet with a diameter of 2 in. flows along the axis. Slower air at 22 ft/s occupies the remainder of the pipe. What is the average velocity for the flow? What is β? What is α?

7.3.1(a) For a water flow with an upstream depth of 20 ft and a downstream depth of 6 ft, what is the force on the gate exerted by the fluid? If the gate is lowered and the downstream depth becomes 5 ft, what would be the force on the gate exerted by the fluid?

7.3.2(a) Let $D_1 = D_2 = D_3 = 1$ ft, $V_1 = 10$ ft/s, and $V_2 = 5$ ft/s. For $p_1 = 80$ psig, determine p_3,

assuming no losses between locations 1 and 3. What is the x component of force on the fluid within the pipe tee? What is the x component of force on the tee exerted by the fluid?

(b) What is the y component of force on the tee? Assume that the losses between locations 1 and 2 are 5 times the kinetic energy term for location 1. (Use the data for V_1, V_2, and the diameters given in Exercise 7.3.2a.)

(c) If D_2 were enlarged to 18 in., what would be the y component of the force on the tee? Use the same losses and information as in Exercise 7.3.2b.

(d) If pipe 2 is enlarged to 18 in. and V_3 is reversed in direction so that the water is flowing toward pipe 2 (the negative x direction) at 11 ft/s, what will be the x component of the force on the tee? (Assume the same energy/mass relationship for the fluid particles arriving at location 2 regardless of their origins. Assume loss terms for the flows from 3 to 2 and from 1 to 2 of 5 times the kinetic energy terms at locations 3 and 1, respectively, in order to estimate the pressures at 3 and 2.) $p_1 = 80$ psig.

(e) Let the flow in pipe 3 be directed away from the tee's junction as in Exercise 7.3.2a but with $D_3 = 3$ in. Draw the momentum and force diagrams for this solution.

7.3.3(a) At what angle θ will the x and z reactions on the vane be the same? (Neglect gravity effects.)

7.3.4(a) For $V_1 = 10$ ft/s and $y_1 = 14$ in., determine y_2.

(b) What is the value of the upstream gravity force? What is the value of the upstream momentum? What is the value of the ratio of this momentum to the gravity force? Determine the values of Fr and its square. How does this value compare with the ratio just found? Can you surmise what the general interpretation of Fr^2 can be in terms of momentum and gravity? (Use the values of V_1 and y_1 of Exercise 7.3.4a.)

(c) Assume that the bottom is inclined at 3° with the horizontal. There would then be a component of gravity in the x direction that is aligned with the bottom. Revise the pressure forces—due to gravity—and include the body force component in a relationship to determine y_2/y_1. What is this ratio for the data given in Exercise 7.3.4a? (Assume $L/y_2 = 5$, where $L/2$ is the length on either side of the disturbance to the place where the flow is uniform.)

7.3.5(a) What is the force on the fluid in the y direction? What is the associated force on the vane?

(b) What is the force on the vane for $\theta = 90°$ and $\theta = 135°$?

(c) For fixed θ, say 135°, what is the force on the vane in the x direction if V_0 is equal to 40, 50, and 60 percent of U_0?

(d) What power is being applied to the vane for the speeds given in Exercise 7.3.5c? What is the optimum value of V_0, based on these results, for maximum power application?

7.6.1(a) Let the outlet blade angle, β_0, be 90° and 80°, successively. What is the output velocity for these radial and forward-turning vanes? Would the latter be suitable for a centrifugal fan for which a significant pressure rise was not important?

(b) Derive an expression for the torque exerted on the impeller for the blades described in Exercise 7.6.1a. (Express the radial output velocity in terms of Q, and not as a numerical value.)

(c) What is the power into the fluid (utilizing the relationship developed in Exercise 7.6.1b) for $\beta_0 = 45°$, 90°, and 110°? Does the power based

on a unit of mass flowing per second increase or decrease with increasing Q for these three blade types?

(d) Assume a configuration such that the exit blade angle, β_0, is 0°. What is V_0 for this case? If the inlet swirl is zero, what is the torque applied to the fluid? What is the torque applied to the impeller?

(e) Exercise 7.6.1d can be considered an introduction to an analysis of a lawn sprinkler. Here the water moves relative to a rotating nozzle. This velocity is the Q that is appropriate to the nozzle divided by the appropriate area of the nozzle. The peripheral velocity of the nozzle and this jet velocity, added vectorially, will give V_0. From this, the frictional torque that keeps the rotating arm at constant speed can be found. For a jet that issues tangentially to an arm that is 18 in. from its rotational axis, find the speed at which the sprinkler arm rotates (in rad/s). Assume that 4.5 gpm of water issues from the nozzle and the nozzle is $\frac{1}{4}$ in. in diameter. The frictional torque arm speed due to the liquid seal is approximately 0.01 rad/s. (450 gpm = 1 cfs.)

7.6.2(a) Determine the angles, inlet and outlet, at the root of the blade.

(b) Determine the change in pressure across the turbine.

(c) If the surface of the lower reservoir in the accompanying diagram is 3/4 m above the exit plane of the turbine blades, what should be the pressure at the exit of the turbine? Is cavitation likely there? What would be the result if the reservoir were below the turbine blades? Assume (1) that both the upper and lower reservoirs vent to the atmosphere, (2) that the losses in the draft tube $= 0.30V_D^2/2g$, and (3) that the losses at point E are $V_E^2/2g$, with $V_E = V_D/5$.

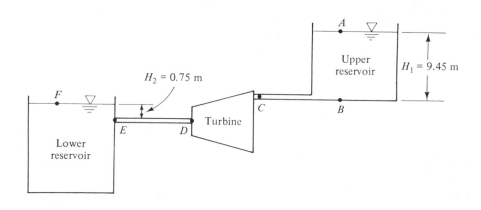

$H_2 = 0.75$ m

$H_1 = 9.45$ m

A

Upper reservoir

C B

Turbine

E D

Lower reservoir

F

PROBLEMS

All pressures are gauge pressure unless stated otherwise.

7.1 The flow in a two-dimensional channel of height b could be $u = (0.5)(y/b)$ if the upper surface is moving at 0.5 m/s. What is the total flux of momentum across a section of height b and 1 m wide? What would be the momentum flux if the average velocity across the area of this slot were constant at 0.25 m/s? What is the momentum correction factor for this case?

7.2 If the flow in a round pipe of radius r_0 is given as $u = u_{max}[1 - (r/r_0)^2]$, what is the momentum correction factor?

7.3 More than likely, the velocity profile in a round pipe would be better described by $u = u_{max}(1 - r/r_0)^{1/7}$ than by the expression in Prob. 7.2. What is the momentum correction factor for this distribution?

7.4 What is the momentum correction factor for the flow described in Fig. P.3.3?

7.5 A horizontal pipe, 6 in. in diameter, terminates in a nozzle that discharges a jet of water 3 in. in diameter with a speed of 30 ft/s. What force is being applied to the nozzle as a result? Are the eight bolts which hold the nozzle to the pipe in tension or compression?

7.6 If the pipe in Prob. 7.5 is installed vertically and the nozzle, whose larger and smaller diameters are 6 and 3 in., is 12 in. long, calculate the force that the retaining bolts must withstand. (*Hint:* Include gravitational effects by estimating the weight of the nozzle as 8 lb.)

7.7 A pipe bend—45° in a horizontal plane—must be supported to relieve the piping system of the induced loads. (See Fig. P.7.7.) The pipe, part of an irrigation system, is 2 m in diameter. The water flows at 150 kPa and 3 m/s. What is the magnitude of the necessary reacting force? Is the support member, located at the inside of the bend, in tension or compression?

Figure P.7.7

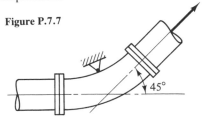

7.8 A reducing elbow, a 90° bend in the horizontal plane, converts a 6-in. pipe, upstream, to a 3-in. pipe downstream. The pressure in the larger section is 50 psi. The losses at the elbow are $KV^2/2g$, in which K is 0.75 and V is the upstream velocity. Determine the forces required to constrain the elbow—magnitudes and directions—when 2 ft³/s of water is flowing. Compare this result with one in which losses are neglected.

7.9 A standard elbow in a 2-in. pipe has losses described by $0.9V^2/2g$. Find the force on this elbow due to the flowing water if the pressure upstream of the elbow is 50 psig and 13 ft/s is the average velocity. (Neglect gravitational effects.)

7.10 A horizontal water jet issues from a two-dimensional slot. It has a height of 200 mm—with a unit width—and a speed of 10 m/s before it strikes a vertical plate. The flow divides evenly, half flowing upward and half downward. What is the force that is exerted on the plate by the jet?

7.11 The horizontal jet referred to in Prob. 7.10 strikes the plate with an angle of 45° between the jet and the surface of a flat plate. Determine the force component in the direction normal to the plate. (Keep in mind that the force component in the tangential direction is nearly zero. This last condition determines the relationship between Q_2 and Q_3, the flows off the two ends of the plate.) What are Q_2 and Q_3?

7.12 The jet of Prob. 7.11 strikes a plate that is not flat (i.e., a surface profile that allows stresses to have a net tangential component). The jets do ultimately leave at 45° with respect to the horizontal. If $Q_3 = 2Q_2$, what are the resulting x and y force components on the vane due to the water? (Take x in the direction of the jet.) The Q_3 jet was diverted 45°.

7.13 A horizontal circular water jet, 3 in. in diameter and moving at 30 ft/s, strikes a vertical flat plate and is deflected radially in all directions. What force is exerted by the jet on the plate? What is the force needed to retain the plate?

Momentum Conservation   Chap. 7

7.14 If the jet in Prob. 7.13 is vertical and the average speed of 30 ft/s is measured very near the nozzle tip, determine the force in the plate as a function of its distance from the end of the nozzle.

7.15 A high-pressure water jet with speed of 2800 ft/s impinges perpendicularly against a thin plate of hard material. In the water there is an abrasive, so that material is removed from the plate. The jet diameter is 0.0152 in., and the nozzle is connected to a tube with a diameter of 0.25 in. Determine the force with which the jet strikes the plate. (Assume $\rho = 2$ slugs/ft^3.) What is the tension in the hose?

7.16 Calculate the force on a vertical flat plate due to a horizontal water jet of width $2b$ for which the velocity is given as $u = u_{max}[1 - (y/b)^2]$ and for which $-b \leq y \leq b$. For such a flow, the average velocity is $2u_{max}/3$. If the speed across the width of the jet were constant at this value, would the force on the plate be greater or less than the value for the original velocity distribution?

7.17 The cart in Fig. P.7.17 expels water through a pump and a horizontal nozzle. The pressure in the reservoir can be maintained at 200 psig, so that the process can be considered steady. What thrust is developed?

Figure P.7.17

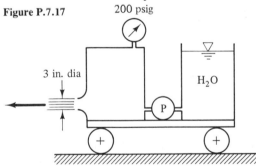

7.18 A standard tee connection is fitted to a horizontal pipe with a diameter of 12 in. in which air is flowing at 100 ft/s. The pressure upstream of the tee is 30 psi, and the flow is divided equally downstream of the tee in each of the 12-in. arms. (The plane of the tee is horizontal.) Assume that the flow is uniform at each section and that the losses in each arm are $0.8V^2/2g$, wherein V is the fluid speed there. Determine the forces exerted by the fluid on the tee. What force must be applied to hold the tee in place?

7.19 If the flow in a round pipe is described as $u = u_{max}[1 - (r/r_0)^2]$, would the force on the tee mentioned in Prob. 7.18 be greater or less than the value calculated there? What would be the numerical value of the ratio?

7.20 A canvas hose 3 in. in diameter is affixed to two connectors on a ceiling. The hose hangs vertically from the ceiling in a loop 20 ft long. Water enters the hose at 40 psi, and there is a 3-psi pressure drop along its length. Assuming that the hose does not begin to swing, what force must the hose fabric be able to withstand? (Assume that the hose weighs 8 oz per foot of length and $Q = 0.95$ cfs.)

7.21 A pipe 1 m in diameter has two lateral branches, as shown in Fig. P.7.21, that are 0.5 m in diameter. The branches have equal flow rates. Oil (SG = 0.87) is flowing at 2 m/s in the largest pipe, where the pressure is 90 kPa at point 1. Determine the forces that act upon the pipe system due to the oil's change of direction. The head loss from 1 to 3 is $0.92V_1^2/2g$. From 1 to 2, it is $0.01V_1^2/2g$, while from 2 to 4 it is $0.01V_4^2/2g$. What effect do these losses have on the force needed to restrain the pipe system? Neglect gravitational effects.

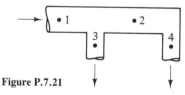

Figure P.7.21

7.22 A conduit 0.5 m in diameter, used in a refinery, makes a horizontal return bend ($\theta = 180°$ in Fig. P.7.22). A liquid hydrocarbon with $\rho = 2500$ kg/m^3 is being pumped at the rate of 0.9 m^3/s. There is a pressure drop in the bend, due to losses, of 2 kPa. Determine the total force that is induced in the bolts in order to hold this U-shaped member in place. The bend is horizontal.

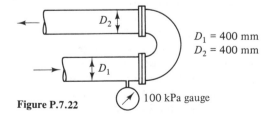

Figure P.7.22

Problems

7.23 A suspension of wood pulp in water is being pumped to two papermaking machines. The flow divides from a 24-in. pipe into two 15-in. pipes (see Fig. P.7.23). Each machine delivers a sheet of mixture to a moving screen (80 ft/s) that is 8 ft wide and $\frac{1}{4}$ in. deep. The pressure in the largest pipe is 10 psi. Assume that the mixture density is 1.9 slugs/ft^3 and that the head loss from 1 to 2 is $0.75V_2^2/2g$. Determine the force on the Y section (magnitude and direction) in order to hold it in place.

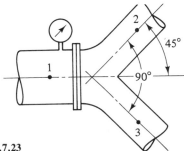

Figure P.7.23

7.24 An insulated pipe 6 in. in diameter is filled with a flowing mixture of water coming from hot and cold water pipes as indicated in Fig. P.7.24. Water at 190°F flows at 2 ft^3/s in one 6-in.-diameter pipe, and water at 50°F flows at 0.5 ft^3/s in the other. Determine the forces on the T section (magnitude and direction). The losses are $h_{3-1} = 0.72V_1^2/2g$ and $h_{2-1} = 0.51V_1^2/2g$. What is the final (mixed) water temperature?

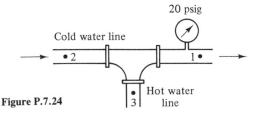

Figure P.7.24

7.25 A hydraulic jump occurs in a river such that the depth goes from 2 to 4 m. What are the upstream and downstream Froude numbers? What is the energy "loss"?

7.26 It is desired to cause an energy "loss" of $\frac{4}{3}$ m (on a N/s basis) in order to change the depth of a stream that is flowing at a depth of 2 m. What is the upstream Froude number? What is the velocity at the 2 m depth?

7.27 A vane on a cart is being moved horizontally at a speed of 20 ft/s. It is being struck by a water jet which is moving at 30 ft/s and which is deflected through 45° by the vane. The area of the jet is 0.2 ft^2. What force is exerted by the water on the vane? What power is produced by the moving vane? If there were a succession of vanes such that *all* of the water in the jet were being effectively utilized, what could be the power output?

7.28 The vane shown in Fig. P.7.28 is diverting a jet that is 0.04 m^2 that has a speed of 60 m/s; 50 percent of the water is being diverted in each direction. The blade is moving in the x direction at 30 m/s. For $\theta_x = 145°$, what will be the power generated by a single vane, a large number of vanes mounted on a rotating disk (i.e., a Pelton turbine), and a system with three jets instead of one (spaced 120° around the turbine wheel)? (Use a model that has the vanes translating in a straight line, and assume that all of the water from the nozzle will be diverted by a series of vanes.)

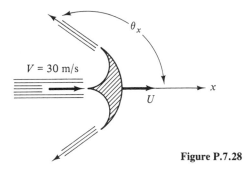

Figure P.7.28

7.29 One way to optimize the design of a Pelton turbine would involve determining the value of the angle θ (see Fig. P.7.28) that would maximize the force on the vane in the x direction. Another optimization would be to find the angle θ and the ratio of blade speed to jet speed that would yield maximum power. Determine θ_x and the speed ratio for the maximum power. How does the force on the vane depend on U/V? (Use a model that has the vanes translating in a straight line, and assume that all of the water from the nozzle will be diverted by a series of vanes.)

7.30 A jet engine is attached to a thrust stand at a test facility. Fuel is added to give $W = 45$ kJ for each 40 lb of air that flows through the engine. Assume that the air enters the engine at atmospheric pressure and negligible velocity through a large intake duct. What exit velocity should be expected in the exhaust section? The jet is 0.5 m in diameter. What thrust would the engine be likely to develop (state your assumption about the exhaust's velocity distribution). Assume $T_{exhaust} = 550°C$.

7.31 A model of a ducted fan engine (i.e., a propeller within a sheet metal duct that changes the momentum of the moving fluid, air, in this case) has an exit diameter of 20 cm. The velocity there is nearly constant over this diameter at 50 m/s. (The entrance momentum is considered to be negligible due to the large entrance region.) What thrust is being produced by the stationary motor? What is the power into the fluid? (Assume standard pressure and temperature conditions.) What reading would a water manometer have if connected to an impact tube that was inserted into the stream? (Assume $T_{exhaust} = 550°C$.)

7.32 The wake behind a flat plate can be described by

$$\frac{u}{U} = \frac{0.664}{\pi}\left(\frac{l}{x}\right)^{1/2}\exp\left(-\frac{y^2 U_0}{4x\nu}\right)$$

in which x is the downstream location where the velocity profile is being examined, l is the length of the plate, y is the ordinate measured normal to the plate; and U_0 is the uniform speed upstream of the plate. (ν is the kinematic viscosity.) What is the flux of momentum across a section at x? Does this value depend upon the location of x? Does this surprise you? Explain.

7.33 If air, for example, flows over a porous flat plate, the velocity profile reaches a constant pattern at a sufficiently large distance x from the leading edge. The x and y components of velocity are, in a particular case, $u = 20[1 - \exp(-y/\delta)]$ and $v = -5$ (both in m/s.), in which δ is a known value at each position along the plate. Apply the momentum theorem to the control volume $OABCO$ by evaluating $\int u(U - u)\,dy$. Relate this integral to the force on the fluid in the x direction, knowing that the pressure is constant along the plate ($\partial p/\partial x = 0$). What is the drag on the plate?

7.34 The air that flows through an axial-flow fan's inlet can be modeled two-dimensionally by air passing through the two-dimensional cascade shown in Fig. P.7.34. The inlet air speed is 20 ft/s, and 120 ft³/min of air is being moved. (Assume standard temperature and pressure.) What is the force that is induced in these stator blades in the z and p (peripheral) directions? (*Note:* Mass conservation requires something about the average value of the z component of velocity.)

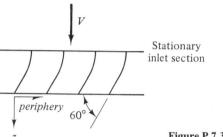

Figure P.7.34

7.35 Air enters a cascade (i.e., a series of vanes) used to straighten a flow modeled two-dimensionally with a speed of 10 m/s so that the air is tangent to the leading edge of the stationary vanes. What force is produced on the cascade if its open area is 0.4 m² (i.e., in the plane perpendicular to the page)? Refer to Fig. P.7.35.

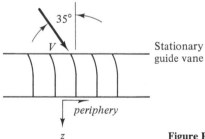

Figure P.7.35

7.36 The blades of the last stage of a multistaged axial-flow turbine can be modeled by the two-dimensional cascade shown in Fig. P.7.36. The absolute velocity V_1 is given as twice the peripheral speed of the blades, U_1, and β_1 is prescribed to be 30°. What will be the angle that V_1 makes with the peripheral direction θ? What is the axial component of velocity ahead of and behind the cascade? For $\beta_2 = 90°$, what must be α_2? What has been the change of

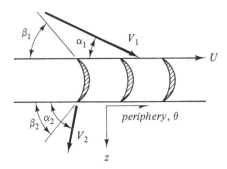

Figure P.7.36

θ-directed momentum per unit mass of fluid flowing through the cascade? What is the force on the vane system per unit mass of flowing fluid? What is the power produced by the vanes per unit mass of flowing fluid, and what is the change of kinetic energy of each unit of mass?

7.37 What should be the ratio of U_1 to V_1 if $\beta_1 = 70°$, $\alpha_1 = 30°$, and $\beta_2 = 25°$ in Fig. P.7.36? What is the expression for the power produced per unit mass of fluid flowing?

7.38 Figure P.7.36 can apply to an axial-flow machine for which the inlet absolute speed is less than the peripheral speed. What is the ratio of U_1 to V_1 for $\beta_1 = 120°$, $\beta_2 = 25°$, and $\alpha_2 = 85°$? Would the power be greater or less if $\alpha_2 = 90°$?

7.39 An axial-flow turbine (a Kaplan turbine) looks like a multibladed propeller. Assume that the cross sections shown in Fig. P.7.39 are taken at a radius of 2 m and that $\beta_1 = 45°$ and $\beta_2 = 30°$ there. What must be the angles β_1 and β_2 at $r = 1$ m if $U_1/V_1 = 1.22$, $\alpha_1 = 15°$, and $\alpha_2 = 0°$ are to be the same at all radii?

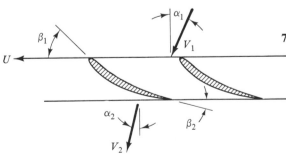

Figure P.7.39

7.40 In Prob. 7.39, the ratio U_1/V_1 may be difficult to maintain constant (α_1, also) in view of the fact that the axial component of V_1 may be constant for all radii but the swirl component of V (in the peripheral direction) may vary as (const)/r. With this information, specify β_1 and β_2 at $r = 1$ m and $r = 2$ m if at $r = 2$ m $\overline{U}/V_z = 2.8$ and $V_\theta/V_z = 0.4$.

7.41 At a constant speed of rotation of the turbine, what must be done to the inlet swirl velocity, by adjusting the inlet wicket gates, if the axial velocity drops 10 percent due to a change in available head?

7.42 Design a set of inlet guide vanes (see Prob. 7.39) for the axial-flow turbine of the type shown in Fig. P.7.39, given $U_1/V_1 = 1.22$, $\beta_1 = 45°$, $\beta_2 = 22°$, and $\alpha_2 = 0°$.

7.43 A radial-flow pump with $D_1 = 2$ in., $D_2 = 7\frac{1}{2}$ in., $\alpha_1 = 90°$, $\beta_1 = 45°$, and $\beta_2 = 90°$ is rotating at 1800 rpm. What is the absolute velocity of the discharging fluid if $Q = 0.2$ ft^3/s? What power is required to pump this quantity of water? Determine the pressure rise $p_2 - p_1$. (Refer to Fig. P.7.43, and $b = \frac{1}{2}$ in.)

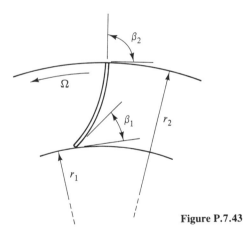

Figure P.7.43

7.44 If, in Fig. P.7.43, β_2 is 90°, the blade is designated as radial; for $\beta_2 > 90°$, the blades are forward-curved. Backward-curved vanes have $\beta_2 < 90°$. For $\beta_2 = 120°$, 90°, and 60°, compare the power and rise in pressure for a pump operating under the conditions of Prob. 7.43. How could one transform the high kinetic energy of the forward-curved blades to produce pressure? Can this be done efficiently and/or easily?

7.45 A lawn sprinkler that has a diameter of 1 ft discharges water at right angles to the rotating arm as shown in Fig. P.7.45. The pressure at the center of the sprinkler is 60 psig. The water leaves at a speed of 25 ft/s with respect to the nozzle, which is rotating at a constant speed of 60 rpm. What torque must the packing gland be exerting on the rotating stem?

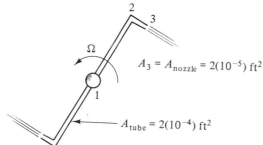

$$A_3 = A_{nozzle} = 2(10^{-5})\ \text{ft}^2$$

$$A_{tube} = 2(10^{-4})\ \text{ft}^2$$

Figure P.7.45

7.46 A radial-inflow turbine with an outside diameter of 1.8 m operates with a head of 182 m at 450 rpm. The inside diameter is 0.9 m. The flow rate is 11.03 m³/s, and 17.82 MW is produced. What torque is produced? What is the change of angular momentum per kilogram of water flowing?

7.47 For the turbine described in Prob. 7.46, the water enters, in effect, only 90% of the peripheral area of the turbine due to the presence of the guide vanes. The height of this opening is 0.25 m. What is α, at $r = 0.9$ m? (see Fig. P.7.39). What should β at the inlet (i.e., $r = 0.9$ m) be? If there is no swirl in the exit flow, what is β_2 at $r = 0.45$ m? What is the ratio U_1/V_1? What is the torque produced (theoretical)? What is the efficiency for this machine that produces 17.82 MW?

7.48 A simple model for the dynamics of the flow through a propeller is associated with Fig. P.7.48. The actual propeller is replaced by a great number of thin blades of the same diameter. The pressure difference across the propeller, acting on the area of this porous *actuator disk*, A_p, gives the propeller thrust. The proposed model assumes further that the fluid is inviscid and incompressible.

Start an analysis of the thrust produced,

$$F = A_p\left(\frac{\rho}{2}\right)(V_3^2 - V_0^2)$$

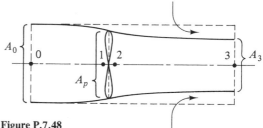

Figure P.7.48

by applying the Bernoulli equation between points on a streamline going through 0 and 1. Do the same for points 2 and 3. If $p_0 = p_3$ and $V_2 \approx V_3$, one can find $p_2 - p_1$ and the associated thrust. Do so. (What was assumed in writing $V_2 \approx V_3$ instead of $V_2 = V_3$?)

Apply the momentum equation in the axial direction to show that the velocity through the actuator disk, V_p, is $(V_0 + V_3)/2$. (Do not neglect the axial momentum of the fluid that crosses the lateral surfaces.)

If the propeller is on an aircraft moving at V_0 in still air, useful work is being produced. Some energy is being added to the fluid, also, by the propeller; the kinetic energy in the slipstream is eventually lost. Show that the efficiency of the propeller can be written as

$$\eta = \frac{\text{work out}}{\text{work in}} = \frac{\text{work out}}{\text{work out} + \text{losses}} = \frac{V_0}{V_0 + V_p}$$

(Note that the kinetic energy of the slipstream per unit mass for a stationary observer is $(V_3 - V_0)^2/2$ if the aircraft is moving at V_0 and if V_3 is measured relative to the aircraft.)

7.49 Mr. Jones states that by assuming no losses in a hydraulic jump even though a small amount could occur, the energy equation gives

$$\frac{V_1^2}{2g} = \frac{h_2^2}{h_2 - h_1}$$

and that when this result is introduced into the momentum equation, the force of the ambient air and bottom on the water, $\mathscr{F}_x$, is of the form

$$\mathscr{F}_x = \frac{\gamma}{2}(h_2^a - h_1^b)\frac{(h_2 - h_1)^c}{(h_2 + h_1)^d}$$

in which a, b, c, and d are integer exponents. Is Mr. Jones (mathematically) correct? (See Fig. P.7.49.) Are any of his results anomalous? Explain.

Problems

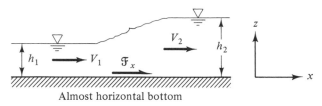

γ = specific weight

Almost horizontal bottom

Figure P.7.49

7.50 A Venturi tube (diameter ratio = 2 : 1, with throat area of 3.14 in.^{2}) is fitted with a tube at its throat in an effort to draw water from a reservoir whose free surface is 20 ft below the tube's horizontal axis. At what pressure on the axis of the throat will water begin to flow from the reservoir into the throat? At what flow rate through the Venturi tube would this pressure be realized? Assume that the pressure in the pipe just upstream of the contraction section in Fig. P.7.50 can be held constant at 8 psig.

7.51 A water jet is used to propel and maneuver a hydrofoil ship. The vanes to guide the jet can be adjusted to slow the ship. A two-dimensional version is shown in Fig. P.7.51 with the expected conditions during braking. (The actual version is axisymmetric.) All jet speeds are 130 ft/s. The velocity into the nozzle is uniform at 50 ft/s. The inlet area is 1.2 ft^2, and the axial jet has an area of 0.25 ft^2. What braking force does the ship experience for the situation shown in the figure? (Consider nozzle and deflector to be vanes.)

7.52 A small water tunnel has been designed to test items under cavitation conditions. Bellows connect an end section as shown in Fig. P.7.52. Will these flexible pieces be in tension or compression? Explain.

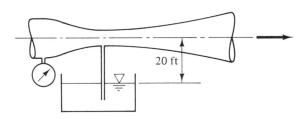

Figure P.7.50

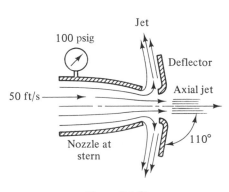

Figure P.7.51

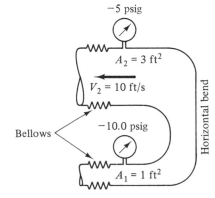

Figure P.7.52

7.53 An early realization of Barker's mill has been reported to be similar to that shown in Fig. P.7.53. (Assume that the jet velocity, which is relative to the nozzle, is independent of the speed of rotation.) What torque would have been produced to mill the grain? (Jets are perpendicular to arms. The jet area is 0.006 m^2, and the jet speed is 11 m/s.) If the torque required for milling were to double because of a new setting of the millstone, what would be the effect on the speed of the mill? Under the two conditions described above, is the grinding power the same?

7.54 A radial-inflow hydrodynamic machine with a diameter of 8 ft rotates at 120 rpm. The absolute velocity of the water entering the unit is 100 ft/s in a direction that is 45° to the radial direction. Specify the angle of the blades at the inlet for "shockless" entry of the liquid onto the impeller.

7.55 A jet of liquid with a velocity profile $w = w_{max}[1 - (2r/D)^2]$ discharges into the air from a horizontal tube of diameter D and gradually acquires a uniform velocity. When this condition is reached, what is the ratio of the jet's diameter to the tube diameter? (Assume the air to be inviscid.)

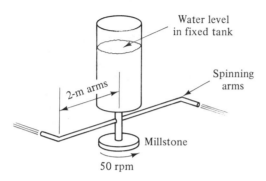

Water level in fixed tank

Spinning arms

2-m arms

Millstone

50 rpm

Figure P.7.53

8

Some Theoretical Predictions of the Effects of Viscosity on Fluid Flow

8.1 INTRODUCTION

The material presented thus far has concentrated on analyses using a control volume of finite size. Integrals were evaluated that gave the net effluxes and temporal changes of finite quantities. This method was found to be useful in the solution of many problems. In some cases, however, additional facts must be specified. There was a hint of this need when we examined the pendant drop in Fig. 1.6.1. The static equilibrium of the drop will depend upon the gauge pressure within the drop, the surface tension, the angle between the tangent to the surface and the vertical direction, the volume of the drop, and its specific weight. Even if the liquid's specific weight were known and the geometry were specified from a photograph, the vertical equilibrium equation for the entire drop would still contain two unknowns. However, Eq. (1.6.1) connected the difference in pressure between the liquid and its surroundings with the surface tension. This was done by considering force equilibrium at a point on the surface and working with infinitesimal distances.

In this chapter we shall look at flows in more detail than before in order to establish differential equations that are relationships between shear stresses and other variables of the flow. The results of our efforts will be the specification of the velocity distribution for the case under examination. One could then integrate this distribution to find the flux of mass, kinetic energy, or momentum leaving a control volume, in the manner of the previous chapters, if that were desired.

Further insight into the aims of the discussion is gained by considering the flow in a horizontal circular conduit like that shown in Fig. 8.1.1. The flow is assumed to be "fully developed." This means that the velocity profile does not change from station to

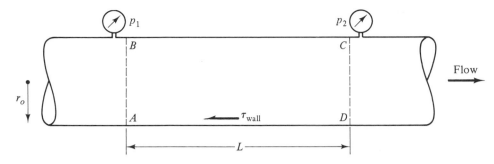

Figure 8.1.1 Finite control surface for analyzing the flow in a horizontal conduit.

station along the pipe.* The energy equation for the control volume $ABCDA$ in Fig. 8.1.1 yields

$$\frac{p_1 - p_2}{\gamma} = (\text{losses})_{1-2} \qquad \begin{pmatrix} \text{per unit weight of fluid} \\ \text{flowing per second} \end{pmatrix} \qquad (8.1.1)$$

The actual velocity profile is of little consequence, because the kinetic energy term is the same at each end of the appropriate control volume.

The momentum equation for the case in Fig. 8.1.1 is not difficult to write, because the net momentum flux is zero. For the present, it will be assumed that gravity acts perpendicular to the page and the centerline of the pipe is horizontal. Also, p_1 and p_2 are centerline pressures. This definition of the problem lets us write the momentum equation as

$$(p_1 - p_2)\pi r_o^2 - 2\pi r_o L \tau_{\text{wall}} = 0$$

These two results leave something to be desired. If one knew the pressure drop $p_1 - p_2$, the losses could be determined and an estimate of the power to pump the fluid could then be made. This pressure drop would be known if τ_{wall} could be found.

In instances where r_o is small, say, 1 cm, and the centerline fluid velocity is not large, say, <10 cm/s, Newton's law of viscosity, $\tau = \mu(du/dy)$, gives a very satisfactory relationship between the shear stresses and the velocity profile. Such a special dependency of the shear stress on the rate of strain will be used in this chapter to discuss some laminar flows. (In Chap. 16, a more general stress relationship will be introduced to treat more complex problems. Equation (4.3.2) is typical of such equations.) We shall take our solutions and compare them with observations to determine how useful the theoretical predictions are. In some cases, the agreement between theory and observation will depend on the magnitude of the result. For example, it might be that, for small velocities, the agreement is quite good, but for large velocities the agreement is poor because the predicted pressure changes are always less than the experimental values. In the course of obtaining solutions for a number of practical situations, we shall discover some of the characteristics of fluid flows. This information will be valuable to us in making estimates of the outcome of cases that are new to us. We can

---

* This would not be true in the region downstream from a sudden expansion or contraction where the flow would be adjusting to the new geometry.

use this information as a guide when discussing somewhat more complicated cases, or when attempting to determine the salient features of a problem so complex that a mathematical analysis would be impossible or impractical.

8.2 FLOW BETWEEN PLANE PARALLEL SURFACES

We now continue our examination of the subject by analyzing the flow shown in Fig. 8.2.1. This is a two-dimensional counterpart of the case shown in Fig. 8.1.1. It is a useful case and somewhat easier than the one for a round conduit, which will be studied subsequently.

Initially the plates will be horizontal, with the lower one stationary and the upper one moving in the positive x direction with a speed of U. It is convenient to assume the plates to be infinitely long. This leads to the idea that there is no beginning or end to the channel, and thus no special beginning and end phenomena need to be considered. Consequently, one *assumes* that the flow at every station in the horizontal direction is the same as at every other station. Finally, the flow is *assumed* to be steady.

The infinitesimal control volume shown in Fig. 8.2.1 is at some elevation y above the bottom plate.* It is natural to assume that the pressure is varying along the length of the fluid layer so that the pressure at the upstream face is simply p, whereas the pressure on the downstream face is $p + (\partial p/\partial x)\,dx$. The magnitude of the shear stress is taken as τ on the lower face. The direction of this stress is assumed without difficulty because the flow is simple and we have certain intuitive notions about the effects that are occurring. Because the flow appears to be moving in horizontal layers (no vertical motions occur, so that $v = 0$), the lower layers tend to retard the upper layers because of the stationary bottom. It is for this reason that the shear stress on the lower face is drawn to indicate a retarding effect. Likewise, the shear stress on the top surface of the fluid element is drawn to the right to indicate that the fluid above this surface is pulling it in the flow direction. The magnitude of this latter shear stress may be different from that on the lower surface, so it is expressed as $\tau + (\partial \tau/\partial y)\,dy$. Placing the shear stress on the top and bottom surfaces in opposite directions is actually a consequence of the

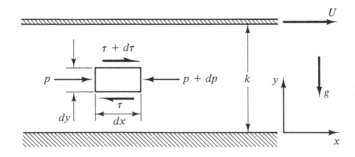

Figure 8.2.1 Infinitesimal control volume for analyzing the flow between horizontal parallel plates. Upper boundary is moving; lower boundary, stationary. Assume unit distance perpendicular to the page.

*In this analysis for a horizontal layer of fluid, gravity will have no effect on the fluid's velocity distribution. This is so because the dynamics that control the horizontal motion of the fluid are independent of the coordinate parallel to gravity. For this reason, y can be used as the vertical coordinate, as is customary for this flow, instead of z as was done heretofore. It is often referred to as *plane Couette flow*. When $\partial p/\partial x = 0$, the velocity distribution is a limiting case of that in the annulus between two concentric cylinders. This circular flow—see Sec. 8.4—was experimentally studied by M. Couette around 1890 while he was working to establish the no-slip condition.

concept of action-reaction. (If it were to occur that our intuition about the problem was faulty and τ proved to be negative in the final solution, we would conclude that the shear stress on the lower surface was directed in the positive x direction. Furthermore, the direction of the stress on the upper surface would also be affected by this change in the sign of τ as well as by the sign and magnitude of $\partial \tau / \partial y$. In Chapter 16 we shall adopt a convention so that the sign of the shear stress is not arbitrary.)

The shear stress and pressure contribute to forces in the x direction. Moreover, there is no acceleration of the material volume, since it maintains its particular speed as it moves along the layer. We note that fluid particles close to the bottom plate move more slowly than those passing through the control surface in Fig. 8.2.1. However, these particles, too, do not accelerate as they move from left to right. Newton's law of motion $\Sigma \mathbf{F} = m\mathbf{a}$ gives for the x direction and for a unit distance perpendicular to the page

$$p\,dy\,(1) - \left(p + \frac{\partial p}{\partial x}\,dx\right)dy\,(1) - \tau\,dx\,(1) + \left(\tau + \frac{\partial \tau}{\partial y}\,dy\right)dx\,(1) = 0$$

This becomes, after dividing by the volume element $(dx)(dy)(1)$,

$$\frac{\partial p}{\partial x} = \frac{\partial \tau}{\partial y} \tag{8.2.1}$$

In Sec. 8.1, we saw that the losses were related to $p_1 - p_2$, which in this case is $-(\partial p / \partial x)\,dx$. Hence, $\partial p / \partial x$ must be negative or zero for nonnegative losses. It is expected that the horizontal pressure gradient does not depend upon the horizontal coordinate x. If it did, we would be treating a case in which the conditions of the problem changed from station to station along the layer, a situation that was excluded at the outset. But how does $\partial p / \partial x$ vary with the vertical coordinate y? This can be examined by referring to Fig. 8.2.2. Here, four points have been chosen at different locations in the channel. Two of the points are directly below two others. If the pressure gradient for the upper points is $(\partial p / \partial x)_u$ and that for the lower points, $(\partial p / \partial x)_l$, one has the following for p_2 and $p_{2'}$:

$$p_2 = p_1 + \left(\frac{\partial p}{\partial x}\right)_u L$$

and

$$p_{2'} = \left\{(p_{1'}) + \left(\frac{\partial p}{\partial x}\right)_l L\right\} = \left\{(p_1 + \gamma K) + \left(\frac{\partial p}{\partial x}\right)_l L\right\} \tag{8.2.2}$$

But p_2 and $p_{2'}$ are also related by the hydrostatic law, since the fluid is not accelerating in the vertical direction. Consequently,

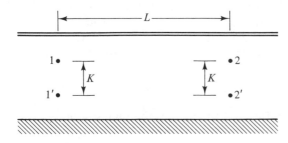

Figure 8.2.2 Pressure at four points in a horizontal flow for which the horizontal pressure gradient is independent of the vertical position.

$$p_{2'} = p_2 + \gamma K = \left[p_1 + \left(\frac{\partial p}{\partial x}\right)_u L\right] + \gamma K$$

Inserting this expression for $p_{2'}$ into Eq. (8.2.2) gives

$$\left[p_1 + \left(\frac{\partial p}{\partial x}\right)_u L + \gamma K\right] = \left\{(p_1 + \gamma K) + \left(\frac{\partial p}{\partial x}\right)_l L\right\} \tag{8.2.3}$$

From a comparison of the terms in brackets, one can conclude that $(\partial p/\partial x)_u = (\partial p/\partial x)_l$; the streamwise pressure gradient is seen to be independent of the vertical coordinate.

Because $\partial p/\partial x$ does not depend on y, as we have just shown, and because it does not depend on x, by virtue of our assumption of a velocity profile that does not change in the direction of flow, we conclude that it is a constant. Equation (8.2.1) then shows that for this situation $\partial \tau/\partial y$ is also a constant. (Note that p varies with x and also y, due to gravity, and so the partial derivative symbol has been retained.) Our velocity profile, which is one that does not vary in the flow direction, is called a *fully developed* profile. Because τ varies only with y, we now write

$$\frac{d\tau}{dy} = \frac{\partial p}{\partial x} = \text{constant} = \frac{p_2 - p_1}{x_2 - x_1} = C \tag{8.2.4}$$

Equation (8.2.4) can be integrated directly to give

$$\tau = Cy + C_1 \tag{8.2.5}$$

The constant C_1 is determined from the boundary conditions.

At this stage, we introduce an important step. Newton's law of viscosity is used to relate the shear stress with the velocity profile. In this way we can proceed to determine that profile. Thus, we write

$$\tau = \mu \frac{du}{dy} \tag{8.2.6}$$

and use this statement in Eq. (8.2.5) with the result that

$$\frac{du}{dy} = \frac{C}{\mu}y + \frac{C_1}{\mu} \tag{8.2.7}$$

This equation for the velocity component u, the only component in this case, can be integrated straightforwardly to give

$$u = \frac{C}{\mu}\frac{y^2}{2} + \frac{C_1}{\mu}y + C_2 \tag{8.2.8}$$

The constant C_2 must also be evaluated from the boundary conditions of the problem.

Often, the only information that is available for the determination of the constants C_1 and C_2 is the velocity or the shear stress at various places in the flow. In the present case, the velocity is prescribed at the upper plate as U. Certainly, we know, the velocity at the lower stationary plate in Fig. 8.2.1 is zero. We can then write the boundary conditions as

$$u_{y=0} = 0 \quad \text{and} \quad u_{y=k} = U \tag{8.2.9}$$

If these values are substituted in Eq. (8.2.8), one has the final solution

$$u = \frac{C}{2\mu}(y^2 - ky) + \frac{U}{k}y \qquad C = \frac{\partial p}{\partial x}$$

and thus
$$u = -\frac{1}{2\mu}\frac{\partial p}{\partial x}(ky - y^2) + \frac{U}{k}y \tag{8.2.10}$$

If one uses the expression for u to determine the quantity of fluid flowing toward the right, one gets

$$Q = \int(u)\,dA = -\frac{\partial p}{\partial x}\left(\frac{k^3}{12\mu}\right) + \frac{Uk}{2} \tag{8.2.11}$$

Finally, Eq. (8.2.6) gives the shear stress on the moving plate as

$$\tau_{y=k} = \left(\frac{\partial p}{\partial x}\right)\left(\frac{k}{2}\right) + \frac{\mu U}{k} \tag{8.2.12}$$

Now that we have studied the flow in a horizontal channel, we shall examine a sloping channel in order to observe the effect of gravity on the flow. In Fig. 8.2.3, the flow between parallel planes is depicted with the x axis making an angle θ with the horizontal plane. This is the same angle that the y axis makes with the h axis, which designates the elevation of a material volume in this special geometry. Gravity acts in the $-h$ direction. If the forces are summed in the x direction, in the same way as previously, one must now include the component of gravitational attraction in the positive x direction. This will be $\rho g\,(dx\,dy)(1)(\sin\theta)$. Then, for this instance of steady flow, we have, instead of Eq. (8.2.1),

$$\left[\left(\frac{\partial \tau}{\partial y}\right)dy\right]dx - \left[\left(\frac{\partial p}{\partial x}\right)dx\right]dy + \rho(dx\,dy)(g)(\sin\theta) = 0$$

which, upon cancellation of the volume $(dx\,dy)(1)$, becomes

$$\frac{d\tau}{dy} = \frac{\partial p}{\partial x} - \gamma\sin\theta \tag{8.2.13}$$

Since positive changes in x are accompanied by negative changes in h, a glance at Fig. 8.2.3 should make it clear that*

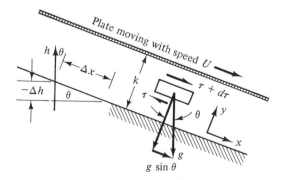

Figure 8.2.3 Infinitesimal control volume for the parallel flow down an inclined plane.

*The partial derivative is necessary because h could vary by changing y for a fixed value of x. This would give $\partial h/\partial y$, which equals $\cos\theta$.

$$\sin \theta = -\frac{\partial h}{\partial x}$$

This relationship for $\sin \theta$ has the consequence that Eq. (8.2.13) can be written as

$$\frac{d\tau}{dy} = \frac{\partial p}{\partial x} + \gamma \frac{\partial h}{\partial x} = \frac{\partial}{\partial x}(p + \gamma h) \qquad (8.2.14)$$

If the y component of the momentum equation is written for the infinitesimal control volume, the result is

$$\tau \, dy - \tau \, dy + \left(p - \frac{\partial p}{\partial y}\frac{dy}{2}\right) dx - \left(p + \frac{\partial p}{\partial y}\frac{dy}{2}\right) dx - \rho g \, \cos \theta \, (dx \, dy) = 0$$

because the fluid is not accelerating in the y direction. The shear stress terms on the vertical surfaces of the fluid element cancel one another, since we expect no variation of the flow—including τ—in the x direction. In view of the fact that $\cos \theta = \partial h / \partial y$, this equation can be reduced to

$$\frac{\partial}{\partial y}[-(p + \gamma h)] = 0 \qquad (8.2.15)$$

The implication of this equation is that $p + \gamma h$ does not vary with y; it is independent of y. We can proceed with the knowledge that $p + \gamma h$ is a function of x—unknown at present—so that the derivative $\partial(p + \gamma h)/\partial x$ depends only on x. Thus the right-hand side of Eq. (8.2.14) cannot depend on y, and this tells us something about the left-hand side of this equation, $d\tau/dy$. Because τ is assumed to vary only with y, $d\tau/dy$ on the left-hand side of Eq. (8.2.14) can only be a function of y, or a constant. However, the right-hand side of this equation has just been shown to be independent of y. This poses no difficulty for us; $d\tau/dy$ must equal a constant. This was the same conclusion that we reached in Eq. (8.2.4), but now the constant is $\partial(p + \gamma h)/\partial x$ instead of simply $\partial p/\partial x$, since $\partial h/\partial x$ was zero in the previous instance. The integration of Eq. (8.2.14) and the application of the boundary conditions are the same as before, so that the result is

$$u = -\frac{\partial}{\partial x}(p + \gamma h)\frac{y}{2\mu}(k - y) + \frac{Uy}{k} \qquad (8.2.16)$$

for the case of a moving upper boundary.

Example 8.2.1

A channel similar to that shown in Fig. 8.2.3 is inclined at an angle of 30° to the horizontal. The upper boundary of the channel is moving with a speed of 5 ft/s in the $-x$ direction. The pressure gradient in the x direction, $\partial p/\partial x$, is 0.3 psi/ft when the channel height k is 0.009 in. The liquid in the channel is an oil with a specific gravity of 0.86 and a viscosity of 120 centistokes (cSt). If the plate is 3 ft long and 2 ft wide, what force is needed to move the plate, if one considers it to be weightless?

Solution One can start with Eq. (8.2.16) and write

$$u = \left(-\frac{\partial p}{\partial x} - \gamma \frac{\partial h}{\partial x}\right)\left(\frac{y}{2\mu}\right)(k - y) + \frac{Uy}{k}$$

with $k = 0.009/12 = 7.5(10^{-4})$ ft and

$$\nu = 1.2 \text{ St} = 1.2 \text{ cm}^2/\text{s} = 1.3(10^{-3}) \text{ ft}^2/\text{s}$$

$$\mu = \rho\nu = [0.86(1.94)](1.3)(10^{-3}) = 2.17(10^{-3}) \text{ slugs/ft-s}$$

Also,
$$\frac{\partial p}{\partial x} = 0.3(144) \text{ (lb/ft}^2)/\text{ft}$$

$$\frac{\partial h}{\partial x} = -\sin 30° = -0.5$$

$$U = -5 \text{ ft/s}$$

Therefore,
$$\tau_{y=k} = \mu \frac{du}{dy} = [-0.3(144) + 0.86(62.4)(0.5)]\left(-\frac{k}{2}\right) - \frac{5\mu}{7.5(10^{-4})}$$

or
$$\tau = +6.14(10^{-3}) - 14.47 = -14.46 \text{ lb/ft}^2$$

so that
$$F = -14.46[2(3)] = -86.8 \text{ lb}$$

Figure 8.2.3 implies that a negative τ is a force per unit area exerted on the plate (by the fluid) that is directed "downhill." Hence, the force found above is also "downhill." This appears to be primarily due to the motion of the plate in the "uphill" direction. An appropriate external force on the plate would be needed to obtain a constant speed for it.

The line of reasoning leading to Eq. (8.2.16) can be followed for the situation shown in Fig. 8.2.4. Here there is a free (gas-liquid) interface. Also, the pressure is constant along the interface. Consequently, $\partial p/\partial x$ must be zero everywhere, since the pressure does not change with x on the free surface and the condition given by Eq. (8.2.15) also applies. (That equation could be differentiated with respect to x, and the result would imply that $\partial(p + \gamma h)/\partial x$ does not vary with y. This is similar to the steps taken previously when discussing the situation depicted in Fig. 8.2.2.

A boundary condition needs to be specified at $y = k$ for this case, since the no-slip condition is not applicable there due to the absence of a rigid boundary. Instead, we shall use a condition on the shear stress at the free surface. This stress on the lower fluid is applied by the fluid above the interface. The lower fluid exerts a stress of equal magnitude, but opposite sense, on the fluid above it. This means that the shear stress varies continuously across the interface (i.e., there is no jump in the magnitude).* If the upper fluid is a gas and the lower one a liquid, one usually assumes that the gas will not exert any significant shear stress on the liquid. Thus, we *assume* that $\tau = 0$ at $y = k$. This condition, and $u = 0$ at $y = 0$, result in the solution [refer to Eqs. (8.2.6) and (8.2.7), with $C = \gamma \, \partial h/\partial x$ now]

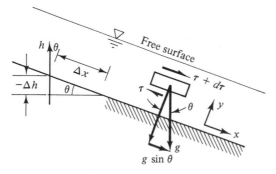

Figure 8.2.4 Infinitesimal control volume for the free-surface flow down an inclined plane.

---
* The velocity also varies continuously across an interface.

$$u = \frac{\gamma}{\mu}(\sin\theta)y\left(k - \frac{y}{2}\right) \qquad (8.2.17)$$

for the flow of a liquid with a free surface down an inclined plane. (The condition $\tau = 0$ at the air-liquid interface would not be valid when modeling the generation of surface waves by wind. Then the wind stresses would be central to the problem.)

8.3 FLOW IN CIRCULAR CONDUITS

With the background that has just been gained for plane flows, it is but a small step to analyze a flow for which cylindrical polar coordinates are natural. For simplicity, the conduit is taken as horizontal in the manner shown in Fig. 8.3.1. The control volume in this figure is a cylindrical shell, chosen because its results are applicable even when considering the flow in the annulus between two concentric tubes. Such a situation is shown in the detail for Fig. 8.3.1.

Again there are forces on the control surface due to the pressures in the fluid that do not cancel when considering the z component.* The shear stress on the inner surface of the cylindrical surface is given as τ, so that the shear stress on the outer surface is written as $\tau + (d\tau/dr)(dr)$. The direction of these stresses indicated in Fig. 8.3.1 was chosen by assuming that they were positive and that the flow near the center of the tube would be moving faster than that near the rigid outer boundary. If this is the case, the fluid nearest the tube's axis would be pulling that surface along in the flow direction. Similarly, the fluid on the outside of the control volume would be retarding the outer surface. (Subsequently, this assignment of directions will be checked to decide whether we are satisfied with the result.) The shear stress τ acts on a surface that has a radius of r with a lateral area of $2\pi r(dz)$. The radius on which $\tau + (d\tau/dr)\,dr$ acts is greater by the amount dr. Keeping this geometric fact in mind, we can write the momentum equation in the axial, or z, direction as

$$2\pi r\,dr\left[p - \left(p + \frac{\partial p}{\partial z}dz\right)\right] + \tau(2\pi r)\,dz - \left(\tau + \frac{d\tau}{dr}dr\right)(2\pi)(r + dr)\,dz = 0 \qquad (8.3.1)$$

If the volume of the fluid element, $2\pi r\,dr\,dz$, is factored out of this result, one obtains

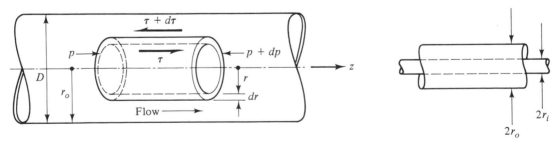

Figure 8.3.1 Infinitesimal control volume for the analysis of a horizontal flow in a circular conduit.

*z is directed along the axis of the pipe regardless of its orientation with the horizontal because cylindrical polar coordinates are being used.

$$-\frac{\partial p}{\partial z} - \frac{\tau}{r} - \frac{d\tau}{dr} - \frac{dr}{r}\left(\frac{d\tau}{dr}\right) = 0 \tag{8.3.2}$$

Finally, dr is allowed to approach zero, so that the last term, the only term in the equation containing this factor, becomes vanishingly small in comparison with the others. This leaves

$$+\frac{\tau}{r} + \frac{d\tau}{dr} = \frac{1}{r}\frac{d(\tau r)}{dr} = -\frac{\partial p}{\partial z} \tag{8.3.3}$$

as our result. This equation can be integrated directly to obtain

$$\tau r = \left(-\frac{\partial p}{\partial z}\right)\frac{r^2}{2} + C_1 \quad \text{or} \quad \tau = \left(-\frac{\partial p}{\partial z}\right)\frac{r}{2} + \frac{C_1}{r} \tag{8.3.4}$$

in which $-\partial p/\partial z$ does not depend on r and is a constant. The constant C_1 must be determined from the information at hand. In the case of a conduit in which fluid is present at $r = 0$, in contrast to the case where the flow is in an annulus between two concentric tubes, the value of C_1 would be zero. Clearly, Eq. (8.3.4) would indicate that τ would be infinite at $r = 0$ if C_1 were not zero. Upon setting C_1 equal to zero, we observe that the shear stress varies linearly with the radius:

$$\tau = \left(-\frac{\partial p}{\partial z}\right)\frac{r}{2} \tag{8.3.5}$$

with the maximum value $\tau_{\text{wall}} = (-\partial p/\partial z)r_o/2$. This result has been obtained with very few assumptions and is thus quite general. The fact that $\partial p/\partial z$ is negative for the flow pictured in Fig. 8.3.1 means that τ will have a positive value and be in the direction that was *assumed* initially.

We now relate the shear stress and the velocity profile in accordance with Newton's *assumption*, $\tau = \mu\, du/dy$. Heretofore we have used this relationship for plane geometries. It is also applicable, however, to this circular cross section with its polar geometry, with a change of sign. (This is due to our choice of the direction of positive τ on the inner section of the control surface. The choice was dictated by our intuitive ideas about τ.) To make our mathematical expression for τ concur with Newton's statement about it, we write

$$\tau = -\mu\frac{dw}{dr} \tag{8.3.6}$$

(In our problem, we expect w, the z component of velocity, to decrease with increasing r; consequently, $dw/dr < 0$ in a region where τ is thought to be positive. Hence, the introduction of the negative sign. This apparently arbitrary "sign convention" begs for a rigorous convention. This is given in Chap. 16.) We now equate Eqs. (8.3.5) and (8.3.6) to get

$$-\mu\frac{dw}{dr} = \left(-\frac{\partial p}{\partial z}\right)\frac{r}{2}$$

An integration of this equation is the next step in the development, and this yields

$$-w = \left(-\frac{\partial p}{\partial z}\right)\frac{r^2}{4\mu} + C_2 \tag{8.3.7}$$

The constant C_2 is evaluated with the knowledge that w is zero at the outer wall, which is stationary. This gives

$$C_2 = -\left(-\frac{\partial p}{\partial z}\frac{r_0^2}{4\mu}\right)$$

This can be reintroduced into Eq. (8.3.6), with result that

$$w = -\frac{1}{4\mu}\frac{\partial p}{\partial z}(r_0^2 - r^2) = \left[\left(-\frac{\partial p}{\partial z}\right)\frac{r_0^2}{4\mu}\right]\left(1 - \frac{r^2}{r_0^2}\right) \tag{8.3.8}$$

The coefficient in square brackets is the value of the velocity on the conduit's centerline, w_o, where $r = 0$. This is

$$w_0 = \left(-\frac{\partial p}{\partial z}\right)\left(\frac{r_0^2}{4\mu}\right)$$

a positive quantity, as expected, for $\partial p/\partial z < 0$. By evaluating

$$Q = \int (w)\, dA = \int_0^{r_0} w_0\left(1 - \frac{r^2}{r_0^2}\right)(2\pi r\, dr)$$

it is possible to show that $V_{avg} = w_0/2$. It is helpful to use this relationship between V_{avg} and w_0. We expand on the definition

$$\frac{w_0}{2} = V_{avg} = \frac{1}{2}\left(-\frac{\partial p}{\partial z}\right)\frac{r_o^2}{4\mu}$$

to obtain

$$V_{avg} = \left(\frac{p_1 - p_2}{L}\right)\frac{D^2}{32\mu} \tag{8.3.9}$$

in which p_2 is the pressure at a point that is a distance L downstream from where p_1 is measured.

Example 8.3.1

A horizontal tube with an internal diameter of 7 mm is required to deliver 10 cm³/s of a chemical with specific gravity $= 2.8$ and viscosity $= 8$ centipoise (cP). What is the difference of pressure in a tube that has a length of 5 m? What power must be expended to deliver the chemical?

Solution   The data are as follows in the SI system:

$$\mu = 0.08\, P = 0.08\ \text{g/cm-s} = \frac{0.08\ (\text{kg}/1000)}{(\text{m}/100)(\text{s})} = 0.008\ \text{kg/m-s}$$

$$\text{Diameter} = 0.007\ \text{m} \quad \text{and} \quad Q = 10(10^{-6})\ \text{m}^3/\text{s}$$

Equation (8.3.9) can be transformed to read

$$Q = \frac{(p_1 - p_2)}{L}\frac{\pi D^4}{128\mu} = 10(10^{-6}) = \frac{(p_1 - p_2)}{5}\frac{\pi(7)^4(10^{-12})}{128(8)(10^{-3})}$$

The result is $p_1 - p_2 = 6.8$ kPa. Also,

$$\text{Losses} = \frac{\Delta p}{\gamma} = \frac{6.8(1000)}{(2.8)[9.8(1000)]} = 0.25 \text{ m}$$

$$= 0.25 \frac{(\text{m-N})/\text{s}}{\text{N/s}} = 0.25 \frac{\text{W}}{\text{N/s}}$$

Hence the minimum power required to deliver the chemical is

$$0.25(Q\gamma) = 0.25(10^{-5})(2.8)(9.8)(10^3)$$

$$= 0.07 \text{ W}$$

One notes that V_{avg}—or also w from Eq. (8.3.8)—is linear in the pressure drop $\Delta p = p_1 - p_2$. (Soon, cases will be discussed for which this linearity is not the result.) Since this pressure drop has been shown to be proportional to the losses, in Sec. 8.1, we have the conclusion that the losses and flow velocity are directly proportional for this flow. J. Poiseuille studied the flow in tubes of small diameter and established the relationship between the flow rate, pressure gradient, and diameter that is given by Eq. (8.3.9). Also, around the middle of the nineteenth century, G. Hagen confirmed these results and observed the effects of temperature on the flow. The work of these two men is commemorated by naming the flow described by either Eq. (8.3.8) or Eq. (8.3.9) *Hagen-Poiseuille flow,* even though G. G. Stokes theoretically derived the result. The steady flow described by these equations, as well as Eq. (8.2.16), has the fluid moving with constant speed along surfaces that are parallel to the walls of the conduit. The flow appears to be moving in sheets, or lamina. Flows of this kind are usually called *laminar flows.* Such flows will be the principal subject of the remaining sections. (Our definition of *laminar flows* will be refined later so that the streamlines need not have constant speed along them.)

Now it would be desirable to verify this theoretical result experimentally. This could be done by plotting Δp as a function of velocity, V_{avg}, for the data from an apparatus of the type sketched in Fig. 8.3.2. Instead of plotting $p_1 - p_2$ directly, this dependent variable will be recast to provide a more useful and general result. Equation (8.3.9) can be solved for $p_1 - p_2$:

$$p_1 - p_2 = 32 V_{\text{avg}} \mu \frac{L}{D^2}$$

and "normalized" by dividing both sides of the equation by an expression that is similar to the pressure rise that occurs at a stagnation point. This was given, in effect,

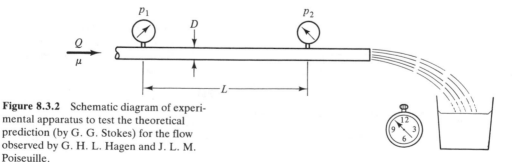

Figure 8.3.2 Schematic diagram of experimental apparatus to test the theoretical prediction (by G. G. Stokes) for the flow observed by G. H. L. Hagen and J. L. M. Poiseuille.

in Example 5.2.4 as $\rho U^2/2$. (One refers to this as the *dynamic pressure*.) In the present case, we shall use $\rho V_{avg}^2/2$ as the value of the normalizing pressure. The result is

$$\frac{p_1 - p_2}{\frac{1}{2}(\gamma/g)V_{avg}^2} = \frac{64}{V_{avg}}\frac{\mu}{\rho}\frac{L}{D^2}$$

or, with a minor rearrangement,

$$\frac{(p_1 - p_2)/\gamma}{(V_{avg}^2/2g)(L/D)} = \frac{64}{(V_{avg}D/\nu)} \tag{8.3.10}$$

Because of Eq. (8.1.1), this can be written as

$$\frac{Losses}{(V_{avg}^2/2g)(L/D)} = \frac{64}{Re}$$

in which Re ($= V_{avg}D\rho/\mu$) is the *Reynolds number* based on the average velocity. (The Reynolds number will be used henceforth to categorize many flows. In Chap. 9, this dimensionless number will be shown to be the ratio of the inertia terms and viscous forces in a flow; however, the very presence of ρ in the numerator and μ in the denominator of the Reynolds number suggests such a conclusion.) The ratio

$$\frac{Losses}{(V_{avg}^2/2g)(L/D)}$$

is defined as f, the *friction factor*. Thus, the experimental procedure would be as follows:

1. Measure Q and find V_{avg}.
2. Measure $p_1 - p_2$, and with the given values of γ, g, L, and D, as well as the determined value of V_{avg}, calculate and locate the ordinate f on a graph.
3. Calculate $V_{avg}D/\nu$, and locate this abscissa value to plot the data point.

Some simulated points have been plotted in Fig. 8.3.4 along with the theoretical prediction, Eq. (8.3.10).

Another experiment to test the merits of the theory that led to Eq. (8.3.10) can be developed. The wall shear stress on the conduit in Fig. 8.3.3 can be measured by the slight movement of the segment of the conduit that is carefully aligned with the system's axis. There is a very small gap at each end of this piece to allow a small displacement due to the force exerted by the wall shear stress. This force is resisted by

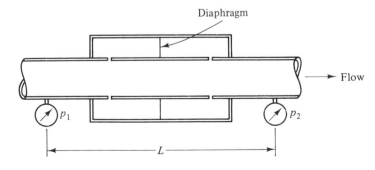

Figure 8.3.3 Schematic drawing of an experimental apparatus to measure the wall shear stress on a circular conduit.

Some Theoretical Predictions   Chap. 8

the elastic deformation of a thin diaphragm or septum, on which are mounted force transducers to sense the deformation. A waterproof housing completes the apparatus. The force on the section of pipe would be inferred from the deflection of the septum and a previous calibration. The shear stress at the wall follows from the value of this force. This shear stress, perhaps normalized, can be plotted against V_{avg} or Q and compared with theory. According to Eq. (8.3.5),

$$\tau = \left(\frac{-\partial p}{\partial z} \right) \frac{r}{2}$$

so that

$$\tau_{wall} = \tau_w = \left(-\frac{\partial p}{\partial z} \right) \frac{r_o}{2} = \left(w_0 \frac{4\mu}{r_o^2} \right) \frac{r_o}{2} \tag{8.3.11}$$

provided one uses the value of w_0 for a laminar flow, from Eq. (8.3.8), instead of the pressure gradient. Again, we shall "normalize" τ_w with a value that has the form of a dynamic pressure, $\frac{1}{2} \rho V_{avg}^2$, to obtain

$$\frac{\tau_w}{\frac{1}{2} \rho V_{avg}^2} = 2V_{avg} \left(\frac{2\mu}{r_o} \right) \left(\frac{1}{\frac{1}{2} \rho V_{avg}^2} \right)$$

which reduces to

$$\frac{\tau_w}{\frac{1}{2} \rho V_{avg}^2} = \frac{16}{V_{avg} D \rho / \mu} \tag{8.3.12a}$$

or

$$\frac{4\tau_w}{\frac{1}{2} \rho V_{avg}^2} = \frac{64}{V_{avg} D / \nu} = f \tag{8.3.12b}$$

The middle term of Eq. (8.3.12b) is the same as that in Eq. (8.3.10). One notes that this term must be nondimensional, since the left-hand side of the equation is a ratio of stresses. The term VD/ν, the Reynolds number, as we have called it, can be interpreted with the aid of Eq. (8.3.12b). It is proportional to the ratio of inertia terms $\rho V^2/2$, and viscous terms τ_w. This is a useful interpretation. Because of it, one says that the viscous effects are large in comparison with inertia effects for low–Reynolds number flows. The opposite would be true for a high-Reynolds number flow. In this chapter, viscosity will usually play an important part in all of the flows. In view of this, we shall be, in effect, treating low–Reynolds number situations.

Figure 8.3.4 contains a set of simulated data that relate $4\tau_w/(\rho V_{avg}^2/2)$, the friction factor f, to the Reynolds number. We wish to determine how well the results of Newton's law of viscosity agree with observations. The hypothetical data are shown for a variety of fluids and conduit materials. What could we conclude from such data? The following deductions are appropriate.

1. Below Re = 2000, Eq. (8.3.10) and (8.3.12b) fit the observations for a broad range of conduits, each having a different surface (Teflon® is not "wetted" by water; cast iron is relatively "rough"; and water "wets" a smooth glass surface).

a. The no-slip boundary condition used in the development of Eq. (8.3.9) is supported by the results.

b. Although the linear relationship $\tau = -\mu \, dw/dr$ might have been introduced for small values of dw/dr, we see that using this relationship can give us realistic predictions even though dw/dr is not necessarily small. (Had the relationship

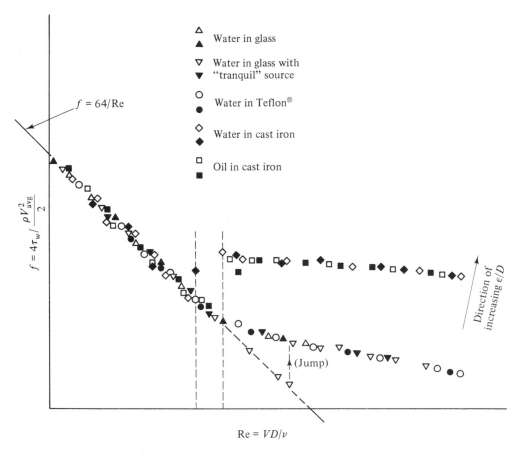

Figure 8.3.4 Plot (on log-log scale) of hypothetical experimental data for the flow in round pipes of different materials using different liquids. Note that ϵ = Size of root-mean-square roughness element. ($\triangle$, $\bigcirc$, $\square$: Increasing Re; $\blacktriangle$, $\bullet$, $\blacksquare$: Decreasing Re.)

between stress and rate of strain been very nonlinear, the agreement for large rates of strain would not have occurred.)

c. In the development of Eq. (8.3.9), the use of $\tau = -\mu(dw/dr)$ is an important step. The fact that the axial velocity w can be obtained through this equation by integration means that it is a "smoothed" result. The average velocity V_{avg} is another "smoothed" or "filtered" result. This latter quantity is used in determining the values of f that appear to have satisfactory agreement between experiment and theory. This indicates that the assumption $\tau = -\mu\, dw/dr$ is not wrong. Is it correct? That question is difficult to answer; nevertheless, the relationship is useful in this and many other flows. The utility of a model is often a significant criterion for employing it. It should again be recalled that the information presented in Fig. 8.3.4 contains V_{avg}. Such integrated quantities can have acceptable values even if the local values are found to be unsatisfactory.

d. Other important assumptions about the nature of flow such as $u_r = 0 = u_\theta$ may also be valid, even though the experiment did not determine their values.

2. When Re is greater than 2000, the correlation between f and Re is more complicated, and Eq. (8.3.9) does not give a reliable estimate of what is observed.

a. The assumptions concerning boundary conditions and the nature of the relationship between the shearing stresses and the velocity distribution, as well as the stipulation that certain of the velocity components are always zero, are all open to question for Re > 2000.

b. There is a "transition" from the type of flow which is described through Eq. (8.3.9) to some other form. The transitions do not always occur at the same value of Re. While it is possible in a carefully controlled experiment to find Eq. (8.3.9) to be applicable as one slowly increases Re beyond 2000, the results of many experiments indicate that f does not usually conform to 64/Re for Re > 2000. For a carefully controlled experiment in which all possible vibrations are eliminated, one will find that suddenly the data points "jump" from the "laminar curve" to an upper curve if a disturbance to the flow occurs. Further increases of Reynolds number in one of these careful experiments would show that the data remain on this upper curve. If the Reynolds number were then slowly reduced, the data points would remain on the upper curve and would not "jump back" to the lower, "laminar" curve as the Reynolds number was reduced toward the value of 2000. Below Re = 2000, the data again correspond to the curve predicted by Eq. (8.3.9) or Eq. (8.3.10). Thus, there seems to be a hysteresis when one goes beyond the value of Re = 2000, and the laminar flow can be considered metastable. This designation can be applied because, if a disturbance with the "correct" frequency and (small) amplitude is present, a laminar flow will undergo the aforementioned transition. The magnitude of the disturbance and its frequency will usually determine the Reynolds number at which the transition occurs.

Once a disturbance of sufficient strength occurs in the fluid, the flow pattern suddenly changes to another state. In this state the velocity components are much different from before—u_r is not identically zero, and u_θ is also not constantly zero. Consequently, the conditions within the fluid that existed at the outset of the experiment are no longer present to maintain a laminar flow beyond Re = 2000. Thus, the flow remains along the upper curve until the Reynolds number is reduced below the value Re = 2000. It appears that a laminar pipe flow, with $u_r = 0$, cannot be transformed into a flow with $u_r \neq 0$ if Re < 2000. In this range of Re a laminar flow is unconditionally stable. (Currently, there is reason to believe that with sufficient care and nearly perfect conditions one can maintain laminar flow at whatever Reynolds number is desired.)

c. The *surface roughness,* denoted by ϵ, and not the chemistry of the surface material influences f at Re > 2000. This roughness is measured by the statistical average of the protrusions on the wall. As ϵ is increased relative to the conduit's diameter, ϵ/D becomes larger and we find that f increases.

d. The wall stress τ_w is usually greater than that given by Eq. (8.3.12b) when the Reynolds number is high.

e. At high values of Re, f is almost independent of Re. This implies that τ_w is almost proportional to the square of V_{avg}. (Note also that Δp is then nearly proportional to V_{avg}^2, whereas for Re < 2000, Δp is proportional to V_{avg}.)

These observations suggest that Eq. (8.3.9), and consequently some of the assumptions leading to this result, are not generally valid for Re > 2000. This indicates that a detailed investigation of the flow near Re = 2000 might prove useful. Osborne Reynolds performed such experiments by letting water flow into a pipe from a reservoir that had no noticeable disturbances. The inlet to the pipe had a gradual reduction in size that could be termed "trumpet-shaped." He inserted a small tube along the axis of the pipe and saw that a streak of dye changed from a thin filament along the axis of the tube to a mixed cloud as the parameter VD/ν was between 1900 and 2000. This is the reason why this parameter has been designated the Reynolds number in his honor. The dye streak shows us that both u_r and u_θ exist when the fine filament breaks up into a sinusoidal curve and finally into a randomly distributed cloud of dye. This means that the velocity components in this case are functions of space and, in a random way, of time. Reynolds recognized this fact and applied his skills to the equations of motion to account for these random fluctuations. We shall discuss his work when turbulent flows are presented in Chap. 13.

However, before leaving Fig. 8.3.4—commonly referred to as the *Moody diagram* in the United States—some comments about its predictions for high–Reynolds number flows would be useful. (L. F. Moody introduced into this country's engineering literature the results of C. F. Colebrook's work in Great Britain on commercial pipe.) By using some additional assumptions about turbulent flows that will be mentioned in Chapter 13, one can deduce a formula for the turbulent flow in pipes that is similar to Colebrook's:

$$\frac{1}{\sqrt{f}} = -0.86 \ln \left(\frac{\epsilon/D}{3.7} + \frac{2.51}{\text{Re} \sqrt{f}} \right) \qquad (8.3.13)$$

A minor rearrangement of this result yields

$$\frac{\epsilon}{D} = 3.7 \exp \left(-\frac{1}{0.86\sqrt{f}} \right) - \frac{9.29}{\text{Re} \sqrt{f}} \qquad (8.3.14)$$

This equation states that ϵ/D is some function of f and Re, or $\epsilon/D = \phi_1(f, \text{Re})$, in which ϕ_1 is the function given in Eq. (8.3.14). In general, one can also infer that, for a given value of ϵ/D and Re, f is uniquely known from Eq. (8.3.14) or Fig. 8.3.4. This means that Eq. (8.3.14) is an implicit relationship for f. One can think of f, determined by ϵ/D and Re, to be specified by a relationship like $f = \phi_2(\epsilon/D, \text{Re})$. As we shall see shortly, the idea of writing a nondimensional variable f as a function of the other nondimensional variables of the problem, ϵ/D and Re, in this case, is an extremely useful approach to solving engineering problems.

The velocity distribution in a horizontal annulus with inner radius r_i and outer radius r_o can be found by starting with Eq. (8.3.4) and using the relationship for shear stress and velocity distribution, $\tau = -\mu \, dw/dr$. The result is

$$w = -\frac{1}{4\mu} \frac{\partial p}{\partial z} \left[(r_o^2 - r^2) - \frac{r_o^2 - r_i^2}{\ln (r_o/r_i)} \ln \left(\frac{r_o}{r} \right) \right] + \frac{W}{\ln (r_o/r_i)} \ln \left(\frac{r_o}{r} \right) \qquad (8.3.15)$$

In this case, the no-slip condition is applied at $r = r_i$ and r_o to evaluate the constants, C_1 and C_2, that occur. The fact that the inner cylinder could be moving in the $+z$ direction with speed W was also introduced. The volume rate of flow associated with the distribution given by this equation is

$$Q = \int w\,(2\pi r)\,dr = -\frac{\pi}{8\mu}\frac{\partial p}{\partial z}\left[r_o^4 - r_i^4 - \frac{(r_o^2 - r_i^2)^2}{\ln{(r_o/r_i)}}\right] + W\pi r_i^2\,\frac{r_o^2/r_i^2 - 1}{2\ln{(r_o/r_i)}} \qquad (8.3.16)$$

For $(r_o - r_i)/r_i \ll 1$, this result reduces to that of Sec. 8.2 where plane flows were analyzed.

If the tube is not horizontal, then the term $\partial p/\partial z$ in the previous equations would be replaced by $\partial(p + \gamma h)/\partial z$. This is the same procedure that was used in Sec. 8.2. (Again, $\partial h/\partial z = -\sin\theta$, in which θ is the inclination of the tube's axis to the horizontal.)

8.4 SWIRLING FLOW

The flow that will be analyzed in this section is *assumed* to be moving with circular streamlines in the manner shown in Fig. 8.4.1. Our goal is to learn how V_θ varies with the coordinates. For convenience we have prescribed an outer boundary of radius r_2 that is rotating at an angular velocity of Ω_2. V_θ is assumed to be the only nonzero velocity component. Because the cylinders are considered to be infinitely long, nothing, including the shear stress, will vary with z. All quantities are assumed to be independent of θ and vary only with the radius. Because there is no variation in the θ direction, the problem is described by the word *axisymmetric*. The pertinent shear is designated by τ; it acts upon a surface area of $2\pi r\,dz$. This shear stress and area lead to a moment about the axis O-O that can be written as

Figure 8.4.1 (a) Schematic drawing for the analysis of the flow between two concentric circular cylinders. V_θ = tangential velocity component. (b) Annulus of length dz. Shaded area is fluid element.

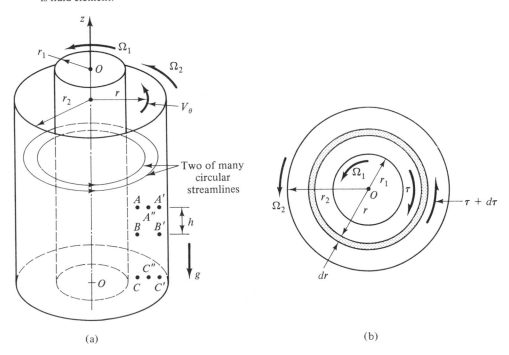

(a)

(b)

$$dM_{O\text{-}O} = (\text{force})(\text{lever arm})$$
$$= -[\tau(2\pi r\,dz)](r) \qquad (8.4.1)$$

because this moment is negative according to the usual sign convention. On the cylindrical surface that is at $r + dr$, the moment is

$$dM_{O\text{-}O} = +(\tau + d\tau)(2\pi\,dz)(r + dr)^2 \qquad (8.4.2)$$

The material within a control volume, consisting of a cylindrical shell between r and $(r + dr)$, is assumed to have a constant angular momentum because no mass crosses its control surface and a steady state is assumed. Hence, the sum of the two moments just determined must be zero according to the principles of angular momentum given in Sec. 7.5. This means that

$$2\pi\left[-\tau r^2 + (\tau + d\tau)(r + dr)^2\right]dz = 0$$

or $\qquad -\tau r^2 + (\tau r^2 + 2\tau r\,dr + \tau\,dr^2 + r^2\,d\tau + 2r\,dr\,d\tau + d\tau\,dr^2) = 0$

so that $\qquad 2\tau r\,dr + r^2\,d\tau + [\tau\,dr^2 + 2r\,dr\,d\tau + d\tau\,dr^2] = 0$

The terms in the square brackets are nonlinear in dr and $d\tau$. As $dr \to 0$, and consequently $d\tau \to 0$, the terms in the square brackets will go to zero faster than those terms that are linear in the differentials and are outside the square brackets. Therefore, in the limit as $dr \to 0$,

$$2\tau r\,dr + r^2\,d\tau = 0$$

This expression can be written as

$$d(\tau r^2) = 0 \qquad (8.4.3)$$

which gives $\tau r^2 = \text{constant} = C_1'$.

In order to determine V_θ, it is necessary to know how it, or its derivative, influences τ. Newton's law of viscosity comes to mind, and it would be natural to write something similar to what was presented in Sec. 8.2. However, one must be cautious about directly adopting information from a Cartesian coordinate system to a polar coordinate system. [We experienced the attendant difficulties in writing Eq. (8.3.6).]

The need for using a valid expression for the shear stress in a variety of coordinate systems points out the desirability of having a general expression for τ that can be applied in the coordinate system at hand. We shall transform the correct expression* for τ in a Cartesian system into cylindrical polar coordinates in Chap. 16.

Here we assume the final result; it will be made plausible later after sufficient preliminary steps have been taken. We write for the shear stress that acts in the θ direction

$$\tau = \mu r\frac{d}{dr}\left(\frac{V_\theta}{r}\right) \qquad (8.4.4)$$

instead of the expression in Sec. 8.2 that applies for a laminar flow in Cartesian coordinates. [Note that Eq. (8.4.4) is dimensionally correct. This need not attest to the validity of the expression, but it is a necessary condition.]

---

* $\tau = (2\mu)(\text{rate of angular deformation})$, or $2\mu[\frac{1}{2}(\partial u/\partial y + \partial v/\partial x)]$ for a two-dimensional flow. This will be discussed further when the Navier-Stokes equations are developed in Chap. 16.

If this expression is used in conjunction with Eq. (8.4.3), one obtains

$$\left[\mu r \frac{d}{dr}\left(\frac{V_\theta}{r}\right)\right]r^2 = C_1'$$

or

$$\frac{d}{dr}\left(\frac{V_\theta}{r}\right) = \frac{C_1'}{\mu}\frac{1}{r^3}$$

This differential equation can be integrated directly to give

$$\frac{V_\theta}{r} = -\frac{C_1'}{2\mu r^2} + C_2'$$

or

$$V_\theta = \frac{C_1'}{r} + C_2' r \tag{8.4.5}$$

The two terms in Eq. (8.4.5) represent an irrotational circular flow (a "bathtub vortex") and a rotational circular flow (rigid-body rotation). The values of C_1' and C_2' are determined from the boundary conditions. If C_2' is zero, the flow will be irrotational.

If we choose the flow regime to be between an inner cylinder of radius r_1 that is rotating at Ω_1 and an outer cylinder of radius r_2 that is at rest, the no-slip condition on the inner and outer cylinders requires that

$$V_\theta = \Omega_1 r_1 = \frac{C_1'}{r_1} + C_2' r_1$$

and

$$0 = \frac{C_1'}{r_2} + C_2' r_2$$

From these two equations, one finds

$$C_1' = \Omega_1 \frac{r_1^2 r_2^2}{r_2^2 - r_1^2} \quad \text{and} \quad C_2' = -\frac{\Omega_1 r_1^2}{r_2^2 - r_1^2}$$

so that Eq. (8.4.5) can be rewritten in terms of these constants as

$$V_\theta = \frac{\Omega_1 r_1^2}{r_2^2 - r_1^2}\left(\frac{r_2^2}{r} - r\right) \tag{8.4.6}$$

or

$$\frac{V_\theta}{r} = \left(\frac{\Omega_1 r_1^2}{r_2^2 - r_1^2}\right)\left(\frac{r_2^2}{r^2} - 1\right)$$

The second expression is more convenient when finding the shear stress on the inner cylinder through Eq. (8.4.4). This expression of Newton's law of viscosity states that for this circular geometry,

$$\tau = \mu r \frac{d}{dr}\left[\left(\frac{\Omega_1 r_1^2}{r_2^2 - r_1^2}\right)\left(\frac{r_2^2}{r^2} - 1\right)\right] \tag{8.4.7a}$$

This expression can be evaluated on the inner cylinder to give

$$(\tau)_{r_1} = \mu(-2)\left(\frac{r_2}{r_1}\right)^2\left(\frac{\Omega_1 r_1^2}{r_2^2 - r_1^2}\right) \tag{8.4.7b}$$

If this shear stress is multiplied by the cylindrical surface area of the inner cylinder that is wetted by the fluid, one has the effective shearing "force" at $r = r_1$ that is creating a moment about the axis O-O. (This is the z axis.) This area is $2\pi r_1 L$, in which L is the length of the cylinder that is immersed in the fluid. The moment about O-O,

$$M_{O\text{-}O} = (\tau)_{r_1}(2\pi r_1 L)r_1$$

is

$$M_{O\text{-}O} = -4\pi\mu r_2^2\left(\frac{r_1^2\Omega_1}{r_2^2 - r_1^2}\right)L \tag{8.4.8}$$

This moment can be compared, perhaps artificially, with another torque magnitude that could be associated with this flow. Such a magnitude can be written as $\rho\Omega_1^2\pi r_1^4 L$. (The reason that such a reference torque is a useful concept, even though it has been introduced in an arbitrary way here, is discussed at great length in Chap. 9.) The ratio of the actual moment, Eq. (8.4.8), to the negative of this reference (pseudo) torque that is designated $T_{O\text{-}O}$, is called a *moment coefficient*. This definition gives

$$\frac{M_{O\text{-}O}}{T_{O\text{-}O}} = C_M = \frac{(-4\pi\mu r_1^2 r_2^2\Omega_1 L)/(r_2^2 - r_1^2)}{-\pi\rho\Omega_1^2 r_1^4 L}$$

so that

$$C_M = 4\left\{\left(\frac{\nu}{\Omega_1 r_1^2}\right)\left[\frac{1}{1 - (r_1/r_2)^2}\right]\right\} \tag{8.4.9}$$

The term $\Omega_1 r_1^2/\nu$ can be thought of as a Reynolds number for the flow. The complete coefficient in the braces has a reciprocal called the *Taylor number,* in honor of Sir Geoffrey Taylor, who worked on rotating flows as part of an enviable scientific career that spanned about 65 years until his death in 1975. For a fixed cylinder geometry, Eq. (8.4.9) indicates that C_M varies inversely with Ω_1. A similar calculation could be executed for the moment coefficient when the inner cylinder is at rest and the outer one is rotating at Ω_2. Let us call this coefficient C_M'. Its exact form need not be derived here, but one might expect that C_M' will also decrease with increasing Ω_2 in this instance. When experiments are conducted, it is found that C_M' is realized over an extensive range of Ω_2. However, C_M, in the form of Eq. (8.4.9), exists over only a small range of values of Ω_1. At higher values of Ω_1, the flow loses its simple, circular streamlines. This will be mentioned again in Chap. 12.

Experimental data for C_M are shown in Fig. 8.4.2. The fact that, according to Eq. (8.4.9), we expect it to vary with $\Omega_1 r_1^2/\nu$ for a given value of r_1/r_2 permits the experimenter to correlate a large range of values of Ω and ν by a *single* curve.

The derivation of Eq. (8.4.6) employed the principle for the conservation of angular momentum. We did not utilize the equations of motion in the radial direction. In Sec. 2.8, the consequences of centripetal acceleration were explored. If we were to write correct expressions for the normal stresses for a laminar flow in cylindrical polar coordinates, the fluid motions would not change the form of the radial equation of motion if only circular streamlines were considered. One would have, as in Eq. (2.8.4), that

$$\frac{\partial p}{\partial r} = \rho\frac{V_\theta^2}{r} \tag{8.4.10}$$

Because we already know V_θ by Eq. (8.4.6), we could determine the manner in which the pressure varies in the fluid. This equation would give the radial dependence of the

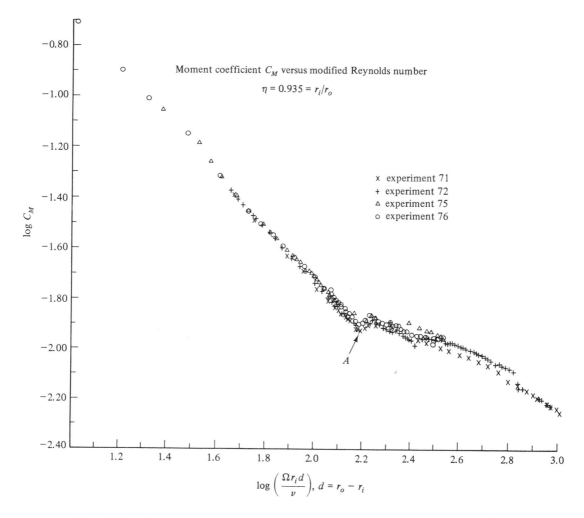

Figure 8.4.2 Plot of data from several experiments for a moment coefficient, $C_M = \dfrac{\text{moment}}{\rho \Omega_i^2 r_1^2 r_2^2 L \pi}$, versus a Reynolds number based on d associated with the flow between concentric cylinders. A indicates onset of secondary flow in annulus. *(From "Proceedings of the 12th International Congress of App. Mechs., 1968," p. 172, Springer-Verlag, by permission)*

pressure, and the hydrostatic law would give the z dependence, provided that the container was not accelerating vertically.

The development of Eq. (8.4.6) also *assumed* that the container was infinitely long. This assumption precluded the existence of end effects and the necessity of applying appropriate boundary conditions. Next, we shall examine some of the consequences of having a fixed bottom in the annulus, such as the bottom of the containing vessel.

In the following discussion, we *assume* initially that V_θ is still the only nonzero velocity component. Because V_θ must decrease as the stationary bottom is approached as a result of applying the no-slip condition there, an anomalous condition will arise. The reason why this situation occurs will be apparent if we observe the conditions at the pairs of points A and A', B and B', and C and C' in Fig. 8.4.1. Each set of points is progressively nearer the bottom. The first set, at least, is sufficiently far that the velocity distribution is effectively given by Eq. (8.4.6) at this location. Equation (8.4.10) tells us that the pressure is greater at A' than at A. Similarly, $p_{B'} > p_B$. From Eqs. (8.4.6) and (8.4.10), the pressure difference $p_{A'} - p_A$ can be calculated. Because of the hydrostatic variation of pressure, $p_B = p_A + h\gamma$ and $p_{B'} = p_{A'} + h\gamma$, so that $p_{B'} - p_B = p_{A'} - p_A$.

A moment's consideration would lead to the conclusion that $p_{C'} - p_C = p_{A'} - p_A$, too, even though the points C' and C are sufficiently close to the bottom that the tangential speed V_θ would be decreased by the presence of the bottom. Hence, at a point C'' the value of V_θ^2/r is less than at point A''. This means that, according to the previous line of argument, the same radial pressure gradient exists at C'' as at A'', but the centripetal acceleration at C'' is less than that at A''. At A'', this acceleration would have to be in balance with the pressure gradient to produce the assumed circular streamlines. The inference is that at C'', the radial pressure gradient and the contribution to the radial acceleration, due to the circular motion, are not in balance. This conclusion is based on our initial assumption about V_θ, and it appears to be in conflict with Newtonian mechanics. A resolution of this paradox might occur if additional nonzero velocity components are considered. This is what we shall now explore.

If the pressure gradient and the radial acceleration for purely circular motion are not equal to one another, then, perhaps, the flow is not purely circular at point C''. The radial pressure gradient that is imposed by the swirling flow, say, at A'', has impressed this pressure gradient, with $p_{C'} > p_C$, on the region near the bottom. In order to have $\mathbf{F} = m\mathbf{a}$ in the radial direction, there must be, at C'', a radial flow component such that there is an additional radial acceleration. This would compensate for the deficient magnitude of V_θ^2/r at C''. Such radially inward motions can be detected when sediment, such as tea leaves or fine sand, falls to the bottom of a stirred vessel. In time, these particles can be observed collecting in the center of the container. There must also be an axial upflow, to satisfy continuity, but the associated shear, similar to that which carried the particles to the center, is insufficient to carry them upward.

Such a small radial inflow on the bottom, with some vertical flow along the walls of the container and along its axis, is an induced flow and is called a *secondary flow*. (It is a different secondary flow from the one responsible for the change in slope in Fig. 8.4.2.) The thin layer of fluid that exhibits the effects of the boundary's motion, or lack thereof, will be encountered again in Secs. 8.5 and 8.6.

8.5 AN UNSTEADY FLOW

The unsteady flow that we shall now examine will be of interest for two reasons. The mere fact that the flow is time-dependent will be a novel feature. But beyond that, the solution will imply that there is only a finite region near the oscillating boundary that is appreciably affected by this motion.

We seek a solution to the problem that is illustrated in Fig. 8.5.1. The plate has been oscillating in simple harmonic motion for a "long" time. (Accordingly, while we seek an unsteady solution, we shall limit ourselves to one in which any transients have died out.)

The plate speed is $U_0 = U(\cos \omega t)$, in which U is the maximum speed. The plate is of infinite length and is located along the x axis. The fluid occupies the entire upper

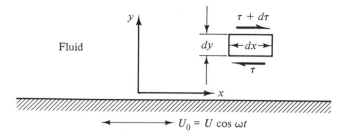

Figure 8.5.1 Infinitesimal control volume used to analyze the flow due to an oscillating plate in a semi-infinite fluid. Assume a unit distance perpendicular to the page.

half plane (i.e., $0 \le r \le \infty$ and $0 \le \theta \le \pi$). Because the fluid is *assumed* to adhere to the plate, it, too, oscillates back and forth. We instinctively believe that the amplitude of this oscillation in the fluid must decay as the coordinate y becomes very large. The solution to our problem will reveal the nature of this decay.

We begin the analysis by examining the small fluid region of size $(dx)(dy)(1)$. For convenience, we have considered the stress on the lower surface to be acting to the left with a magnitude of $\tau\, dx\,(1)$, and we have assumed that on the upper surface there is a shear force acting in the positive x direction of $(\tau + d\tau)(dx)(1)$. Furthermore, the fluid is *assumed* to be without a pressure gradient in the x direction, and all of the motions are assumed to be parallel to the x axis (i.e., $v = 0$). This means that the sum of the x-directed forces, $[-\tau\, dx + (\tau + d\tau)\, dx]$, is equal to ρa_x, in which

$$a_x = \frac{Du}{Dt} = \frac{\partial u}{\partial t} + u\frac{\partial u}{\partial x} + v\frac{\partial u}{\partial y} \tag{8.5.1}$$

for a two-dimensional flow, according to Eq. (4.6.6b). The last term in this acceleration term is zero, since v is assumed to be zero everywhere.

The equation for mass conservation, Eq. (4.8.6), states that for an incompressible two-dimensional flow,

$$\frac{\partial u}{\partial x} + \frac{\partial v}{\partial y} = 0$$

With v *assumed* to be identically zero everywhere, we have no spatial changes in v. Hence, $\partial v/\partial x = 0 = \partial v/\partial y$. This fact makes the previous equation reduce to $\partial u/\partial x = 0$. From this, we can conclude that u cannot change with x. This is exactly what we wish! (Moreover, it is what we expect for a problem for which there are no special geometric features to distinguish one x position from another.)

With the fact that $v = 0 = \partial u/\partial x$, Eq. (8.5.1) reduces to $a_x = \partial u/\partial t$. This says that the acceleration of the material volume, a_x, is equal to the time rate of change of u at the appropriate location. We can write Newton's law of motion for the x direction as

$$\rho\left(\frac{\partial u}{\partial t}\right)(dx\, dy) = d\tau\,(dx) \tag{8.5.2}$$

Now τ varies only with y; everything is assumed to be independent of x because of the infinite length of the plate. Hence, $d\tau = (d\tau/dy)\, dy$. This prescription for $d\tau$ permits Eq. (8.5.2) to be written as

$$\rho\frac{\partial u}{\partial t} = \frac{d\tau}{dy} \tag{8.5.3}$$

The *assumption* of $v = 0$ implies an orderly flow, and we now explicitly state that we seek a laminar solution so that $\tau = \mu\, \partial u/\partial y$.* When this constitutive equation is utilized in conjunction with Eq. (8.5.3), one has the result

$$\rho \frac{\partial u}{\partial t} = \mu \frac{\partial^2 u}{\partial y^2} \tag{8.5.4}$$

This equation can be solved by a variety of methods. Here we shall employ the inelegant, but effective, method of guessing a plausible solution and inserting it into Eq. (8.5.4) to see whether our intuition is correct.

We expect the solution to decay with y. Furthermore, since the boundary is forcing the fluid through a simple harmonic motion, we would expect the fluid motion to be of the same form and frequency—after starting transients decay—but with a differing amplitude and phase due to the distance from the plate. For this reason, we try as a possible solution

$$u = [f(y)][\cos(\omega t - \beta y)]$$

The amplitude variation has been accounted for by the unknown function $f(y)$, and a phase change that is linear in y has been selected. We know that at the plate, where $y = 0$, the fluid speed u must be $U_0 = U \cos \omega t$. This indicates that

$$f(0) = U$$

If our assumed solution is inserted into Eq. (8.5.4), we find that

$$f(y) = U \exp(-y \sqrt{\omega/2\nu}) \quad \text{and} \quad \beta = \sqrt{\frac{\omega}{2\nu}}$$

so that our solution is

$$u = U \exp(-y \sqrt{\omega/2\nu}) \left[\cos\left(\omega t - y \sqrt{\frac{\omega}{2\nu}} \right) \right] \tag{8.5.5}$$

because the differential equation and the boundary conditions have been satisfied.

We have now solved this time-dependent problem. In doing so, one may have noticed that Eq. (8.5.4) has a second derivative with respect to y. Thus we should expect that to determine our solution we would need more than the one boundary condition, $f(0) = U$, that was employed. Actually, $f(y)$ contained two exponentials, but one had a positive exponent so that it would tend to grow to infinity at large distances y. The requirement that the solution must remain finite for all values of y dictates that the constant that multiplies this particular undesirable exponential must be set to zero. This consideration leads to the form of $f(y)$ and Eq. (8.5.5).

It is useful to examine this equation. There will be some values of y that will make the argument of the cosine such that there is a $180°$ phase shift between the local velocity and the velocity of the oscillating plate. At any value of y, the amplitude of the velocity will be diminished due to the coefficient $\exp(-y\sqrt{\omega/2\nu})$. It is interesting to note that $e^{-5} \approx 0.01$, so that wherever $y\sqrt{\omega/2\nu} = 5$, the magnitude of u will be about 1 percent of the motion of the plate. We shall designate δ as the value of y for which the fluid motion is 1 percent of the boundary motion.† Thus, for this problem,

---

* While u does not vary with x, the partial derivative is used because u does change with time and y.
† The 1 percent value is arbitrary.

$$\delta \approx \frac{5}{\sqrt{\omega/2\nu}} \qquad (8.5.6)$$

Example 8.5.1

What is δ for $\omega = 60(2\pi)$ and $\nu = 10^{-4}$ m²/s, a value appropriate for water? Is δ larger or smaller for air at $\omega = 120\pi$?

Solution

$$\delta_{H_2O} = \frac{5}{\sqrt{120\pi/2(10^{-4})}}$$

$$= \frac{5}{10^2(13.7)} = 0.0036 \text{ m} = 3.6 \text{ mm}$$

Since $\nu_{air} > \nu_{H_2O}$,

$$\delta_{air} > \delta_{H_2O}$$

The result of this effort shows that, at high frequency and low kinematic viscosity, the distance δ becomes very small. For these conditions there is only a small layer of fluid within which there is appreciable motion due to the oscillating plate. This layer, near the boundary, where the motion of the fluid is affected by the viscous effects from the motion—or lack of it—is called the *boundary layer*.

In this instance, the existence of a boundary layer was inferred from the solution to the problem that was posed. In other situations, we shall assume the presence of such a boundary layer in order to formulate the solution. We do this in Sec. 8.6.

Had one used a control volume for the analysis that led to Eq. (8.5.3), one would have utilized Eq. (7.8.1). Because the flow is assumed to have only an x component of velocity, which does not vary with x, the surface integral in that equation will be zero. The volume integral, $\partial(\int \rho u \, d\forall)/\partial t$, would be $\partial(\rho u \, dx \, dy \, dz)/\partial t$ because of the infinitesimal control volume for which u and ρ are, in effect, the average values. This expression can be written as $(\partial u/\partial t)\rho \, dx \, dy \, dz$ for constant density. The net x-directed surface force on the infinitesimal control volume is $[(\partial \tau/\partial y) \, dy] \, dx \, dz$. These force and momentum change rates can be equated, as required by the momentum equation, to produce Eq. (8.5.3).

8.6 LAMINAR BOUNDARY LAYERS

The case of the oscillating plate in a semi-infinite fluid that was treated in Sec. 8.5 has an interesting feature. The solution to the problem shows that there is a *limited* region near the plate where the fluid velocity goes from the value prescribed by the plate's motion to the value that is associated with positions far from the plate. We have called this distance the *boundary-layer thickness*. The idea that there could be only a small distance from the solid boundary where the effects of shear would be important in high–Reynolds number flows was proposed by Ludwig Prandtl in 1904. (In assuming that the viscous effects are confined to a boundary layer, we are also allowing for the great importance of inertia effects in the dynamics of the problem at distances greater than δ from the boundary.)

For the oscillating plate treated in Sec. 8.5, the distance δ is independent of the position along the infinitely long surface. A different situation occurs for the semi-infinite plate whose momentum equation was discussed in Sec. 7.4. Figure 8.6.1 summarizes what we have already learned about this problem. We can make additional progress on the understanding of this problem once we remember that an integration of the shear stress along the plate, τ_w, should also yield the shearing force on the plate, which is called the *drag*. Thus, for unit width and for a plate of length X,

$$\text{Drag} = \int_0^X [\tau_w(x)]\, dx \tag{8.6.1}$$

The rate of change of this drag with distance along the plate is $d(\text{Drag})/dx$, and differentiation of Eq. (8.6.1) gives

$$\frac{d(\text{Drag})}{dx} = \frac{d}{dx}\left\{\int_0^X [\tau_w(x)]\, dx\right\} = \tau_w(x) \tag{8.6.2}$$

This result is useful because we also know from Fig. 8.6.1 that

$$\text{Drag} = -\mathscr{F}_x = \rho U^2 \int_0^\delta \frac{u}{U}\left(1 - \frac{u}{U}\right) dy = \rho U^2 \delta\left[\int_0^1 \frac{u}{U}\left(1 - \frac{u}{U}\right) d\left(\frac{y}{\delta}\right)\right] \tag{8.6.3}$$

in which the term in square brackets is a constant once a suitable expression for u/U is substituted in the integral. (The most suitable expression is obviously the one that satisfies the governing differential equation. This may be difficult to determine; often, however, an expression for the velocity profile that is a good approximation and meets the boundary conditions can be used with surprisingly satisfactory results.) If Eq. (8.6.3) is differentiated in conformity with Eq. (8.6.2), one obtains

$$\frac{d(\text{Drag})}{dx} = \frac{d}{dx}(\rho U^2 \delta)\left[\int_0^1 \frac{u}{U}\left(1 - \frac{u}{U}\right) d\eta\right] = \frac{d\delta}{dx}\rho U^2\left[\int_0^1 \frac{u}{U}\left(1 - \frac{u}{U}\right) d\eta\right]$$

with $\eta = y/\delta$. The quantities ρ and U have been considered to be independent of x, and the term δ has been differentiated because we know that it does in fact depend on x.

Figure 8.6.1 Finite control volume for analysis of boundary-layer flow past a flat plate aligned with a uniform flow. Outflow $= \rho \int_0^\delta (U - u)\, dy$; volume efflux across $CD = \int_0^\delta u\, dy$; shaded area $= \int_0^\delta (U - u)\, dy$). Plate is of unit width.

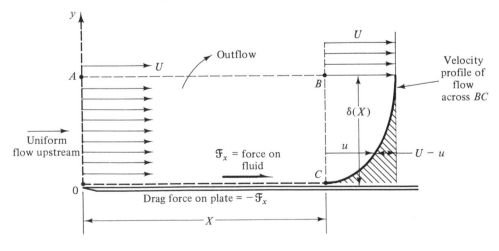

Some Theoretical Predictions   Chap. 8

Now that the integral expression for the drag has been differentiated, we can combine Eqs. (8.6.2) and (8.6.3) as

$$\tau_w = \frac{d\delta}{dx}\rho U^2\left[\int_0^1 \frac{u}{U}\left(1 - \frac{u}{U}\right)d\eta\right] = \frac{d\delta}{dx}\rho U^2 K_1 \tag{8.6.4}$$

in which the value of K_1 will depend upon the function u/U that is substituted in the integral. But we can extend this result, since for laminar flow

$$\tau_w = \mu\left(\frac{du}{dy}\right)_{y=0} = \mu\left[\left(\frac{du}{d\eta}\right)_{\eta=0}\right]\frac{d\eta}{dy} = \mu\left[\left(\frac{du}{d\eta}\right)_{\eta=0}\right]\frac{1}{\delta} \tag{8.6.5}$$

because the substitution $\eta = y/\delta$ implies $d\eta/dy = 1/\delta$. We shall see shortly an example of a reasonable mathematical function for the velocity distribution, u. This function will have η as the variable, so that we write*

$$\frac{u}{U} = f(\eta) \quad \text{and} \quad \frac{du}{d\eta} = Uf'(\eta)$$

With this expression for $du/d\eta$, we can rewrite Eq. (8.6.5) as

$$\tau_w = \mu U[f'(0)]\frac{1}{\delta} = \frac{\mu U K_2}{\delta} \tag{8.6.6}$$

Example 8.6.1

For the approximate velocity distribution

$$\frac{u}{U} = \sin\left(\frac{\pi y}{2\delta}\right) = \sin\left(\frac{\pi\eta}{2}\right)$$

find K_1 in Eq. (8.6.4) as well as $f'(\eta)$ at $\eta = 0$, which is K_2 in Eq. (8.6.6).

Solution

$$K_1 = \int_0^1 \sin\left(\frac{\pi\eta}{2}\right)\left[1 - \sin\left(\frac{\pi\eta}{2}\right)\right]d\eta$$

$$= \left(-\frac{2}{\pi}\cos\frac{\pi\eta}{2}\right)\Big|_0^1 - \frac{1}{2}\left(\eta - \frac{1}{\pi}\sin\pi\eta\right)\Big|_0^1$$

$$= \frac{2}{\pi} - \frac{1}{2}$$

For $f = \sin(\pi\eta/2)$,

$$f' = \frac{d}{d\eta}\left[\sin\left(\frac{\pi\eta}{2}\right)\right] = \frac{\pi}{2}\cos\left(\frac{\pi\eta}{2}\right)$$

$$f'(0) = \frac{\pi}{2} = K_2$$

If the expressions for τ_w in Eqs. (8.6.4) and (8.6.6) are equated, the result is

$$\frac{\mu U K_2}{\delta} = \rho U^2 K_1\frac{d\delta}{dx}$$

---

*The symbol $f(\eta)$ denotes a function of η; it has no relationship with the friction factor f that was introduced in Sec. 8.3.

so that
$$\frac{d(\delta^2/2)}{dx} = \frac{\mu}{\rho}\frac{1}{U}\frac{K_2}{K_1}$$

The right-hand side is a constant, and so

$$\frac{\delta^2}{2} = \frac{\nu}{U}\frac{K_2}{K_1}x + C_1 \qquad (8.6.7)$$

Because δ approaches zero as x approaches 0, $C_1 = 0$. With C_1 evaluated, we see that δ varies as the square root of x for laminar flow.

Example 8.6.2

Determine the boundary-layer thickness for air at 20°C and 760 mm Hg flowing over a plate 3 m long at 10 m/s. Use the values of K_1 and K_2 in Example 8.6.1.

Solution From Eq. (8.6.7),

$$\delta^2 = \frac{2K_2}{K_1}\frac{3}{10}[1.5(10^{-5})] \qquad \rightarrow \delta = 10.2 \text{ mm}$$

Verify this result by checking the values for K_1, K_2, and ν.

It is convenient now to introduce the value of δ in Eq. (8.6.7) into the expression for τ_w, Eq. (8.6.6). This leads to

$$\tau_w = \frac{\mu U K_2}{\nu x / U}\sqrt{\frac{K_1}{2K_2}} \qquad (8.6.8)$$

This last result is useful because it shows the manner in which τ_w varies with distance along the plate. (As x increases, the slope of the velocity profile at $y = 0$ decreases.) An average value for the surface stress can be obtained by dividing the total drag by the area of the plate, $(X)(1)$, in which X is the length of the plate. Using Eqs. (8.6.3) and (8.6.7), this would be

$$\frac{\text{Drag}}{X(1)} = \rho U^2 \frac{\delta}{X}K_1 = \rho U^2 \frac{K_1}{X}\sqrt{\frac{\nu X}{U}}\sqrt{\frac{2K_2}{K_1}} = \rho U^2\sqrt{\frac{\nu}{UX}}\sqrt{2K_1 K_2}$$

which can be written as

$$\text{Average shear stress} = \frac{\text{Drag}}{X(1)} = \rho U^2 \frac{1}{(\text{Re}_x)^{1/2}}\sqrt{2K_1 K_2}$$

in which Re_x is a Reynolds number based on the length of the plate up to the position X. (When $X = L$, the length of the plate, one writes Re_L.). It is subsequently useful to form the ratio

$$\frac{\text{Average shear stress}}{\text{A convenient dynamic pressure}} = \frac{\text{Drag}/(X)(1)}{\rho U^2/2}$$

just as we did when we were concerned with the friction factor, f. The form of this ratio is

$$\frac{\text{Drag}/X}{\rho U^2/2} = \frac{2\sqrt{2K_1 K_2}}{(\text{Re}_x)^{1/2}} \qquad (8.6.9)$$

The term $\text{Drag}/[\frac{1}{2}\rho U^2(\text{area})]$ for the flat plate is called the *skin friction coefficient*. Later, we shall examine bluff bodies (plates, spheres, cylinders, and so on) and note that the shape of the body introduces "form drag." The conclusion of this development is that the skin friction coefficient, a nondimensional term, depends on the remaining physical parameters of the problem—such as U, ρ, μ, and the distance along the plate. The solution of the differential equation for $f(\eta)$ is obtained numerically with an electronic computer. (Chap. 16 discusses the procedure.) The solution provides

$$C_f = \frac{1.328}{(\text{Re}_X)^{1/2}} \qquad (8.6.10)$$

and

$$\delta = 5.0\sqrt{\frac{\nu X}{U}} \qquad (8.6.11)$$

Figure 8.6.2 shows the way in which the skin friction coefficient C_f varies with Re_X. Again, we see that our laminar-flow "model," Eq. (8.6.9), is useful up to a point. Also, when there is a change to another type of flow (i.e., a flow transition), the drag is increased over the laminar value, indicating an increase in the shear stress at the plate. This occurs because the flow becomes turbulent.

The curve labeled "transition" is obtained as a composite of the expected results for laminar and turbulent boundary layers. In Chap. 13, we shall see that a turbulent boundary layer on a flat plate is thicker, but the velocity gradient at the wall is larger, than for the laminar boundary layer. This leads to higher shear stresses, greater drag, and a larger skin friction coefficient for turbulent flows.

A laminar boundary layer will become more and more sensitive to disturbances the longer it flows along a plate. At some point along the plate depending upon the Reynolds number that is based upon the distance from the leading edge, some disturbances in the flow will not be damped out by the viscous friction. Viscosity will begin to play an additional role. It will also provide the means for transferring energy from the main flow to turbulent fluctuations. As the (plate) Reynolds number is increased, conditions are enhanced so that the laminar boundary layer changes to a turbulent one ever closer to the leading edge. This means that the fraction of the plate's length experiencing turbulent shear stresses becomes greater as the Reynolds

Figure 8.6.2 Drag law for smooth plates.

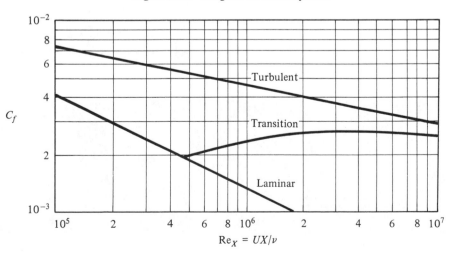

number becomes larger. Consequently, the value of C_f becomes asymptotically close to the "turbulent" curve in Fig. 8.6.2 for high (plate) Reynolds numbers.

Figure 8.6.3 shows the nature of the flow past a flat plate for various values of the plate Reynolds number. The reader should refer to Fig. 8.6.2 to appreciate the effect of the length of laminar boundary layer on the overall skin friction coefficient.

Figure 8.6.3a illustrates the laminar boundary layer on a flat plate aligned with the flow, for $Re_L \approx 2(10^4)$. Figure 8.6.3b shows the advent of boundary-layer instability when $Re_L \approx 7.5(10^4)$. Two-dimensional waves, called *Tollmien-Schlichting waves,* first appear. Subsequently, three-dimensional disturbances grow into a random, turbulent boundary layer.

Figure 8.6.3c shows a boundary-layer flow over a flat plate with an angle of attack of 1° to 2°. This small angle of attack can prevent the oncoming flow from immediately adjusting to the contour of the plate along its upper surface. The point (or line, on a

Figure 8.6.3(a)–(f) Laminar boundary layer flow over a flat plate for various values of the plate Reynolds number. *(Figures (a)–(b), (d)–(f) courtesy of H. Werlé, O.N.E.R.A., Paris, France)*

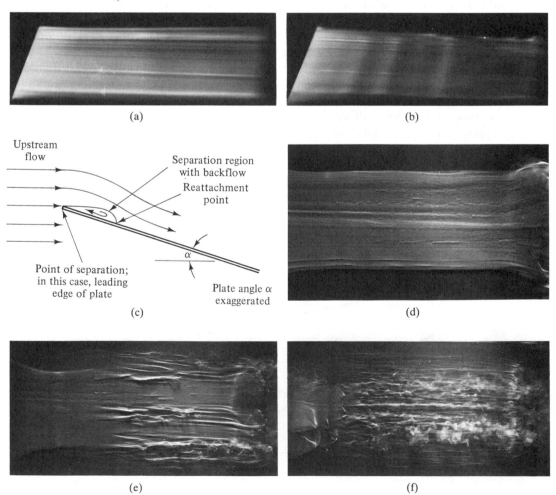

(a)

(b)

Upstream flow

Separation region with backflow

Reattachment point

Point of separation; in this case, leading edge of plate

α

Plate angle α exaggerated

(c)

(d)

(e)

(f)

Some Theoretical Predictions   Chap. 8

two-dimensional surface) at which the streamline departs from the surface of the plate is called the *separation position*. Downstream of the place of separation, the fluid moves in a direction that is opposite to that existing just upstream of the separation point. In this figure, the separation zone is small and the pressure gradients near the plate are sufficient to cause the streamline that initially separated from the body to reattach itself to the surface.

Figures 8.6.3d through f show a boundary-layer flow on a flat plate with a 1° to 2° attack angle producing a pressure gradient and a small separation bubble at the leading edge. "Separation" occurs when the upstream flow is unable to move in its original direction in close proximity with the boundary. The fluid moves in a direction that is away from the boundary. Aft of the separation line, the fluid is moving in the upstream direction along the boundary. Under the proper condition, the fluid may "reattach" to the boundary. (The increasing Reynolds number affects the state of the flow after it "reattaches.")

In Fig. 8.6.3d, separation and reattachment are laminar at $Re_L \approx 2.5(10^5)$. In Fig. 8.6.3e, where $Re_L \approx 4(10^4)$, the reattachment shows disturbances that grow (i.e., transition to turbulence); a spanwise structure in the turbulent flow is evident. At higher plate Reynolds numbers (Fig. 8.6.3f), the flow reattachment initially has a discernible structure that grows and becomes random farther downstream; thus, the skin friction coefficient is the integrated composite of these different flow regimes along the plate.

The transition of the boundary layer from a laminar to a turbulent state will be affected by the plate's roughness and the degree of turbulence in the ambient fluid that flows over the plate. Both of these factors will determine the level of the disturbances which are being imposed upon a laminar boundary layer and which tend to destabilize it. Under carefully controlled conditions in a laboratory, transition to a turbulent boundary layer can be delayed until Reynolds numbers of about $5(10^6)$ are reached. However, for the levels of turbulence that can be expected in most applications, the value of about $5(10^5)$, as shown in Fig. 8.6.2, is typically the transition Reynolds number.

It must be emphasized that the parameter called the *Reynolds number* will occur many times in the study of fluid mechanics. It will be used in conjunction with the flow in pipes, over plates, and past spheres, for example. In each case, a velocity, length, and kinematic viscosity will be used. This length can be different in each case. It is a characteristic length for the problem at hand. For the pipe, it is the inside diameter; and for a sphere, it is the outside diameter. For a plate, one uses the length from the leading edge to the station in question—often the end of the plate—to define the Reynolds number. For a Reynolds number that can be used to characterize a boundary layer, the boundary-layer thickness might be used as the length scale. In view of this it is not surprising that each flow phenomenon, with its different length scales and different physical features, will have different "critical" Reynolds numbers. For the internal flow in a pipe, transition to turbulence usually occurs around Re = 2000, but it can be delayed to Re ≈ 50,000 under special conditions. For a flat plate, the critical Re is between $5(10^5)$ and $5(10^6)$, depending upon the conditions.

The skin friction coefficient shown in Fig. 8.6.2 could be used to estimate the force needed to tow a plate (or a platelike structure) through a fluid. The submerged sides and bottom of a ship can be thought of as sections of long plates, and the curves shown in Fig. 8.6.2 might be used to obtain an estimate of the viscous drag. A ship also generates waves that decay. Thus the ship creates some potential energy in the fluid in the form of a wave train which is dissipated (through viscosity). The energy to create these waves comes from the power input to the ship, which in turn is the product of the ship's speed, U, and the wave drag force. Since the naval architect is interested in the power that is required to propel ships, there is an interest in the product (U)(wave drag + viscous drag). It is not obvious, or necessarily true, that these two types of drag are independent of one another. (The wave motion may be accompanied by a flow field which itself influences the viscous drag.)

The amount of viscous drag on plates was measured by William Froude in a towing tank built by the British Admiralty. He varied the speed, length, and surface finish of the test plates. Even with this pioneering work on viscous effects (skin friction), Froude's name is linked with wave drag, a subject that he advanced by championing the results of ship model testing. In Chap. 9, the use and significance of the nondimensional term that we now call the *Froude number* will be studied.

Example 8.6.3

If a flow enters a two-dimensional channel from a reservoir, the flow is initially almost uniform in velocity. Gradually, the parabolic profile described in Sec. 8.2 results as the fluid flows down the channel.* If Q is the volume rate of flow per unit width, the uniform inlet velocity is Q/h. The maximum velocity downstream where the laminar profile is fully developed is $\frac{3}{2}V_{avg}$, or $\frac{3}{2}Q/h$. The downstream flow has a flux of kinetic energy flowing across the channel's cross section that is greater than that at the entrance. This gives, from the energy equation, a decrease in pressure even if viscous effects are not considered. Estimate the length of the conduit over which this change in the velocity profile takes place.

Solution   The development of the profile is due to viscosity and is akin to the development of the boundary layer over a flat plate for an *external* flow (i.e., one that has some fluid regions that are unconstrained by some external geometric boundaries). The flow in the channel under discussion is an *internal* flow, because the flow is within lateral walls that affect the type of flow at all points between them. In this case, the centerline speed increases as the fluid enters the channel because of the no-slip condition on the wall and the requirement that the volume flow rate remain the same at each cross section.

Even though there are some differences between the external boundary layer flow discussed in this section and the internal flow now being considered, an initial estimate of the entrance length L_t can be obtained by determining the requirements for a laminar boundary layer to grow to a height half the channel's height k. Equation (8.6.11) would give $k/2 = 5\sqrt{\nu L_t/U}$, so that $L_t/k = 0.01\, Uk/\nu$. One might be inclined to use a value for U intermediate to the centerline value at the inlet and at the final station. A detailed computation for this two-dimensional case shows that the bulk of the flow has effectively reached its final distribution by $L_t/k = (0.04)(Uk/\nu)$, whereas the centerline velocity is nearly equal to $\frac{3}{2}V_{avg}$ at a distance about 60 percent farther down the channel. Thus, Eq. (8.6.11) would give the order of magnitude of the more complete result.

Example 8.6.4

Determine the viscous drag due to the installation of a keel that is 4 m long in the fore-aft direction and extends 3 m below the hull. The estimate should be made for a speed of 7

---

* This change was neglected in Secs. 8.2 and 8.3 because the channels or pipes were assumed to be very long, compared with their height.

knots; the boat will be moving downwind at this speed. (*Assume* that there are no three-dimensional flow effects at either the keel's end or its junction with the hull.)

Solution   The water will be *assumed* to be at 10°C, so that $v = 1.5(10^{-6})$ m²/s. Since 1 knot ($= 6076$ ft/h) corresponds to 6076/3600 ft/s (or 0.514 m/s per knot), 7 knots is equivalent to 7(0.514) m/s. Consequently, the Reynolds number for the plate is

$$\text{Re}_L = \frac{3.6(4)}{1.5(10^{-6})} = 9.6(10^6)$$

For Reynolds numbers greater than 500,000, turbulent flow exists over most of the plate's length. Thus, one would not use the laminar curve in Fig. 8.6.2. If the plate were not extremely rough, one would use a curve that accounted for the change of the laminar boundary-layer flow to a turbulent flow. The curve labeled "transition" would give a C_f of about $2.4(10^{-3})$ at $\text{Re}_x = 10^7$. From a definition of the skin friction coefficient, the shear force on the plate is

$$\text{Drag} = \frac{(3.0)(10^{-3})(1000)(3.6)^2[(4)(3)]}{2} \quad (2) \quad \text{N}$$

The last factor of 2 is due to the two sides of the keel that are submerged in the flow. What is the numerical value of the drag? Is it reasonable?

★ 8.7 LAMINAR BOUNDARY LAYERS OVER TWO-DIMENSIONAL BODIES

The flow over a flat plate that is aligned with the flow is the simplest case of the development of a boundary layer. The external flow was *assumed* to have a constant speed U, and the pressure gradient in the flow direction was zero. Neither of these requirements is generally the case, a fact that is apparent if one considers the flow in a two-dimensional channel with plane upper and lower walls that are inclined to one another. If the channel narrows in the flow direction, the fluid must speed up, and $dp/dx < 0$. The opposite is true when channel walls are set so that the cross section increases in area in the flow direction. Moreover, when a fluid flows past a body, the pressure on the surface of the body will depend upon the location. At the most upstream part of a symmetric body aligned with the flow, the *nose*, the pressure will correspond to the stagnation pressure. As the flow moves over the body, its speed increases and eventually decreases to zero at the rear of the body. At the point of maximum velocity, the pressure would be predicted by the Bernoulli equation to be a minimum. Boundary-layer flows with $dp/dx \neq 0$ will be discussed in this section, and the work will be extended to the flow over bodies in which dp/dx varies with x, the distance along the surface.

The equations that are applicable to boundary-layer flows are differential equations, just as was true of the flows treated in most of the previous sections. All of the differential equations in those sections were developed on a case-by-case basis. Because boundary layers are quite general, the equations that describe the fluid motion in them are best developed from the general equations for fluid motion, the Navier-Stokes equations. This is done in Chap. 16; here we shall follow the pattern of the previous section and utilize integrated results to show the overall behavior of boundary layers. Velocity profiles that are useful and generally correct will be used in doing this, even though they might not satisfy the governing differential equations at specific points within the boundary layer. While one would wish that the equations of motion were satisfied everywhere, their satisfaction in an integral sense can provide useful results. We can derive such tools by extending the work of Sec. 8.6 by examining the situation shown in Fig. 8.7.1.

This figure shows the fluid flowing over a flat plate for which $dp/dx \neq 0$. (The fact that a flat plate is chosen does not limit our results, because we shall require that a body's radius of curvature be sufficiently large that the body can be considered to be *locally* flat. Then Cartesian coordinates can be used with the y axis perpendicular to a surface that might not be plane.) The coordinate system is located at some designated position along the surface, possibly at the nose, and the velocity within the boundary layer is prescribed there as $u_{x=0}$, which is a function of the coordinate y, the distance perpendicular to the surface of the body. The control volume is chosen so that the distance h is at least as great as the boundary-layer thickness δ at the location X along the body that marks the length of the control volume.

We begin by writing an expression for the momentum efflux from the control volume shown in Fig. 8.7.1:

$$\int u\rho \mathbf{V} \cdot \mathbf{n} \, dA = \int_0^h \rho u_{x=X}^2 \, dy - \int_0^h \rho u_{x=0}^2 \, dy + \int_0^X \rho v U \, dx$$

The second integral on the right-hand side is a prescribed value, while the first and last ones will depend on the location X where the integrals are evaluated.

The forces that act upon the control volume will be assessed next:

$$\Sigma F_x = -\left(\int_0^X \frac{\partial p}{\partial x} \, dx \right) h - \int_0^X \tau_w \, dx$$

in which τ_w is the local shear stress *on* the plate's surface. These forces can now be related to the momentum efflux terms in the same way that has been used previously:

$$\int_0^h \rho u^2 \, dy - \text{constant} + \int_0^X \rho v U \, dx = -h \int_0^X \frac{\partial p}{\partial x} \, dx - \int_0^X \tau_w \, dx$$

For constant density, ρ can be factored out of the momentum integrals. Moreover, these integrals can be differentiated with respect to x to give

Figure 8.7.1 Finite control volume for analyzing growth of the boundary layer along a flat plate.
On OA: $p = p_{x=0}$, $u = u_{x=0}$, $\mathbf{V} \cdot \mathbf{n} = -u$.
On AB: $u_{y=h} = U$, from Bernoulli equation, $\mathbf{V} \cdot \mathbf{n} = +v$.
On BC: $p = p_{x=0} + \int_0^X \frac{\partial p}{\partial x} dx$, $u = u_{x=X}$, $\mathbf{V} \cdot \mathbf{n} = +u$.

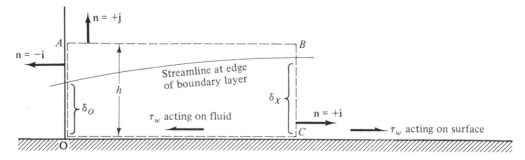

$$\rho \frac{d}{dx}\left(\int_0^h u^2 \, dy\right) + \rho(vU) = \left(-\frac{\partial p}{\partial x}\right)h - \tau_w \qquad (8.7.1a)$$

if the force terms in the momentum equation are also differentiated. The v term in this equation can be rewritten by integrating the continuity equation for an incompressible flow, $\partial v/\partial y = -\partial u/\partial x$, to give $v = -\int(\partial u/\partial x)\,dy$. Because the velocity at the outer edge of the boundary layer depends only on x, we write $U = U(x)$, and it can be treated as a constant when integrating with respect to y at some station along the plate, X. This means that

$$vU = -U\int_0^h \frac{\partial u}{\partial x}\,dy = -\int_0^h U\frac{\partial u}{\partial x}\,dy$$

Moreover, the differentiation of the first term in Eq. (8.7.1a) produces $\int_0^h (2u\,\partial u/\partial x)\,dy$. These steps permit Eq. (8.7.1a) to be rewritten as

$$\int_0^h \left(2u\frac{\partial u}{\partial x}\right)dy - \int_0^h U\frac{\partial u}{\partial x}\,dy - \left(-\frac{1}{\rho}\frac{\partial p}{\partial x}\right)h = -\frac{\tau_w}{\rho} \qquad (8.7.1b)$$

The last term in the left-hand side can be written as

$$\int_0^h \left(-\frac{1}{\rho}\frac{dp}{dx}\right)dy$$

because we have assumed that the pressure gradient parallel to the surface is independent of y. If this is the case, it is equal to the value at the edge of the boundary layer. We *assume* that the Bernoulli equation is applicable along the streamline that bounds the boundary layer. (Viscous effects are absent at the outer edge of the boundary layer.) Along this streamline, $p/\rho + U^2/2 =$ constant, so that $dp/dx = -\rho U\,dU/dx$. This important step, and its underlying assumption, which will be substantiated in Chap. 16, allows all of the integrals on the left-hand side of Eq. (8.7.1b) to be collected together for convenience as

$$\int_0^h \left(\underset{a}{2u\frac{\partial u}{\partial x}} - \underset{b}{U\frac{\partial u}{\partial x}} - \underset{c}{U\frac{dU}{dx}}\right)dy = -\frac{\tau_w}{\rho}$$

The terms in the expression have been identified with letters so that they can be recognized in an equivalent expression,

$$\int_0^h \left(\underset{a}{2u\frac{\partial u}{\partial x}} - \underset{b}{U\frac{\partial u}{\partial x}} - \underset{d}{u\frac{dU}{dx}}\right)dy - \int_0^h \left(\underset{c}{\frac{dU}{dx}U} - \underset{d}{\frac{dU}{dx}u}\right)dy = -\frac{\tau_w}{\rho}$$

if the term designated "d" is added and subtracted from the equation. The "b" and "d" terms in the first integral are $-d(uU)/dx$, and the derivative term is a common factor in the second integral. Therefore

$$-\frac{d}{dx}\int_0^h (uU - u^2)\,dy - \frac{dU}{dx}\int_0^h (U - u)\,dy = -\frac{\tau_w}{\rho}$$

The negative signs can now be eliminated to give the *momentum-integral equation for a boundary layer:*

$$\frac{d}{dx}\left[\int_0^h u(U - u)\, dy\right] + \frac{dU}{dx}\left[\int_0^h (U - u)\, dy\right] = \frac{\tau_w}{\rho} \qquad (8.7.2)$$

These integrals are usually rewritten in a form that requires U^2 to be factored out of the first integral and U out of the second. This produces

$$\frac{d}{dx}\left[U^2 \int_0^h \frac{u}{U}\left(1 - \frac{u}{U}\right) dy\right] + U\frac{dU}{dx}\left[\int_0^h \left(1 - \frac{u}{U}\right) dy\right] = \frac{\tau_w}{\rho} \qquad (8.7.3)$$

Each of the resulting integrals has units of length, and the notation

$$\delta_1 = \int_0^h \left(1 - \frac{u}{U}\right) dy \qquad (8.7.4)$$

$$\delta_2 = \int_0^h \frac{u}{U}\left(1 - \frac{u}{U}\right) dy \qquad (8.7.5)$$

is commonly used although one often sees δ^* used as an alternative designation for δ_1 and θ for δ_2. These lengths are called the *displacement* and the *momentum thickness*, respectively. With this notation, Eq. (8.7.3) is expressed as

$$\frac{d}{dx}(U^2 \delta_2) + U\frac{dU}{dx}\delta_1 = \frac{\tau_w}{\rho} \qquad (8.7.6)$$

While h, the upper limit in the integrals that define δ_1 and δ_2, was set at some value that was sufficiently large, these integrals would not change their values if h were the boundary-layer thickness, δ. This is because for $y \geq \delta$, the viscous effects are negligible.

Example 8.7.1

In Sec. 8.6, the flow past a flat plate with $\partial p / \partial x = 0$ was discussed. A boundary-layer profile that closely approximates a theoretical solution to the problem is $u/U = \sin(\pi y / 2\delta)$, with $U = $ constant. This constant U allows Eq. (8.7.6) to be written as

$$\frac{d}{dx}(\delta_2) = \frac{\tau_w}{\rho U^2}$$

The momentum thickness depends on x through the growth of δ; this fact follows from evaluating Eq. (8.7.5) as follows:

$$\delta_2 = \int_0^\delta \left(\sin\frac{\pi y}{2\delta}\right)\left(1 - \sin\frac{\pi y}{2\delta}\right)(dy)$$

$$= \delta \int_0^1 \left(\sin\frac{\pi\eta}{2}\right)\left(1 - \sin\frac{\pi\eta}{2}\right)(d\eta) \qquad \text{if } \eta = \frac{y}{\delta}$$

$$= \delta\left[\int_0^1 \left(\sin\frac{\pi\eta}{2}\right) - \frac{1}{2}\int_0^1 \left(1 - \cos 2\frac{\pi\eta}{2}\right)(d\eta)\right]$$

$$= \delta\left\{-\frac{2}{\pi}\left(\cos\frac{\pi\eta}{2}\right)\Big|_0^1 - \frac{1}{2}\left[(1 - 0) - \frac{1}{\pi}(\sin\pi\eta)\Big|_0^1\right]\right\}$$

$$= \delta\left(\frac{2}{\pi} - \frac{1}{2}\right) = \delta\left(\frac{4 - \pi}{2\pi}\right)$$

Equation (8.7.6) can be expanded to read

$$2U\frac{dU}{dx}\delta_2 + U^2\frac{d\delta_2}{dx} + U\frac{dU}{dx}\delta_1 = \frac{\tau_w}{\rho}$$

and this result can then be written as

$$U(\delta_2)' + \delta_2 U'\left(2 + \frac{\delta_1}{\delta_2}\right) = \frac{\tau_w}{\rho U}$$

or, by multiplying this equation by δ_2/v, as

$$\frac{U\delta_2\delta_2'}{v} + \left(2 + \frac{\delta_1}{\delta_2}\right)\frac{U'\delta_2^2}{v} = \frac{\tau_w\delta_2}{\mu U} \qquad (8.7.7)$$

This form of the momentum equation for a boundary layer is useful for predicting the evolution of the velocity profile in the boundary layer next to a stationary surface. We shall use this result in conjunction with an assumed velocity profile that has many of the characteristics that would be expected of the actual profile.

The function used for u/U in Example 8.7.1 could also have been a polynomial meeting the boundary conditions that were selected as the most important ones to be applied. The function $u/U = 2\eta - 2\eta^3 + \eta^4$ satisfies the following conditions:

$$\eta = 0: u = 0 \quad \text{and} \quad \frac{\partial^2 u}{\partial y^2} = 0$$

$$\eta = 1: u = U \quad \text{and} \quad \frac{\partial u}{\partial y} = 0 \qquad (8.7.8)$$

The sine function that was used in the example meets the same conditions. The value of the second derivative at $y = 0$, the solid boundary, is zero in order to comply with a requirement imposed by the equation of motion. If, in Eq. (8.7.1a), h were nearly zero, the forces on the control volume would be, with $dy = h$,

$$-dy\left[\left(\frac{\partial p}{\partial x}\right)dx\right] - \int_0^X (\tau_w)\, dx + \int_0^X \left[\tau_w + \left(\frac{\partial\tau}{\partial y}\right)dy\right]dx$$

in which the last integral expresses the shear stress on the layer at the small distance dy from the boundary. This means that the net force on the fluid layer of height dy above the surface reduces to

$$-dy\left[\left(\frac{\partial p}{\partial x}\right)dx\right] + \int_0^X\left[\left(\frac{\partial\tau}{\partial y}\right)dy\right]dx$$

The momentum flux terms will vanish as $dy \to 0$, since both u and v vanish at $y = 0$. Thus, if one differentiates the momentum equation for this thin layer, just as was done when h was greater than δ, one obtains

$$0 = -\left(\frac{\partial p}{\partial x}\right)dy + \left(\frac{\partial\tau}{\partial y}\right)dy$$

or

$$\frac{\partial\tau}{\partial y} = \frac{\partial p}{\partial x} \quad \text{at } y = 0 \qquad (8.7.9)$$

Because $\tau = \mu\, \partial u/\partial y$, the boundary condition at the wall is written as

$$\frac{\partial}{\partial y}\left(\mu \frac{\partial u}{\partial y}\right) = \frac{\partial p}{\partial x} = \mu \frac{\partial^2 u}{\partial y^2} \qquad (8.7.10)$$

This result requires that the polynomial expression for u/U take the pressure gradient into account. The function

$$\frac{u}{U} = (2\eta - 2\eta^3 + \eta^4) + \frac{\lambda}{6}(\eta - 3\eta^2 + 3\eta^3 - \eta^4) \qquad \text{with } \eta = \frac{y}{\delta} \qquad (8.7.11)$$

can provide for the pressure gradient through the parameter λ. When the boundary condition given by Eq. (8.7.9) is satisfied by the velocity profile specified by Eq. (8.7.11), one finds that

$$\lambda = \frac{U'\delta^2}{\nu} \qquad (8.7.12)$$

if it is recalled that $dp/dx = -\rho U U'$. (Higher-order polynomials could also be used if additional conditions are imposed at $\eta = 0$ or 1. For example, one can justify setting the third derivative of u with respect to y equal to zero at $\eta = 0$ and/or 1.)

The case treated previously, $U = $ constant, would correspond to $\lambda = 0$. A limiting value of λ can be gotten from the derivative of the velocity. If $\partial u/\partial y > 0$ at $y = 0$, the velocity increases as one moves away from the wall. But $\partial u/\partial y < 0$ means that the flow would increase in magnitude in the *negative x* direction, as one proceeded away from the boundary. Such a flow reversal would be a *backflow,* and this is what is encountered when the flow separates from a body. Figure 1.1.3 shows the region of flow separation with the associated backflow. We shall define the *inception of separation* to occur when $\partial u/\partial y = 0$ at $y = 0$. The value of λ for this case can be deduced from Eq. (8.7.11) to be -12. If $\partial u/\partial y \geq 0$ in the boundary layer and u is never greater than U, then Eq. (8.7.11) requires that λ not be greater than 12. Thus the shape of the boundary-layer velocity profile is prescribed by λ, the *shape factor,* and its value should be between -12 and $+12$, if no other constraints are placed upon the flow. Subsequently, however, reasons will be given to reduce the positive limit.

When the velocity profile given by Eq. (8.7.11) is introduced into the integrals that define δ_1 and δ_2, the results are

$$\frac{\delta_1}{\delta} = \frac{3}{10} - \frac{\lambda}{120} \qquad (8.7.13)$$

and
$$\frac{\delta_2}{\delta} = \frac{37/5 - \lambda/15 - \lambda^2/144}{63} \qquad (8.7.14)$$

(For $\lambda = 0$, $\delta_2/\delta = 0.1175$ for the fourth-order polynomial, whereas the value was 0.1366 in Example 8.7.1, in which the sine function was used.)

Equation (8.7.7) describes the change of δ_2 in terms of δ_1, τ_w and other parameters. The wall shear stress can be related to λ through $\tau_w = \mu(\partial u/\partial y)_{y=0}$ and Eq. (8.7.11). The result is

$$\tau_w = \frac{\mu U}{\delta(2 + \lambda/6)} \qquad (8.7.15)$$

If one could solve Eq. (8.7.7) so that τ_w is known along the surface, one could integrate it and find the drag due to skin friction. Equally well, one could determine the evolution of the velocity profile from the nose of a symmetric airfoil up to the place where separation occurs, $\lambda = -12$. But what is the value of λ at the nose, the stagnation point? We need to know its value in order to begin the calculations. Equation (8.7.7) will aid us in our quest for this initial value, which will be designated λ_0 and found to be 7.072 *if* the body has an upstream stagnation point. Therefore, the range of the shape parameter is usually from -12 to 7.052, instead of $+12$, the value obtained if one arbitrarily controls the magnitude of u/U in the boundary layer.

This value of λ_0 follows directly if Eq. (8.7.7) is rewritten by incorporating the definitions of δ_2 and τ_w that are given by Eqs. (8.7.14) and (8.7.15). The steps follow below. First,

$$\frac{U\delta_2\delta_2'}{\nu} + \left(2 + \frac{\delta_1}{\delta_2}\right)\frac{U'\delta_2^2}{\nu} = \frac{\tau_w\delta}{\mu U}\left(\frac{\delta_2}{\delta}\right)$$

or

$$\frac{U\delta_2\delta_2'}{\nu} + \left(2 + \frac{\delta_1}{\delta_2}\right)(\lambda)\left(\frac{\delta_2}{\delta}\right)^2 = \left(2 + \frac{\lambda}{6}\right)\left(\frac{\delta_2}{\delta}\right) \tag{8.7.16}$$

At the stagnation point, U will be zero but dU/dx, or simply U', will be positive. Thus, the first term in Eq. (8.7.16) is zero at the stagnation point, and the value of $\lambda_0 = 7.052$ is the lowest positive value of λ that satisfies the relationship for the remaining terms in Eq. (8.7.16). (The remaining solutions give λ values that are outside the range -12 to $+12$.)

With Eq. (8.7.16) as the means of predicting shape factors λ for the velocity as a function of x, and with bounding values for this factor available to us, it is natural to turn to the solution of this problem. This was first done by K. Pohlhausen in 1921 and T. von Kármán in 1921, for a related case, with the result that the method that uses the polynomial representation in the momentum equation for the boundary layer is usually called the *Kármán-Pohlhausen method*. When the method was originally used, and, subsequently, when a considerable amount of experience with it had been gained, analog and digital computers were virtually unknown. To provide a systematic and direct method of solution, researchers adopted techniques such as splitting the equation appropriately and solving it by the "method of isoclines." Others found that one of the functions of λ that arose in the course of solving the problem varied almost linearly for values of λ between 0 and -12. (The value of zero occurs when U reaches its maximum value along the boundary.) The *assumption* of a linear variation of this function over the *entire* range of λ permits an equation for δ_2 to be integrated without difficulty. This long series of improvements to aid in the solution of Eq. (8.7.16) will not be discussed, because we now have digital computers that can rapidly solve the problem, even if they were to have only a modest amount of available memory. We shall pursue a nontraditional manner of utilizing a numerical solution now.

We begin, just as did some of the researchers alluded to previously, by making a change of variable. We let

$$Y = U\frac{\delta_2^2}{2\nu} = \frac{\lambda}{2}\frac{U}{U'}\left(\frac{\delta_2}{\delta}\right)^2 \tag{8.7.17}$$

A differentiation of Y with respect to x gives

$$\frac{dY}{dx} = \frac{U\delta_2\delta_2'}{\nu} + \frac{U'\delta_2^2}{2\nu}$$

The first term on the right-hand side of this expression is the first term on the left-hand side of Eq. (8.7.16), while the second term in this last equation is $\lambda(\delta_2/\delta)^2/2$, a function of λ only. These observations lead one to write Eq. (8.7.16) as

$$\frac{dY}{dx} - \frac{\lambda}{2}\left(\frac{\delta_2}{\delta}\right)^2 + \left(2 + \frac{\delta_1}{\delta_2}\right)\lambda\left(\frac{\delta_2}{\delta}\right)^2 = \left(2 + \frac{\lambda}{6}\right)\left(\frac{\delta_2}{\delta}\right)$$

or

$$\frac{dY}{dx} = \left(2 + \frac{\lambda}{6}\right)\left(\frac{\delta_2}{\delta}\right) - \lambda\left[\left(2 + \frac{\delta_1}{\delta_2}\right) - \frac{1}{2}\right]\left(\frac{\delta_2}{\delta}\right)^2 \qquad (8.7.18)$$

The right-hand side of Eq. (8.7.18) is solely a function of λ. Y is zero, from its definition that includes U, when x is zero. Thus one can find the change of Y at $x = 0$ over a chosen interval dx and subsequently the value of Y itself at this value of $x = 0 + dx$. But at this value of x the functions U and U' are known because they are prescribed by the shape of the body being considered. Consequently, λ can be found at $x = (1)(dx)$ from the definition of Y given in Eq. (8.7.17). With this value of λ, a number less than λ_0, one can again evaluate the right-hand side of Eq. (8.7.18) to determine dY/dx at the first step along the surface, $x = (1)(dx)$. In principle, this process can be continued for $x = (2)(dx)$ until $x = (N)(dx)$, in which N is the number of the increment that makes $\lambda < -12$. The point on the surface where the -12 value is realized will be the location where separation is predicted for the two-dimensional body under examination.

The numerical procedure that has just been described is called *Euler's method*. It is a simple scheme but can give poor results if the changes in slope are very rapid. To account for such changes, one could first calculate the value of Y and λ at $x = (1)(dx)$. This would be the first trial at this point. With these values, one could estimate the slope at this point from Eq. (8.7.18), just as before. But now, instead of proceeding to $x = (2)(dx)$, one could make an improved estimate of dY/dx in the interval between $(0)(dx)$ and $(1)(dx)$. Such an estimate could be

$$\frac{[\text{Slope at } x = (0)(dx)] + [\text{slope at } x = (1)(dx)]}{2}$$

With this "weighted average slope" one could estimate anew the values of Y and λ at the end of the interval $x = (1)(dx)$. This procedure, the simplest form of *Euler's improved method*, can be continued to $x = (2)(dx)$ and then onto $x = (N)(dx)$. Figure 8.7.2 is intended to give a qualitative justification for using more than one derivative in an interval. The reader can refer to Carnahan, Luther, and Wilkes' book *Applied Numerical Methods*, or similar books on numerical analysis, to find even better algorithms.*

Appendix G gives some results for the flow past a circular body for which $U = 2U_\infty \sin x$ if the radius is 1 so that $\theta = x/r = x$. The numerical procedures give a solution that indicates separation around $108°$ from the stagnation point. In Schlichting's book *Boundary Layer Theory*,† a more exact formulation of the problem

---

* Brice Carnahan, H. A. Luther, James O. Wilkes, *Applied Numerical Methods*, John Wiley & Sons, 1969.

† Herman Schlichting, *Boundary Layer Theory*, 7th ed, McGraw-Hill Book Co., 1979.

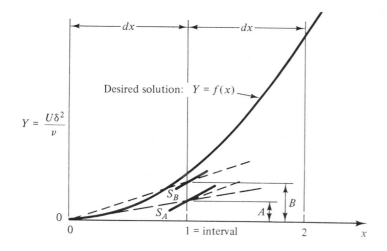

1. Slope given at origin, $(dY/dx)_0$.

2. A: First estimate of $Y(1)$ using slope given in Step 1.

3. S_A: Slope given by expression for $dY/dx = f(x, Y)$ using
   $x = x(1)$ and $Y = Y(1)$ from first estimate, Step 2.

4. Average slope from Steps 1 and 3.

5. B: Second estimate of $Y(1)$ using slope of Step 4.

6. S_B: Slope calculated from $dY/dx = f(x, Y)$ using
   $x = x(1)$ and $Y = Y(1)$ from second estimate, Step 5.

7. To extend the figure, return to Step 1, using slope S_B; procedure
   repeated in next interval of x to obtain $Y(2)$ at $x(2)$.

Figure 8.7.2 Representation of stepwise improvement in estimate of slope in Euler method for solving a differential equation.

is presented with a solution of 110°. (The polynomial approximation to the velocity distribution is not used in this "exact" solution.) However, this disagreement of a few degrees is not the main difficulty. It is experimentally observed that when the boundary layer is laminar on a circular cylinder, the flow separates around 82°.

The calculations show that λ decreases slowly until its value is zero and then, after $U' = 0$, it decreases rapidly toward -12. Thus, if the position of $U' = 0$ that is specified *a priori* is different from that which occurs on the actual body, then the position for separation that is predicted will disagree with observations. We know that a description of U that is proportional to the sine of θ cannot be reliable after 90°, because the actual flow is not symmetric about the y axis. Thus, a more realistic estimate of the point of separation would follow from using a function for U that predicted the observed pressure distribution on the body. K. Hiemenz provided Pohlhausen with such a polynomial for an example in his paper. The predicted result is 81°! That polynomial, scaled for a circle of unit radius, is $U/U_\infty = 1.815684x - 0.2713598x^3 - 0.0473245x^5$.

The Kármán-Pohlhausen method can be applied to thin bodies such as struts or wings and gives good results. This can be expected because such bodies have relatively small separation regions aft of the body, so that the expression for U that determines the pressure distribution used in the method is often very close to the one that can be predicted theoretically by means of potential flow.

If Eq. (8.2.11) is examined, it will be noticed that when Q and U are held constant, then $\partial p/\partial x$ will vary with the height of the channel, k. This becomes quite evident if the equation is written as

$$Q - \frac{Uk}{2} = \frac{(-\partial p/\partial x)k^3}{12\mu}$$

and plotted in the manner of Fig. 8.8.1. There, the left-hand side of the equation is plotted as a function of the constant spacing within the channel, k (curve 1), and plots of the right-hand side of this equation (curves 2 and 3) are superimposed. The pressure gradient associated with curve 2 is greater than that for curve 3. The intersection of either of the curves 2 or 3 with curve 1 forms a solution. This figure makes it evident that in the solution for the larger k (that is, k_3), the magnitude of the pressure gradient is the smaller.

In this section we shall examine the effect of a gradually changing distance k on the pressure distribution in the channel. Because k will be assumed to change only slightly in the direction of flow, there will be a small change in the value of the velocity, u, from section to section. That such a change takes place means that u varies with x in Fig. 8.8.2, as does k, so that $\partial u/\partial x$ is not exactly zero. Because of this, the fluid must be accelerating as it moves through the channel, positively or negatively, depending on the sign of θ. The magnitude of this acceleration is obtained by examining Eq. (4.6.6b), which states

$$a_x = \frac{Du}{Dt} = \frac{\partial u}{\partial t} + u\frac{\partial u}{\partial x} + v\frac{\partial u}{\partial y} \tag{8.8.1}$$

for a two-dimensional flow. The flow in our converging channel is considered to be *steady*, so that $\partial u/\partial t = 0$, but the other terms in Eq. (8.8.1) are neither individually nor collectively zero.

Consequently, the equation governing the motion of the infinitesimal material volume shown in Fig. 8.8.2 is

$$\text{Shear force + pressure force} = (\text{mass})a_x$$

or

$$\left(\frac{\partial \tau}{\partial y}dy\right)dx - \left(\frac{\partial p}{\partial x}dx\right)dy = (\rho\,dx\,dy)a_x \tag{8.8.2}$$

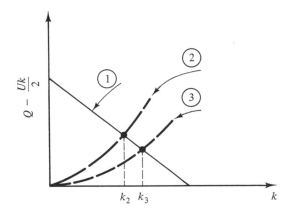

Figure 8.8.1 Solutions of the equation

$$Q - \frac{Uk}{2} = \frac{(-\partial p/\partial x)\,k^3}{12\mu}$$

(i.e., intersection of curve 1 with curves 2 and 3) for values of k as a function of the pressure gradient.

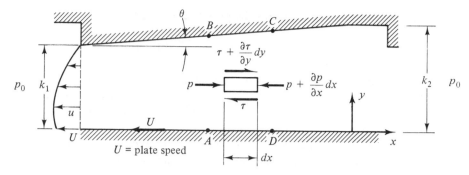

Figure 8.8.2 Infinitesimal control volume for the analysis of the flow between almost parallel plates.

Whereas the right-hand side of this equation was exactly zero in Sec. 8.2 where parallel plates were discussed, we shall attempt to formulate the problem for plates that are slightly nonparallel. For this case, a_x will not be exactly zero; however, we hope to show that it is small in comparison with the other terms that remain in the equation. These are $\partial\tau/\partial y$ and $\partial p/\partial x$. If the ratio of a_x to one of these terms, say, $\partial\tau/\partial y$, proves to be much less than unity, we may reasonably assume that the acceleration term is small in comparison with the viscous term. More important, we shall attempt to solve the equation of motion with the *assumption* that the acceleration terms can be neglected.

In order to make such an order-of-magnitude analysis in this case, it is convenient to apply the momentum equation to the control volume $ABCDA$ in Fig. 8.8.3. The shear stresses are *assumed* to be approximately the same on the upper and lower surfaces; then the forces in the x direction are

$$-2(\tau_{\text{wall}})(dx) + \left[p - \frac{\partial p}{\partial x}\left(\frac{dx}{2}\right)\right]k - \left[p + \frac{\partial p}{\partial x}\left(\frac{dx}{2}\right)\right](k + dk) + p(dk)$$

in which the last term is the x component of the force due to the average pressure, p, over the length dx on the sloping upper surface. The slope of this surface is dk/dx. The other pressure terms are, of course, those applied at either end of the control volume.

Figure 8.8.3 Control volume with infinitesimal length and finite height.

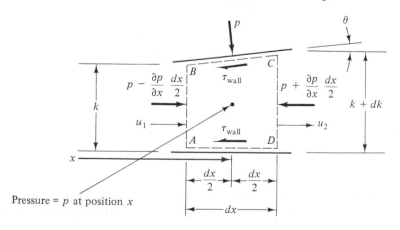

When one simplifies the terms for the pressure, one gets

$$-2\tau_{\text{wall}}(dx) - \left(\frac{\partial p}{\partial x}\right)k(dx)$$

for the x component of the forces on the fluid in the control volume. The integration of the momentum efflux terms for the entire control surface gives us $\beta\rho Q(\bar{u}_2 - \bar{u}_1)$, in which β is a momentum correction factor and the overbar denotes an average value. This average value is *assumed* to be nearly the same at the two locations because of the gradual change of k. The average velocity can be expressed in terms of Q and the width, k or $k + dk$. Then the momentum efflux can be written as

$$\beta\rho Q^2\left(\frac{1}{k + dk} - \frac{1}{k}\right) \quad \text{or} \quad \beta\rho Q^2\left(-\frac{dk}{k^2}\right)$$

if $dk/k \ll 1$ in the denominator. Equating the forces applied to the fluid in the control volume to the momentum efflux produces, after dividing by dx,

$$-2\tau_{\text{wall}} - \left(\frac{\partial p}{\partial x}\right)k = -\beta\left(\frac{\rho Q^2}{k^2}\right)(\tan\theta) \tag{8.8.3}$$

This equation gives us an explicit expression for the momentum, or inertia, terms in our equation of motion. To get an estimate of the wall shear stress, we *assume* that τ_{wall} is some number F times the product of μ and the average velocity divided by half the channel width. Accordingly,

$$\tau_{\text{wall}} = F\left(\frac{\mu Q/k}{0.5k}\right)$$

Then the ratio of the inertia term to the wall shear in our momentum equation, Eq. (8.8.3), is

$$\frac{(\beta\rho Q^2/k^2)(-\tan\theta)}{2F\mu Q/k^2}$$

This reduces to

$$\left(\frac{\rho Q}{\mu}\right)(-\tan\theta)\left(\frac{\beta}{2F}\right)$$

Now, this ratio can be made as small as desired, independent of the value of the factor $\beta/2F$, by having the angle θ or $\rho Q/\mu$ (a Reynolds number, in effect) small. This can be achieved by a small value of Q or a large value for the kinematic viscosity $v = \mu/\rho$. Either, or both, conditions will be *assumed* in what follows. This will allow us to neglect the acceleration terms with respect to the other terms in Eq. (8.8.2). The result is

$$\frac{\partial\tau}{\partial y} - \frac{\partial p}{\partial x} = 0$$

with the consequence that

$$\mu\frac{\partial^2 u}{\partial y^2} = \frac{\partial p}{\partial x} \tag{8.8.4}$$

With the adoption of Eq. (8.8.4) as our mathematical model, we have a result similar to Eq. (8.2.10) as the appropriate solution. This equation will now be applied to the configuration shown in Fig. 8.8.2, in which the lower plate is moving to the left with a magnitude of U at $y = 0$. In this case,

$$u = \frac{\partial p}{\partial x}\frac{1}{2}\mu(y^2 - ky) + \frac{U}{k}(y - k)$$

and
$$Q = -\frac{\partial p}{\partial x}\frac{1}{12\mu}k^3 - \frac{Uk}{2} \tag{8.8.5}$$

by starting with Eq. (8.2.8). Even though k varies with x, continuity requires that Q be the same at each station along the channel.

If, in Eq. (8.8.5), one solves for dp/dx, the result is

$$-\left(\frac{Uk}{2} + Q\right)(12)\frac{\mu}{k^3} = \frac{dp}{dx} \tag{8.8.6}$$

Because the slope of the channel can be written as dk/dx, it proves to be convenient to solve for the variation of p in Eq. (8.8.6) not as an explicit function x but as a function of k. This is accomplished by setting $dp/dx = (dp/dk)(dk/dx)$. The term dk/dx is the slope of the channel and has been defined as $\tan\theta$ in Fig. 8.8.3. Equation (8.8.6) will be solved by using $(dp/dk)(\tan\theta)$ instead of dp/dx. The result is

$$\frac{dp}{dk} = \frac{-1}{\tan\theta}\frac{(Uk/2 + Q)12\mu}{k^3}$$

which integrates to

$$p = \frac{12U\mu}{\tan\theta}\left(\frac{1}{2k} + \frac{1}{2}\frac{Q/U}{k^2}\right) + C_1 \tag{8.8.7}$$

The constant C_1 must now be determined. Furthermore, Q must be specified, since it will be determined through a pressure gradient in accordance with Eq. (8.8.5) and not by a single pressure. What should that pressure gradient be?

At the inlet to the channel, the height is k_2 and the pressure is p_0, as shown at the right of Fig. 8.8.2. The same pressure is assumed to occur at the exit of the channel, where the height is k_1. With these things in mind, we can determine C_1 and Q in Eq. (8.8.7). This equation gives p at k_1 and k_2 as

$$p_0 = \frac{6U\mu}{\tan\theta}\left(\frac{1}{k_1} + \frac{Q/U}{k_1^2}\right) + C_1 \tag{8.8.8}$$

as well as

$$p_0 = \frac{6U\mu}{\tan\theta}\left(\frac{1}{k_2} + \frac{Q/U}{k_2^2}\right) + C_1$$

A subtraction of these equations from one another gives

$$Q = -\frac{Uk_1k_2}{k_1 + k_2} \tag{8.8.9}$$

subsequent to a rearrangement of the terms.* If Eq. (8.8.8) is subtracted from Eq. (8.8.7), the result that was sought is

---

* The minus sign indicates that the net flow is from the right to the left.

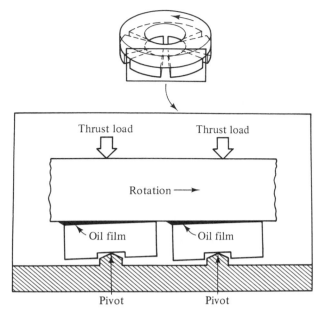

Thrust load   Thrust load

Rotation →

Oil film   Oil film

Pivot   Pivot

Figure 8.8.4 Hydrodynamic bearing. (*Courtesy of Kingsbury, Inc.*)

$$p - p_0 = \frac{6U\mu}{\tan\theta}\left[+\frac{1}{k} - \frac{1}{k_1} + \frac{Q}{U}\left(\frac{1}{k^2} - \frac{1}{k_1^2}\right)\right]$$

The coefficient Q/U can be eliminated through Eq. (8.8.9), and after several lines of algebraic steps, this relationship for the pressure can be reduced to

$$p - p_0 = \frac{6U\mu}{\tan\theta}\left(\frac{k - k_1}{k^2}\right)\left(\frac{k_2 - k}{k_1 + k_2}\right) \tag{8.8.10}$$

All of the terms in the expression are positive. Therefore, $p - p_0$ will be positive, so that the pressure within the channel will be greater than that existing in the spaces ahead of and behind it. Consequently, the integration of this pressure distribution along the length of the upper and lower surfaces of the channel will yield a resultant force that acts on the surface in a direction away from the channel itself. This tapered channel could occur between a stationary surface that is in close proximity to a rotating disk, similar to the oil film shown in Fig. 8.8.4. The combination of the stationary surface and the rotating disk would form a thrust bearing.*

To find the resultant force on the lower surface of the bearing, one determines $\int_0^L (p - p_0)(dx)(1)$, in which L is the bearing's length. However, x is not a convenient variable of integration, because $p - p_0$ does not contain x explicitly. But dx and dk are interconnected by the slope of the channel, $\tan\theta$, as before. Therefore, we write

$$F = \frac{1}{\tan\theta}\int_{k_1}^{k_2}(p - p_0)\,dk$$

$$= \frac{6U\mu}{(\tan\theta)^2}\left(\frac{1}{k_1 + k_2}\right)\left[(k_1 + k_2)\ln\frac{k_2}{k_1} - (k_2 - k_1) + k_1 k_2\left(\frac{1}{k_2} - \frac{1}{k_1}\right)\right]$$

---

*The motion picture "Low Reynolds Number Flows," by Sir Geoffrey Taylor, produced by the National Committee for Fluid Mechanics and distributed by the Encyclopedia Britannica Educational Corporation, treats this and similar material.

This expression can be simplified. Because of the geometry of the bearing, one can deduce that

$$(\tan \theta)^2 = \left(\frac{k_2 - k_1}{L}\right)^2 \tag{8.8.11}$$

With this fact, the force on the bearing, per unit width, can be written as

$$F = \frac{6U\mu L^2}{(k_2 - k_1)^2}\left(\ln\frac{k_2}{k_1} - 2\frac{k_2 - k_1}{k_1 + k_2}\right)$$

which can be simplified and rewritten effectively with k_2/k_1 appearing as a design parameter. The final equation is

$$F = \frac{6U\mu L^2}{k_1^2(k_2/k_1 - 1)^2}\left(\ln\frac{k_2}{k_1} - 2\frac{k_2/k_1 - 1}{k_2/k_1 + 1}\right) \tag{8.8.12}$$

This equation can be used to determine the value of k_1/k_2 that would produce maximum thrust; the result is $k_2/k_1 \approx 2.2$. However, the bearing design would not be complete then, because the shear stress on the bearing surfaces would have to be evaluated in order to determine the power needed to operate the bearing.* A good design criterion would minimize this power while maximizing the available thrust.

The viscous drag on the upper surface of the bearing can be obtained through a simple extension of the work already accomplished. In a manner similar to Eq. (8.2.12), the shear stress can be expressed as $-(dp/dx)k/2 + \mu U/k$. This means that the bearing drag, D, is

$$D = \int_0^L \left[-\frac{\partial p}{\partial x}\frac{k}{2} + \frac{\mu U}{k}\right](dx)(1)$$

The pressure gradient in this expression is given by Eq. (8.8.6), which contains Q. This flow is given by Eq. (8.8.9) as $Q = -Uk_1 k_2/(k_1 + k_2)$. Again, because the shear stress—the term in square brackets in the last integral—is implicit in x, through $k = k_1 + x \tan \theta$, the area element $(dx)(1)$ is replaced by $dk/\tan \theta$ in executing the integration. The result for a unit width is

$$|D| = \frac{2U\mu L}{k_2 - k_1}\left(2 \ln\frac{k_2}{k_1} - 3\frac{k_2 - k_1}{k_2 + k_1}\right) \tag{8.8.13}$$

Figure 8.8.5 shows the predicted velocity and pressure distributions in a tapered channel.

Thrust bearings similar to the one whose characteristics are given in Eqs. (8.8.12) and (8.8.13) were patented around 1900 by A. Kingsbury in the United States and by A. G. M. Michell in Australia. These designs usually have stationary bearing pads that are segments of a circular ring. The pads have a layer of soft material in case contact is made between them and the hardened disk that is attached to the rotating shaft. It is this disk and the inclined pads that generate the pressure distribution illustrated in Fig. 8.8.5. This is the hydrodynamic pressure that supports the load. (The hydrodynamic pressure is in addition to the support provided by the hydrostatic pressure, which in this kind of bearing is not the primary agent for the support of the shaft.) For some of

---

* The pressure distribution in the sloping upper surface would produce a force in the x direction. However, for small angles, this will be a minor effect in comparison with the shear stresses.

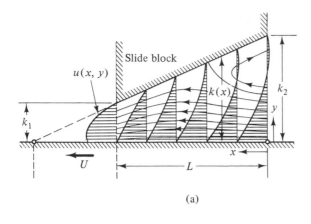

(a)

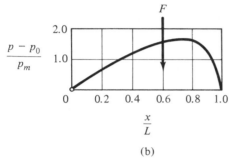

(b)

Figure 8.8.5 Hydrodynamic lubrication in a bearing: (a) flow in the wedge between the slide block and a plane guide surface; (b) pressure distribution. *(From H. Schlichting's Boundary–Layer Theory, 7th ed., McGraw-Hill Book Co., 1979)*

these designs, the bearing can tip about an axis and change the angle θ with respect to the plane of the rotating disk. In this way, for a fixed value of k_1, the ratio k_2/k_1 can change with the load. The minimum practical value of k_1 is about 0.01 mm—about one-tenth of the thickness of this page. Figure 8.8.4 shows the features of such a bearing.

The propeller thrust of a ship moving at 28 knots provided by a turbine producing 16,560 kW can be about 1 meganewton (MN). The bearing might have an outside diameter of 1050 mm and an inside diameter of 830 mm. The ring could be divided into 10 bearing pads. In design calculations, a surface speed of 8.86 m/s at the mean diameter of the pads could be used for a shaft speed of 180 rpm. In the bearing used by the *Bremen,* built around 1930, an oil with a viscosity of 40 centipoise (cP) at 50°C and parts with the dimensions just given were used. About 24 kW was expended by bearing friction. One can show that Eqs. (8.8.12) and (8.8.13) generally substantiate these thrust and drag figures.

Before closing this particular section, it is useful to review the thought process that has been used. We were interested in mathematically modeling the flow in a bearing. It was idealized in the manner of Figs. 8.8.2 and 8.8.3. There were simplifications because we assumed that there were not any significant end effects at the entrance and exit of the tapered region and that there was uniformity in the direction perpendicular to the printed page. This representation of the problem led us to consider the dynamics of the fluid flow. Even though it was recognized that some fluid acceleration would be taking place, we found it expedient to assess the magnitude of the inertia effect and subsequently neglect it in comparison with the other terms associated with Newton's laws of motion. We then formed a differential equation that we could solve. This process of neglecting terms that are small when the other terms in the equation are much larger is done whenever it is necessary to find a realistic solution to an otherwise intractable problem. This solution must be assessed to see whether, in fact, the assumed insignificance of the neglected term can be justified. One designates mathematical solutions that neglect the inertia terms *"creeping flows"* or *Stokes flows.* George Stokes made such an assumption when he solved the problem for the slow motion of a viscous fluid about a sphere (about 1850).* Experiments show that his results are confirmed when the ratio UD/ν is less than 0.1. This ratio, which contains the speed of the sphere in a quiescent fluid, its diameter, and the kinematic viscosity of the fluid, is similar to the ratio Q/ν that was encountered in this section.

Sometimes, however, all the terms in a theoretical model can be significant and should be included. Although the resulting equation may be too complicated to solve, one can learn a great deal by treating some limiting cases. This was the case for the analysis of a hydraulic jump in Example 7.3.4. The inertia and gravity forces were assumed to dominate the viscous terms. In effect, any momentum changes due to entrainment in the disturbed free surface of the "roller" were also neglected. The effects of fluid mixing would also be neglected if a hydraulic jump were to occur when water separated two immiscible hydrocarbons. Then, however, the viscous shear stresses would be important in the analysis of the phenomenon. If, however, one were treating the intrusion of a layer of salt water between two other layers of different salinity, the viscous stresses and the momentum interchanges at the surfaces of the intruding liquid ought to be included in an analysis of the problem. From these comments one can deduce that while viscosity is ever-present, it might not be of paramount importance in some problems. It was very important in the cases treated in this chapter where inertia effects were either identically zero or found to be small by comparison. These are low–Reynolds number flows. In Sec. 16.3, high–Reynolds number flows will be discussed, and there significant results will ensue from neglecting a small portion of the shear stresses near the boundary of a rigid body and all of the shear stresses at points that are remote from it.

The inclusion and exclusion of some forces in an analysis, after reaching a conclusion about their relative magnitudes, implies that force ratios should be examined with care when one models a problem. Such force ratios, as well as geometric, speed, and energy ratios, are part of the topic to which we turn in the next chapter.

8.9 SUMMARY

The analysis of the flow situations presented in this chapter all began with a free-body diagram of an infinitesimal material volume of fluid. The momentum changes of this material volume and the forces applied to it were related to yield a differential equation that could be integrated analytically or numerically. All of the cases treated were limited to the regime of laminar flow, so that Newton's law of viscosity was employed to relate the shear stresses with the velocity derivatives.

The case of the oscillating plate gave a result that supports Prandtl's assumption of the existence of a boundary layer. The region of fluid near the plate that is affected

---

*However, it was Osborne Reynolds who investigated the hydrodynamics of lubrication, about 35 years after this work of Stokes.

by the motion of the plate becomes thinner as the fluid's viscosity decreases and/or the frequency of oscillation increases. The latter condition is equivalent to saying that the maximum speed of the plate increases. Prandtl had assumed that the boundary layer would decrease as the Reynolds number for the flow past a flat plate increased. However, a shear stress would continue to exist, even as the viscosity became very small, because the no-slip condition would be enforced for all real fluids.

In Sec. 8.7, we were interested in the effects of viscosity within the thin layer near a solid boundary. This layer is thin when the Reynolds number is large. By using an integral equation for the flow in a boundary layer, Eq. (8.7.2), and a reasonable polynomial representation of the velocity distribution in a boundary layer, the concepts presented for the flow past a flat plate in Sec. 8.6 can be extended to the flow over a surface whose curvature is relatively large. The steps associated with such an analysis culminated in Eq. (8.7.18).

In Sec. 8.8, the flow must vary from point to point within a tapered channel. The complete differential equation that describes the flow must include acceleration terms. However, these terms make a simple solution impossible, even though such a solution gives useful results. Without knowing what will be the outcome, the probable magnitude of the terms in the (differential) equation of motion can be compared. In the event that a particular term is found to be very small with respect to another, it would be reasonable to neglect that term if the resulting equation could be solved readily. By assuming a low–Reynolds number flow, the acceleration effects were neglected, and a solution that predicts the pressures in a hydrodynamic bearing was developed. Such low-Re flows are often called *Stokes*, or *creeping*, *flows*. Their differential equations are also referred to as the *lubrication equations*.

EXERCISES

8.2.1(a) The shear force in the example is -86.8 lb. Why is the force of the fluid on the plate directed downhill?

(b) Determine whether the magnitude and sign of the shear stress on the lower plate are the same as on the upper plate.

(c) Is the liquid flowing uphill or downhill under the influence of the plate's motion and the pressure gradient? What is the net flow?

8.3.1(a) What would be the volume rate of flow if the pressure difference were 100 kPa?

(b) What is the centerline velocity of the liquid for the condition of the example?

(c) What is the wall shear stress for the 6.8 kPa pressure difference in the example?

8.6.1(a) What are K_1 and K_2 if $u/U = y/\delta$?

(b) Determine whether $u/U = 2\eta - 2\eta^3 + \eta^4$ satisfies the desired boundary conditions for u. What other boundary condition does it satisfy at $\eta = 1$? Is this last condition reasonable?

(c) For $f(\eta)$ to be of the form $A\eta + B\eta^3$, what must be the value of the constants A and B in order to satisfy the most obvious boundary conditions on u?

8.6.2(a) What are the values K_1 and K_2 in the example?

(b) What would be the boundary-layer thickness if water at the same conditions had been used?

(c) How long would the plate be if the boundary-layer thickness at its end were twice the value in the example, for the same conditions for flow?

8.6.3(a) For a pressure gradient of -0.3 psi/ft and a plate separation k of 0.09 in., what would be the expected transition length at the inlet of a horizontal channel?

(b) The constant in Eq. (8.6.11) is 5. Had this constant been 2.5, what would have been the initial estimate of the transition length for the flow in the channel?

(c) Estimate the difference in pressure from the change in kinetic energy associated with a uniform and a parabolic velocity profile.

(d) Assume a value of the channel Reynolds number, Uh/ν, to be 1000. ($U = 10$ cm/s; $h = 1$ cm; $\nu = 1$ centistokes.) Estimate the Δp associated with the shear stress that acts over L_t with the aid of Fig. 8.6.2.

8.6.4(a) What would be the drag on the keel at 20 knots?

(b) What would be the shear drag on the keel in a wind of 20 knots if the boat is stored out of the water?

(c) What would be the expected drag on a 3-m addition to the height of a vertical stabilizer of an aircraft's tail that is 4 m long in the flow direction? A flight speed of 370 km/h is anticipated in air at 20°C.

(d) On the basis of the results of Exercises 8.6.4b and c, could the drag coefficient for the aircraft's tail stabilizer be determined by tests in water? Explain.

PROBLEMS

8.1 If the plate velocity in Example 8.2.1 were changed so that it had the opposite sign, what would the thickness of a steel plate have to be so that no force other than its gravitational attraction would be needed to obtain the same speed at the same pressure gradient?

8.2 Let water flow down an inclined plane similar to that shown in Fig. 8.2.4. The depth is 0.11 in. What are the maximum speed of the water, the average flow rate, and the shear stress on the plane due to the water? What is the pressure on the plane?

8.3 A film of water in a rainstorm is driven upward along a surface inclined at 45° to the horizontal. Derive an expression relating the local wind shear stress, the thickness of the film, and the average speed of film advance.

8.4 A disk rotates about a vertical axis at the rate of 100 rpm. It is submerged in an oil with a viscosity of 50 cP, and there is a vertical gap of 2 mm between it and a comparable stationary disk with a diameter of 15 cm. What torque is required to achieve this motion? Assume that the analysis for the flow between parallel plates can be applied, i.e., that streamline curvature does not play a significant role.

8.5 A thin plate (0.3 mm thick) with a surface area (one side) of 0.05 m^2 and an effective weight of 7.5 N (buoyancy effects have been accounted for) is supported as shown in Fig. P.8.5 so that a constant distance is maintained from the parallel surfaces. Determine the speed of descent of the plate if $\mu = 0.3$ kg/m-s.

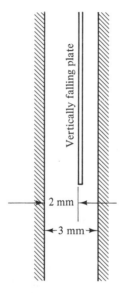

Vertically falling plate

2 mm

←3 mm→

Figure P.8.5

8.6 A horizontal flat plate 2 ft long and 1.2 ft wide supports a total load of 5000 lb that is equally distributed over its surface (see Fig. P.8.6). Between the plate and the floor there is a uniform oil film with a thickness of 0.010 in. Seals prevent the oil from flowing out of the 2-ft edges of the plate. Oil enters into the gap under pressure along a groove that is parallel to the 1.2-ft sides. The groove is midway between these sides so that one can assume that half of the oil moves toward each of them if the plate is at rest.

(a) For a viscosity of 0.0002 slugs/ft-s, determine the pressure in the vicinity of the oil supply groove. (Neglect entrance effects.)

(b) If the load is increased at the same time the

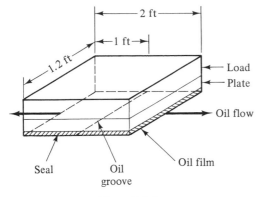

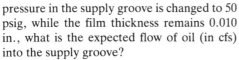

Figure P.8.6

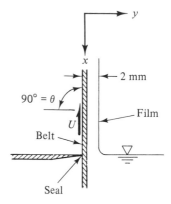

Figure P.8.7

pressure in the supply groove is changed to 50 psig, while the film thickness remains 0.010 in., what is the expected flow of oil (in cfs) into the supply groove?

(c) If the load is again increased so that the gap is reduced by half but the oil flow remains unchanged, what will be the pressure near the oil groove that supplies the oil?

(d) If the plate and its load are translated across the floor at 0.6 ft/s in a direction parallel to the 2-ft side, draw representative velocity profiles in the oil.

(e) For the case mentioned in part (d), will the oil flow symmetrically under the plate? Explain.

(f) What is the horizontal force needed on the plate-load assembly to achieve the translation given in part (d)?

8.7 A 1.5-m-wide belt moves vertically upward out of a vat of liquid ($\mu = 0.02$ kg/m-s; $SG = 0.88$) at a speed of 2.1 m/s. (See Fig. P.8.7.) A film of liquid 2 mm thick is on the belt. What shear stress is induced on the belt? Is the vat being filled or emptied by the action of the belt? (The former could certainly be the case if the belt were at rest.)

8.8 What must be the relationship between the velocity of the upper plate, U, and the pressure gradient, $\partial p/\partial x = c$, for $Q = 0$ in Eq. (8.2.11)?

8.9 What is the shear stress on the surface of the vertical shaft in Fig. P.8.9? What is the speed of descent of the shaft?

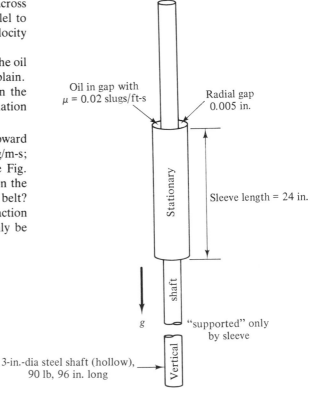

Figure P.8.9

8.10 A piston moves horizontally to the left at a speed of 0.5 in./s in a passage that has an internal diameter of 2 in. The piston's diameter is 1.5 in. and its length is 8 in. The pressure just ahead of the piston is 75 psig, while it is 25 psig just behind it. The viscosity of the liquid in the annulus is $2(10^{-5})$ slugs/ft-s. What are the shear stress and the total force on the piston? What force is on the piston when it is at rest but with the same pressure difference? Compare the shear stress on a stationary piston in a cylinder using Eqs. (8.3.15) and (8.2.10), the latter of which neglects the curvature of the surfaces.

8.11 A horizontal channel formed by two parallel plates has a pressure gradient in the $+x$ direction of -300 kPa/m. The channel is k millimeters high. What are the volume rate of flow per unit width, the average velocity, and the shear stress on the walls? Express this stress as a function of the average velocity. Is this stress greater or less than an estimate using the average velocity and $k/2$, half the channel width, in Newton's law of viscosity?

8.12 A round tube is inclined to the horizontal at an angle θ. Each end of it is attached to a reservoir at atmospheric pressure. Estimate the laminar flow rate of a liquid with viscosity μ in a tube of diameter D and length L. If $L = 50$ mm, $D = 0.3$ mm, $Q = 1$ cm³/min, and $\theta = \pi/2$, what is the kinematic viscosity of the liquid? The difference of the liquid levels in the two reservoirs is 75 mm.

8.13 A two-dimensional horizontal channel has water in its lower half with oil [SG = 0.9 and $\mu = 2(10^{-3})$ kg/m-s] in its upper half. A pressure gradient of -6 kPa/m in the flow direction is measured in the water layer. What is the rate of flow of oil and water in the channel if it is 12 mm high?

8.14 A piston (length = 4 in.) is slowly descending with speed U in a vertical tube whose diameter is only slightly larger than its own. ($D_{tube} = 2.00$ in. and $D_{piston} = 1.96$ in.) There is an upward flow of fluid in the gap between the piston and cylinder which can be modeled as the laminar flow between two parallel plates along which there is a pressure difference. (Assume a viscosity that is 5 times that of water and a specific gravity of 0.9.)

(a) For given values of Q and U, what is the relationship between the pressure gradient along the gap and the shear stress on the cylinder?

(b) What is the relationship between the value of Q and the pressure gradient for the prescribed value of U?

(c) What is Q in terms of U and the geometric data that are given?

(d) Draw a free-body diagram of the descending piston and write the equation that governs its motion.

(e) By referring to the previous parts of this problem, tell whether any additional information must be given to determine the weight of the piston for a given speed of descent, U.

8.15 Assume that the flow between parallel plates can model the flow in the clearance space between the plunger and cylinder of a diesel fuel injector. The inside diameter of the cylinder is 0.500 in., and the clearance is 0.0003 in. The plunger stroke is 0.5 in., and when the engine is rotating at 200 rpm, the injector sprays fuel into the cylinder 100 times per minute. Use the average piston speed as the speed of the moving surface, U. If the pressure in the cylinder is 1500 psi, what is the expected leakage past the piston (length = 1.5 in.) due to this pressure at one end, if atmospheric pressure is assumed to exit at the other end of the piston? Assume that the viscosity of the fuel is 3 times that of water.

8.16 Under different operating conditions, the piston described in Prob. 8.15 travels along its axis at 20 ft/s. What force must be applied to the piston to move it at this speed while there is no pressure difference between its ends?

8.17 A tube with an internal diameter of 0.015 in. and length 2 in. is part of a fuel-metering system. What is the maximum pressure drop that can be expected to produce laminar flow of a fuel with a viscosity of $5(10^{-5})$ slugs/ft-s and a specific gravity of 0.8? (Assume the flow to be fully developed in the tube.)

8.18 Use Eq. (8.3.13) to estimate the wall shear stress on a smooth tube with a diameter of 4 in. if the Reynolds numbers are, successively, 10,000 and 50,000 when water is flowing. What are the average velocities and pressure gradients for these two cases? (Note that f and τ_w are related.)

8.19 A hypodermic needle with an internal diameter of 0.25 mm and a length 50 mm should deliver 0.05 cm³/s of a serum with a viscosity 10 times that of water. The needle is connected to a syringe with an internal diameter of 15 mm. What force (in N and lb) must be applied to the piston in the syringe to achieve this flow? (Assume that flow is fully developed in the needle.)

8.20 A tube 3.4 m long with an internal diameter of 1 mm is used to deliver water from a reservoir that is used for intravenous feeding. The reservoir is 2 m above the terminus of the tube, which is at atmospheric pressure. What is the expected flow rate? What would be the flow rate if the water at the end of the tube were at a pressure of 10 kPa above atmospheric pressure?

8.21 A heat transfer device consists of two concentric tubes which have inside diameters of 1 and 2 mm. Their outside diameters are 1.6 and 2.6 mm, respectively. If water flows horizontally through both the inner tube and the annulus between the tubes, what is the ratio of the pressure gradients for the same rate of laminar flow? For this condition, what is the ratio of the maximum fluid speeds in the two flow cross sections? What is the maximum wall shear stress?

8.22 A sleeve bearing with an internal diameter of 1.346 in. supports a journal with a diameter of 1.340 in. The oil has a viscosity of 10 cP. What is the friction torque necessary to rotate the journal at 500 rpm if the bearing is 5 in. long? What is the corresponding power? If water is used to cool this bearing, what quantity must be supplied, per unit time, in order to keep the oil temperature in the bearing constant?

8.23 The sleeve bearing and journal of Prob. 8.22 are to operate at 100 rpm. It is suggested that the frictional torque could be significantly reduced if the clearance between the bearing and the journal were enlarged by removing material from the journal. What are the ratios of the torques to the original torque for clearances that are double and triple the original clearance?

8.24 Two concentric cylinders have an annular gap of 1 in. The outer cylinder has an internal diameter of $2\frac{7}{8}$ in. and rotates at 1250 rpm by means of external ball bearings while the inner one is stationary. Compare the shear stresses on the inner and outer cylinder. What torques do these shear stress induce if a liquid with a viscosity of 7 cP is in the annulus? (Assume that the annulus is 8 in. long and full of this liquid.)

8.25 Use Fig. 8.4.2 to estimate the viscosity of an oil in an annulus 3.5 in. in outside diameter and 10 in. long. The inner cylinder rotates at 10 rev/s when a torque of 0.05 lb-ft is applied to it. The specific gravity of the liquid is 0.85.

8.26 For the situation described in Prob. 8.25, what would be the torque necessary to drive the cylinder at half the speed given? What would be the torque if the abscissa value in Fig. 8.4.2 were 3.0?

8.27 An oil with a viscosity of 0.002 slugs/ft-s and a specific gravity of 0.9 is in the gap between two cylinders with a length of 9 in. and external and internal diameters of 5 and 5.4 in. What torque is needed to turn the inner cylinder at 40 rpm while the outer cylinder is stationary? If the inner cylinder is restrained and the outer cylinder is driven at 40 rpm, what is the required torque and power?

8.28 In Eq. (8.4.6), let $r = r_2 - y$, in which y is measured radially inward from the outer cylinder. Expand the solution in terms of the variable y/r_2. Does the result compare with Eq. (8.2.10) if the ratio r_1/r_2 is near 1? What would be an acceptable value for this ratio in order to assume that V_θ varies linearly with y?

8.29 Verify that Eq. (8.5.5) is the solution to Eq. (8.5.4) for the boundary conditions discussed in Sec. 8.5.

8.30 Does Eq. (8.5.5) imply that portions of the fluid are moving in the direction opposite that of the plate? If this is so, identify the location of this fluid.

8.31 A plate 20 in. long and 15 in. wide oscillates in the bottom of a tank containing glycerine with a 1.2 in. amplitude. The plate moves at a maximum of 5 cycles per second. What is the maximum force that will be induced in the cable system that causes the motion? If it is desired to have the free surface oscillate with an amplitude that is one one-hundredth of the plate's, what should be the depth of liquid?

For this criterion, what should be the depth of water used to perform initial tests on the apparatus?

8.32 What are the expected drag and skin friction drag coefficient for the case in Example 8.6.2? How does the value of C_f compare with the result obtained from Eq. (8.6.10) and Fig. 8.6.2?

8.33 How much should the fore-aft distance in Example 8.6.4 be changed to reduce the keel's drag 10 percent?

8.34 Examples 8.6.1 and 8.6.2 contained an assumed velocity profile. For air at standard conditions and $U = 3$ ft/s, determine the boundary-layer thickness at $x = 10$ ft and 15 ft. What is the volume rate of flow parallel to the plane wall that is within the boundary layer at these two stations? If the two flow rates are not equal, explain the difference.

8.35 For the velocity profile given in Example 8.6.1, find the distance δ^* such that $\delta^* U = \int(U - u)\,dy$ is the volume rate of flow that does not flow parallel to the boundary. The limits on the integral are 0 and δ, and the distance δ^* is called the *displacement thickness*.

8.36 Equation (8.7.16), which describes the evolution of the boundary layer, can also be applied to the flow over a flat plate. For the flow without a pressure gradient, λ is zero at the leading edge for two reasons, by virtue of the definition given by Eq. (8.7.12). (The boundary-layer thickness is zero there, and U' is zero everywhere.) Equations (8.7.13) and (8.7.14) show that if λ is zero, there is a (constant) proportionality between the various δ's. Use these observations to show that Eq. (8.7.16) predicts a boundary-layer growth that is proportional to $x^{0.5}$ in this case. Does the result of this problem agree with that of Eq. 8.6.7?

8.37 If a flat plate is aligned with an oncoming flow of air with a constant pressure gradient, Eq. (8.7.16) can be used to determine the boundary-layer growth along the plate. Because the boundary-layer thickness is zero at the leading edge, λ is also zero there, according to Eq. (8.7.12). This would imply that $Y = 0$ at the leading edge. Consequently, Eq. (8.7.18) can be used to find the value of Y at

station 1, $0 + dx$. Use these observations to determine the boundary-layer growth along a plate that is aligned with the plane of symmetry between two flat, diverging walls with an angle of 6° between them. The speed of the air is 3 ft/s just upstream of the plate's leading edge, and the walls are 2 ft apart at this point. Terminate the calculations when the plates are 3 ft apart. Plot the boundary-layer growth with distance along the plate.

8.38 Refer to Prob. 8.37 and determine the boundary-layer growth for a favorable pressure gradient, one for which the pressure decreases in the streamwise direction, by having the walls converge at 6° from a width of 2 ft at the plate's leading edge. Terminate the calculations when the walls are 1 ft apart. Plot the boundary-layer growth with distance along the plate that is on the plane of symmetry.

8.39 Use Eq. (8.6.9) to determine the skin friction coefficient C_f for the approximate solution $f(\eta) = \sin(\pi\eta/2)$. Compare the result with Eq. 8.6.10. What can you conclude?

8.40 The program in Appendix G can be modified to change the number of steps along the surface of the cylinder. How does the step size affect the point of separation when using $U = 2U_\infty \sin(x/1)$?

8.41 Use the program in Appendix G, or a variation thereof, to verify that the Kármán-Pohlhausen method can predict well the laminar separation point.

8.42 Equation (8.8.12) gives the theoretical thrust force with a simple model that does not include the curved streamlines associated with a rotating device. Let $k_2/k_1 = 2.2$, $k_1 = 0.01$ mm, $U = 9$ m/s, and $\mu = 40$ cP. Assume that the width of the bearing is 110 mm and its (effective) length is 3000 mm. What thrust could be developed?

8.43 What is the drag on the bearing that is modeled in Prob. 8.42? What power is expended in the bearing?

8.44 What is the proportion of the drag on a bearing that is due to the pressure gradient $-\partial p/\partial x$?

8.45 Compare the thrust-carrying capacity of bearings with $k_2/k_1 = 2$, 2.2, and 2.4.

9

The Use of Dimensionless Parameters in Fluid Mechanics

9.1 INTRODUCTION

This chapter will examine some of the specific results that have already been developed, with a view toward generalizing them. In this way, the assembled information will be more useful. At the same time, it should become clear how to gather other information expeditiously.

A particular example that can introduce our discussion appears in Section 8.6, where the viscous drag on a flat plate was treated. The ratio of the average shear stress on the plate—shear force divided by plate area—to the dynamic pressure, $\rho U^2/2$, was defined as the skin friction coefficient C_f. It was pointed out that drag coefficients have also been determined experimentally for a host of two- and three-dimensional bodies. This collection of data is useful in situations for which the resistance effects of the fluid flowing past bodies are of concern.

Example 9.1.1

Determine the drag per unit width of a plate 400 mm long that is one of the control surfaces of a rocket moving at 950 km/h in air at −20°C.

Solution The kinematic viscosity of air at this temperature is $1.1(10^{-5})$ m²/s, and 950 km/h corresponds to 264 m/s. The Reynolds number for the plate is

$$\mathrm{Re}_L = \frac{264(0.40)}{1.1(10^{-5})} = 9.6(10^6)$$

Thus, the drag coefficient will be the same as in Example 8.6.4, and the drag can be calculated by using the appropriate speed, density, and area.

This example is in principle no different from Example 8.6.4, but we now recognize that the drag coefficient for the plate in question could have been found by doing a test on a body of different size, and possibly in a different medium. This idea is often essential to the process by which experimental modeling is conducted in mechanics. The nondimensional data from one experiment are useful in another situation, provided both applications have conditions for which the appropriate nondimensional, independent parameters are the same. This is a most useful result and one upon which we shall expand in this chapter.

(We shall see that not all of the parameters need be the same in two situations in order to have the results of one applicable to the other. Usually, it is impossible to have all the parameters equal, but certainly the most important ones should be the same. We shall try to put substance into the phrase *most important ones* in Sec. 9.3. At the present, it is observed that if one were to conduct tests on the oscillations of a liquid column in a U tube, surface tension would have to be taken into account for tubes of small bore. However, in similar tests with surge tanks of large diameter, the effects of surface tensions would not be noticeable in the free-surface motions.)

The existence of the functional relationship between C_f and Re is but one instance in which a dependent variable, C_f in this case, is a function of an independent variable such as Re. Often the exact relationship between the two is unknown over a broad range of the independent variable yet we are confident that as long as the same physical phenomena are present in two situations, then the same nondimensional outcome of an experiment will result for the same dimensionless operation conditions. (A case in which shock waves exist and one in which there are no shock waves would not be corresponding situations.) Note carefully that we would not require that V, L, and ν in the Reynolds number be the same in both situations. Only the ratio VL/ν must be the same.

In a sense, we are postulating a deterministic system, such that if the input conditions are the same from time to time, the outputs will be the same, even allowing for statistical variation. In effect, the physical process contains a "transfer function" that converts the information about the input conditions into an output that we call the *solution*. If the input data to this transfer function are properly scaled in the form of nondimensional variables, the output will be a nondimensional variable that has great utility.

The notion of a transfer function for a deterministic physical process is reasonable, but it is not obvious how one chooses the dimensionless parameters in a particular situation. Surely, there must be some basis for the process. The development of the equation that relates the height ratio for a hydraulic jump provides some useful insight. It was found in Chap. 7 that

$$\frac{V_1^2}{gy_1} = \frac{1}{2}\left(\frac{y_2}{y_1}\right)\left(\frac{y_2}{y_1} + 1\right) \qquad (9.1.1)$$

This result can be solved for y_2/y_1, and it is evident that this ratio will be a function of V_1^2/gy_1 and some constants. What that function happens to be is not important now. What matters is the process by which we came to that conclusion. The problem was analyzed by means of the momentum equation. The kinds of forces that were acting in the flow direction were considered along with the physical quantities that influenced those forces; the momenta were also considered and their ingredients properly inserted into the equation.

Two processes were at work. The first one was the selection of the important physical characteristics of the problem that were relevant to the solution. The second step was the inclusion of these physical parameters into the appropriate mathematical statement of a physical "law," momentum conservation, in this case. The result of the analysis is Eq. (9.1.1).

When a complete and/or convenient mathematical formulation of the physical problem cannot be found, the second part of the process mentioned before is not possible, but the first one still is. This step is choosing, to the best of one's ability, the physical parameters that are deemed to affect the desired result. The selection of the important physical parameters of a problem—e.g., lengths, physical constants, and boundary conditions—is not an easy task, but it is an important one. Experience and insight will be useful in making the proper choices. The material in Chap. 8 was presented to provide a measure of these two assets. For every case that is undertaken, the investigator should start by inquiring into the interrelationships that are associated with the "conservation" of mass, energy, and momentum. The thoughts that stem from these considerations should spawn other useful ideas.

If the hydraulic jump were to be examined for the first time, one might ask about the parameters that determine the downstream height, y_2. A guide in the process of selection would be some general notions about the laws of nature that influence the phenomenon. For the case of the hydraulic jump in a channel of flowing water, one might speculate that y_2 would depend upon the upstream velocity and height. Since the jump is something like a wave at an interface, gravity would play a role because the potential energy would depend upon the mass of the medium and gravity. Thus γ enters the problem. Since the density of water is about 850 times that of the air in contact with the water, perhaps only the density of water would be significant in the problem. Momentum effects must also be assessed for the problem. The momentum will be determined by the depth of the fluid, its speed, and its density. The shear forces on the bottom of the flow channel where the hydraulic jump occurs could contribute to the change of momentum. Such shear stresses would be affected by the fluid's viscosity μ and by the local unsteadiness of the flow, which in turn would be influenced by the fluid's density. The roughness of the channel bottom, as measured by a representative roughness height denoted by ϵ, could alter y_2. Further deliberation might result in the justifiable inclusion of additional physical parameters in the problem (e.g., the slope of the channel). However, those parameters that have already been mentioned may be considered sufficient for now to delineate the problem.

We can summarize our conclusions so far by writing

$$y_2 = \phi(y_1, V_1, \rho, \mu, \gamma, \epsilon) \tag{9.1.2}$$

in which the symbol ϕ denotes the *functional relationship* that is sought. An exploratory study of the magnitude of the shear forces to be expected in the bed of the channel—perhaps one similar to that which led to Eq. (7.4.7)—could suggest that viscous forces play a secondary role in this problem. This would indicate that μ would not be essential to a preliminary examination of the phenomenon. If this were the case, one could infer from the balance of the forces—only hydrostatic forces remain—and the momentum changes that γ and ρ are involved in this balance in a special way. It is only their ratio, g, that matters. This would lead us to write

$$y_2 = \hat{\phi}(V_1, y_1, g) \tag{9.1.3}$$

if ϵ is also considered to have only a secondary influence on y_2. At this stage, the function $\hat{\phi}$ would still be unknown to us* but we would be in a position to determine it experimentally through a series of good tests. The sound planning of those tests would already have been started through the thought process that resulted in Eq. (9.1.3). However, one is able to extend and improve upon this thinking through additional reflection upon the roles played by the physical variables. We now turn to this matter.

9.2 THE Π THEOREM

It is convenient to continue to use the hydraulic jump as an aid in the development of our thinking. We postulated initially that the equation

$$y_2 = \phi(y_1, V_1, \rho, \mu, g, \epsilon) \tag{9.2.1}$$

is a statement for a deterministic process which will yield a given height for y_2 each time the same values of y_1, V_1, ρ, μ, g, and ϵ occur. This height for y_2 will happen regardless of the system of units in which y_1, V_1, ρ, and so on, are measured. The elevation y_1 could be given in feet or fathoms, for example. Then, if a consistent set of measurement units are employed for V_1, ρ, μ, and so on, in Eq. (9.2.1), a value of y_2 will result in the same set of units. This appears to be obvious because of our belief that the "transfer function," ϕ, is independent of the system of measurement. In what follows we shall use two reasonably common measuring systems and relate the results obtained from them. This will provide us with the pattern for converting from one system to another. Then we shall introduce a special system of measurement and thereby gain a result of great generality.

Suppose that y_2, y_1, V_1, and so on, are measured in a foot-hour-slug measurement system—one possibility, albeit a nonstandard one. Our seven variables y_2, y_1, V_1, ρ, μ, g, and ϵ contain all of these basic measurement units. Also, for the present, let $\hat{y}_2$, $\hat{y}_1$, $\hat{V}_1$, and so on, be values that are measured in a meter-second-kilogram system. Thus, to change from one system to another, one would set

$$\begin{aligned} \hat{y}_2 &= \alpha_1 y_2 \\ \hat{V}_1 &= \alpha_2 V_1 \\ \hat{\rho} &= \alpha_3 \rho \end{aligned} \tag{9.2.2}$$

and

in which α_1, α_2, and α_3 are scale factors to convert measurements in the foot-hour-slug system to those of the meter-second-kilogram system. By using the conversion factor for meters to feet, one can show that $\alpha_1 = 0.3048$ m/ft. Similarly, $\alpha_3 = 515.4$ (kg/m³)/(slugs/ft³). The factor α_2 converts speed units in one system to speed units in another, and thus α_2 implicitly contains a time scale. This is apparent from

$$\alpha_2 V_1 = \hat{V}_1 \quad \text{or} \quad \alpha_2 \frac{L}{T} = \frac{\hat{L}}{\hat{T}}$$

in which L is in feet and T is in hours, while $\hat{L}$ is in meters and $\hat{T}$ is in seconds for this presentation. But $\alpha_1 L = \hat{L}$ from Eq. (9.2.2), so that

---

* It is hoped that the experiments would show the function to be similar to $(-1 + \sqrt{1 + 8V_1^2/gy_1})(y_1/2)$ and that they would substantiate the analysis of Chap. 7 that yielded Eq. (9.1.1).

$$\alpha_2 \frac{L}{T} = \alpha_1 \frac{L}{\hat{T}} \quad \text{or} \quad T = \frac{\alpha_2}{\alpha_1} \hat{T}$$

If T is in hours and $\hat{T}$ in seconds, then $\alpha_2/\alpha_1 = 1/3600$ h/s. The α's are the scale factors for the dimensions of the system, perhaps implicitly, as in the case of the present α_2. *There are as many α's as there are basic dimensions* of the problem. In some problems, there are only two dimensions (e.g., for a pendulum, there are length and time). The factor α_3 is a scale factor for the density (i.e., mass) of the system:

$$\alpha_3 \rho = \hat{\rho} \qquad \alpha_3 \frac{M}{L^3} = \frac{\hat{M}}{\hat{L}^3}$$

or

$$M = \left(\frac{L^3}{\hat{L}^3}\right)\left(\frac{1}{\alpha_3}\right)\hat{M} = \left(\frac{1}{\alpha_1^3 \alpha_3}\right)\hat{M}$$

If M is specified in slugs and $\hat{M}$ in kilograms, the numerical value of α_3 becomes 14.594 kg/slug.

Our discussion of Eq. (9.2.1) implies that in our second system of units,

$$\hat{y}_2 = \phi(\hat{y}_1, \hat{V}_1, \hat{\rho}, \hat{\mu}, \hat{g}, \hat{\epsilon}) \tag{9.2.3}$$

since the function ϕ is not dependent upon the units of measurement. But the inter-relationships given by Eq. (9.2.2) state that $\hat{y}_1 = \alpha_1 y_1$, $\hat{V}_1 = \alpha_2 V_1$, and so on, so that Eq. (9.2.3) can be written as*

$$\hat{y}_2 = \alpha_1 y_2 = \phi\left(\alpha_1 y_1, \alpha_2 V_1, \alpha_3 \rho, \alpha_1 \alpha_2 \alpha_3 \mu, \frac{\alpha_2^2 g}{\alpha_1}, \alpha_1 \epsilon\right) \tag{9.2.4}$$

Equation (9.2.4) is a minor, but important, restatement of Eq. (9.2.3). It literally says that the values $\hat{y}_1$, $\hat{V}_1$, and so on, in one system can be scaled to give y_2 and the appropriate scale factor, α_1. Why is this important? Simply because α_1, α_2, and α_3 need not have been scale factors associated with a conversion to the metric system even though we used such a "transformation" to facilitate the discussion.[†]

Assume for the present that the length scale factor α_1 could be $1/y_1$, in the case of a hydraulic jump. The velocity scale factor, α_2, also might be set to $1/V_1$; the density scale factor α_3 could be equated to $1/\rho$. This would mean that Eq. (9.2.4) would be

$$\alpha_1 y_2 = \frac{y_2}{y_1} = \phi\left(1, 1, 1, \frac{\mu}{V_1 y_1 \rho}, \frac{g y_1}{V_1^2}, \frac{\epsilon}{y_1}\right) \tag{9.2.5}$$

---

* Note that

$$\hat{\mu} = \frac{\hat{M}}{\hat{L}\hat{T}} = \frac{\alpha_1^3 \alpha_3}{\alpha_1(\alpha_1/\alpha_2)} \frac{M}{LT} = \alpha_1 \alpha_2 \alpha_3 \frac{M}{LT} = \alpha_1 \alpha_2 \alpha_3 \mu$$

A similar pattern can be used to verify that $\hat{g} = (\alpha_2^2/\alpha_1) g$.

[†] We note that there are other, perhaps unconventional, measurement systems, based on different standards. The human frame is drawn about 9 times the length of the head as a standard for good proportion. The distance ahead of an object where something significant occurs may not be best specified in feet or meters. Instead, the number of diameters of the sphere in an experiment, for example, could be used to give the location. Equally important, for some phenomena the critical unit of time may not be the second (i.e., 1/31,556,025.9747 of the time for the earth to orbit the sun as measured at noon on January 1, 1900). A unit of time that is particular to the problem, say, the time for the fluid to move one diameter of a sphere, may be more meaningful.

Each factor in the relationship of Eq. (9.2.5) is nondimensional. Moreover, this last equation has three fewer variables than Eq. (9.2.4) because three constants have been introduced. There are three constants because we had three α's; one for each of the characteristic scales associated with the problem. Now we shall justify setting $\alpha_1 = 1/y_1$.

Let us not measure y_1 in feet or meters. Instead, assume that we observe that the depth of water upstream of the jump, y_1, exactly corresponds to the distance between the author's elbow and the tip of his extended hand. We choose to call this length a *del** and observe incidentally that it also corresponds to 1.5 ft. Hence α_1 would be $\frac{2}{3}$ del/ft. We would have $\tilde{y}_1 = \alpha_1 y_1$, $\tilde{y}_2 = \alpha_1 y_2$, and so on. (The tildes signify quantities in this third measurement system.) A measurement of time could be made with a pendulum that was deemed to be the standard for the realm. A special vessel, filled with water, could be used to specify $\tilde{\rho}$ in "jugs." Then, instead of Eq. (9.2.3), we would have

$$\tilde{y}_2 = \phi(\tilde{y}_1, \tilde{V}_1, \tilde{\rho}, \tilde{\mu}, \tilde{g}, \tilde{\epsilon})$$

for our new system of measurement. If we chose to convert that system to a foot-slug-hour system, we would have something like Eq. (9.2.4),

$$\alpha_1 y_2 = \phi(\alpha_1 y_1, \alpha_2 V_1, \alpha_3 \rho, \ldots, \ldots, \ldots) \tag{9.2.6}$$

which has been abbreviated, since we wish to concentrate on the first term of the function ϕ. What is that term? It is the product $\alpha_1 y_1$ in which y_1 is the numerical value for the depth in the foot system that is associated with the observed value of $\tilde{y}_1$. We took the upstream depth, $\tilde{y}_1$, to be the standard of length, in dels, of 1 so that this product has a numerical value of exactly *unity* in this case. Consequently, $\alpha_1 = 1/y_1$. This is the consequence of measuring everything with respect to the upstream depth. The upstream depth was used as the reference length.

Similarly, we can use the upstream velocity as the standard and say, for example, that the downstream velocity is 12 percent of the upstream velocity. This would lead to $\alpha_2 = 1/V_1$. The justification for setting $\alpha_3 = 1/\rho$ should now be apparent. Thus, if we use reference, or characteristic, scales that are associated with the problem at hand, we are able to recast the problem in terms of nondimensional variables.

The number of these variables, also called *pi parameters*, will be less than in the original variable set. Usually, the number is less by the number of significant dimensions of the problem. In our example with the hydraulic jump, we originally had chosen seven variables—six independent ones and one dependent one—and there were three significant dimensions (mass, length, and time). Thus, we would expect four $(7 - 3)$ dimensionless pi parameters to be associated with the problem. If we define, in Eq. (9.2.5),

$$\pi_1 = \frac{y_2}{y_1} = \text{height ratio}$$

$$\pi_2 = \frac{V_1 \rho y_1}{\mu} = \text{Reynolds number}$$

$$\pi_3 = \frac{V_1^2}{g y_1} = (\text{Froude number})^2$$

and

$$\pi_4 = \frac{\epsilon}{y_1} = \text{relative roughness}$$

---

* A measurement standard called the *ell* did exist, but the del is Debler's elbow.

we can write symbolically

$$\pi_1 = \phi(\pi_2, \pi_3, \pi_4)$$

In this way we have demonstrated the truth of Buckingham's pi theorem. This states that

> If one has a system of physical parameters that are n in number and these parameters contain m independent dimensions, then the system can be described by $n - m$ independent, dimensionless parameters.

We shall devote much of the next section to selecting the n parameters and finding the $n - m$ dimensionless groups.

9.3 CHOOSING THE EXPERIMENTAL PARAMETERS

The previous discussion concerning the hydraulic jump was intended to introduce the reader to the utility of dimensional analysis. It is now profitable to extend these notions to other types of flow. The major handicap that the reader may have in choosing the significant parameters in a new situation will be the lack of familiarity with a particular phenomenon. But a little bit of common sense will be extremely valuable. Also, the fundamental laws of physics will provide the framework for picking the salient features of the problem under examination. In this section, we shall discuss several flow phenomena and make statements about the important factors in each situation. Those factors that determine the outcome of an experiment become the independent parameters of the problem. A particular result of the experiment becomes the dependent parameter.

9.3.1 Flow in Pipes

It is useful to recall our consideration of the factors that would influence the pressure drop in a round horizontal conduit. The term *pressure drop* stimulates thoughts about forces. The ideas about the momentum equation are easily part of the same thought process. One can now focus on the elements in such an equation, even though it might not be possible to write down a complete description of them. What are the forces? For a round pipe and a viscous fluid, shear stresses should be considered. These shear stresses will be influenced by the viscosity, μ. The velocity profile itself will be a factor in the magnitude of the shear stresses; consequently, the velocity in the pipe should be a variable that would influence the pressure changes along its length. For a given type of flow (laminar or turbulent), the maximum or average velocity could be used as the variable, since the two are interrelated. The size of the pipe will determine the velocity gradient for a given discharge. Thus, the diameter of the pipe plays a role. The roughness of the inside surface of the pipe will affect the momentum transfer near the wall and, consequently, should be included in the discussion. Is the fluid density important? It is a part of the total change of momentum. Sometimes this change is called the change of the inertia terms. Are such inertia terms important for the flow in a pipe of constant cross section? Our analysis for the steady, laminar flow did not include ρ. The constant discharge of a turbulent liquid from a pipe could be another matter! Such flows can be steady, with respect to time averages, but are associated with a great number of high-frequency fluctuations of the flow variables (such as velocity and pressure). Thus, in a turbulent flow there are accelerations of the fluid particles, so it is likely that ρ should enter the picture. Experiments will show that, in a certain regime of flows, ρ is indeed important. In the laminar regime, ρ is of no consequence, a fact that supports the theoretical analysis of Sec. 8.3. Finally, we might expect that the change in pressure will depend on the length of the pipe. But for fully developed flow, the pressure changes should be proportional to L. Thus, our investigation will be more general if we focus on the change of pressure per unit length, $\Delta p/L$.

A consideration of the variables in the problem of pipe flow such as this leads us to write

$$\frac{\Delta p}{L} = \overline{\phi}_1^*(\mu, V, D, \epsilon, \rho) \tag{9.3.1a}$$

in which $\overline{\phi}_1^*$ is a function that must be determined. With this expression as a beginning, it is useful to ask whether any additional effects are important. We can only use our best judgment in augmenting the list of important variables. Sometimes terms that do not have a significant influence on the results will be naively included. Our subsequent experimental observations should show us that this is the case. This may prove to be an expensive piece of information, but it is a new experience and may be of advantage in the future in a different situation. Alternatively, we might fail to include an important parameter, with the result that a strong correlation of the results with the parameters that have been selected will not occur. Whenever experimental results contain anomalies, one should become suspicious that the initial notions about the phenomenon are incomplete or incorrect. In such cases, we are forced to reexamine the problem with hope of discovering the deficiencies in our original premise. For example, compressibility may be an important feature in some conduit flows. (The transport of natural gas in pipelines is an example for which compressibility can be significant.) In Chap. 14, we shall see that under certain conditions, the pressure drop may be sufficient to cause a considerable expansion in volume of the gas. This, in turn, greatly affects how much fluid can be moved along the conduit. For such flows, the correlation expressed by Eq. (9.3.1a) is insufficient and some parameter that is a good indicator of the influence of compressibility must be chosen. The *speed of sound, c,* is one such indicator as well as k, the ratio of specific heats. Thus, when concerned with compressible fluids, we write, instead of Eq. (9.3.1a),

$$\frac{\Delta p}{L} = \overline{\phi}_1(\mu, V, D, \epsilon, \rho, c, k) \tag{9.3.1b}$$

We shall return to this statement when we proceed to discuss the nondimensional parameters in Sec. 9.5. We next turn to the flow past a circular cylinder in order to expand our horizons.

9.3.2 Pressure Distribution on a Circular Cylinder

If we wish to know something about the pressure distribution on a circular cylinder when a fluid is streaming over it, we might ask, "What is the difference in pressure between a point on the surface of the cylinder and the stagnation point?" Call this difference Δp. We begin by assuming the fluid to be viscous. Then we draw upon all the experience that we have assembled so far. It would be reasonable to conjecture that the pressure at a given location would depend upon the point on the surface of the cylinder as determined by the angle θ. Should the diameter of the cylinder affect the phenomenon? We might seek to answer the question on the basis of the experience gained with the flow past the flat plate. Shear stress effects can influence pressure. The actual physical distance from the stagnation point is dependent upon both the angle θ and the radius of the cylinder. Thus it is natural to include the diameter D as a variable that could affect the pressure. Does density enter the picture? The fluid is moving quite slowly near the stagnation point. (By definition, the velocity is zero there.) As it moves away from the stagnation point and passes close to the cylinder, the fluid will be accelerating. Thus, inertia terms are important. The magnitude of these inertia terms will be affected by both the density ρ and the local velocity. The latter will be a function of the upstream velocity, U_∞. Because the shear stresses may affect the pressure, through the momentum principle, the viscosity μ may be a factor in the problem. What about the velocity gradient? Again, the velocity gradient at the surface of the cylinder will be influenced by the position along the surface. (This seems to have been specified already through our inclusion of both the diameter and the angle in our considera-tion.*) Compressibility may be important; for now, however, we shall restrict our-selves to cases for which we can consider the fluid to be incompressible. Thus the speed of sound will not be included as a parameter, because of our decision to limit the scope of our experimental analysis. What about roughness of the cylinder's surface? The roughness may influence the shear, and we have suspected that the shear may influ-ence the pressure. In view of this, it would seem natural to include a roughness parameter, ϵ, which might be the average—or some other statistical measure—of the height of the protrusions. These considerations lead us to write

$$p - p_\infty = \overline{\phi}_2(\theta, D, U_\infty, \rho, \mu, \epsilon) \qquad (9.3.2)$$

in which p_∞ is the ambient pressure and $\overline{\phi}_2$ is the functional relationship that is sought. (The associated nondimensional parameters will be determined in subsequent sections after we consider some more flow cases.) Information about this phenomenon was given in terms of nondimensional variables in Fig. 5.4.3.

9.3.3 Drag on a Cylinder

Often, the local pressure on a cylinder may not be of interest. Instead, the total drag on the object is of concern. If we look back upon the last expression for the pressure at various positions on the cylinder, it is reasonable to conclude that the net force in the direction of the airflow acting upon the cylinder would be influenced by this pressure distribution as well as by the shear stresses. The total force would be obtained by correctly integrating the shear stresses and the pressure distribution, completely around the cylinder. A moment's examination of Eq. (9.3.2) would lead us to conclude that the drag force on the cylinder would be influenced by all of the elements except the angle θ; the angle would be eliminated by the integration process. Hence, we would write for a cylinder

---

* We can recall from Sec. 8.7 that as one moves along a surface, the velocity gradient changes.

$$\text{Drag} = \overline{\phi}_3(D, U_\infty, \rho, \mu, \epsilon) \tag{9.3.3}$$

excluding, of course, the possibility of compressibility effects, with associated shock waves or cavitation effects on the body.

9.3.4 Lift on a Body

Many bodies have, in addition to drag, a force component that is perpendicular to the direction of the oncoming fluid stream. This is called *lift*. What are the physical parameters in a flow that would influence the lift on a body? We might speculate about the elements that would fit into an equation for forces. Then, too, we think of lifting surfaces as being winglike.* Since wings come in various shapes, we must now decide on a particular shape to facilitate our present discussion. We now contemplate the lift due to a curved shape such as that shown in Fig. 9.3.1. This is an airfoil with a given relationship between the chord c and the curvature of the foil as specified by the maximum camber, e, located at f. We assume that we are interested in members of a particular family of *geometrically similar* foils. This means that, for all of the foils, the ratios e/c and f/c are constants. If this is the case, the only adjustable dimension is the length of the chord, c. Once this is specified, the remaining geometric features such as camber are determined, because all of the foils are geometrically similar. Clearly, since the lift will be determined by the size of the lifting surface, we shall limit ourselves to considering the lift per unit length of span (i.e., the span of the foil will not enter the discussion). What else is there to consider? Integration of the vertical components of the pressure and shear forces is needed to get the total force on the foil in the vertical direction, so that the items that influence pressure and shear stress should also influence the lift. The parameters ϵ, ρ, μ, and U_∞ are natural to consider in this problem also. But one other element that was not included in the previous drag calculation will be important in our present consideration. (It will also influence the drag on this particular body.) It is the *angle of attack,* α. This is the angle made by the chord line with respect to the direction of the oncoming fluid stream. (If you reflect upon your observations of birds, you may have noted that as they approach their perch they seem to be nearly standing still. Their act of landing is accompanied by an apparent high angle of attack of their extended wings. While not all features of bird and aircraft flight

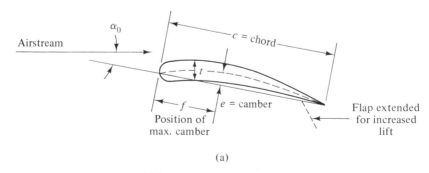

(a)

Figure 9.3.1 (a) Schematic drawing of airfoil to illustrate lift.

* However, when a wind blows past angle-iron members of a structure, lift forces are induced on them as well as drag forces.

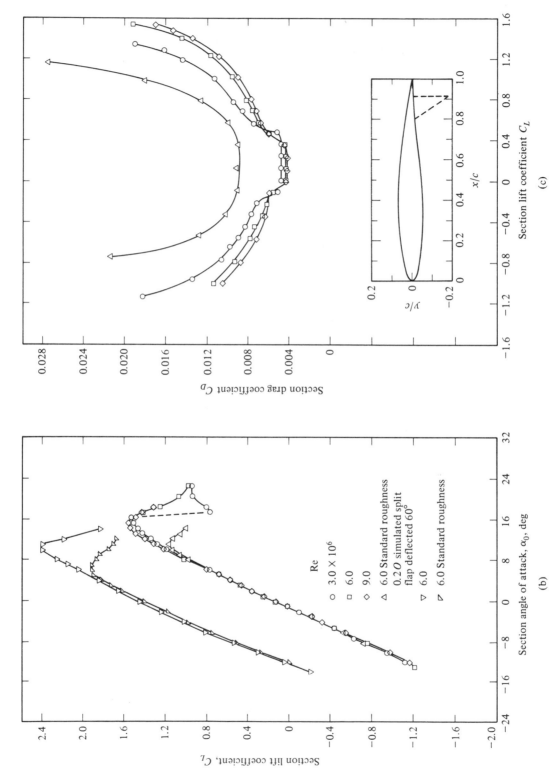

Figure 9.3.1 (b) Lift coefficient versus angle of attack, $\alpha_{\circ}$, for NACA 64_1-222 wing section. (c) Drag coefficient versus angle of attack, $\alpha_{\circ}$, for a NACA 64_1-222 wing section. ((b) and (c) From "Theory of Wing Sections" by Abbott and von Doenhoff, Dover Publications, Inc., by permission.)

are similar, one might be inclined to suspect that the angle of attack will play a role in the determination of the lift. Our experience with kites, which have nearly flat lifting surfaces, also reinforces our supposition that the angle of attack will be important.)

In view of this, it seems reasonable to write

$$\text{Lift} = \overline{\phi}_4(c,\, U_\infty, \rho, \mu, \epsilon, \alpha) \tag{9.3.4}$$

in which c, in this case, is the characteristic dimension of the body.

9.3.5 Wake Shedding

Another phenomenon can be mentioned if we return to the flow past a cylinder. When a fluid stream passes over the cylinder, a flow results that is two-dimensional near the upstream portions of the body. As the fluid flows over the body, it is retarded by it. After the fluid flows past the body, one can detect fluid particles that are slower than the free-stream velocity U_∞. The region these particles occupy is called the *wake*. The wake behind a bluff body is often similar to those which are shown in Fig. 9.3.2.* An examination of this wake will show that it is unsteady even though the oncoming stream is steady. This unsteadiness is associated with a periodic change of the boundary layer along the upper and lower surfaces of the cylinder. The fluid does not flow in only one direction along the entire length of most bodies. At some point, the streamline departs from the body's surface. This location has been called the *separation point*. Beyond this point, the fluid flows toward it from the rearward part of the body. Vortices, or swirls, appear and shed alternately from separation points on the upper and lower portions of the cylinder. This means that the actual pressure distribution on the cylinder also varies periodically in synchronization with the shifting location of the wake. There are, then, varying forces both in the direction of the fluid stream and transverse to it. The transverse forces cause the cylinder to vibrate; depending upon the geometry and material of the cylinder, this vibration could be at a resonant frequency. The result of this forcing function can be the initiation of large-amplitude vibrations that can cause damage to the structure—a bridge or a Venetian blind— through fatigue or large-amplitude deformations. Boiler tubes as well as aircraft struts are subject to such oscillations.

The list of applications could go on, and consequently a study of the factors that affect the frequency of such oscillations is of interest. It appears plausible that the frequency of shedding can be associated with the oscillations of the boundary layer. To understand what influences these oscillations, we should know about all of the features that could affect separation. We think that separation is enhanced whenever a fluid passes into a region where the pressure increases with distance in the direction of the fluid motion. (Such a situation would occur in a Venturi tube when the fluid moves from the narrow section, the throat, to a position farther down in the diverging section of the device. If the rate of this pressure rise is not too large, the flow may not separate; but it appears to have a tendency to separate if the angle of divergence becomes too large, say, greater than 7°.)

---

* At even higher Reynolds numbers than those shown, the pattern of flow is more disorganized, but the overall structure persists. It is called the *von Kármán vortex street* (probably because the vortices resemble trees lining a street on an architect's drawing).

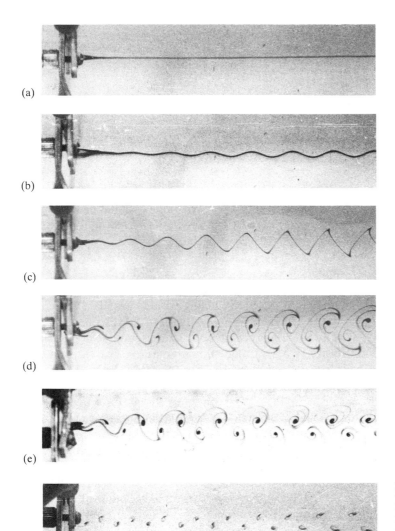

(a)

(b)

(c)

(d)

(e)

(f)

Figure 9.3.2 Wake patterns behind a circular cylinder: (a) Re ≈ 30; (b) Re ≈ 40; (c) Re ≈ 47; (d) Re ≈ 55; (e) Re ≈ 67; (f) Re ≈ 100. *(From "Physical Fluid Dynamics" by David Tritton, Oxford University Press, 1988, by permission.)*

In the case of the circular cylinder, the fluid particles passing along the uppermost or lowermost portions of the surface must decelerate if they are to continue along the surface to the rear stagnation point. The magnitude of the deceleration will depend on the fluid density ρ and the speed of the oncoming flow, U. Consequently, these variables should influence separation and its frequency, f. The diameter of the cylinder will determine the length of the path of the particles that pass near the cylinder and consequently will specify the time for the velocity changes (i.e., deceleration) to occur. This deceleration will be affected by the forces on the fluid particles. The fluid stresses will depend on U and μ as well as ρ. (Note that p depends on U and ρ, from the Bernoulli equation, which is a good model to predict the streamwise pressure gradient along the unseparated portion of the body's surface.) If we restrict our discussion to smooth cylinders, we would conclude that the frequency of wake shedding could depend on the following variables:

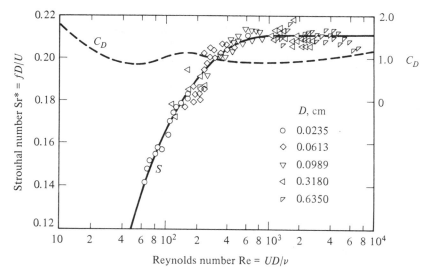

Figure 9.3.3 The Strouhal number versus the Reynolds number for the flow past a circular cylinder as measured by A. Roshko. *(N.A.C.A. Report 1191)*

$$f = \overline{\phi}_5(U, \rho, D, \mu) \tag{9.3.5}$$

This frequency has been nondimensionalized with U and D in Fig. 9.3.3, which presents experimental results for three decades of the Reynolds number.

It is hoped that sufficient examples have been given to provide the reader with an inkling of how one could proceed in a new situation by the recollection of the applicable physical laws.

9.4 CALCULATION OF THE Π PARAMETERS

Once the n dimensional parameters* for a particular problem have been selected, one chooses m of them to serve as scaling parameters. If we assume that the dimensions of length, time, and mass are present in the dimensional parameters, we will have

1. A characteristic length scale. (Usually, this is an important size dimension of the system, the upstream depth y_1, in our example of the hydraulic jump.)
2. A characteristic time scale. (Sometimes, this can be a characteristic velocity which, with the characteristic length, implicitly gives a characteristic time; e.g., V_1 and y_1 define a characteristic time scale.)
3. A characteristic scale for mass. (Often, this is the medium's density, ρ.)

These m parameters are combined with the remaining $n - m$ variables, which have dimensions, to form the $n - m$ dimensionless groups. The characteristic scales that are selected are necessary to achieve the dimensionless character of the combination.

---

* Be sure to include the dependent variable, too, in the number n.

For our example of the hydraulic jump that was discussed in Sec. 9.1, we had, in Eq. (9.1.2), $y_2 = \phi(y_1, V_1, \rho, \mu, g, \epsilon)$. The characteristic scales could be chosen to be y_1, V_1, and ρ. We can form π_1 by combining y_2 with the characteristic scales of length, time, and mass to form a dimensionless parameter as follows:

$$\pi_1 = y_2(y_1)^{a_1}(V_1)^{b_1}(\rho)^{c_1} \qquad (\text{dimensionless: } M^0 L^0 T^0)$$

$$M^0 L^0 T^0 = (L)(L)^{a_1}\left(\frac{L}{T}\right)^{b_1}\left(\frac{M}{L^3}\right)^{c_1}$$

By equating the exponents on M, L, and T, one can solve for the values of a_1, b_1, and c_1.

$$M: c_1 = 0$$
$$L: 1 + a_1 + b_1 - 3c_1 = 0$$
$$T: -b_1 = 0$$

Hence, $a_1 = -1$, so that $\pi_1 = y_2/y_1$.

In the same way, we combine y_1, V_1, ρ, and μ to form π_2:

$$\pi_2 = \mu(y_1)^{a_2}(V_1)^{b_2}(\rho)^{c_2}$$

$$M^0 L^0 T^0 = \frac{M}{LT}(L)^{a_2}\left(\frac{L}{T}\right)^{b_2}\left(\frac{M}{L^3}\right)^{c_2}$$

We can continue the procedure and show that $a_2 = -1 = b_2 = c_2$.

Starting with

$$\pi_3 = g(y_1)^{a_3}(V_1)^{b_3}(\rho)^{c_3}$$

we are led to the values of a_3, b_3, and c_3, with the result that

$$\pi_3 = g(y_1)^1(V_1)^{-2}\rho^0$$

Finally, π_4 becomes ϵ/y_1. The result is that an equivalent form of Eq. (9.1.2) is

$$0 = \tilde{\phi}(\pi_1, \pi_2, \pi_3, \pi_4) = \tilde{\phi}\left(\frac{y_2}{y_1}, \frac{\mu}{\rho V_1 y_1}, \frac{gy_1}{V_1^2}, \frac{\epsilon}{y_1}\right)$$

Example 9.4.1

Equation (9.3.3) can be given a nondimensional form by choosing D, U, and ρ as the scales for length, time, and mass. What is this form?

Solution There are six variables in this equation, and the three dimensions of M, L, and T occur. Hence, $n = 6$ and $m = 3$, so that there are $6 - 3$, or 3, dimensionless π parameters to be formed. These are

$$\pi_1 = (\text{Drag})(\rho)^{a_1}(D)^{b_1}(U_\infty)^{c_1} = \left(\frac{ML}{T^2}\right)\left(\frac{M}{L^3}\right)^{a_1}(L)^{b_1}\left(\frac{L}{T}\right)^{c_1}$$

$$= M^0 L^0 T^0$$

$$\pi_2 = (\mu)(\rho)^{a_2}(D)^{b_2}(U_\infty)^{c_2} = \left(\frac{ML}{T^2}\right)\left(\frac{M}{L^3}\right)^{a_2}(L)^{b_2}\left(\frac{L}{T}\right)^{c_2}$$

$$= M^0 L^0 T^0$$

and

$$\pi_3 = (\epsilon)(\rho)^{a_3}(D)^{b_3}(U_\infty)^{c_3} = (L)\left(\frac{M}{L^3}\right)^{a_3}(L)^{b_3}\left(\frac{L}{T}\right)^{c_3}$$

$$= M^0 L^0 T^0$$

To solve for a_1, b_1, and c_1, we write equations in which the sums of the exponents of M, L, and T are zero. We do this in a formal way by writing for the terms associated with π_1 the following:

$$M: 1 + 1(a_1) + 0(b_1) + 0(c_1) = 0$$
$$L: 1 - 3(a_1) + 1(b_1) + 1(c_1) = 0$$
$$T: -2 + 0(a_1) + 0(b_1) - 1(c_1) = 0$$

This system of equations will have a solution provided that the determinant of the coefficients of the unknowns is nonzero. Similarly, the solution for a_2, b_2, and c_2 results from the system of equations that will make the net exponent of M, L, and T zero for the terms belonging to π_2

$$M: 1 + 1(a_2) + 0(b_2) + 0(c_2) = 0$$
$$L: -1 - 3(a_2) + 1(b_2) + 1(c_2) = 0$$
$$T: -1 + 0(a_2) + 0(b_2) - 1(c_2) = 0$$

The exponents a_3, b_3, and c_3 necessary for π_3 are obtained in the same way:

$$M: 0 + 1(a_3) + 0(b_3) + 0(c_3) = 0$$
$$L: 1 - 3(a_3) + 1(b_3) + 1(c_3) = 0$$
$$T: 0 + 0(a_3) + 0(b_3) - 1(c_3) = 0$$

The nonzero determinants that are necessary for a solution to these systems of equations could be formed by writing down an array with m (= the number of dimensions) rows and n (= the number of variables) columns. Such an array has as its elements the exponent of the particular dimension that is contained in the particular variable. In our case, we would have

	Drag	ρ	D	U_∞	μ	ϵ
M	1	1	0	0	1	0
L	1	-3	1	1	-1	1
T	-2	0	0	-1	-1	0

The determinant of the terms outlined with the dashed line was required to be nonzero for a solution of a_1, b_1, and c_1.* The result of the algebra associated with the dimension-variable matrix gives us a_1, b_1, and so on, so that

$$\pi_1 = \frac{\text{Drag}}{\rho U_\infty^2 D^2} \qquad \pi_2 = \frac{\mu}{\rho U D} \qquad \pi_3 = \frac{\epsilon}{D}$$

The first of these ratios is proportional to the drag coefficient, the second is the reciprocal of the Reynolds number, and the third is a relative roughness. In view of these results and Buckingham's pi theorem, we now write

$$\frac{\text{Drag}}{(\rho U_\infty^2/2)(\text{frontal area})} = \phi\left(\text{Re}, \frac{\epsilon}{D}\right) \tag{9.4.1}$$

Experimental data for cylinders and spheres are given in the form of this equation in Fig. 9.4.1. The locus of experimental points determines the function ϕ for each case. The data for which compressibility effects are important are shown by the inclusion of the Mach number as a parameter.

---

*This system for a_2, b_2, and c_2 depended on the same determinant.

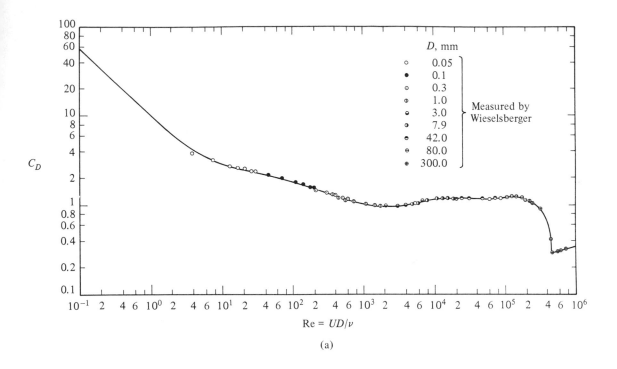

(a)

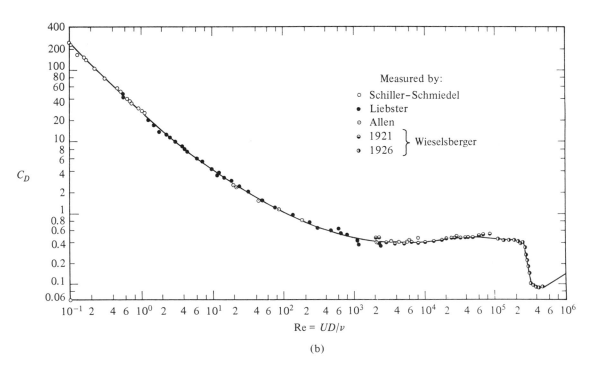

(b)

Figure 9.4.1 (a) Drag coefficient for a circular cylinder versus Reynolds number. (b) Drag coefficient for a sphere versus Reynolds number. *(From* Boundary-Layer Theory, *by H. Schlichting, McGraw-Hill Book Company, by permission.)*

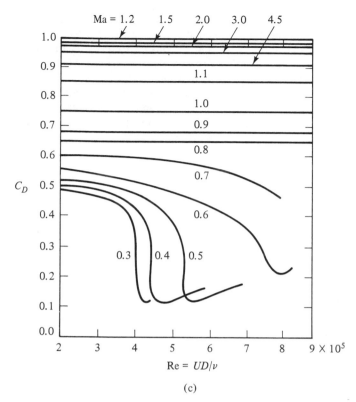

Ma = 1.2   1.5   2.0   3.0   4.5

C_D

Re = UD/ν

(c)

Figure 9.4.1 (c) Effect of the Mach number on the drag coefficient of a sphere for a range of Reynolds numbers, as measured by A. Naumann. *(From* Boundary-Layer Theory, *by H. Schlichting, McGraw-Hill Book Company, by permission.)*

The π_1 term would be equivalent to the one just found if we had sought to non-dimensionalize the product

$$(U_\infty)(\text{drag})^{a_1'}(\rho)^{b_1'}(D)^{c_1'}$$

If we had solved for a_1', b_1', and c_1', we would have required the determinant outlined by the dotted line to be nonzero. In general, for $(n - m)$ pi parameters to exist, all of the $m \times m$ determinants that can be formed by elements of the $m \times n$ matrix must be nonzero. Then the rank of the matrix will be m. (The *rank* of the matrix, r, is equal to the order of the largest nonzero determinant that can be formed by the matrix.)

In some special cases—usually those wherein a casual determination of the primary dimensions misses some special feature of the problem—the rank of the dimension-variable matrix can be less than m. This gives us the rule that *the number of π parameters is $(n - r)$,* in which r is usually equal to m, the number of dimensions of the system. Sometimes, in choosing these dimensions, say, in a statics problem, we make an unwise choice. This would be true if we choose to nondimensionalize such a problem with M, L, and T when only forces and lengths (or, equivalently, distances) are at issue. Then we would have erroneously—but seemingly necessarily, through $F = ML/T^2$—introduced the mass and time dimensions. The former is usually associated with a dynamic case. However, the next example shows that this is of no difficulty.

Example 9.4.2

A pendant drop such as that shown in Fig. 1.6.1 has a maximum diameter D that depends on the tube diameter d. What other variables are appropriate to the determination of D? Determine the form of the relationship between the dimensionless variables.

Solution A glance at Fig. 1.6.1 should remind us of the force balance that must be satisfied. The surface tension σ and the specific weight γ are appropriate to include. Consequently, we are assuming that $D = \phi(d, \sigma, \gamma)$. If M, L, and T are chosen as principal dimensions, as is often the case, these four variables of the problem should give $1 (= 4 - 3)$ pi parameter. The dependent variable will not be chosen as a characteristic parameter, in accordance with our usual desire to have an explicit relationship for D. Besides that, the variable d can adequately fulfill the role of a length scale. Mass and time are contained in σ and γ, so that we seem to have sufficient means to find the exponents in $D(d)^a(\gamma)^b(\sigma)^c$ to yield $M^0 L^0 T^0$. If the dimensions of the variables are substituted in this expression, we have

$$L(L)^a \left(\frac{M}{T^2}\right)^b \left(\frac{M/T^2}{L^2}\right)^c = M^0 L^0 T^0 \tag{9.4.2}$$

The system of equations for the solution of the exponents is

$$M: 0 + 0(a) + 1(b) + 1(c) = 0$$
$$L: 1 + 1(a) + 0(b) - 2(c) = 0$$
$$T: 0 + 0(a) - 2(b) - 2(c) = 0$$

However, the first and third equations are not independent. Thus, any value of c will give an associated value of b in the first equation and a value of a in the second. The value of c could be zero, so that $a = -1$ and the π term—the only π term we expect if the $(n - m)$ rule is adhered to—is D/d. Then

$$\frac{D}{d} = \phi(\text{nothing but constants})$$

We have lost the influence of σ and γ on the problem.

This seemingly awkward result serves to alert us that, upon an examination of the dimension-variable matrix, it can be shown to be zero for at least one 3×3 determinant, but is nonzero for all 2×2 determinants. This means that the rank of the matrix is 2 and we should, in reality, expect $(4 - 2)$ as the number of dimensionless ratios. Now we are inclined to look at Eq. (9.4.2) anew and observe that the dimensions of the problem are, effectively, L and M/T^2. The latter is the dimension of σ. Then we can form two π parameters, with d and σ as the two characteristic scales. The result is $D/d = \phi(\sigma/\gamma d^2)$.

In general, one can always proceed by checking initially the dimension-variable matrix to see whether $m = r$. However, this seems to the author to be a considerable effort that is usually unnecessary. Rather, when one uses dimensional analysis, it is suggested that the problem be analyzed from the point of view of one attempting to model the phenomenon. Include all of the salient agents and processes that are involved. This would be identical to the procedure that one would use if we could write the appropriate algebraic or differential equations. If, after choosing the appropriate scales, the steps to determine the exponents in the π parameters uncover a case for which $r \neq m$, then one should look to see how the choice of dimensions should be revised. Thus, for the case in Example 9.4.2, the static phenomenon would make the most logical primary dimensions force and length. A final example should round out this discussion.

Example 9.4.3

The excess pressure sensed at the tip of an impact tube depends on the velocity of the fluid and its density. What dimensionless variables are associated with this problem?

Solution It seems clear that ρ contains mass and length, while V is length/time. Thus the three variables in three dimensions should produce $(3-3)$, or zero, π parameters. After reflecting upon the phenomenon, we are still quite sure that only the three variables that have been chosen are of importance to the problem. The dimension-variable matrix is

$$
\begin{array}{cccc}
 & \Delta p & \rho & V \\
M: & 1 & 1 & 0 \\
L: & -1 & -3 & 1 \\
T: & -2 & 0 & -1
\end{array}
$$

and the 3×3 determinant is zero. (Verify this!) This means that there may be only two principal dimensions. We choose ρ and V as the characteristic scales, since we can show that the dimensions of Δp can be obtained from ρV^2—i.e., $(M/L^3)(L^2/T^2)$. Consequently,

$$
\frac{\Delta p}{\rho V^2} = \pi_1 \quad \text{and} \quad \pi_1 = \phi(\text{constant})
$$

is the result, so that π_1 is a constant. (Is $\pi_1 = 0.5$?)

The procedure for calculating the π parameters should now be clear. Still, there are some questions to be answered about the selection of the variables that lead to these parameters:

1. In the hydraulic jump case that was treated previously, it is natural to ask whether the π parameters would have been different if μ had been chosen for the characteristic mass (the reason for doing this may not be obvious), or if ϵ had been chosen as the characteristic length.
2. Also, the π parameters may change depending upon which variable is selected as the dependent variable in a particular application.

We shall answer the first question by choosing ϵ as the characteristic length* in the case of a hydraulic jump, for which we earlier argued that

$$
y_2 = \phi(y_1, V_1, \rho, \mu, g, \epsilon)
$$

If once again V_1 were selected for the time scale and ρ for the standard of mass in the problem, the π parameters would become

$$
\pi_1' = \frac{y_2}{\epsilon} \qquad \pi_2' = \frac{V_1 \rho \epsilon}{\mu} \qquad \pi_3' = \frac{V_1}{g \epsilon} \qquad \pi_4' = \frac{y_1}{\epsilon}
$$

by means of the method that was previously worked out in detail. From this result, it is clear that

$$
\pi_1' = \frac{y_2}{y_1}\left(\frac{y_1}{\epsilon}\right) = \frac{\pi_1}{\pi_4} \qquad \pi_2' = \frac{V_1 \rho y_1}{\mu}\left(\frac{\epsilon}{y_1}\right) = \pi_2 \pi_4
$$

and so on. The unprimed π's are as given in Eq. (9.2.5). The consequence of this is that in choosing different characteristic scales for a given problem, different π parameters are obtained. However, the parameters that come from one selection of scales are directly related to those that are the result of a different choice of scales.

---

* This choice is neither an obvious nor a reasonable one, but it does allow a point to be made.

Often it is not only the choice of the characteristic scales that affects the dimensionless ratios, but also the specific choice of the dependent variable. To illustrate this point, we shall consider the flow in a circular conduit. One rewrites Eq. (9.3.1b), the dependency of $\Delta p/L$ on the other flow parameters, as

$$\frac{\Delta p}{L} = \tilde{\phi}(Q, \rho, \mu, D, \epsilon) \tag{9.4.3}$$

in which $\tilde{\phi}$ denotes the relationship, regardless of whether it is a mathematical function, a series of charts, or tabular data. Once the terms on the right-hand side of Eq. (9.4.3) are specified, the left-hand side can be assigned a value. One can write Eq. (9.4.3) also as

$$0 = \tilde{\phi}(Q, \rho, \mu, D, \epsilon) - \frac{\Delta p}{L}$$

The entire right-hand side of this equation can be thought of as a new function ϕ of the variables Q, ρ, μ, D, ϵ, and $\Delta p/L$. This can be written as

$$0 = \phi\left(Q, \rho, \mu, D, \epsilon, \frac{\Delta p}{L}\right) \tag{9.4.4}$$

The advantage of Eq. (9.4.4) is its treatment of the variables in a uniform way. Although in some cases $\Delta p/L$ may be the quantity that must be determined, in other cases Q, or perhaps D, may be unknown and depend upon the other parameters.

If information is desired about $\Delta p/L$, one proceeds to nondimensionalize Eq. (9.4.4) by choosing three physical parameters of the problem that scale its length, time, and mass variables. If D, Q, and ρ are chosen as these characteristic scaling factors and they are combined with the remaining physical parameters, one at a time, the following nondimensional parameters are obtained:

$$\pi_1 = \frac{\Delta p}{L}\left(\frac{D^5}{Q^2\rho}\right) \qquad \pi_2 = \frac{Q}{vD} \tag{9.4.5}$$

and

$$\pi_3 = \frac{\epsilon}{D}$$

The introduction of a few constants for convenience permits one to write*

$$\frac{\Delta p/L}{(Q^2/\pi^2)(16/D^5)(\rho/2)} = \phi_1\left(\frac{4Q}{\pi v/D}, \frac{\epsilon}{D}\right) \tag{9.4.6}$$

or

$$f = \phi_1\left(\text{Re}, \frac{\epsilon}{D}\right) \tag{9.4.7}$$

in which Re is the Reynolds number and

$$f = \frac{[\Delta p/\gamma]/L}{(Q^2/D^5)(16/\pi^2)(1/2g)}$$

is the friction factor. (The specific weight, γ, has been introduced for subsequent convenience.) The term in the square brackets is commonly called the *head loss* h_f, due to pipe friction. Figure 9.4.2, the Moody diagram, presents the best experimental correlation of the dimensionless parameters of Eq. (9.4.7).

* The tildes, the subscripts, and the like indicate that the various ϕ's are different functions.

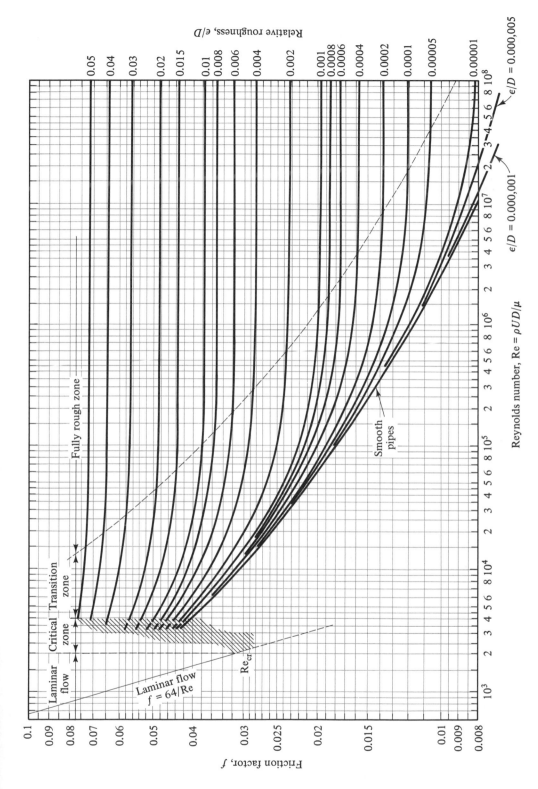

Figure 9.4.2 (a) Moody diagram: friction factor for fully developed flow in circular pipes versus Reynolds number with relative roughness as a parameter. (*From R. Giles's "Fluid Mechanics & Hydraulics," Schaum's Outline Series in Engineering, McGraw-Hill Book Company, 1962.*)

Kind of Pipe or Lining (New)	Values of ϵ in ft	
	Range	Design Value
Brass	0.000005	0.000005
Copper	0.000005	0.000005
Concrete	0.001–0.01	0.004
Cast iron – uncoated	0.0004–0.002	0.0008
Cast iron – asphalt dipped	0.0002–0.0006	0.0004
Cast iron – cement lined	0.000008	0.000008
Cast iron – bituminous lined	0.000008	0.000008
Cast iron – centrifugally spun	0.00001	0.00001
Galvanized iron	0.0002–0.0008	0.0005
Wrought iron	0.0001–0.0003	0.0002
Comm. and welded steel	0.0001–0.0003	0.0002
Riveted steel	0.003–0.03	0.006
Transite	0.000008	0.000008
Wood stave	0.0006–0.003	0.002

(b)

Figure 9.4.2 (b) Effective roughness height ϵ for various pipe materials. *(From R. Giles's "Fluid Mechanics & Hydraulics," Schaum's Outline Series in Engineering, McGraw-Hill Book Company, 1962.)*

But one may not be interested in $\Delta p/L$ in a particular engineering problem. The flow rate Q may be desired as a function of the other variables. One would then mentally rewrite Eq. (9.4.4) as

$$Q = \overline{\Phi}_2\left(\frac{\Delta p}{L}, \mu, \rho, \epsilon, D\right) \tag{9.4.8}$$

If one now chooses D, μ, and ρ as the scaling parameters for length, time, and mass in Eq. (9.4.8), the following result is obtained* if factors such as 2, 4, and π are introduced for convenience:

$$\frac{4Q}{D\nu\pi} = \phi_2\left(\frac{D^{3/2}}{\nu}\left[2g\frac{\Delta p}{L}\frac{1}{\gamma}\right]^{1/2}, \frac{\epsilon}{D}\right) \tag{9.4.9}$$

or

$$\text{Re} = \phi_2\left(\text{Re } f^{1/2}, \frac{\epsilon}{D}\right) \tag{9.4.10}$$

Thus, by choosing a different independent variable and different scaling factors in Eq. (9.4.4), a different set of nondimensional parameters has resulted. The function ϕ_2 in Eq. (9.4.9) gives the value of Q when the remaining variables of the problem are specified. Figure 9.4.3 presents a useful form of the relationship for ϕ_2.

---

*It can be shown that in some transient-flow problems, $D^2/(\mu/\rho)$ serves as a characteristic time for decay.

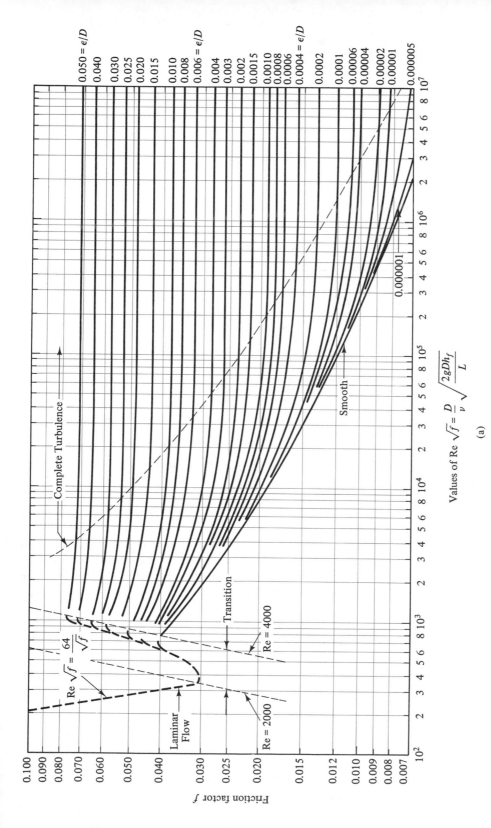

Figure 9.4.3 (a) Modified Moody diagram to determine pipe flow rate when head loss due to friction is specified. ϵ = size of surface imperfections, ft. d = actual inside diameter, ft. *(From R. Giles's "Fluid Mechanics & Hydraulics," Schaum's Outline Series in Engineering, McGraw-Hill Book Company, 1962.)*

Kind of Pipe or Lining (New)	Values of ϵ in ft	
	Range	Design Value
Brass	0.000005	0.000005
Copper	0.000005	0.000005
Concrete	0.001–0.01	0.004
Cast iron – uncoated	0.0004–0.002	0.0008
Cast iron – asphalt dipped	0.0002–0.0006	0.0004
Cast iron – cement lined	0.000008	0.000008
Cast iron – bituminous lined	0.000008	0.000008
Cast iron – centrifugally spun	0.00001	0.00001
Galvanized iron	0.0002–0.0008	0.0005
Wrought iron	0.0001–0.0003	0.0002
Comm. and welded steel	0.0001–0.0003	0.0002
Riveted steel	0.003–0.03	0.006
Transite	0.000008	0.000008
Wood stave	0.0006–0.003	0.002

(b)

Figure 9.4.3 (b) Effective roughness heights ϵ for various pipe materials. *(From R. Giles's "Fluid Mechanics & Hydraulics," Schaum's Outline Series in Engineering, McGraw-Hill Book Company, 1962.)*

Also, Eq. (9.4.4) could be rewritten for D in terms of the other variables. This would give

$$D = \hat{\phi}_3\left(Q, \frac{\Delta p}{L}, \mu, \rho, \epsilon\right) \qquad (9.4.11)$$

in which $\hat{\phi}_3$ is the appropriate relationship, albeit a function that may be unknown at present. Because D is the dependent variable, it should not be chosen as a scaling factor. Instead, the physical parameters Q, μ, and ρ are chosen. One has, after the necessary dimensionless parameters are determined, in accordance with the procedure of this section,

$$\text{Re} = \overline{\phi}_3\left(\frac{\Delta p}{L}\frac{Q^3}{\nu^5}\frac{1}{\rho}, \frac{\epsilon\nu}{Q}\right)$$

The numbers 4, π, and $g/8$ are introduced as a matter of convenience to change the function $\overline{\phi}_3$ to $\hat{\phi}_3$, which appears as

$$\text{Re} = \hat{\phi}_3\left(\frac{g}{g}\frac{\Delta p}{\rho}\frac{1}{L}\frac{2}{4^2}\left[\frac{Q}{\pi}\right]^3\left[\frac{4}{\nu}\right]^5, \epsilon\frac{\pi}{Q}\frac{\nu}{4}\right)$$

or

$$\text{Re} = \phi_3\left(\frac{h_f}{L}\frac{g}{8}\left[\frac{Q}{\pi}\right]^3\left[\frac{4}{\nu}\right]^5, \epsilon\frac{\pi}{Q}\frac{\nu}{4}\right) \qquad (9.4.12)$$

The first parameter on the right-hand side can be rewritten as

$$\frac{h_f}{16Q^2/(2g\,\pi^2 D^4)(L/D)}\left(\frac{4Q}{\pi D v}\right)^5$$

or $\qquad\qquad\qquad\qquad f\ \mathrm{Re}^5$

This implies that Eq. (9.4.12) can be written as

$$\mathrm{Re} = \hat{\phi}_3\!\left(\mathrm{Re}^5\, f,\ \frac{\epsilon \pi v}{4Q}\right) \qquad\qquad (9.4.13)$$

or as

$$\mathrm{Re} = \phi_3\!\left(\mathrm{Re}\, f^{1/5},\ \frac{\epsilon \pi v}{4Q}\right) = \phi_3\!\left(\mathrm{Re}\, f^{1/5},\ \frac{\epsilon/D}{R}\right) \qquad\qquad (9.4.14)$$

Such a relationship is shown in Fig. 9.4.4. The use of this form of the relationship for ϕ_3 is illustrated in the following example.

Example 9.4.4

Consider a situation involving 1000 ft (305 m) of cast iron pipe through which water at 68°F (20°C) is pumped at the rate of 12 cfs (0.336 m³/s). The "head loss" due to friction must be limited to 2 ft (0.6 m) because of the power available to the pump. Tests show that for this kind of pipe, $\epsilon \approx 0.001$ ft. What diameter of pipe is required?

Solution   Equation (9.4.11) is our starting point, but we shall, in effect, use Eq. (9.4.14).
   Using English units for this problem, $4/v = 3.77(10^5)$ s/ft², $h_f/L = 2 \times 10^{-3}$, $Q/\pi = 3.82$ cfs, $g/8 = 4.03$ ft/s² and $\epsilon = 0.001$ ft. Thus, the parameter at the top of Fig. 9.4.4 is

$$\mathrm{Re}\, f^{1/5} = 3.77(10^5)\{4.03[2(10^{-3})](3.82^3)\}^{1/5}$$
$$= 3.21(10^5)$$

and the parameter at the right in Fig. 9.4.4 becomes

$$\frac{\epsilon \pi}{Q}\frac{v}{4} = 0.001\left(\frac{1}{3.82}\right)\left[\frac{1}{3.77(10^5)}\right] = 0.69(10^{-9})$$

These values of $\mathrm{Re}\, f^{1/5}$ and $\epsilon \pi v/4Q$ also give a value for f, on an expanded scale. One prefers to use this value of f, 0.0175, and find D from

$$f = \frac{h_f/L}{(Q^2/\pi^2)(16/2g)(1/D^5)}$$

in which only D is unknown. This is because the percent error in the result due to inaccuracies in reading f is less than the percent errors for inaccurate readings of Re. In this case, D equals 2.00 ft (0.6 m).

We have seen that the choice of the characteristic scales for a given problem may not be obvious. With the information available from Eq. (9.2.1), someone might choose y_1 and someone else ϵ. Often the choice may depend on precedent, but the choice should always result in the most illuminating presentation of the results. Generally, one would not choose the desired dependent variable for a scaling factor or repeating variable, since then the dependent variable would appear in many of the dimensionless parameters. The result would be an implicit relationship for the selected dependent variable, instead of a more desirable explicit relationship.
   In a particular hydraulics problem, free-surface conditions cause the list of variables in Eq. (9.4.3) to be increased by surface tension σ and gravity, g. Finally,

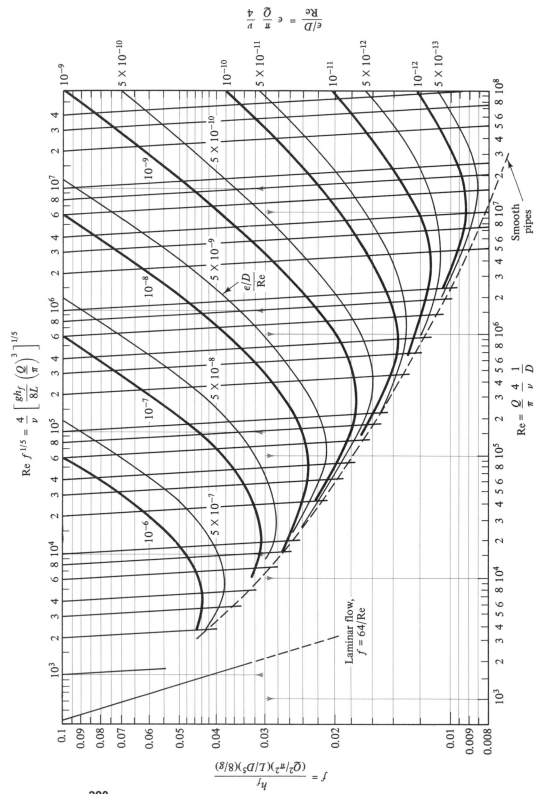

Figure 9.4.4 Modified Moody diagram to determine the pipe size when the flow rate and head loss are specified. *(Author's paper, by permission of the A.S.C.E.)*

pressure transients could be important, so that the speed of sound c in the liquid would affect the results because these transients are propagated by the elasticity of the medium. Then, instead of Eq. (9.4.3), we would have, since $k = 1$ for liquids,

$$\frac{\Delta p}{L} = \phi(V, \rho, \mu, D, \epsilon, \sigma, g, c)$$

The choice of ρ, D, and V as the characteristic scales for mass, length, and time gives the following dimensionless parameters:

$$\pi_1 = \frac{\Delta p}{\rho V^2} \qquad \text{pressure coefficient}$$

$$\pi_2 = \frac{VD\rho}{\mu} \qquad \text{Reynolds number, Re}$$

$$\pi_3 = \frac{\epsilon}{D} \qquad \text{relative roughness}$$

$$\pi_4 = \frac{V^2 D\rho}{\sigma} \qquad \text{Weber number, We}$$

$$\pi_5 = \frac{V^2}{gD} \qquad \text{square of the Froude number, } \text{Fr}^2$$

and
$$\pi_6 = \frac{V}{c} \qquad \text{the Mach number,* Ma}$$

such that
$$\hat{\phi}(\pi_1, \pi_2, \pi_3, \pi_4, \pi_5, \pi_6) = 0 \qquad (9.4.15)$$

It should be emphasized that the dimensional analysis simplifies and classifies an experimental procedure. The dimensional analysis does not, however, usually give the form of the unknown function $\hat{\phi}$. It will remain for an experimental program to establish the relationship, or correlation, between the π parameters. In Sec. 9.5, we shall examine the nature of some correlations for specific situations.

9.5 INTERPRETATION OF THE DIMENSIONLESS PARAMETERS

In repeating some of Stokes' analytical steps, we concluded in Eq. (8.3.12b) that

$$\frac{1}{f} = \frac{\rho V_{\text{avg}}^2/2}{4\tau_0} = \frac{\text{Re}}{64}$$

This would indicate that the magnitude of the ratio of the inertia terms, ρV_{avg}^2, to the viscous stresses, exemplified by τ_0, is determined by the value of Re. Such an analysis helps us to interpret the π parameter $\rho VD/\mu$, the Reynolds number, as the ratio of inertia forces to viscous forces in Eq. (9.4.15). Figure 9.5.1 shows the empirical relationship between f and Re for values of Re beyond which the laminar model no longer appears valid. In this range, the roughness parameter becomes important (the research presented in Fig. 9.5.1 is the forerunner of the Moody diagram in Fig. 9.4.2).

Other factors could also enter the picture, depending on the particulars of the situation. If the medium is a gas, compressibility of the medium has been mentioned as

---

*In liquids, $c^2 = (\text{bulk modulus of elasticity})/\rho$, and in these variables, π_6 is termed the *Cauchy number*.

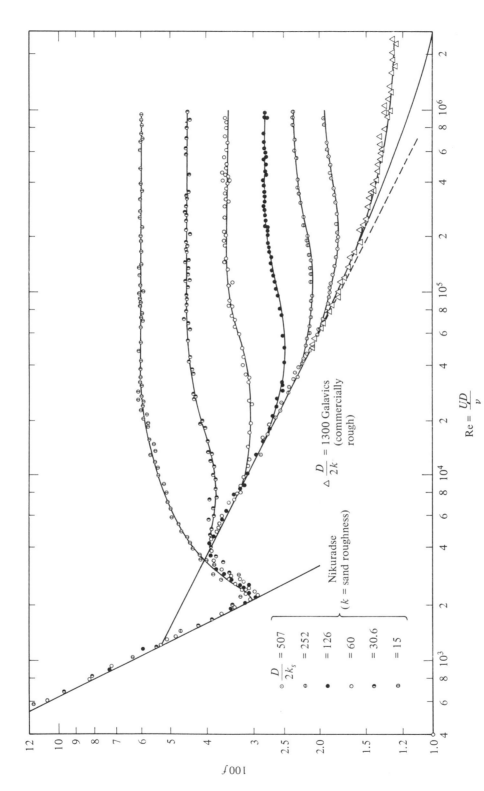

Figure 9.5.1 Friction factor versus Reynolds number for various pipes that were roughened by bonding sand grains of different sizes, k_s, to the inside surface. *(After Nikuradse in H. Schlichting's Boundary-Layer Theory, McGraw-Hill Book Co., by permission)*

a property of importance. This effect can be accounted for by the π parameter V/c, in which c is the speed of sound in the medium. This V/c, called the *Mach number*, represents the ratio of speed of translation of the fluid with respect to the speed of propagation of elastic waves. Although this is not a "force ratio" as is f or Re, one could also view the Mach number as the ratio of inertia forces to elastic forces.

One final remark concerning the types of parameters that are associated with conduit flow: Figure 9.5.1 indicates that if one were to conduct an experiment in a horizontal conduit with water, the pressure differential would increase with increasing V^2, because f is nearly constant at high values of Re for rough pipes. If the experiment were conducted at a contant upstream pressure, one would find the downstream pressure decreasing nearly as the square of V_{avg}. Ultimately, the pressure at the end of the conduit would be reduced to a point near the vapor pressure of water, for the operating temperature. This new water phase, steam, would alter the physical situations in the conduit. Further reductions of pressure at the downstream station where vapor has formed might not result in a continuation of the relationships between the Δp and Q that existed before cavitation commenced. Such an occurrence will affect the performance of pumps and other equipment that are located downstream of the point of cavitation.

Fluid vapor may occur locally to some measurable degree even before vapor pressure is reached throughout the fluid. Local accelerations in the system may result in local "boiling." This phenomenon can happen at the tips of a propeller or at the inlet of a centrifugal pump. (See Figs. 1.8.2 and 1.8.3.) A parameter that is a measure of the possibility that a flow will produce vapor at some location is the *cavitation coefficient K*, expressed by

$$K = \frac{p_0 - p_v}{\rho V_0^2} \tag{9.5.1}$$

where
p_0 = absolute pressure at a reference location
V_0 = reference velocity
p_v = vapor pressure (absolute)
ρ = fluid density at reference point*

The numerator of K is the pressure excess above vapor pressure, and the denominator is proportional to the local dynamic pressure. This cavitation index is a special case of the pressure coefficient, $\Delta p/\rho V^2$, that relates pressure forces and inertia forces. In Fig. 9.5.2, we can see how K affects the drag on bodies. The associated vapor cavities are shown in Fig. 9.5.3.

When the hydraulic jump was analyzed in Example 7.3.4, the nondimensional ratio V_1^2/gy_1 was developed. This is commonly designated Fr^2, the square of the *Froude number*. When the momentum equation was used to construct the Froude number, it was evident that this term was proportional to the ratio of the inertia and gravity forces in the flow. Subsequently, in Chap. 7, the speed of a small disturbance on a free surface was determined to be $\sqrt{gy}$. Thus the Froude number can also be interpreted as the ratio of the fluid speed to the speed of propagation of a small disturbance wave.

---

*This coefficient is sometimes denoted by σ; however, this symbol has been used already to denote surface tension, in accordance with widespread custom.

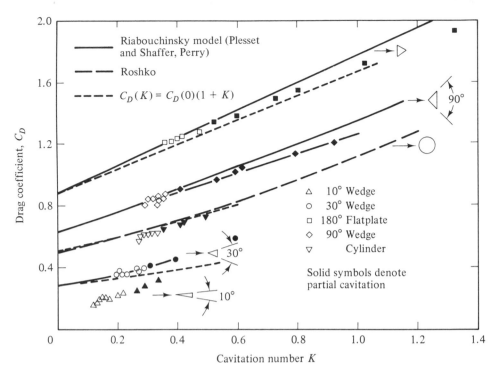

Figure 9.5.2 Effect of cavitation number on the drag coefficient of circular and noncircular cylinders. *(C.I.T. Hydro. Lab. Rep. E-73.6, 1957, by R. L. Waid)*

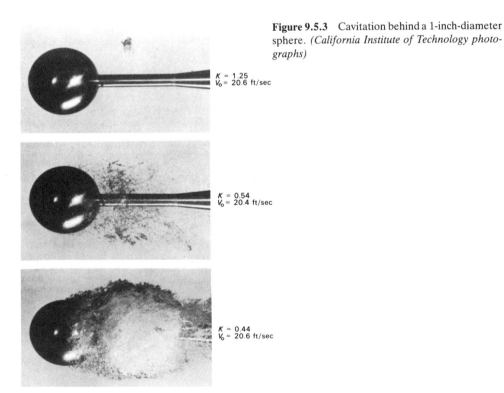

Figure 9.5.3 Cavitation behind a 1-inch-diameter sphere. *(California Institute of Technology photographs)*

$K = 1.25$
$V_0 = 20.6$ ft/sec

$K = 0.54$
$V_0 = 20.4$ ft/sec

$K = 0.44$
$V_0 = 20.6$ ft/sec

The ratio between inertia forces and surface tension is called the *Weber number* We = $\rho V^2 L/\sigma$, in which L is a characteristic length and σ is the surface tension of the liquid. This ratio, too, can be interpreted as the square of a ratio—the ratio between the speed of the fluid and the speed of propagation of capillary waves.

The speed of propagation (i.e., celerity) of some waves depends on their wavelength. Surface waves that are dominated by the effects of inertia and gravity have the longest wavelengths and propagate the fastest. The celerity decreases with wavelength for waves for which surface tension replaces gravity as one of the main agents for motion. (The last vestiges of waves in a coffee cup or teacup are capillary waves.) Hence, if inertia, gravity, and surface tension are all present in the proper proportions, the wave speed decreases initially with wavelength and ultimately begins to increase. The analysis that establishes this result predicts the wavelength at which the minimum celerity occurs. We shall call it λ_{min}; it equals

$$2\pi \left[\frac{\sigma}{g\rho(1-S)} \right]^{1/2}$$

The symbol S in this expression is the ratio of the density of the upper and lower fluid at the interface where the waves occur (e.g., about 1/850 for an air-water interface). At λ_{min}, the analysis gives

$$c_{min} = \left\{ \frac{2[(1-S)g\sigma/\rho]^{1/2}}{1+S} \right\}^{1/2}$$

As a consequence of the different force ratios that the Reynolds, Froude, and Mach numbers signify, it is natural that these numbers will appear as π parameters when the particular problem deals with inertia, viscous, gravity, and elastic forces. The utility of these dimensionless parameters is the subject of the next section.

9.6 MODELING

In the introductory section of this chapter, the drag exerted on two different plates was mentioned. We saw there that if these two plates were submerged in two different fluids and subjected to two different free-stream velocities, the same drag coefficient would result *if* the Reynolds numbers for the two plates were made the *same*. The italicized words in the preceding sentence provide the key ideas concerning the techniques of modeling that will now be discussed. In fact, these ideas provide a justification for the presentation of this chapter.

In the course of the work in this chapter, we have seen that one can examine a particular flow phenomenon, say, the flow past a bluff body, and conclude that

$$\frac{\text{Drag}}{\frac{1}{2}\rho V^2 A} = C_D = \phi\left(\text{Re}, \text{Fr}, \text{Ma}, \text{We}, \frac{\epsilon}{D} \right) \qquad (9.6.1)$$

in which Re is the Reynolds number and the other abbreviations represent parameters listed at the end of Sec. 9.4. Similar relationships have been described through Eqs. (8.6.9) and (9.4.1), among others. The associated functional relationships are described in Fig. 9.4.1a and b. However, Fig. 9.4.1 does not contain all of the parameters that are included in Eq. (9.6.1). Why is this?

The reason is that Eq. (9.6.1) provides a general expression that covers all of the cases associated with its development. But sometimes, in a particular study of a bluff body, gravitational effects are of minor importance. For example, if the flow velocities are quite large and the excess temperature of the body is quite small, the oncoming fluid will carry the warm, buoyant fluid along with it, rather than letting it rise appreciably near the body. From this, one would infer that the buoyancy effects on the body associated with heating would be minor. Then, Fr in Eq. (9.6.1) would be an unimportant parameter (i.e., would not strongly affect the results). Yet the speed of the flow could be slow enough that compressibility effects were also minor. Accordingly, the Mach number, which describes the magnitude of the ratio between the inertia and compressibility effects in a given situation, is of little consequence. Also, if the body is completely submerged, surface tension effects would disappear, so that the effect of We in the problem would disappear. In view of these particular aspects of the problem, Eq. (9.6.1) would reduce to a simpler form:

$$C_D = \hat{\phi}\left(\text{Re}, \frac{\epsilon}{D}, \ldots\right) \qquad (9.6.2)$$

wherein any terms after ϵ/D would be geometric ratios appropriate to the body (e.g., the ratios of the three axes of an ellipsoid). Experience may show that for a particular ellipsoid, C_D is not sensitive to Re for a particular range of Re and ϵ/D. For example, a sphere has a drag coefficient of 0.4 for $2(10^5) > \text{Re} > 2(10^3)$ in spite of a large variation of ϵ/D. This situation might not be anticipated initially, but tests like those that led to Fig. 9.4.1b give us this result. But Fig. 9.6.1 shows that in a small range of Re, the value of ϵ/D is an important factor in C_D.

These observations lead us to conclude that in a particular situation, one should analyze the problem by considering the important physical variables while excluding those that appear to be unimportant in the case under study. When a scale model is

Figure 9.6.1 Effect of ambient turbulence on the drag coefficient for spheres. The significant influence is limited to only a narrow range of Reynolds numbers. (*From S. Goldstein's "Modern Developments in Fluid Dynamics," Dover Publications, Inc., by permission.*)

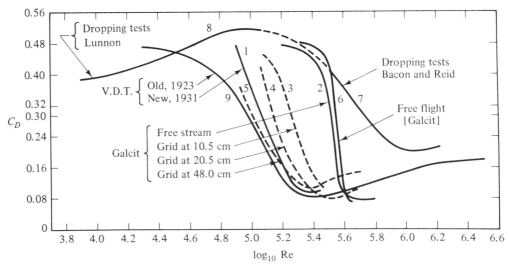

Dimensionless Parameters in Fluid Mechanics   Chap. 9

used in an experimental study, the various geometric ratios between the model and a full-scale prototype will necessarily be the same. Then, the various dimensionless parameters in tests of the model and its full-scale counterpart should be the same. One should expect the same value of the drag coefficient; or a pressure coefficient, $\Delta p/(\frac{1}{2}\rho V^2)$; or a moment coefficient, $M/[\frac{1}{2}\rho V^2 A\,(\text{length})]$. Once we know the dimensionless drag coefficient, or any other appropriate coefficient, we can calculate the prototype effect—drag, in this case—on the basis of measurements with the model.

Example 9.6.1

A sphere with a diameter of 20 mm is submerged in a stream of water ($\nu = 10^{-6}$ m²/s) during a model test. The water approaches the sphere at 1.1 m/s. Determine the force on a sphere with a diameter of 500 mm that is in an airstream with a speed for which the Reynolds number is the *same* as that found in the model test.

Solution The constancy of the Reynolds number between the model and prototype tests is important. Because the larger sphere is a scaled-up version of the model sphere, the drag coefficient of the model test should be realized when operating the prototype sphere in air instead of water at the appropriate speed, provided the relative roughnesses is the same. This drag coefficient could be determined from the force measurements taken during the model test. The resulting data point should agree with the information of Fig. 9.4.1b. We see that the value will depend upon the Reynolds number. The model tests were conducted at

$$\text{Re} = \frac{.02}{10^{-6}}(1.1) = 2.2(10^4)$$

Figure 9.4.1b shows that for this value of the Reynolds number, $C_D = 0.4$. We now recognize that the drag coefficient on the larger sphere will be the same if it is associated with the same Reynolds number. Then

$$(C_D)_{\text{model}} = (C_D)_{\text{prototype}} = 0.4 \tag{9.6.3}$$

so that

$$(\text{Drag})_{\text{prototype}} = \frac{0.4\rho V^2(\text{area})}{2} \tag{9.6.4}$$

Once V is known, the drag is determined. This velocity is a special value. It is the velocity that is prescribed by the Reynolds number which, in turn, specified the drag coefficient at 0.4. The two tests were to be conducted so that

$$\text{Re}_{\text{model}} = \text{Re}_{\text{prototype}} \tag{9.6.5}$$

giving the requirement that

$$V_{\text{prototype}} = \left(\frac{VD}{\nu}\right)_{\text{model}}\left(\frac{\nu}{D}\right)_{\text{prototype}}$$

$$= V_m\left(\frac{D_m}{D_p}\right)\left(\frac{\nu_p}{\nu_m}\right)$$

in which the subscripts m and p denote *model* and *prototype*, respectively.

The ratio of the diameters is called the *scale ratio* of the model. The area of the prototype in Eq. (9.6.4) is related to the scale ratio associated with a scale model. We would have

$$\frac{A_p}{A_m} = \left(\frac{D_p}{D_m}\right)^2$$

$$= (50/2)^2 \tag{9.6.6}$$

For the data given in this example, one can determine that the force on the model was 0.06 N by using

$$C_D = \frac{\text{Drag}}{\frac{1}{2}\rho V^2(\pi D^2/4)}$$

Sometimes, however, it is not possible to obtain conditions so that the arguments in a physical relationship, such as Eq. (9.6.1), are the same in the model and the prototype environments. For example, it may not be possible to obtain surface finishes for small and large spheres that would yield the same value of ϵ/D. The surface finish of mechanical parts, such as pumps, is controlled by the grain size of the sand used in the casting process that produces the pieces of the machine. This size is not subject to convenient modification. The consequence of this is that some tests must be conducted, of necessity, without complete similarity between model and prototype conditions. In the case of pump testing, one assumes that the efficiencies that are obtained in model tests will be exceeded in full-scale equipment, because the effects of roughness, ϵ/D, while not directly comparable between model and prototype, will be less important in the larger piece of equipment. The greater relative roughness in models should increase the observed effect of losses in the model tests.

There are other cases of "incomplete" modeling. When conducting tests on a submarine, viscous effects are dominant. For a surface vessel, wave generation produces resistance, as do the surface shear stresses. However, if one attempts to conduct tests with Eq. (9.6.1) in mind such that

$$\text{Re}_\text{model} = \text{Re}_\text{prototype}$$

and

$$\text{Fr}_\text{model} = \text{Fr}_\text{prototype}$$

it will be required that the fluids used in the model and prototype tests have a specialized relationship between their viscosities. This may put a virtually impossible demand upon the model tests, because water is the material in which most ships move. If this is the case, what can be done?

The practical solution to this problem was proposed by Froude, who divided the total actual ship resistance R_T by $(1/2)(\rho SV^2)$, in which S is the wetted area of the ship, to obtain a total drag coefficient C_T. He then proposed to deal with frictional and wave-making resistance separately by setting C_T equal to the sum of two different coefficients that are determined separately. With this *assumption,* one would have

$$C_T = C_\text{Fr} + C_\text{Re} \tag{9.6.7}$$

with the *frictional coefficient* C_Re a function of Reynolds number only and C_Fr, the *wave drag coefficient,* solely a function of Froude number. This latter design parameter was already in use by others when Froude broadly advocated its merit. The coefficients in Eq. (9.6.7) are then determined as follows:

1. Operate the model according to Froude scaling, so that

$$V_m = V_s\left(\frac{L_m}{L_s}\right)^{1/2} \tag{9.6.8}$$

in which the subscript s denotes the ship and m, the model.

2. The *total model resistance,* R_{Tm}, is determined in the laboratory by towing tests.

3. The *frictional resistance* of the *model,* $R_{Re,m}$, is estimated from "smooth, flat plank" data,* assuming a plate of equal length and area to those of the model. (The model, while operating at a lower Reynolds number than the prototype, is made to have a turbulent boundary layer by the addition of a "trip wire" or similar device.)

4. The wave resistance of the model—largely the consequence of wave making with some wake generation—is found by subtraction, so that

$$R_{Fr,m} = R_{Tm} - R_{Re,m} \qquad (9.6.9)$$

5. Froude scaling is then used to determine the wave resistance of the ship by

$$C_{Fr,s} = C_{Fr,m} \qquad (9.6.10)$$

or

$$R_{Fr,s} = R_{Fr,m} \left(\frac{V_s}{V_m}\right)^2 \left(\frac{S_s}{S_m}\right)$$

from the definition of C_{Fr}. Using Eq. (9.6.8), one has

$$R_{Fr,s} = R_{Fr,m} \left[\left(\frac{L_s}{V_m}\right)^{1/2}\right]^2 \left(\frac{L_s}{L_m}\right)^2$$

since the areas will be proportional to the square of the scale ratio, $L_s/L_m = \lambda$. This yields

$$R_{Fr,s} = R_{Fr,m} \lambda^3 \qquad (9.6.11)$$

6. The frictional resistance of the ship, $R_{Re,s}$, is again estimated from the flat-plate data. The design speed of the ship, V_s, is used in computing the Reynolds number that is used to obtain $C_{Re,s}$ from an appropriate design curve such as Fig. 13.3.3. This value of $C_{Re,s}$ is equal to

$$\frac{R_{Re,s}}{\frac{1}{2}\rho V_s^2 S} \qquad (9.6.12)$$

7. The total ship resistance is then

$$R_{Ts} = R_{Re,s} + R_{Fr,s} \qquad (9.6.13)$$

in which $R_{Re,s}$ was estimated in step 6 and $R_{Fr,s}$ comes from step 5.

But ship model testing is not the only field that must abandon the idea of complete similitude. The testing of hydraulic models of rivers and lakes is another instance. In order to have the watercourse conveniently modeled within a laboratory, the length would have to be drastically reduced, say, to $L_m = L_p/31{,}500$ or L_p/x, in which x is the *length scale.* The *depth scale* might be chosen to be different, say, $H_m = H_p/960$ or $H_m = H_p/y$, in order that the model depth not be too shallow. If this were the case, one might expect significant capillary effects to appear in the wave patterns, which, of

---

* Froude towed many kinds of planks at his test facility. Karl Schoenherr compiled a great body of flat-plate drag data and presented it in a form similar to Fig. 8.6.2. His curve is widely used in evaluating ship model test data and is presented as Fig. 13.3.3.

necessity, would have to be of small amplitude. Because waves are of general interest in such studies, Froude number scaling is considered to be of primary importance. This gives

$$\frac{V_m}{V_p} = \sqrt{\frac{H_m}{H_p}} \tag{9.6.14}$$

The ratio of the speed of a small wave in the model to the speed of a wave occurring in nature would be $\sqrt{H_m/H_p}$ in accordance with the discussion about small surface disturbances in Sec. 7.7. The time for these waves to travel the length of their respective basins would be

$$T_m = \frac{L_m}{\sqrt{gH_m}}$$

and

$$T_p = \frac{L_p}{\sqrt{gH_p}}$$

so that

$$\frac{T_m}{T_p} = \frac{\sqrt{y}}{x} \tag{9.6.15}$$

For the scale ratios $x = 31,500$ and $y = 960$, Eq. (9.6.15) gives

$$T_m = T_p\left(\frac{30.98}{31,500}\right) = 9.7(10^{-4})T_p \tag{9.6.16}$$

If the wave were due to a tide with a period of 12.4 h, T_p would be 12.4 h, or 44,640 s. This gives $T_m = 43.5$ s. Osborne Reynolds operated a model with such a distorted geometry to study the action of the tides on sand deposition in the estuary of the Mersey River (the tides in this region are quite large, with a change of about 20 ft in the water's elevation between low and high tide).

When it was first obvious that a *distorted scale*—where, by definition, the vertical scale is x/y larger than the horizontal scale—was useful, it was noted that care was necessary to ensure that the model depth would not be too shallow. If this is the case, the proper inertia effects will not be present to give waves of sufficient amplitude to be relatively free of the influence of surface tension. An overly shallow model can also induce unwanted viscous effects, in a manner that will now be outlined.

Waves are an oscillatory motion in which inertia effects are usually important. The fluid moves slightly back and forth. If one moves with a particle of fluid, the bottom of the model or lake apears to move to and fro. Such a situation was presented in Sec. 8.5, where an oscillating plate was analyzed. It was learned there that the effect of the moving plate decayed away from the plate according to the relation $e^{-y/\sqrt{2\nu T}}$, in which T is the period of the oscillating plate and ν is the kinematic viscosity. When $y/\sqrt{2\nu T}$ is about 5, the result, e^{-5}, is so small that the influence of the bottom on the velocity is small. We have called this associated value of y the *laminar boundary-layer thickness*. Hence, we define

$$\delta = 5\sqrt{2\nu T} \tag{9.6.17}$$

It is apparent that δ will be small if ν is small or T is small (i.e., at high frequency). If water is used in the model and the prototype, the value of δ_m becomes, from Eqs. (9.6.17) and (9.6.15),

$$\delta_m = 5\sqrt{2\nu T_m} = 5\sqrt{2\nu T_p}\left(\frac{y^{1/4}}{x^{1/2}}\right)$$

We shall insert H_m and H_p successively into this result for convenience, so that

$$\frac{\delta_m}{H_m} = \frac{5\sqrt{2\nu T_p}(y^{1/4}/x^{1/2})}{H_m}$$

and

$$\frac{\delta_m}{H_m} = \frac{5}{H_p}\frac{H_p}{H_m}\sqrt{2\nu T_p}\frac{y^{1/4}}{x^{1/2}}$$

Using $H_p/H_m = y$, we obtain

$$\frac{\delta_m}{H_m} = \frac{5}{H_p}y\frac{y^{1/4}}{x^{1/2}}\sqrt{2\nu T_p}$$

If we require that $\delta_m/H_m \le 1$, which would make for a very small viscous boundary layer compared with the total depth, we will have

$$\frac{5}{H_p}\sqrt{2\nu T_p}\frac{y^{5/4}}{x^{1/2}} \le 1 \tag{9.6.18}$$

With $\nu = 10^{-5}$ ft^2/s for water, and using the scaling of Osborne Reynolds that has been referred to previously, we require that the actual depth of the Mersey River estuary, H_p, be such that

$$H_p \ge 7.07\sqrt{0.446}\frac{(960)^{5/4}}{(31,500)^{1/2}} = 7.07\frac{(0.67)}{177.5}(5344) = 142.6 \text{ ft}^*$$

Equation (9.6.18) can also show us that for a given horizontal scale ratio x, there is a bound on the vertical scale ratio y. We see also that

$$\frac{y^{5/4}}{x^{1/2}} \le \frac{H_p}{5\sqrt{2\nu T_p}} \tag{9.6.19}$$

This discussion points up the fact that while *complete similitude* (i.e., identical independent π parameters in model and prototype) is desirable to obtain the same dependent π parameters in model and prototype, sometimes this goal is not practical to attain. Different π parameters, representing different force ratios, may demand awkward or unobtainable test requirements. Then experimenters must choose the most important parameters in accordance with the appropriate scaling laws and account for the remaining terms as well as possible, or in accordance with the existing practice of their specialty in the profession.

---

*The Mersey River does not seem to comply with this criterion. The factor 5 in Eq. (9.6.18) is too large as a gauge for the extent of viscous effects in this case. The flow will most likely not be laminar. In a turbulent flow, the boundary layer is thicker than for a laminar flow. However, the distance from the boundary of intense streamwise shear stresses that would cause sand movement on the bottom is a smaller fraction of the boundary-layer thickness for a turbulent flow. This fact could account for the apparent success of Reynolds' experiments in predicting silting patterns in the river. Experiments in 1939 with a larger model ($x = 800$, $y = 120$) would have $H_p \ge 66$ ft according to Eq. (9.6.18), a figure closer to the actual mean depth.

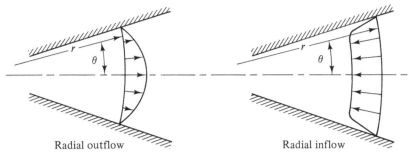

Figure 9.7.1 Symmetric radial outflow and inflow between two nonparallel plates.

9.7 SELF-SIMILAR FLUID MOTIONS

When one produces an analytical model for a particular fluid flow, the situation may sometimes be idealized to facilitate the analysis. We have done this ourselves for the case of the flow between parallel plates. In Sec. 8.2, we assumed that the plates were infinitely long in the direction of flow and that there was no variation with respect to the transverse direction, the width. Let us consider a seemingly related case, the flow between radially diverging walls. This case is illustrated in Fig. 9.7.1. If we assume that the flow is purely radial (either the inflow or the outflow), the continuity equation is more conveniently employed when written in cylindrical polar coordinates (see Eq. 4.8.8).

$$\frac{1}{r}\frac{\partial(ru_r)}{\partial r} + \frac{1}{r}\frac{\partial u_\theta}{\partial \theta} + \frac{\partial u_z}{\partial z} = 0 \tag{9.7.1}$$

Since u_r is the only nonzero component of velocity, the continuity equation reduces to*

$$\frac{1}{r}\frac{\partial}{\partial r}(ru_r) = 0 \quad \text{or} \quad u_r = \frac{f(\theta)}{r} \tag{9.7.2}$$

in which the function $f(\theta)$ is unknown. (It can be determined with the use of the equations of motion.) At $r = 1$, $u_r = f(\theta)$, and at $r = \frac{1}{2}$, u_r is twice the value at $r = 1$. This means that the velocity profile, $f(\theta)$, will be "stretched" according to the factor $1/r$ to give the value of the velocity at a particular θ for some chosen value of r. We say that u_r has a *similarity solution,* since once $f(\theta)$ is known at $r = 1$, u_r is known at all r by a simple scaling factor. Two solutions for $f(\theta)$ are shown in Fig. 9.7.1. One notes that the results for radial outflow and inflow are significantly different. The velocity profile for the radial inflow exhibits boundary-layer characteristics (i.e., there is a very thin layer in which the flow is retarded by the wall). For this flow, the pressure is decreasing in the direction of flow. (Such a pressure gradient is expected from the Bernoulli equation.) The pressure will be increasing in the flow direction for the case of radial outflow. When this latter situation occurs, flow separation from the boundary becomes a distinct possibility. If the included angle between the walls exceeds 7°, there is a tendency in experiments for the flow to become unsymmetrical with respect to a line bisecting the angle between the walls. The outflowing fluid tends to flow along one wall, and there is

_____
* The origin, $r = 0$, is excluded from consideration.

something approximating a weak radial inflow along the other wall. This inflow is turned aside near the narrowest part of the flow channel so that, in actuality, a weak swirling motion is generated and maintained by the discharging fluid.

Example 9.7.1

Point source is the term that is used to describe a flow which radiates in all directions from the origin. There is only a radial component of velocity, u_R. [It was introduced by Eq. (5.4.14).] If one imagines a spherical control surface with the origin at the center, it has a surface area of $4\pi R^2$. If at each point on this surface the radial velocity has the same value (i.e., if u_R is a function of R only), the volume flux across the control surface is

$$Q = 4\pi R^2 u_R \qquad (9.7.3)$$

or

$$u_R = \frac{Q}{4\pi R^2} = \frac{\text{const}}{R^2}$$

This is a similarity velocity profile, with the magnitude having a scaling factor of $1/R^2$.

Example 9.7.2

A line source is the two dimensional counterpart of the point source. [See Eq. (5.4.11).] If one thinks of a line, such as a porous pipe, drawn perpendicular to the page in Fig. 9.7.2 and the fluid moves radially away in planes that are perpendicular to this line, one has a line source.* The radial velocity, u_r, is independent of z, the direction of the line, unless specifically stated to the contrary in a particular application.

If we use a circular cylindrical control surface, with an axis of *unit length,* the volume flux will be

$$Q = \int u_r\, dA = u_r \int dA = u_r (2\pi r)(1)$$

if u_r is independent of θ, as assumed. Hence,

$$u_r = \frac{Q}{2\pi r (1)} = \frac{\text{constant}}{r} \qquad (9.7.4)$$

which is another similarity profile. In fact, it is a particular case of $u_r = f(\theta)/r$, in which $f(\theta) = $ constant. Now Fig. 9.7.1 shows $f(\theta)$ to be nearly constant over a major portion of the flow region for radial inflow. We might conclude from this that the walls, with their no-slip condition, influence the general flow region in only a small region near the boundary for the particular case shown in the second diagram in Fig. 9.7.1.

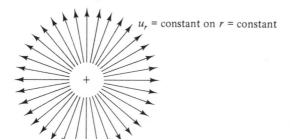 $u_r = $ constant on $r = $ constant

Figure 9.7.2 Velocity distribution in the *xy* plane for a line source directed along the *z* axis.

*One lawn-watering device is a hose with many small perforations that spray the water perpendicular to the hose; the flow, when viewed from the end of such a hose, would look similar to Fig. 9.7.2.

The laminar flow over a flat plate that was introduced in Sec. 8.6 is another situation for which a self-similar velocity profile is possible. In carrying out our discussion on the boundary layer in Sec. 8.6, we *assumed* the existence of a velocity profile in the form

$$\frac{u}{U} = f\left(\frac{y}{\delta}\right)$$

in which δ is a function of x. We found that δ is equal to

$$(\text{constant})\left(\sqrt{\frac{\nu x}{U}}\right)$$

so that we can write the velocity profile as

$$\frac{u}{U} = f\left([\text{constant}]\left[\frac{y}{\sqrt{\nu x/U}}\right]\right)$$

Such an independent variable as $(y/\sqrt{\nu x/U})$ is an example of a similarity variable. The same value for this dimensionless parameter will result for a host of values of x and y.

A two-dimensional jet also exhibits the possibility of a similarity solution. Such a case is illustrated in Fig. 9.7.3. This figure shows that the velocity in a jet changes greatly in the y direction, while the change in the x direction is gradual. This is one of the characteristics of a boundary-layer flow. Figure 9.7.4 shows some profiles predicted by theory using dimensionless variables.

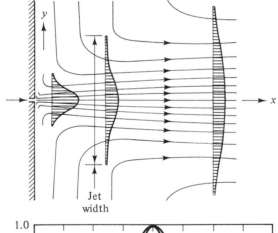

Figure 9.7.3 Streamlines and velocity profiles for a two-dimensional jet issuing out of a wall. *(From H. Schlichting's* Boundary-Layer Theory, *7th edition, McGraw-Hill Book Company, 1979)*

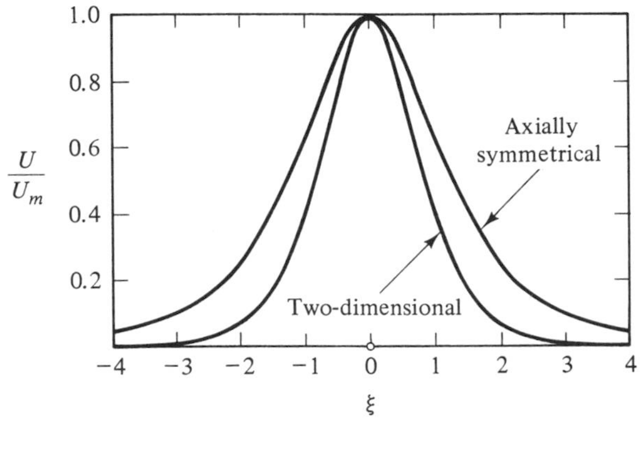

Figure 9.7.4 Theoretical velocity distribution in a two-dimensional jet, similar to that in Fig. 9.7.3. Dimensionless variable $\xi = 0.275\, K^{1/3} y/(\nu x)^{2/3}$, where K is (jet momentum flux)/ρ. Theoretical results are also presented for a circular jet with $\xi = 0.244(K')^{1/2} y/\nu x$. K' is three-dimensional equivalent of K. *(From H. Schlichting's* Boundary-Layer Theory, *7th edition, McGraw-Hill Book Company, 1979)*

The "defect" in velocity profile that occurs in the wake behind a cylinder displays a similarity profile at large distances from the body. Some results are shown in Fig. 9.7.5.

A consequence of being able to find a "similarity" solution for a particular problem is the considerable simplification that results. The introduction of a similarity variable, where one exists, can simplify the system of differential equations that must be solved. The solution to the problem, if it can be obtained, is also made increasingly valuable because of the compact way a great deal of information is presented. Figure 9.7.5 is an illustration of experimental efforts to disclose the existence of a similarity solution.

Figure 9.7.5 (a) Wake velocity profiles $\hat{u} = U - u$, (b) Measured and theoretical velocity distribution in a two-dimensional wake behind circular cylinders. Solid line is for laminar wake and dashed line for turbulent. *(From H. Schlichting's* Boundary-Layer Theory, *McGraw-Hill Book Company, by permission)*

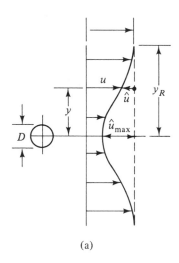

(a)

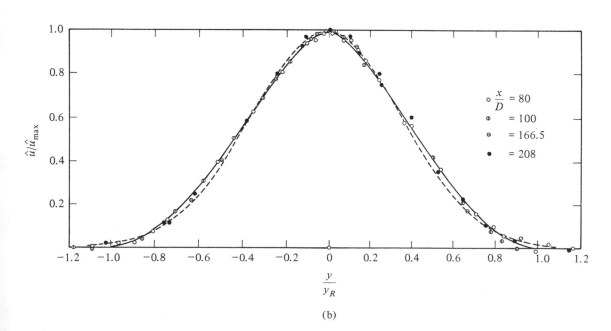

(b)

9.8 SUMMARY

In this chapter, the concept of similarity has been discussed because of its importance in experimental research. We have also seen that experimental studies can be effectively planned and summarized by using dimensionless variables. The results of the laboratory work can be extended to other situations to give meaningful predictions of prototype performance using the data from models in dimensionless form. Lastly, similarity can also be used in the context of "self-similar" velocity profiles, which are functions of a "similarity variable." This similarity can result in the transformation of the partial differential equations of motion to ordinary differential equations. One then has an analytical model of the flow which may be more easily solved. We shall experience this simplification in Chap. 16. Equally well, these theoretical concepts may suggest means of presenting the experimental data, such as in Fig. 9.7.5, to extend their usefulness.

It is necessary to realize that when one writes $C_D = \phi(\text{Re}, \epsilon/D)$, for example, the function ϕ connotes a "transfer function" that gives the same value of C_D whenever the same values of Re and ϵ/D are used. This is the basis for assuming that the drag coefficient on a model will be equal to a prototype's drag coefficient, provided both are operating at the same Re and ϵ/D and no extraneous physical processes are present. The "transfer function" is the physical phenomenon, and its natural constraints are the subject of our interest.

EXERCISES

9.1.1(a) What is the drag (per unit width) of a plate 1 m long that is part of a control surface for an airplane flying at 380 km/h in air at $-20°C$?

(b) What is the drag (per unit width) of a plate 3 m long that is part of a drone aircraft flying at 237.5 mph in air at $-4°F$?

(c) At what speed must one tow a 10-ft-long plate in water at 68°F to obtain the same drag coefficient as for a plate 100 ft long—a section of an airplane—flying at 400 mph in air at $-10°F$?

9.4.1(a) Instead of choosing D, U, and ρ as repeating variables, choose D, U, and μ and determine the dimensionless parameters therewith.

(b) Expand the possibilities given by Eq. (9.3.3) by including the effects of compressibility. These are thought to include the speed of sound c and the ratio of the specific heats k. Now, the drag depends on seven variables, so that there will be $5 (= 8 - 3)$ dimensionless variables. Find them by using D, U, and ρ as scaling variables.

(c) Equation (9.3.3) interrelates six variables. Another form of this expression could be $U_\infty = \hat{\phi}_3(\text{drag}, D, \rho, \mu, \epsilon)$. Find the dimensionless parameters by using D, ρ, and μ as scaling variables.

9.4.2(a) Determine the dimensionless parameters using d and γ as scaling variables.

(b) If it is recognized that the dimensions of a static pendant drop are controlled by forces in static equilibrium, one might select at the outset a system of dimensions that is based on F, L, and T instead of M, L, T. Hence, $\gamma = F/L^3$ and $\sigma = F/L$. Determine the dimensional parameters for the static pendant drop using an F, L, T system. How do these parameters compare with those in Example 9.4.2?

(c) The diameter of a liquid drop on a horizontal, nonwetting surface would depend on the drop's volume, σ, and γ. Choose the volume and σ as scaling parameters to determine the dimensionless ratios for the problem.

9.4.3(a) If the flow is compressible, the excess pressure at the tip of an impact tube also depends upon the speed of sound c and the ratio of specific heats k. Determine the appropriate dimensionless parameters for this compressible situation.

(b) In a compressible flow, the temperature at the tip of a slender body, T_s, is called the *stagnation temperature*. It depends upon the speed of the gas flow, V; the static temperature (as measured by a sensor moving with the gas) of the gas, T; the gas constant, R; and the ratio of the specific heats, k. Determine the dimensionless parameters for this case by choosing V and T as scaling parameters.

9.4.4(a) What diameter of pipe could be used if the head loss due to friction could be 4 ft of water?

(b) What diameter of pipe could be used if the flow rate were 18 cfs and the head loss could be 4 ft of water?

(c) What length of 2-ft-diameter pipe ($\epsilon = 0.001$ ft) should be used if the losses must be limited to 3 ft at a flow rate of 12 cfs?

9.6.1(a) What should be the speed of a water test with the 500-mm sphere at the same Reynolds number as in Example 9.6.1?

(b) What should be the speed of a water test in mph for a sphere with an 18-in. diameter for the same Reynolds number as in Example 9.6.1? What is the drag force on this sphere?

(c) Between Re = 10^3 and Re = 10^5, the drag coefficient varies little for a sphere. (The same is true for a circular disk, over which the flow

pattern also does not change significantly with Reynolds number.) But such constancy of C_D is often not the case. Compare the drag on the 500-mm sphere in water at 0.2 and 2.0 m/s.

9.7.1(a) For a point sink flow, a radial velocity of 5 m/s was measured at $R = 2.1$ m. What velocity should exist at $R = 4.5$ m?

(b) A point sink is used to model the flow into the inlet of a pipe for an underground water well. Water flows up this pipe at 5 gallons per minute (448 gpm = 1 cfs). What should be the radial velocity of the water in the well at $R = 1$ ft and 5 ft?

9.7.2(a) A porous pipe discharges cooling water uniformly (radially) into a large vessel. The pressure needed to pump this water leads one to believe that the radial velocity on the outer surface of the pipe (outside diameter = 100 mm) is 0.01 m/s. Where would you expect the radial velocity to be 1 percent of the value at the pipe wall?

(b) If the pipe in Exercise 9.7.2a is 1.6 m long, what is the volume rate of flow through the porous wall?

PROBLEMS

9.1 The amplitude of a light wave, A_s, that is scattered by a particle depends upon the amplitude and wavelength of the incident wave, A_i and λ. In addition, the size of the particle, D, and the distance from it to the observer, L, are important. What are the dimensionless variables? Use the incident parameters as some of the scale values.

9.2 The deflection of a cantilever beam depends upon the load at its end, its length, and the depth of the particular cross section. However, an aluminum beam will deflect differently from a steel beam. What material constant should be included in the analysis? (Neglect gravity.) Form the dimensionless parameters for this problem, and from them predict how the deflection will depend upon the depth of the section if all the other parameters are held constant.

9.3 The lateral vibrations of a stretched wire will exhibit a frequency that depends upon its length, diameter, density, and tension.

Develop the associated dimensionless parameters.

9.4 The period of longitudinal vibrations in a rod depends upon its length, density, and modulus of elasticity. What dimensionless parameters characterize this phenomenon?

9.5 It is found that the angle of twist of a circular rod under pure torsion depends upon the applied torque, the length of the rod, its diameter, and an elastic constant. Use the length and modulus as characteristic scale values to develop the relative dimensionless parameters. Can you anticipate any special role for the length in this dimensionless expression? How does the angle of twist per unit length vary with the diameter of the rod?

9.6 A steel rod vibrates at a frequency of 256 hertz (Hz). What would be the frequency of geometrically similar rods of aluminum that were of the same length? Of rods twice the length? (Refer to an appropriate handbook for any necessary physical constants.)

9.7 The flow over a dam's spillway will depend upon the fluid's density and viscosity, as well as the height of the water above the spillway's crest. Do you expect the gravitational constant or surface tension to be important? What are the dimensionless variables needed to express the flow rate over the spillway, Q?

9.8 A submerged hump that spans a watercourse can control the flow. If the length of a horizontal portion of such a broad-crested weir is B, determine the dimensionless parameters that can be used to correlate the flow rate. (Refer to Prob. 9.7 for some additional parameters.) For tests with fresh water with a prototype exposed to sea water, what should be the factor that relates the model's viscosity if the model's length scale is 1/15? (Use an SG of 1.03 for sea water.) What should be the ratio of the Q's for the model and prototype for the same value of the dimensionless variables? If the prototype is to operate at 10°C, what should be the water temperature for the model tests?

9.9 A spherically symmetric explosion propagates a shock wave whose radius will depend on time and the energy at the instant of release, E. The pressure and density of the ambient medium, p_0 and ρ_0, should be factors in the analysis, in addition to the ratio of the specific heats at constant pressure and volume, k. Use E, t, and ρ as repeating variables to develop a dimensionless relationship for the radius of the shock front.

9.10 The Stokes number is the ratio of pressure forces to viscous forces. Determine its form by combining the pressure p, a length scale L, a characteristic speed U, and the viscosity μ into a dimensionless parameter. Show that it is the product of the pressure coefficient and Reynolds number.

9.11 A 575-ft-long ship is designed to travel at 30 knots. At what speed should an 8-ft-long model be towed?

9.12 What is the ratio of the drag forces that should be encountered by an aircraft flying at sea level and at 25,000 ft under similar conditions? (Use standard atmosphere tables to estimate the conditions, and Appendix A as a guide to the other fluid properties.) What is the ratio of the two engine powers for these two altitudes?

9.13 A 1/10 model of a submersible used for televising the ocean floor is to be tested in a wind tunnel that can pressurize the air to a specific weight of 0.7 lb/ft³, at 50°F. The drag on the model was 2.1 lb when the air speed was 90 ft/s. What is the expected drag that the prototype will experience at the operating condition that corresponds to the model test? What power will this require of the prototype's motor? (The viscosity will not depend upon the pressure in the wind tunnel, but the kinematic viscosity will, because of the density that is part of it.)

9.14 The ship and model described in Prob. 9.11 should be analyzed by including the effects of shear stresses on the drag of the vessel. With this additional feature of the modeling process, what is the expected drag (viscous and wave) on the prototype? Use a draft of 25 ft and a beam of 50 ft.

9.15 The data from a pipe network consisting of bends and valves are to be used to estimate the losses in a system that is twice as large in scale. The original system was tested at a Reynolds number of 10^5 while the water temperature was 65°F. Assume that pipe roughness is zero in the determination of the ratio of losses for the two systems at the same Reynolds number. What is that ratio?

9.16 Under conditions of dynamic and geometric similarity, tests are to be conducted in a 2-in. pipe with water and a 4-in. pipe with air. The tests are to be conducted at 14.7 psia and 70°F. If the air velocity is 48 ft/s, what is the corresponding speed for the water? What is the ratio of the pressure differentials at corresponding points of the two pipes? What are the corresponding pressure gradients?

9.17 A rectangular wing with a chord of 8 in. and a length of 48 in. is tested in a wind tunnel at 70°F and a speed of 280 ft/s. The chord line is at an angle of attack of 10°, and the resultant force, measured normal to the chord, is 83 lb. What are the drag and lift coefficients for this wing? If this is a model of a wing that has a chord of 12 ft, what should be the prototype speed that corresponds to this model test? What are the expected prototype lift and drag for this model test?

9.18 An aircraft is tested at 890 mph when flying at sea level on a day when the temperature corresponds to that of the standard atmosphere. What is the Mach number for this test? What Mach number can be expected at 10,000 and 40,000 ft at the same speed?

9.19 The aircraft in Prob. 9.18 was developed from tests in a wind tunnel where the pressure was 200 psia at 40°F. What was the speed of the air in the tunnel?

9.20 A ship's propeller develops a thrust F that depends on its forward speed U, its diameter D, and a rotational speed N. The fluid's density and viscosity are also important parameters. Because such a propeller will create some waves, gravitational attraction must be included in any analysis. (Is the depth of submergence important?) What are the dimensional parameters that will determine the thrust coefficient if U, D, and ρ are selected as scaling factors? What additional dimensionless parameter will occur if cavitation is a consideration? (The significant dimensional parameter is the vapor pressure.) How can these dimensionless ratios serve to describe the thrust of an aircraft's propeller?

9.21 The thrust that a propeller produces is equal to the drag on the craft at constant speed. The power to produce this speed is the product of this thrust and the vehicle's speed. The power applied to the propeller is the product of the torque and the angular velocity. This torque can be expressed as the product of a moment (torque) coefficient and $\rho N^2 D^5$. (See Sec. 8.4.) Thus, the efficiency of a propeller depends on the same variables as the thrust, with the inclusion of C_M. Develop an expression for this efficiency that depends on dimensionless variables.

9.22 The drag on a smooth sphere will depend only on inertia and viscous terms in a horizontal flow of a turbulence-free fluid. (Any normal stress will be dependent on these same variables and, thus, does not constitute another independent variable.) Choose the speed of the flow U, the diameter of the sphere D, and the density ρ, to form the dimensionless variables. If in very slow motion the functional dependence of C_D on the Reynolds number is found to be 24/Re, develop an expression for the drag on a sphere for such slow motions.

9.23 The flow of water in a heat transfer device is

to be modeled with air in a 1/10 scale model. The prototype velocity is to be 24 ft/s. What speed will occur in the model for similar conditions? What is the ratio of the pressure differentials to be expected at comparable locations in the model and the prototype?

9.24 A centrifugal pump is a hydrodynamic machine that is similar to the propeller discussed in Prob. 9.20. However, the parameter U would be replaced by Q/D^2 when treating pumps and turbines, where Q is the volume rate of flow. The design of a pump which is to deliver 4000 gpm at a head of 240 ft is to be examined by a 1/3 scale model. The prototype pump conveys an oil with a viscosity of 1.8 poise (P) and a specific gravity of 0.92 while rotating at 1375 rpm. The model tests are to be conducted with water. What will be the model's speed to test the design parameters? What are the corresponding model flow and head?

9.25 A fluid coupling is another type of hydrodynamic machine. Its ability to transmit torque will depend upon the usual parameters and the ratio of the speeds of the output and input shafts. This influence is usually expressed by the slip ratio S, which is $(1 - \text{speed}$ ratio). Develop a set of dimensionless parameters that characterize the torque of a fluid coupling by using D, ρ, and N_{input} as characteristic scales. (If the volume of the fluid in the coupling can be changed, it will also be a parameter in the expression for the torque. What is the corresponding dimensionless parameter?) Experience shows that the Reynolds number influences the results only slightly and can be used to modify the initial calculation at a subsequent stage in the design process. For the same slip ratio and fluid volume, how will the torque change for a 30 percent decrease in the input speed to a test unit?

9.26 The buoyant force per unit mass that arises from a heated surface with temperature $Temp_w$ can be characterized as $g\alpha(Temp_w - Temp)$, in which α is the coefficient of thermal expansion and has dimensions of $(\text{degrees})^{-1}$. A measure of the inertia force per unit mass is given by W^2/L, in which W is a characteristic speed and L is a scale length. Determine the ratio of these two forces. If this ratio is equated to the ratio Gr/Re^2, determine the form of the Grashof number, Gr, by introducing the parameters for the Reynolds number, Re.

9.27 The force of buoyancy is mentioned in Prob. 9.26. The viscous force in a flow would be proportional to the product of the shear stress and an area that can be represented by $(\mu W/L)L^2$. This force would be divided by ρL^3 if the shear force per unit mass is wanted. The Rayleigh number Ra is the ratio of the buoyant force to the viscous force. What is its form?

9.28 The energy transfer associated with conduction in a fluid can be represented by $k(T_w - T)/\rho L^2$. T_w is a surface temperature, while T is the temperature within the fluid. L is a scale length, and k is the thermal conductivity of the fluid. The energy transfer due to convection can be scaled by $c_p U(T_w - T)/L$, in which c_p is the specific heat at constant pressure and U is a reference velocity. The ratio of convection to conduction energies is the Péclet number Pe. What is its form? Equate the Péclet number to the product of the Reynolds number and another dimensionless term, the Prandtl number Pr. What is the form of the Prandtl number?

9.29 If a heat transfer process is unsteady, the convection energy must include a term that characterizes the temporal part. The product $c_p(T_w - T)/t$, in which t is time, does this. The ratio of this unsteady energy transfer to the conduction heat transfer is called the *Fourier number* Fo. Find its form after referring to Prob. 9.28.

9.30 Refer to Fig. 5.4.3 to determine the velocity in air and water (70°F) that would be necessary to obtain a pressure coefficient of about -0.2 on the rearward part of a circular cylinder. What would be the pressure there for these two cases if the ambient pressure is 14.7 psia?

9.31 If the wing characteristics shown in Fig. 9.3.2 were obtained on a 1/10 scale model, what should be the ratio of the lift and drag forces on the model and prototype?

9.32 If a wing with a chord of 10 ft and a length of 50 ft were of the same shape as the wing whose characteristics are given in Fig. 9.3.2,

what would be the maximum weight of the aircraft when flying at 250 mph? What would be the most probable angle of attack of the wing? (Use Table A.3 at about 10,000 ft and two wings per aircraft.) What power would be generated by the engines for a 100 percent efficient propeller system?

9.33 A circular cylinder ($D = 4$ mm) flutters in a cross wind (air at 101.1 kPa and 20°C). If the Strouhal number Sr is 0.212, determine the wind speed. What would be the wind speed if this Strouhal number were measured on a wire that is 0.4 mm in diameter?

9.34 Compare the drag on a cylinder having a diameter of 20 mm when it is being towed in water and in air (101.1 kPa and 20°C) at $Re = 10^5$. Make a similar comparison if the cylinder were towed in air at one-tenth and at 10 times the velocity in water when $Re = 10^5$ for the water tests. Do the same drag forces result if the cylinder is towed at the same speed in air and water?

9.35 If a sphere is towed under *all* the conditions of Prob. 9.34, what is the ratio of the drag forces in each case, using the drag in water at $Re = 10^5$ as the reference value.

9.36 Compressibility will affect the drag on a body. Determine the drag on a sphere with an air speed of (about) 500, 700, 1100, and 1400 ft/s. Do this when the Reynolds number is 10^5, 10^6, and 10^7. (Assume 14.7 psia and 60°F.)

9.37 What is the pressure drop per unit length in a cast iron pipe 12 in. in diameter if water (68°F) is flowing at an average velocity of 9 ft/s?

9.38 For the Reynolds number that is implicit in Prob. 9.37, what is the speed of air in a duct of the same size whose effective roughness is comparable with that of a galvanized iron pipe? What pressure drop would be expected under these conditions?

9.39 Two pipes of the same length and diameter exhibit a friction factor of 0.01 at $Re = 10^6$. What is the ratio of their relative roughnesses? If the kinematic viscosities differ in the two pipes by a factor of 10, how would the average fluid speeds differ?

9.40 Two identical pipes have flows that exhibit the same friction factor, 0.01 at $Re = 10^6$, but the head losses are in the ratio of 2/1. What is the ratio of the kinematic viscosities of the fluids for the same Q?

9.41 Two pipes with the same nominal geometric values convey water at the same speed and Reynolds number, 10^6, so that the measured friction factor, 0.02, is also the same. If one pipe is commercial steel, estimate the value of the roughness of the other pipe.

9.42 It is proposed to repeat the original experiments that were performed by Nikuradse in pipes. If the data at $\text{Re} = 2(10^4)$ are to be checked in a 3-in. diameter pipe, what must be the size of the sand grains if a friction factor of 0.04 is to be realized? Is sand of this average size usually found in the world?

9.43 Arrangements are to be completed to test a sphere at a constant drag coefficient of 0.3 under controlled conditions of free-stream turbulence. What range of air speeds (80°F and 14 psia) should be anticipated for a 1-ft-diameter sphere? What range of drag forces should be encountered?

9.44 Data for axisymmetric jets in water and air reveal different locations where $u/u_m = 1$. For the same axial distance x, what is the ratio of the radial distances y for two such jets, one in air and one in water, that have the same kinematic momentum K' ($=$ momentum/ρ)? If, instead, the jets had the same momentum, what would be the ratio of the radial distances? The two-dimensional jet described in Fig. 9.7.4 appears to have an effective width of 2ξ for ξ between 2.2 and 3. How does such a jet widen as a function of x only? What is the rate of growth with x of an axisymmetric jet?

9.45 For a two-dimensional jet, the centerline velocity decreases with $x^{-1/3}$. (See Fig. 9.7.4.) The similarity parameter for the width varies proportionately to $y/(\nu x^2)^{1/3}$. Compare the momentum and mass rate of flow across two planes normal to the x axis for which x varies by a factor of 2. Comment on the momentum and flow rate ratios that you obtain, and give any theoretical justification for the values you obtain.

9.46 The two-dimensional wake shown in Fig. 9.7.5 has a maximum (i.e., centerline) speed defect that decreases with $x^{-1/2}$, and a width b

that increases with $x^{1/2}$. Is this speed and width variation in concert with a constant (wake) momentum in the direction of the free-stream flow? Why should the wake momentum be a constant in this case if a uniform flow approaches the body?

9.47 Refer to Prob. 9.46 and predict whether the mass flow rate in the direction of the free stream is the same at the two x locations.

9.48 In one process to recover oil from porous rock strata, the volume per second of oil, Q, that can be obtained depends upon the rock's average pore size d, the viscosity μ, the surface tension σ between the oil and the water that is pumped into the ground to replace the oil, and the pressure drop per unit distance that can be achieved by the water that is pumped to replace the oil, $\Delta p/L$. In addition, the curvature of surfaces of the oil droplets in the rock can be characterized by an average radius of curvature of the pores, $\bar{r}$. Develop the associated nondimensional parameters for this problem as

$$\frac{Q\mu}{d^4(\Delta p/L)} = \phi\left(\frac{\sigma}{\Delta p/L}, \frac{\bar{r}}{d}\right)$$

9.49 The results of Prob. 9.48 could be applied to a laboratory study of the situation. Assume that the rock is to be modeled by a very long tube of diameter d. Initially, only oil fills the tube so that σ, the interfacial tension, is zero because no water is in the tube. Also, the ratio $\bar{r}/d$ is taken as $(d/2)/d$, since the tube's radius is the only radius of curvature. To what value do you think the function

$$\phi\left(\frac{\sigma}{\Delta p/L} = 0, \frac{\bar{r}}{d} = \frac{1}{2}\right)$$

should reach? (*Hint:* Would the experiment simulate Poiseuille flow?)

9.50 In the situation discussed in Prob. 9.49, assume that $\bar{r}$ is directly proportional to d in a particular type of rock. Assume that certain chemicals in the water used to flood the oil field decrease σ by a factor of 2. If, under these conditions, the pressure gradient is also cut in half, would you expect the volume rate of oil recovery to increase or decrease? If, in a second oil field, the pore size in the rock is one-half that in the first field but the relationship $\bar{r}/d$ does not change, what will be the expected Q in the second field? Assume that the same $\Delta p/L$ will be applied to both oil fields.

9.51 When a light fluid (e.g., sewage, with density ρ_0) is released from a porous conduit at the bottom of a lake or other body of water of depth d and colder temperature, and therefore higher density, it rises toward the surface. If the medium has a density of ρ_b at the bottom and this density is the same at all elevations, then the light fluid rises to the surface as a jet whose cross-sectional area increases with elevation. This broadening of the jet is due to the entrainment of the denser medium into the jet, which has the lower density. However, the body of water may have a density gradient due to temperature. It may be that the bottom temperature hardly varies but the free-surface temperature is higher due to the absorption of solar energy, with the result that the surface density ρ_s is less than ρ_b. Then, for $\rho_b < \rho_0 < \rho_s$, the light fluid that is discharged at the bottom will not reach the free surface but will rise to a level z_∞. Assume that z_∞ depends on the density gradient $(\rho_b - \rho_s)/d$; the initial excess of buoyant force to drive the fluid upward, $g' = g(\rho_b - \rho_0)/\rho_b$; and the initial vertical momentum that is determined by ρ_0 and the vertical velocity w_0. Derive a set of π parameters relating z_∞, $(\rho_b - \rho_s)/d$, g, w_0, and ρ_0 giving

$$\frac{z_\infty g'}{w_0^2} = \phi\left(\frac{[\Delta\rho/d]w_0^2}{g'\rho_0}\right) \quad \text{where} \quad \frac{\Delta\rho}{d} = \frac{\rho_b - \rho_s}{d}$$

Was z_∞ chosen as a characteristic length in the above result? Explain why one would, or would not, do this,

9.52 If, in Prob. 9.51, g' does not change from winter to summer and effluent and lake bottom temperature do not change, how must w_0 be changed from winter to summer in order that $z_\infty g'/w_0^2$ remain constant if $\Delta\rho/d$ in the summer is twice $\Delta\rho/d$ in the winter? If z_∞ is 88 ft in the summer, what will it be in the winter?

9.53 Refer to Prob. 9.51 and assume that w_0 is held constant throughout the year; also, that ρ_0 is less in the summer than in the winter but affects the results principally through g',

which is considered here to be twice as large in the summer as in the winter. What must be the value of $\Delta\rho/d$ in the summer, with respect to the winter value, for the same value of $z_\infty g'/w_0^2$ to be expected? What, then, is z_∞ in the winter if a value of 88 ft is found in the summer?

9.54 If there are two horizontal layers of liquids of different densities—the denser liquid constituting the lower layer—then waves can be created at the interface if it is disturbed. The amplitude of the waves, a, depends upon the difference in the specific weights of the two liquids, $g\,\Delta\rho$; the surface tension σ; a length parameter L, which can be taken as the depth of the layers (assumed to be equal); and the wavelength λ. Find the dimensionless parameter(s). (It is reasonable that L enters the problem, since if the layers are bounded above and below, these boundaries should influence the vertical extent of the oscillations of the intermediate interface. Can you justify the selection of $g\,\Delta\rho$ and σ as important parameters in this case? The wavelength will affect the curvature of the interface. Should this be important?)

9.55 The force on a sluice gate, R_x, depends on the upstream velocity U_1; the upstream and downstream fluid depths, designated h_1 and h_2, respectively; the fluid's density; and its specific weight (i.e., gravity). (Viscosity should play a minor role.) Determine the appropriate dimensionless parameters.

9.56 The situation described in Prob. 9.55 is discussed in Example 7.3.1. Show that

$$\frac{R_x}{\rho U_1^2 h_1} = \frac{1}{2}\frac{gh_1}{U^2}\left(1 - \frac{h_2^2}{h_1^2}\right) + \left(1 - \frac{h_1}{h_2}\right)$$

Note that $h_1^2\gamma/2$ and $\rho U_1^2 h_1$ in Example 7.3.1 are representative of the upstream forces due to gravity and inertia. Is the equation above in dimensionless form? Could this equation be plotted so that $R_x/(\rho U_1^2 h_1)$ would be a function of U_1^2/gh_1, with h_2/h_1 as a parameter? If this is the case, what would be the connection between the resulting curve and the curves obtained experimentally by applying the results of Prob. 9.55?

9.57 If a vertical rectangular channel of width L (unit distance perpendicular to the page) is filled with a liquid with density ρ and the channel is submerged in a vessel with a liquid that is less dense, as shown in Fig. P.9.57, the heavier fluid need not flow downward if the surface tension at the interface is sufficiently large. If the interface is displaced vertically, a pressure difference will be induced due to the difference of densities. This pressure difference will be balanced by the surface tension and the curvature of the interface. A limiting condition is also shown in the figure. Carry out the analysis for this problem, assuming static equilibrium of the surface AB, and write the result in nondimensional form.

9.58 Apply dimensional analysis to the situation described in Prob. 9.57. Assume that the maximum displacement η_{max} is a function of the difference in specific weight, $g\,\Delta\rho$; the surface tension σ; and the width L. Develop the appropriate parameters.

9.59 The rise of the heavier fluid between the two parallel walls in Fig. P.9.59 is a function of $g\,\Delta\rho$, b, σ, and the angle of contact with the wall, α. Develop the nondimensional parameters upon which h/b depends.

9.60 Review the choice of variables and the logic that led to Eq. (9.4.6). Then choose Q, μ, and D as the repeating variables and show that this choice would lead to

$$\left(\frac{\Delta p}{L}\right)\left(\frac{D^4}{Q\mu}\right) = \phi_1^*\left(\text{Re}, \frac{\epsilon}{D}\right)$$

in which ϕ_1^* is a function that must be determined. Rewrite the left-hand side of the expression above in terms of familiar variables to obtain.

$$f\,\text{Re} = \phi_1^*\left(\text{Re}, \frac{\epsilon}{D}\right)$$

9.61 What is the force on the prototype sphere in Example 9.6.1? At what speed would this force be realized?

9.62 If the Froude and Reynolds numbers are to be kept constant, what must be the ratio of the viscosities of the fluids used in the model and prototype tests? Is it practical to require such a ratio if one of the fluids is water?

9.63 Verify the values of α_2 and α_3 in Eq. (9.2.2). In the case of α_3, note that water has a density of 1.932 slugs/ft³ or 1000 kg/m³ under standard gravitational conditions. Use this fact and α_1 to verify the value of the mass scale factor, $\alpha_3(\alpha_1)^3$.

9.64 Estimate the pressure drop in psi between two points 15 ft apart in a horizontal pipe with an internal diameter of $\frac{3}{4}$ in. if the relative roughness ϵ/D is 0.015 when water at 60°F is flowing with an average speed of 2.5 ft/s. What would be the expected pressure drop if the flow (i.e., Q) were reduced by a factor of 10?

9.65 The pressure developed in a centrifugal pump is related to the speed of the pump, the volumetric discharge rate, the impeller diameter, and the pertinent physical constants of the fluid being conveyed. What are the dimensionless groups that are appropriate to correlate the power delivered to the fluid?

9.66 A crank of radius A is attached to a connecting rod of length L. The latter is fastened to a slider that moves to and fro as the crank turns at an angular velocity of ω. If x is the displacement of the slider, upon what dimensionless variables does x/L depend? Carry out a kinematic analysis of this case to show that the solution $\pi_1 = \phi(\pi_2, \pi_3)$ is *not* of the form $\pi_1 = C\pi_2\pi_3$. Under what condition would the displacement x vary (almost) sinusoidally?

$\rho_1 > \rho_0$ **Figure P.9.57**

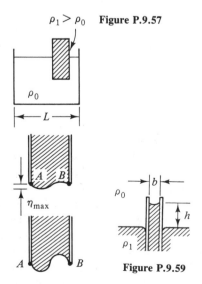

Figure P.9.59

10

Flow in Conduits (Internal Flows)

10.1 INTRODUCTION

In Chap. 6, where the energy equation was developed, we used the flow in conduits in examples of the application of the theory. We shall now discuss this kind of flow in further detail. One form of energy equation is

$$\dot{W} + \dot{2} + \dot{m}\left(\frac{\alpha_1 V_1^2}{2} + \frac{p_1}{\rho_1} + gz_1\right) = \dot{m}\left(\frac{\alpha_2 V_2^2}{2} + \frac{p_2}{\rho_2} + gz_2\right) + \dot{m}(e_2 - e_1)$$

for the flow that enters and leaves a section of pipe of length L. If one divides each term of this equation by the weight rate of flow through the pipe, $g\dot{m}$, the following equation results:

$$\frac{\dot{W}}{\gamma Q} + \frac{\dot{2}}{\gamma Q} + \left[\frac{\alpha_1 V_1^2}{2g} + \frac{p_1}{\gamma_1} + z_1\right] = \left[\frac{\alpha_2 V_2^2}{2g} + \frac{p_2}{\gamma_2} + z_2\right] + \frac{e_2 - e_1}{g} \qquad (10.1.1)$$

For the flow of liquids, the term $(e_2 - e_1)/g$ is commonly considered to be a loss of the available (mechanical) energy to the flow. It is usual to designate this term h_L, the *head loss*. For a horizontal pipe of constant diameter with fully developed flow and no energy input in the form of $\dot{W}$ or $\dot{2}$, Eq. (10.1.1) indicates that $h_L = \Delta p/\gamma$. Usually one designates the terms in the square brackets in Eq. (10.1.1) the *total head* for the flow. The individual terms are, respectively, the *kinetic energy head*, the *pressure head*, and the *elevation head*. A plot of the total head along the length of the pipe is called the *energy grade line*, and the corresponding plot of the pressure and elevation heads is called the *hydraulic grade line*. The elevation of the hydraulic grade line, relative to the pipe's centerline, would be the position to which a column of fluid would rise in a vertical tube attached to a pressure tap on the pipe. The energy grade line will decrease in elevation along the pipe as a result of the mechanical energy dissipation that occurs. There will be increases wherever there is an input of energy.

The presence of shear stresses within a fluid inevitably leads to a conversion of mechanical energy terms into thermal quantities. Pipe friction is one manifestation of this process. Energy dissipation also results when the conduit contains an abrupt change in cross section. Thus, we shall reserve the designation h_L for the *entire loss* of head for the entire pipe system. The designation h_f will denote that part of h_L due to *pipe friction*. Our discussion of the pressure drop per unit length of pipe, $\Delta p/L$, in Chap. 9 led us to conclude that

$$\frac{\Delta p/L}{\rho V^2/D} = \phi\left(\text{Re}, \frac{\epsilon}{D}\right)$$

This dimensionless unit pressure drop was designated f, the *friction factor*. For a horizontal pipe, it follows that

$$f = \frac{\Delta p/\gamma}{(V^2/2g)(L/D)} = \frac{h_f}{(V^2/2g)(L/D)} \tag{10.1.2}$$

This f is plotted in Figs. 9.4.2 to 9.4.4. They show how f varies with Re, for instance. From these curves one can deduce how h_f changes as a function of the flow parameters.

While h_f has been considered until now in connection with a horizontal pipe, should it not be the same for a vertical pipe? In such a pipe the same dissipative process would be at work. It is for this reason that we can broaden the definition of f to determine the head loss due to pipe friction, h_f, for a straight pipe of *any* slope. After h_f has been determined for a prescribed flow situation and all the other elements of energy dissipation along the flow path have been assessed, they can be summed together to yield h_L. This term can then be inserted into Eq. (10.1.1) to solve for $\dot{W}$, Δp, or any other unknown.

Such solutions will be the major emphasis of this chapter. We shall also include additional terms that are part of h_L similar to those mentioned in the development of Eqs. (7.7.9) and (7.7.10).

10.2 PIPES IN SERIES AND PARALLEL

We now turn to applying Fig. 9.4.2, the Moody diagram, in some examples to gain some experience with its use.

Example 10.2.1

A horizontal cast iron pipeline has a 1500-ft length with a diameter of 24 in. Water at 68°F must be pumped through this pipe at 12 cfs. What pressure drop can be expected? [Remember to use only units consistent with slugs, feet, and seconds (and pounds).]

Solution   The information that is given allows us to find ϵ from the table provided by Fig. 9.4.2b so that $\epsilon/D = 0.001/2 = 0.0005$ and

$$\text{Re} = \frac{VD}{\nu} = \frac{Q}{\pi D^2/4}\left(\frac{D}{\nu}\right) = \frac{(4/\pi)(12/2)}{1.06(10^{-5})} = 7.2(10^5)$$

Figure 9.4.2 can be used, as resketched in Fig. 10.2.1, to yield $f = 0.0175$. This means that

$$h_f = f\frac{L}{D}\frac{V^2}{2g} = 0.0175\left(\frac{1500}{2}\right)\left(\frac{V^2}{2g}\right)$$

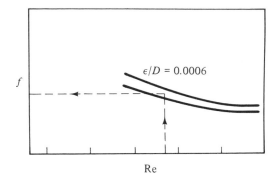

$\epsilon/D = 0.0006$

f

Re

Figure 10.2.1 Simple graph extracted from Moody diagram of Fig. 9.4.2 to show its utilization.

with $V = 3.82$ ft/s from Q/A. This gives

$$h_f = 2.97 \approx 3 \text{ ft}$$

As a result, $\Delta p = \gamma(2.97) = 186$ lb/ft^2, from Eq. (10.1.1).

Example 10.2.2

If, over the 1500 ft of pipeline in Example 10.2.1, there had been an increase of elevation of 15 ft, what would the expected change in pressure be for the data given in the example?

Solution Because the flow (i.e., Re and ϵ/D) has not changed, h_f will be the same as in Example 10.2.1. Since $\dot{W}$ and $\mathcal{Q}$ are zero, we have, from Eq. (10.1.1),

$$0 + 0 + \left(\frac{\alpha_1 V_1^2}{2g} + \frac{p_1}{\gamma} + 0 \right) = \left(\frac{\alpha_2 V_2^2}{2g} + \frac{p_2}{\gamma} + 15 \right) + 2.97$$

or

$$\frac{p_1 - p_2}{\gamma} = 15 + 2.97 = 17.97 \text{ ft}$$

Then

$$\Delta p = \gamma(17.97) \approx 1121 \text{ lb/ft}^2$$

Had the pipeline in Example 10.2.1 been twice as long, one would expect the losses to have been twice as large. In a sense, the losses in the second 1500 ft are added to those in the first 1500 ft. *The losses are additive in a series pipe system.* In such a series arrangement, the volume rate of flow is the same in each pipe. Therefore, if there are n pipes in *series,* one has

$$V_i = \frac{Q}{A_i} \quad \text{and} \quad h_f = \sum_{i=1}^{n} f_i \frac{L_i}{D_i} \frac{V_i^2}{2g} \qquad (10.2.1)$$

This implies that in order to get each of the n parts of the losses, one needs to use the Moody diagram, or its equivalent, n times. Finally, one must use Eq. (10.1.1) to solve for Δp, or perhaps $\dot{W}$. (There will be regions, downstream of the coupling of one pipe to another, in which the flow changes to meet the requirements of a different size or different surface properties. Any associated energy dissipation there will be examined in Sec. 10.3, where minor losses are discussed.)

If the two 1500-ft lengths of pipe were connected in parallel, part of the total discharge would pass through each pipe. Regardless of which pipe a fluid particle passes through, we postulate that, when the streams converge, the energy per unit mass of fluid and the pressure are the same. (This implies that the total head will be independent of the path through parallel pipes.) Thus, the *loss in all pipes in parallel is the same*. Thus, we have assumed for n pipes in *parallel* that

$$h_f = \left(\begin{array}{c}\text{constant}\\\text{independent}\\\text{of } i\end{array}\right) = f_i \frac{L_i}{D_i} \frac{V_i^2}{2g} \quad \text{and} \quad Q = \sum_{i=1}^{n} Q_i \qquad (10.2.2)$$

Example 10.2.3

A 1500-ft length of cast iron pipe with a diameter of 1 ft is in parallel with a 2-ft-diameter pipe made of concrete the same length. The pipe friction loss is prescribed at 3 ft. What is the total flow through the two pipes? The pipes are horizontal.

Solution   Since we are given h_f and must determine V, or Q, we can effectively use Fig. 9.4.3.* Here, h_L is $\Delta p/\rho g$, because of Eq. (10.1.1). The value of the abscissa in Fig. 9.4.3 becomes, for the cast iron pipe,

$$\frac{1}{1.06(10^{-5})} \sqrt{\frac{2(32.2)(1)(3)}{1500}} = \frac{0.358}{1.06(10^{-5})} = 3.4(10^4)$$

If we use uncoated cast iron and assume it to be old, $\epsilon \approx 0.002$. For $\epsilon/D = 0.002/1$, Fig. 9.4.3a yields $f \approx 0.025$. From the definition of f, we have

$$0.025 \frac{L}{D} \frac{V^2}{2g} = h_f = 3 \text{ ft}$$

This gives

$$V = \frac{(2g)(1)(3)}{(0.025)(1500)} = 2.27 \text{ ft/s}$$

The corresponding Q is $2.27(\pi)(1)^2/4 = 1.78$ cfs. For the concrete pipe, the value of the abscissa to use in Fig. 9.4.3 is $9.6(10^4)$ for $\nu = 10^{-5}$ ft^2/s. If we have reason to assume that the pipe is newly installed and smooth, we can choose ϵ as 0.001. Then $\epsilon/D = 0.001/2 = 0.0005$ and $f \approx 0.017$, from Fig. 9.4.3. In a similar fashion to the result for the other pipe,

$$0.017 \frac{L}{D} \frac{V^2}{2g} = h_L = 3 \text{ ft}$$

This definition gives

$$V = \frac{2g(2 \text{ ft})(3 \text{ ft})}{0.017(1500 \text{ ft})} = 3.89 \text{ ft/s}$$

with
$$Q = VA = 3.89(\pi)(2)^2/4 = 12.23 \text{ cfs}$$

The sum of the flows through both parts of the parallel network is $1.78 + 12.23$, or 14.01 cfs.

Although we have used the Moody diagram, in various forms, in these examples, it might be possible to obtain some information from one equation that related f, Re, and ϵ/D. The Colebrook equation, given as Eq. (8.3.13), is one such relationship.

---

* If only Fig. 9.4.2 were available, one could assume that the Reynolds number was sufficiently large that f would depend only on ϵ/D. One could then solve the problem for the appropriate f from Fig. 9.4.2, using the asymptotic value at large Re for the given ϵ/D. Using this solution, one would check to see that, indeed, the value of Re was large enough to use the asymptotic value of f. On the basis of this result, a new value of f might be chosen and a new calculation performed. (Example 9.4.4 provides a case for which D must be found when Q and h_f are known.)

However, it suffers in that it is implicit in f. For a given value of Re and ϵ/D, one would have to iterate to determine f. On a PC (personal computer), this is not significantly difficult, but it might be more inconvenient for a hand-held calculator without considerable programmable steps and memory capacity. Explicit formulations, which are approximations to the Colebrook equation, are available. One relationship, due to S. W. Churchill,[*] is

$$f = 8\left[\left(\frac{8}{\text{Re}}\right)^{12} + \frac{1}{(A+B)^{3/2}}\right]^{1/12} \tag{10.2.3}$$

where

$$A = \left[2.457 \ln \frac{1}{(7/\text{Re})^{0.9} + 0.27\epsilon/D}\right]^{16}$$

$$B = \left(\frac{37{,}530}{\text{Re}}\right)^{16}$$

10.3 MINOR LOSSES

The term *minor losses* usually refers to the energy dissipation associated with the flow of fluids through passages that have an abruptly changing geometry. However, what is minor in one case may not be so in another. Because of variations in the geometry of flow passages, random, large-scale eddies are usually generated in the flow. Such turbulence is associated with a pressure drop that varies approximately as the square of the average velocity. In the examples of Sec. 10.2, one can also infer that most of the flows in conduits will be in the turbulent range of the appropriate Moody-like diagram. Thus, even when abrupt changes in the geometry of the flow passage are not necessarily present, say, in a well-designed 90° elbow, the increased pressure changes due to induced secondary flows would probably be turbulent rather than laminar. These induced flows are similar to the flow mentioned at the close of Sec. 8.4. Figure 10.3.1 shows a time-dependent secondary flow that can be generated in a pipe tee.

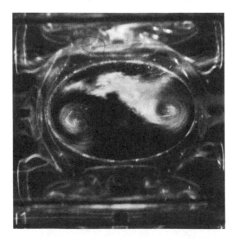

Figure 10.3.1 Secondary swirl patterns in center portion of a pipe tee which has flow entering from each of the lateral arms.

[*] S. W. Churchill, "Friction-factor equation spans all fluid-flow regimes," *Chem. Eng.*, 84:No. 2, pp. 91–92, 1977. Churchill uses a different definition of the friction factor, and this fact requires the factor 8 in Eq. (10.2.3).

It is common to designate the losses due to valves, elbows, transitions, and so on, as $K(V^2/2g)$, in which V is the average velocity in the pipe. K will have a value that is determined by the particular configuration but is almost independent of the Reynolds number, provided that Re is above 10^4. This topic was broached in Examples 6.3.4 to 6.3.6, with no mention of the turbulent flow that dictated the form of the relationships. Table 10.3.1 gives values of K for some pipeline transition pieces. A second method for determining minor losses, which allows for some effect of the Reynolds number on the losses, is also used. For this case, the minor loss is written as $f(L_e/D)V^2/2g$, in which L_e is some *equivalent length* of pipe that is tabulated. The value of f is taken for the appropriate conduit and flow conditions. Some values of L_e/D are also listed in Table 10.3.1. In Table 10.3.2, coefficients are listed to calculate the loss coefficients for branching and merging pipes. In the branching case, the parent pipe is labeled 1 and the daughter pipes are designated 2 and 3. This leads to two coefficients, K_{12} and K_{13}.

TABLE 10.3.1 FRICTION OF VALVES AND FITTINGS IN TERMS OF K OR L_e/D*

Head loss $= K(V^2/2g)$ **or** $f(L_e/D)V^2/2g$

Item	Description	K
Sudden enlargement†	$D_1/D_2 = 0.0$	1.0
	0.33	0.8
	0.5	0.56
	0.7	0.26
	0.9	0.04
Sudden contraction	$D_2/D_1 = 0.9$	0.04
	0.7	0.25
	0.5	0.5
	0.0	1.0
Slightly rounded entrance from reservoir		0.25
Rounded flow nozzle entrance		0.04

Pipe Fitting	Description	L_e/D
Globe valve	Fully open	340
Gate valve	Fully open	13
	3/4 open	35
	1/2 open	160
	1/4 open	900
Elbows	45° standard elbow	16
	90° standard elbow	30
	90° long-radius elbow	20
Standard tee	With flow through run	20
	With flow through branch	60
180° bend	Close pattern return	50
90° bends	Bend radius = 1 pipe diameter	20
	= 2 pipe diameters	12
	= 3 pipe diameters	16
	= 4 pipe diameters	24

* Courtesy of Crane Co.
† Refer to Eq. 7.7.9.
D_1 = upstream diameter; D_2 = downstream diameter.

TABLE 10.3.2 PARABOLIC FIT FOR MINOR HEAD LOSS COEFFICIENTS h_{ml} FOR BRANCHING AND MERGING PIPES OF THE SAME DIAMETER*

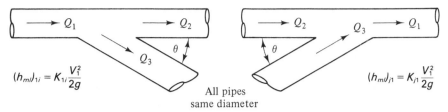

All pipes
same diameter

For branching pipes, $(h_{ml})_{1i} = K_{1i}V_1^2/2g$. For merging pipes, $(h_{ml})_{j1} = K_{j1}V_1^2/2g$.

$$K = A + B\,(Q_3/Q_1) + C\,(Q_3/Q_1)^2$$

(Two significant decimal places for K)

	Q	A	B	C
K_{12}	$\left.\begin{array}{l}45°\\90°\end{array}\right\}$	0.121425	−0.243212	0.580355
K_{13}	45°	0.9175	−1.50768	1.04911
	90°	0.945357	−0.450179	0.790179
K_{21}	45°	0.0353574	1.03768	−1.62053
	90°	0.037143	0.723571	−0.160714
K_{31}	45°	−0.908214	2.8816	−1.6116
	90°	−1.16107	3.79874	−1.75445

Values of A, B, C yield two significant decimal places for K.
* Adapted from figures in H. Press and R. Schröder, *Hydromechanik in Wasserbau* (Wilhelm Ernst & Sohn, Berlin, 1966).

Both are used with the upstream velocity V_1. For the merging flow, K_{21} and K_{31} are used with the downstream velocity which is also, but unmistakedly, denoted as V_1 in the definition sketch for Table 10.3.2.

While it is often the case that the losses associated with bends and valves are small in comparision with the losses in a pipeline, such a situation is not universally true, as indicated by the example that follows. However, it is not uncommon in an initial analysis of a pipeline system to disregard the losses in bends and valves in making a preliminary estimate of the flow. Subsequent refinements of the analysis should take such minor losses into account.

Example 10.3.1

A hydraulic model of the duct that connects the boiler flue to the chimney of a power plant is to be tested. This duct is 30 ft high and 20 ft wide and is to be modeled by a duct that is 6 in. by 4 in. A water reservoir with a free surface 20 ft above the horizontal centerline of the model is to be used in the tests. The Bernoulli equation would indicate that sufficient velocity should result so that the prototype's Reynolds number of $8(10^5)$ could be achieved in the model. Determine whether this conclusion is corrrect when a preliminary trial is made with a plastic tube 4.25 in. in diameter and 3 ft long. This tube would be connected to the reservoir with a rubber hose 3 in. in diameter and 30 ft long. A gate valve, a 90° long-radius elbow, and a slightly rounded entrance at the reservoir complete the circuit. These dimensions are shown in Fig. 10.3.2.

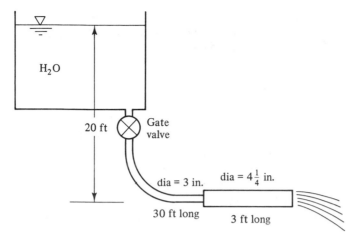

Figure 10.3.2 Schematic diagram for Example 10.3.1.

Solution An energy equation for an adiabatic system without the addition of mechanical energy can be written in terms of the weight rate of flow between the free surface and the exit of the plastic pipe. The result is

$$\frac{0^2}{2g} + \frac{0}{\gamma} + 20 = \frac{V_{exit}^2}{2g} + \frac{0}{\gamma} + 0 + h_L \qquad V_{exit} = V_{tube}$$

Using K values from Table 10.3.1 for the entrance, the long-radius elbow, the valve, and the sudden expansion between the rubber and plastic tube of 0.25, 20, 13, and 0.26, respectively, one has

$$h_L = \frac{V_{rubber}^2}{2g}\left[0.25 + 20 + 13 + \left(\frac{fL}{D}\right)_{rubber} + 0.26\right] + \frac{V_{tube}^2}{2g}\left(\frac{fL}{D}\right)_{tube}$$

It will initially be assumed that the Reynolds number for the prototype will be achieved in the model. Then, for smooth pipe such as a rubber hose and plastic pipe, f can be taken as 0.012 from the Moody diagram. The expression for the head loss becomes

$$h_L = \frac{V_{tube}^2}{2g}\left\{\left(\frac{4.25}{3}\right)^4\left[33.51 + 0.012\left(\frac{30}{3/12}\right)\right] + \frac{0.012(3)}{4.25/12}\right\}$$

with

$$V_{rubber} = V_{tube}\left(\frac{4.25}{3}\right)^2$$

If this result is inserted into the energy equation that was written at the outset, the expression for the velocity becomes

$$20 = \frac{V_{tube}^2}{2g}[4.03(33.51 + 1.44) + 0.102 + 1]$$

$$V_{tube} = 3.01 \text{ ft/s}$$

The Reynolds number for the plastic pipe, assuming the water to be at 70°F, can be calculated to be

$$Re = \frac{3.01(4.25/12)}{1.06(10^{-5})} = 10^5$$

It appears that the head loss terms were sufficiently large to render the prediction based upon the Bernoulli equation too optimistic. Moreover, one can see that the minor losses through the elbow and valve were the major contributors in producing the head loss. It would be instructive to determine the length of rubber tubing necessary to give a head loss equivalent to that of the elbow and valve. (*Ans.:* 685 ft)

10.4 TRANSITION TO FULLY DEVELOPED PIPE FLOW

In Table 10.3.1, the loss coefficients for sudden expansions and sudden contractions are listed. The values for the former case can be estimated by the use of the momentum and energy principles in a manner similar to that in Sec.7.7. The losses associated with a sudden contraction are due in part to the change of the velocity profile in the conduit subsequent to a contraction. It was observed in Sec. 8.6 that when a fluid with nearly uniform velocity begins to pass over a solid boundary, a boundary layer will develop. In a fashion similar to that discussed in Example 8.6.3, this boundary layer thickens as the distance downstream from the start of the sudden contraction is increased. Ultimately, a point is reached along the conduit where the boundary layers from the side walls converge. After this point, the velocity profile does not change as long as the conduit's geometry does not change. In this region of the conduit, one has a *fully developed velocity profile* with associated wall shear stresses that are balanced by the pressure gradient. This pressure gradient is related to h_f, as we have seen. But prior to the establishment of the fully developed profile, the velocity gradient at the wall is greater than it is subsequently. This means that the wall shear stresses are highest at the entrance to the conduit near the reservoir and slowly decrease to the value for fully developed flow. This suggests that over the length of the pipe in which the velocity profile makes its transition to the fully developed profile, the (h_f/length) will be larger than after the *transition length* has been exceeded. One additional factor contributes to the increased h_f at the inlet to a conduit after a sudden contraction. The almost uniform velocity profile at the conduit's inlet has a lower kinetic energy and momentum flux, when integrated over the conduit's cross section, than the velocity profile that exists when fully developed flow has been achieved. If this latter location is labeled point 2 and the former designated point 1, application of the energy and momentum equations will show for a horizontal conduit that over the transition length from point 1 to 2,

$$h_L = [2\beta - (\alpha + 1)]\left(\frac{V^2}{2g}\right) + \left(\frac{4}{\rho g}\right)\left(\frac{L}{D}\right)\tau_{avg}$$

in which β and α are the momentum and kinetic energy correction factors for the fully developed flow. (The last term in this expression would be $\Delta p/\gamma$ if $\alpha = 1 = \beta$.) Because this τ_{avg} will be larger than for fully developed flow, we shall write the second term in the equation for h_L as

$$(\bar{f} + f')\left(\frac{L}{D}\right)\left(\frac{V^2}{2g}\right)$$

in accordance with the definition of $4\tau_w$ contained in Eq. 8.3.12b. In this special case, the subdivision has $\bar{f}$ as the usual friction factor for fully developed flow and f' as an additional contribution because the flow is, in fact, not fully developed. This would mean that the increase over the usual h_f for a developing flow is

$$\Delta h_L = \left[2\beta + \frac{f'L}{D} - (\alpha + 1) \right]\left(\frac{V^2}{2g}\right) \qquad (10.4.1)$$

The coefficient of the average kinetic energy term in the square brackets would be the value of K equal to 0.25 (see Table 10.3.1) for a smooth entrance. Additional losses would occur for an abrupt contraction due to the separation that would be inevitable as the flow moved around the sharp corner. This separation would be accompanied by energy losses in the vortices that would result. The length L in Eq. (10.4.1) is the *distance over which the velocity profile is in transition* to a fully developed profile. It will subsequently be *designated L_t*. The factor f' in Eq. (10.4.1) is an integrated value over the length L_t that accounts for the varying wall shear stress within this distance.

It is necessary to have an estimate of the transition length in a pipe when installing pressure taps to monitor the flow and when locating a flow-metering device whose calibration is dependent upon having a fully developed flow upstream. In Example 8.6.3, an estimate was made of this length, L_t. The ratio of L_t to the channel's width was seen to depend upon the Reynolds number of the flow, in which the width was used along with the average velocity and the kinematic viscosity. For circular pipes in *laminar flow,* a similar relationship exists. It is

$$\frac{L_t}{D} = 0.06\left(\frac{VD}{\nu}\right) \qquad (10.4.2)$$

in which V is the average velocity in the pipe of diameter D. In laminar flow, the influence of the wall on fluid motion proceeds layer by layer into the fluid away from the wall. This is similar to the way thermal energy is transferred through conduction in a fluid at rest. But if there is a considerable amount of mixing, such as occurs in a turbulent flow, then the heat energy will be conveyed from place to place in the fluid by mass transport also. This fluid motion, called *convection*—and, in some contexts, *advection*—is a very effective way to transfer heat (i.e., energy) as well as other quantities.

In turbulent motion, the retardation effect of the wall is, in effect, transferred into regions remote from the wall more efficiently than in laminar flow. The result is that the turbulent boundary layers are usually thicker than laminar layers that might exist under similar conditions. Hence, it is understandable that the transition length for pipes is less in turbulent flow than in laminar flow. The lateral motions in such flows appear to promote the fully developed velocity distribution so effectively that the dependency of L_t on the Reynolds number almost vanishes. A conservative rule of thumb for *turbulent pipe flows* is that *transition* is completed in 50 diameters. For a turbulent flow containing large eddies, L_t/D may be as low as 30.

10.5 NONCIRCULAR CONDUITS

Ventilation ducts are often rectangular in cross section. Similarly, the coolant passages in internal-combustion engines are noncircular in many cases. Because the variety of possible cross sections in which a fluid can be flowing is so large, it would be helpful if there were a useful method for transforming the information for circular conduits, for which much is known, to the particular case at hand. Richard Von Mises considered this problem (about 1910) and proposed that the geometric properties of the cross section be used to define an equivalent diameter for a comparable circular conduit. With this accomplished, the Moody diagram could be used to predict the losses. His reasoning was based upon a proper scaling of the equations of motion. He pointed out that the diameter of a circle was the ratio between the area and the circumference. This gave rise to the concept of the *hydraulic diameter** D_h:

$$D_h = 4 \frac{\left(\begin{array}{c}\text{flow area of conduit's}\\ \text{cross section}\end{array}\right)}{\left(\begin{array}{c}\text{corresponding wetted perimeter}\\ \text{on stationary boundaries}\end{array}\right)} \qquad (10.5.1)$$

Tests were made on triangular and rectangular conduits and conduits of other shapes nearly a decade later, and the results appeared to confirm von Mises' conjecture. This seems to be surprising when one sees the results of some of the tests conducted in the early 1930s. They indicated that there is a considerable amount of secondary flow in such conduits. These flows have velocity components in the plane of the cross section that bring the high axial velocity existing along the centerline of the conduit toward the corners of the polygon forming the cross section. Slow-moving fluid is also transported from the middle of the sides of the polygon toward the center of the conduit. (See Fig. 10.5.1.) The result of these secondary motions is to change the velocity profile in a manner that von Mises would have had difficulty imagining. Subsequent tests have shown that the use of Eq. (10.5.1) and the Moody diagram for turbulent flows yields discrepancies with experimental values that are surprisingly small. Consequently, one can use Eq. (10.5.1) without hesitation in most cases, especially if the conduit's cross section is not greatly different from circular (i.e., hexagonal or octagonal).

The secondary flows mentioned above are particularly unwanted in wind tunnels, which are often square or rectangular in cross section. It is common practice to put a small fillet in the corners to convert the section to an eight-sided polygon with two sets of unequal segments for the sides. This reduces the secondary flow appreciably.

Example 10.5.1

A ventilation duct 1 ft by 2 ft in cross section should deliver 3600 cfm of air at 68°F. What pressure drop, in inches of water, can be expected in a horizontal run of 200 ft?

Solution The area of the duct is 2 ft², and its perimeter is 6 ft. These values give $D_h = 4(2)/6 = 1.33$ ft. The average velocity in the actual duct is

$$V = \frac{Q}{A} = \frac{(3600/60)}{2} = 30 \text{ ft/s}$$

 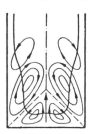

Figure 10.5.1 Schematic representation of secondary (turbulent) flows in noncircular pipes. *(From S. Goldstein's "Modern Developments in Fluid Mechanics," Dover Publications, Inc., by permission.)*

* *Hydraulic radius,* which is defined quite differently, is another term used.

With $\nu = 1.7(10^{-4})$, Re $= 2.3(10^5)$. The walls of the duct are assumed to be smooth, so that using the curve $\epsilon/D = 0$ in the Moody diagram (Fig. 9.4.2) gives $f = 0.0155$. This gives a head loss of

$$\frac{0.0155(200/1.33)(30)^2}{2g}$$

in feet of air. To express the pressure differential in lb/ft^2, one must next multiply this number by the specific weight of air, 0.0745 lb/ft^3. The answer to the question, in inches of water, is

$$(\Delta h)_{\text{in. H}_2\text{O}} = (12)\frac{\{(0.0155)(200/1.33)(30^2/2g)(0.0745)\}}{62.4}$$

$\uparrow$ conversion factor, feet to inches

10.6 SEPARATION IN PASSAGES

At several places in the preceding discussions, the possibility that a flow will not follow the shape of a solid boundary was mentioned. The separation from the edge of the flat plate in Fig. 1.8.9 is a case in point. When there is a sudden contraction in a round pipe, the fluid must move toward the axis of the pipe before proceeding parallel to the axis after it goes through the opening. Such abrupt changes of direction must be associated with the presence of large pressure gradients. If the local gradient is not large enough, the flow will not change its direction to keep it attached to a sudden change of boundary contour. One can observe this when watching water flow out of a hole in a thin-walled tank. The width of the jet is less than the diameter of the hole.

If there is a sudden expansion, a similar situation occurs. The fluid does not immediately accommodate itself to the larger diameter. As it moves out of the smaller conduit into the larger, it tends to continue in its original direction parallel to the axis of the system. The stream of flow separates from the edge of the smaller conduit and moves into the larger cross section in the manner of a jet. It slowly expands to fill the cross section, and in the region between this jet and the wall of the larger conduit there is an extensive mixing region. In the accompanying eddies and swirls, a fraction of the fluid's available pressure head is converted into thermal energy. Separation can also occur when the conduit expands gradually. This was mentioned in Sec. 9.7, where the topic of similarity solutions was broached. Such a separation is not due to a lack of sufficient local pressure gradients to change the momentum of the fluid. It is due to the localized shear stresses in the boundary layer.

Figure 10.6.1 shows a diverging passage between plane walls. (A circular cross section could also be treated, but the additional complexity is not warranted here.) Radial streamlines are also depicted, along with a velocity profile that is a simple model of a laminar flow in the passage. The model shows that the radial velocity along the centerline has been diminished in accordance with the widening flow passage and the requirements of mass conservation. The figure also shows the effect of the no-slip condition at the wall and the presence of a boundary layer. The flow is retarded only in the neighborhood of the stationary wall by its presence and the action of viscosity. While the model that is illustrated shows the viscous effects to be localized between two streamlines, *BC* and *EF,* it is understandable that, as the flow continues down the diverging passage, the effects of viscosity will find their way toward the center of the

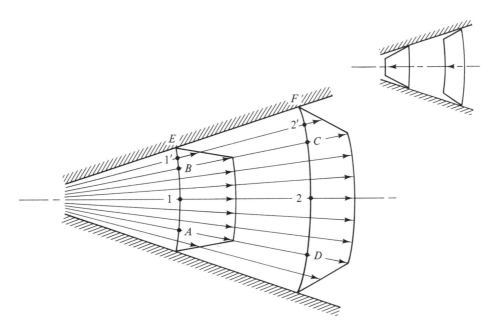

Figure 10.6.1 Model of velocity distribution for radial outflow between nonparallel walls, with inset showing radial-flow case.

conduit. However, as the following discussion will show, the diverging flow shown in Fig. 10.6.1 contains an anomaly. It is not likely to proceed along the diverging passage without changing its character. In practice, one finds that the flow separates from the walls of the conduit if their included angle is more than 7°.

In the region that has not been affected by the viscosity, *ABCDA*, one can write the Bernoulli equation,

$$\frac{\Delta p}{\gamma} = \frac{p_2 - p_1}{\gamma} = \frac{V_1^2 - V_2^2}{2g}$$

In the region where viscosity is assumed to act, *BEFCB*, the energy equation—with suitable kinetic energy correction factors—becomes

$$\frac{\Delta p'}{\gamma} = \frac{p_{2'} - p_{1'}}{\gamma} = \frac{\alpha_1 V_{avg,1}^2 - \alpha_2 V_{avg,2}^2}{2g} - h_L$$

A review of Sec. 6.2 will show that for the linearly varying velocity profile, α is equal to 2. Then one has

$$\frac{\Delta p'}{\gamma} = \frac{p_{2'} - p_{1'}}{\gamma} = \frac{2(V_1/2)^2 - 2(V_2/2)^2}{2g} - h_L$$

In using the particular streamline configuration shown in Fig. 10.6.1, there is no need to be concerned about mass conservation, since there is no flow across streamlines.

The pressure difference between points *B* and *C* is the same as between points 1 and 2 for a channel that has gravity acting perpendicular to the plane *ABDCA*. Moreover, the pressure variation along the streamlines near the wall will be the same as along those near the channel's centerline, if the velocity varies nearly linearly from the wall. This is a conclusion from the differential equations for the flow; they are presented in Chap. 16. (The nearly linear change of velocity is an ingredient in our model. In Chap. 16 boundary layers and their governing equations are discussed. It is concluded there that the pressure gradient in the streamwise direction does not vary significantly across the boundary layer.) The assumption of constant Δp along a streamline regardless of its position from the wall would undoubtedly occur between the ends of a vertical diverging channel that connects one larger reservoir with another. Such a situation would have $\Delta p'$ and Δp equal in the previous expressions. Equating these pressure differences yields

$$\frac{V_1^2 - V_2^2}{2g} = \frac{1}{2}\frac{V_1^2 - V_2^2}{2g} - h_L$$

or

$$h_L = \frac{1}{2}\frac{V_1^2 - V_2^2}{2g}$$

From this negative head loss, one must conclude that the flow pattern drawn in Fig. 10.6.1 is not likely to occur. Moreover, the pattern would also not occur if $|\Delta p'| > |\Delta p|$. (Such a pressure condition would be present if the velocity profile near the wall had an appreciable amount of curvature that extended all the way to the passage's centerline.) This "negative loss" is in contrast to the loss for the radial-inflow case shown in the inset in this figure. For this situation, the head loss will be positive, as required.

If the diverging flow of Fig. 10.6.1 is improbable, what *might* occur? Figure 9.7.1 shows a predicted result. This prediction is based upon the equations of Chap. 16 that have $\Delta p'$ less than Δp for small volume rates of flow. This $\Delta p'$ permits the head loss to be positive for such small flow rates. As the quantity of flow increases, the pressure changes along the wall increase, so that $\Delta p' \geq \Delta p$. Then the fluid no longer flows outward along the wall. The pattern for outflow in Fig. 9.7.1 does not exist; a flow that is (usually) asymmetric with respect to the channel's centerline occurs (in theory and practice). A significantly different flow pattern ensues from that of Fig. 10.6.1 or Fig. 9.7.1. See Fig. 10.6.2.

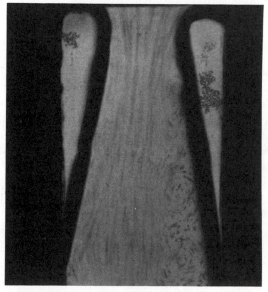

Figure 10.6.2 Radial outflow between nonparallel walls whose included angle is large enough to cause an asymmetric flow. The stream adheres to the wall at the left, and this stream drives a vortex in the right-hand portion of the channel.

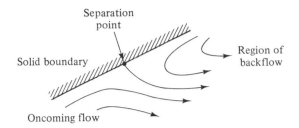

Figure 10.6.3 Schematic drawing of flow separation along a rigid boundary in a viscous flow.

From this example, one can draw the conclusion that when a viscous flow near a wall encounters an increase of pressure with distance, the streamline near the wall "prefers" to depart from the wall. (This departure has been previously referred to as *separation*.) Figure 7.6.2 shows what happens when the flow near a plate approaches the stagnation point at the base of a cylinder that is attached to the plate. The result is a separated flow that is called a *horseshoe vortex*.

This conclusion about the separation of the flow from a boundary when it moves toward a region of higher average pressure was contained in Prandtl's first paper on boundary-layer theory. He showed that there would be flow in the opposite direction just downstream of the separation point similar to that shown in Fig. 10.6.3. The eddies in this reverse-flow region would contribute to the energy losses for a diverging conduit. If the conduit's angle of divergence is small enough, the flow rate low enough, and the viscosity high enough, then purely radial outflow is possible. However, then there is no boundary layer with a small thickness. The effects of viscosity are important across the entire channel.

Prandtl's arguments about separation were made when he was discussing the flow past circular cylinders and not flows within conduits.* The flow in Fig. 10.6.1 was convenient for introducing the connection between positive pressure gradients—also called *adverse pressure gradients*—and separation. In this case, the value of h_L was determined and found to be of the wrong sign. In Chap. 11, we turn to the flow past bodies, for which separation is also an important phenomenon.

10.7 FLOW THROUGH ORIFICES

When a liquid flows through an orifice, it can produce a jet that is submerged in the same liquid. This would occur if there is an orifice in a section of a pipeline and it is used to measure the volume rate of flow. It could also occur if an orifice or nozzle is at the terminus of a conduit that enters submerged into a reservoir. An example of this would be the warm-water discharge port of a power station into a river. Such jets are *submerged jets* and are in contradistinction to the *free jet* that occurs if the liquid is discharged into the atmosphere (or another gas.) The water jet from a hose nozzle would be an example of this. The distinction between the two is primarily associated with the entrainment of the ambient fluid that occurs with the submerged jet and the relative importance of surface tension.

When such jets—either from a sharp-edged orifice or a short nozzle—leave the opening, the liquid's cross-sectional area is usually less than that of the aperture. A

---

* It is said, however, that flow in ventilation ducts whetted his interest in fluid mechanics.

minimum cross-sectional area would then ensue. This occurs at the *vena contracta* that is located about two orifice diameters from the exit plane. This contraction is largely due to inertia effects. The upstream liquid moves toward the centerline of the orifice in order to pass through it. Consequently, it has radial inertia. As the liquid passes through the orifice, it may still have some of this radial inertia. The amount of it will depend upon the nozzle or orifice design. (One can expect that inertia at the edge would not be negligible for a sharp-edged orifice plate, with the sudden change of flow direction at the edge of the orifice.) Thus, the radially inflowing fluid just upstream of the orifice will tend to continue to move radially inward as the jet is formed. Separation at the orifice's edge occurs. The pressure distribution within the jet will alter this radial momentum, but a noticeable effect will occur in the form of the vena contracta.

In calculating the flow associated with free jets, one should use the energy equation appropriately and apply a *contraction coefficient* to the orifice's area to calculate the discharge. If the ratio of the conduit's diameter to that of the orifice is very large, say, greater than 100, then the irrotational flow theory of Chap. 5 predicts that the ratio $A_{\text{jet}}/A_{\text{orifice}}$, the contraction coefficient, is 0.611. Figures 10.7.1 to 10.7.3 present a variety of experimentally determined values. These are presented in the form of a discharge coefficient, which is the product of the contraction coefficient and a velocity coefficient. The latter number accounts for viscous effects in the orifice or nozzle and is usually between 0.95 and 0.99. Thus $C_d = C_c C_v$.

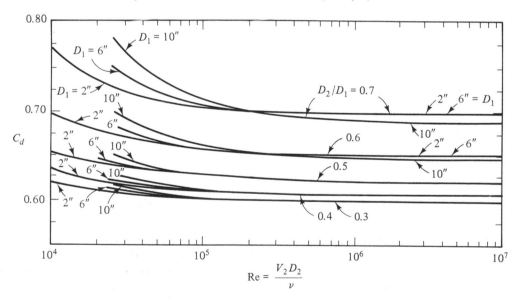

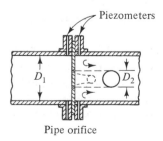

Pipe orifice

Figure 10.7.1 Discharge coefficients for square-edged orifices with flange taps. *(From H. Rouse's "Engineering Hydraulics," by permission of the Iowa Institute for Hydraulic Research)*

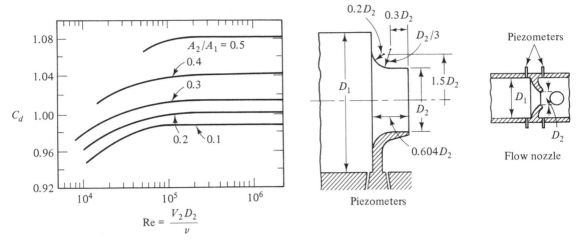

Figure 10.7.2 Discharge coefficients for short flow nozzles. *(From H. Rouse's "Engineering Hydraulics," by permission of the Iowa Institute for Hydraulic Research)*

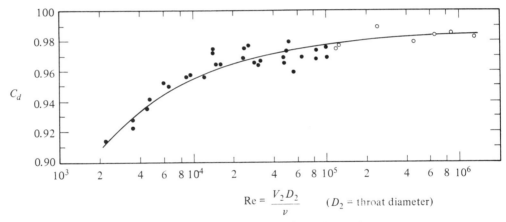

Figure 10.7.3 Discharge coefficients for Venturi meters having a diameter ratio of 0.5. *(From H. Rouse's "Engineering Hydraulics," by permission of the Iowa Institute for Hydraulic Research)*

For liquid jets of small diameter, the surface tension of the liquid can also reduce the diameter. This will occur when the liquid discharges from a long tube for which there should be no contraction because of the absence of inertia effects. (Strictly speaking, the liquid in the jet will not be at atmospheric pressure, but under some "hoop stress" associated with the surface tension model.) Such surface tension contraction can produce jet diameters that are approximately 60 percent of the tube diameter.

The surface tension also usually produces undulations on the liquid jet's liquid-air interface. These grow in amplitude with distance from the opening, and drops are formed. Under certain conditions, such as very low Reynolds numbers, observations have been made that show jet diameters that are slightly larger than the diameter of the forming opening. (One observes this in some non-Newtonian fluids as well.) This phenomenon is termed *die-swell*.

EXERCISES

10.2.1(a) If concrete were used instead of cast iron, what would be the pressure drop in 1500 ft of pipe?

(b) What would the value of h_f have been if the flow had been 18 cfs?

(c) Calculate the value of h_f for 6 cfs. Plot h_f versus Q, and infer the head loss at 15 cfs from the diagram.

10.2.2(a) What would be the pressure change if the elevation had decreased by 15 ft over the 1500-ft length?

(b) If there were a pump between the ends of the pipe with the 15-ft increase in elevation, what would be the pressure change over the 1000-ft length? The pump adds 10 ft-lb/s per pound of fluid flowing per second. What should be the power supplied to the pump if it is 85 percent efficient?

(c) If the pipe is vertical and there is no pressure change between two points A and B, what must the length of the pipe between these points be if 12 cfs is flowing?

10.2.3(a) What would be the flow if the concrete pipe were 2000 ft in length?

(b) Determine the total flow if the head loss across the parallel circuit is 4 ft.

(c) Now find the total flow if the head loss across the parallel circuit is 6 ft. Then plot the total flow versus h_f of 5 ft using this diagram. What is the flow in each pipe under these conditions?

10.3.1(a) If the valve is half open, what is the expected flow? (Assume the plastic pipe is attached as given.)

(b) What would the elevation of the reservoir have to be if the desired Reynolds number is to be achieved? (Assume that the length of the hose is this elevation plus 10 ft.)

(c) What is the expected speed of the flue gases in the actual duct? Base the Reynolds number on a length that is 4 times the value of (area)/(duct perimeter). Assume the flue gases to have the properties of air at 250°F.

10.5.1a What is the answer to the example?

(b) What is the pressure drop if the duct is square with sides that give a cross-sectional area of 2 ft^2?

(c) What is the pressure drop if the duct is square with sides that give a perimeter of 6 ft?

PROBLEMS

10.1 A pipeline is constructed from commercial steel pipe (8 in. in diameter) along level terrain. It is desired that it carry 400 gpm of an oil with viscosity SAE 30 at 32°F. What is the pressure drop in 1 mile of pipeline? What power is necessary to pump the oil over a distance of 1 mile? What power would be required during weather for which the average temperature is 70°F?

10.2 If the pipeline of Prob. 10.1 must climb a hill of 326 ft at one location and the pressure is −2 psig there, what pressure must be in the pipeline at a location 1 mile upstream and on the level plain for a flow rate of 400 gpm at 32°F?

10.3 If it is desired that the pipeline described in Prob. 10.1 never have a negative gauge pressure while it is horizontal, what is the maximum distance between pumping stations that can be allowed for such a horizontal run

if at the exit of a pump a gauge has pressure reading of 90 psi when a flow rate of 400 gpm occurs at 32°F? What would be the minimum wall thickness of the pipe to keep the circumferential (hoop) stresses less than 3000 psi?

10.4 Plastic pipe with a diameter of 4 in. is used to draw water from an irrigation ditch. The inlet of a pump with a discharge of 250 gpm is connected to this pipe. The pump is used to supply the water to the pipes and spray nozzles of the irrigation system and is 3 ft above the surface of the water in the ditch. The operation of the pump ceases to be satisfactory if the inlet pressure falls below −8 psig. What is the maximum horizontal distance that the pump can be placed from the irrigation ditch to ensure satisfactory performance?

10.5 A cottage gets its water from an artesian well

that spills into an open pool. A galvanized iron pipe with a 2 in. diameter has been laid between this pool and the cottage, which is 480 ft away. A pump has been purchased that can supposedly deliver 10 ft-lb of energy to each pound of water flowing (i.e., 10 ft of head). Can this pump deliver 20 gpm through the horizontal pipe to the cottage? If the pump is located on the ground floor of the cottage, what pressure would you expect at the inlet to the pump?

10.6 A cast iron pipe with a diameter of 10 in. has a length of 1035 ft and has been extended another 800 ft with 8-in. cast iron pipe because of economic considerations. If the pipeline is horizontal, what is the pressure differential in the 10- and 8-in. sections for a flow of 0.6 cfs of water? What must the elevation difference be between the ends of the pipeline if the pressure difference over the entire 1835 ft is to be zero?

10.7 Under a particular set of circumstances, the system described in Prob. 10.6 cannot deliver the desired 0.6 cfs of water. Pressure measurements are made at the beginning and end of the section of 10-in. horizontal pipe. The readings are 60 and 48 psig. What should the flow rate be? What should the pressure be at the end of the 8-in. section?

10.8 The elevation difference between a mountain (water) reservoir and a power station is 1000 m. A welded steel pipe with a diameter of 6 m has been laid between the reservoir and the station over a distance of 1.2 km. A single jet of water with a diameter of 300 mm leaves the pipe through a nozzle. What is the expected rate of flow? What is the kinetic energy of the water in the jet? How much power could be developed if the Pelton turbine is 90% efficient?

10.9 The capacity of the station described in Prob. 10.8 is to be increased. A second pipeline to the reservoir is planned. Either 6- or 8-m pipe will be used. What increase in flow rate should be expected if the 6-m pipe is used? The 8-m pipe?

10.10 Would the volume rate of flow through six parallel plastic pipes (use the roughness corresponding to drawn tubing) that are 135 ft long with a diameter of 2 in. be greater or less than through eight plastic pipes with a 1.5-in. diameter and the same length?

10.11. Lubricating oil ($\mu = 4$ P and SG $= 0.9$ at 70°F) is pumped horizontally 1000 yd in a 2-in. commercial steel pipe at the rate of 20 gpm. The line is scheduled for replacement with a 4-in. commercial steel line of the same length. What should be the new flow rate for the same pressure drop?

10.12 If, prior to the construction of the larger pipeline in Prob. 10.11, the design is reconsidered and the smaller line is not to be removed, what pump head should be required to flow the oil through both pipes at a rate of 40 gpm?

10.13 The flow rate of the lubricating oil in Prob. 10.11 was first increased in the existing line by keeping the oil at 100°F so that the viscosity decreased by a factor of 4 and the density decreased by 5 percent. Could the hope of a twofold increase in flow be realized for the original pressure differential?

10.14 A tank with a pressure of 65 psig contains water and is connected to an open tank whose free surface is above that in the pressurized tank. This elevation difference was originally 100 ft and subsequently changed to 250 ft. What would have been the flow rate in a 300-ft length of commercial steel pipe with a diameter of 6 in. for the two elevation differences? What power must be supplied to an 80 percent efficient pump to have equal flows before and after the elevation change?

10.15 A pump supplies a vessel that is at 90 psi with water from a 12-in. cast iron pipe 1000 ft long. The pump's discharge pressure is 110 psi when the water surface in the vessel is 7 ft above the pump's centerline. Determine the volume rate of flow to the pressure vessel. Five 6-in. galvanized pipes 500 ft long connect this tank with an open reservoir. What must be the difference of elevation of the free surfaces in the tank and the reservoir so that the level in the pressure vessel remains constant.

10.16 If 30 gpm of a chemical (viscosity $= 10$ cP and SG $= 0.95$) will flow from an open storage tank to a chemical vat at 10 psi, the existing 1-in. commercial steel line will not have to be replaced. The bottom of the tank is 35 ft above the surface of the liquid in the vat. There is a gate valve, and there are five standard elbows in the line that is currently 50 ft long. If the existing installation does not meet the flow requirement, suggest a cost-effective remedy.

10.17 A 4-in. commercial steel pipeline runs horizontally from a pump for 800 yd and then increases elevation 100 ft through two long-radius elbows. A 600-yd horizontal length of pipe completes the run from the pump to a plant that processes a chemical that flows in the pipeline at the rate of 0.5 cfs. Assume a kinematic viscosity of $2(10^{-5})$ ft²/s and a specific gravity of 1.7. What is the pressure at the entrance to the plant if the pump's discharge pressure is 120 psig? What would the pressure be if the effect of the pipe bends were neglected?

10.18 Water at 60°F is being pumped at the rate of 0.5 cfs from one basin to another that is 2000 ft higher. The pipeline is 3 in. in diameter and 6000 ft long and made of galvanized iron. The overall efficiency of the pump and motor is 75 percent. If the cost of electricity in the locality is 10 cents per kilowatt-hour, what is the hourly cost of the energy to pump the water?

10.19 A pump has suction and discharge pipes with diameters of 60 and 30 cm. The suction gauge is 1 m below the pump's centerline, while the pressure gauge is the same distance above it. These gauges read 50 kPa and 150 kPa, respectively. What power is being added to the water if the flow rate is 0.25 m³/s? What power must be added to the pump's rotor if it is 85 percent efficient? If the pump's seals and bearings require 3 percent of the power supplied to the fluid, what power must the motor supply to the unit? What is the overall efficiency of the pump?

10.20 A horizontal ventilation duct, 1 ft × 2 ft, is 100 ft long and should supply air at an average speed of 5 ft/s at 70°F. What pressure drop in psi should be expected in this length? Express the answer in (inches of water)(specific weight of water).

10.21 Is it likely that the shear stress at the wall of the duct in Prob. 10.20 could overcome the static friction that a grain of sand would experience if it were resting on the wall of the duct? Would the airflow "sweep" the duct clean? (Assume a coefficient of friction of 0.3, a $\frac{1}{64}$ in. cube for the sand, and that τ_w acts on the lateral surface of the cubical grain.)

10.22 A square wind tunnel, 8 ft on a side, is capable of reaching air speeds of 120 mph for automotive testing. If a circuit is 210 ft long, what fan horsepower is needed? (In an actual installation, there would be portions of larger cross section to provide settling chambers to reduce the turbulence level in the air. However, use the 8 ft × 8 ft figure for a preliminary estimate.)

10.23 A resort furnishes water for fire protection by pumping water from a lake to a tank whose water level is 75 ft above the lake level. A 4-in.-diameter pipe, 75 ft long and having two elbows, connects the inlet of the pump to the lake. The discharge line of the pump is 225 ft long and 2 in. in diameter. An open gate valve and three elbows are in this line. The system is designed to deliver 0.2 ft³/s. Determine the horsepower required of a 72 percent efficient pump.

10.24 A 12-in. pipeline of commercial steel is proposed to carry 28,888 barrels per day (1 bbl = 5.61 ft³) of oil from a refinery to a seaport terminal that is 350 miles distant. The fuel has a viscosity of 2 cP, and its specific gravity is 0.73. The minimum pressure in the line should be 8 psia because of the fuel's vapor pressure at an average temperature of 100°F. If the pressure in the line at the refinery is 110 psia, at what distance from the refinery should the first pump be placed? Estimate the number of identical pumping units that should be installed along the pipeline if there is a restriction that no pump should increase the pressure more than 120 psi.

10.25 The San Giacomo power-generating system takes water that has come from an Alpine reservoir and connects it to a reservoir at the power station that is equipped with Pelton turbines. This is done with about 14.3 km of pipe ($\epsilon = 4$ mm) that is 3.5 m in diameter. (There is an additional 658-m length of

2.5-m pipe to change the elevation of the water, but do not consider this.) What is the pressure drop at the rated flow of 37.5 m³/s? ($v = 1.5(10^{-6})$ m²/s.) What power loss does this represent? (Rated power = 183 MW.)

10.26 Program Eq. (10.2.3) on a digital computer and compute the friction factor at Reynolds numbers of 10^3, 10^4, 10^6, and 10^8 for relative roughnesses of 0.05, 0.002, and 0.00001. How do these values compare with the values from Fig. 9.4.2?

10.27 Use Table 10.3.2 to determine K_{12} and K_{13} for a 90° branching pipe tee. Determine K_{21} and K_{31} for a 90° merging pipe tee. Do these values correspond, approximately, to the fL_e/D values given in Table 10.3.1 for Q_2 or $Q_3 = 0$? (Assume $f = 0.025$ in making this comparison.)

10.28 A sharp-edged orifice with a diameter of 0.02 in. is installed at the terminus of a pipeline that is 0.4 in. in diameter to form a fuel spray nozzle. (Assume that the fuel is kerosene at 32°F.) If the average velocity through the orifice is 50 ft/s, what is the pressure difference across the orifice?

10.29 The flow of air into a 1600-cm³ engine that is operating at 3000 rpm is being measured in a duct 300 mm in diameter by a short flow nozzle 100 mm in diameter. What is the expected reading of a water manometer that is connected across the nozzle? (A four-cycle engine is being tested so that 1600 cm³ of air is being ingested during each two revolutions of the engine.)

10.30 If the manometer reading in Example 5.2.2 is 9 in. of water, and air at 70°F is flowing in the meter, what is the volume rate of flow if viscous effects are taken into account? Compare this answer with that predicted by Eq. (5.2.8). What is the percent error? (Assume $D_1 = 6$ in. and $D_2 = 3$ in.)

10.31 Refer to Fig. 10.7.1 and discuss what the effect on C_d should be for an orifice in a horizontal pipeline that carries sediment which deposits itself at the bottom of the upstream face of the orifice.

10.32 A large tank is to be pressurized so that it can force a coolant (assume water) through a pipe system that is the equivalent of 1630 ft of horizontal 1-in. galvanized pipe. Determine the pipe friction loss through such a system for flows from 10 to 150 gpm. Plot the results as a pipe system curve, which has the "head required" as the ordinate and Q as the abscissa. What will be the flow if the tank is pressurized to 60 psig?

10.33 A water reservoir with its free surface at an elevation of 90 ft is directly connected to another whose free surface is at an elevation of 8 ft. 1800 feet of 6-in. welded steel pipe is used. A third reservoir, with free-surface elevation of 130 ft, connects with this pipe by means of a 6-in. steel conduit 720 ft long and a wye connection that makes a 45° angle with the 1800-ft run of pipe at its midpoint. Determine the total flow of water into the lowest reservoir.

10.34 The flow into and out of the three reservoirs shown in Fig. P.10.34 can be determined in a variety of ways. An efficient and general way is given in Appendices D and E. But because the system is a relatively simple one, we shall solve it here without resorting to the appendix. (Afterward, the reader would be well-advised to do so in order to gain increased abilities.)

The head is known at points A, B, and C. Also, $Q_{AJ} + Q_{BJ} = Q_{JC}$, if all of the flow from the higher reservoirs enters the lowest reservoir during steady-state operation. With this assumed set of flow directions, the head at the junction, point J, is also assumed. The value should be between 100 and 300 ft in this case. ($H_J = V_J^2/2 + p_J/\gamma + h_J$.

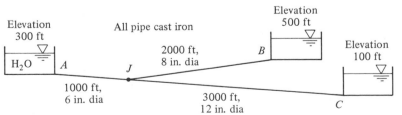

Figure P.10.34

Elevation 300 ft

All pipe cast iron

Elevation 500 ft

H_2O A

J

2000 ft, 8 in. dia

B

Elevation 100 ft

1000 ft, 6 in. dia

3000 ft, 12 in. dia

C

Thus, the elevation of J is unimportant.) Now the energy equation can be written from A to J, B to J, and J to C. The head loss term in each of these equations will give V_{AJ}, V_{BJ}, and V_{JC}. The constraint on the assumed value of H_J is flow continuity at J. If it is not satisfied, a new value of H_J must be assumed and the process repeated. A digital computer makes this process almost effortless. Values of f used at first can be those for $Re = 10^8$, but these must subsequently be refined. What is Q_{JC}?

10.35 In Prob. 10.34, the flow directions were assumed and the problem was solved for a succession of assumed H_J. However, this direct course of action does not always lead to a solution. The reason is quite simple. The direction of flow in the various conduits is not always obvious. Because the energy equation assumes that the losses are positive *downstream*, and because V^2 is always positive, regardless of the algebraic sign of V, there may not be a solution with the initially assumed directions. If this occurs, new directions must be assumed—as logically as possible—and the process started over. Determine Q_{JC} for the configuration shown in Prob. 10.34, but change the 300-ft elevation to 150 ft and the 12-in. pipe diameter to 4 in.?

10.36 Solve Prob. 10.34 by the method of Appendix E.

10.37 Solve Prob. 10.35 by the method of Appendix E.

10.38 Solve for the flow between the two reservoirs in Fig. P.10.38 using the procedure outlined in Prob. 10.35.

10.39 Solve Prob. 10.38 by drawing the "system curve" between the two reservoirs as discussed in Appendix D.

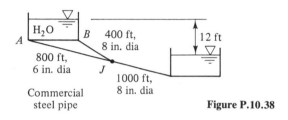

Figure P.10.38

11

Flow Around Bodies (External Flows)

11.1 INTRODUCTION

The flows of Chap. 10 were contained within finite regions. The fluid velocity was zero at the wall of the conduit and increased toward its centerline. In Chap. 11, the no-slip condition will be applied at the boundaries of the body and the velocity will approach that of the oncoming fluid stream at distances that are "far" from the body. The dimension that constitutes "far" has been clarified by Prandtl's *boundary-layer theory*. If the Reynolds number based on some characteristic of the body—its length or width, for example—is very large, say, 10^6, the free-stream velocity is approached at points that are very close to the surface of the body. As the Reynolds number decreases, the distance at which the fluid velocity approximates that of the *free stream* becomes greater.* For Reynolds numbers near unity, the extent of the influence of the boundary on the velocity field becomes very great and must be reckoned with in analyzing such flows and utilizing experimental results associated therewith.

The principal emphasis of this chapter will be on the forces that are induced on the submerged body. The slow motion of a viscous fluid past a sphere provides a specific case to illuminate the present discussion. Figure 11.1.1 is a schematic drawing showing the coordinate system and other details. In drawing Fig. 11.1.1, it was *assumed* that the fluid velocity and the pressure are symmetric with the z axis, which is aligned with the flow. (Hence there is no variation of any quantity in the ϕ direction.) For this reason, the shaded area element can be used. On this element the pressure p, and the shear stress τ, do not vary. They do change, however, if the angle θ is changed. Each of

---

* This distance, δ, was introduced in Sec. 8.6 for a laminar boundary layer on a flat plate. It varies as $\mathrm{Re}_x^{-0.5}$ in this case.

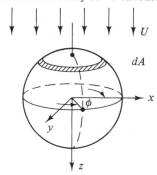

Uniform fluid velocity in $+z$ direction

dA

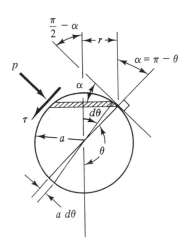

Figure 11.1.1 Axisymmetric flow past a stationary sphere. $dA = 2\pi r(R\ d\theta)$, $r = a \sin \alpha$, $\alpha = \pi - \theta$, ϕ = azimuthal angle (longitude). Note that z, in this case, is *not* necessarily related to the direction of gravity.

these stresses contributes to the force that the fluid exerts on the sphere in the $+z$ direction. We call the sum of all the fluid's forces on a body, in the direction of oncoming flow, the *drag* on the body. We call the force perpendicular to the direction of motion the *lift*. (This would be in the vertical direction when the motion is horizontal.) In the present case, there is no lift.

Example 11.1.1

We can calculate the shear force component in the z direction due to the shear stress by using the implications contained in Eq. (2.4.1). (We must keep the directions of the stresses—shear stresses, in this case—in mind, however.) The magnitude of the shear force on the shaded area is

$$|d\mathbf{F}_{\text{shear}}| = \tau |d\mathbf{A}|$$

and

$$|d\mathbf{A}| = 2\pi(R\ \sin \alpha)R\ d\theta \qquad \text{with } R = a \tag{11.1.1}$$

because the circumference of the sector of the sphere is $2\pi a \sin \alpha$ at the latitude determined by θ. But this force is tangent to the surface, and the tangent makes an angle of $(\pi/2 - \alpha)$ with the z direction. Hence, the z component of the elemental shear force is

$$(dF_z)_{\text{shear}} = +\tau[2\pi a^2(\sin \alpha)\ d\theta] \cos\left(\frac{\pi}{2} - \alpha\right) \tag{11.1.2}$$

in which the positive sign preceding τ is associated with the direction of τ and the z axis. To obtain the total effect of the shear stresses on the body, Eq. (11.1.2) needs to be integrated. This can be done if τ and p are known from experiments or suitable theory. Such a theory for creeping flow is beyond the scope of this book, but G. G. Stokes in 1851 gave the reasons for writing

$$p = -\frac{3}{2}\frac{U}{a}\mu \cos \theta \tag{11.1.3}$$

and

$$\tau = +\frac{3}{2}\frac{U}{a}\mu \sin \theta \tag{11.1.4}$$

for this case. An integration of Eq. (11.1.2) then yields:

$$(F_z)_{\text{shear}} = 4\pi\mu Ua \qquad (11.1.5)$$

The pressure on the sphere, p, will produce an additional force in the z direction,*

$$(F_z)_{\text{pressure}} = 2\pi\mu Ua \qquad (11.1.6)$$

Equation (11.1.6) is an example of the integrated force called the *pressure drag*. Because this second part of the total drag can be altered significantly by minor changes of the body's shape, it is also called the *form drag*. Such alterations in geometry may not change the associated shear stresses significantly if the Reynolds number is large, with the result that the total drag can be decreased. Accordingly, the drag coefficient would be reduced. This situation occurs if the change of shape—or some parameter associated with the flow—affects the point of flow separation from the body. We encountered this when Figs. 1.1.3 and 1.1.4 were discussed. The change in the point of separation on a sphere can reduce the drag tenfold. In this chapter, it will be seen that transforming a sphere into a "teardrop" can have the same effect on reducing the drag. Figure 11.1.2 shows how streamlining can affect the wheel loading of a motorcycle.

Some bodies with sharp corners will induce flow separation at well-defined positions that are almost independent of changes in the flow parameters. Such bodies,

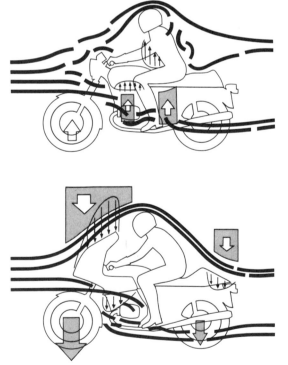

Figure 11.1.2 Effect of modification to a motorcycle's design on the airflow and pressure distribution that are experienced by rider and machine. *(Courtesy of the Bavarian Motor Works, Munich, W. Germany)*

---

*In this case, the total normal stress on the sphere includes a contribution that is due to the fluid motion. However, this part of the normal stress is symmetric with respect to the xy plane and contributes nothing to F_z for the *particular* case mentioned here.

often termed *blunt bodies,* will have drag coefficients that are independent of the Reynolds number—and other parameters—over a wide range of values.

Separation was discussed at the close of Chap. 10 with respect to internal flows. Prandtl first shed significant light on this phenomenon in his 1904 paper on the theory of the boundary layer. He did so by analyzing the flow past bodies, particularly a circular cylinder. (Prandtl, a professor of mathematics, had made experimental observations in a small apparatus. These may have helped in the formulation of his ideas and buttressed his conclusions.) Flow separation affects the lift as well as the drag on a body. The pressure distribution on a wing will be influenced by the orientation of the wing with respect to the relative wind (i.e., the vector sum of the wind and wing velocities). As the angle between the relative wind and the wing's *chord* (a line drawn between its leading and trailing edges) is increased, the flow will separate from the upper surface of the wing. (The angle is called the *angle of attack.*) This is shown in Fig. 11.1.3a to d. The eddy—with its associated backflow—that forms of the lifting surface of the wing can have a dramatic effect on the lift when the angle of attack reaches a critical value. This value will differ for each wing design. Fig. 9.3.1b presents

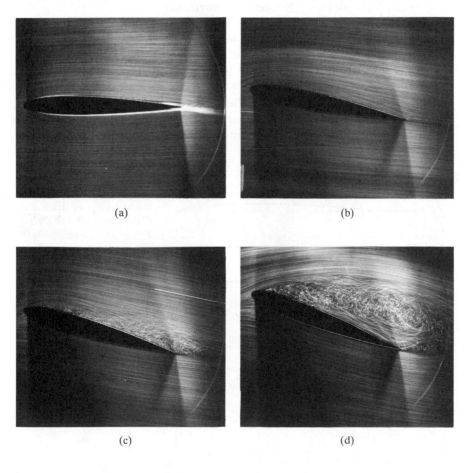

(a) (b)

(c) (d)

Figure 11.1.3 High-speed flow around a wing profile with an angle of attack (a) $\alpha = 0°$, (b) $\alpha \approx 10°$, (c) $\alpha \approx 15°$, and (d) $\alpha \approx 20°$. *(Courtesy of H. Werlé, O.N.E.R.A., Paris, France)*

Sec. 11.1   Introduction   **429**

data that relate the lift coefficient C_L with the angle of attack. This lift coefficient has the same form as the drag coefficient (the lift force replaces the drag force). *Stalling* of the wing occurs when the peak of the C_L curve is reached and the lift drops drastically due to separation.

It seems natural to say that the lift on a wing is due to the integrated effect of the pressure distribution on its surfaces. But how does this distribution occur? What are the dynamics of the fluid that cause the pressure variations? The pressure is relatively high at the "nose" of the airfoil due to stagnation effects. This pressure decreases somewhat as the fluid flows along one side of the wing—usually the lower surface—but still remains positive with respect to the ambient pressure. In contrast, the effective pressures on the upper side become negative. A "suction" is induced. The combination of the high pressure on the underside and the low pressures on the upper side results in the lifting force. Figure 11.1.4 shows a representative pressure distribution on an airfoil. (It is such a pressure distribution that is drastically disturbed when separation occurs.) A discussion of the factors inducing these pressures will be given in Sec. 11.3, but now we shall see that this pressure differential between the surfaces of a wing can have another effect on its performance.

If the wing is on an airplane, there will also be a tip to it. The trajectory of the air flowing along the underside of the wing near the tip will be influenced by the negative pressure on the wing's upper side. There is a pressure gradient within the three-dimensional fluid region from the lower side of the wing to the upper side. (For a two-dimensional model of the wing, there are no ends to the wing. The wing, itself, forms a barrier to this gradient.) This gradient will be strong near the wing tip and will affect the flow. Fluid from the lower side of the wing will flow around the wing tip to the upper side. This creates a *wing-tip vortex*. This is illustrated in Figs. 11.1.5 and 11.1.6.

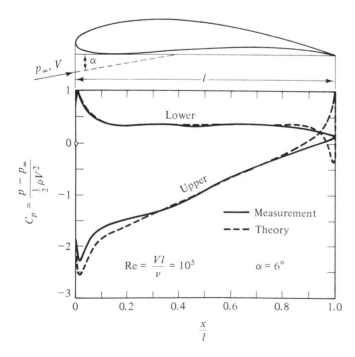

Figure 11.1.4 Pressure coefficient variation along the upper and lower surfaces of an airfoil with angle of attack of 6°. *(From Goldstein's "Modern Development in Fluid Dynamics," Dover Publications, Inc., by permission)*

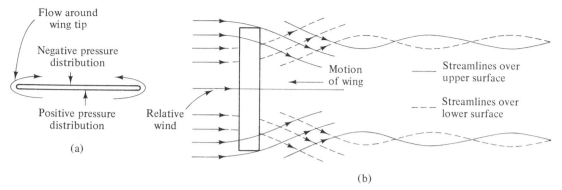

(a)

(b)

Figure 11.1.5 Formation of a wing-tip vortex from a wing of finite span with positive lift. (a) front view and (b) plan view of wing. *(From "Mechanics in Fluids" by Duncan, Thom, and Young, Edward Arnold Ltd., London)*

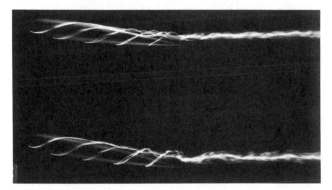

(a)

(b)

Figure 11.1.6 Effect of angle of attack on tip vortex of a rectangular wing at a (chord) Reynolds number of 50,000. (a) $\alpha = 7.3°$; (b) $\alpha = 10°$. *(Courtesy of H. Werlé, O.N.E.R.A., Paris, France)*

For such a finite-length wing, the induced flow due to the wing-tip vortices—in the opposite rotational sense at the two ends of the wing—will be a net downward flow component. This is called a *downwash*. Small aircraft following behind a much larger craft can experience destabilizing effects due to such a downwash. The result of changing the momentum of a portion of the fluid flowing toward a stationary wing is the introduction of additional force components upon the wing due to its finite length. The associated force in the flow direction is termed the *induced drag*. Such a contribution to the total drag of a lifting surface is a momentum effect and not a viscous one.

The influence of the ends of the lifting surface must be taken into account when conducting wind tunnel tests. In order to exclude end effects, baffles are used. These are planes set normal to the axis of the foil or cylinder that prevent the flow that would go from one side to the other around the end of the foil. They are thin enough so as not to disturb the flow, to which they are aligned, but they could, nevertheless, induce new flow phenomena where they intersect the surface of the foil. (See Fig. 7.6.2.)

11.2 DRAG ON BODIES

There have been an extensive number of fruitful investigations concerning the flow past two- and three-dimensional bodies. The design rules that are derived from the application of the boundary-layer theory and Prandtl's ideas about separation have produced

TABLE 11.2.1 DRAG COEFFICIENTS FOR CYLINDERS OF VARIOUS SHAPES

Form	Drag coefficient		
	$L/D = \infty$	$L/D = 5$	$L/D = 1$
$U_\infty \rightarrow$ (circle)	1.20 $\quad(10^3 < \mathrm{Re} < 10^5)$ 0.33 $\quad(\mathrm{Re} > 5(10^5))$	0.74 0.35	0.63
(half cylinder)	1.20		
(half cylinder, curved)	1.20		
(curved)	2.30		
(square)	2.05		1.05
(diamond) $\updownarrow D$	1.55		0.8
(triangle)	1.55		
(triangle)	2.0		
(flat plate)	1.9, Re $\approx 10^5$	1.20	1.16
(ellipse) 3:1	0.1, Re $\approx 10^5$		
(airfoil) 3:1	0.06		
(T-section)	2.0*		

*A lift coefficient of considerable magnitude can be induced also.

our present, highly developed state of technology, which can predict the flow and drag characteristics of surfaces whose shapes gradually change. However, many blunt and angular geometric shapes are fabricated for use, or are found in nature. There is at present no theoretical way to predict satisfactorily the shear stress and pressure distributions around these shapes.* In order to gain information about the drag forces that

*Computer programs to predict the flow around complex bodies cannot yet do this in a realistic fashion when there is a significant amount of separation. It is then necessary to "fine-tune" the program with the results of applicable experiments.

are induced by a fluid flow, recourse to data gathered in experiments is necessary. Table 11.2.1 shows a collection of drag coefficients for long (i.e., almost two-dimensional) bodies. Table 11.2.2 presents similar information for three-dimensional bodies.

If a body is adjacent to a plane surface, as is a building protruding above the ground, one must take into account the peculiar flows and stresses that are induced in this instance. (Refer to Fig. 7.6.2.) The consequences of the associated "interference" effects are drag coefficients that are different from those which would be calculated if one simply took into account the fraction of the total body that was on one side of the plane and submerged in the flow. Thus, a hemisphere projecting above a plane that is aligned with the flow would have a higher drag coefficient than half of the coefficient for a sphere that is far from any boundaries. If a sphere falls within a tube, one will observe that the drag coefficient will depend on D_{sphere}/D_{tube}. In the case of bodies that are submerged below a liquid-liquid or liquid-gas interface, there is a different type of interaction with the free surface, if it is in close proximity. Under such conditions it is found that the local elevation of the free surface will be affected by the movement of—or flow past—the body. The usual result is that waves are generated; however, the specially designed bulbous appendage put on the bow of a ship (see Fig. 1.8.6) serves to suppress bow waves. The data in Tables 11.2.1 and 11.2.2 augment the data contained in Fig. 9.4.1 for a circular cylinder and a sphere, as well as in Fig. 8.6.2 for flat plates aligned with the flow.

TABLE 11.2.2 DRAG COEFFICIENTS FOR THREE-DIMENSIONAL BODIES WITH THE AXIS OF SYMMETRY PARALLEL TO THE ONCOMING FLOW*

Form	Drag Coefficient
Sphere	$10^3 < \text{Re} < 10^5$ 0.4–0.5
	$\text{Re} = 6(10^6)$ 0.2
Hemisphere:	
Curved surface upstream	0.42
Flat surface upstream	1.17
Hemispherical shell	
Hollow downstream	0.36
Hollow upstream	1.40
60° cone pointed upstream	0.50
Ellipsoid (1 : 2 with major axis parallel to flow)	$\text{Re} > 2(10^5)$ 0.07
Tear drop	$\text{Re} > 2(10^5)$ 0.05
Disk	
Single	1.15
Two or tandem	
$L/D = 1$	0.93
$L/D = 2$	1.04
$L/D = 3$	1.54
$L/D = 7$	2.30
Circular cylinder	
$L/D = 0$	1.15
$L/D = 1$	0.91
$L/D = 2$	0.85
$L/D = 4$	0.99

* See Table 11.2.1 for short-cylinder data.

Figure 11.2.1 is called a *lift-drag polar diagram*. It is for a complete hydrofoil boat, so that it includes wave drag, aerodynamic hull drag, and hydrodynamic foil drag. One would use it in conjunction with Fig. 9.3.1c for a particular foil shape to design a vehicle that is to be fitted with such a wing.

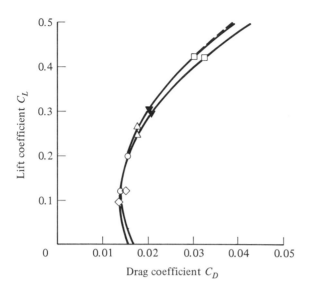

Figure 11.2.1 Lift-drag polar diagram for a complete hydrofoil boat. *(From Wright and Otto, "J. Hydronautics," Oct. 1980. Reprinted with permission of the American Institute of Aeronautics and Astronautics.)*

Example 11.2.1

A raindrop may fall slowly enough during its initial stage of formation to obey Stokes' law for the drag on a sphere. (Surface tension at low speeds can be sufficient to keep the drop nearly spherical if its diameter is less than 4 mm.) If, in the present case, the drop is 0.1 mm in diameter and falls at (almost) a constant rate, what should this speed be?

Solution Since the speed is to be assumed constant, there are no unbalanced external forces on the drop. Hence, the drag on the drop will be equal in magnitude to its effective weight (i.e., its weight in air):

$$\text{Weight} = \frac{\gamma(\pi D^3)}{6} = \frac{9.81(1000)(\pi)(0.0001)^3}{6}$$
$$= 5.14(10^{-9}) \text{ N} = \text{drag}$$

One of the variables in Fig. 9.4.1b is the drag coefficient C_D, or

$$\frac{\text{Drag}}{\frac{1}{2}(\rho V^2 \pi D^2/4)}$$

Thus, with the result just found for the drag,

$$C_D = \frac{\gamma \pi D^3/6}{\rho \pi V^2 D^2/8} = \frac{9.81(0.0001)(4)}{3V^2}$$

The variable against which C_D is plotted in Fig. 9.4.1b is the Reynolds number. The viscosity of air at about 0°C will be used: $1.3(10^{-5})$ m²/s. Thus

$$\text{Re} = \frac{0.0001V}{1.3(10^{-5})}$$

A solution could ensue if one tried several assumed values of V in the expression of the

Reynolds number used in Fig. 9.4.1a and determined the corresponding value of C_D, and then checked to see whether that value agreed with the $9.81(0.0001)(4)/3V^2$ established previously. If one were to solve many such problems, it would be useful to replot the data of Fig. 9.4.1b, which relates

$$\frac{\text{Drag}}{(\rho V^2/2)(\pi D^2/4)}$$

and $VD\rho/\mu$ as $C_D = \phi(\text{Re})$. It follows from this that

$$C_D \, \text{Re}^2 = \text{Re}^2 \, \phi(\text{Re}) = \hat{\phi}(\text{Re})$$

and $C_D \, \text{Re}^2$ is proportional to $(\text{Drag})\rho/\mu^2$. The transformation of the data in Fig. 9.4.1b can easily be done to give a graph with $(\text{Drag})\rho/\mu^2$ and Re as coordinates.

In this case, however, the solution procedure can make use of the fact that the points in Fig. 9.4.1b that came into consideration for this problem are located at the far left of the figure. This is the Stokes flow regime and has $C_D = 24/\text{Re}$. The value of the drag coefficient is the same as before, so that it is possible to write an explicit expression for V by relating $24/\text{Re}$ and the known drag as

$$\frac{9.81(0.0001)(4)}{3V^2} = \frac{24}{0.0001V/1.3(10^{-5})}$$

After solving for V, one would have to be sure that the associated Reynolds number was in the range below 0.1 for which the formula for the drag coefficient is applicable. Does the solution for V in this example meet this criterion?

Example 11.2.2

A 50-ft-tall mast has a circular cross section that varies from 24 in. at the base to 12 in. at its tip. What horizontal force must its support (cantilever) sustain in a 60-mph wind? (Neglect any effects of ground interactions.) What is the maximum bending moment induced in the mast?

Solution   It will be assumed that the force is the sum of many forces on short cylinders of varying diameter. (As there are no "end effects" of this contiguous group, the drag coefficient for infinitely long cylinders will be utilized.)

An estimate of the Reynolds number for the cylinders in question is

$$\text{Re} \approx \frac{1.5(88)}{10^{-4}} \approx 1.32(10^6)$$

in which an average diameter has been used. This Reynolds number is beyond the boundary-layer transition point, $\text{Re} \approx 5(10^5)$. For $\text{Re} > 10^6$, very little experimental data are available from which to infer C_D. Nevertheless, a value of $C_D = 0.35$ is reasonable in this case.

The incremental force on a cylinder of length dz is

$$d(\text{Drag}) = \left(\frac{C_D \rho V^2}{2}\right)[D \, dz]$$

in which the term in square brackets is the frontal area in question. The diameter D can be expressed as

$$D = 2 - \frac{z}{50}$$

Hence,

Figure 11.2.2 Wind tunnel tests of the space shuttle and its launching equipment to determine wind loads. *(Courtesy of Cermak, Peterka, and Petersen, Inc., by permission of Martin Marietta Corp.)*

$$\text{Drag} = \frac{0.35(0.0024)(88)^2}{2} \int_0^{50} \left(2 - \frac{z}{50}\right) dz$$

$$= 3.25 \left(2z - \frac{z^2}{100}\right)\Bigg|_0^{50}$$

$$= 3.25(50)(3)/2 = 244 \text{ lb}$$

The moment induced at the base is

$$M = \left(\frac{C_D \rho V^2}{2}\right) \int_0^{50} z \left(2 - \frac{z}{50}\right) dz$$

$$= 5417 \text{ lb-ft}$$

Figure 11.2.2 shows an example of wind tunnel research to determine the wind forces on tall buildings or structures and the associated flow patterns.

11.3 LIFT

The question concerning the origin of the pressure distributions that cause lift has not been addressed as yet. An insight into a mechanism that explains the phenomenon can be gotten from Figs. 11.1.3a–d. In Fig. 11.1.3a, one can see the stagnation point at the foremost tip of the leading edge. As the angle of attack increases, the stagnation point moves slightly to the underside of the wing. At the trailing edge of the wing, one can see the fluid that has traveled past the upper surface merging with the fluid going past the lower side. This is true even as the angle of attack is increased up to the advent of the stall condition. These observations will now be coupled with the results of irrotational flow theory. This discussion is based upon the work in Sec. 5.4.

The stream function $\psi = Uy[1 - a^2/(x^2 + y^2)]$ describes the flow past a circular cylinder of radius a in the absence of viscous effects and separation. If an irrotational vortex is added to this flow, the circle $r = a$ can still remain a streamline. The result is

$$\psi = Uy\left(1 - \frac{a^2}{x^2 + y^2}\right) - \left(\frac{\Gamma}{2\pi}\right) \ln\left(\frac{r}{a}\right) \tag{11.3.1}$$

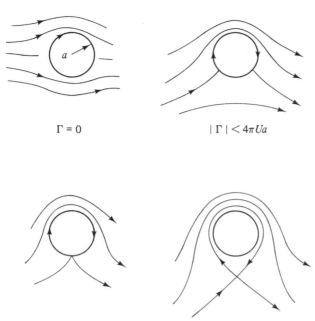

$\Gamma = 0$

$|\Gamma| < 4\pi Ua$

$|\Gamma| = 4\pi Ua$

$|\Gamma| > 4\pi Ua$

Figure 11.3.1 Theoretical flow patterns associated with increasing the circulation about the body.

The constant Γ is called the *circulation*, because it is the value of the integral $\int \mathbf{V} \cdot d\mathbf{s}$ around a circle of arbitrary radius r. * ($|d\mathbf{s}|$ in this case would be $r\,d\theta$.) Figure 11.3.1 shows the effect of increasing Γ on the flow past the circular cylinder. These flow patterns can be realized in experiments by rotating the circular cylinder. The wake behind the cylinder is greatly suppressed because the separation process is impeded by moving the boundary. The effect of increasing Γ is to move the forward stagnation point downward and also displace the rearward stagnation point.

It is possible to map the points on the circle $r = a$ onto a straight line by a process called *conformal mapping*. (Such a procedure is similar to transforming the surface of a globe onto a map by some method of projection—perhaps a Mercator projection.) Each point (x, y) on the circle is associated with a point on the (X, Y) line. The equations needed to make the transformation are of the form $(X, Y) = (2a \cos \theta, 0)$, wherein a is the radius of the circle and θ is the polar angle to the point on the circle.† This is not an arbitrary relationship, but one that provides the means to study the flow about a flat plate as a function of the angle of attack, α. One does this by studying the flow about a circle with $x'y'$ axes that are inclined to the flow direction, far from the plate, by the angle α. Figure 11.3.2 illustrates this. The flat plate at the upper right in this figure is a rudimentary airfoil, perhaps a simple kite. (The shape of this "foil" can be changed, if necessary, by further conformal transformations to resemble closely a variety of airfoils, such as the one shown at the lower right of Fig. 11.3.2.) One feature

---

* This is generally true. In this case, it would be easy to demonstrate for large values of r. In particular, if U were very small, v_θ would be primarily $\Gamma/2\pi r$. Then the integral $\int \mathbf{V} \cdot d\mathbf{s}$ would approach $\int (\Gamma/2\pi)\,d\theta$ with limits of 0 and 2π.

† The reader may wish to show that this case is a limiting case of $(X, Y) = [a(1 + b^2/a^2) \cos \theta, a(1 - b^2/a^2) \sin \theta]$ and $b = a$. When $b < a$, the transformation maps the points on the circle onto an ellipse.

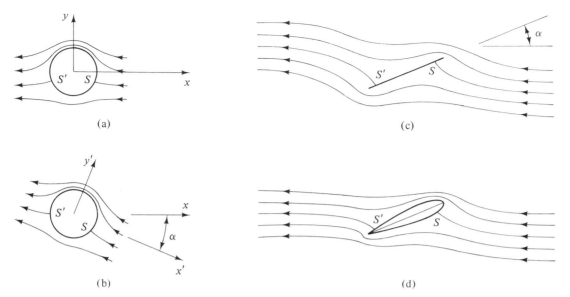

Figure 11.3.2 (a) Flow past a circular cylinder with circulation for a flow parallel to the $+x$ axis as $x \rightarrow \infty$. (b) Flow as $x \rightarrow \infty$ makes an angle of $-\alpha$ with the $+x$ axis. (c) Flow past a flat plate at an angle of attack α. Circle in part (b) is transformed into straight line of length $4a$ positioned along the x axis. Then entire picture—flow pattern and plate—is rotated counterclockwise through α so that flow appears parallel to x axis far from plate. (d) Flow past foil-shaped cylinder at angle of attack α. Shape of body is generated from a different transformation of the circle slightly more complex than the one that produced the flat plate. *(From "Mechanics of Fluids" by Duncan, Thom, and Young, Edward Arnold Ltd., London)*

of the flow pattern shown in this figure is unsatisfactory, however, when one recalls the observations made concerning Fig. 11.1.3. The flow at the trailing edge in the current model, shown in Fig. 11.3.3, does not show a smooth confluence of the fluid streams passing above and below the plate. In fact, there is a sharp reversal of the flow at the trailing edge, which would imply that there was an infinite acceleration—an infinite force, too—at this point. (The leading edge will not concern us further, as it is usually well rounded.) This unsatisfactory condition could be eliminated if the circulation Γ were adjusted to move point S' to the trailing edge of the plate as in Fig. 11.3.4. F. W. Lanchester postulated that a flow must accompany the means to produce the necessary circulation to cause a smooth merging of the streams. (In spite of Lanchester's contribution, this requirement is called the *Kutta condition.**) The consequence of this circulation is to speed up the flow on one side of the airfoil and retard it on the other, with both speeds approaching the same value at the trailing edge. A further consequence, and the one that we are looking for, is that the pressures on one side of the wing are reduced with respect to those on the other side. A lifting force is the result, just as a ball—a baseball, tennis ball, Ping-Pong ball, or golf ball—would

---

* Lanchester formulated a complete theory that enables one to calculate the lift on infinitely long airfoils. Prandtl also worked on this topic, and the theory today honors both of their names. M. W. Kutta was a German mathematician. (The Runge-Kutta method of solution of differential equations also bears his name.) He developed, almost simultaneously with the Russian N. E. Joukowsky, the transformations needed to create airfoil shapes that could be studied with the aid of the theory associated with irrotational flows.

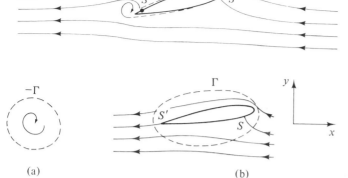

Figure 11.3.3 Formation of eddy (starting vortex) during starting stages of flow past an airfoil in a viscous fluid. *(From "Mechanics of Fluids" by Duncan, Thom, and Young, Edward Arnold Ltd., London)*

Figure 11.3.4 Generation of circulation Γ about foil (b) is accompanied by equal, but opposite, circulation about the trailing vortex that drifts off in the $-x$ direction to infinity (a), leaving net circulation equal to zero. *(From "Mechanics of Fluids" by Duncan, Thom, and Young, Edward Arnold Ltd., London)*

(a)   (b)

experience a force normal to its path of motion if it were caused to spin while translating.

This postulate of Lanchester's, with its requirement for the development of circulation by the lifting surface, provides the explanation that is used to account for the generation of lift. A result of this theory is that C_L for a flat plate is $2\pi \sin \alpha$. This result indicates that C_L increases with α, but it does not give any suggestion about the possibility of "stall," because the basis for, and the concept of, flow separation is not inherent in the theory. Despite this shortcoming, the Lanchester-Prandtl theory is obviously useful in its range of applicability.

The manner in which the necessary circulation is generated to satisfy the Kutta condition is explained by the presence of a *starting vortex*. As the flow commences, there is separation at the trailing edge. (There is no upwind flow from the trailing edge along the upper surface of the foil.) This separation induces a vortex at the trailing edge (see Fig. 11.3.4) that is swept off the foil as a starting vortex. One such vortex is shown in Fig. 11.3.5. Note also in Fig. 11.3.6 the *stopping vortex* that occurs when the flow ceases. This vortex cancels the starting vortex's circulation. That vortex has in the interim moved to infinity.

Figure 11.3.5 Formation of counterclockwise eddy at the trailing edge, producing a clockwise circulation around the airfoil as it starts to move.

Figure 11.3.6 Formation of a clockwise eddy as airfoil stops.

Example 11.3.1

The hydrofoil boat shown in Fig. C.2a in the appendices has the following characteristics:

Weight (out of water): 246,400 lb
Type of hydrofoil section: NACA 16 series
Average area of hydrofoil: 180 ft^2
Available engine power: 7600 hp

The maximum lift coefficient can be taken as 0.37 (based on average foil area) and can be used to calculate the boat's speed when it is foilborne. (This is the moment when the foils can support the hull's weight completely. Then the hull need no longer be in the water and utilize Archimedes' principle to support itself. In preliminary calculations, the buoyancy of the foils can be neglected.) What is the power needed to attain this foilborne speed, assuming 100 percent efficiency? What should be the maximum speed of the boat if then the drag is 21,850 lb and the overall propulsive efficiency is 38 percent?*

Solution   We start by using the definition of the lift coefficient to find

$$V^2 = \frac{\text{lift}}{\frac{1}{2}\rho A C_L} = \frac{246,400}{\frac{1}{2}(1.932)(180)(0.37)}$$

$$V = 61.88 \text{ ft/s} = 36.6 \text{ knots}$$

The lift-drag polar diagram, Fig. 11.2.1, is a source for a drag coefficient. A value of 0.025 is reasonable for the specified lift coefficient:

$$\text{Drag} = 0.025 \tfrac{1}{2}(1.932)(180)(61.88)^2, \text{ or}$$

$$= \frac{C_D}{C_L}(\text{lift}) = \frac{0.025}{0.37}(246,400) = 16,649 \text{ lb}$$

Minimum liftoff power $= V(\text{drag}) = 1873$ hp

$$\text{Maximum effective power} = 7600(0.38)(550)$$
$$= 1.5884(10^6) \text{ ft-lb/s}$$

$$V_{\text{max}} = \frac{\text{maximum power}}{\text{drag at } V_{\text{max}}} = \frac{1.5884(10^6)}{2.185(10^4)}$$
$$= 72.7 \text{ ft/s} = 43 \text{ knots}$$

---

* Values of 50 to 60 percent can be obtained.

(a)

(b)

(c)

Figure 11.4.1 Effect of small changes of shape on flow past a body illustrated by three steps in development of Scirocco stern. *(Courtesy of the Volkswagen Corporation, Wolfsburg, W. Germany)*

11.4 SEPARATION AND BOUNDARY-LAYER CONTROL

The phrase *boundary-layer control* (BLC) came into usage after 1945, when there was a new effort to increase the lift on wings by increasing the angle at which stall occurred. This was done by various means. One way that became practical was the suction of air near the trailing edges of the wing. This forestalled separation and inhibited the occurrence of stall.*

The application of "tripping wires" to bodies to induce a turbulent boundary layer, and to change the point of flow separation, is also a form of boundary-layer control. Moreover, the alteration of the surface of a sphere to form a "teardrop" can be thought of as a type of BLC. The form of this "afterbody" has been a subject of research for many years in order to develop profiles with a low drag or a desired pressure gradient along the surface to inhibit separation. The popular term *streamlining* is, in effect, a form of BLC. Efforts in this direction need not only have the effect of smoothing the surface profile of the body. Sometimes, the addition of a "spoiler" can change the flow characteristics in the rear of a bluff body and hence change the drag coefficient. Figure 11.4.1 illustrates this through three steps in the design of the stern of the Scirocco:

1. With the first design (Fig. 11.4.1a), the flow separates at the roof's edge. The rear window is dirtied by the wake, which entrains road dust. To prevent this from occurring, the lower edge of the rear hatch was changed.

2. In Fig. 11.4.1b, the separation line is now below the window, which should now remain clean longer. However, the force dynamometer shows that a higher air resistance is also present.

3. The problem was finally solved with the spoiler configuration shown in Fig. 11.4.1c. It keeps the window clean *and* reduces air resistance. This design gave the additional benefit of increased vehicle stability at high speeds.

---

*The jet engines that power today's aircraft ingest large quantities of air to produce the thrust. Some of it can come from a set of numerous holes near the trailing edge of the wing via connecting ducts.

Momentum can be added to the fluid in the boundary layer of wings by providing slots near the leading edge or flaps at the trailing edge. Fig. 11.4.2 shows some of the combinations. Air that has not been exposed to the wing previously can be forced along the boundary layer to mix with air that has flowed along the lifting surface from

Figure 11.4.2 Typical high-lift devices. (a) Characteristics of several flaps with NACA 23012 airfoil. *(From NACA Tech. Rep. 664)* (b) Characteristics of Clark Y wing with various slots and flaps. *(From NACA Tech. Note 459)*

Designation	Diagram	$C_{L,max}$	α at $C_{L,max}$
Basic airfoil		1.54	15.5°
With 0.2C split flap deflected 60°		2.53	12°
With 0.2C plain flap deflected 60°		2.38	12.5°
With 0.2C slotted flap deflected 50°		2.76	13.5°
With 0.27C Fowler flap deflected 30°		2.90	10.5°

(a)

Designation	Diagram	$C_{L,max}$	α at $C_{L,max}$
Basic airfoil		1.27	14°
With Handley–Page slot		1.84	28°
With Handley–Page slot and Fowler flap		3.37	16°

(b)

the outset.[*] The lift enhancement that is possible with a combination of slot, flap, and suction is shown in Fig. 11.4.3. The flow coefficient in this figure is $Q_{suction}/V$ (wing area).

The two (noncircular) cylinders on the *Alcyone,* shown in Fig. 11.4.4, serve to assist the diesel motor to propel the ship. They are hollow and have two sets of holes along the length of the surface through which air is sucked by a fan that exhausts through an opening at the top. A movable flap controls the amount of suction through each of the two sets of holes. It will reduce the aspiration area on the windward side of the cylinder, which is effectively a thick wing profile. These cylinders experience a thrust load perpendicular to the relative wind in the same way as an aircraft's wing or a well-set sail. The plate at the top of the cylinder reduces the induced drag.

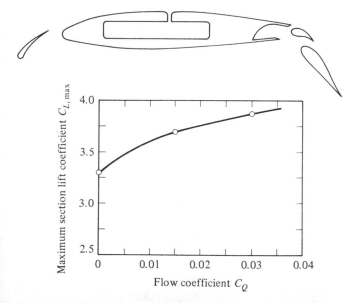

Figure 11.4.3 Configuration and maximum lift characteristics for NACA 64₁-212 wing section with leading-edge double-slotted flap and boundary-layer control. *(From "Theory of Wing Sections" by I. H. Abbott and A. E. Von Doenhoff, Dover Publications, Inc., by permission)*

Figure 11.4.4 The *Alcyone,* developed by the Foundation Cousteau and the Cousteau Society to use effectively the wind as a supplemental energy source. *(Photo courtesy of The Cousteau Society, a membership-supported environmental organization)*

[*] Wilbur and Orville Wright obtained patents on such devices, but the Handley-Page designs proved to be superior.

11.5 SUMMARY

The flow past most bodies is accompanied by *flow separation*. The pressure on the body increases aft of the locus of points comprising the separation line. The integrated effect of the pressure on the body's surface yields a force—a drag—in the direction of the upstream flow. This *form drag* is in addition to the viscous drag that the body experiences. In the case of bluff bodies, such as a sphere, the form drag is many times larger than the drag due to shear stresses for high Reynolds numbers. Through a modification of the body's shape, it is possible to move the region of separation far aft on the body and greatly reduce the form drag. This process is called *streamlining* the body. The fluid motions near the separation line are generally unsteady, so that unsteady forces of considerable magnitude can be induced on the body.

A component of the net force on the body that is perpendicular to the direction of the ambient flow is called the *lift*. For bodies that are used as wings or other lifting surfaces, one can change the lift by changing *the angle of attack* of the body. (The drag may also change simultaneously by an acceptable amount.) As the angle of attack increases, so does the lift, until a critical angle is reached. At that point, there is a flow separation over a major portion of the body's surface, most usually the upper surface of a wing, and a radically different pressure distribution on that surface ensues. Whereas formerly the pressures on those surfaces were very low when compared with other portions of the complete body surface, now the pressures are much higher, so that the lift is less. When this occurs on an aircraft, it cannot maintain its direction of flight and rapidly loses altitude. This is called *stalling* of the wing, or simply *stall*.

The pressure distribution on a wing also induces *wing-tip vortices* that create a *downwash* aft of the wing. Currently, the tips of subsonic aircraft are the subject of engineering development to provide high lift under the conditions of high angle of attack and low speeds. *Tipsails,* short sections of wing tip that are almost at right angles to the main wing, appear to be useful in this regard. See Fig. 11.5.1. As mentioned previously, one attempts to alter the pressure distribution on a wing so as to prevent separation, and, consequently, stall, by shaping the foil; however, blowing and sucking devices, as well as slots and flaps, are also part of *boundary-layer control*.

Figure 11.5.1 Tipsails at the tip of the wings of a jet airplane for corporate use. (*Courtesy of Beech Aircraft Corporation, Wichita, Kansas*)

EXERCISES

11.2.1(a) What is the expected terminal velocity of a spherical hailstone (use SG = 0.92) 1 cm in diameter?

(b) What would be the diameter of a sphere that has a terminal velocity of 3 m/s in air at 0°C if it has a specific gravity of 7.9? What is the answer for water at 20°C?

(c) An ellipsoid of revolution has a volume of $4\pi r^2 L/6$, in which L is the major axis and r the maximum radius of revolution. Find the expected terminal speed of an ellipsoidal hailstone ($r = 1$ cm, $L = 2$ cm) that falls so that its major axis is vertical. (sq = 0.92, again.)

11.2.2(a) What would be the force on the mast if it were of a constant diameter equal to 1.5 ft? What would be the moment induced at the base of the mast?

(b) If at the tip of a pole 1.5 ft in diameter a sphere with a diameter of 4 ft were mounted, what would be the moment induced at the base of the pole?

(c) The natural frequency of vibration of a steel mast such as the one in Exercise 11.2.2a is $0.56\sqrt{EIg/Ad/50^2}$, in which E is the elastic modulus, I the moment of inertia of the cross section of area A, and d the weight per unit volume of the pole's material. (g is introduced to give the mass per unit volume of the pole.) What is the natural frequency in this case? How does this frequency compare with the shedding (Strouhal) frequency?

11.3.1(a) What power must be expended by the water jet that drives the boat (assume 100 percent efficiency) at liftoff?

(b) On the basis of the drag at maximum speed, estimate the foil's overall drag coefficient at this speed.

(c) Since drag $= C_D/(C_L/\text{lift}) = \text{lift}/(C_L/C_D)$, the minimum drag will be associated with the maximum value of C_L/C_D. Estimate this value from Fig. 11.2.1. (*Hint:* Is the ratio C_L/C_D the slope of the ray drawn from the origin to any point on the curve in Fig. 11.2.1?)

PROBLEMS

11.1 Using direct integration, determine the force due to the pressure distribution that is associated with a flow over a sphere (Re < 0.1). What is the total drag force on the sphere? (In low-Re flows, the form drag is of the same order of magnitude as the drag due to viscous shear. This is not so for high-Re cases.)

11.2 What is the form drag on the sphere whose pressure distribution is approximated in Fig. P.11.2? What is the associated drag coefficient? If the measured (total) drag coeffi-

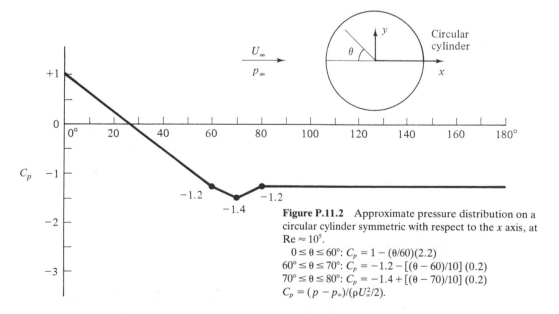

Figure P.11.2 Approximate pressure distribution on a circular cylinder symmetric with respect to the x axis, at Re ≈ 10^5.

$0 \le \theta \le 60°$: $C_p = 1 - (\theta/60)(2.2)$
$60° \le \theta \le 70°$: $C_p = -1.2 - [(\theta - 60)/10] (0.2)$
$70° \le \theta \le 80°$: $C_p = -1.4 + [(\theta - 70)/10] (0.2)$
$C_p = (p - p_\infty)/(\rho U_\infty^2/2)$.

cient were 0.6, what could you conclude about the order of magnitude of the viscous shear drag?

11.3 What is the form drag coefficient for a circular cylinder whose pressure distribution is approximated in Fig. P.11.3? What is the (total) measured drag coefficient for a cylinder at the same Reynolds number? Are vis-

cous effects on the drag of a cylinder very important at this Reynolds number?

11.4 Figure 11.1.4 is roughly approximated in Fig. P.11.4. What is the lift coefficient for the airfoil? How does your calculation using pressure measurements agree with the lift coefficient that was measured with a force transducer, 1.23?

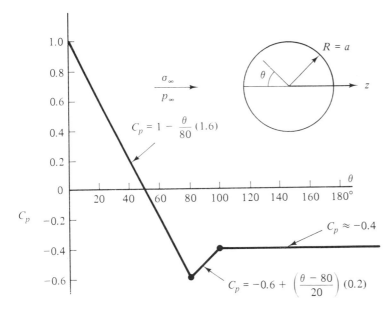

$C_p = 1 - \dfrac{\theta}{80}$ (1.6)

$C_p = -0.6 + \left(\dfrac{\theta - 80}{20}\right)$ (0.2)

$C_p \approx -0.4$

Figure P.11.3 Approximate pressure distribution around a sphere with a laminar boundary layer, Re ≈ 1.6(10⁵). $C_p = (p - p_\infty)/(\rho U_\infty^2/2)$.

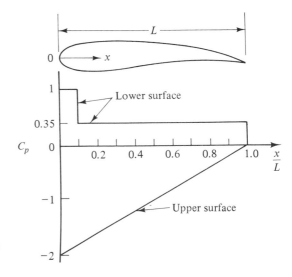

Figure P.11.4 Approximate pressure distribution on upper and lower surfaces. $C_p = (p - p_\infty)/(\rho U_\infty^2/2)$.

11.5 For the pressure distribution given in Prob. 11.4, what is the induced moment about the z axis, which is normal to the page at the origin? At what point along the chord does the resultant lift force tend to act?

11.6 Any point on the lift-drag polar diagram in Fig. 11.2.1 will give both the lift and the drag on the boat, normalized in the same way. Hence, C_L/C_D = lift/drag. Is this ratio the slope of the line from the origin to a point in question? Describe a simple (geometric) means to find the point on the curve that maximizes lift/drag.

11.7 Draw a lift-drag polar diagram for the airfoil whose characteristics are shown in Fig. 9.3.1. What is the angle of attack for the maximum lift/drag ratio? (Refer to the discussion at the beginning of Prob. 11.6.)

11.8 Compare the drag force on a streamlined, foil-shaped cylinder (1×3) and a circular cylinder at the same speed in the same fluid, if the thickness of the foil is the same as the diameter of the cylinder. For an equal drag on the two types of cylinders, what would be the ratio of the foil thickness to the cylinder diameter for $10^3 < \mathrm{Re} < 10^5$?

11.9 An airfoil can have a drag coefficient of about 0.005. If such a foil were 1 ft thick, what would be the diameter of a circular rod that had the same drag for $10^3 < \mathrm{Re} < 10^5$?

11.10 A 30-ft-tall mast has the form of an airfoil (chord/thickness = 3). It is 6 in. thick. Three guy wires, 3/16 in. in diameter and 37 ft long, support the mast. What is the total drag on the mast support system in a 40-mph wind that is in the direction of the mast's chord line? What fraction of this drag is due to the wires?

11.11 Assume that the mast that is described in Prob. 11.10 is cantilevered at its base. What bending moment is exerted there by the 40-mph wind? Assume here that the guy wires are not attached.

11.12 When a flat plate is aligned with a flow, it experiences a shear drag and no lift. At small angles of inclination to the oncoming flow, a lift force is induced. This force varies nearly linearly for small angles of attack α. The mathematical solution for this lift, per unit length of (infinite) span, is

$\rho U^2(\text{chord})\pi \sin(\alpha + \beta)$, in which β is zero for a flat plate. Assume that such a plate with a chord and span of 8 and 16 ft, respectively, could serve as a model for a hang glider. If the angle of attack is limited to 10° to avoid stalling the wing, what must be the minimum airspeed to support a person who weighs 140 lb? (Assume that the 2/1 aspect ratio of this wing does not seriously affect the lift coefficient. This is nearly true for aspect ratios >4.) If the glider's wing is slightly cambered, β could be 0.09 radians. How would such a camber affect the minimum speed of the glider to support the 140-lb person?

11.13 If a flat brass plate 2 in. wide and 12 in. long is supported in cantilever fashion at its base, its end will deflect $wl^4/8EI$, in which w is the load per unit length, l is the length of the beam, E is Young's modulus, and I is the planar moment of inertia of the cross section. What must be the thickness of the plate if the maximum deflection is to be 0.5 in. in a water stream of 20 ft/s?

11.14 Two people are carrying a 4-ft × 8-ft sheet of plywood vertically when they are overtaken by a 40-mph gust of wind. If the wind is perpendicular to the sheet, estimate the force that must be resisted by the transporters.

11.15 If an oil tanker 280 m long, 75 m wide, and with a draft of 25 m is moving at 15 knots, estimate the drag due to viscous effects on the sides and bottom. (1 knot is about 6080 ft/h.) What power is being expended to overcome the water's friction?

11.16 If the tanker in Prob. 11.15 is unloaded, its draft could be 10 m. This would give an additional 15 m of freeboard in the unloaded condition over the 5 m for the loaded state. If the unloaded tanker were nearly at rest in a harbor but in a storm with 80-mph winds from the side, what force would be exerted on the tanker by the air? (Neglect any interactions between the air and the water surface. Also assume that the superstructure—ship's bridge, and so forth—is 10 m high and 25 m wide.) At what (constant) speed would the tanker be drifting sideways?

11.17 An automobile is traveling at 40 mph as it emerges from a tunnel into a crosswind of 60

mph. It is 15 ft long, 5 ft wide, and 5 ft high and weighs 2200 lb. If on wet pavement the static friction coefficient of the tires is 0.65 and the sliding coefficient is 0.6, would you expect the car to slide sideways in the cross-wind? (Assume a drag coefficient of 1.25.)

11.18 The side rails in the design of a bridge are to be poured concrete slabs. Originally, they were to be plane surfaces that were perpendicular to the roadway. If they are 8 ft high and each section of the bridge between expansion joints is 100 ft, what lateral force should a wind of 70 mph exert on the expansion joint supports? (Consider the two side rails of the bridge to be acting independently.) If these side rails were triangular in cross section with the vertical side toward the roadway, what would be the difference in the lateral wind loading?

11.19 The two curved semicircular surfaces in Fig. P.11.19 are 8 in. long and have a diameter of 1 in. They are attached to a vertical axle by spokes of radius $r_1 = 6$ in. What torque is produced when a 20-mph wind blows perpendicular to the spokes? If this torque is twice the average torque when turning at constant speed, how much greater is the speed of rotation when the wind is 60 mph if the friction torque varies linearly with the speed?

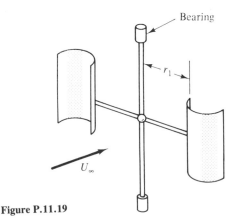

Figure P.11.19

11.20 Two hemispherical shells 50 mm in diameter are located symmetrically about a vertical axis so that their centers are 180 mm apart. (One shell has its curved surface upstream, while the other has it downstream.) The de-

vice is placed in an airstream of 5 m/s and the air is directed normal to the plane of the opening in the shell. What is the torque induced about the vertical axis for such a flow configuration? Hollow cones with a vertex angle of 60° now replace the hemispherical shells. (The cones and the hemispheres have the same diameter.) If the friction torque in the bearings that support the vertical rod is linearly proportional to speed, which device would rotate the faster, and by what factor?

11.21 A tower is to be constructed of square rods that will form a series of triangular grids. The plan is to weld 1100 ft of 2-in.-square bars into one of the plane faces of this tower. What difference will there be in the force of the wind on such a face if the bars are welded (1) with a flat surface facing the wind and (2) with an edge windward? Assume a wind speed of 60 mph, and ignore any effects due to air interaction of the wind with the ground.

11.22 At what distance of separation L do two disks of diameter D in tandem, i.e., one directly behind the other, each experience same drag as a single disk? Now consider two disks that are separated by a distance 3 times their diameter. What would be the percent decrease in axial drag if a cylindrical (lateral) surface of the same diameter were to connect them?

11.23 A short circular cylinder ($L/D = 1$) is to be replaced by a cylinder of the same diameter but with $L/D = 2$. What will be the percent change in drag?

11.24 What diameter teardrop would have the same drag coefficient as a sphere with a 2-in. diameter in subcritical flow?

11.25 What power would be saved when propelling an aerodynamic body in the form of a teardrop instead of a 2/1 ellipsoid? The diameter of both should be 0.8 m, and the air speed is 200 m/s.

11.26 An ellipsoid of revolution with the major axis twice the minor axis has a volume of $4\pi r^2 L$, with $L = 2(2r)$. How does the drag on such an ellipsoid compare with that of a sphere with the same volume?

11.27 If a slender circular rod is dropped so that its axis remains vertical, the drag will vary linearly with the speed of fall if the Reynolds number is very small. Using the subscript 1 to denote this case, one has $\mathrm{drag}_1 = K(u_1)$. If the same rod is dropped with its axis horizontal and the Reynolds number is again very small, the drag can once again be expressed linearly as $\mathrm{drag}_2 = C(u_2)$. However, the speed of descent in the second case is measured as one-half that in the first. Because the weight of the rod is the drag in each case, this implies that $K = C/2$. If the rod is dropped with its axis at an angle with the horizontal, it will not fall vertically. It will drift to one side. Show this by using the facts just given, Fig. P.11.27, and the force component equilibriums that exist for terminal velocity. In the motion picture "Low Reynolds Number Flows,"* G. I. Taylor uses a clever geometric construction to show that the maximum angle of drift, α, is about 19°, which gives an accompanying angle of δ near 34°. Check this result by writing an appropriate program for a digital computer. Does the numerical solution give angles close to 19.4° and 37.7°? The equilibrium for the forces in the vertical direction shows that the speed along the line of drift is dependent upon the drift angle, for a fixed weight. What is that speed at the maximum angle of drift?

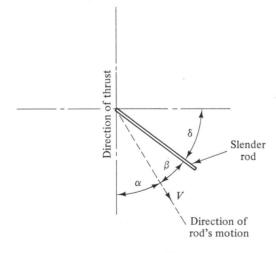

Figure P.11.27

* See motion picture references.

11.28 The rod in Prob. 11.27 could equally well be moving in a horizontal plane if a thrust were applied to it along the line PQ. The solution of the force equation for terminal speed shows that the thrust increases with α for a fixed speed along the drift angle. This indicates that the maximum thrust will occur at the maximum value of α. What is the maximum thrust for $V = 1$? Does the thrust vary greatly with the angle of drift? (Taylor goes on to talk about the motion of spermatozoa, whose motion would be Re $\ll 1$. If a spermatozoon were to orient its tail at the angle $(\alpha + \beta)$ and move it in the direction α, it would receive the maximum thrust needed to overcome fluid friction and move its main body in the direction PQ shown in Fig. P.11.27.)

11.29 A circular cylinder with $L/D = 5$ is released in air and could fall with its axis vertical, horizontal, or somewhere in between. Assume a drag coefficient of 1.2 for subcritical flow when the axis is vertical, and develop the equations for the motion of the cylinder, provided it does not tumble, in the manner outlined in Prob. 11.27.

11.30 If an aircraft were fitted with the NACA 64_1-212 wing described in Fig. 9.3.1 and were just lifting off from a runway (at a 0° angle of attack) at 150 knots with a weight of 150,000 lb, what would be the effective wing area if the wing had a flap deflected at 60°? If 45,000 lb of thrust were available at the time, what would have been the distance covered along the runway from start to takeoff? What speed would the aircraft need to have if it did not have the flaps? What would the corresponding runway length have to be? If the aircraft had been fitted with the devices mentioned in Fig. 11.4.3, a lift coefficient of 2.4 could be expected at zero angle of attack. What would be the ratio of the required runway lengths for the standard wing, a wing with flap, and a wing with leading-edge slot, flaps, and suction?

11.31 An aircraft that is 150 ft long has a wingspan of 108 ft. The wings have a chord that varies from 20 ft to 8 ft. The fuselage has an effective diameter of 14 ft. At the cruising altitude of 31,000 ft, a maximum load of 190,000 lb can be carried at 450 knots if there is no wind. What is the effective lift coefficient of the wings? Estimate the skin friction drag when cruising at 31,000 ft at 450 knots. (Neglect the effects of the empennage in this calculation.) If each of the three engines produces 15,000 lb of thrust at sea level—use the ratio of air densities to estimate the performance at altitude—what fracton of the maximum available power is being used under cruise conditions to overcome skin friction?

11.32 If the aircraft described in Prob. 11.31 can take off at 150 knots, what is its effective lift coefficient at takeoff? What is the ratio of lift to drag?

11.33 What is the reason that the drag coefficient on a circular cylinder is drastically smaller at $Re > 10^5$? Why is the drag coefficient for a body with sharp corners usually independent of the Reynolds number? The drag coefficient for a streamlined shape is listed in Table 11.2.1 as 0.06. Do you expect this value to be independent of Reynolds number? Explain.

In the following problems concerned with raindrops, consider the mass to be constant unless otherwise specified. This assumes no evaporation or condensation. Also assume the drop to remain spherical because of surface tension, even though for large raindrops this may not be the case.

11.34 In a nonprecipitating stratus cloud, the droplets might have a diameter less than 10 μm. What updraft velocity would be necessary to keep them from falling?

11.35 If someone says that the terminal velocity of a raindrop with a diameter of 3 mm is about 8 m/s, would you consider this statement to be reasonable? Explain.

11.36 If a raindrop with a diameter of 0.5 mm is 10 m directly below a drop that is 1.0 mm in diameter, what distance will it travel before the drops collide? Estimate the speed of the composite (spherical) drop.

11.37 During a modeled rainstorm, all of the drops are initially falling so that one is 1 m directly below another. All, except one, are 0.5 mm in diameter. The exception is 1.0 mm in size. Estimate its size (due to collisions with drops directly below it) after it has fallen $\frac{1}{2}$ km.

11.38 With what force will a raindrop with a diameter of 5.8 mm strike a horizontal surface at its terminal velocity?

11.39 The largest measured hailstone had a circumference of 17.5 in. and weighed 1.67 lb. Assume it to have been spherical and estimate its terminal velocity and the force with which it struck Coffeyville, Kansas.

11.40 With what kinetic energy would a hailstone that is 1 in. in diameter strike the ground? Assume terminal velocity and an average specific gravity of 0.92 because of the porosity in the ice crystals.

11.41 Does a 2-mm water drop fall in air with the same terminal speed with which an air bubble of the same size would rise in water? If the ratio of speeds is not unity, what is it?

12

Secondary Flows and Flow Stability

12.1 INTRODUCTION

The horseshoe vortex generated near the base of the cylinder attached perpendicular to the flat surface in Fig. 7.6.2 is a consequence of the dynamics within the fluid flowing near the intersection of the surfaces. (The primary flow is parallel to the plane surface, in this case.) The induced horseshoe vortex is an example of a secondary flow, because locally the primary flow cannot satisfy the dynamic relationships and the boundary conditions. An additional flow pattern must be present. The "teacup" flow discussed at the end of Sec. 8.4 is another case with a secondary flow. The dynamics in this case are relatively simple, and one can readily see how the presence of the rotating flow far from the bottom of the container will necessarily induce a radial inflow near the bottom. (This also occurs in a cyclonic storm due to the interaction of the rotating wind system and the ground.) In this chapter, some other situations will be explored in which secondary flows occur. The purpose of the work is to gain new insights into these flow phenomena and to provide some additional background for the important case in which an orderly flow is transformed into a random flow. That will be the focus of Chap. 13.

12.2 BASIC IDEAS

When analyzing a particular physical situation, such as the "teacup" in Sec. 8.4, it is possible that a paradox will occur when certain initial assumptions about the flow are carried to their logical conclusions. The paradox can sometimes be resolved by the existence of a secondary flow. Figure 12.2.1 presents an example of such an occurrence. Let us initially *assume* that the liquid is at rest everywhere in the container.

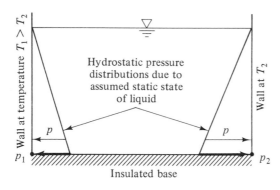

Figure 12.2.1 Rectangular tank with opposite vertical walls at different temperatures creating an unconditionally unstable situation.

Then the pressure must increase from the top of the liquid to the bottom. However, because of the temperatures that are assumed to exist along the walls, the fluid density along the left wall should be less than that along the right. Consequently, the pressure at point 2 will exceed that at point 1, so that $p_2 - p_1 > 0$. But if this is so, the fluid along the bottom cannot be at rest; a weak motion must exist so that the associated shear stresses can balance the horizontal pressure gradient. This small motion can alter the hydrostatic pressure distribution in the tank somewhat, but $p_2 - p_1$ will remain positive.

One could say that the primary flow was $\mathbf{V} = 0$ and the secondary flow was the thermally induced convection. Others might say that, because of the convection, $\mathbf{V}$ is not zero at any time or anywhere, so that the division between primary and secondary flow made here is artificial. Such an argument would miss the mark here, because it is only intended to emphasize the dynamics that can cause such a flow to be necessarily present for all nonzero values of $T_2 - T_1$. The reader can sense that there are other phenomena that can be analyzed by making some assumptions and following logically the deductions that can be made from them. We turn now to another example.

Figure 12.2.2 shows an idealization of the temperature and salinity conditions in a primitive model of water near the equator. The absorbed solar radiation warms the surface layer. The related evaporation increases the salinity of the water there. This increase in the water's density due to the salt concentration could be compensated for by a decrease in the density of the water due to temperature. One can conceive of a

Figure 12.2.2 Two horizontal layers of water with different temperatures and salt concentrations.

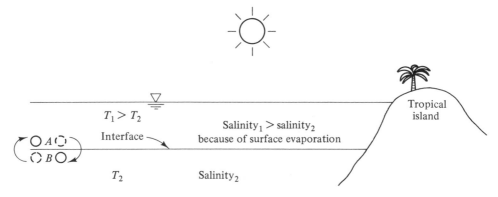

case for which this compensation is complete. Then the solution's density would be constant with depth. Imagine now a situation wherein packets of fluid at points A and B are interchanged. As long as there is no interaction of these packets with their surroundings, the density of each packet in its new environment would be the same as that of its surroundings. There would be no unbalanced forces on them, and the packets would remain where they were. However, the transplanted particles have different temperatures and salinities from those of their surroundings. What happens if there is diffusion of heat and mass between the packets and their surroundings because of these gradients?

If there is heat transfer but no salinity transfer, the packet that is now at A will be warmed. Since the particles were less saline than their neighbors and this condition has not changed, they will now be less dense than particles at the same elevation. They will be lighter and will begin to rise, like a submerged cork in water. For this process, in which only heat has been transferred, the occupants at position B that have come from A will be cooled and will begin to sink. The result is a continuation of the displacement that may have been initiated by some random means. The fluid, once disturbed a small amount, continues in motion. This would be designated as a situation that is *unstable to small disturbances*.

But if the physical characteristics of the water are such that salinity diffusion exists while there is no sensible heat transfer, then an analysis would show a stable situation for small disturbances. The displaced particles would tend to be driven back to their original positions as they lose or gain salt from their neighbors immediately after being transplanted.

From this discussion, it can be inferred that the stability of this system, in which two diffusion processes occur, will be influenced by the ratio of diffusion of heat to diffusion of salt. (The reader is encouraged to examine handbooks to verify that this ratio is greater than 1. If a layer of salt water is floated over a horizontal layer of fresh water at the same temperature, would the two layers be unstable to small disturbances?)

The sharp line of demarcation between warm and cool water, as well as high and low salinity, is a simple model that might be improved upon. Figure 12.2.3 presents an attempt at a better simulation for this particular case of *double diffusion*. Depending upon the slopes of the temperature and salinity distributions, the water could be in a situation wherein its density increases with depth, or is constant. (Experience indicates

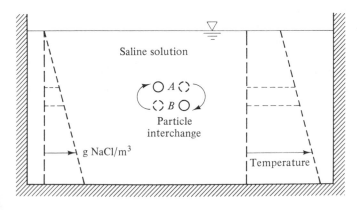

Figure 12.2.3 Vessel with stable salinity gradient and unstable temperature gradient.

that the least dense liquid should not be the deepest.) We now turn our attention to the consequences of interchanging particles at points A and B for the case in which the effects of salinity and temperature compensate one another to give a constant-density medium.

Once again, if there is thermal diffusion without salinity diffusion, the packet of transplanted fluid particles will rise. The situation would be unstable. If only salinity diffusion took place, the case would be a stable one. However, the displacement of particles takes a finite interval of time in actuality. What could occur in that interval? Could it happen that, during the excursion of the packet from A in Fig. 12.2.3 to point B, the passage will occur extremely slowly? In transit, the particle could transfer enough heat and salt that its density is nearly the same as its neighbors' at the same elevation. If this were so, the particle would not have very much excess buoyancy. The "driving force" would be small and might be balanced by the presence of small shear forces that would be induced by the motion in a sufficiently viscous fluid. Consequently, the particle could come to rest. It would neither continue to move away from its origin nor tend to return there. One would judge the situation to be *neutrally stable to the small disturbance*. By altering the physical constants or the temperature and salinity gradients in the fluid, one could have either a stable or an unstable case.

Of course, one could cause the interchange to occur suddenly and not give time for the diffusion processes to take place completely. One would then expect that the dominant diffusion process would dictate the outcome in a fashion similar to that associated with Fig. 12.2.2. Thus, some physical cases can be unstable to large disturbances even though they might be stable to small disturbances. (It has been suggested by oceanographers that one can create a "fountain" to harvest some of the ocean's food products by creating a large disturbance through a submerged pump. This would provide the initial interchange of fluid needed to destabilize a situation such as that depicted in Fig. 12.2.3.)

There has been great interest in problems for which the flow is unstable to small disturbances. Some problems have been amenable to analysis, and solutions have been obtained that are definitive and in concert with experimental observations. (This was true in the case of the results of Tollmien and Schlichting for the instability of the boundary-layer flow over a flat plate. See Fig. 8.6.3b.) The flow between two concentric rotating cylinders that was studied in Sec. 8.4 provides the basis for another example of flow instability. We shall examine it next because of its importance in the development of the subject, as well as the insight that it brings.

The flow within the annulus was shown by Eq. (8.4.5) to be the sum of circular motions due to a rigid-body rotation Ar and an irrotational vortex B/r. The values of the constants A and B depend upon the radii and the speeds of the outer and inner cylinders. For the present, we shall deal with the case that B is zero. Furthermore, it will be *assumed* that the fluid is inviscid. Then, according to Eq. (2.8.4), the radial pressure gradient and the purely rotational flow are related by Newton's laws of motion:

$$\frac{\partial p}{\partial r} = \rho \frac{V_\theta^2}{r}$$

If $V_\theta = Ar$, the pressure gradient near the inside cylinder is

$$\left(\frac{\partial p}{\partial r}\right)_{r_i} = \rho A^2 r_i$$

and similarly, near the outside cylinder,

$$\left(\frac{\partial p}{\partial r}\right)_{r_o} = \rho A^2 r_o$$

We shall now assume that a particle near the outside, at r_o, interchanges its position with a particle at r_i. Because of the inviscid *assumption,* the interchange would occur without any applied torques, so that the angular momentum per unit mass, rV_θ, of the particles would not be changed. The fluid particles at r_o have an angular momentum of $r_o(Ar_o)$, so that a particle that was transplanted to r_i without changing its angular momentum would have to have a θ-directed velocity of $V_\theta = Ar_o^2/r_i$ when it arrived at r_i. Such a displaced particle would have a radial acceleration of $-V_\theta^2/r$ ($= -A^2 r_o^4/r_i^3$) at r_i if there were no further radial motion. What forces are exerted on this displaced particle at r_i to account for this acceleration? The ambient radial pressure gradient will provide one such force. The gradient at r_i was just determined to be $\rho A^2 r_i$. If the possibility of some radial motions is not excluded, the equation of motion in the r direction at r_i is $-\partial p/\partial r = \rho a_r$, or

$$-\rho A^2 r_i = \rho a_r = \rho\left(-\frac{A^2 r_o^4}{r_i^3} + a_r \text{ associated with radial motions}\right)$$

A revision of this statement would read:

$$1 = \frac{(Ar_o^2)^2}{(Ar_i^2)^2} - \frac{1}{A^2 r_i}(a_r \text{ associated with radial motions})$$

Because $r_o > r_i$, this equation demands that there be some radial motion to give additional terms to the radial acceleration. The conclusion is that the particle that was brought toward the center without loss of angular momentum must have a positive contribution to a_r due to a radial motion. It should accelerate outward. This means that the transplanted particle would be returning to its original position. The end result would be that the disturbance associated with the radial displacement of the particle has been annulled. A stable condition exists.

Had the velocity distribution in the annulus been the other extreme, B/r, then the ambient pressure gradient near the inside cylinder would be

$$\left(\frac{\partial p}{\partial r}\right)_{r_i} = \rho\frac{V_\theta^2}{r} = \rho\frac{B^2}{r_i^3}$$

The angular momentum per unit mass of a particle at r_o ($>r_i$) would be $(rV_\theta)_o$, and this would be the constant B. If the angular momentum of such a particle were preserved upon moving to r_i, its θ velocity would be obtained from $B = (rV_\theta)_i$. The result for the displaced particle at r_i is $V_\theta = B/r_i$. This is the same θ velocity as that of the neighboring particles at r_i that were not displaced and which determine the ambient pressure gradient. One would conclude from this that the particle that just arrived at r_i from r_o would be in dynamic equilibrium there. The ambient pressure gradient and its own radial acceleration, due only to the θ motion, would be in balance. There would be no need to require any radial acceleration contributions due to a radial motion to balance the equation. The particle would remain in the new position. The disturbance would neither grow nor decay. One would have a case of *neutral stability for the inviscid fluid.*

In the first case of flow in the circular annulus that was just discussed, the square

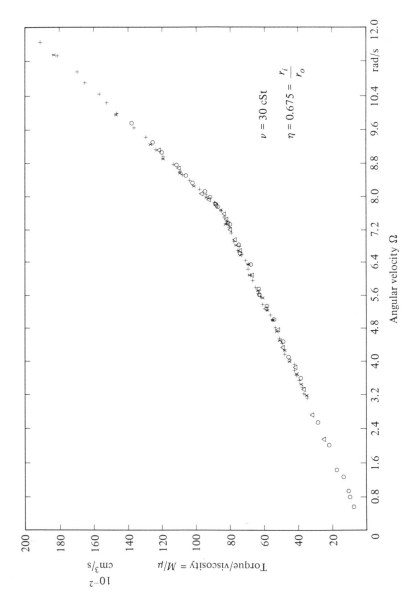

Figure 12.2.4 Normalized torque versus speed data for the flow in an annulus with $\eta = 0.675$. (The change of slope implies a change in the flow pattern within the annulus.) (From "Proceedings of the Twelfth International Congress of App. Mech., 1968," p. 172, Springer-Verlag, by permission.)

$\nu = 30 \text{ cSt}$

$\eta = 0.675 = \dfrac{r_i}{r_o}$

Angular velocity Ω

Torque/viscosity $= M/\mu$
10^{-2}
cm^3/s

of the angular momentum increased with increasing radius and the motion was stable. In the second case, the angular momentum was constant and neutral stability occurred. It would take the same pattern of argumentation to show that if the square of the angular momentum decreased with increasing radius, then any interchange that preserved angular momentum of particles of different radii would be unstable for an inviscid fluid.

One can anticipate that viscosity, if present in the analysis, would introduce shear stresses that would oppose any radial motions. This would mean that viscosity would increase the stability of flows which were already stable when the fluid was considered to be inviscid. In the case where the square of the angular momentum decreases outward, viscosity could provide a reason why some of these flows are stable, even though they would be unstable if inviscid.

Figure 12.2.4 illustrates the relationship between a measure of the torque M and the angular velocity of the inner cylinder Ω. The outer cylinder is at rest. (This means that neither of the constants A and B, mentioned above, is zero. The angular momentum, rV_θ, obviously decreases as r increases, since V_θ is zero at the outer cylinder.) The initial straight-line portion of the curve in the figure is exactly the same as that predicted by the results for a V_θ obtained with the appropriate values of A and B. The radial velocity would be zero. The viscous flow appears to be stable to disturbances, despite the existence of a criterion, for an inviscid fluid, that predicts the contrary. However, a change in the slope of the curve does occur if the angular velocity of the inner cylinder is sufficiently high. This change is associated with a different flow within the annulus. The θ velocity will change, and radial and axial velocity components will then be present. This flow is pictured in Fig. 12.2.5. It was made visible by adding finely divided aluminum flakes (e.g., aluminum paint pigment) to silicone oil. The bands are visible because the orientation of the flakes in adjacent layers is different. They reflect the light differently. The particles in each layer are moving on helical paths. Their motion is similar to that of a wooden bead that moves along a coiled spring whose axis is bent into a circular shape. In each of the successive layers, this helix changes its direction. One band has a "right-handed" helix, while the adjacent bands are "left-handed." This permits the fluid at the interface between the bands to have a common velocity. (Additional instabilities occur as the inner cylinder is slowly

Figure 12.2.5 Flow pattern in an annulus for a flow regime that is comparable to the high-slope region in Fig. 12.2.4.

speeded up. The bands become wavy. If only the outer cylinder is rotated, as illustrated in Fig. 1.7.1, these instabilities do not occur.)

How can we rationalize the results of this viscous case when an inviscid criterion predicts unconditional instability? One can sense the role of viscosity with a simple analysis, although a more complete analysis would be needed to give a quantitative criterion. (This was done by G. I. Taylor in 1923.) If the outer cylinder is at rest, then the angular momentum of the fluid particles near it is nearly zero. If one of these particles moved radially inward as a result of some random disturbance, viscous interactions with neighboring particles could give it a small θ velocity component. The displaced particle would arrive at some location near the inner cylinder with an angular velocity that undoubtedly would be less than that of the particles that were always there. Thus, the new particle would be in a pressure gradient that would tend to force it further inward, if it were not for the presence of viscous shear stresses. It is conceivable that during this period of time these stresses could be sufficient to arrest the motion of the particle so that it would acquire the θ velocity of its neighbors. This would occur through viscous effects in the θ direction. This explanation requires that viscosity alter the θ velocity component and act to suppress the radial component of any randomly moving particles.

If the equations of motion are written to include viscous effects, this speculation about the effect of viscosity can be confirmed. To do this, one writes each of the velocity components as the sum of a term due to the basic state of motion plus a term due to the perturbed motion. For example,

$$u_r = U_r + \tilde{u}_r$$
$$u_\theta = U_\theta + \tilde{u}_\theta$$
$$u_z = U_z + \tilde{u}_z$$
$$p = P + \tilde{p}$$

with the capital letter denoting the basic state and the tilde a perturbation to that motion. This solution requires for our annular flow that $U_r = 0$, $U_\theta = Ar + B/r$, and $U_z = 0$ for the unperturbed flow state. For such a state, the disturbances $(\tilde{u}_r, \tilde{u}_\theta, \tilde{u}_z)$ are zero. Equation (8.4.5) can be interpreted to mean that the U_r, U_θ, and U_z values just given satisfy the equations of motion, since u_r and u_z were implicitly assumed to be zero in that equation. Because the complete equations of motion given in Chap. 16 allow for radial and axial components, they permit small values of $\tilde{u}_r$, $\tilde{u}_\theta$, $\tilde{u}_z$, and $\tilde{p}$ to exist in a steady way in this case, provided that the parameter called the *Taylor number*,

$$T = -\left(\frac{\Omega}{\nu}\right)^2 \frac{(D_2 - D_1)^4}{4} \frac{(D_2/D_1)^2}{1 - (D_2/D_1)^2} \tag{12.2.1}$$

is greater than a certain, critical value. A result of Taylor's work is shown in Fig. 12.2.6. It shows the inviscid criterion for neutral stability and the modification (experimental and theoretical) due to viscosity.

The instability of the rotating flow within an annulus provides a useful example for a discussion, because only momentum changes and the associated momentum diffusion are important for the case. There are no rates of thermal or material diffusion in the problem.

The last example of instability that will be discussed includes thermal diffusion. A layer of fluid in a gravity field can be heated from below to cause the fluid at the

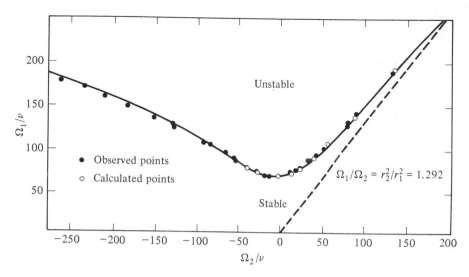

Figure 12.2.6 Taylor's theoretical and experimental results for $R_1 = 3.55$ cm, $R_2 = 4.035$ cm. The coordinates are in units of cm^{-2}. *(From "Hydrodynamics Stability" by P.G. Orazin and W.H. Reid, Cambridge University Press, by permission.)*

lowest point in the test chamber to be the least dense. Our experience tells us that the densest fluids are to be found at the bottom of a static container so that the act of creating the least dense particles there may produce some interesting results. It is one of the oldest instability problems that has been examined. Figure 12.2.7 shows an interferogram—the black and white lines represent surfaces of constant temperature—of two-dimensional convection between rigid horizontal surfaces. In this case, two-dimensional rolls are generated, with the axis of the rolls parallel to the short side of the rectangular horizontal surface that forms the floor (or ceiling) of the test chamber. (Compare this with the three-dimensional cells of Fig. 1.6.4 for a different, but related, problem. The instability that may have received the first serious study is that due to the variation of surface tension on the free surface of a liquid. This can occur if there are differences in temperature along the surface that could be induced by random disturbances. Surface tension is reduced by increasing temperature, a fact that allows cooking oil to spread on the bottom of a warm skillet. Figure 1.6.4 shows a laboratory experiment involving this phenomenon. Because the convection cells for a surface tension–driven instability are similar to those in which a fluid layer is heated from below, these two cases were often confused with one another in the first 50 years of the study of flow instabilities.)

A question that is prompted by the discussion of the case presented in Fig. 12.2.7 is whether or not the case of heating a layer of fluid from below so that $(T_{bottom} - T_{top}) > 0$ constitutes a case of necessary secondary flow (i.e., unconditionally unstable case) or one for which a neutral stability criterion can be established. If the latter case were true, then there should be a parameter whose magnitude would indicate whether or not a secondary motion (i.e., an instability of the primary motion) should be expected. Experimental evidence is sought to decide this question, and Fig. 12.2.8 provides the answer.

Figure 12.2.7 Buoyancy-driven convection rolls. The interferogram presents a side view of convection instability of silicone oil in a rectangular box that has a width about 11 times its height. Fluid is heated from below and produces rolls that are parallel to the shorter side of the test chamber's base. *(Courtesy of Dr. H. Oertel, Jr., in* Thermal Instabilities *in* Convective Transport and Phenomena, *J. Zierep and H. Oertel eds, Braun Verlag, Karlsruhe, 1982)*

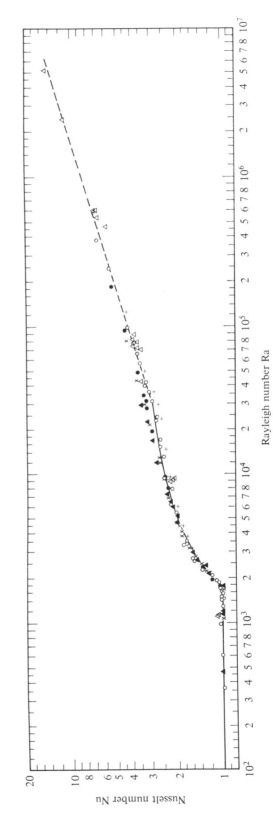

Figure 12.2.8 Silveston's experimental results on the heat transfer in various liquids showing Nusselt number versus Rayleigh number. *(From "Hydrodynamic and Hydromagnetic Stability" by S. Chandrasekhar, Dover Publications, Inc., by permission)*

The abscissa in Fig. 12.2.8 is a dimensionless number called the *Rayleigh number*. It is defined as

$$\mathrm{Ra} = \frac{g\alpha(T_{\mathrm{bottom}} - T_{\mathrm{top}})h^3}{\kappa\nu} \tag{12.2.2}$$

in which Ra = Rayleigh number
  g = gravitational constant
  α = coefficient of volume expansion (dimensions: 1/temperature)
  h = layer thickness
  κ = coefficient of thermal diffusivity (dimensions: L^2/time)
  ν = kinematic viscosity (dimensions: L^2/time)

The terms in the numerator of the Rayleigh number are the agents that force the instability, while the terms in the denominator are those that restrain it. Thus, the Rayleigh number represents the ratio of the factors that tend to cause the motion to the factors that would annul it. It is clear that Ra must be greater than 1 for an instability to occur. The ordinate in the figure, the Nusselt number, is the ratio of the actual heat transferred to the heat that would be transferred if only heat conduction were to occur in the layer for the imposed temperature difference between the upper and lower surfaces. As long as conduction does, in fact, exist in the layer, the ordinate should be equal to 1. However, one observes that after a certain value of the Rayleigh number, the ordinate is greater than 1. The conclusion is that some other heat transfer process than pure conduction must be taking place when the Rayleigh number exceeds a critical value. The new state of heat transfer is laminar, natural convection. If one uses a different representation of the experimental data, it becomes obvious that this convection will transfer much more heat across the layer than conduction for the same temperature difference.

Figure 12.2.8 provides the answer to our question about the existence of neutral stability for this case. For a Rayleigh number less than 1708, one can expect that conduction is the mode of heat transfer. There would be no fluid motions in the layer. For values of the Rayleigh number above this critical value, one should expect convection, with its attendant velocities. It is noteworthy that the critical value of the Rayleigh number that was just given comes from a theoretical analysis of the problem. Because this theoretical value is close to that observed experimentally, it is believed that the theoretical model embodies most of the important features of the problem. (If the upper surface is free—not in contact with a solid surface—then the theoretical critical Rayleigh number is near 1100.)

But despite some successes, stability theory currently cannot usefully predict the transition of a laminar flow in a round pipe to a new state. (That state should be turbulent pipe flow.) The results for the flow between parallel plates are no better. The problem of turbulence is a difficult one. Some of the best minds of recent decades have worked on it. There are scientists who have despaired of ever understanding it, even scientists who never despaired of comprehending, say, nuclear structure. Turbulence may be a subject that is as difficult to solve as the origin and future of our universe. However, you can be assured that turbulence will continue to occupy the thoughts and energies of creative people in the future. But even if one is not now able to give a definitive model of the turbulence process, one can provide a great deal of computational and experimental experience upon which good estimates can be made about new turbulent situations. We can hope that the future will bring additional insight into these "rules" for turbulent flow.

12.3 SUMMARY

If one were to require that the volume of a liquid in a segment of a circular jet be the same as that in a spherical drop, and also require that the pressure within the liquid jet and the drop be the same for the same ambient surroundings, one could conclude that the segment must have a length that is 1.7 times its circumference. Equation (1.6.1) would be the source of this conclusion. Such a calculation was the basis for one of the first analyses of the breakdown of a jet into drops, which is pictured in Fig. 1.6.5. A current analysis of this problem to determine whether a cylindrical or spherical surface should occur takes into account the fact that the processes involved are dynamic and not static ones. This theory predicts that the fastest growing disturbance has a wave length that is 1.47 times the jet's circumference.

While surface tension is an important mechanism in determining the breakdown of jets, the examples in this chapter have pointed out that density and angular momentum distributions can also alter a flow pattern if it is sufficiently disturbed. Various physical properties of the fluid such as viscosity, mass and heat diffusivities, and the dimensions of the flow regime can affect the transition from one flow state to another. This geometric effect is evident in Eqs. (12.2.1) and (12.2.2).

Even though the flow through a pipe can develop a swirling motion due to centrifugal effects that occur in a curved, or bent, portion—see Fig. 12.3.1—not all flows must develop a laminar secondary flow if conditions are changed. However, all flows will gain one characteristic as their speed is increased or some particular parameters are changed. Their motions will become random and appear chaotic. This change in the type of flow was noticed by Osborne Reynolds and mentioned in Sec. 8.3. Chapter 13 deals with this subject more fully.

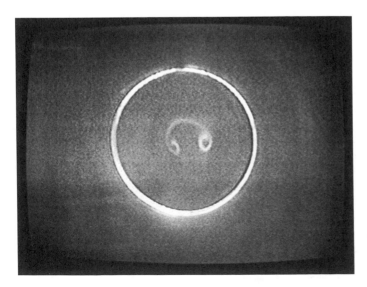

Figure 12.3.1 Television image of secondary flow in a branch of a pipe-wye. Flow pattern was made visible by the addition of a mixture of water vapor and CO_2 from "dry ice."

PROBLEMS

12.1 Is the "teacup" secondary flow associated with an instability caused by a small disturbance? Justify your point of view.

12.2 Is "double diffusion" possible if one had layers of sugar and salt water? Would it be necessary that one of the solutions be above the other?

12.3 When a melted alloy begins to solidify, the resulting crystals form in a solution of constantly varying composition. Does this mean that the crystals will have a chemical distribution that will depend on the cooling rate because of "double diffusion"?

12.4 For the data given in Fig. 12.2.4, what is the critical Taylor number given by Eq. (12.2.1)? Does this value agree with the results given in Fig. 12.2.6?

12.5 What is the significance of the dotted (straight) line in Fig. 12.2.6?

12.6 The data in Fig. 12.2.8 show that the critical Rayleigh number is about 1708. What would be the minimum temperature difference to induce convection in a layer of air or water that is bounded by rigid isothermal boundaries? Assume that the layer depth is 10 mm. $(\nu/\kappa)_{air} = 0.7$, $(\nu/\kappa)_{H_2O} = 7$.

12.7 If a fluid layer has rigid isothermal boundaries above and below it, the critical Rayleigh number is about 1708. However, if the upper surface is an isothermal free surface, the number is around 1100. Explain why the critical Rayleigh number should be less in the latter case.

12.8 What could be the reason for the change in the shape of the curve for a Rayleigh number of about 30,000?

12.9 Does the formation of drops in a liquid jet provide some grounds for the observation that in fires near high-voltage lines, the fire-fighters handling the hose are not harmed when water strikes the wires?

12.10 A water jet issues from an orifice 2 mm in diameter at a speed of 1 m/s. At what frequency should the orifice by disturbed to enhance the formation of drops? What would be the frequency for 10 m/s?

13

Turbulent Flow

13.1 INTRODUCTION

In Secs. 8.3 and 8.6, the laminar flows in round conduits and over flat plates were discussed. In both cases, it was found that the laminar solution gave results that agreed with experiments up to a limiting value of the Reynolds number, Re_{cr}. (This parameter had a length scale that was appropriate to the particular case. The limiting values are approximately 2000 and 500,000 for round pipes and flat plates, respectively.)

The actual flow in a pipe that replaces the laminar flow requires a pressure gradient considerably greater than that of its predecessor for the same average velocity. Consider now a laminar flow that is characterized by stage 1 in Table 13.1.1, and observe how it progresses to stages 2 to 5 and beyond as Re_{cr} is slightly exceeded. An oscillation in the flow as one goes from one flow state to another in Table 13.1.1 would be manifested by a pulsating discharge from the end of the pipe.

The fact that a doubling of a small pressure gradient could, in some cases, double the quantity of fluid flowing per unit of time was known from Poiseuille's experiments. (His experiments were confined to the laminar regime of flow. Hagen, in Berlin, had conducted experiments over a wide range of flows while trying to show the effect of temperature on discharge. His findings in 1854 showed that the relationship between discharge and pressure gradient was not always linear.) George Stokes' analysis in 1848 had predicted the proportionality between flow and pressure gradient. To check his work, he compared the results of his equation with data that had been gathered for a flow that could not have been laminar. He found the correspondence disappointing. Between 1850 and 1880, people continued to seek a rational explanation for the observed fact that, after a certain pressure gradient was reached, the relationship between Δp and the volume rate of flow ceased to be linear. The pressure gradient

TABLE 13.1.1 LAMINAR FLOW TRANSITION IN PIPES

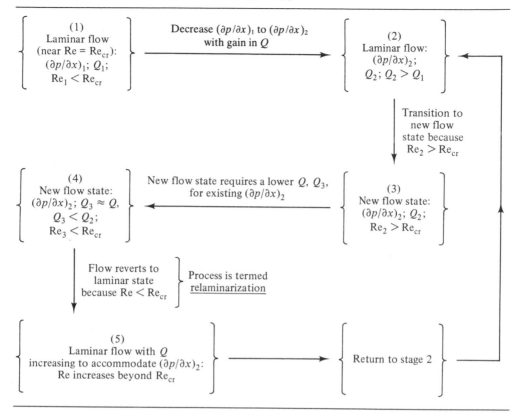

seemed to vary as the average velocity to the 1.75 power. (Data had been collected in France as early as 1732 by the junior Couplet that indicated such an exponent.) An explanation that was accepted in many circles was the existence of a second viscosity coefficient that came into effect differently in different experiments.

Osborne Reynolds was unsatisfied with this resolution of the paradox. He conducted experiments that identified the abrupt occurrence of the transition from laminar flow as the pressure gradient was increased. He observed the linear variation of the volume flow rate with the pressure gradient before the transition, as well as the 1.75 power law subsequent to the change in flow. He verified that a change indeed took place by inserting dye through a fine tube aligned along the axis of a larger glass tube. When the flow versus pressure gradient curve changed slope, he found that the thread of dye also changed appearance. During the laminar flow, the dye flowed almost unaltered down the pipe as a thin thread of colored water. But as the pressure gradient was increased, a point was reached at which the straight line of dye became "wavy." This waviness increased greatly in amplitude with slight increases in the pressure gradient until the line of dye no longer appeared regular. Rather, it assumed random positions, and the existence of strong radial velocity components became evident from the rapid diffusion of the dye into the flowing water.

Reynolds did many things that are of interest to engineering and science. Of particular significance to us here is that he identified a critical parameter for the experi-

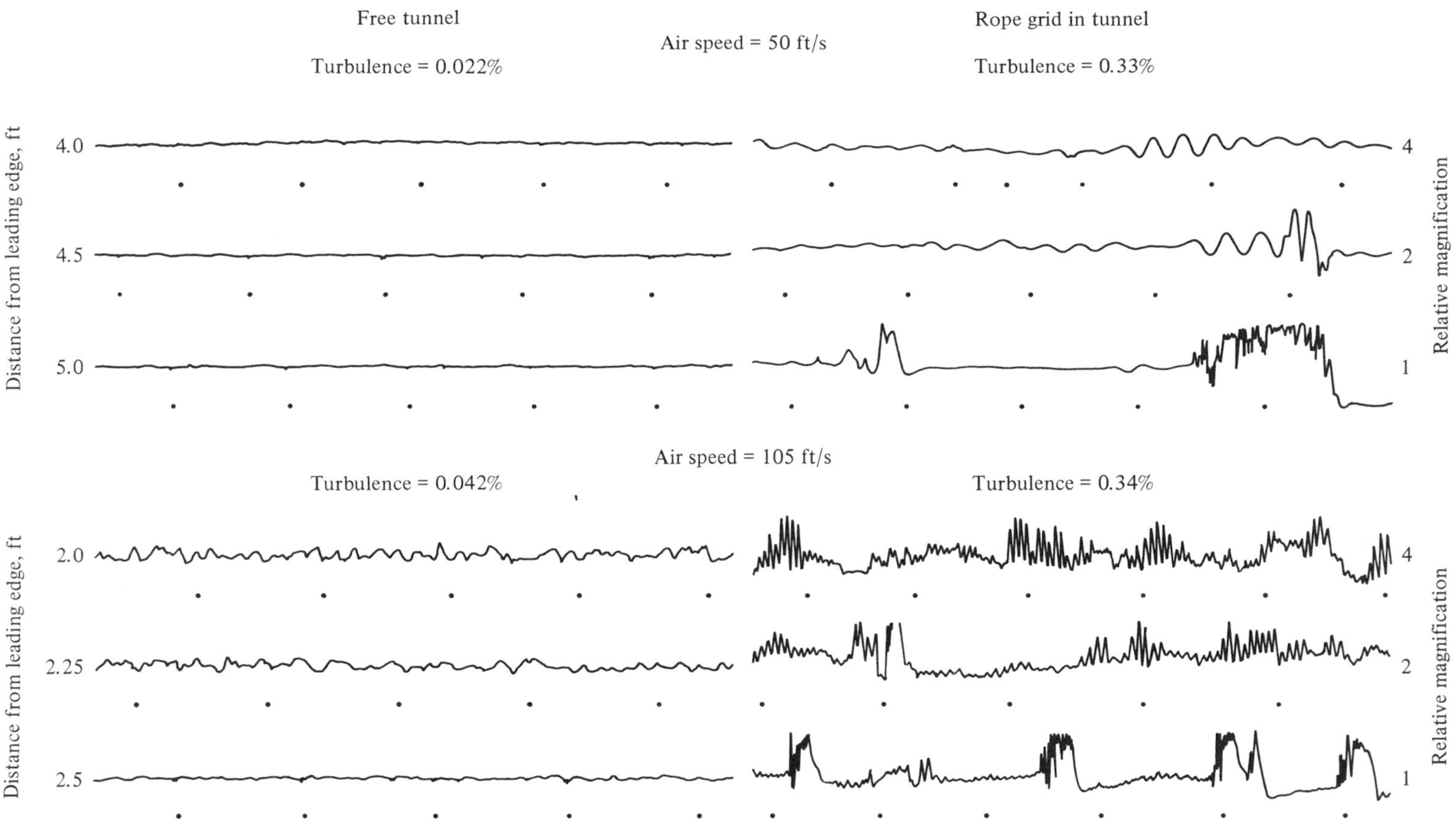

Figure 13.1.1 Oscillograms of fluctuations in streamwise velocity component in boundary layer showing effect of increasing the turbulence in the stream from 0.022 to 0.042 percent. Distance from surface = 0.004 in. Time interval between dots = 1/30 s. *(Courtesy of the National Institute of Standards and Technology)*

ment, the product of the fluid speed, the size of the pipe, the density of the fluid, and the reciprocal of the viscosity. (It was Arnold Sommerfeld, a German physicist who worked in topics from fluid mechanics to atomic theory, who proposed that this parameter is called the *Reynolds number.*) We now call the flow that replaces the laminar flow in pipes after the critical Reynolds number is exceeded a *turbulent flow.*

It was observed by Reynolds that turbulent flows have local velocity fluctuations even though the quantity of flow is ostensibly constant with time. These fluctuations appeared to be random, and this randomness is often used as a hallmark for turbulent flows. Figure 13.1.1 shows some electrical signals that can be correlated with a velocity component measured in a turbulent flow. The signal comes from a very thin heated wire (about 50 μm in diameter). As the speed of the air passing over the wire changes, so does its cooling effect. Thus, the temperature of the wire, and its resistance, are constantly changing. This change of resistance gives rise to the signal shown in the figure. (The wire probe and the associated electronic circuits are called a *hot-wire anemometer.*) Experiments show that when a fluid is being discharged from a pipe in a steady, but turbulent, manner, there are locally rapid oscillations that appear to be centered about some average value of the local velocity.

Even though Reynolds did not have such electric signals to analyze, he proposed that the velocity components at a point in a flow be subdivided into an average and a fluctuating component. Thus, for the x direction, the decomposition is $u = \bar{u} + u'$. In this notation, the barred quantity is the *average value,* and the primed value is the instantaneous value that is fluctuating around the average value. It is called the *fluctuating,* or *turbulent,* velocity component. For the *temporal mean component,* we shall also employ the convention $\bar{u} = U$. This subdivision is shown in Fig. 13.1.2. Similar subdivisions are used for all three of the velocity components in the equations of motion. The equation for a_x is written here for convenience; the others are given by Eqs. (4.6.7) and (4.6.8).

$$\frac{Du}{Dt} = \frac{\partial u}{\partial t} + u\frac{\partial u}{\partial x} + v\frac{\partial u}{\partial y} + w\frac{\partial u}{\partial z}$$

If one assumes an incompressible fluid, with the associated consequence that

$$\frac{\partial u}{\partial x} + \frac{\partial v}{\partial y} + \frac{\partial w}{\partial z} = 0$$

Figure 13.1.2 Definition drawing for average and fluctuating velocity portions of a velocity component. For long periods of time T, area under $\bar{u}$ curve (dashed line) is equal to area under u curve (solid line); i.e., $\int_0^T u \, dt = \bar{u}T$.

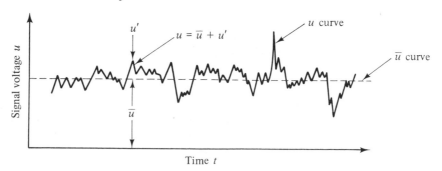

then the x-directed acceleration equation can be written as

$$\frac{Du}{Dt} = \frac{\partial u}{\partial t} + \frac{\partial}{\partial x}(u^2) + \frac{\partial}{\partial y}(vu) + \frac{\partial}{\partial z}(wu) \tag{13.1.1}$$

Thus, it is not surprising that, when $u = U + u'$, $v = V + v'$, and so on, are substituted in the equations of motion, terms such as $\partial(u'v')/\partial y$ occur. These are in addition to the terms associated with the main motion, for example, $\partial(UV)/\partial y$. Reynolds reasoned that if one knew the average value of $\partial(u'v')/\partial y$, denoted by $\overline{\partial(u'v')}/\partial y$, it would be possible to solve for the mean values of the velocity components. He derived equations that were the temporal averages of the fluctuating terms; such terms are identified with an overbar. Such an average can be thought of as the result of integrating the instantaneous value of the derivative $\partial(u'v')/\partial y$ over a long period of time to obtain the area under a curve such as the one shown as an example in Fig. 13.1.2 and then dividing by the time period T to obtain an average ordinate that gives the same area. We write this mathematical process as

$$\text{Time average of } u'v' = \overline{u'v'} = \frac{1}{T}\int_0^T (u'v')\,dt \tag{13.1.2}$$

and

$$\text{Time average of } \frac{\partial(u'v')}{\partial y} = \frac{\overline{\partial(u'v')}}{\partial y} = \frac{1}{T}\int_0^T \frac{\partial(u'v')}{\partial y}\,dt$$

Also,

$$\frac{1}{T}\int_0^T \frac{\partial(u'v')}{\partial y}\,dt = \frac{\partial}{\partial y}\left[\frac{1}{T}\int_0^T (u'v')\,dt\right] = \frac{\partial}{\partial y}\,\overline{(u'v')}$$

This indicates that

$$\frac{\overline{\partial(u'u')}}{\partial y} = \frac{\partial(\overline{u'u'})}{\partial y} \tag{13.1.3}$$

Consequently, the average of the derivative is equal to the derivative of the average.

We have averaged (i.e., integrated) over a long time period. The frequence of the varying signal that is to be averaged will influence what this time period should be. The next example indicates this.

Example 13.1.1

Find the time average of the square of a voltage signal given by $100(\sin \omega t)^2$.

Solution   Let $A = 100 \sin^2 \omega t$. Then

$$\overline{A} = \text{time average of } 100 \sin^2 \omega t$$

$$= \frac{1}{T}\int_0^T 100 \sin^2 \omega t \, dt = \frac{100}{T}\int_0^T \frac{1 - \cos 2\omega t}{2}\,dt$$

$$= \frac{50}{T}\left(t - \frac{1}{2\omega}\sin 2\omega t\right)\Big|_0^T = 50\left(1 - \frac{1}{2\omega T}\sin 2\omega T\right)$$

If $\omega T \gg 1$, $(1/\omega T)\sin 2\omega T \approx 0$, regardless of the value of T. In this limit, $T \gg 1/\omega$, $\overline{A} = 50$.

It must be emphasized that electrical signals from transducers in turbulent flow are not regular like the one used in this example. The function that was used exhibited bounded changes with the parameter t and was easily integrated. In practice, with random oscillations, the integration would be performed numerically or, more usually, elec-

Turbulent Flow   Chap. 13

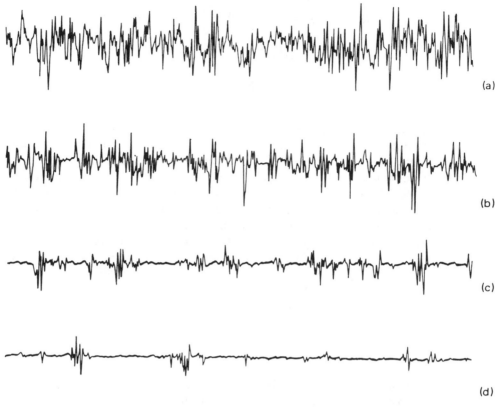

(a)

(b)

(c)

(d)

Figure 13.1.3 Oscillograms in turbulent wake at $x/d = 28$, showing increasing intermittency at edge. $y/d =$ (a) 0.87, (b) 2.25, (c) 3.4, (d) 4.2. (*Note:* These traces were obtained with electronic boosting of high frequencies; this improves the contrast between laminar and turbulent periods by showing a quantity related to the vorticity.) *(From Dr. D. Tritton's "Phy·ical Fluid Dynamics," 1988, Oxford University Press, by permission)*

tronically. Figure 13.1.3 shows the shape of electronic signals from different locations in a turbulent wake. The fraction of the time in which the turbulent fluctuations are greater than a predetermined value is called the *intermittency*. This value is often denoted by γ.

Today, with available solid-state electrical components, the u signal with its fluctuations can be recorded and the time average of the signal up to the instant of the measurement can be determined. On the assumption that it will continue to occur, this average value can be subtracted from the detected u signal. This gives the instantaneous fluctuating value of u, which is u'. The same could be done for the v signal. Moreover, it can be done at two locations that are close together. Then, the product $u'v'$ can be formed for the signal at each location. One of the two products can also be subtracted from the other to form, in effect, the derivative after the separation distance is used as a divisor. All this can be done at almost the speed of light. Next, the instantaneous derivative of the product of fluctuations can be integrated by a suitable electronic circuit. The result is one of the terms arising from the fluctuations that would be applicable to the acceleration term in a Cartesian system of coordinates:

$$\frac{\partial}{\partial x}\,(\overline{u'u'}), \frac{\partial}{\partial y}\,(\overline{v'u'}), \frac{\partial}{\partial z}\,(\overline{w'u'}) \qquad \text{for the } x \text{ direction}$$

$$\frac{\partial}{\partial x}\,(\overline{u'v'}), \frac{\partial}{\partial y}\,(\overline{v'v'}), \frac{\partial}{\partial z}\,(\overline{w'v'}) \qquad \text{for the } y \text{ direction} \qquad (13.1.4)$$

$$\frac{\partial}{\partial x}\,(\overline{u'w'}), \frac{\partial}{\partial y}\,(\overline{v'w'}), \frac{\partial}{\partial z}\,(\overline{w'w'}) \qquad \text{for the } z \text{ direction}$$

Although the complete equations of motion, the Navier-Stokes equations, are not presented until Chap. 16, there is no reason why the interested student cannot read that material next and understand the general way in which the shear stresses enter the expressions in the equation $\mathbf{F} = m\mathbf{a}$. However, it is also apparent from a review of Sec. 8.2 that the y derivative of the shear stress is used in the equations of motion along with the pressure gradient. In an extension to a three-dimensional Cartesian system, one would have x, y, and z derivatives of the shear stresses. It has become customary since the time of Reynolds to include the density with the arguments of the derivatives given above, the derivatives arising from momentum fluctuations, and call the resulting terms—$\rho\overline{u'u'}$, $\rho\overline{u'v'}$, and so on—*Reynolds stresses*. This designation is, strictly speaking, a misnomer, because the origin of the terms stems from momentum and not stress considerations. For a two-dimensional channel such as the one considered in Sec. 8.2, the turbulent equation of motion is

$$\rho\left(U\frac{\partial U}{\partial x} + V\frac{\partial U}{\partial y}\right) = -\frac{\partial p}{\partial x} + \frac{\partial}{\partial y}[\tau - \rho(\overline{v'u'})] \qquad (13.1.5)$$

The left-hand side is zero, as before, for the parallel-wall channel. $\partial(\overline{u'u'})/\partial x$ is assumed to be zero, even though $\overline{u'u'}$ is not. If the Reynolds stress terms were known, one could use an expression such as Newton's law of viscosity for the viscous stress τ and solve for the mean motion. This quest for the mean motion in a turbulent flow is the motivation for combining the momentum fluctuations with the viscous stress terms. The combination can be viewed as a stresslike effect.

There is, however, a snag in the procedure to determine the mean motion in this way. The Reynolds stresses are not known at the outset. They contain the unknown velocity fluctuations (u', v', w'). In the case of the viscous effects, one relates the shear stresses to the mean motion by means of Newton's law of viscosity. But what is the relationship between u' and U, for example? This is a question that has occupied researchers since the time of Reynolds. It is called the *closure* issue because additional interrelationships between the unknown quantities are required in order to have the number of equations equal the number of variables for the problem.

13.2 TURBULENT FLOW IN PIPES

Because the flow in pipes has so many applications, it is not surprising that this case has been investigated extensively. In 1920, Stanton, Marshall, and Bryant from Great Britain published "On the Conditions at the Boundary of a Fluid in Turbulent Motion." They wrote:

> The general conclusions which are arrived at from the whole series of experiments here described are that in turbulent motion there exists at the boundary a layer of fluid of finite thickness, which is in laminar motion, and which has zero velocity at the boundary.

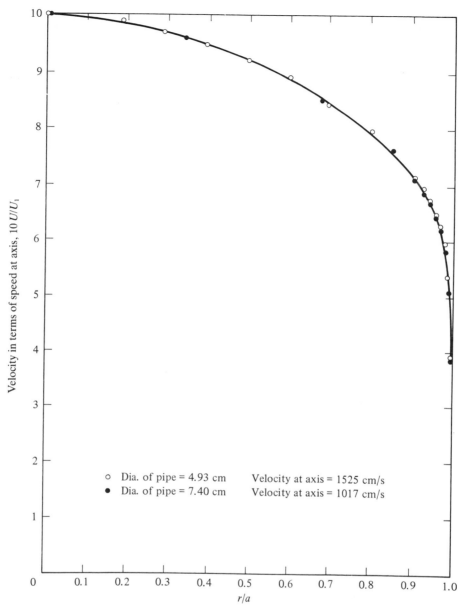

Figure 13.2.1 Velocity distribution in two smooth pipes in which $U_1 D$ is constant *(After Stanton, in "The Mechanical Viscosity of Fluids," Proc. of the Roy. Soc. of London, v. 85 (1911), by permission)*

Today, it is believed that their statement need only be modified slightly. The layer near the wall gives some of the appearances of laminar motion; however, it has other characteristics that belie such a state. (This current view will be explored later.) Stanton, Marshall, and Bryant went on to say, in effect, that the wall shear stress in turbulent flow can be obtained from Newton's law of viscosity by using the derivative of the mean velocity and evaluating it at the wall. (The zero velocity at the wall observed, or inferred, by these researchers indicates that all of the fluctuating velocity components also vanish there.) Some of the velocity profiles measured by Stanton in 1911 are shown in Fig. 13.2.1. One sees the rapid increase in the average velocity as one moves away from the wall, and then a more gradual change over the remainder of the conduit's cross section.

One of the first attempts to account for the increased pressure gradients needed for turbulent pipe flow was the introduction by Boussinesq of a *turbulent exchange coefficient* ϵ in 1877. Then $\overline{u'v'}$ was identified with $-(\epsilon \, dU/dy)$. This model is still used, with ϵ commonly called the *eddy viscosity*, but the proposal has drawbacks. One does not know how ϵ varies with position in a particular problem. Moreover, its form appears to be different for circular and noncircular pipes.

In Germany, around 1910, Blasius correlated the results of many measurements made in the United States and deduced that for smooth pipe and turbulent flow the friction factor could be expressed as

$$f = \frac{0.316}{\text{Re}^{1/4}} \qquad (13.2.1)$$

This correlation was used by Prandtl, who took an active role in turbulence research, a field that profited from his fertile mind. In the 1920s he proposed:

1. Near the wall, the velocity is not determined by the size of the pipe. The velocity depends upon the distance from the wall, y; the density ρ; the *wall shear stress* τ_w; and the viscosity, μ.
2. The velocity distribution is similar regardless of the centerline speed. (In what follows, the centerline speed will be designated U_1.)

This second assumption can be expressed as

$$\frac{U}{U_1} = f\left(\frac{y}{r_o}\right)$$

in which y is the distance from the wall of the conduit of radius r_o and f is a function that must be determined in some manner. He went on to specialize this relationship to

$$\frac{U}{U_1} = \left(\frac{y}{r_o}\right)^q \qquad (13.2.2)$$

in which q must be found from experiments. Prandtl's reason for assuming a *"power law"* might have been suggested by an examination of Stanton's data in Fig. 13.2.1. However, Prandtl concluded that $q = \frac{1}{7}$ by matching his assumed velocity distribution with the correlation of Blasius and required that his assumption 1, above, must be realized by the proper value of q. Such power laws are still used today because of their simplicity and convenience of use. Subsequent work by L. Prandtl, T. von Kármán, and G. I. Taylor led to the establishment of logarithmic laws for the mean velocity. (In the material that follows, only an outline of the research that spanned approximately 50 years will be given. For a more detailed discussion, the reader should refer to more specialized or advanced texts that treat turbulent flow.)

The logarithmic velocity distribution was the outcome of several different models of the flow. Prandtl developed the *mixing-length theory,* which has similarities with Maxwell's kinetic theory of gases with its mean free path between collisions of small particles. His model of the transverse movement of packets of fluid with different mean velocities led to the definition of a mixing length that was determined by the existing Reynolds stress and the mean velocity derivative:

$$l^2 = \frac{\overline{u'v'}}{|dU/dy|^2}$$

$$= \frac{\tau/\rho}{|dU/dy|^2} \tag{13.2.3}$$

A plausible assumption about the way l varies with position was then made. Since the turbulent fluctuations should vanish at the wall, it was assumed that l equals a constant times the distance from the wall. Next, a statement about the shear stress was needed to utilize Eq. (13.2.3). For a laminar channel flow, Eq. (8.2.5) states that the shear stress decreases linearly with the distance from the wall. However, very near the wall in a turbulent flow, Prandtl assumed that the turbulent shear stress is nearly constant. If y now represents the distance from the wall, Eq. (13.2.3) and the previous assumptions yield

$$[(\text{Const})_1(y)]^2 = \frac{(\text{const})_2/\rho}{|dU/dy|^2}$$

This equation is satisfied if U varies logarithmically with y.

It is important to remember that subsequent measurements failed to reveal many of the ingredients implicit in the *mixing-length* model. Thus, in this case, a flawed concept was capable of providing some observable results. (Experience has shown that the clumps of turbulent fluid cannot be considered small in comparison with some defined "mixing length" and that the efffects on such packets of fluid should not be modeled by a discrete collision process. Furthermore, Prandtl excluded pressure fluctuations as agents for affecting the momentum fluctuations.)

Van Kármán was a colleague of Prandtl at Göttingen and subsequently headed an institute in Aachen before coming to the United States and directing aerodynamics research at California Institute of Technology. He proposed another model that was based upon the similarity of the equations for the velocity fluctuations. Starting with the observation of Stanton for the measurements summarized in Fig. 13.2.1 that $U_1 - U = C(r/r_0)^2$ near the center of the pipe, he expressed the velocity at a point in terms of the velocity, and its first and second derivatives, at a reference point in the pipe. He used these expressions when he applied the equations of motion to the region near the center of a pipe. Then length scales were introduced so that the equations of motion for the fluctuating motions did not explicitly depend upon the position of the reference point or the location of the random fluid motion. This means that the turbulence was *homogeneous*. The conclusion that an important length scale was the ratio of the first derivative of the velocity profile to its second derivative followed from his analyses:

$$\left(\begin{matrix} \text{Characteristic length} \\ \text{for turbulence} \end{matrix} \right) = \bar{l} = \frac{|dU/dy|}{|d^2\,U/dy^2|} \tag{13.2.4}$$

However, some of the assumptions inherent in this theory, including the two-dimensionality of the turbulence, have also not borne the test of time.

If one identifies this length scale with Prandtl's definition of the *mixing length l* for Reynolds stresses, one obtains a logarithmic velocity profile. (One introduces a constant to relate $\bar{l}$ and l; $l = \kappa\bar{l}$.)

Taylor did much of his work at Cambridge University and had a productive career that spanned 70 years and many subjects. His model of turbulence had vortices as a key element. These vortices preserve their angular momentum in regions where μ was thought to play a minor role in the dynamics of the flow. This is the case near the center of the pipe. While the acceptance of this hypothesis was limited, Taylor's results were often in concert with the theories of Prandtl and von Kármán and, more important, in some experiments correlated the data in a superior way. Taylor's development of many facets of turbulence theory has been a monumental achievement in itself, but that was only one part of his total contribution to mechanics.

Despite the ultimate inadequacies of these theories, they did provide many of the seminal ideas about turbulence, and they stimulated the experimental research that was to prove some of them insufficient. (The histories of astronomy and atomic physics, to mention but two cases, contain similar tales.) Experiments did show, however, that when the axial velocity in a pipe (or in the boundary layer of a flat plate) was normalized suitably, a logarithmic relationship would correlate the experimental data. The scaling for the velocity is the *friction velocity*, $U_\tau = \sqrt{\tau_w/\rho}$, in which τ_w is the shear stress at the wall of the conduit, or the flat plate. The other variable in the correlation was yU_τ/ν. These variables are, in fact, the ones suggested by Prandtl as those which are important for describing turbulence near a wall.

Information from measurements taken in pipes of city water systems or from laboratory experiments has always guided the significant steps in the process of understanding turbulent flows. Each theoretical step to develop a model of these flows was examined in the light of the available data. Reynolds, himself, compared his tests using lead pipes with those of Darcy, who experimented with the pipes for Dijon's municipal water system. Subsequent tests showed that Prandtl's one-seventh power law failed to give agreement with the observed friction factor if the Reynolds number, VD/ν, was greater than 100,000. As measurements for higher and higher Reynolds numbers were recorded, new power laws were introduced. For flows with Re $\approx$ 400,000, a one-eighth law was used. Prandtl remarked that the development of the logarithmic law should be useful for all Reynolds numbers and that there was no need to continue to build new test facilities capable of reaching ever higher values of Re. But despite the progress that was made in measuring average velocity profiles with Pitot-static tubes—perfected by Darcy around 1850—little was known about the random motions associated with turbulent flows. A crucial step in developing the instrumentation that could portray the velocity fluctuations was the invention of the radio and the subsequent developments in the field of electronics. Amplifiers could be built so that the electrical signals generated by a hot-wire anemometer could be recorded and analyzed.

The work of Laufer, published in 1953, serves well to identify most of the hallmarks of turbulent flows. The most obvious independent variable is y/r_o, just as it was in Prandtl's power laws. Some of Laufer's results are shown in Figs. 13.2.2 and 13.2.3. It should be mentioned that his data not only give us insight into the details of the turbulent flow, but they also are of the correct magnitude and sign to satisfy the theoretical equations of motion. We see that his average velocity profiles in Fig. 13.2.2 have slopes near the wall that give values of the shear stress there, τ_w, that are in concert with the pressure gradients, given in Fig. 13.2.3. [In Sec. 8.3 we observed that in a pipe of radius a, $\tau_w = (\partial p/\partial z)(a/2)$.] Because the slopes are in the ratio of $6.21:1$, one can conclude that the friction velocities, $\sqrt{\tau_w/\rho}$, will be approximately in the ratio of $2.5:1$. We shall refer to this fact shortly.

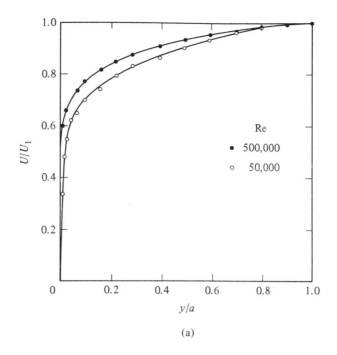

(a)

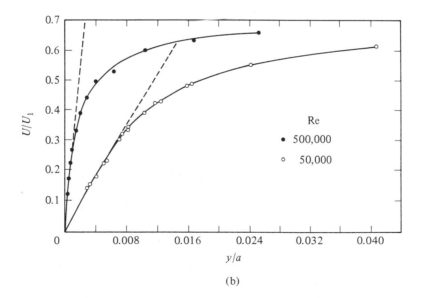

(b)

Figure 13.2.2 (a) Mean-velocity distribution in a circular duct. (b) Mean-velocity distribution near wall of duct. Dashed lines computed from pressure drop measurements. *(From J. Laufer, Natl. Advisory Comm. Aeron. Tech. Note 2954, 1953)*

Sec. 13.2   Turbulent Flow in Pipes

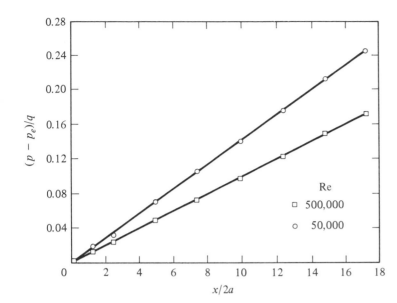

Figure 13.2.3 Mean-pressure distribution along pipe axis nondimensionalized with $q = \rho U_1^2 / 2$. p_e = pipe exit pressure; x = distance upstream from exit. *(From J. Laufer, Nat'l Advisory Comm. Aeron. Tech. Note 2954, 1953)*

Figure 13.2.4a shows the decay of axial turbulent fluctuations near the center of the pipe. The symbol u'' denotes the square root of $\overline{u'u'}$. This term is commonly called the *root-mean-square* (rms), because it is the square root of a "mean-squared" quantity. In terms of Example 13.1.1, this is $\sqrt{50}$ for a signal given by $B = 10 \sin \omega t$, since the mean of the square of this signal is $B^2 = A = 50$. One observes in Fig. 13.2.4a that, as the center of the tube is approached, the values of u'' are the same for the two Reynolds numbers. This suggests that near the center of the pipe, a Reynolds number based on the average velocity is not a significant parameter, because u'' is independent of it. Because Re is the ratio of inertia terms to viscous terms and the inertia in the fluctuations present at the tube's core is significant, one can conclude that the effect of viscosity near the tube's axis is minor. This conclusion is reinforced by the lack of dependence on Re in the curves in Fig. 13.2.5 of the radial and angular velocity components.

However, Fig. 13.2.4b shows that the u'' data are not correlated for the tenfold change in Re as one approaches the wall. A similar conclusion can be made about the poor correlation of the data in Fig. 13.2.5 for small values of y/a.

Because $\overline{u'v'}$ is a significant Reynolds stress, one can see in Fig. 13.2.6 that a major effect of turbulence vanishes as one approaches the tube's axis. [This is satisfying, because in the development of the governing relationships, Eq. (8.3.5), the shear stress vanishes on the axis for axisymmetric flow.] This figure again indicates that, at locations away from the wall, the parameter y/a serves to correlate the data for the two values of Re.

These data suggest that there is a region next to the wall wherein y/a does not serve as a useful parameter for displaying the mean and fluctuating parts of the velocity. But farther from the wall ($y/a > 0.3$, as one estimate), there is good correlation with this parameter. Nevertheless, U/U_1 in Fig. 13.2.2 does not seem to be so favorably correlated. This "blemish" would disappear if the data were correlated by U/U_τ. (One can test the utility of this proposal by using the 2.5 : 1 ratio for U_τ, for the two Reynolds

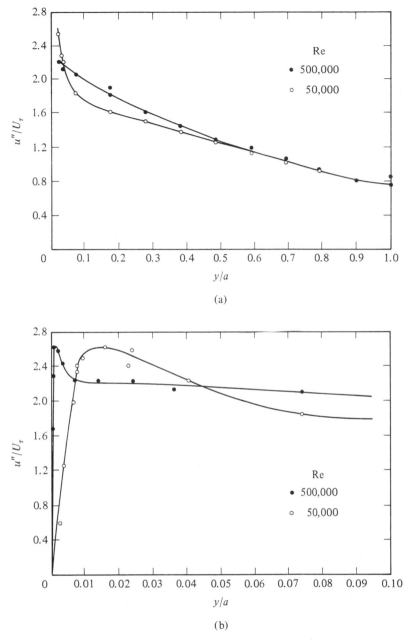

Figure 13.2.4 (a) Distribution of rms value u'' of axial fluctuation u'. (b) u'' distribution near wall. *(From J. Laufer, Natl. Advisory Comm. Aeron. Tech. Note 2954, 1953)*

numbers mentioned previously, and plotting the results.) The region of the pipe where y/a correlates the turbulent flow characteristics is called the *core-flow* region. Nearer the wall, one must use a length scale other than the pipe radius a to correlate the data. This region near the wall is called the *wall layer*. Within this wall layer there appear to

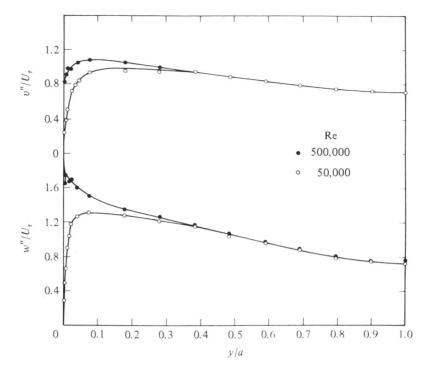

Figure 13.2.5 Distribution of rms values for radial and tangential velocity fluctuations, denoted by v'' and w'', respectively. *(From J. Laufer, Natl. Advisory Comm. Aeron. Tech. Note 2954, 1953)*

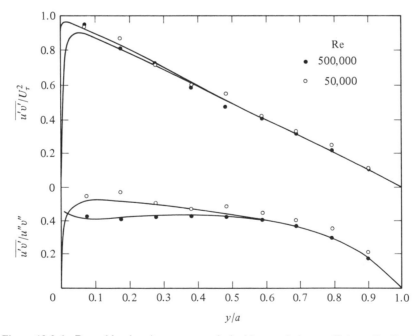

Figure 13.2.6 Reynolds shearing stress and double-correlation-coefficient distributions. Curves are calculated from measured dU/dr, u'', and v''. *(From J. Laufer, Natl. Advisory Comm. Aeron. Tech. Note 2954, 1953)*

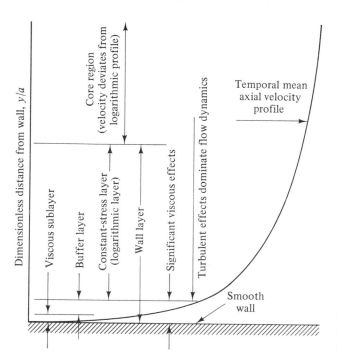

Figure 13.2.7 Designations for portions of the mean-velocity profile in a smooth walled pipe of radius a.

be *three zones*. These are identified in Fig. 13.2.7. Laufer's data, presented in Fig. 13.2.8, show the characteristics of these three regions.

1. Nearest the wall, u''/U_τ increases steadily; v''/U_τ and w''/U_τ behave similarly. The scale length is ν/U_τ, so that the independent parameter is yU_τ/ν, *which is often denoted as y^+*. This length scale is not unexpected in direct proximity to the wall. At the wall itself, $\tau_w = \mu \, dU/dy$, which implies that for an approximately constant slope at the wall,

$$U = \frac{y\tau_w}{\mu} = \frac{y(\tau_w/\rho)}{\nu}$$

The definition of $U_\tau = \sqrt{\tau_w/\rho}$ permits this linear variation of U with y at the wall to be written as

$$\frac{U}{U_\tau} = \frac{yU_\tau}{\nu} = y^+ \tag{13.2.5}$$

[The use of y^+ is implicit in Prandtl's assumptions that were employed to develop his "power" laws, Eq. (13.2.2).] An appreciation for the large values of u'' that can occur is obtained by noting that at $y^+ = 10$, $u''/U_\tau = 2.5$ in Fig. 13.2.8, and that U/U_τ in Fig. 13.2.9 is about 9. Hence, $u''/U \approx 2.5/9 \approx 28$ percent. Thus, the rms value of a fluctuation is about one-fourth of the local mean velocity at a point that is very near to the wall. While viscosity is very important at such locations, the region near the solid boundary is definitely not laminar. For this reason, the region is now called the *viscous sublayer,* in contrast to the previous designation, *laminar sublayer*. Figure 13.2.8 shows that the turbulent shear stresses, represented by $\overline{u'v'}/U_\tau^2$ for one example, have a region of slow growth near the wall where $y^+ \leq 5$. The *viscous sublayer* is usually restricted to $y^+ \leq 5$.

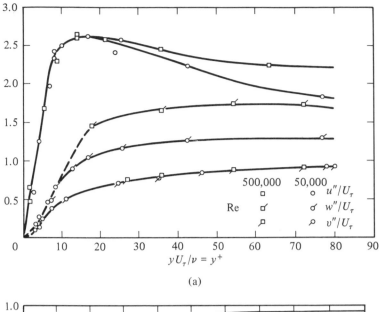

(a)

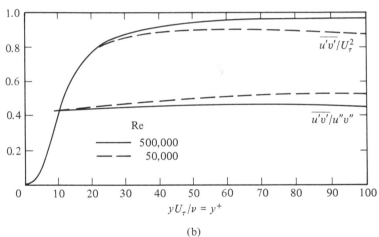

(b)

Figure 13.2.8 (a) u', v', and w' distributions versus y^+, the nondimensional wall coordinate. (b) Turbulent shearing stress and double-correlation coefficient near the wall. *(From J. Laufer, Natl. Advisory Comm. Aeron. Tech. Note 2954, 1953)*

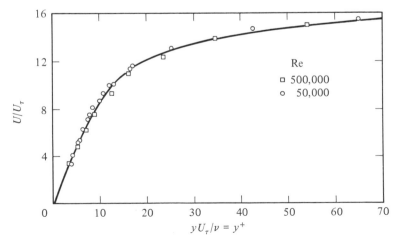

Figure 13.2.9 Velocity distribution in a smooth duct at different Reynolds numbers versus the nondimensional wall coordinate y^+

2. The *buffer zone* is for $5 < y^+ < 30$–50, approximately. The Reynolds stresses increase in an almost linear fashion here. The change of the turbulent stresses with distance is markedly different for y^+ between 30 and 50. In the viscous sublayer range of y^+, $\overline{u'v'}/U_\tau^2 \approx 10$ percent, so that the turbulence stresses are a small fraction of the wall shear stress for $y^+ < 5$. In the buffer zone, the turbulent effects on the flow grow from 10 to 90 percent of the wall shear stress. The conclusion is that in the range of y^+ between 5 and 50, both viscous and turbulent stress are important. One notes that u''/U_τ reaches its maximum in this zone. This prompts some investigators to believe that the term *buffer zone* is misleading. They prefer *production* or *generation zone* for $5 < y^+ < 30$.

3. The *constant-stress* region applies to $y^+ > 50$. (It will also be the region where the mean velocity profile is correlated by a logarithmic function.)

Refer to Fig. 13.2.10 to observe the variation of U/U_τ with $\log y^+$ in these three zones.

Example 13.3.1

For the flow of water in a smooth pipe with an 8-in. diameter at a Reynolds number of $5(10^5)$, estimate the distance y at $y^+ = 15$ (where the largest values of $\overline{u'u'}/U_\tau$ are found).

Solution The Moody diagram, Fig. 9.4.2, provides us with $f = 0.013$ at Re = $5(10^5)$. Since, according to Eq. (8.3.12b), $f = 4\tau_w/(\rho U_{avg}^2/2)$,

$$\frac{\tau_w}{\rho} = 0.125 U_{avg}^2 f \quad \text{or} \quad U_\tau = 0.04 U_{avg}$$

Figure 13.2.10 Mean-velocity profile correlated to show a logarithmic characteristic away from the wall. *(From H. Schlichting's* Boundary-Layer Theory, *7th edition, McGraw-Hill Book Company, by permission)*

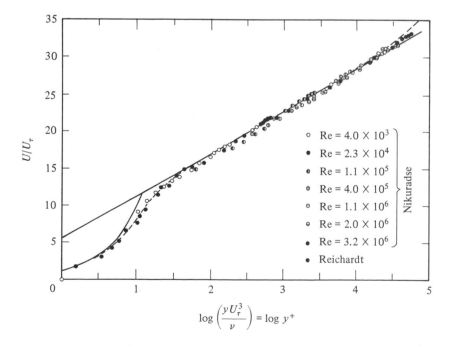

$$U_{avg} = \frac{Re\,\nu}{D} = \frac{5}{8/12} \quad \text{if} \quad \nu \approx 10^{-5} \text{ ft}^2/\text{s}$$

$$= 7.5 \text{ ft/s}$$

$$U_\tau = 0.302 \text{ ft/s}$$

Then
$$15 = y^+ = \frac{yU_\tau}{\nu} = y(3.02)(10^4)$$

$$y = 4.96(10^{-4}) \text{ ft} = 0.006 \text{ in.}$$

or about twice the thickness of a sheet of paper.

After reviewing Laufer's graphic representations, one can be reassured that two length ratios, y/a and y^+, are useful to relate the experimental results. One parameter, a, correlates the data near the axis of the pipe, and the other, U_τ/ν, is pertinent near the wall. A summary of the previous observations would include the proposition that there is a region correlated by y^+ that contains a viscous sublayer, a buffer layer, and a constant-stress layer. We shall now examine these regimes further.

We begin by giving two displays of data for the mean velocity based on the two length scales just mentioned. These are shown in Figs. 13.2.10 and 13.2.11. While these two displays have different ordinates, it appears, just as in the case of Laufer's work, that there is a range of pipe locations in which either abscissa variable would

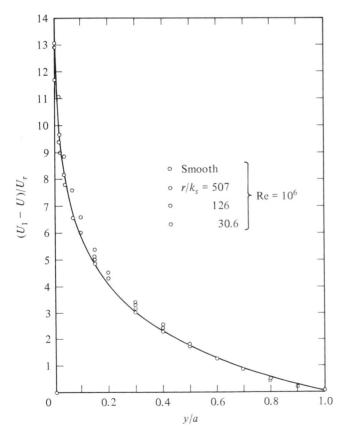

Figure 13.2.11 Mean-velocity profile displayed in the form of a "velocity defect" correlation after J. Nikuradse in N.A.C.A. TM 1292.

correlate the data from various experiments. This region is not directly adjacent to the wall; in fact, the region has Reynolds stresses large in comparison with the viscous stresses. (Nevertheless, viscosity may not be unimportant. It will be the agent for transferring energy from the mean flow to the fluctuations.) In this region, the mean velocity U is thought to depend upon the distance from the wall, y; the pipe radius a; the viscosity and density; and the wall shear stress τ_w. This means that a relationship between these variables can be written in the form of a function ϕ,

$$\phi(U, y, a, \mu, \rho, \tau_w) = 0$$

in accordance with the work in Chap. 9. The choice of ρ, a, and τ_w as scales for mass, length, and time—recall that the frictional velocity $U_\tau = \sqrt{\tau_w/\rho}$—leads to

$$\frac{U}{U_\tau} = \hat{\phi}\left(\frac{U_\tau a}{\nu}, \frac{y}{a}\right) \tag{13.2.6}$$

in which $\hat{\phi}$ is a function of the dimensionless terms $U_\tau a/\nu$ and y/a. We shall designate the former as Re^+, a *turbulence Reynolds number*, and the latter as η. A relationship could equally well be presented as

$$\frac{U}{U_\tau} = \tilde{\phi}\left(\frac{U_\tau y}{\nu}, \eta\right) \tag{13.2.7}$$

by combining the dimensionless independent variables in a different way. This second presentation of U/U_τ, and the experience gained from a review of Fig. 13.2.8, would indicate that as $y \to 0$, $\tilde{\phi}$ depends primarily on $U_\tau y/\nu \, (= y^+)$; whereas near the center of the pipe, where $y \to a$ (or $\eta \to 1$), the function $\tilde{\phi}$ depends primarily on η. The general expression $\hat{\phi}$ or $\tilde{\phi}$ will now be compared with an additional experimental observation that was made by Fritsch, who worked with von Kármán. Fritsch observed that his velocity profiles for different degrees of wall roughness were well correlated by considering the difference between the maximum speed and the local mean speed of flow as well as the frictional velocity U_τ. He found that if U_1 is written as the maximum or centerline speed, all of his velocity profiles, expressed as $(U_1 - U)/U_\tau$, could be superimposed on one another for the same value of η, as long as η was not too small. (It should be mentioned that Stanton, too, saw the utility of $U_1 - U$, but he did not propose the use of U_τ as a normalizing factor.) This means that, remote from the wall, a *"velocity defect"* law has the form

$$\frac{U_1 - U}{U_\tau} = g(\eta) \qquad \text{for } \eta \to 1 \tag{13.2.8}$$

or

$$\frac{U}{U_\tau} = \frac{U_1}{U_\tau} - g(\eta) \qquad \text{for } \eta \to 1 \tag{13.2.9}$$

Another limiting case was indicated by the experimental literature. As mentioned previously, for $\eta \to 0$, Eq. (13.2.7) can be written as $U/U_\tau = \tilde{\phi}(U_\tau y/\nu, 0)$. The function $\tilde{\phi}$ can contain constants, and we shall write it as

$$\frac{U}{U_\tau} = Af\left(\frac{U_\tau y}{\nu}\right) + \text{const}$$

or

$$\frac{U}{U_\tau} = Af(y^+) + \text{const} \tag{13.2.10}$$

In 1938, C. Millikan observed that the experimental existence of the functions g and f in Eq. (13.2.9) and (13.2.10) was a necessary and sufficient reason for the mean velocities in a common region to be correlated by a logarithmic law. One need not assume any theoretical models such as those of Prandtl, Taylor, and von Kármán.

If a general equation is to be applicable over a range of yU_τ/ν and y/a ($= \eta$) that is common to both (i.e., for η around a value η_1 that must be determined), then Eq. (13.2.7) must have the form

$$\frac{U}{U_\tau} = A \ln \frac{yU_\tau}{\eta} + \overline{\phi}(\eta) = \tilde{\phi}\left(\frac{yU_\tau}{\nu}, \eta\right) \tag{13.2.11}$$

so that as $\eta \to 0$, but is not equal to zero,

$$\frac{U}{U_\tau} = A \ln \frac{yU_\tau}{\nu} + \overline{\phi}(\approx 0) \tag{13.2.12}$$

By comparing this result with Eq. (13.2.10), it follows that Millikan had found that $f(y^+)$ was necessarily logarithmic. He determined the constant A and $\overline{\phi}(0)$ from the results of several experimenters and deduced that Eq. (13.2.10) should be written as

$$\left. \begin{array}{l} \dfrac{U}{U_\tau} = 2.5 \ln y^+ + 5.0 \\[3mm] \dfrac{U}{U_\tau} = 5.75 \log_{10} y^+ + 5.0 \end{array} \right\} \tag{13.2.13}$$

or

for very small values of y, but usually $y^+ > 30$. (As $y^+ \to 0$ and viscous effects dominate, one would expect $U/U_\tau = yU_\tau/\nu = y^+$, which is a linear variation with y^+.) Experiments conducted in the years subsequent to Millikan's analysis have not significantly changed the constants. Either form of Eq. (13.2.13) is usually called the *"law of the wall."*

If Eq. (13.2.11) were to be valid on the pipe's centerline, where $U = U_1$, then

$$\frac{U_1}{U_\tau} = A \ln \frac{aU_\tau}{\nu} + \overline{\phi}(1) \tag{13.2.14}$$

A combination of Eqs. (13.2.11) and (13.2.14) gives the "velocity defect" law as

$$\frac{U_1 - U}{U_\tau} = -A \ln \eta - [\overline{\phi}(\eta) - \overline{\phi}(1)]$$

This equation would be written more explicitly in terms of η if the function $\overline{\phi}(\eta)$ could be inferred from the experimental data. For smooth pipe and large η, $\overline{\phi}(\eta)$ is approximated by $\overline{\phi}(1)$. This yields a "velocity defect" law in the form

$$\left. \begin{array}{l} \dfrac{U_1 - U}{U_\tau} = -2.5 \ln \eta \\[3mm] = -5.75 \log_{10} \eta \end{array} \right\} \tag{13.2.15}$$

Millikan's analysis helped him to deduce the value of the constants from the experimental data in a way that was superior to what others before him had done. He concluded that data over the entire range of η could be useful to determine the form of $\overline{\phi}(\eta)$ and the constants that result from it.

In deriving the "velocity defect" law, it was assumed that the logarithmic law, Eq. (13.2.11), was valid at $\eta = 1$. This was the limiting point where this law would be true. An examination of Figs. 13.2.10 and 13.2.11 generally supports this assumption (there is, however, a slight deviation at the highest values of y^+—where $\eta = 1$). At this point, the "law of the wall" predicts values of U/U_τ that are less than the observed values. This is not a serious caveat for Eq. (13.2.13) (Subsequently, we shall examine a deduction from using it at $y = a$.); therefore

$$\frac{U_1}{U_\tau} = 2.5 \ln \frac{aU_\tau}{\nu} + 5.0$$

or

$$\frac{U_1}{\sqrt{\tau_w/\rho}} = 2.5 \ln \left(\frac{D}{2\nu} \sqrt{\frac{\tau_w}{\rho}} \right) + 5$$

Equation (8.3.13) defines τ_w in terms of the friction factor for pipes, f, as

$$\tau_w = \frac{f(\rho U_{avg}^2)}{8}$$

If this relationship is used in the previous expression for U_1, one obtains

$$\sqrt{8} \left(\frac{U_1}{U_{avg} \sqrt{f}} \right) = 2.5 \ln \left(\frac{U_{avg} D}{2\nu} \sqrt{\frac{f}{8}} \right) + 5$$

$$= 2.5 \ln (\text{Re} \sqrt{f}) + (5 - 2.5 \ln 4\sqrt{2})$$

One can speculate that the ratio U_1/U_{avg} is a fixed number. Experimentally, this ratio appears to be close to 1.23. If this value is used, one finds that

$$\frac{1}{\sqrt{f}} = 0.72[\ln (\text{Re} \sqrt{f}) + 0.216] \tag{13.2.16}$$

This result compares favorably with a general expression developed by Colebrook from experiments for pipes over a wide range of Re and ϵ/D. If $\epsilon/D = 0$ is used in Eq. (8.3.13), the Reynolds number from that empirical correlation would be slightly higher than that from Eq. (13.2.16) when comparing them for the same value of f.

The theoretical models to correlate the velocity profiles in smooth pipes have been extended to cases with rough boundaries. Once again, experimental results provided the standards by which the models were judged. J. Nikuradse provided the data shown in Fig. 9.5.1, as well as much of the data used to evaluate the constants for the "law of the wall." Figure 13.2.12 shows some of his velocity profiles. They show that as the Reynolds number increases, the profile becomes "fuller" (i.e., more nearly constant). A similar trend appears as one increases the smoothness of the surface. Stanton's data were principally for smooth pipes. He displayed his results in a manner similar to that used subsequently by Nikuradse and stated that he followed a suggestion by Rayleigh in doing so. Colebrook performed a great number of tests on commercial pipe and employed Nikuradse's variables to present his findings. He found that the pipe he tested displayed slightly different flow characteristics from Nikuradse's in the Reynolds number range where the flow becomes turbulent. He ascribed this to the uniformity of Nikuradse's roughness. The form of Eq. (8.3.13) was selected by Colebrook to correlate his experimental results. L. F. Moody presented Colebrook's contribution in a chart that could be easily used by engineers.

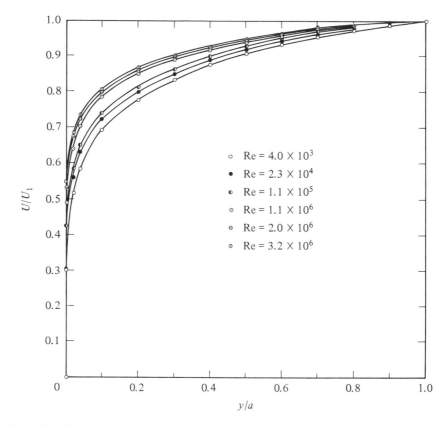

Figure 13.2.12 Velocity distribution in a smooth pipe of radius a for varying Reynolds number *(After Nikuradse in H. Schlichting's* Boundary-Layer Theory, *7th edition, McGraw-Hill Book Company, by permission)*

13.3 FLOW OVER FLAT PLATES

The velocity profiles in fully developed pipe flow have the characteristic that the extent of detectable velocity variation extends from the wall to the pipe's centerline. The boundary layer grows from the beginning of the pipe until it reaches the radius. This occurs over a distance called the *transition length*. For a flat plate, the boundary-layer thickness δ continues to grow along the length of the plate. This suggests that δ should replace the pipe radius a as the characteristic scale length. However, much of the physics that was assumed to be associated with the turbulence near the wall of a pipe is also considered to prevail near the surface of a flat plate. Also, the mechanics that exists far from the plate or pipe wall should be similar. Because the boundary layer on a plate—even a turbulent one—will be developing, τ_w will depend on the location along the plate. When discussing flat-plate boundary layers in Sec. 8.6, the pressure gradient $\partial p / \partial x$ was taken to be zero. However, the proper use of the momentum equation used to obtain the associated results makes it clear that τ_w could depend upon the pressure gradient, as Sec. 8.7 makes clear.

Figure 13.3.1 shows some mean velocity profiles taken along a flat plate. The curve represented by $\gamma = 1$ corresponds to turbulent flow 100 percent of the time. This

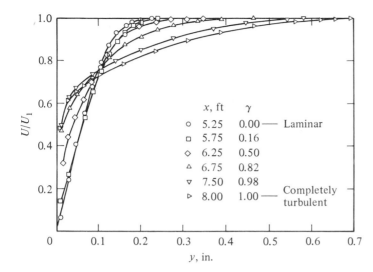

 is represented by the plot with the following legend.

	x, ft	γ	
○	5.25	0.00	Laminar
□	5.75	0.16	
◇	6.25	0.50	
△	6.75	0.82	
▽	7.50	0.98	
▷	8.00	1.00	Completely turbulent

Figure 13.3.1 Mean-velocity profiles through transition region. $U_1 = 80$ ft/s; free-stream turbulence of 0.03 percent. γ = fraction of test duration for significant turbulent intensity in flow. *(From "Contributions on the mechanics of boundary-layer transition," by G. B. Schubauer and P. S. Klebanoff, N.A.C.A. Report 1289.)*

profile shows higher τ_w than the profile for a laminar flow ($\gamma = 0$). The turbulent profile shows U/U_1 approaching unity more gradually than the completely laminar profile. This leads to a thicker boundary layer after a transition to a turbulent boundary layer has occurred.

Any useful theory about turbulent boundary layers should predict local values of τ_w that can be integrated to give the total drag on the plate. Many experiments were performed in flumes in which water flowed at constant depth. In such experiments, the pressure along the bottom of the channel would be constant. (A large part of the data that form the basis for the "law of the wall" was gathered in this way.) Another case in which τ_w could vary while the pressure gradient along the plate was zero occurs when submerged flat plates are towed in long tanks. (W. Froude, whose name is associated with wave phenomena because he advocated the correct model laws for performing such tests, provided much of the early towing data. In his laboratory he measured the drag on a variety of plates that would be similar to those on a ship's hull for a variety of rust and fouling conditions.)

Figure 13.3.2 shows some drag comparisons between the results of theory and experiment for smooth plates. (The theory is semiempirical, since the "law of the wall" has constants obtained from experiments.) The friction velocity in the "law of the wall" contains τ_w, and it is this value that must be integrated when making the comparison in Fig. 13.3.2. Some of Froude's results are assembled with the tests of others on rough plates in Fig. 13.3.3. There is a trend that indicates that the rougher the plate, the higher the drag. This implies a higher total skin friction coefficient from correspondingly higher local values of τ_w. Figure 13.3.4 presents the expected effect on the total skin friction coefficient $C_f = \text{Drag}/(\rho U^2 A/2)$ for a plate of length l and an equivalent sand roughness of k_s. (In the definition of C_f just given, A is the wetted surface area of the plate.) The points in this figure were inferred from Nikuradse's data that were gathered with artificially roughened pipes, Fig. 9.5.1.

A special feature of the turbulence found near the surface of a flat plate is the transition from a laminar boundary layer that occurs immediately downstream from the plate's leading edge. One can infer from Figs. 13.3.2 and 13.3.3 that this occurs for

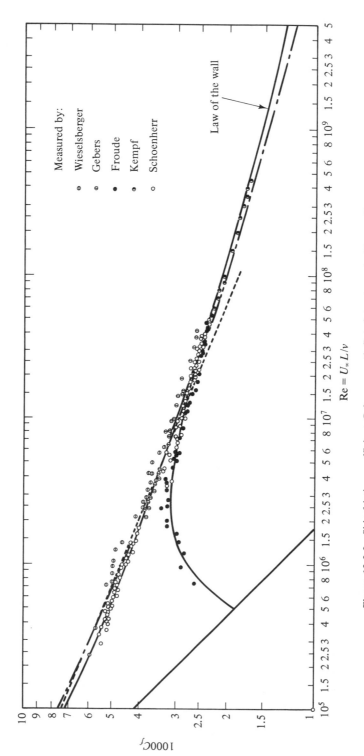

Figure 13.3.2 Skin friction coefficient C_f for a smooth flat plate at zero incidence. Curve at lower left is for a laminary boundary layer. (*From H. Schlichting's Boundary-Layer Theory, 7th edition, McGraw-Hill Book Company, by permission.*)

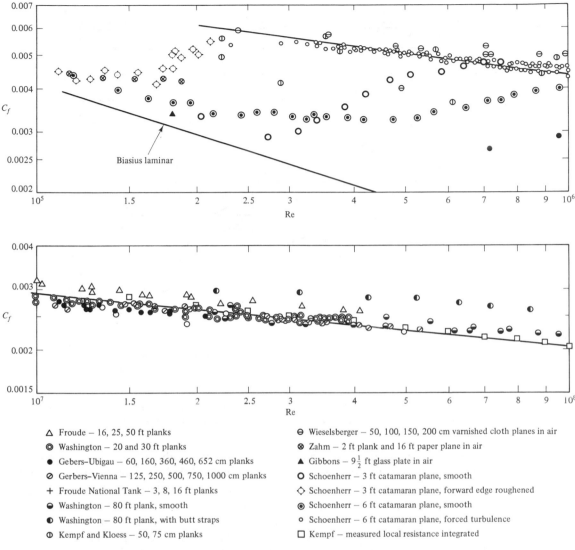

Figure 13.3.3 Flat-plate skin friction coefficient for materials used in ship construction, after Schoenherr. *(Copyright of the Society of Naval Architects and Marine Engineers, reproduced with permission)*

The legend for Figure 13.3.3:

△ Froude – 16, 25, 50 ft planks
◎ Washington – 20 and 30 ft planks
● Gebers–Ubigau – 60, 160, 360, 460, 652 cm planks
⊘ Gerbers–Vienna – 125, 250, 500, 750, 1000 cm planks
+ Froude National Tank – 3, 8, 16 ft planks
⊖ Washington – 80 ft plank, smooth
◑ Washington – 80 ft plank, with butt straps
① Kempf and Kloess – 50, 75 cm planks

⊖ Wieselsberger – 50, 100, 150, 200 cm varnished cloth planes in air
⊗ Zahm – 2 ft plank and 16 ft paper plane in air
▲ Gibbons – $9\frac{1}{2}$ ft glass plate in air
○ Schoenherr – 3 ft catamaran plane, smooth
◇ Schoenherr – 3 ft catamaran plane, forward edge roughened
◉ Schoenherr – 6 ft catamaran plane, smooth
∘ Schoenherr – 6 ft catamaran plane, forced turbulence
□ Kempf – measured local resistance integrated

a smooth plate near a Reynolds number, based on the plate's length, of $5(10^5)$. The influence of plate roughness on the position of transition is implicit in Fig. 13.3.3. For a given value of Re_L below $5(10^5)$, and at prescribed values of fluid speed, viscosity, and plate length, the drag coefficient increases with plate roughness. This could be due to increased τ_w, but it also results from a greater fraction of the plate being exposed to a turbulent boundary layer. The shifting of the value of Re_L at which transition occurs supports this inference.

The effect of surface roughness on the transition to a turbulent boundary layer was also observed by researchers studying the drag on cylinders and spheres. (Refer to the dips in the drag coefficient data in Fig. 9.4.1.) Moreover, they found that the

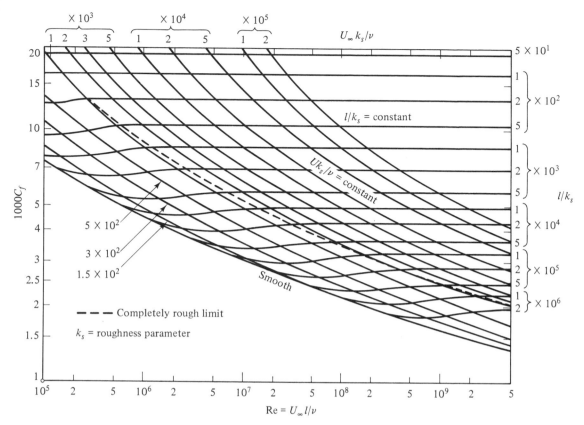

Figure 13.3.4 Estimate of skin friction coefficient C_f for sand-roughened plates based on data for sand-roughened pipes. *(From H. Schlichting's* Boundary-Layer Theory, *7th edition, McGraw-Hill Book Company, by permission)*

turbulence intensity in the air flowing far away from the body influences the drag on a sphere. The results in well-designed wind tunnels were different from those carried out with drop towers in open air. Accordingly, the drag on a standardized sphere was used as a criterion to gauge the turbulence level in early wind tunnels. This means that, in order to conduct tests that would reveal the nature of the interactions between the plate and the fluid in the boundary layer, great attention should be given to the testing facility itself. In the 1930s, a wind tunnel was built at the National Bureau of Standards in Washington, D.C., that had the necessary low free-stream turbulence level. (The velocity signals shown in Fig. 13.1.1 were recorded while that facility was being calibrated.) G. Schubauer and H. Skramstad studied the mechanics of transition of laminar to turbulent flow over a flat plate with this new tool. Their work was motivated, in part, by the results of a theory that had been developed by H. Schlichting, who was also a student of Prandtl. According to this theory, if one had a fluid without any disturbances and it flowed over a plate, one could disturb it at a critical frequency that would cause the resultant fluid motions to grow in an unbounded manner. If this theory were correct, one would expect that the final fluid state would be turbulent. (This would be in accord with the state that is observed.) However, it is conceivable that the transition would be to a second, more complicated laminar flow. This does occur in some cases, as was shown in Chapter 12.

Schubauer and Skramstad let their undisturbed air move past a steel ribbon that was positioned in the boundary layer of a flat plate. They oscillated the ribbon with a magnetic field and caused a disturbance to the flow that was of known frequency. For the conditions of their experiment (i.e., air speed, measurement location, physical constants), they found that velocity waves were induced in the boundary layer by the vibrating ribbon. Sometimes the oscillations were damped, and at other times these motions were amplified. The frequency of ribbon excitation appeared to be the determining factor. The motions became amplified, and subsequently turbulent, at frequencies that were in accord with Schlichting's calculations. This experimental result promised to unlock some of the secrets of turbulence and its transition from the laminar state. Despite considerable progress that has been made on other transitions of laminar flow, the transition to turbulence remains a clouded issue. (Nearly all the *successful* theoretical predictions are those that treat the transition between one laminar flow and a second, more complicated, laminar flow.)

Some of the velocity signals recorded by Schubauer and Skramstad are presented in Figs. 13.3.5 and 13.3.6. In Fig. 13.3.5, one sees that the recorded signals change from (almost) sinusoidal to random as one moves downstream along the plate. Figure 13.3.6 shows that under certain conditions the correlation between the u signals at two locations goes from positive (i.e., the signals are, effectively, in phase) to negative (i.e., out of phase) as one moves down the plate. This observation was predicted by Schlichting's theory.

On the basis of the aforementioned experiments, and numerous subsequent ones such as those shown in Fig. 8.6.3, some of the processes associated with the flow transition along a flat plate can be explained. While a laminar flow has (by definition) zero free-stream turbulence, it could nevertheless contain small, periodic oscillations that have a wide range of frequencies. Some of these vibrations decay, but others are amplified when the fluid flows over the plate. Waves (called *Tollmien-Schlichting waves*) appear in the flow, and these, in turn, grow into nonperiodic, random motions. In this process, one can observe small regions of fluid, nearly at the surface of the plate and called *turbulent spots,* that seem to suddenly explode into turbulent motion. These spread downstream and interact with more and more of the fluid in the boundary layer. Moreover, motions are induced in the boundary layer periodically across the width or span of the plate. Initially, these motions show a regular pattern across the width of the plate. Ultimately, the motions become chaotic.

Recent experiments have identified *hairpin vortices* (for example, swirling fluid motions about a U-shaped axis) that appear to interact with the fluid in contact with a plate. As these vortices approach the plate, they tend to lift out the fluid next to the plate that resides between the arms of the vortex. This process occurs in an explosive way. Some calculations based upon the equations for fluid motions show similar behavior.

Random hairpin vortices can be studied effectively by producing a similar structure in the wake of a hemispherical protrusion that is submerged in the boundary layer of a flat plate. Figure 13.3.7a illustrates the features of a vortex formed in a water channel. Figure 13.3.7b shows one view of a hairpin vortex that was formed during an experiment. The flow has been made visible by means of the hydrogen bubbles that are generated on wires parallel to the plate and perpendicular to the mean velocity in the channel.

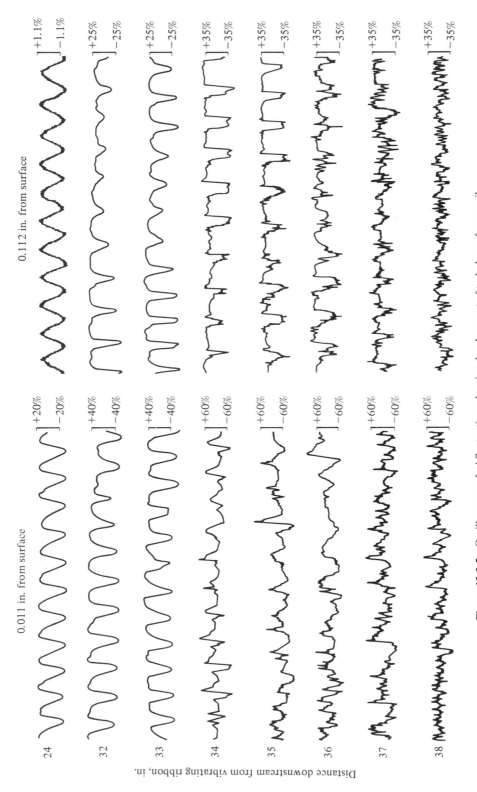

Figure 13.3.5 Oscillograms of u' fluctuations showing development of turbulence from oscillations produced by vibrating ribbon placed 3 ft from leading edge of plate. (*Courtesy of the National Institute of Standards and Technology*)

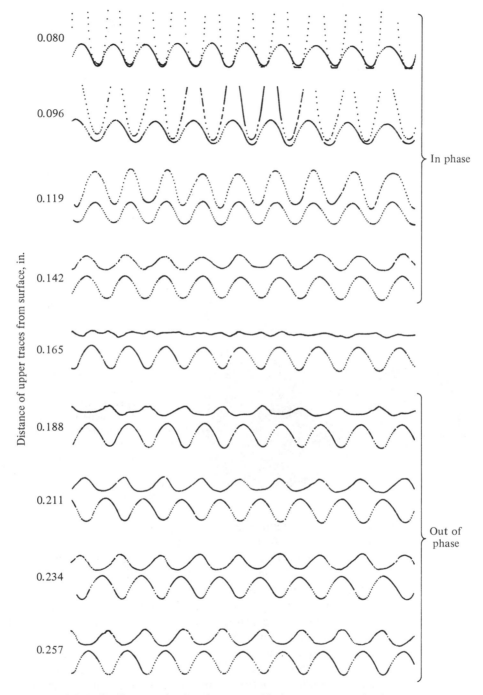

Figure 13.3.6 Oscillograms showing phase reversal in the *u* component of oscillations excited by a vibrating ribbon. *(Courtesy of the National Institute of Standards and Technology)*

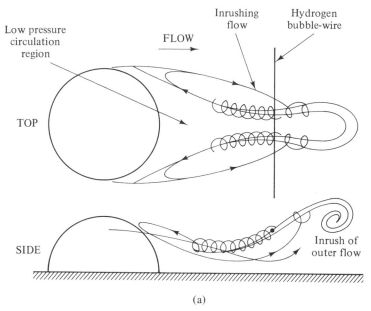

(a)

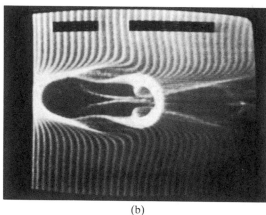

(b)

Figure 13.3.7 (a) Schematic drawing of formation of hairpin vortex on lee side of a hemispherical protrusion on a flat plate. (b) Top view of hydrogen bubble line pattern showing formation of hairpin vortex. (Upstream bubble wire located at $y/R = 0.6$, and downstream wire at $x/R = 6$. x and y directions are aligned with the flow and the normal to the plate, respectively. Flow Reynolds number based on the protrusion's radius $= 750$. Boundary layer is 120 percent of the protrusion's radius R.) (Reproduced from M. S. Ascarlar and C. R. Smith's article in *J. Fluid Mechanics*, v. 175 (1987), by permission)

Some turbulent disturbances have velocity fluctuations that can be correlated over distances that are very much larger than the length scale of the disturbance itself. The correlation between a velocity component at two different points, a *spatial correlation,* can be a significant characteristic of a turbulent flow.

The interface between a turbulent jet and the quiescent ambient fluid is termed a *shear layer,* because of the strong velocity gradients there. The fluid mixing that occurs there sometimes contains *coherent structures* which maintain some identifiable characteristics or properties over a considerable period of time. Figure 13.3.8a shows the changes in the vortices that originate at the beginning of a jet. (See Fig. 1.1.5.) Two vortices can be seen to form a pair. If the jet flow is disturbed periodically at the correct frequency, no pairing occurs. Moreover, pairing can be enhanced by other special disturbance frequencies. This is shown in Fig. 13.3.8b and c.

Despite the success of sophisticated flow visualization techniques that have rendered boundary-layer motions visible, the quest continues to understand turbulent

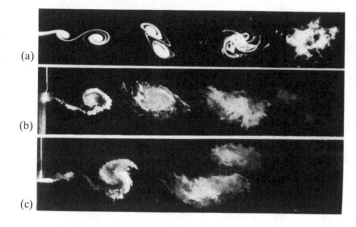

(a)

(b)

(c)

Figure 13.3.8 (a) Laminar separated boundary layer with vortex roll-up, pairing, and ultimate transition to turbulence at Re of about 10^4, with length scale based upon flow section shown in photograph. *(Courtesy of Prof. P. Freymuth, University of Colorado)* (b) Periodic excited turbulent mixing layer without pairing at Re of about 10^5, with length scale based upon flow section shown in the photograph. (c) Turbulent mixing layer with artificially excited pairing condition at Re $\approx 10^5$. *(b) and (c) courtesy of P. Mensing and H. Fiedler, T. U. Berlin)*

boundary-layer flows completely. A large body of experimental data, including the "law of the wall," allow one to make good predictions about simple new cases, but a workable general model for such boundary-layer flows is not currently available. The development of the digital computer, with its ability to manipulate large quantities of experimental data—some of these experiments are computational—has certainly enlarged the professional community's ability to work on the question of turbulence. However, a century after the identification of the problem by Reynolds, one still wishes for better models for the turbulence phenomena.

13.4 SUMMARY

This chapter is in large part descriptive. Turbulent flows have been studied for about 100 years, but a satisfactory composite description of the various phenomena that occur is still lacking. However, a great deal is known about such flows, and this information has been used to assess the various theories that have been proposed. The ideas associated with eddy viscosity, mixing length, and a similarity hypothesis appear useful in certain special cases, but lack generality because they are each at odds with the experimental observations in some way.

Our current capability of performing numerical experiments on large digital computers has greatly expanded our ability to model turbulent flows. New statistical ideas are being explored. While many of the advanced numerical schemes use new theoretical insights into turbulence, many numerical codes for useful engineering problems are currently based upon relatively simple notions, similar to the ideas discussed in this chapter.

Some of the theories presented in this chapter predict a logarithmic variation of the mean velocity with respect to location, something that is observed over a major portion of a pipe's diameter. (The experimental trends of the average axial velocity toward the center and wall of pipes have been shown to dictate a logarithmic variation in the region between.)

The measurements show a wall layer and a core region in which the data can be correlated by means of different length scales. Near the wall, one must use a length scale composed of the friction velocity U_τ and the kinematic viscosity. The resulting dimensionless length ratio is y^+, whereas in the *core region* one uses y divided by the pipe radius with success to correlate the data. Between these two regions is the *buffer layer,* where the Reynolds stresses reach their highest values. They tend to approach a nearly constant value as y^+ becomes greater than 50, so that one terms this portion of the pipe the *constant-stress region.* (Recall that while the same kind of viscous stresses occur in laminar and turbulent flows, in the latter case the Reynold stresses are much larger whenever one has $y^+ > 10$.)

While pipe (and channel) flows have been the subject of many of the studies of fully developed turbulent flows, one is also interested in the development of turbulent flows from an initially laminar flow along a flat plate. An examination of the experimental data shows that a turbulent boundary will be thicker than a laminar boundary layer at the same plate Reynolds number, even though the mean velocity gradient measured at the plate itself is the larger for the turbulent case. (This accounts for the increased local skin friction in turbulent flow.)

The development of the large-scale (turbulent) fluctuations along a flat plate led Tollmien and Schlichting to examine the situation from a theoretical point of view. They succeeded remarkably with a relatively simple theory to predict many of the observations. This theory was an outgrowth of the young field of study called *hydrodynamic stability.* Persons working in this area had been able to successfully predict the occurrence of various secondary laminar flows.

While it is expected that the reader will have assimilated a number of results, concepts, and definitions, it is hoped that this review of the development of our understanding of turbulence will be seen as an example of the general evolution of science. Rational theories are proposed and tested. The weak portions are abandoned and replaced with the results of new insights. Sometimes an existing theory can be patched to make it more suitable or general. Sometimes a completely new approach needs to be adopted in order to make fresh progress toward an understanding of the world around us. This pattern has many examples: our concepts of the atom and our universe come quickly to mind, but one should not forget the controversies that preceded the definitive works of Euler and Newton. And has not Einstein rethought some of the latter's work?

EXERCISES

13.1.1(a) A signal is measured as $110 \sin 120\pi t$ volts. What is its time average? What is the time average of the square of this signal? What is the value of the square root of the second answer?

(b) The electrical signal $(V_0 + A \sin \omega t)$ is measured. What is the average value of the square of this signal? What is the value of the square root of this signal?

(c) An electrical signal can also have a triangular wave form. Suppose such a signal is positive for one-half of the cycle and negative for the remaining half. Draw such a wave and let it have a frequency of ω and a maximum amplitude of 2. What is the time average of the square of this signal? (*Hint:* Can you perform the integration geometrically?)

13.2.1(a) Find y at $y^+ = 15$ for $\mathrm{Re} = 5(10^4)$ in water. (Laufer's experiments for this Reynolds number were in air.)

(b) Where were Laufer's probes at $y^+ = 15$ for $\mathrm{Re} = 5(10^4)$ and $5(10^5)$? He had air flowing in a 9.72-in. I.D. brass tube.

(c) Where does the constant-stress region begin in a smooth 8-in. duct in which water flows at $\mathrm{Re} = 5(10^4)$ and $5(10^5)$?

Turbulent Flow   Chap. 13

PROBLEMS

13.1 Show that, if A and B are two turbulent quantities, $\overline{A + B} = \overline{A} + \overline{B}$. (This would imply that the average of a sum is the sum of the averages.)

13.2 Since $U = \frac{1}{T}\int u\,dt$, a substitution of $u = U + u'$ in the integral would imply that $\frac{1}{T}\int u'\,dt = 0$. Consequently, $\overline{u'} = 0$. Verify this result.

13.3 If the second term of the right-hand side in Eq. (13.1.1) is averaged according to the procedure defined by Eq. (13.1.2), one can show that, except for a factor of z,

$$\frac{\partial \overline{(u^2)}}{\partial x} = \frac{\partial (U^2)}{\partial x} + \frac{\partial \overline{(u'u')}}{\partial x} + \overline{\frac{u'\,\partial U}{\partial x}} + \overline{\frac{U\,\partial u'}{\partial x}}$$

Verify this statement, and also show that the last two terms are zero because of the conclusions made in Prob. 13.2.

13.4 For a voltage given by $0.3 \sin(1000t)$, determine the average value as well as a suitable integration interval T. (The variable t is in seconds.)

13.5 What is the rms value of the voltage in Prob. 13.4?

13.6 For the data in Fig. 13.2.1, estimate the wall shear stress for the curve that summarizes the two cases. (Assume 65°F.) From these data estimate the pressure gradients in the experiments. Check to see whether these gradients agree with Blasius' formula, Eq. (13.2.1). The fluid is air.

13.7 Use the one-seventh power law, Eq. (13.2.2), to estimate the average velocity in terms of U_1.

13.8 Equation (13.2.3) is the result of von Kármán's theory. It can be amplified by assuming a relationship for the length scale, Eq. (13.2.4), from Prandtl's hypothesis.

(a) Show that

$$\frac{\tau}{\rho} = (\kappa)^2 \frac{(U')^4}{(U'')^2}$$

(b) For the flow between parallel plates, τ varies linearly as $\tau_0 y/h$. (The plate separation is $2h$, and y is measured from the cen-

ter.) In such a channel, U'' is negative for $y > 0$. Show that a velocity distribution given by

$$\frac{U_1 - U}{U_\tau} = -\frac{1}{\kappa}\left[\ln\left(1 - \sqrt{\frac{y}{h}}\right) + \sqrt{\frac{y}{h}}\right]$$

satisfies the equation deduced in part (a) by performing the indicated differentiations.

13.9 If $l = Ky$, Prandtl's mixing-length definition, given by Eq. (13.2.4), would yield $U/U_\tau = (1/\kappa)\ln y + C$. Verify this statement. (y is measured from the wall.)

13.10 Use the friction velocity and the results of Prob. 13.6 to plot U/U_τ and yU_τ/ν for Stanton's data in Fig. 13.2.1.

13.11 Use the 2.5:1 ratio of U_τ for Laufer's two Reynolds numbers and plot U/U_τ versus yU_τ/ν for these data. Does Fig. 13.2.9 result?

13.12 Laufer took measurements of the mean velocity very close to the wall of his conduit. What was the proximity of the nearest measurement? Do these measurements lend support to the assumption of the no-slip condition? (I.D. = 9.72 in.)

13.13 Use the criterion $y^+ > 50$ to define the constant-stress region where the logarithmic law should apply. What is this distance, in inches, for Laufer's two different values of Re in a 9.72-in. I.D. tube?

13.14 While Eq. (13.2.13) does not model the flow at $y = 0$, its integral over nearly the entire pipe's cross section would give a good estimate of the volume rate of flow. Use this equation down to $y^+ = 1$, which is somewhat beyond its range of applicability, according to Fig. 13.2.10, to estimate the average velocity in a round duct. (For $y^+ < 10$, the fluid velocities are very small and contribute a negligible amount to the flow rate.)

13.15 If Eq. (13.2.13) were valid near the centerline (and indeed over most of the pipe), that is, if

$$\frac{U}{U_\tau} = 2.5 \ln\frac{yU_\tau}{\nu} + 5.0$$

in which

$$U_\tau = \sqrt{\frac{\tau_w}{\rho}} = \sqrt{\frac{f U_{avg}^2}{8}}$$

then one could integrate the equation and find an expression for U_{avg} in a turbulent flow. What is this expression? (The integration could go from the equivalent of $y^+ = 10$ to $y = a$.) Can you use this expression to find the "constant" value for the ratio U_1/U_{avg}?

13.16 Compare U_1/U_{avg} for the two flows in Fig. 13.2.2a.

13.17 Compare the value of Re for $f = 0.03$ and 0.02 using Colebrook's correlation [Eq. (8.3.13)] and Blasius' much earlier correlation for smooth pipe, $f = 0.316/\mathrm{Re}^{0.25}$.

13.18 Do the data in Figs. 13.2.1 and 13.2.12, from Stanton and Nikuradse, respectively, agree for the comparable Reynolds numbers?

14

Compressible Flows

14.1 INTRODUCTION

The development of the conservation laws for mass, energy, and momentum in Chaps. 3, 6, and 7 was not restricted to incompressible fluids. Although we have concentrated on this kind of medium until now, the general results of those chapters are equally applicable to gases. Such substances have a great number of characteristics in common with liquids, even though a few phenomena such as cavitation—with its associated relationship with the gaseous phase—are of concern solely with the flow of liquids.† We recognize that liquids are compressible to a slight degree, but that gases exhibit this property in a significant way. This trait of a gas is in an equation of state that inter-relates the pressure, density, and temperature of the gas. Any two of these variables prescribe the thermodynamic state of the gas. Although there are many equations of state, each for a specific substance, we shall employ the ideal gas law in this chapter. This law is based upon experimental observations for gases of low density. For this condition, it is observed that the pressure changes in proportion to the density if the temperature is held constant (Boyle's law); and if the pressure remains constant, the density is inversely proportional to the temperature (Charles' law and Gay-Lussac's law). These observations can be combined to form a simple equation of state for a gas:

$$p/\rho = RT \qquad (14.1.1)$$

in which R is a constant for a particular gas and T is the absolute temperature.

The influence on the flow of a gas due to changes of density with pressure and temperature will be studied in this chapter. Such influences will be manifested in the

† The flow of superheated steam in turbines can be described in terms of the material in this chapter.

results that follow from the combined use of the following *steady-state* forms of the basic equations.

Mass conservation:

$$\int \rho \mathbf{V} \cdot \mathbf{n} \, dA = 0 \qquad (14.1.2)$$

in which constant velocity and physical properties are *assumed* to exist at the various segments of the control surface.

Energy conservation: †

$$\frac{\dot{W}}{\rho Q} + \frac{\dot{\mathcal{Q}}}{\rho Q} + \left(\frac{V_1^2}{2} + \frac{p_1}{\rho_1} + gz_1 + e_1 \right) = \left(\frac{V_2^2}{2} + \frac{p_2}{\rho_2} + gz_2 + e_2 \right) \qquad (14.1.3)$$

Momentum conservation:

$$\Sigma F_x = \int \rho u \, \mathbf{V} \cdot \mathbf{n} \, dA \qquad (14.1.4)$$

and, of course, the equation of state, Eq. (14.1.1).

In constrast to our previous work, we will not lump all of the losses into the difference $e_2 - e_1$. Instead, we shall examine the portion of the energy that is transformed into internal energy and hence into temperature. This is possible for gases because of the extensive knowledge that has been established about the thermodynamics of these substances. (In Appendix F, some salient features of thermodynamic processes are reviewed.)

14.2 SOUND WAVES

A simple case of compressible flow is associated with the transmission of sound. For many centuries, it has been recognized that sounds are produced by vibrating objects. A bell becomes mute if one grasps its rim and eliminates the sensible vibrations. Sounds reach our ears over great distances. In the early seventeenth century, Robert Boyle used a vacuum pump and with great care showed that sound was not transmitted through a vacuum. A medium was necessary for the propagation to occur. Philosophers pondered the possible ways that a medium could serve to bring sounds to our ears. Measurements were made that recorded the fact that sound reached the ears quicker if it appeared to travel through water, or a cast iron pipe, rather than air. A theory that accounted for the motion of a gas during the transmission process, which did not require any translation of the medium, was related to the study of ripples on the surface of water. This is the wave theory of sound. The theory postulates that the passage of a surface of increased pressure is followed by the passage of one of decreased pressure, just as the crest of a surface wave is followed by a trough. Sound waves, or acoustic waves, have subsequently been shown to obey the same differential equation as surface waves. It is only natural that one calls it the wave equation.

The pitch and the intensity of a sound do not appear to affect its speed of propagation. Moreover, the speed of the medium does not seem to play a role in the phenomenon either. But the situation is different if one moves toward or away from a

---

† Note that $\dot{W}$ and $\dot{\mathcal{Q}}$ are positive if the energy in the form of heat and work is transferred *to* the system of fluid particles.

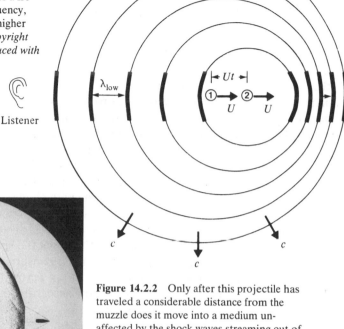

Figure 14.2.1 Effect of moving source on wavelength (and hence frequency) of emitted sound wave. Outermost circular wave front was emitted when source was at position 1, and innermost, at position 2. (Wave fronts are circular because sound propagates in all directions from source at same velocity.) Stationary listener on left hears a frequency lower than the source frequency, and a listener on the right hears a higher frequency than the source. *(© Copyright 1986 Hewlett-Packard Co. Reproduced with permission.)*

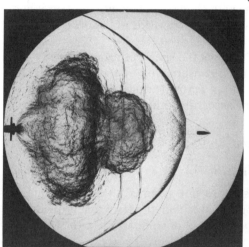

Figure 14.2.2 Only after this projectile has traveled a considerable distance from the muzzle does it move into a medium unaffected by the shock waves streaming out of the rifle barrel. Surfaces of steep pressure gradient were caused by the moving projectile within the barrel. Photograph also shows rapidly discharging smoke from projectile's propellent. *(By G. Klingenberg of the Fraunhofer Institute for High speed Dynamics, Ernst Mach Institute, Freiburg, F.R.G., with permission)*

source of sound with speed U. The speed of the sound is denoted by c, and the wavelength of the acoustical wave is abbreviated by the symbol λ. If the observer is stationary, the successive waves are found to occur with a period $T = \lambda/c$. When the observer moves with speed U toward the source of the sound, the sound waves appear to be approaching at a speed of $U + c$. The waves are still the same distance apart, so that the time period between successive waves is sensed by the ear to be $\lambda/(U + c)$. From this shorter period, the observer has the sensation of a tone with a higher frequency, or pitch. This is the effect that C. J. Doppler discussed in 1842. A moving source of sound produces the same effect. This is illustrated in Fig. 14.2.1. If the emitter is moving at a speed greater than the speed of sound, it leaves the sound wave behind. Then the sound wave reaches the observer after the source has already passed by.

Figure 14.2.2 shows a curved pressure wave behind a projectile that is moving faster than the speed of sound. This wave has a large change of pressure across it. The change of density across the wave affects the speed of light through it, and this produces the image shown in the photograph. Initially, this pressure wave moved faster than the projectile, but as the wave expanded it slowed down and the projectile passed through it. An observer stationed at the right of the photograph would experience the Doppler effect of the sound wave and experience the sharp report due to it after the missile had passed.

14.3 THE SPEED OF A SOUND WAVE

If the source of sound is quite small, it can be considered to be a point source which radiates its energy in spherical waves. Far from the source these waves will have a large radius of curvature; locally, they can be thought of as plane waves. Such a plane wave is shown in Fig. 14.3.1, traveling from the right to the left with a speed c. If this speed is constant, one can adopt a moving frame of reference that translates with the moving wave. Then, just as for the moving hydraulic jump of Chap. 7, a steady-state problem ensues for which all of the fluid velocities are taken relative to the moving frame. These velocities are shown in Fig. 14.3.1 along with changes of velocity, pressure, and density that could occur as the molecules of gas pass from one side of the pressure wave to the other.

We shall determine a relationship that governs the speed of propagation of the pressure wave, c, by using the equations of mass conservation and momentum. The former relationship relates the rate of mass influx across surface AB with the mass efflux across surface CD, since there is no flow across the other surfaces. With the aid of Fig. 14.3.1, this concept can be written as

$$\rho VA = (\rho + d\rho)(V + dV)A \qquad (14.3.1)$$

so that
$$0 = \rho \, dV + V \, d\rho + d\rho \, dV \qquad (14.3.2)$$

The changes of pressure, velocity, and density are not assumed to be small in Eq. (14.3.2). If they are small, as they are for a sound wave, then the term that is quadratic in the changes, $d\rho \, dV$, can be neglected as being small in comparison with the terms

Figure 14.3.1 (a) Planar sound wave moving to left with speed c through a medium at rest. (b) Coordinate frame associated with moving wave. Fluid speeds are relative to this frame.

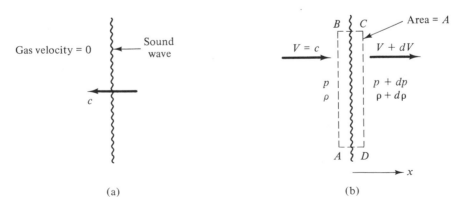

(a)

(b)

Compressible Flows   Chap. 14

$\rho \, dV$ and $V \, d\rho$, which contain the changes in V and ρ to the first power. We *assume* now a small disturbance, so that the product $d\rho \, dV \approx 0$ and

$$\rho \, dV = -V \, d\rho \tag{14.3.3}$$

The momentum equation can be applied to our control volume. The summation of forces is $pA - (p + dp)A$. The final momentum rate, per unit mass, is the final x component of velocity, $V + dV$. The initial momentum rate, per unit mass in the x direction, is simply V. The mass rate of flow is given by either the right- or the left-hand side of Eq. (14.3.1). The latter is the more convenient form to use in this case. Thus, we equate the net force on the fluid in the control volume to the net change of momentum efflux across the control surfaces:

$$pA - (p + dp)A = \rho VA[(V + dV) - V]$$

or

$$-dp = \rho V \, dV \tag{14.3.4}$$

(This momentum equation would be correct even if dp and dV were not infinitesimal, because no terms were neglected in developing it.) This equation contains $\rho \, dV$, as does Eq. (14.3.3). If Eq. (14.3.3) is substituted in Eq. (14.3.4), one has

$$-dp = V(-V \, d\rho)$$

so that

$$V^2 = \frac{dp}{d\rho} = c^2 \tag{14.3.5}$$

since the speed of the gas to the left of the control volume is c, with respect to the moving frame of reference.† Equation (14.3.5) would give the value of c if $dp/d\rho$ were known. How does the pressure change with density for this case of a sound wave? We must inquire into the process by which the gas undergoes thermodynamic changes of state in going across the pressure wave.

If the upstream pressure and density are p_u and ρ_u, then the equation of state is $p_u / \rho_u = RT_u$. Similarly, for a downstream pressure and density of p_d and ρ_d, the downstream temperature would be $RT_d = p_d / \rho_d$. One *plausible assumption* about the passage of the acoustic wave would be that the gas temperature remains constant. This seems reasonable, because it implies that speech travels from the speaker to the listener at the ambient temperature. If this is the case, $T_u = T_d$, so that

$$RT_u = \frac{p_u}{\rho_u} = \frac{p_d}{\rho_d} = RT_d$$

or

$$\frac{p}{\rho} = RT = \text{constant} = B_1 \tag{14.3.6}$$

for an isothermal process. This is *not* an equation of state, but an equation that describes the assumed thermodynamic process by which the gas would change its state while the pressure wave passes. From this relationship for p and ρ with constant T, one can calculate $dp/d\rho$. This is made convenient by first taking the logarithm of both sides of Eq. (14.3.6), so that $\log p - \log \rho = \log B_1$. Both sides of this equation can be differentiated to give $dp/p = d\rho/\rho$, so that

---

† This frame of reference is translating with the sound wave at speed c in a gas that is at rest far from the wave.

$$\left(\frac{dp}{d\rho}\right)_{\text{isothermal}} = \frac{p}{\rho} = RT \qquad (14.3.7)$$

With $dp/d\rho$ determined for an assumed isothermal process, one can now calculate the speed of the sound wave from Eq. (14.3.5):

$$c_{\text{isothermal}} = \sqrt{RT}$$

For air at 68°F (20°C),

$$c_{\text{isothermal}} = \sqrt{(1716)(460 + 68)} = 954 \text{ ft/s}$$

This answer is less than the observed speed of sound by such an amount as to raise doubts about the validity of the steps that led to its derivation. These steps involved Eqs. (14.3.3), (14.3.4), and (14.3.6). The first two results are basic principles of physics, while the last is the result of assuming an isothermal process. This *assumption* may be *incorrect*. If the actual thermodynamic process is not isothermal, what is a reasonable alternative to Eq. (14.3.6)?†

We *assume* now that the process proceeds without a transfer of heat (i.e., adiabatically) to the gas, which undergoes compression, and that this process is reversible. This latter facet of the assumed process can be justified by the relatively small decay of sound energy in air. The fact that one can hear sounds over long distances, when no background noise is present, would indicate that there is only a minor amount of irreversibility (i.e., losses) present in such situations. A general expression from thermodynamics for entropy changes between two states, developed in Appendix F, is

$$s_2 - s_1 = c_v \ln\left[\frac{p_2}{p_1}\left(\frac{\rho_1}{\rho_2}\right)^k\right] \qquad (14.3.8)$$

For a reversible adiabatic process, $s_1 = s_2$. We shall assume this constancy of entropy in our work. A thermodynamic process for which this is true is called an *isentropic process*. Equation (14.3.8) provides a zero difference in entropy if the argument of the logarithm is unity. This would require that $(p_2/p_1)/(\rho_1/\rho_2)^k = 1$, which is equivalent to $p/\rho^k = \text{constant} = B_2$. For such an isentropic process, $dp/d\rho$ can be calculated and used in Eq. (14.3.5) to give the value of c. This derivative can be found by again taking the logarithm of p/ρ^k. This yields $\ln p - k(\ln \rho) = \ln B_2$ with the result that $dp/p = k\, d\rho/\rho$. Solving for $dp/d\rho$ for an isentropic process,

$$\left(\frac{dp}{d\rho}\right)_{\text{isentropic}} = \frac{kp}{\rho} = kRT$$

This relationship, along with Eq. (14.3.5), states that

$$c = \sqrt{kRT} \qquad (14.3.9)$$

which for air at 68°F (20°C) gives

$$c = \sqrt{(1.4)(1716)(528)} = 1126 \text{ ft/s } (343.3 \text{ m/s})$$

This result is in good agreement with observation, a fact that gives us some confidence in using an isentropic process to model other gas flows where no dissipative processes appear to be present.

---

† Newton and others assumed an isothermal process when they derived the sonic speed. It was not until Laplace's time, about 150 years later, that a better hypothesis was advanced.

14.4 AN ENERGY EQUATION FOR GASES

In Appendix F, the internal energy per unit mass of an ideal gas is written as $(c_v T + e_0)$. With this physical quantity expressed in this way, we have a property of the fluid which can be used in the general energy equation, an expression of the first law of thermodynamics, which was discussed in Chap. 6. If we follow the same pattern which was used to derive Eq. (6.3.3), we can verify that

$$\frac{\dot{W}}{\rho Q} + \frac{\dot{2}}{\rho Q} + gz_1 + \frac{V_1^2}{2} + \frac{p_1}{\rho_1} + (c_v T_1 + e_0) = gz_2 + \frac{V_2^2}{2} + \frac{p_2}{\rho_2} + (c_v T_2 + e_0)$$

In this equation e_0 is a reference internal energy that can be cancelled subsequently. Moreover, the subscripts 1 and 2 denote any two positions along the axis of a duct or nozzle in which a gas is flowing. The equation of state is assumed to be $p/\rho = RT$, so that we can write

$$\frac{\dot{W}}{\rho Q} + \frac{\dot{2}}{\rho Q} + gz_1 + \frac{V_1^2}{2} + T_1(R + c_v) = gz_2 + \frac{V_2^2}{2} + T_2(R + c_v) \qquad (14.4.1)$$

This form of the energy equation, which is appropriate for an ideal gas, can be simplified for many cases. When changes in elevations are small, the z terms, which are related to the changes of potential energy, can be neglected, since they are minor with respect to the other terms in the equation. Moreover, only cases for which $\dot{W}$ and $\dot{2}$ are exactly zero will be studied at the present time. The value of $R + c_v$ is c_p, the specific heat at constant pressure, which can be substituted in Eq. (14.4.1) to yield

$$\frac{V_1^2}{2} + c_p T_1 = \frac{V_2^2}{2} + c_p T_2 \qquad (14.4.2)$$

when changes in elevation are neglected. This result is also equivalent to Eq. (6.3.6a) if the relationship $\Delta h = c_p \Delta T$ is included. Finally, the value of c_p from Eq. (F.8) allows Eq. (14.4.2) to be written as

$$\frac{V_1^2}{2} + \left(\frac{k}{k-1}\right) R T_1 = \frac{V_2^2}{2} + \left(\frac{k}{k-1}\right) R T_2 \qquad (14.4.3)$$

Note carefully that the gas constant R in Eq. (14.4.3) has units of ft-lb/slug-°R or J/kg-K. If the gas constant is given in terms of ft-lb/lb$_m$-°R, one must convert it to the appropriate units. The latter constant is here designated $\hat{R}$ instead of R.

Example 14.4.1

Use Eq. (14.4.3) to estimate the temperature of air which is leaving from a large reservoir where $p_0 = 500$ psia and $T_0 = (460 + 70)$°R.† The orifice through which the gas flows has an opening such that the velocity is 583 ft/s there. (For air, $\hat{R} = 53.3$ ft-lb/lb$_m$-°R. Since 1 lb$_m$ = 1/32.2 slug, a factor of 32.2 will be needed.)

Solution   Because the reservoir is "large," V_0 is zero there. This means that Eq. (14.4.3) reduces to

---

† Reservoir conditions will be denoted by the subscript 0. In this example the flow will be considered to be between locations 0 and 2; point 1 is an arbitrary place between points 0 and 2 but does not enter this particular discussion.

$$\left(\frac{1.4}{1.4-1}\right)[32.2(53.3)](460+70) = \left(\frac{583}{2}\right)^2 + \left(\frac{1.4}{1.4-1}\right)[32.2(53.3)]T_2$$

Solve for T_2.

We must now consider the value of T_2. Does the fact that this example implies that $T_2 < T_0$ come as a surprise to you? Does a CO_2 fire extinguisher produce "dry" ice when it is discharged? Of course, the temperature of solid CO_2 is less than normal room temperature. ($T_2 = 501.7°R$ in the example.)

Another question might be, "What is the pressure at the orifice?" In this case, Eq. (14.4.3) is insufficient—even if some state relationship between p and ρ could be introduced for T. For now, we *assume* that the ideal gas equation is applicable, so that Eq. (14.4.3) can be rewritten to read

$$\frac{V_2^2}{2} = \frac{V_0^2}{2} + \frac{k}{k-1}\left(\frac{p_0}{\rho_0} - \frac{p_2}{\rho_2}\right) = \frac{V_0^2}{2} + \frac{k}{k-1}\frac{p_0}{\rho_0}\left(1 - \frac{p_2}{p_0}\frac{\rho_0}{\rho_2}\right) \tag{14.4.4}$$

If we refer to Example 14.4.1, we note that $V_0 = 0$ in the reservoir, and V_2 was given in terms of T_2. Thus, V_2 is in terms of p_2/ρ_2. This means that we have one equation, Eq. (14.4.4), to solve for p_2 and ρ_2 in terms of the ratio p_0/ρ_0. This is awkward, because we have one equation and two unknowns. What is the way out of this difficulty? Is it possible that p_2 and ρ_2 are interrelated in some additional way? If this relationship were known, then one would have the additional result to combine with Eq. (14.4.4) to give the needed two equations for the two unknowns, p_2 and ρ_2. What is the process by which the gas undergoes changes in its pressure and density from the conditions in the reservoir, point 0, to the orifice, point 2? We shall *assume* that the changes occur (1) *reversibly* and (2) *adiabatically,* just as in the case of an acoustic wave. These assumptions allow us to write $p_0/p_2 = (\rho_0/\rho_2)^k$ as the relationship, according to Eq. (F.15), which describes the *process* by which the gas has its thermodynamic variables altered from state (p_0, ρ_0) to (p_2, ρ_2). The consequence of this relationship for the thermodynamic process is that Eq. (14.4.4) can be written as

$$\frac{V_2^2}{2} = \frac{V_0^2}{2} + \frac{k}{k-1}\left(\frac{p_0}{\rho_0}\right)\left[1 - \left(\frac{p_2}{p_0}\right)^{(k-1)/k}\right] \tag{14.4.5}$$

We have just *assumed* a specific thermodynamic process by which the gas properties change from one thermodynamic state to another. This assumption and the reasons for making it may not be immediately apparent. One might ask whether or not an isothermal process could be employed. The obvious, although perhaps incomplete, answer to this alternative is that many everyday experiences argue that the process is not isothermal. (See Example 14.4.1 for the flow out of a reservoir, as well as the determination of the acoustic speed in Sec. 14.2.)

If a gas has a low thermal conductivity, which is actually the case, then there should be a very limited amount of conductive heat transfer between the small volumes of gas which constitute the region of flow. Thus the idea of an adiabatic process is not unrealistic. The fact that the process is reversible is also associated with the types of problems which we shall consider. We shall assume that the flows are in regions whose geometry makes possible minute and continuous stages in the transition from one thermodynamic state to another. We shall assume no viscosity, so that the major dissipative mechanism is absent. As a result, the assumption of an isentropic flow is not unreasonable. It greatly simplifies our calculations and does, in fact, provide answers that are in concert with many observations.

Before leaving the subject of the thermodynamic process by which the gas expands (or compresses), we should mention that if the gas used in Example 14.4.1 is pursued further, an interesting result appears. After it has passed through the orifice in the nozzle that is connected to the reservoir, the gas comes to rest in the space outside the reservoir. Even though the changes of the thermodynamic states of the outflowing gas were considered isentropic, there will be some change in entropy for the gas itself after it has left the orifice and come to rest in the surrounding atmosphere. Consequently, a sudden expansion of the gas beyond the nozzle and its coming to *rest* would not be reversible.

If the gas flow emanates from a reservoir, V_0 can be taken to be negligible compared with the other terms in Eq. (14.4.5). For the present, we shall let p_2 be located just inside the nozzle at its exit. The density can then be obtained from Eq. (F.15), $p/\rho^k = \text{const}$, so that the mass conservation equation, Eq. (14.1.2), would be written as

$$\dot{m} = \rho V A = A_{\text{exit}} \left[\rho_0 \left(\frac{p_2}{p_0} \right)^{1/k} \right] \sqrt{\left(\frac{2k}{k-1} \right) \left(\frac{p_0}{\rho_0} \right) \left[1 - \left(\frac{p_2}{p_0} \right)^{(k-1)/k} \right]} \qquad (14.4.6)$$

From this equation, we can show that $\dot{m} = 0$ for $p_2/p_0 = 1$. Also, $\dot{m}$ vanishes for $p_2/p_0 = 0$. This is because ρ_2 would be zero for such a situation. It seems natural that $\dot{m}$ is nonzero in the interval $0 < p_2/p_0 < 1$. The plot of Eq. (14.4.6) as the dashed line in Fig. 14.4.1 bears this out.

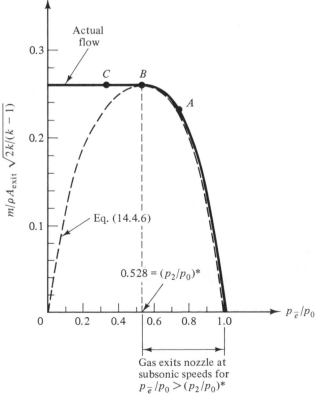

Figure 14.4.1 Simulated experimental data (solid curve) for flow in a converging nozzle, with results predicted from Eq. (14.4.6), for $k = 1.4$.

Example 14.4.2

For a nozzle with an exit area of 0.0625 in.^{2}, what is the mass rate of flow for the conditions given in Example 14.4.1?

Solution We know that $\dot{m} = \rho V A = \rho_2 V_2 A_2$. Only ρ_2 needs to be determined. We further have $T_2 = 501.7°R$ and $p_2/p_0 = (\rho_2/\rho_0)^k$. The equation of state of the gas gives us $p_2/p_0 = \rho_2 R T_2/\rho_0 R T_0$. Then we can simply couple the indicated substitutions to give

$$\frac{p_2}{p_0} = \frac{\rho_2 T_2}{\rho_0 T_0} = \left(\frac{\rho_2}{\rho_0}\right)^k$$

or

$$\frac{\rho_2}{\rho_0} = \left(\frac{T_2}{T_0}\right)^{1/(k-1)} = \left(\frac{501.7}{530}\right)^{1/0.4} = 0.8718$$

The equation of state shows that the reservoir density is 0.07916 slug/ft^3, requiring that $\rho_2 = 0.0690$ slug/ft^3. Then

$$\dot{m} = 0.069(583)\left(\frac{0.0625}{144}\right) = 0.0175 \text{ slug/s}$$

[Show that a direct application of Eq. (14.4.6) with $p_2/p_0 = 0.8252$ yields the same value for $\dot{m}$. Verify that $p_2 = 413$ psia.]

The mass rate of flow out of a converging nozzle, according to Eq. (14.4.6), would appear to follow the dashed curve in Fig. 14.4.1. If one lets the variable $p_2/p_0 = P$ and attempts to find the maximum flow rate $\dot{m}_{max}$ as a function of P, one must solve the equation $d\dot{m}/dP = 0$. This calculation yields

$$\left(\frac{p_2}{p_0}\right)_{\dot{m}\,max} = \left(\frac{2}{k+1}\right)^{k/(k-1)} \tag{14.4.7}$$

The temperature and density ratios can then be found to be

$$\left(\frac{T_2}{T_0}\right)_{\dot{m}\,max} = \frac{2}{k+1} \tag{14.4.8}$$

and

$$\left(\frac{\rho_2}{\rho_0}\right)_{\dot{m}\,max} = \left(\frac{2}{k+1}\right)^{1/(k-1)} \tag{14.4.9}$$

Example 14.4.3

The reservoir in Examples 14.4.1 and 14.4.2 contained air at 500 psia and 70°F. The exit area of the nozzle in Example 14.4.2 was 0.0625 in^2. What should the pressure just inside the nozzle exit, location 2, be for the maximum rate of mass flow? What is the corresponding $\dot{m}$?

Solution Equation (14.4.7) gives

$$p_2 = 500\left(\frac{2}{1.4+1}\right)^{1.4/0.4} = 500(0.528) = 264 \text{ psia}$$

If the receiver pressure (just outside the nozzle) is this pressure, then the gas would flow without further expansion—hence reversibly—into the receiver chamber.

The mass rate of flow is

$$\dot{m} = \rho_2 V_2 A_2 = \rho_0\left(\frac{2}{2.4}\right)^{1/0.4} V_2 A_2$$

V_2 follows from the energy equation, Eq. (14.4.3), with T_2 coming from Eq. (14.4.8). These expressions give

$$V_2^2 = \frac{2kR}{k-1}\left(T_0 - \frac{2}{k+1}T_0\right) = \frac{2kR}{k+1}T_0 = 2\left[\frac{(1.4)(32.2)(53.3)}{2.4}\right]T_0 = 1.4(32.2)(53.3)(442)$$

The result is $V_2 = 1030.5$ ft/s. Equation (14.4.8) shows that $T_2 = 442°R$. From an examination of the value 442 in the last relationship for V^2, one infers that $V_2^2 = kRT_2$. According to Eq. (14.4.8), one can see that V_2 is equal to the speed of sound, Eq. (14.3.9), at the exit of the nozzle for the maximum $\dot{m}$. Now, we can complete the calculations for $\dot{m}$, using

$$\rho_0 = \frac{(500 \times 144)}{(32.2)(53.3)(530)}$$

In *actual* practice, the mass flow rate follows the solid line in Fig. 14.4.1, instead of the dashed curve. This deviation from the results of Eq. (14.4.6) can be explained with the aid of Fig. 14.4.2. As the ratio of the receiver pressure to the reservoir pressure, $p_{\bar{e}}/p_0$, is reduced, the mass rate of flow is affected by the pressure $p_{\bar{e}}$. (This occurs near point A of Fig. 14.4.1.) Changes of pressure downstream of the exit of the nozzle are propagated to a point just upstream of the nozzle's exit, point 2, at the speed of sound, so that the appropriate thermodynamic relationships (e.g., energy equation, isentropic flow) must be satisfied within the nozzle. When the pressure in the receiver, $p_{\bar{e}}$, is reduced to the point that p_2 satisfies Eq. (14.4.7) *and* the maximum flow rate is realized—point B of Fig. 14.4.1—the velocity of the gas is sonic (i.e., equal to the local speed of sound) at the exit of the nozzle (see Example 14.4.3). This shows that the gas is moving *out of* the nozzle at the same rate at which the pressure wave,[†] an acoustic wave for small pressure changes, would move *into* the nozzle. The net effect is that any reduction in pressure in the reservoir below the reduction given by Eq. (14.4.7), $(p_2/p_0)^*$ in Fig. 14.4.1, is not communicated into the nozzle. Hence, nothing changes *within* the nozzle. Then the pressure just inside the nozzle, p_2, no longer equals $p_{\bar{e}}$. The mass rate of flow remains constant. This independence of $\dot{m}$ with respect to $p_{\bar{e}}/p_0$ is called *choking*. The difference between the flow predicted by Eq. (14.4.6) and the choked flow is shown in Fig. 14.4.1. This phenomenon can also occur in steam valves and other passages through which a compressible fluid flows.

For $p_{\bar{e}}/p_0 < (p_2/p_0)^*$, there is a locus of points that form a surface across which there is a pressure change so that the gas just inside the nozzle can adjust to the lower pressure just outside the nozzle. Since these pressure gradients are large, one calls the region where they occur a *shock wave*. Because the gas is moving into a region of rapidly decreasing pressure, one speaks of an "*expansion*" shock wave, in this case. An entropy change occurs across this shock wave. If the exit, or receiver, pressure $p_{\bar{e}}$ is such that $p_{\bar{e}}/p_0 < (p_2/p_0)^*$, the exit pressure does not affect the pressure at point 2 or at any other point within the nozzle shown in Fig. 14.4.2. The lowest receiver pressure for which $p_{\bar{e}} = p_2$ (i.e., in which case the flow is sonic at the exit, point 2) will be designated $p_{\bar{e}}^*$. At this pressure, the sonic flow can continue into the receiver from the converging nozzle isentropically. (It is customary to introduce the asterisk to denote quantities

---

† This pressure wave is analogous to changing the depth of flowing water in a horizontal channel by lowering a vertical gate. A surge wave will travel upstream from the gate in the manner discussed at the beginning of Sec. 7.7.

Sec. 14.4   An Energy Equation for Gases

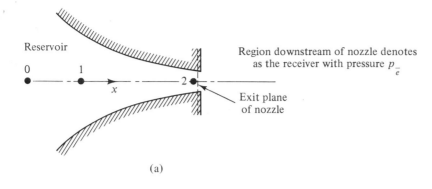

Reservoir

0   1

Region downstream of nozzle denotes
as the receiver with pressure $p_{\bar{e}}$

Exit plane
of nozzle

(a)

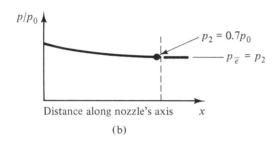

p/p_0

$p_2 = 0.7p_0$

$p_{\bar{e}} = p_2$

Distance along nozzle's axis x

(b)

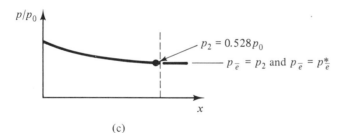

p/p_0

$p_2 = 0.528p_0$

$p_{\bar{e}} = p_2$ and $p_{\bar{e}} = p_{\bar{e}}^*$

x

(c)

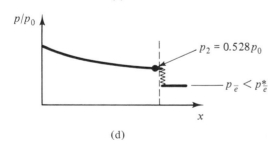

p/p_0

$p_2 = 0.528p_0$

$p_{\bar{e}} < p_{\bar{e}}^*$

x

(d)

Figure 14.4.2 Pressure distributions in a converging nozzle for conditions marked in Fig. 14.4.1: (a) Definition sketch. (b) Receiver pressure at 70 percent of reservoir's (condition A, Fig. 14.4.1). (c) Receiver pressure at critical value, 52.8 percent of reservoir's for $k = 1.4$ (condition B, Fig. 14.4.1). This receiver pressure is denoted $p_{\bar{e}}^*$. (d) Receiver pressure $p_{\bar{e}} < p_{\bar{e}}^*$ (condition C, Fig. 14.4.1).

that are associated with a gas flow at the speed of sound at the location of the minimum area in a nozzle.)

It is now reasonable to ask what would occur once the receiver pressure was reduced below $p_{\bar{e}}^*$—after which further reductions of pressure could not affect the flow—and then raised to a value that was above $p_{\bar{e}}^*$. Would the fact that the flow was sonic at the exit mean that receiver pressures that are subsequently in excess of $p_{\bar{e}}^*$ could not be communicated to the nozzle and affect the flow rate? One would think that this should not be the case and that the mass rate of flow should decrease as $p_{\bar{e}}$ is increased in the range $p_e^* \leq p_{\bar{e}} \leq p_0$. One can provide a reasonable argument that this would be the case. However, the steps in the argument rely on the material in Sec. 14.6, where the relationship between nozzle shape and Mach number is discussed.

14.5 THE EFFECT OF MACH NUMBER ON A FLOW

We now continue the case of the flow issuing out of a reservoir, with point 0 in the reservoir and point 1 at some arbitrary section in a converging nozzle, so that $V_0 = 0$ and from Eq. (14.4.4),

$$\frac{V_1^2}{2} = \left(\frac{k}{k-1}\right)\frac{p_1}{\rho_1}\left(\frac{p_0}{p_1}\frac{\rho_1}{\rho_0} - 1\right) = \left(\frac{k}{k-1}\right)\frac{p_1}{\rho_1}\left[\left(\frac{p_0}{p_1}\right)^{(k-1)/k} - 1\right] \quad (14.5.1)$$

which is equivalent to Eq. (14.4.5). It is instructive to solve for the ratio of p_0/p_1 in this equation:

$$\frac{p_0}{p_1} = \left[1 + \left(\frac{V_1^2}{2}\frac{k-1}{kp_1/\rho_1}\right)\right]^{k/(k-1)}$$

If the equation of state is used, one can substitute for p_1/ρ_1 so that

$$\frac{p_0}{p_1} = \left(1 + \frac{V_1^2}{2}\frac{k-1}{kRT_1}\right)^{k/(k-1)} \quad (14.5.2)$$

This equation would give us p_0/p_1, and hence p_1, if V_1^2/kRT_1 were known. Indeed, this term is the square of the *Mach number* which is denoted by Ma. (Recall that Ma = V/c.) Consequently, the Mach number at the exit plane of the nozzle would specify the pressure there, if point 1 were associated with the exit plane. Some additional insight into the effect of the Mach number on the pressure difference $p_0 - p_1$ can be had by expanding the right-hand side of Eq. (14.5.2) by means of the binomial theorem.† The result is

$$\frac{p_0}{p_1} = 1 + \left(\frac{k}{k-1}\right)\left(\frac{V_1^2}{2}\frac{k-1}{kRT_1}\right) + \frac{1}{2!}\left(\frac{k}{k-1}\right)\left(\frac{1}{k-1}\right)\left(\frac{V_1^2}{2}\frac{k-1}{kRT_1}\right)^2 + \cdots$$

so that

$$p_0 = p_1 + \frac{V_1^2}{2}\frac{p_1}{RT_1} + \frac{1}{2}\frac{V_1^2}{2}\frac{p_1}{RT_1}\left(\frac{V_1^2}{2}\frac{1}{kRT_1}\right) + \cdots \quad (14.5.3)$$

The factor $(V_1^2/2)(p_1/RT_1)$ is common to all terms in the series for $(p_0 - p_1)$ and can be written as $V_1^2\rho_1/2$ because of the ideal gas law. This permits Eq. (14.5.3) to be rewritten as

$$\frac{p_0 - p_1}{\rho_1 V_1^2/2} = 1 + \left(\frac{\text{Ma}_1^2}{4} + \frac{2-k}{24}\text{Ma}_1^4 + \cdots\right) \quad (14.5.4)$$

The terms without parentheses would be the result if the Bernoulli equation for an incompressible fluid were used between the reservoir and the location 1. Thus, the terms in parentheses give the contribution to $p_0 - p_1$ that is due to the compressibility effects associated with the flow.

---

† $(a + x)^n = a^n + na^{n-1}x + \frac{n(n-1)}{2!}a^{n-2}x^2 + \cdots$ for $x^2 < a^2$.

We see that for small values of Ma_1, say, 0.2, the correction to the pressure difference would be minor. For $Ma_1 > 0.5$, the actual pressure differences between the reservoir and the location 1 must include a great number of the terms in Eq. (14.5.4) that are contained in the parentheses. In such a case, one should use Eq. (14.5.1) at the outset. This would be a prudent step in *any* compressible-flow problem. Equation (14.5.4) was developed solely to display clearly the effects of compressibility. The series representation does not lend itself to convenient computation.

14.6 NOZZLE GEOMETRY AND MACH NUMBER VARIATION

The equation of mass conservation, augmented by Newton's laws of motion in the direction of the nozzle's axis, can be rearranged to predict the change of the Mach number that accompanies the change of area in converging and diverging nozzles. The mass rate of flow at *any* section in a nozzle is $\dot{m} = \rho V A$. For a given flow, $\dot{m}$ is constant along the nozzle even though ρ, V, and A change. Hence,

$$d(\dot{m}) = 0 = d(\rho V A)$$

This can be expanded to read $0 = d\rho \ (VA) + dV \ (\rho A) + dA \ (\rho V)$; after dividing through by ρVA, one has

$$0 = \frac{d\rho}{\rho} + \frac{dV}{V} + \frac{dA}{A} \tag{14.6.1}$$

Newton's laws and the associated Euler equation given by Eq. (5.6.5) apply to our situation. This means that

$$\rho V \frac{dV}{dx} = -\frac{dp}{dx} \tag{14.6.2}$$

By solving for ρ in this equation and substituting the result in Eq. (14.6.1), one finds

$$0 = -\frac{d\rho}{dp} V \, dV + \frac{dV}{V} + \frac{dA}{A}$$

so that

$$\frac{dV}{dA}\left(\frac{1}{dp/d\rho} - \frac{1}{V^2}\right) = \frac{1}{VA}$$

This result can be rearranged by factoring $1/V^2$ to give

$$\frac{dV}{dA} = \frac{1/VA}{\frac{1}{V^2}\left(\frac{V^2}{dp/d\rho} - 1\right)}$$

If the flow process occurs isentropically, $dp/d\rho = kRT$, so that

$$\frac{dV}{dA} = \frac{V/A}{\dfrac{V^2}{kRT} - 1} = \frac{V/A}{Ma^2 - 1} \tag{14.6.3}$$

Equation (14.6.3) will always have a positive numerator on the right-hand side. The value of Ma in the denominator will determine the sign of the complete right-hand side. For a contracting section, dA is negative (i.e., $dA/dx < 0$). The value of V

increases from zero as the fluid moves from the reservoir. Consequently, $dV > 0$ and $dV/dA < 0$ in a converging nozzle attached to a reservoir. It follows that Eq. (14.6.3) requires that Ma < 1 in such a converging section. In a diverging section, $dA > 0$. If the speed decreases in this section, as it does in the expanding section of a Venturi tube, dV (i.e., dV/dx) will be negative. Equation (14.6.3) indicates Ma < 1 for this case also; if Ma > 1 in a diverging section, the sign of dV must be positive. This indicates that V is increasing when Ma > 1 in a diverging section. The only remaining question posed by Eq. (14.6.3) is the value of Ma at the throat of a converging-diverging nozzle where $dA = 0$. To answer this question, we write Eq. (14.6.3) as

$$\frac{V}{A}\frac{dV}{dx} = \frac{dA/dx}{\text{Ma}^2 - 1}$$

The right-hand side is 0/0 at the throat with Ma = 1. This indeterminate form can be resolved by the application of L'Hôpital's rule. Then

$$\frac{V}{A}\frac{dV}{dx} = \frac{d^2A/dx^2}{2\text{Ma}\left[d(\text{Ma})/dx\right]} \tag{14.6.4}$$

The numerator of the right-hand side is not zero; its value is determined by the curvature of the throat section. Thus, dV/dx will not be zero, and it will depend upon whether the Mach number is increasing or decreasing in the diverging portion of the nozzle.

We now return to answer the question raised at the end of Sec. 14.4. There we encountered the choking phenomenon in a converging nozzle. As the pressure in the receiver is reduced to the value needed to obtain sonic flow (Ma = 1) at the exit, the exiting gas stream, a jet, should have $dA/dx = 0$ there if the stream acts in accordance with Eq. (14.6.3). The flow in the nozzle is not affected with Ma = 1 at the exit. The flow rate would continue at a maximum (see Fig. 14.4.1) as the receiver pressure is reduced further. At the exit, the jet would be at a higher pressure than the receiver pressure $p_{\bar{e}}$, and the jet would expand. However, dA/dx could still be zero at the nozzle's exit.

If the pressure is now slowly increased above that needed for sonic exit conditions, what will occur? It can be argued that the higher pressure in the receiver causes the exiting gas jet to contract. This implies that $dA/dx < 0$. In turn, from Eq. (14.6.3), since dA/dx is not zero, Ma^2 is not 1. In fact, it must be less than unity. Then the pressure disturbances could enter the nozzle and alter the conditions there. The result would be a reduction in the mass rate of flow. (This explanation of the effect of the receiver pressure on the discharging jet's geometry also supports the occurrence of choking. Upon leaving the nozzle, the jet could expand if the pressure in the receiver were below that needed to obtain Ma = 1 at the exit. This expansion would be consistent with the needed value for dA/dx for the jet.)

14.7 SUPERSONIC FLOW

If in a converging nozzle the maximum gas velocity is sonic, how can supersonic speeds be achieved? A converging-diverging (de Laval) nozzle is the answer. Such a nozzle is shown in Fig. 14.7.1. Point 0 is in the reservoir, and point e is just upstream of the

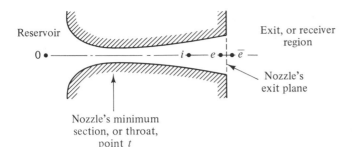

Reservoir

0 •

Nozzle's minimum
section, or throat,
point t

Exit, or receiver
region

Nozzle's
exit plane

Figure 14.7.1 Schematic diagram of
converging-diverging nozzle.

nozzle's exit. The subscript i will denote *any* point within the nozzle. Once again, we shall employ the energy equation,

$$\frac{V_0^2}{2} + \frac{p_0}{\rho_0}\frac{k}{k-1} = \frac{V_i^2}{2} + \frac{p_i}{\rho_i}\frac{k}{k-1} \tag{14.7.1}$$

the mass conservation equation,

$$\dot{m} = \rho_i V_i A_i \tag{14.7.2}$$

an equation of state,

$$\frac{p}{\rho} = RT \tag{14.7.3}$$

and a "process" equation, isentropic flow, in this case,

$$\frac{p_0}{p_i} = \left(\frac{\rho_0}{\rho_i}\right)^k \tag{14.7.4}$$

The momentum equation will be employed later in Sec. 14.8 where a nonisentropic condition is discussed and Eq. (14.7.4) cannot be used.

Because point 0 is within the reservoir, $V_0 \approx 0$. This means that a combination of Eqs. (14.7.1) through (14.7.4) will yield a result similar to Eq. (14.4.6) for the mass rate of flow at the nozzle exit,

$$\dot{m} = A_e \rho_0 \left(\frac{p_e}{p_0}\right)^{1/k} \sqrt{\frac{2k}{k-1}\frac{p_0}{\rho_0}\left[1 - \left(\frac{p_e}{p_0}\right)^{(k-1)/k}\right]} \tag{14.7.5}$$

Thus, for given nozzle design, reservoir conditions, and exit pressure, one can solve for $\dot{m}$ in Eq. (14.7.5). If, for the given A_e and p_e, one specifies A_i at various places along the axis, the associated values of p_i can be calculated, since $\dot{m}$ is now known. To find p_i, Eq. (14.7.5) is written more generally for the ith station as

$$\dot{m} = A_i \rho_0 \left(\frac{p_i}{p_0}\right)^{1/k} \sqrt{\frac{2k}{k-1}\frac{p_0}{\rho_0}\left[1 - \left(\frac{p_i}{p_0}\right)^{(k-1)/k}\right]} \tag{14.7.6}$$

Now p_i can be found at any location within the nozzle for which the flow has been isentropic. Consequently, a complete distribution of pressure through the nozzle can be given once the exit pressure p_e in Eq. (14.7.5) is specified. Figure 14.7.2 shows the variation in the half width of a planar nozzle.

Example 14.7.1

Air at 20°C and 101 kPa from the atmosphere is drawn through a converging-diverging nozzle. The reduced pressure at the nozzle's exit is 91.4 kPa. The flow is isentropic

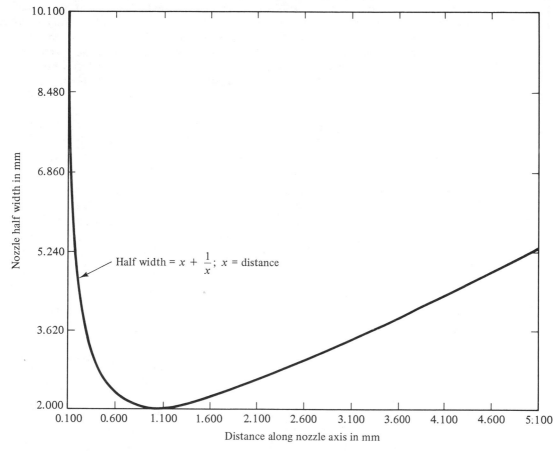

Figure 14.7.2 Shape of planar converging-diverging nozzle used to calculate flow variables in Figs. 14.7.3–14.7.6 given by half width $= 1/x + x$ with x = axial distance.

throughout the nozzle. Determine the mass rate of flow if the exit area is 0.033 m². What is the value of the pressure at the location where the area is 0.022 m²?

Solution   The pressure ratio is $p_e/p_0 = 0.905$ for the given conditions, and $\rho_0 = 101(1000)/286.6(293)$ from the equation of state. Substituting these values in Eq. (14.7.5) yields

$$\dot{m} = 0.033(1.203)(0.905)^{0.714}V_e$$

in which

$$V_e = \sqrt{\frac{2.8}{0.4}(286.6)(293)[1-(0.905)^{0.286}]}$$

and so

$$\dot{m} = 4.76 \text{ kg/s}$$

At the section where the area is 0.022 m², the mass rate of flow is also 4.76 kg/s. Equation (14.7.6) becomes

$$4.76 = 0.022(1.203)\left(\frac{p_i}{p_0}\right)^{0.714}V_i$$

with

$$V_i = \sqrt{\frac{2.8}{0.4}(286.6)(293)\left[1-\left(\frac{p_i}{p_0}\right)^{0.286}\right]}$$

and must now be solved for p_i/p_0. An iterative solution method is needed. An initial estimate follows after it is noted that the flow is subsonic at the nozzle's exit. Hence, the flow is behaving somewhat like that of an incompressible fluid, according to Sec. 14.6. This indicates that the pressure at the smaller section is lower than at the larger. An initial choice of $p_i/p_0 = 0.75$ is made. The value of the right-hand side of Eq. (14.7.6) is 4.64 kg/s for the pressure ratio of 0.729. Thus, the pressure at the smaller section is about 73.6 kPa [= 0.729(101)].

The upper curve in Fig. 14.7.3 is another example of the use of Eq. (14.7.6). Once p_i is known, V_i, ρ_i, and T_i can be found. These variables are plotted in Figs. 14.7.4 to 14.7.6. The upper curve in Fig. 14.7.3 is labeled "subsonic" because the gas speed is, in fact, subsonic in the converging and diverging sections. The *particular* value of p_e that was selected—along with the prescribed nozzle geometry of Fig. 14.7.2—gives a speed at the minimum section of the nozzle—the throat—that is sonic. This situation, Ma = 1, occurs only at the throat, in agreement with the development of Sec. 14.6.

Figure 14.7.3, which results from having the proper exit pressure to obtain sonic speed at the throat, has a second branch, labeled "supersonic." This path would not exist if sonic conditions had not been reached at the throat. A gas with Ma = 1 at the

Figure 14.7.3 Pressure variation within a converging-diverging nozzle with sonic flow at the throat and isentropic flow throughout the nozzle.

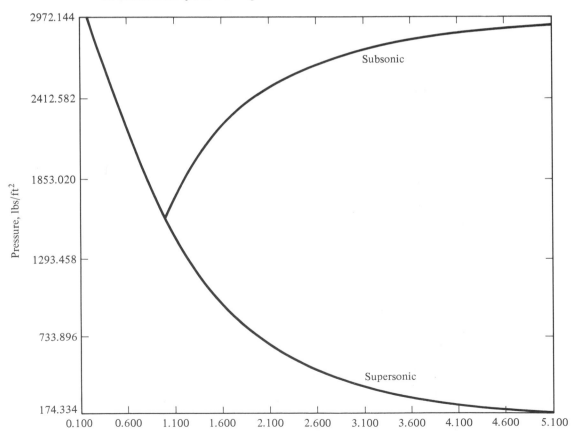

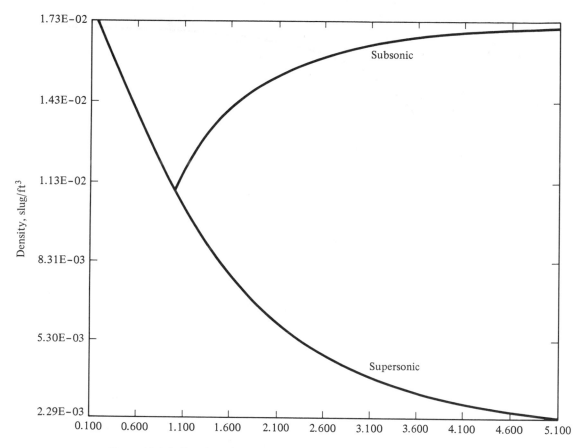

Figure 14.7.4 Density variation within a converging-diverging nozzle with sonic flow at the throat and isentropic flow throughout the nozzle.

throat can also have $dV > 0$ there. The flow becomes supersonic in the diverging section in accordance with Eq. (14.6.6), which related the Mach number to the nozzle's geometry. But in order for the flow to continue isentropically throughout the nozzle, the exit pressure must be uniquely determined. This value would come from Eq. (14.7.5), by solving for p_e with the value of $\dot{m}$ such that there is sonic flow at the throat.

The mass rate of flow in the nozzle for sonic flow at the throat can be derived easily. Equations (14.7.1) through (14.7.4) can be used as follows to give, for Ma = 1 at the throat, the values of the velocity, $V_t = V^*$; pressure, $p_t = p^*$; temperature, $T_t = T^*$; and density, $\rho_t = \rho^*$ (the asterisk signals that sonic conditions are being given):

$$\text{Ma}^* = \frac{V^*}{\sqrt{kRT^*}} = 1 \tag{14.7.7}$$

and using this velocity in the right-hand side of Eq. (14.7.1), one has

$$T^* = T_0 \left(\frac{2}{k+1}\right) \tag{14.7.8}$$

From Eq. (14.7.8) and the equation for an isentropic process, one obtains

Sec. 14.7   Supersonic Flow

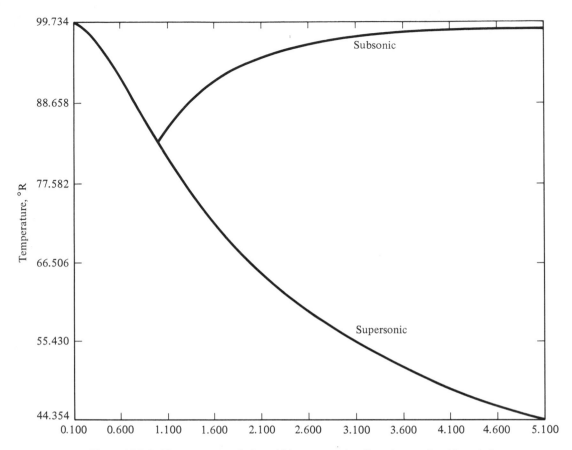

Figure 14.7.5 Temperature variation within a converging-diverging nozzle with sonic flow at the throat and isentropic flow throughout the nozzle.

$$p^* = p_0 \left(\frac{2}{k+1}\right)^{k/(k-1)} \tag{14.7.9}$$

and

$$\rho^* = \rho_0 \left(\frac{2}{k+1}\right)^{1/(k-1)} \tag{14.7.10}$$

Then one can write $\dot{m} = A\rho V$ as

$$\dot{m}^* = A_t \left[\left(\frac{2}{k+1}\right)^{1/(k-1)} \rho_0\right]\left[kR\left(\frac{2}{k+1}\right)T_0\right]^{1/2} \tag{14.7.11}$$

When Ma = 1 at the throat, Eq. (14.7.5) gives two real values for p_e, the pressure just upstream of the nozzle's exit. One of these we shall call p_e'. This value yields sonic flow at the throat and subsonic flow throughout the diverging section. The other root, p_e''', gives the exit pressure for isentropic and supersonic flow throughout the diverging section.† When $p_{\bar{e}}$, the pressure in the receiver, is less than the value of p_e' but greater

---

† Note that the pressure has been assigned special values, p_e' and p_e'''. When the subscript on p received an overbar, the pressure at a point different from e is being considered. Whereas e is just inside the nozzle, $\bar{e}$ is just outside it.

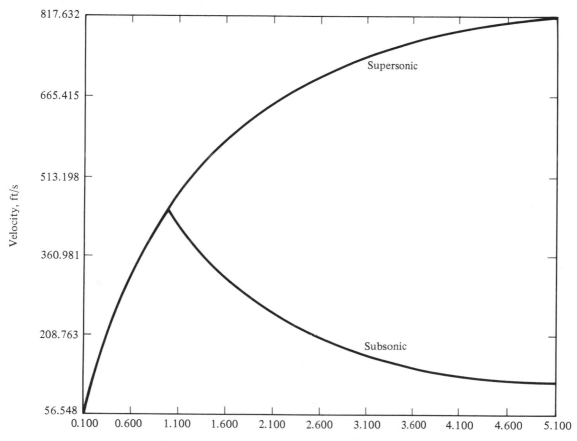

Figure 14.7.6 Fluid speed variation within a converging-diverging nozzle with sonic flow at the throat and isentropic flow throughout the nozzle.

than p_e''', a difficulty arises. Equation (14.7.5) does not permit both sonic flow at the throat and values of the nozzle exit pressure p_e in the range $p_e''' < p_e < p_e'$. This equation allows only the two pressures, p_e' and p_e'''. One can argue again that as the receiver pressure $p_{\bar{e}}$ is lowered to just below the value of p_e', the pressures in the diverging section of the nozzle are affected because the flow is subsonic there. However, the associated sonic flow at the throat prevents the downstream pressure changes from affecting the pressure in the converging section that occurs upstream of the throat. Thus, the thermodynamics in the converging part, and hence the mass flow rate, are not influenced by letting $p_{\bar{e}} < p_e'$.

It appears reasonable that receiver pressures between p_e' and p_e''' can be specified in an experiment. If they are, what will be the flow within the nozzle? One would like to model this flow mathematically and use the applicable statements from among those given by Eqs. (14.4.1) to (14.7.4). Of these, only the isentropic relationship, Eq. (14.7.4), lacks the needed generality. It is applicable only to a reversible process, and such a process, by itself, cannot model the exit pressures in the range between p_e' and p_e'''. The process that does occur is accompanied by a sudden change of pressure and density of the gas and results in an increase of entropy. The location where the large

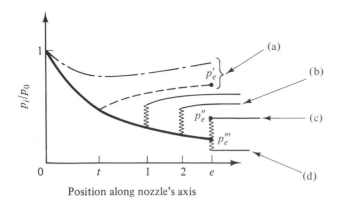

Figure 14.7.7 Pressure distributions within converging-diverging nozzle for various receiver pressures, showing kinds of shock waves that can be produced as receiver pressure $p_{\bar{z}}$ is varied. (a) Range of $p_{\bar{z}}$ for subsonic flow in nozzle. (b) $p_{\bar{z}}$ such that normal shock wave occurs at section 2. (c) maximum $p_{\bar{z}}$ for supersonic flow in entire divergent section; $p_{\bar{z}} = p_e''$. (d) $p_{\bar{z}} < p_e'''$.

changes of thermodynamic properties occur is again called a *shock front,* or *shock wave.* With Eq. (14.7.4) now no longer assumed to apply for $p_e''' < p_e < p_e'$, we clearly need a relationship to take its place in order to have the needed four equations for the four unknowns, V, p, ρ, T.

One possibility is to *assume* that the irreversibilities occur at *one* place in the expanding section, *with* the flow reversible up to, and after, this point. The assumed flow sequence is shown in Fig. 14.7.7 by one of the solid lines; which line is followed depends on the place in the nozzle where the shock wave occurs (e.g., station 1 or 2 in the figure) or, equivalently, on the value of p_e that is prescribed. If the shock wave occurs within the nozzle—location 1 or 2 in Fig. 14.7.7, for example—the pressure discontinuity is perpendicular to the nozzle's axis, except very near to the wall, because of viscous effects. This is the reason that such phenomena are called *normal shock waves.* If the shock wave occurs at the exit, point e in Fig. 14.7.7, it is usually not a plane. Furthermore, for pressures $p_{\bar{z}}$ that are below p_e''', the gas will expand. The shock wave in this case is called an *expansion shock wave.* If the reservoir pressure is above p_e''', the shock waves are *compression* shock waves. The only such wave that can be a plane wave would occur if $p_{\bar{z}} = p_e''$. This fact will be demonstrated in the next section. Figure 14.7.8 shows the oblique shock waves from a supersonic jet during an engine

Figure 14.7.8 Jet engine exhaust showing three-dimensional shock wave pattern during static test. *(Courtesy of United Technologies, Pratt & Whitney)*

test. Figures 14.7.9 and 14.7.10 confirm the flows that are expected when $p_{\bar{e}} = p_e'''$ and $p_e''' < p_{\bar{e}} < p_e''$.

By making the assumption of isentropic flow previously, we added another equation to the collection that governs the types of flow under consideration. However, until now we have not employed the momentum equation. If only pressure forces are considered, one can write for the momentum equation in the direction of the nozzle's axis

$$-\frac{dp}{dx} = \rho V \frac{dV}{dx} \tag{14.7.12}$$

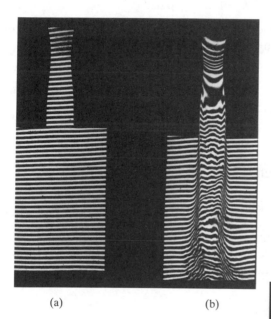

(a) (b)

Figure 14.7.9 (a) Interferogram of converging-diverging nozzle prior to beginning of flow. Lines correspond to surfaces of constant density. (b) Isentropic gas flow in the nozzle with the receiver pressure adjusted to have supersonic flow without shock waves. Some distortion of the density surfaces is seen in the mixing region between the jet and the ambient gas in the receiver. Exit Ma ≈ 1.5. *(Photograph courtesy of Prof. Daniel Bershader, Stanford University.)*

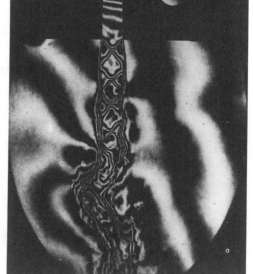

Figure 14.7.10 Oblique shock waves in oblique receiver for receiver pressure above that needed for full expansion, $p_{\bar{e}}'''$, but less than pressures producing planar shock waves within the nozzle. Exit Ma ≈ 1.5. *(Photograph courtesy of Prof. Daniel Bershader, Stanford University.)*

We shall now show that the conditions of *energy and entropy conservation,* along with the ideal gas law, imply a relationship that is the same as the momentum equation. Thus, for isentropic flows, the momentum equation is redundant. One can show this by starting with the energy equation,

$$\frac{V^2}{2} + \frac{c_p}{R}\frac{p}{\rho} = \text{constant}$$

This equation can be differentiated with respect to x to give

$$\rho V\left(\frac{dV}{dx}\right) + \frac{k}{k-1}\left(\frac{dp}{dx} - \frac{p}{\rho}\frac{d\rho}{dx}\right) = 0$$

The $d\rho/dx$ in this equation will come from the isentropic relationship, $p/\rho^k = \text{constant}$. A differentiation of this expression gives

$$\frac{d\rho}{dx} = \frac{1}{kRT}\frac{dp}{dx}$$

The differentiated energy equation can then be written as

$$\rho V\left(\frac{dV}{dx}\right) + \frac{k}{k-1}\left(\frac{dp}{dx} - \frac{RT}{kRT}\frac{dp}{dx}\right) = 0$$

After factoring out dp/dx, the result is

$$\rho V\frac{dV}{dx} + \frac{\partial p}{\partial x} = 0$$

which proves the redundancy of the momentum equation for isentropic flow. However, in Sec. 14.8, where nonisentropic states are considered, recourse to the momentum equation will (and must) be made.

14.8 SHOCK WAVES IN NOZZLES

The momentum equation can be applied to the control volume shown in Fig. 14.8.1 to analyze the flow across a plane shock wave that could occur in the nozzle. The result is

$$p_u A_u - p_d A_d = \rho_d A_d V_d^2 - \rho_u A_u V_u^2 = \dot{m}(V_d - V_u) \tag{14.8.1}$$

in which the subscripts u and d denote "upstream" and "downstream." Also, the pressure rise is assumed to occur at one place, and so $A_u = A_d$. The shock wave is

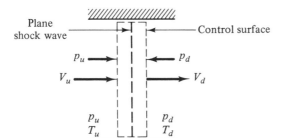

Figure 14.8.1 Control volume for studying shock waves in a nozzle. Control surface has upstream area A_u and downstream area A_d. In this case, $A_u = A_d$.

assumed to be perpendicular to the axis of the nozzle for this analysis. Equation (14.8.1), along with the energy equation,

$$\frac{V_u^2}{2} + \frac{k}{k-1}\frac{p_u}{\rho_u} = \frac{V_t^2}{2} + \frac{k}{k-1}RT_t = \frac{k}{k-1}RT_0$$

or

$$\frac{V_u^2}{2} + \frac{k}{k-1}\frac{p_u}{\rho_u} = \frac{V_d^2}{2} + \frac{k}{k-1}\frac{p_d}{\rho_d} \qquad (14.8.2)$$

is coupled with the mass conservation equation,

$$V_u\rho_u = V_d\rho_d \qquad (14.8.3)$$

to permit a solution for V_d, ρ_d, and p_d if V_u, ρ_u, and p_u are specified. The value of p_u needed in Eqs. (14.8.1) and (14.8.2) follows from Eq. (14.7.6), once the value of A_u is given. The density ρ_u can be found by using the reservoir conditions to evaluate the constant for the isentropic flow. V_u is determined by using the energy equation between the throat and the location just ahead of the shock wave. This relationship is given in the equation preceding Eq. (14.8.2). Because the flow is isentropic up to the shock wave, p_u can replace p_i in Eq. (14.7.6). The section A_u where the shock wave occurs is specified for A_i in this equation.

With the subsequently computed values of the conditions downstream of the shock, one can add the isentropic relationship to the energy equation and mass conservation equation to solve for conditions between the downstream side of the shock wave and the nozzle exit. In this way, values of the pressure just inside the nozzle exit, p_e, can be found. The velocity at the exit will be subsonic, so that we can expect that the receiver pressure $p_{\bar{e}}$ equals p_e. This assumed model of the flow is borne out by observation. When $p_{\bar{e}} < p_e'''$, the flow passes isentropically through the entire nozzle and there is an expansion (nonplanar) shock wave at the exit. There is also a range of pressures greater than p_e''' for which a compression shock wave (also nonplanar) occurs at the nozzle exit. This situation is discussed further after the next example.

Example 14.8.1

Air at 20°C and an atmospheric pressure of 101 kPa is drawn through a convergent-divergent nozzle by a reduced pressure in the receiver section at the nozzle exit. Determine the thermodynamic conditions upstream and downstream of a normal shock wave that occurs exactly at the nozzle's exit. The throat of the nozzle is 0.02 m², and the exit area is 0.033 m².

Solution   The fact that a shock wave occurs at the exit—or at any location downstream of the throat—means that in Fig. 14.7.7 the lowermost curve was followed after the throat. Any of the dashed curves would not result in a steep pressure gradient in the nozzle. Hence, sonic flow must exist at the throat of the nozzle, and the flow is supersonic up to the point where the shock wave occurs. In the present case, this is at the nozzle exit. This does not limit the steps that follow, except that there is no thermodynamic path *within* the nozzle *after* the shock wave for this particular problem. (We shall comment upon this downstream thermodynamic path after solving the problem at hand.) Equations (14.7.7) to (14.7.10) give the conditions at the throat corresponding to Ma = 1 there, with $k = 1.4$, as

$$T_t^* = T_0\left(\frac{2}{k+1}\right) = (273 + 20)(0.833) = 244 \text{ K}$$

$$p_t^* = 101\left(\frac{2}{k+1}\right)^{k(k-1)} = 101(0.528) = 53.4 \text{ kPa}$$

and

$$\rho_t^* = \frac{p_t^*}{RT_t^*} = \frac{(53.4)(10^3)}{(287)(244)} = 0.76 \text{ kg/m}^3$$

Furthermore

$$V_t^* = \sqrt{kRT_t^*} = \sqrt{1.4(287)(244)} = 313 \text{ m/s}$$

and

$$\dot{m}^* = V_t^* \, \rho_t^* \, A_t = 4.76 \text{ kg/s}$$

The last result could have been obtained by using Eq. (14.7.11) directly. With this mass rate of flow, the upstream pressure p_u, where the shock wave occurs, can be found from Eq. (14.7.6), since the flow is isentropic up to the upstream station. This solution, through trial and error or graphic means, gives two values of p_u, one of which is for subsonic conditions. The answer for supersonic conditions is†

$$p_u = 0.133 p_0 = 13.43 \text{ kPa}$$

The density can be determined from Eq. (14.7.4):

$$\rho_u = \rho_0 \left(\frac{p_u}{p_0}\right)^{1/k} = \rho_0(0.236) = 0.283 \text{ kg/m}^3$$

and T_u, from the equation of state, is 168 K.

Because $\dot{m} = \rho V A$, we have

$$V_u = \frac{\dot{m}}{\rho_u A_u}$$

or

$$V_u = \frac{4.76}{(0.283)(0.033)} = 509 \text{ m/s}$$

We now proceed to calculate the downstream variables. To do this, we shall borrow a result that will be developed later in this chapter. We shall see that at a plane shock wave in a nozzle, $V_u V_d = c_t^2$, in which c_t is the sonic speed at the throat. This gives us $V_d = 313^2/509 = 192$ m/s.

The remaining steps are easy. Mass conservation gives

$$\rho_d = \frac{V_u \rho_u}{V_d} = \frac{509(0.283)}{192} = 0.750 \text{ kg/m}^3$$

Use either the momentum or the energy equation to show that

$$p_d = 59 \text{ kPa} \quad \text{or} \quad \frac{p_d}{p_0} = 0.584$$

The downstream temperature can be found from the equation of state. The entropy change is, according to a variation of Eq. (A.F.13),

$$s_d - s_u = \Delta s = c_v \ln\left[\frac{p_d}{p_u}\left(\frac{\rho_u}{\rho_d}\right)^k\right]$$

$$= c_v \ln[4.4261(0.256)]$$

$$= 717(0.1230) = 88.22 \frac{\text{J}}{\text{kg} - \text{K}}$$

† This value can be verified by substituting in Eq. (14.7.6) to see whether it gives the value of $\dot{m}$ that was found previously.

Up to this point, all of the calculations are applicable to determining the conditions ahead and behind a normal shock wave at any point within the diverging portion of the nozzle. If this discontinuity had occurred ahead of the exit, one would have to determine the path of thermodynamic states that the gas traversed in flowing to the exit. What these states are considered to be will be discussed momentarily.

Before continuing, it is worthwhile to make some final remarks about the example. The p_d that was calculated was the downstream pressure for a plane shock wave that occurred in the exit plane of the converging-diverging nozzle. Hence, p_d is the exit (receiver) pressure for this case. (We have $p_u = p_e$, the pressure just inside the nozzle, and $p_d = p_{\bar{e}}$, the pressure just outside the nozzle.) We shall designate the *maximum* receiver pressure for which a plane shock wave occurs at the nozzle exit as p_e''. For a receiver pressure $p_{\bar{e}} = p_e''$, a *plane* shock wave stands at the nozzle exit. If the receiver pressure is reduced to p_e''', then there is supersonic flow coming out of the nozzle—$p_{\bar{e}} = p_e = p_e'''$—without any shock waves being formed. But for a receiver pressure between these two values, $p_e''' < p_{\bar{e}} < p_e''$, *two-dimensional* compression shock waves are formed at the nozzle exit. The analysis of such waves is beyond the scope of the present discussion. Figure 14.7.7 includes a summary of the phenomena that occur at the nozzle exit for $p_{\bar{e}} < p_e''$.

In Example 14.8.1 we presented the interrelationships between the various thermodynamic variables ahead of, and behind, a plane shock wave in a step-by-step way. However, in completing the solution it was necessary to use the expression $V_u V_d = c_t^2$. This result can be derived by using the equations of mass conservation and energy conservation and the momentum principle—that is Eqs. (14.8.1) to (14.8.3)—in the following manner.

We start with the conservation of energy across the plane shock wave. This, according to Eq. (14.8.2), is

$$\frac{V_u^2}{2} + \frac{k}{k-1}\frac{p_u}{\rho_u} = \frac{V_d^2}{2} + \frac{k}{k-1}\frac{p_d}{\rho_d}$$

For ρ_d we use $\rho_u V_u/V_d$ from the mass conservation equation, and for p_d we introduce an expression for p_u from the momentum equation that is given by Eq. (14.8.1). This gives us

$$\frac{V_u^2}{2} + \frac{k}{k-1}\frac{p_u}{\rho_u} = \frac{V_d^2}{2} + \frac{k}{k-1}\left(\frac{V_d}{V_u}\right)\left(\frac{1}{\rho_u}\right)\left[\frac{-\dot{m}}{A}(V_d - V_u) + p_u\right]$$

If this equation is rearranged, we can write

$$0 = \frac{V_d^2 - V_u^2}{2} + \frac{k}{k-1}\left\{-\frac{p_u}{\rho_u} - V_d\left[\frac{\dot{m}}{V_u \rho_u A_u}\right](V_d - V_u) + \frac{V_d}{V_u}\left(\frac{p_u}{\rho_u}\right)\right\}$$

The term in square brackets is unity (because of mass conservation), so this expression can be simplified to

$$0 = \frac{V_u^2}{2}\left[\left(\frac{V_d}{V_u}\right)^2 - 1\right] + \frac{k}{k-1}\left[\frac{p_u}{\rho_u}\left(\frac{V_d}{V_u} - 1\right) - V_d V_u\left(\frac{V_d}{V_u} - 1\right)\right]$$

If there were no shock wave discontinuity at the particular position in the nozzle, V_u would equal V_d and the equation would be satisfied. There is another possibility, however, which is of greater interest. We can find it by factoring out $(V_d/V_u - 1)$ from this equation and canceling it. What remains is

$$\frac{V_u^2}{2}\left(\frac{V_d}{V_u} + 1\right) + \frac{k}{k-1}\left(\frac{p_u}{\rho_u} - V_d V_u\right) = 0$$

which can also be written as

$$\frac{V_u^2}{2} + \frac{k}{k-1}\frac{p_u}{\rho_u} = V_u V_d\left(\frac{k}{k-1} - \frac{1}{2}\right)$$

The left-hand side of this equation is $c_t^2(k+1)/2(k-1)$, according to Eq. (14.8.2), so that we can write, after adding the fractions on the right-hand side,

$$\frac{c_t^2(k+1)}{2(k-1)} = \frac{V_u V_d(k+1)}{2(k-1)}$$

Clearly, this equation is satisfied by the condition

$$V_u V_d = c_t^2 = (V^*)^2 \tag{14.8.4}$$

It is also possible to combine Eqs. (14.8.1) to (14.8.3) so that separate equations result for the pressure and for the density or velocity occurring downstream of the shock wave. These equations are called the *Rankine-Hugoniot equations* and are as follows:

$$\frac{p_d}{p_u} = \frac{[(k+1)/(k-1)](\rho_d/\rho_u) - 1}{[(k+1)/(k-1)] - (\rho_d/\rho_u)} \tag{14.8.5}$$

and

$$\frac{V_u}{V_d} = \frac{\rho_d}{\rho_u} = \frac{1 + [(k+1)/(k-1)](p_d/p_u)}{[(k+1)/(k-1)] + (p_d/p_u)} \tag{14.8.6}$$

In Example 14.8.1, we saw how one can calculate the downstream conditions associated with a normal shock wave. It is assumed, and confirmed by experiment, that when the shock wave occurs within the nozzle, the gas subsequently flows to the nozzle's exit in a reversible manner. Hence, p/ρ^k is again constant, and there is no change of entropy for this portion of the flow. True, there was a change of entropy across the shock wave, and we shall call this value Δs. Because downstream of the normal shock there is assumed to be no further change of entropy, at all points downstream (designated i) a relationship to the reservoir conditions is given by

$$\frac{p_i}{p_0}\left(\frac{\rho_0}{\rho_i}\right)^k = e^{\Delta s/c_v} = S \tag{14.8.7}$$

with the value of S depending upon the value of Δs that has occurred across the shock front. This is a general result from the thermodynamics, and can be derived from Eq. (F.12). If we use Eq. (14.8.7) instead of the isentropic relationships, for which $S = 1$, the mass rate of flow in the nozzle after the plane shock wave is

$$\dot{m} = A_i \rho_0\left(\frac{p_i}{p_0}\right)^{1/k}\frac{1}{S^{1/k}}\sqrt{\frac{2k}{k-1}\frac{p_0}{\rho_0}\left[1 - \left(\frac{p_i}{p_0}\right)^{(k-1)/k}S^{1/k}\right]} \tag{14.8.8}$$

[This result should be compared with Eq. (14.7.6), for which S was taken as unity.] Note that the value of S in this last equation can be obtained by letting *position i*—*somewhere downstream of the shock wave*—be the exit position or a point just downstream of the shock front. The pressures associated with these two locations are p_e and p_d. (Remember that the shock wave has occurred at, or previous to, A_i.) Then one would have

$$\frac{p_e}{p_0}\left(\frac{p_0}{p_e}\right)^k = S = \frac{p_d}{p_0}\left(\frac{p_0}{p_d}\right)^k = \frac{p_i}{p_0}\left(\frac{p_0}{p_i}\right)^k \qquad (14.8.9)$$

It is useful to note that Eq. (14.7.11) also gives $\dot{m}$, because the flow is sonic at the throat if a plane shock wave is to exist farther downstream. This mass flow rate can be used in Eq. (14.8.8). When the appropriate terms are canceled, one has

$$A_i\left(\frac{p_i}{p_0}\right)^{1/k}\frac{1}{S^{1/k}}\sqrt{\left[1-\left(\frac{p_i}{p_0}\right)^{(k-1)/k}S^{1/k}\right]/(k-1)} = A_t\left(\frac{2}{k+1}\right)^{1/(k-1)}\sqrt{\frac{1}{k+1}}$$

or $$\frac{1}{S^{1/k}}\sqrt{1-\left(\frac{p_i}{p_0}\right)^{(k-1)/k}S^{1/k}} = \left[\frac{A_t}{A_i}\left(\frac{2}{k+1}\right)^{1/(k-1)}\sqrt{\frac{k-1}{k+1}}\right]\left(\frac{p_i}{p_0}\right)^{-1/k} \qquad (14.8.10)$$

The right-hand side of Eq. (14.8.10) can be plotted on log-log paper versus p_i/p_0 to give a straight line. For the prescribed value of S for the shock wave, the left-hand side of Eq. (14.8.10) can be plotted as a curve. Where the curve and the straight line intersect, one has a value of p_i/p_0 for which the right- and left-hand sides have a common value. A solution has been achieved. Such solutions can be replotted in the form of Fig. 14.8.2. The current state and availability of digital computers make the solution of Eq. (14.8.10) a seemingly effortless task. Figure 14.8.2 includes points like A that are part of the solutions to Examples 14.8.1 and 14.8.2.

Example 14.8.2

An extension is placed in the nozzle treated in Example 14.8.1 so that now $A_e = 0.04$ m². The conditions in the receiver are adjusted so that a normal shock wave occurs at the point where the original nozzle ended. ($A_{\text{throat}}/A_{\text{shock}} = 0.606$.) What is the exit pressure?

Solution   The solution to this new situation can utilize much of what we already know from Example 14.8.1. For instance, with the shock wave still located where the area is 0.033 m², we can conclude that p_u, p_0, and the other upstream quantities are the same as in Example 14.8.1 and

$$S = e^{\Delta s/c_v} = e^{0.1230} = 1.13093$$

With this value, we can try for a solution of Eq. (14.8.10) in the form that is pictured in Fig. 14.8.2. It is associated with the $S = 1.13093$ curve and the vertical line associated with 0.02/0.04 for A_t/A_i, or A_t/A_e in this case. Point C in Fig. 14.8.2 is the state point for the exit. Hence, for p_i taken as p_e, the ordinate in the figure is

$$\frac{p_e}{p_0} = 0.645 \quad \text{or} \quad p_e = 65.1 \text{ kPa}$$

Because the shock wave occurs ahead of the exit, the flow leaving the nozzle is subsonic and p_e is also the receiver pressure, $p_{\bar{e}}$.

We can now summarize the previous procedure. If we know the value of A_i where a shock wave occurs, the method of Example 14.8.2 allows us to calculate S and p_d,

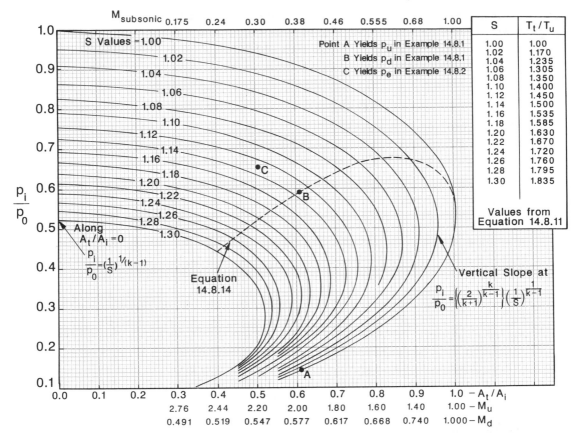

Figure 14.8.2 Entropy changes associated with normal shock waves in converging-diverging nozzles, obtained from Eq. (14.8.10).

among other things. With a knowledge of A_e and the use of Eq. (14.8.10), we can find p_e, because the value of S is the same as before.

If one is required to find the position within the nozzle where a normal shock wave occurs, then p_e/p_0 and A_e/A_t must be specified. Then, using the subscript e for the ith station to denote the exit, Eq. (14.8.10) is solved for S. We shall use this value of S shortly.

Equation (14.8.1), for the momentum relationship across the shock wave, can be written as

$$p_u - p_d = \frac{\dot{m}}{A_u}(V_d - V_u) = \frac{\dot{m}}{A_u}\frac{c_t^2 - V_u^2}{V_u}$$

after using Eq. (14.8.4). By dividing through by ρ_u, we obtain

$$\frac{p_u}{\rho_u} - \frac{p_d}{\rho_u} = c_t^2 - V_u^2 = 2kR\frac{T_u - T_t}{k-1}$$

if the energy equation is also used between the throat and the position just upstream of the shock wave. Equation (14.8.9) will now be used for its value of p_d in terms of the value of S that has just been determined. The last equality now becomes

$$RT_u - S \frac{p_u}{\rho_u}\left(\frac{\rho_d}{\rho_u}\right)^k = 2kR \frac{T_u - T_t}{k - 1}$$

The ratio ρ_d/ρ_u can be replaced through Eq. (14.8.3), which describes mass conservation across the shock wave. We then have

$$T_u\left[1 - S\left(\frac{V_u}{V_d}\right)^k\right] = 2k\frac{T_u - T_t}{k - 1}$$

Equation (14.8.4) can be usefully employed again to give

$$1 - S\left(\frac{V_u^2}{c_t^2}\right)^k = 2k\frac{1 - T_t/T_u}{k - 1}$$

Again, the energy equation can be introduced to give V_u and c_t in terms of the appropriate temperatures. Thus, we can write

$$1 - S\left(\frac{T_0/T_t - T_u/T}{T_0/T_t - 1}\right)^k = 2k\frac{1 - T_t/T_u}{k - 1}$$

Because the ratio T_0/T_t is $(k + 1)/2$, for sonic flow at the throat—see Eq. (14.7.8)—a final form can be written as

$$1 - S\left(\frac{(k + 1)/2 - T_u/T_t}{(k - 1)/2}\right)^k = 2k\frac{1 - T_t/T_u}{k - 1} \tag{14.8.11}$$

This equation is conveniently solved for S by prescribing values of T_u/T_t. The interpolated results are listed in Fig. 14.8.2 for convenience.

To find the position of a normal shock wave within a nozzle when the exit pressure is known, the table in Fig. 14.8.2 can be used to determine the associated value of S and T_u/T_t. (Recall that T_t is known for prescribed values of T_0.) From T_u and T_0, the value of V_u is known from the energy equation. Isentropic flow from the reservoir to just upstream of the shock wave prescribes the value of ρ_u. The mass rate of flow at the throat equals $V_u\rho_u A_u$, so that A_u, and thus the location in the nozzle, are known.

Example 14.8.3

A convergent-divergent nozzle has $A_t/A_e = 0.02 \text{ m}^2/0.04 \text{ m}^2$ and $p_e/p_0 = 0.645$. The reservoir conditions are 20°C and 101 kPa. Determine the thermodynamic conditions and area where a normal shock wave occurs.

Solution Figure 14.8.2, with an ordinate of 0.645 and an abscissa of 0.5 for the exit conditions, yields a S value of about 1.13, at point C. (The value is 1.13093, according to Example 14.8.2.) The value of 1.480 for T_t/T_u is obtained from the S values in the table shown as part of Fig. 14.8.2. Therefore, from Eq. (14.7.9),

$$T_t = 293\left(\frac{2}{2.4}\right) = 244 \text{ K}$$

so that

$$T_u = \frac{244}{1.480} = 165 \text{ K}$$

$V_u(= 508.5 \text{ m/s})$ follows from Eq. (14.8.2). The upstream density ρ_u equals $\rho_0(T_u/T_0)^{1/(k-1)}$ by using Eq. (14.7.4) and the equation of state to eliminate p. Its value is 0.282 kg/m³. The mass rate of flow through the throat and the shock cross section is the same. Thus,

$$\frac{A_t}{A_u} = \frac{\rho_u V_u}{\rho_t V_t}$$

$$= \frac{V_u}{V_t} \left(\frac{T_u}{T_t}\right)^{1/(k-1)}$$

$$= \left(\frac{1}{1.48}\right)^{2.5} \left[\frac{508.5}{1.4(287)(244)}\right]$$

$$= 0.3753(508.5)/313.1$$

$$= 0.61$$

$$A_u = 0.02/0.61 = 0.033 \text{ m}^2$$

This result is associated with point B in Figure 14.8.2 as the following discussion makes evident.

In this example we found A_t/A_u as a function of T_u/T_t and V_u/V_t. Because the energy equation relates V_u and V_t through T_u and T_t, one can anticipate that A_t/A_u is a function of T_u/T_t (and hence S) alone. We can verify this surmise by starting with Eq. (14.8.2),

$$\frac{V_u^2}{2} + c_p T_u = \frac{V_t^2}{2} + c_p T_t$$

which can be rewritten as

$$\frac{V_u^2}{2c_p T_t} + \left(\frac{T_u}{T_t} - 1\right) = \frac{V_t^2}{2c_p T_t} = \frac{V_t^2}{kRT_t}\left(\frac{k-1}{2}\right) = (1)\left(\frac{k-1}{2}\right)$$

In writing this result, we employed $V_t = \sqrt{kRT_t}$ because $\text{Ma}_t = 1$ when there is a normal shock wave in the nozzle. These expressions for V_u and V_t lead to

$$\frac{V_u}{V_t} = \left[1 + \frac{2}{k-1}\left(1 - \frac{T_u}{T_t}\right)\right]^{1/2} \tag{14.8.12}$$

Therefore, from $\dot{m} = \text{constant}$,

$$\frac{A_t}{A_u} = \frac{V_u}{V_t}\left(\frac{T_u}{T_t}\right)^{1/(k-1)} \tag{14.8.13}$$

from which one can conclude that

$$\frac{A_t}{A_u} = \left(\frac{T_u}{T_t}\right)^{1/(k-1)}\left[1 + \frac{2}{k-1}\left(1 - \frac{T_u}{T_t}\right)\right]^{1/2} \tag{14.8.14}$$

This expression is evaluated in Table 14.8.1 and can be used to determine the location of the shock wave in divergent portions of the nozzle.

The area ratios of Eq. (14.8.14) are also plotted as the dashed curve in Fig. 14.8.2. This curve represents the momentum condition across the normal shock wave, because Eq. (14.8.11) is at its root. It forms a constraint on the possible exit pressures and normal shock waves within the nozzle. Figure 14.8.3 details this application of Eq. (14.8.14). Hence point B in Fig. 14.8.2 is a solution to Example 14.8.3

If the receiver pressure is less than p_e', corresponding perhaps to point E_1 in Fig. 14.8.3a, the path will be at constant S from the reservoir past the throat, point T, to U_1.

TABLE 14.8.1 THERMODYNAMIC CONDITIONS UPSTREAM OR DOWNSTREAM OF A NORMAL SHOCK WAVE WITH ENTROPY CHANGE $(\Delta s)c_v$ In (S) IN TERMS OF CONDITIONS AT THE THROAT

T_t/T_u	T_t/T_d	S	Ma_u	Ma_d	A_t/A_u	T_t/T_u	T_t/T_d	S	Ma_u	Ma_d	A_t/A_u
1.0	1.0	1.0	1.0	1.0	1.0	1.5	0.889	1.13987	2.0	0.577	0.592593
1.02	0.982	1.00009	1.058	0.946	0.997261	1.52	0.888	1.1488	2.03	0.572	0.577987
1.04	0.969	1.00061	1.114	0.901	0.989945	1.54	0.887	1.15783	2.059	0.568	0.563795
1.06	0.958	1.00176	1.166	0.864	0.979157	1.56	0.886	1.16697	2.088	0.563	0.550008
1.08	0.949	1.00358	1.217	0.832	0.965739	1.58	0.885	1.1762	2.117	0.559	0.536621
1.1	0.941	1.00608	1.265	0.804	0.950347	1.6	0.885	1.18552	2.145	0.555	0.523623
1.12	0.935	1.00921	1.311	0.78	0.933492	1.62	0.884	1.19493	2.173	0.551	0.511006
1.14	0.929	1.01295	1.356	0.759	0.91557	1.64	0.883	1.20442	2.2	0.547	0.49876
1.16	0.925	1.01723	1.4	0.74	0.896921	1.66	0.883	1.21398	2.227	0.543	0.486874
1.18	0.920	1.02201	1.442	0.723	0.87778	1.68	0.882	1.22362	2.254	0.540	0.47534
1.2	0.917	1.02724	1.483	0.707	0.858356	1.7	0.881	1.23332	2.28	0.537	0.464146
1.22	0.913	1.03289	1.523	0.693	0.838812	1.72	0.881	1.24308	2.307	0.534	0.453284
1.24	0.910	1.03891	1.562	0.68	0.819276	1.74	0.880	1.25291	2.332	0.531	0.442743
1.26	0.908	1.04528	1.6	0.668	0.79985	1.76	0.879	1.26279	2.358	0.528	0.432513
1.28	0.905	1.05195	1.637	0.658	0.780616	1.78	0.879	1.27272	2.383	0.525	0.422585
1.3	0.903	1.05891	1.673	0.648	0.761639	1.8	0.879	1.2827	2.408	0.522	0.412949
1.32	0.901	1.06614	1.709	0.638	0.742968	1.82	0.878	1.29273	2.433	0.52	0.403596
1.34	0.899	1.07359	1.74	0.63	0.72464	1.84	0.878	1.30281	2.458	0.517	0.394516
1.36	0.898	1.08127	1.778	0.622	0.706687	1.86	0.877	1.31293	2.482	0.515	0.385702
1.38	0.896	1.08915	1.811	0.614	0.689128	1.88	0.877	1.32309	2.506	0.512	0.377143
1.4	0.895	1.09722	1.844	0.607	0.671979	1.9	0.877	1.33329	2.53	0.51	0.368832
1.42	0.893	1.10545	1.876	0.6	0.655249	1.92	0.876	1.34353	2.553	0.508	0.360762
1.44	0.892	1.11385	1.908	0.594	0.638945	1.94	0.876	1.35381	2.577	0.506	0.352922
1.46	0.891	1.12239	1.939	0.588	0.623068	1.96	0.876	1.36411	2.6	0.504	0.345307
1.48	0.89	1.13107	1.97	0.583	0.607618	1.98	0.875	1.37445	2.623	0.502	0.337908
						2.0	0.875	1.38482	2.646	0.5	0.330719

T_t = throat temperature

T_u = temperature at upstream surface of normal shock wave

T_d = temperature at downstream surface of normal shock wave

At the station in the nozzle for U_1 the curve of constant entropy that is associated with E_1 will also require that the shock wave occur at D_1, so that the momentum equation can be satisfied across the shock wave. This fixes the position of the shock front and, consequently, the upstream condition U_1. A similar situation occurs for E_2, E_3, and E_4. When the receiver pressure is reduced to E_e, point D_e coincides with E_e. The normal shock wave is at the exit plane. Receiver pressures lower than the pressure p_e'' corresponding to point E_e cannot result in a normal shock wave at the exit. The surface of pressure discontinuity must be more complex than just a plane.

Receiver pressures lower than p_e''' must again fail to satisfy the momentum equation for a plane shock front. Nonplanar expansion shock waves are the result.

It should be apparent to the reader that Examples 14.8.1 to 14.8.3 are closely related. Whereas in the first example we examined conditions in the nozzle from the reservoir to the shock wave at a prescribed location, the second example concentrated on conditions from the shock wave to the nozzle exit. In the third example, we worked from the exit upstream to the shock wave at an (initially) unknown location. We also needed information about the mass flow rate. This came from the sonic condition at

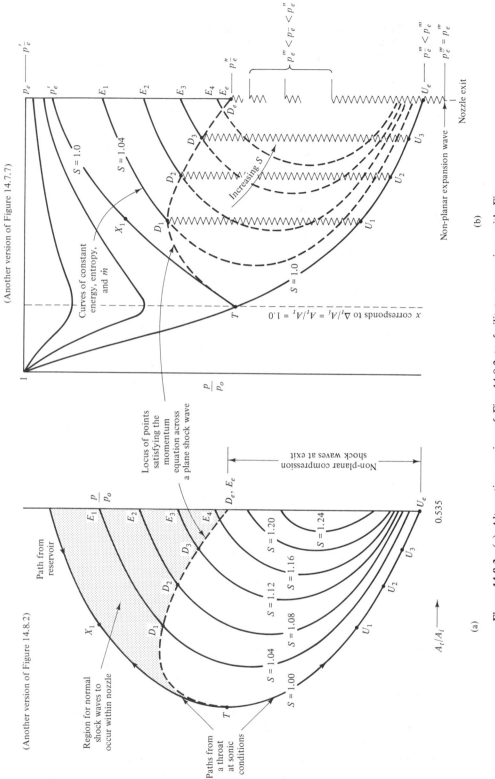

Figure 14.8.3 (a) Alternative version of Fig. 14.8.2 to facilitate comparison with Fig. 14.8.3b. (b) Inclusion of entropy lines into a plot of pressure distributions within a converging-diverging nozzle with corresponding points for Fig. 14.4.3a.

(Another version of Figure 14.7.7)

$\dfrac{p}{p_o}$

Curves of constant energy, entropy, and $\dot{m}$

x corresponds to $\Delta_i/A_i = A_i/A_t = 1.0$

Increasing S

$S = 1.0$

$S = 1.04$

$S = 1.0$

$p_e''' < p_e'' < p_e'$

$p_e'' < p_e''$

$p_e''' < p_e''$
$p_e''' = p_e'''$

Nozzle exit

Non-planar expansion wave

(b)

(Another version of Figure 14.8.2)

$\dfrac{p}{p_o}$

Path from reservoir

Locus of points satisfying the momentum equation across a plane shock wave

Region for normal shock waves to occur within nozzle

Non-planar compression shock waves at exit

Paths from a throat at sonic conditions

$S = 1.00$
$S = 1.04$
$S = 1.08$
$S = 1.12$
$S = 1.16$
$S = 1.20$
$S = 1.24$

$A_t/A_i \longrightarrow$

0.535

(a)

the throat. In all the cases, the same principles were used, but in each, the "conservation equations" and other thermodynamic relationships were combined slightly differently to meet the task at hand. Figure 14.8.2 does, however, combine these relationships.

*14.9 VISCOSITY AND COMPRESSIBLE FLOW (ADIABATIC FLOW IN PIPES)

In the previous sections of this chapter, the flowing gas was assumed to be inviscid. Viscosity, however, will lead to thermodynamic irreversibilities and makes isentropic flow a fragile idealization of reality. In this section, we shall deal with one aspect of the effects of viscosity. We shall learn how it can vary the Mach number along a duct of constant area. The concept of losses in a pipe was mentioned when the flow of incompressible fluids was treated in Chaps. 6 and 8. The friction factor

$$f = \frac{\text{losses}}{(V^2/2g)(L/D)}$$

was a way of nondimensionalizing these losses. We recognized at that time that the term *losses* was not to be taken literally, because, in fact, this quantity of energy was being transformed through the fluid's viscosity into internal energy. A very small temperature increase of the liquid would occur. But the increased internal energy could not be converted *practically* to useful work. It is in this sense that the energy is not recoverable, or is lost. However, when the fluid is compressible, some of the viscous dissipation can be converted into changes of internal energy, pressure, or density. The manner in which this redistribution occurs is dictated by the type of thermodynamic process that accompanies the flow. Then the entropy change can be determined. Subsequently, using the thermodynamic definition of the entropy change, we could measure the extent of the irreversibilities, or losses.

In this section, we shall examine the effect of fluid friction in a conduit on the changes of the thermodynamic variables p and ρ or p and T. Because of the associated dissipation of energy, the flow cannot be reversible. The analysis that ensues prescribes the flow variables along the pipe so that, in effect, we can develop the process by which these variables change. Initially, the goal is to learn how the fluid speed u changes along the length of the pipe.

This analysis for a compressible flow will use the friction factor f in the context of a wall drag coefficient. This concept was developed in writing Eq. (8.3.12b). It is useful here because the starting point is the momentum equation for the control volume shown in Fig. 14.9.1. The wall shear stress τ_0 ($\tau_0 = \tau_w$ in Chap. 13) is included in the figure, and we can write the momentum equation for the x direction as follows:

$$\pi r_0^2 [\rho u]\{(u + du) - u\} = \pi r_0^2 \left\{ p - \left(p + \frac{\partial p}{\partial x} dx \right) \right\} - 2\pi r_0\, dx\, \tau_0 \qquad (14.9.1)$$

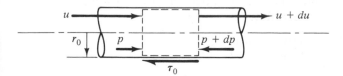

Figure 14.9.1 Control volume for analyzing the compressible flow in a pipe with a wall shear stress.

The term in square brackets in Eq. (14.9.1) is related to the mass rate of flow at any section:

$$[u\rho](\text{Area}) = \text{constant} \quad \text{or} \quad u\rho = C_1 \qquad (14.9.2)$$

since we are concerned with a conduit that has a constant area. For such a conduit, we have

$$u\rho = (u + du)(\rho + d\rho) = C_1 \qquad (14.9.3)$$

Equation (14.9.1) simplifies readily to read

$$\rho u \frac{du}{dx} = -\frac{dp}{dx} - \frac{L\tau_0}{r_0}$$

By using Eq. (8.3.12b), $4\tau_0/(\rho u^2/2) = f$, we can extend this result to give

$$\rho u \frac{du}{dx} + \frac{dp}{dx} = -\frac{4}{D}\left[\frac{f}{4}\left(\rho \frac{u^2}{2}\right)\right]$$

provided that the substitution $2r_0 = D$ is used. If instead of $u(du/dx)$ we write $\frac{1}{2}(du^2/dx)$, we have the formulation

$$\frac{1}{u^2}\frac{d(u^2)}{dx} + \frac{dp/dx}{\frac{1}{2}\rho u^2} = -\frac{f}{D} \qquad (14.9.4)$$

An expression for dp from the energy equation can be inserted constructively into this equation for u^2. By *assuming* no energy transfers from external sources to the fluid in the conduit as heat or work, we can write the energy equation as

$$\frac{u^2}{2} + c_p T = \text{constant}$$

Since $u = \dot{m}/A\rho$, this equation can be written as $(\dot{m}/A\rho)^2/2 + c_p T = \text{constant}$. This is an equation that relates two thermodynamic variables, $1/\rho$ and T, because $\dot{m}$ and A are constant in a duct. This relationship is called a *Fanno line*.

One can show that curves of constant s have the form $(T^{c_v/R})/\rho = \text{constant}$. This result follows directly from $ds = dh - dp/\rho$ with $ds = 0$. The curve with the greatest entropy will intersect a Fanno line at one point. If one equates the slope of a curve of constant s and the slope of a Fanno line, one finds the point of tangency to be at Ma = 1. Figure 14.9.2 illustrates this relationship.

From the fact that $c^2 = kRT$, the energy equation can also be written as

$$\frac{u^2}{2} + c_p\left(\frac{c^2}{kR}\right) = \text{constant} \qquad (14.9.5)$$

or

$$\frac{u^2}{2} + c_p\left(\frac{p}{R\rho}\right) = \text{constant} \qquad (14.9.6)$$

The last of these expressions will be used immediately, but Eq. (14.9.5) will be useful at a subsequent stage. The differentiation of Eq. (14.9.6) gives

$$\frac{d(u^2)}{2} + \frac{c_p}{R}\left(\frac{dp}{\rho} - p\frac{d\rho}{\rho^2}\right) = 0$$

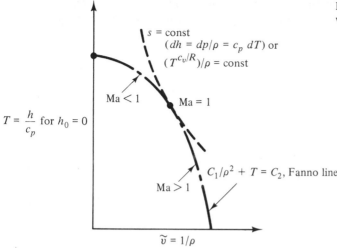

Figure 14.9.2 Intersection of a Fanno line with a curve of constant entropy at Ma = 1.

or

$$dp = p\frac{d\rho}{\rho} - \frac{\rho}{2}\frac{R}{c_p}d(u^2)$$

so that

$$\frac{dp}{dx} = \frac{p}{\rho}\frac{d\rho}{dx} - \frac{\rho}{2}\frac{R}{c_p}\frac{d(u^2)}{dx}$$

This equation can be combined with Eq. (14.9.4), with the result that

$$\frac{1}{u^2}\frac{d(u^2)}{dx} + \frac{\frac{p}{\rho}\frac{d\rho}{dx} - \frac{\rho}{2}\frac{R}{c_p}\frac{d(u^2)}{dx}}{\frac{1}{2}\rho u^2} = -\frac{f}{D} \qquad (14.9.7)$$

The expression $(1/\rho)(d\rho/dx)$ is related to the term $(du^2/dx)/u^2$ if we add the equation for mass conservation to our analysis. This is Eq. (14.9.3). If this expression is simplified, one obtains

$$\frac{d\rho}{\rho} = -\frac{du}{u}$$

or

$$\frac{1}{\rho}\frac{d\rho}{dx} = -\frac{1}{u}\frac{du}{dx}$$

with the result that

$$\frac{1}{\rho}\frac{d\rho}{dx} = -\frac{1}{u^2}\left(u\frac{du}{dx}\right) = -\frac{1}{2u^2}\frac{d(u^2)}{dx}$$

This expression will be inserted into Eq. (14.9.7) along with a substitution for p from the equation of state for an ideal gas, $p = \rho RT$. The result is that the term $(du^2/dx)/u^2$ becomes a common factor to several of the terms in the equation. We can then write

$$\frac{1}{u^2}\frac{d(u^2)}{dx}\left(1 - \frac{RT}{u^2} - \frac{R}{c_p}\right) = -\frac{f}{D} \qquad (14.9.8)$$

This result is not quite sufficient for our purposes. If the right-hand side were known, the differential equation could be solved for u^2 if T were prescribed as a function of x.

Instead of concerning ourselves with T, however, we will concentrate on c, the speed of sound, which is related to T. This implies that Eq. (14.9.8) should be written as

$$\frac{1}{u^2}\frac{d(u^2)}{dx}\left(1 - \frac{c^2}{ku^2} - \frac{k-1}{k}\right) = -\frac{f}{D}$$

or

$$\frac{1}{u^2}\frac{d(u^2)}{dx}\left(1 - \frac{1}{k\,\mathrm{Ma}^2} - 1 + \frac{1}{k}\right) = -\frac{f}{D}$$

and finally as

$$\frac{1}{u^2}\frac{d(u^2)}{dx}\frac{\mathrm{Ma}^2 - 1}{k\,\mathrm{Ma}^2} = \frac{f}{D} \tag{14.9.9}$$

if the definition of the Mach number, $\mathrm{Ma} = u/c$, is used. Moreover, if $\mathrm{Ma}^2 = u^2/c^2$ is differentiated, we can write du^2/dx in terms of $d(\mathrm{Ma}^2)/dx$, with the result that the entire left-hand side of Eq. (14.9.9) contains only functions of Ma, its derivatives, and constants. This situation will lead to a solution for Ma as a function of x. The needed steps of differentiation are

$$d(\mathrm{Ma}^2) = d\left(\frac{u^2}{c^2}\right) = \frac{d(u^2)}{c^2} - \frac{u^2}{c^2}\frac{d(c^2)}{c^2}$$
$$= \frac{1}{c^2}\left[d(u^2) - \mathrm{Ma}^2\,d(c^2)\right] \tag{14.9.10}$$

The appearance of dc^2 causes no difficulties, because it can be eliminated by differentiation of Eq. (14.9.5); this step gives

$$dc^2 = -\frac{Rk}{2c_p}du^2$$

The consequence of this is that Eq. (14.9.10) can be written as

$$d(\mathrm{Ma}^2) = d(u^2)\left(1 + \mathrm{Ma}^2\frac{Rk}{2c_p}\right)\Big/c^2$$

or

$$\frac{1}{u^2}\frac{d(u^2)}{dx} = \frac{\dfrac{c^2}{u^2}\dfrac{d(\mathrm{Ma}^2)}{dx}}{1 + \mathrm{Ma}^2(k-1)/2}$$

so that Eq. (14.9.9) can be written in a useful form as follows:

$$\frac{1}{k\,\mathrm{Ma}^4}\frac{d(\mathrm{Ma}^2)}{dx}\left\{\frac{\mathrm{Ma}^2 - 1}{1 + \mathrm{Ma}^2(k-1)/2}\right\} = -\frac{f}{D} \tag{14.9.11}$$

The friction factor in Eq. (14.9.11) will vary along the pipe, but its value will be positive. This means that the right-hand side of the equation is negative. For $\mathrm{Ma} < 1$, the term in curly brackets is negative, so that $d(\mathrm{Ma}^2)/dx$ must be positive. That means that the Mach number will increase as the fluid moves along the conduit. But if $\mathrm{Ma} > 1$ in the pipe, the term in curly brackets would be positive, so that $d(\mathrm{Ma}^2)/dx$ must be negative. The result is that Ma for a supersonic flow is reduced along the pipe. Hence, the flow, at whatever Mach number occurs, tends toward $\mathrm{Ma} = 1$. The position where $\mathrm{Ma} = 1$ is labeled L_{max}.

The reason for the L_{max} designation was contained in the full paragraph preceding Eq. (14.9.5), in which the Fanno line is mentioned. Since the pipe friction induces an irreversible effect, $ds/dx > 0$. Therefore, the maximum entropy occurs at the end of the pipe. This also implies that Ma = 1 at the end of the pipe if the pipe is sufficiently long. Thus, a pipe whose length is L_{max} will choke the flow in a manner similar to a converging nozzle. The exit Mach number will be unity. The inlet Mach number must then be determined by suitable integrations of Eq. (14.9.11). This calculation will be considered next.

The manner in which f varies with x could present some minor difficulties in the integration of Eq. (14.9.11). However, if an average value of f is used, say

$$\bar{f} = \frac{1}{L_{max}} \int_0^{L_{max}} f \, dx$$

then Eq. (14.9.11) contains only polynomials of Ma^2. This use of $\bar{f}$ facilitates the integration between $x = 0$, where Ma = Ma_0, and $x = L_{max}$, where Ma = 1. Otherwise, one would need to relate f through the changing value of u. This velocity would affect the Reynolds number, and f would change somewhat in accordance with the curves of the Moody diagram (Fig. 9.4.2). One can verify that the result with $\bar{f}$ is

$$\bar{f}\left(\frac{L_{max}}{D}\right) = \frac{1 - Ma_0^2}{k\, Ma_0^2} + \frac{k+1}{2k} \ln \frac{(k+1)Ma_0^2/2}{1 + (k-1)Ma_0^2/2} \qquad (14.9.12)$$

(Remember that in this expression Ma_0 is the Mach number at the inlet of the duct, the location from which L_{max} is measured.)

If air is flowing in a smooth conduit with a diameter of 100 mm and Ma = 1 at 20°C, the velocity would be 343 m/s. The viscosity of air would give an associated Reynolds number of 1.7×10^6. By referring to the Moody diagram, one obtains a value of f which is near 0.01. This would indicate the approximate rule: For flows with $Ma_0 < 1$, $\bar{f}$ should be greater than 0.01; and for $Ma_0 > 1$, $\bar{f}$ should be less than 0.01.

Example 14.9.1

An airstream flows from a convergent nozzle at 583 ft/s at a temperature of 566°R. The Mach number for these conditions is 0.5. If a smooth insulated pipe with a diameter of 0.75 in. were added to the nozzle, find the length at which the flow would become sonic.

Solution   Equation (14.9.12) gives, assuming $\bar{f} = 0.02$,

$$\frac{\bar{f}}{D}L_{max} = 2.14 + 0.857 \ln 0.2857$$

or

$$L_{max} = 0.75\left(\frac{1.97}{0.02}\right) = 40 \text{ in.}$$

If the energy equation is used between the inlet and the outlet of the pipe, the result is

$$\frac{583^2}{2} + \frac{kR}{k-1}(566) = \frac{kRT_{L,\,max}}{2} + \frac{kR}{k-1}T_{L,\,max}$$

This expression gives $T_{L,\,max}$, from which the sonic (fluid) speed at L_{max} can be determined. Because the flow in the convergent nozzle ahead of the pipe is assumed to be isentropic, the pressure and density at the entrance to the pipe, point 2, can be found from the reservoir conditions, point 0. (They are 500 psia and 593°R.) The conditions at the entrance of the pipe, are found, just as in Example 14.4.2, as

$$p_2 = 500 \left(\frac{566}{593}\right)^{1.4/0.4} = 425 \text{ psia}$$

and

$$p_2 = \rho_0 \left(\frac{566}{593}\right)^{0.4} = 0.982 \, \rho_0$$

This latter figure can be used to find the mass rate of flow per unit area, which in turn prescribes the density at L_{max} in the pipe. With the density and temperature known there, it is possible to calculate the pressure in the tube at L_{max}.

If the length of pipe at the exit of a converging nozzle is specified, one can proceed as follows. By first assuming choked flow, one can calculate the Mach number at the inlet to the pipe. As this point is at the exit of the converging nozzle, this Mach number determines all of the thermodynamic conditions there if isentropic nozzle flow is assumed. This information in turn determines $\dot{m}$. The energy equation will specify the temperature at the exit of the duct. The requirement that Ma = 1 at L_{max} permits the determination of the velocity from the temperature there. The mass rate of flow can be used to calculate the density of the gas, which, along with the temperature and the equation of state, establishes the pressure just inside the end of the duct. This pressure will be greater than the exit pressure for choked flow.

It is useful to recall that in Chap. 10, where internal flows were considered, the flow of gases was not excluded from the discussion. At low Mach numbers, say, 0.01, Eq. (14.9.12) would indicate values of L_{max}/D of the order of magnitude of $5(10^5)$. This fact might be important in planning gas pipelines but would be relatively unimportant in forced-air heating systems.

14.10 SUMMARY

The fundamental ideas that are relative to compressible flows should be familiar by now. Mass conservation, energy conservation, and momentum conservation must be considered just as in previous chapters. But now an equation of state is needed. To this list of physical laws, we have added the second law of thermodynamics. Related to this is the concept of entropy; and for the flow in converging nozzles, isentropic flow is utilized. But constancy of entropy cannot exist in convergent-divergent nozzles if all exit pressures are to be experienced. In order to model these processes, we are guided by experiment. A sharp change of the pressure, temperature, and velocity is assumed to occur at one location along the axis of the convergent-divergent nozzle. These changes will be accompanied by an increase in entropy. We call this discontinuity in the thermodynamic variables a *plane shock wave*. Just ahead of the shock front, the experimental flow is supersonic, and behind it the flow is subsonic. Thus, the thermodynamic path $TX_1 D_1 E_1$ in Fig. 14.8.3b is precluded as a possibility.

An almost anomalous situation occurs in the diverging portion of such a converging-diverging nozzle. A supersonic flow speeds up as the area increases. This is because for our compressible medium, the density can also change, depending upon the flow conditions. We saw that by developing the equation relative to the flow from a reservoir of gas, the Bernoulli equation is the first approximation. The correction terms are functions of the square of the Mach number. These corrections are small for Ma < 0.3, about 105 m/s. Hence, in ventilating systems and many other applications, air could be considered an incompressible medium during a flow process.

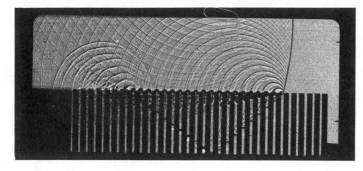

Figure 14.10.1 Shock waves can be damped by an energy dissipater, which consists of a matrix with many narrow open spaces through which the wave must pass. The comb model shown could represent houses and streets in a city or a protective structure that surrounds a fragile object. *(By H. Reichenbach of the Fraunhofer Institute for High-Speed Dynamics, Ernst Mach Institute, Freiburg, F.R.G., with permission.)*

Only in Sec. 14.9 did we discuss an effect on the flow that would be due to fluid viscosity. This effect can be a minor factor in some gas flows, such as nozzle flows, in making predictions about overall effects. Still, viscosity will leave its telltale signs. The shock waves will not be infinitesimally thin. Their thickness will depend on the viscosity. In the diverging passage of the nozzle, flow separation is likely to occur. This changes the effective passage area. Viscosity also enters the question of whether or not plane shock waves exist, instead of two-dimensional waves, within the nozzle. To answer this question, and many others, one would have to turn to experimental evidence and more encompassing theories. (Figure 1.8.4 shows the effects of boundary layers and separation in a converging-diverging nozzle.)

The work in this chapter has been only the beginning. (Goethe, the German writer and intellect, said, "Every beginning is difficult!") Nevertheless, we have made a good start. We have been able to address many questions and have formed a sound basis for further study. An understanding of the two-dimensional phenomena shown in Fig. 14.10.1 would be part of that further study.

EXERCISES

14.4.1(a) What is the speed of sound at the exit for this example?

(b) What would be the reservoir temperature if Ma = 1 at the exit for T_2 given in the example?

14.4.2(a) What is $\dot{m}$ for $p_2/p_1 = 0.242$?

(b) What is p_2/p_1 if $\dot{m} = 0.0073$ slug/s but $A_2 = 0.05$ in^2?

14.4.3(a) What is $\dot{m}$? (What are the units of A_2?)

(b) Is the Mach number always equal to 1 at the nozzle exit when the maximum mass rate of flow occurs in a converging nozzle?

14.7.1(a) What is the temperature in the nozzle where the area is 0.022 m^2, if the flow is sonic at the throat and subsonic subsequently?

(b) What should be the area of the throat if the flow is sonic there?

(c) For another installation, the jet's exit pressure is 101 kPa and the temperature is 10°C. (Exit area = 0.033 m^2.) If $\dot{m}$ is again 4.76 kg/s, what

are the reservoir pressure and temperature if the flow is sonic at the throat and subsonic thereafter?

14.8.1(a) What is $\dot{m}$ if T_0 is 10°C but all the other given data are the same as in the example?

(b) What is the density just upstream of the nozzle's exit where the shock wave occurs? ($T_0 = 10$°C once again.)

(c) What is the entropy change across the shock wave at the nozzle's exit if the reservoir gas temperature is 10°C?

14.8.2(a) Determine the pressure within the nozzle upstream of the shock wave that is located at $A_i/A_i = 0.55$ by referring to Fig. 14.8.2. (Use the nozzle and reservoir conditions of the example.) Use the appropriate equations from among Eqs. (14.8.1) to (14.8.7). Find ρ_u, T_u, and Ma$_u$, also.

(b) Should the result for p_d/p_0 lie along the dashed curve in Figure 14.8.2 for a shock wave at $A_t/A_i = 0.55$?

(c) If the receiver pressure is increased to $p_0/p_{\bar{e}} = 0.655$, determine the area ratio where the normal shock wave appears and the thermodynamic conditions upstream and downstream of the shock wave. What are Ma_u and Ma_d?

14.8.3(a) What are the conditions in the nozzle just downstream of the shock wave?

(b) What are the conditions in the nozzle where the area is 0.0364 m²?

(c) Use Fig. 14.8.2 and Table 14.8.1 to obtain the area where the shock wave should occur in this example.

(d) For the same reservoir conditions but $A_t/A_e =$ 0.45 and $p_e/p_0 = 0.65$, determine the location of any shock wave that might occur by using Figure 14.8.2.

14.9.1(a) Verify that the reservoir temperature is 593°R, and determine the mass rate of flow in the pipe.

(b) What are the temperature, velocity, density, and pressure of air at L_{max}?

(c) For the same value of inlet temperature and pressure, but a Mach number of 0.1, find L_{max}/D. Compare this value with that of the example (use $\bar{f} = 0.02$ for convenience, but comment upon the value of f that would be more appropriate).

PROBLEMS

14.1 Air at 20°C and 180 kPa flows in a pipe with an area of $16(10^{-4})$ m². What is the velocity in the pipe at a location, say, A_2, if the air originated in a reservoir where the pressure was 300 kPa and the temperature 77°C? What is $\dot{m}$?

14.2 Assume that all of the conditions in Prob. 14.1 are the same except that the reservoir temperature is unknown. If the flow is isentropic between the reservoir and A_2, what should the reservoir temperature be?

14.3 Beginning with the universal gas constant 8317J/kmol/K, determine the value of c_p for helium, nitrogen, and carbon dioxide. The values of k for these gases are 1.66, 1.4, and 1.29, respectively.

14.4 Determine the change in entropy for an isothermal expansion of air in a closed container for which the volume changes fivefold.

14.5 What is the entropy change during an isobaric expansion for which the volume increases fivefold?

14.6 What is the change of entropy for Prob. 14.1?

14.7 If Laplace saw the flashes from Napoleon's cannons from a distance of 2 km, what could have been the time delay before he heard their report? Assume a temperature of 16°C.

14.8 A bolt of lightning is seen, and the thunderclap is heard 6 s subsequently. Estimate the distance from the observer to the location where the lightning has struck if the temperature is 80°F.

14.9 What is $\dot{m}$ for a case similar to that treated in Examples 14.4.1 and 14.4.2, if $A_2 = 0.0625$ in² but $V_2 = 500$ ft/s?

14.10 If $\dot{Q} = 0$ and $\dot{W} = 0$, one can use the energy equation to analyze aspects of the conditions for the compressor, combustor, and turbine of a gas turbine. The heating value of the fuel in such a turbine is about 20,000 Btu/lb, which is equivalent to $5(10^8)$ ft-lb/slug. About 5 percent, by weight, of the air from the compressor would be routed to the combustor. The rest is later mixed with the heated portion and sent to the turbine. If the air and fuel mixture is originally 1050°R, 12 atmospheres, and Ma = 0.25, what should be the values of the temperature, pressure, and Mach number at the exit of the combustor? If these combustion products (assume that they have the properties of air) are now mixed with the remaining 95 percent of the compressor's discharge, what should be the temperature and pressure of the mixture? (Assume that the combustion takes place at constant pressure within a duct of constant area with fuel/air = 1/14.)

14.11 The mass flow in a wind tunnel is 2 kg/s when the reservoir conditions are 1 MPa and 500°C. What mass rate of flow can be expected if the reservoir pressure is changed to

2 MPa and its temperature remains the same?

14.12 (a) For the data in Example 14.4.1, what must be the pressure at point 2 (refer to Fig. 14.4.2) for $Ma_2 = 1$?

(b) What is T_2 for part (a)?

(c) What is $\dot{m}$ if $A = 0.05$ in^2 for part (a)?

(d) What is $\dot{m}$ if the pressure in the reservoir is reduced to give a value of p_2 that is 90 percent of the value calculated in part (a)? (The temperature in the reservoir remains constant.)

14.13 If the airflow in Prob. 14.2 continues adiabatically to a location in the pipe where $V_3/\sqrt{kRT_3} = 0.9$, what are the temperature and velocity there? What are the density and pressure there?

14.14 What should be the axial force between stations 2 and 3? (Refer to Probs. 14.2 and 14.13.)

14.15 Assume that air flows at a speed of V_1 with a pressure and density of p_1 and ρ_1. A streamlined body is placed in the flow and the pressure p_s is measured at the foremost tip of the body. This pressure and p_1 can be used to infer V_1. What percent of error in V_1 will result if the flow is treated as incompressible but in reality the Mach number is 0.1, 0.3, 0.5, or 0.7?

14.16 An aircraft is moving at an elevation of 2000 m in a standard atmosphere at $Ma = 0.7$. What is the speed of the aircraft? What is the temperature sensed by a thermocouple at the nose of the craft that is directed into the airstream? What pressure would be present at the tip of an impact tube that is used as an airspeed indicator?

14.17 An airplane flying at 3000 m has an impact tube as an airspeed indicator that senses a stagnation pressure 52% higher than the local ambient pressure. At what Mach number is the airplane flying? What could be the highest temperature of the airplane's surface (due to aerodynamic heating)?

14.18 A poor connection in a CO_2 pipeline allows the gas at 200 psia and 70°F to leak to the atmosphere (14.7 psia and 70°F). What is the velocity of the escaping jet? What is its temperature?

14.19 If an earth orbiter at 200,000 ft in preparation for landing is reported to be at $Ma = 8$, what should be its speed in mph? What is its speed at 50,000 ft if a Mach number of 3 is recorded there?

14.20 Assume that air decreases its velocity isentropically in a diffuser from 600 to 300 ft/s. At the location of the higher speed, the temperature and pressure are 40°F and 10 psia, and the area is 1.1 ft^2. What is the mass rate of flow? What are the temperature and pressure at the downstream location? What is the area there?

14.21 A turbine is operating at 25,000 ft where the conditions are described by the standard atmosphere. The exhaust gases, largely air, enter a converging-diverging nozzle at 300°F and 40 psia. If the gases pass through and out of the nozzle isentropically (full expansion), what are the exit speed and Mach number? Entrance $Ma = 0.6$.

14.22 Assume that the combustion product from a rocket is CO_2. The gas leaves the rocket supersonically from a (reservoir) chamber at a temperature and pressure of 2600 K and 3 MPa. The nozzle's exit pressure is 25 kPa. Find the theoretical exit velocity and temperature. If the exit area is 1 cm^2, what is the mass rate of flow? What is the area of the throat?

14.23 The nozzle on a rocket is designed to operate with p_e''' as its back pressure when at an altitude of 20,000 m. (Standard atmosphere: $T = 216.7$ K and $p = 0.05457p_0$, with $p_0 = 101.33$ kPa.) What should be the velocity of the exiting jet if the pressure in the reservoir ahead of the nozzle is 100 kPa? Determine the pressure at the nozzle's throat under these conditions. Find the temperature in the exiting jet if the reservoir temperature is 220 K. Determine the exit Mach number. What is $\dot{m}$? If A_{exit} is 0.005 m^2, estimate the area of the throat. The gas is helium.

14.24 Explain why the intersection of the $A_e/A_i = 0.606$ line and the $S = 1.00$ curve at $p_i/p_0 \approx 0.9$ in Fig. 14.8.2 is not a solution for Exercise 14.8.1. Describe the flow in the entire nozzle for this value of p_i/p_0.

14.25 Eliminate $\dot{m}$ between Eqs. (14.7.6) and (14.7.11) and solve for p_i/p_0 as a function of

A_i/A_{throat} for $k = 1.4$. (Use a microcomputer and employ an iterative solution, if necessary.) Find the two exit pressures for isentropic flow with $A_{exit}/A_{throat} = 1.2$.

14.26 If in a nozzle and reservoir system a shock wave occurs where the area is 0.025 m^2, what is the Mach number just upstream of the shock front? (Reservoir temperature and pressure are 20°C and 202 kPa. Throat area $= 0.02 \text{ m}^2$.) What is the Mach number just downstream of the shock discontinuity? Determine the strength of the shock wave by assessing the change of entropy across it.

14.27 What is the pressure change across the shock wave in Prob. 14.26?

14.28 The pressure in the receiver is 70 percent of the reservoir pressure of 300 kPa. The area ratio between the throat and the exit is 0.5. Is the flow isentropic? Justify your answer.

14.29 The nozzle in a rocket was designed to operate with p_e as its back pressure when at an altitude of 20,000 m. (Refer to Table A.3 for thermodynamic data and to Prob. 14.23 for pressure and geometric information.) Where is the shock wave with respect to the nozzle when the rocket is at 15,000 m?

14.30 Verify Eq. (14.8.11) by completing all the intermediate steps beginning with $p_u/\rho_u - p_d/\rho_d = c_t^2 - V_u^2$.

14.31 If the length of the tube in Example 14.9.1 is doubled but the same reservoir system is used, determine the pressure and temperature at the end of the tube.

14.32 A converging-diverging tube has exit and throat areas of 2 ft^2 and 1 ft^2, respectively. Atmospheric air enters the tube and is exhausted from it at 8.5 psia. What is the air temperature leaving the tube if it enters at 68°F?

14.33 If the Mach number upstream of a normal shock wave in air ($p = 101.3$ kPa and $\rho = 1.225 \text{ kg/m}^3$) is 2, what are the Mach number, pressure, and temperature just downstream of the shock wave?

14.34 A normal shock wave occurs in a converging-diverging nozzle such that the static pressure downstream of the shock wave is the same as that of the throat. What is the area of the duct at the shock location with respect to the throat area?

14.35 A large pressurized tank of nitrogen (1500 psia and 70°F) is equipped with a converging-diverging nozzle that exhausts into air at 14.7 psia and 70°F to simulate a rocket motor. What should be the ratio of the areas at the exhaust and throat portions of the nozzle to develop the maximum gas velocity at the exit of the nozzle? What should be the area ratio for maximum thrust? What is the magnitude of this thrust?

14.36 A supersonic wind tunnel has a test section with an area that is twice that of the throat. Downstream of the test section, the duct contracts to a control point where the area is 1.5 times that of the throat, after which it expands again into some vacuum tanks where its area is 4 times that of the throat. When the tunnel has atmospheric air (101.3 kPa) as its reservoir, a normal shock wave stands at the reduced control section downstream of the test section. What is the air's speed and Mach number inside the test section? What are the temperature and pressure of the air as it enters the vacuum tanks?

14.37 For a wind tunnel of the type described in Prob. 14.36, what should be the ratio of the test section to throat area to obtain $Ma = 2$ in the tunnel? If a normal shock wave is now made to stand at the control location downstream of the test section, again with an area 1.5 times that of the throat, what are the pressure and temperature at the entrance to the vacuum tank ($A_{tank}/A_{throat} = 4$)?

14.38 A normal shock wave stands in a converging-diverging nozzle. The conditions just upstream of the shock wave are $T_u = (273 + 10)$ K, $p_u = 200$ kPa, and $Ma_u = 2.0$. Find the corresponding variables just downstream of the shock wave.

14.39 Nitrogen flows from one large tank to another through a converging-diverging nozzle. The gas is to flow sonically at the nozzle's throat at the rate of 1 kg/s when the supply pressure is 700 kPa and 40°C. What should be the area of the nozzle's throat and exit if the gas is to flow isentropically into the receiving tank at a Mach number of 2.2? (Assume that the reservoir conditions remain constant.)

14.40 A design for a jet engine's exhaust nozzle

assumes that it has an inlet area of 0.95 m². Gas that has been expanded through the turbine enters the nozzle at 40 m/s, 800 K, and 280 kPa. The nozzle is intended to fully expand the exhaust down to the ambient pressure that corresponds to the altitude at which it is flying. Determine the area ratio ($A_{\text{exit}}/A_{\text{throat}}$) for altitudes of 0, 5000, and 10,000 m. What is the corresponding thrust developed by the nozzle at these altitudes? What is the exhaust thrust without the nozzle? Does the inlet kinetic energy significantly affect the nozzle's thrust?

14.41 An aircraft flying at an altitude of 20 km on a standard day ingests air at Ma = 2.2 and slows it to Ma = 1.4 in a supersonic diffuser. A normal shock wave is the cause for the change in Ma, and subsequently the air further decreases its Mach number due to the duct's geometry. Determine the pressure and temperature at the point where Ma = 0.49 if the subsonic process is isentropic. How has the duct changed area as the Mach number goes from 2.2 to 1.4 to 0.49? [*Note:* Figure 14.8.2 is not limited to flows beginning from rest. A supersonic flow entering a supersonic diffuser can "move" isentropically along the lower portion of the S = 0 curve until a shock wave occurs. The post-shock state point is along the dashed line that represents Eq. (14.8.14). In order to further increase the pressure as the area of the supersonic diffuser decreases, the state points of the gas are associated with the S value determined by the entropy increase across the shock wave. The value of Ma$_d$ is associated with the area of the location in the diffuser. The value of p_0 in this case is a reference value that corresponds to the pressure in a hypothetical reservoir.]

14.42 In a converging-diverging nozzle an air flow with Ma = 2 near the exit is produced. This flow is sensed by a Pitot-static tube in the nozzle. Downstream of the shock wave, the flow comes to rest at the tip of the probe. What should be the pressure sensed by the probe if the pressure just upstream of the shock wave is 0.07(101.3) kPa at 203K? What should be the pressure in the reservoir? (If the probe is aligned with the flow, the shock wave will be normal to the stag-

nation streamline, since it will be on the axis of symmetry of the three-dimensional shock wave. This will allow one to use the one-dimensional equations and Fig. 14.8.2.)

14.43 A supersonic aircraft is undergoing tests at 15,000 m. A Pitot-static tube transmits a pressure of 60 kPa above the ambient pressure. What is the (subsonic) Mach number behind the shock wave that is thought to exist ahead of the Pitot-static tube? What is the Mach number of the aircraft? (Note that the shock wave's surface will likely be a surface of revolution around the probe, and only on the centerline, the line of symmetry, are the "normal" shock wave equations of this chapter applicable. Elsewhere, one needs to use multidimensional equations to describe the shock wave parameters.)

14.44 A converging-diverging nozzle is attached to a reservoir where the temperature and pressure are 1000°F and 100 psia. A normal shock wave occurs at a point where the area is 2.5 times that at the throat. (Throat area = 2 in².) The exit area is 3 times that of the throat. What are the nozzle exit temperature and pressure? What would the pressure have to be downstream of the shock wave if the wave were to stand at the nozzle's exit? What would be the exit pressure for supersonic flow throughout the nozzle and no shock wave at the exit?

14.45 A nozzle is operating with air with a Mach number of 2.5 at the exit. If the reservoir temperature and pressure are 50°C and 600 kPa, what is the receiver pressure $p_{\bar{e}}$ such that no shock waves occur? What must be the receiver pressure for a normal shock wave to occur at the nozzle's exit? What is the maximum pressure in the receiver for Ma = 1 at the throat? What are the temperatures corresponding to these pressures?

14.46 The conditions upstream of a shock wave and those at the throat of a converging-diverging nozzle are related by the equation of mass conservation, $V_u \rho_u A_u = V_t \rho_t A_t$.
(a) Show that this equation can be rewritten as

$$\left(\frac{T_u}{T_t}\right)^p \text{Ma}_u = \frac{A_t}{A_u}$$

with $p = (k + 1)/2(k - 1)$.

(b) Use the energy equation between A_t and A_u to solve for T_u/T_t and obtain

$$\frac{A_t}{A_u} = \left[\frac{1}{\text{Ma}_u^2(k-1)/2 + 1}\right]^p (\text{Ma}_u)\left(\frac{k+1}{2}\right)^p$$

(c) What is the area ratio A_t/A_u for $\text{Ma}_u = 0.35$, and is this result in agreement with one of the abscissa scales of Fig. 14.8.2?

14.47 Air flows in a constant-area duct. At one point, the temperature, pressure, and Mach number are 1000°R, 99 psia, and 0.15, respectively, while farther downstream these variables have the value 1800°R, 90 psi, and 0.25. What is the rate of heat transferred between these two stations for these data? For the given upstream conditions, what would be the downstream velocity for $\text{Ma} = 0.25$ and $\dot{\mathcal{Q}} = 0$? What would be the pressure for isentropic conditions?

14.48 Air flows into a smooth, insulated constant-area tube at $\text{Ma} = 0.2$, 20 psia, and 600°R from a reservoir. (Diameter = 1 in.) Assume an average friction coefficient of 0.02, and find the distance at which $\text{Ma} = 0.7$. At what distance would $\text{Ma} = 1$? What is the mass flow rate? Now add 3 ft to the tube and determine the mass flow rate.

14.49 Nitrogen is to be pumped into a 20-cm smooth, insulated pipe at 1 MPa (absolute), 40°C, and 15 m/s. What pressure is to be expected 2000 m downstream? What should be the velocity and Mach number there?

14.50 A supersonic wind tunnel can be operated by reducing the pressure in vacuum tanks downstream of the test section and using atmospheric air as the reservoir. If these vacuum tanks are connected to the tunnel by a 6-in. smooth, insulated pipe, how long can this pipe be if the Mach number within it is to remain greater than or equal to 1 while the test section is at $\text{Ma} = 2$? (Assume atmospheric air at 20°C and 101.3 kPa.)

14.51 A wind tunnel has air at 1000 psia and 80°F in high-pressure containers that are remote from the reservoir chamber, to which is attached a converging-diverging nozzle. A smooth, insulated pipe 4 in. in diameter is used to interconnect the high-pressure and reservoir vessels, the latter of which should be at 100 psia. What is the maximum length

of 4-in. pipe that can be used without choking the flow?

14.52 After having located the wind tunnel's stagnation chamber and the high-pressure tanks a distance of 50 ft apart, a need arises to replace the insulated connecting pipe. What should be the minimum diameter to prevent choking for the conditions given in Prob. 14.51?

14.53 If, instead of a short converging nozzle, the air from a reservoir (300°C and 0.9 MPa) is passed through a 2.5-cm-diameter insulated tube that is 30 cm long, what is the expected flow rate if the exit pressure is 101.3 kPa? How does this flow rate compare with that obtained with a short nozzle?

14.54 Exhaust blowers to remove sawdust transport air with the suspended particles through ducts for discharge into a separator. If the pressure and temperature in such a separator are about 14.7 psia and 70°F, what must be the pressure and velocity of the air entering the duct in order to have a discharge velocity of 165 ft/s in an 8-in.-diameter duct that is 500 ft long? What power must be supplied to the blower? (Assume that the concentration of solids is sufficiently low that the physical properties of air can be used for these calculations.) Is there an optimum length of 8-in. duct for the given exit conditions?

14.55 Air is used to transport solids for drying. If the inlet to a drying column is a tube 15 ft long and the tube is to transport 0.5 lb of air per second—neglect the presence of the solid in this calculation—what should be its diameter if the pressure drop is to be less than 5 psi when the discharge pressure and temperature are 14.7 psia and 200°F?

14.56 Determine L_{max}/D for the viscous flow of air in an insulated duct ($D = 1$ ft). Assume $T = 80°F$ and $p = 20$ psia, and an initial air speed of 3 ft/s. Assume $\bar{f} = 0.015, 0.025$.

14.57 Assume that at the inlet of an insulated duct $\text{Ma} = 0.1$, determine the fraction of L_{max}/D for the location where $\text{Ma} = 0.5$. Assume $f = \bar{f} = 0.02$.

14.58 Using a microcomputer, numerically integrate Eq. (14.9.11) with an appropriate (variable) f for the case of Example 14.9.1. Compare the result with using $\bar{f} = 0.02$.

15

Open Channel Flow

15.1 INTRODUCTION

We encountered open channel flow when the Bernoulli equation was first introduced in Chap. 5. The flow under a sluice gate was treated because in the usual cases, it occurs without an appreciable loss of energy. Subsequently, when we studied the momentum principle, we were able to determine the force that would be induced on a sluice gate as a consequence of a flow. In examining the interconnection between the momentum and energy principles in Chap. 7, we analyzed the loss of mechanical energy, $V^2/2g + z + p/\gamma$, that must accompany an increase in the free-surface elevation in a channel with a horizontal bottom. This would be the result for a surface wave or a hydraulic jump.

Example 5.2.3 treated the problem that is again illustrated in Fig. 15.1.1. When using the energy equation to solve for the elevation of the water over the hump, we discovered that two solutions appeared to be possible. Because the study of open channel flows is often concerned with the elevation of the free surface in various parts

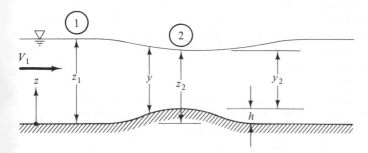

Figure 15.1.1 Definition drawing for two-dimensional flow in channel having change in bottom elevation.

of the channel, it is appropriate that we begin this chapter by seeking a means to predict which of the two solutions appropriate to Fig. 15.1.1 is the more likely to occur. Following the approach in Example 5.2.3, but including losses, we can write the energy equation as

$$\frac{V_1^2}{2g} + z_1 = \frac{V_2^2}{2g} + z_2 + \text{losses}_{1\text{-}2}$$

The volume rate of flow is the same at each section, so that

$$V_1 z_1 = V_2(z_2 - h) = V_2 y_2 = Q$$

If $V_1 z_1/(z_2 - h)$ is used instead of V_2 in the energy equation, it becomes

$$\frac{1}{z_1}\frac{V_1^2}{2g} + 1 = \frac{1}{z_1}\frac{V_1^2 z_1^2}{2g(z_2 - h)^2} + \frac{z_2}{z_1} + \frac{\text{losses}_{1\text{-}2}}{z_1}$$

which can be revised as

$$\frac{1}{2}\frac{V_1^2}{gz_1} + 1 = \frac{1}{2}\frac{V_1^2}{gz_1}\frac{1}{(z_2/z_1 - h/z_1)^2} + \frac{z_2}{z_1} + \frac{\text{losses}_{1\text{-}2}}{z_1}$$

In Example 5.2.5, z_2 was either 1.71 or 4.94, assuming that the losses were zero. (If losses were present, the dotted horizontal line in Fig. 5.2.9 would intersect the curve in such a way that the two solutions would be slightly closer together.) It is useful for further discussion to write the energy equation as

$$\frac{1}{2}\frac{V_1^2}{gz_1}\left[1 - \frac{1}{(z_2/z_1 - h/z_1)^2}\right] + \left(1 - \frac{z_2}{z_1}\right) = \frac{\text{losses}_{1\text{-}2}}{z_1} \qquad (15.1.1a)$$

These losses should be nonnegative. What is the relationship between V_1^2/gz_1, which is Fr_1^2, and z_2/z_1 so that this energy condition is realized? As $z_2/z_1 \rightarrow h/z_1$, the term in square brackets approaches $-\infty$. Because of this, Fr_1 must approach zero—it cannot be negative—so that the positive "sum" within parentheses can influence the algebraic sign of the losses. (The value of this parentheses term approaches $1 - h/z_1$.) The result will be a positive, finite value on the right-hand side of Eq. (15.1.1a). As z_2/z_1 increases beyond h/z_1, the term in square brackets becomes less negative. Fr_1 can increase to give a negative product that must be compensated by $1 - z_2/z_1$, which concurrently becomes decreasingly positive. This smaller positive value of $1 - z_2/z_1$ leads to the result that Fr_1 cannot continue to increase as z_2/z_1 approaches a value of 1. It is clear that for $z_2/z_1 = 1$, and no losses, $\text{Fr}_1 = 0$ in Eq. (15.1.1a). (A zero Froude number could imply $V_1 = 0$, which would be consistent with a system without losses.) In the range of z_2/z_1 between h/z_1 and 1, the maximum value of Fr_1 proves to be less than 1 if $h/z_1 < 1$.

Thus, when the upstream Froude number is less than 1, there is a decrease in the fluid's elevation in going from position 1 to position 2. This is shown in Fig. 15.1.2. The curves show the value of Fr_1 for values of z_2/z_1, under the assumption that there are no losses. If losses are indeed present, the value of Fr_1 for a particular value of z_2/z_1 will be within the shaded region of the particular curve.

Figure 15.1.2 indicates that if $\text{Fr}_1 > 1$, then z_2/z_1 would be greater than 1. This means that there would be an increase in the free surface's elevation. The shape of the diagram given in Fig. 15.1.2 can be comprehended by referring to Eq. (15.1.1a) once again. As $z_2/z_1 \rightarrow (1 + h/z_1)$, the term in square brackets begins to vanish, but $1 - z_2/z_1$

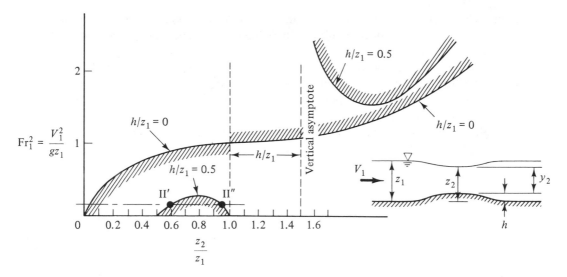

Figure 15.1.2 Solutions for the ratio of surface elevations as a function of upstream Froude number and bottom elevations.

concurrently becomes negative. For positive, or even zero, losses to occur, Fr_1 would have to become very large. Again, as $z_2/z_1 \to \infty$, $1 - z_2/z_1$ becomes very negative. However, the term within the square brackets becomes nearly 1 for such large z_2/z_1 values. The consequence is that $Fr_1 \to \infty$ as $z_2/z_1 \to \infty$. Equation (15.1.1a) implies that Fr_1 would approach $2(z_2 + \text{losses})/z_1$ for large z_2/z_1. For intermediate values of z_2/z_1, Fr_1 should reach a minimum. Because the losses must be positive in a realistic case, the admissible values of Fr_1 are within the shaded region of Fig. 15.1.2; the exact point therein is determined by the magnitude of the losses.

Example 15.1.1

Water is flowing in a horizontal channel at a depth of 0.8 m and passes over a hump with a height of 0.4 m. At the crest of the hump, a measurement indicates that $z_2/z_1 = 1.9$. What is the volume rate of flow in the channel per unit width? Assume no fluid losses in passing over the hump.

Solution Figure 15.1.2 indicates that the flow upstream of the hump has a Froude number greater than unity. Equation (15.1.1a) can be written as

$$\frac{1}{2}Fr_1^2\left[1 - \frac{1}{(1.9 - 0.5)^2}\right] = -(1 - 1.9)$$

$$Fr_1 = 1.92$$

Therefore, $\qquad V_1 = 1.92\sqrt{9.8(0.8)} = 1.92(2.8) = 5.37 \text{ m/s}$

and the flow rate is $\quad V_1 y_1 = 5.37(0.8) = 4.29 \text{ m}^3/\text{s}$

for each meter of channel width.

With this discussion of the *subcritical* ($Fr_1 < 1$) and *supercritical* ($Fr_1 > 1$) upstream flow, we now turn, as before, to specifying which of the two values in Fig. 5.2.9

is likely to occur. The two solutions for z_2/z_1 are also indicated as II′ and II″ in Fig. 15.1.2. We can see that both solutions satisfy the requirement that $z_2/z_1 < 1$, since Fr_1 is less than 1.

We can observe that the smaller value of z_2 gives a value for Fr_2, $V_2/\sqrt{g(z_2 - h)}$, that is greater than unity, while the larger z_2 in the solution yields $Fr_2 < 1$. This means that our solution has given the possibility of a supercritical and a subcritical flow at the apex of the hump.

An answer is facilitated by recasting the energy equation in terms of the depth of liquid, $y_2 (= z_2 - h)$. This leads to writing Eq. (15.1.1a) as

$$\frac{V_1^2}{2g} + z_1 = \left(\frac{V_2^2}{2g} + y_2\right) + h + \text{losses}_{1\text{-}2} \qquad (15.1.1b)$$

or, with $V_2 = \tilde{Q}/y_2$,

$$\frac{V_1^2}{2g} + z_1 = \left(\frac{\tilde{Q}^2}{2gy_2^2} + y_2\right) + h + \text{losses}_{1\text{-}2} \qquad (15.1.2)$$

The terms grouped in parentheses in these equations are termed the *specific energy* E_2, and $\tilde{Q}$ is designated here the *volume rate of flow per unit width*. (In the subsequent paragraphs, the loss term will be *assumed* to be zero. Later, we shall comment upon the effect of the losses on the results.) Consequently, Eq. (15.1.2) is

$$\frac{\tilde{Q}^2}{2gy_1^2} + y_1 = \frac{\tilde{Q}^2}{2gy_2^2} + y_2 + h$$

since $z_1 = y_1$. This means, in terms of specific energy, that

$$E_2 = E_1 - h \qquad (15.1.3)$$

If $\tilde{Q}$ and y_2 were known, E_2 would be a definite number. But for the stated value of $\tilde{Q}$, there are other y_2's which would give this same value of E_2. This is shown through the drawing in Fig. 15.1.3. The curve through point C in this figure is attained by starting with the equation defining specific energy,

$$E_1 = \frac{\tilde{Q}^2}{2gy_1^2} + y_1 \qquad (15.1.4)$$

which is recast as

$$y_1^2(E_1 - y_1) = \frac{\tilde{Q}^2}{2g} \qquad (15.1.5)$$

(Statements similar to the last two could be written at any station by using the appropriate subscripts.) For small values of $\tilde{Q}$, say, $\tilde{Q} = 0.5$ cfs/ft, $E - y \to 0$ in Eq. (15.1.5), so that the 45° asymptote is apparent. Similarly, for large values of E, $y \to 0$.

For a specified flow rate $\tilde{Q}$, the right-hand side of Eq. (15.1.5) is a constant. With E_1 specified, two positive values of y ensue, as Fig. 15.1.3 indicates. For a given value of the upstream specific energy E_1, E_2 need not remain fixed, since it will vary with h, according to Eq. (15.1.3). Other curves could also be drawn for different flow rates, using Eq. (15.1.5), and these are shown as dashed lines in Fig. 15.1.3. This figure is helpful in understanding open channel flows.

In Example 5.2.3, with $\tilde{Q} = 10$ m²/s, $z_1 = y_1 = 5$ m, and $h = 0.5$ m, E_1 equals $10^2/[(2)(9.8)(25)] + 5$, which is 5.2041. This value corresponds to $E_2 + h$, according to

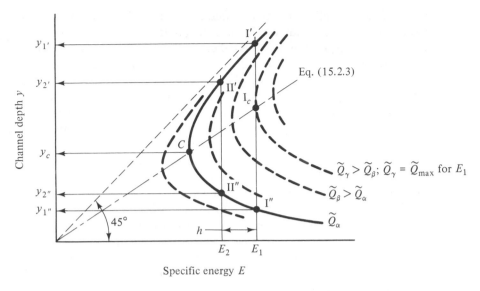

Figure 15.1.3 Channel depth as a function of specific energy of the flow.

Eq. (15.1.3). We shall now designate y as the water's depth somewhere between the point where the hump begins to rise and the highest point of the bottom, where $h = 0.5$. The value E_1 has been marked in Fig. 15.1.3, and we shall now observe what must happen to E, and the depth y, as one proceeds from the foot of the profile in Fig. 15.1.1 to the crest, where $E = E_2 = E_1 - h$ and $y = y_2$. In going to this latter point, the bottom's elevation changes from zero to h, so that the specific energy of the fluid will decrease as position 2 in Fig. 15.1.1 is approached. This is in concert with Eq. (15.1.3). The decrease in E is obtained by subtracting h from the value of E_1 that prescribed point 1 and recording the value on the abscissa of Fig. 15.1.3 as E_2. There are two values for y associated with E_2. These are denoted as y_2' and y_2''. For the particular data selected for the example that has served us in this discussion (Example 5.2.3), the value of y_2' associated with the upper branch of the yE curve is 4.44, and along the lower branch, $y_2'' = 1.21$. These two depths are termed *alternate depths*; one is the alternate of the other for a given value of the specific energy E_1. Again, we note that point II′ corresponds to subcritical *(tranquil)* flow and II″ to supercritical *(rapid* or *shooting)* flow.

While proceeding from the base of the hump to its crest, the liquid depth has gone from the value associated with E_1 to that of E_2. As the specific energy is decreased, one moves from point I′, which is determined by the upstream conditions, toward II′ in Fig. 15.1.3. This is the first "solution" for station 2 that is encountered while continuously moving from point I′ along the solid curve in this figure.

One cannot proceed directly from solution II′ to II″ for the channel's bottom profile that is given by h. If one were to prescribe a locus of vertical points between II′ and II″, one would be requiring values of $\tilde{Q}$ that are different from that initially prescribed. (This could be the case if the flow should change sufficiently due to a transient condition, or, more important, if the channel width were decreasing—since $\tilde{Q} = Q/\text{width}$.)

If h were sufficiently large that a solution to Eq. (15.1.3) would be associated with point C in Fig. 15.1.3, one would have the minimum specific energy for the given value of $\tilde{Q}$. The corresponding depth is called the *critical depth* y_c. For a bottom profile with a maximum elevation that is sufficient to obtain the critical depth, one could approach point C from either point (E_1, y_1') or point (E_1, y_1''), depending on the upstream depth. From point C, associated with the crest of a very special hump for the given $\tilde{Q}$, it would appear that one could then proceed along either the upper (subcritical) or lower (supercritical) portion of the curve Fig. 15.1.3 to obtain a downstream depth at a prescribed location. The actual path, however, is not ambiguous. It will be determined by conditions downstream of the hump in the channel.

One can see from Fig. 15.1.3 that for large values of E_1 that are accompanied by the large values of y_1 of subcritical flow, the change in surface elevation will be of the order of magnitude of the height of the disturbance, h. Similarly, if the flow is supercritical upstream, the changes in surface elevation will be much smaller than h if the initial specific energy is large. However, for values of the specific energy E_1 near the "nose" of the curve, at point C, a small disturbance—due to a small value of h—will cause a relatively large change in the surface elevation. After the small hump, the free surface will move toward its original elevation, but waves that are slow to damp would be generated.

15.2 CRITICAL CONDITIONS

The flow conditions that lead to point C in Fig. 15.1.3 are of interest because this point divides the Ey curve into subcritical and supercritical flows. Starting with $E = y + \tilde{Q}^2/2gy^2$, one can find the manner in which E varies with y, much as the qualitative results were found at the close of the last section, as

$$\frac{dE}{dy} = 1 - \frac{\tilde{Q}^2}{gy^3} = 1 - \frac{V^2}{gy} = 1 - \text{Fr}^2 \qquad (15.2.1)$$

At point C, E does not vary with y, and so here $dE/dy = 0$. Consequently,

$$0 = 1 - \text{Fr}^2 \quad \text{and} \quad V_c^2 = gy_c$$

or

$$y_c = \sqrt[3]{\frac{\tilde{Q}^2}{g}} \qquad (15.2.2)$$

Then E_c, $(y_c + V_c^2/2g)$, becomes equal to $\frac{3}{2}y_c$, so that

$$y_c = \tfrac{2}{3}E_c \qquad (15.2.3)$$

These results indicate that, for a prescribed $\tilde{Q}$, y_c and E_c can be found. In fact, Eq. (15.2.3) would plot as a straight line with a slope of $\frac{2}{3}$. This line is the locus of critical conditions for which the specific energy is a minimum for a given discharge per unit width, $\tilde{Q}$.

If we examine Fig. 15.1.3, we note that as $\tilde{Q}$ increases from $\tilde{Q}_\alpha$ through $\tilde{Q}_\gamma$, the curves for $\tilde{Q}$ (shown dotted) move to the right of the diagram. For a given value of E_1, one could draw a vertical line in the figure and find all of the curves for constant $\tilde{Q}$ which intersect this line. These curves, with their associated values of $\tilde{Q}$, would give the appropriate alternate depths y' and y'' that would be expected for this E_1. One curve, with a maximum value of $\tilde{Q}$, would be tangent to the vertical line drawn through E_1. The alternate depths would be equal, and one would have "critical" conditions again. This would correspond to the maximum discharge per unit width, $\tilde{Q}$, for a given initial value of the specific energy.

This discussion could also have proceeded by writing E in terms of $\tilde{Q}$ and y by considering E as a constant, E_1, and then finding $d\tilde{Q}/dy$. This derivative will be zero for a maximum value of E between zero and the prescribed value E_1. The solution to the associated equations yield, as we now have cause to expect, a result similar to Eqs. (15.2.2) or (15.2.3).

Example 15.2.1

The example that we have treated heretofore (Example 5.2.3) has a $\tilde{Q}$ of 10 m²/s and $y_1 = 5$ m, so that $E_1 = 5.2041$ m. If this channel is gradually narrowed while the bottom remains at a constant elevation, what is the maximum value of $\tilde{Q}$ that can be achieved?

Solution   In view of what has been said, we seek a critical condition, since we want $\tilde{Q}_{max}$ for E_{given}. This means that the associated critical depth would be, after Eq. (15.2.3),

$$y_c = \tfrac{2}{3}(5.2041) = 3.47 \text{ m}$$

and Eq. (15.2.2) would give

$$\tilde{Q}_c = \{[\tfrac{2}{3}(5.2041)]^3(9.8)\}^{1/2}$$
$$= 20.25 \text{ m}^2/\text{s}$$

The original discharge was 10 m³/s per meter of width, but the channel has been reduced in cross section and the same total flow occurs from section to section. If the width at the narrowest section were called w_{min}, then

$$Q = (w_{min})(20.25) = (w_{original})(10)$$

This means that w_{min} is 49 percent of the original channel width. We shall next inquire into the consequences of having made the channel narrower (e.g., to minimize the length of a bridge that spans a river).

It appears clear, after some thought, that in solving Example 15.2.1 we have moved in the diagram, depicted in Fig. 15.1.3, vertically downward from I′ to some intermediate point, I_c, where the flow is "critical." If we examine Eq. (15.2.2), which applies to this condition, we have $V_c^2/gy_c = 1$ at all the points given by critical conditions. Hence, when one has narrowed the channel and increased $\tilde{Q}$ to $\tilde{Q}_{max}$, the conditions that occur subsequently in the channel—either supercritical or subcritical—will depend upon the conditions downstream of the contraction. This is similar to the result that was found when the bottom elevation was such that y_c was reached.

Example 15.2.2

In solving for the water's elevation z_2 at the crest of the hump, we found in Example 5.2.3 for $z_1 = y_1 = 5$ m, $\tilde{Q} = 10$ m²/s, and $h = 0.5$ m that $z_2 = (1.21 + 0.5)$ or $(4.44 + 0.5)$ when the losses were neglected. These correspond to

$$(\text{Fr}_2')^2 = \frac{10^2}{(9.8)(4.94 - 0.5)^3} = 0.12$$

and

$$(\text{Fr}_2'')^2 = \frac{10^2}{(9.8)(1.21)^3} = 5.76$$

We now know that since the original flow was subcritical and the hump did not produce "critical" flow conditions, the flow will remain subcritical. The elevation is thus 4.94 m, and the fluid depth above the hump, y_2, is 4.44 m.

15.3 EFFECT OF CHANNEL CROSS SECTION ON A CONSTANT FLOW

We shall now examine the effect of channel cross section, specifically, by assuming a rectangular channel and investigating the interrelationship between the bottom elevation, or the channel width, on the local Froude number. This parameter will serve to characterize the situation. The channel is nominally horizontal, but local changes of height h can occur.

If the channel has a constant width and the bottom elevation changes, then $\tilde{Q}$ remains constant along it. The total energy at some reference location upstream where the bottom serves as the elevation datum ($z_0 = y_0$) is E_0. At subsequent downstream positions,

$$E_0 = h + y + \frac{\tilde{Q}^2}{2gy^2} = h + E \tag{15.3.1}$$

in which h is the change of the bottom's height and y is the depth of the liquid. At various downstream places, y and h may be different. We now ask how the variables change and how the changes interrelate. (It has been seen in Chap. 14, on compressible flows, that Mach number and nozzle shape are also interrelated.) We shall seek to know how the quantities change with the coordinate x, which is taken as positive in the direction of flow. For this reason, we differentiate Eq. (15.3.1) with respect to x:

$$0 = \frac{dh}{dx} + \frac{d}{dx}\left(y + \frac{\tilde{Q}^2}{2gy^2}\right)$$

or

$$0 = \frac{dh}{dx} + \left(1 - \frac{\tilde{Q}^2}{gy^3}\right)\frac{dy}{dx}$$

since E_0 and $\tilde{Q}^2$ are constants. However, as we noted in writing Eq. (15.2.1), $\tilde{Q}^2/gy^3$ is Fr^2, and so our result can be written as

$$\frac{dh}{dx} = \frac{dy}{dx}(\text{Fr}^2 - 1) \tag{15.3.2}$$

If the bottom is rising in the direction of the stream, dh/dx is positive. Hence, the depth will increase—$dy/dx > 0$—or decrease accordingly as Fr is greater or less than 1. Clearly, we can also make deductions about a receding bottom. These relationships are summarized in Table 15.3.1.

TABLE 15.3.1

$\dfrac{dy}{dx} > 0$	for	$\dfrac{dh}{dx} > 0$	Fr > 1
		$\dfrac{dh}{dx} < 0$	Fr < 1
$\dfrac{dy}{dx} < 0$	for	$\dfrac{dh}{dx} > 0$	Fr < 1
		$\dfrac{dh}{dx} < 0$	Fr > 1

What is the condition when $dh/dx = 0$ in a horizontal channel bed? This would be at the crest of a hump along the bottom. Then Eq. (15.3.2) would require that $dy/dx = 0$ and/or Fr = 1. When $dy/dx = 0$, the liquid's depth will be either maximum or minimum. Since this occurs where $dh/dx = 0$, one has also a maximum or a minimum of the free surface's elevation, z.

In Section 15.6 we shall utilize the possibility that $dy/dx \neq 0$ but Fr = 1. This situation could occur when a liquid spills over the top of a broad-crested protrusion. This is usually called a *broad-crested weir,* and it can serve to control the flow. As the liquid cascades over the hump where $dh/dx = 0$, the free surface is constantly falling in such a way that the depth is decreasing. Hence, $dy/dx \neq 0$. This would require the Froude number to be unity at this location of maximum obstruction.

The *many assumptions* that have been introduced are apparent. Some of them are a uniform flow, no losses, and a hydrostatic pressure distribution. The last requirement would not occur if the submerged hump caused the flow to change speed significantly or to produce streamlines with a small radius of curvature. (The flow over a spillway or weir that is thin in the flow direction would have such streamlines.) Hence, it would be advisable to calibrate a specific installation after the appropriate principles have been utilized in its design.

It is possible that the hump will be higher than the height required to produce critical conditions. (That is, one will have moved to the left from point E_1 in Fig. 15.1.3 by an amount h so that one is now to the left of the "knee" in the curve.) Then this hump would serve as a control device and cause a change of the upstream conditions. Perhaps the surface elevation would be changed. (Weirs and gates are used to control the flow rate in a channel and are often referred to simply as *controls.*) The specific energy ahead of the control would be increased. The change of the reference specific energy would have to be sufficient for the upstream and downstream specific energy points both to be on a single Ey curve.

Example 15.3.1

If it is known that the bottom of a channel has risen 0.1 m due to a hump while the free surface has lowered 0.115 m in the same 2 m of channel length, what Froude number can be expected in this section of the channel?

Solution For this case, Eq. (15.3.2) would read

$$\frac{0.1}{2} = -\frac{0.115}{2}(\mathrm{Fr}^2 - 1) \qquad \text{with Fr} = 0.36$$

Similar considerations apply to a horizontal channel of varying width but of rectangular cross section. If the width is w, we know that

$$w\tilde{Q} = Q = \text{constant} = Vyw \tag{15.3.3}$$

and the energy equation for a condition without losses would be written as

$$\frac{V_1^2}{2g} + y_1 = \frac{V_2^2}{2g} + y_2 = \frac{V_i^2}{2g} + y_i \tag{15.3.4}$$

in which there is no reason to explicitly write the value 1 or 2 for i. This equation becomes

$$\frac{\tilde{Q}^2}{2gy_i^2} + y_i = \text{constant} \tag{15.3.5}$$

since $V_i = \tilde{Q}/y_i$. $\tilde{Q}$ and y_i are both variables, and for the expression of volume conservation given by Eq. (15.3.3),

$$d(w\tilde{Q}) = 0$$

or

$$w\frac{d\tilde{Q}}{dx} + \tilde{Q}\frac{dw}{dx} = 0$$

Moreover, the way in which the specific energy will change with x is obtained by differentiating Eq. (15.3.5):

$$\frac{\tilde{Q}}{gy^2}\frac{d\tilde{Q}}{dx} - \frac{\tilde{Q}^2}{gy^3}\frac{dy}{dx} + \frac{dy}{dx} = 0 = \frac{\tilde{Q}}{gy^2}\frac{d\tilde{Q}}{dx} + \frac{dy}{dx}(1 - \text{Fr}^2)$$

$d\tilde{Q}/dx$ can be eliminated between this equation and the previous one, yielding

$$\frac{y}{w}\frac{dw}{dx} = \frac{dy}{dx}\left(\frac{1}{\text{Fr}^2} - 1\right) \tag{15.3.6}$$

One can interpret this result by referring to Table 15.3.2.

TABLE 15.3.2

$\frac{dy}{dx} > 0$	for	$\text{Fr} < 1$	$\frac{dw}{dx} > 0$
		$\text{Fr} > 1$	$\frac{dw}{dx} < 0$
$\frac{dy}{dx} < 0$	for	$\text{Fr} > 1$	$\frac{dw}{dx} > 0$
		$\text{Fr} < 1$	$\frac{dw}{dx} < 0$

Since, in Eq. (15.3.6), $y/w > 0$, we can conclude for the narrowest parts of the channel, where $dw/dx = 0$, that $dy/dx = 0$ and/or $\text{Fr} = 1$. If the critical condition is reached (i.e., by altering conditions so that the operating condition, in effect, moves vertically along a given E line and finds the $\tilde{Q}$ curve that is tangent to it in the Ey diagram—as in Fig. 15.1.3), the flow would then be able to proceed either subcritically or supercritically in the widening portion of the channel. The manner of the flow would again be determined by the downstream conditions. If the channel is too narrow and $\tilde{Q}$ is so high there that no Ey curve will intersect the prescribed E_1, the channel is designated "choked." The result must be a change in the upstream total head, which in this case is also the specific energy.

15.4 NONRECTANGULAR CHANNEL SECTION

Most natural watercourses do not have a rectangular cross section. In many populated areas, the riverbeds and shorelines are given a concrete facing. The cross section that results is nearly trapezoidal. Fortunately, our previous work for a rectangular channel is useful for these cases too. If A is the local cross-sectional area and B is the local breadth of the water surface, then in the energy equation, say, Eq. (15.3.4), $V = Q/A$. Small changes in the channel's area, due to increases in liquid depth, are $dA = B \, dy$, which is equivalent to $dA/dy = B$. This result would be useful for finding the values of y for a minimum specific energy, just as we did in determining the conditions given by Eq. (15.2.2). It is left as an exercise for the reader to show that

$$Q_c^2 = \frac{gA^3}{B} \quad \text{or} \quad \frac{Q_c^2}{B^2} = \tilde{Q}_c^2 = g\left(\frac{A}{B}\right)^3 \tag{15.4.1}$$

for critical conditions. [*Hint:* Start with a form of the energy equation that employs V, and then substitute $V = Q/A$. Follow the pattern that led to Eq. (15.2.2).]

By comparing this result for $\tilde{Q}_c$ with Eq. (15.2.2), we can see that A/B is a generalized "depth." A further examination of the situation would show the advantage of writing the local Froude number—previously $\tilde{Q}^2/gy^3$—as $\tilde{Q}^2/g(A/B)^3$ or $(Q^2/B^2)/g(A/B)^3$, which can be simplified to Q^2B/gA^3. It is now tempting to see whether, by starting from first principles, a result that is analogous to Eq. (15.3.2) would follow, keeping in mind the definition of the local Froude number that we have just given. Such a result would occur for a triangular cross section if the revised definition of Fr were kept in mind and the change in h were due to a change of elevation of the trough and not due to a hump. This would keep the cross section congruent, something that would not be the case if a horizontal channel's width varied alone.

15.5 MOMENTUM CONSIDERATIONS IN OPEN CHANNEL FLOWS

A review of the analysis for determining the force on the fluid due to a sluice gate, $\mathscr{F}_x$, that is given in Chap. 7 would show that for the rectangular channel in Figure 15.5.1,

$$\frac{\gamma_1 y_1^2}{2} - \frac{\gamma_2 y_2^2}{2} + \mathscr{F}_x = \rho \tilde{Q}(V_2 - V_1) \tag{15.5.1}$$

for the situation shown. This expression, which separates the total forces and momentum changes, is ideal for analyzing the problem initially. Now we shall rewrite it to meet our present needs:

$$\frac{\gamma y_1^2}{2} + \rho \tilde{Q}V_1 = \frac{\gamma y_2^2}{2} + \rho \tilde{Q}V_2 + (-\mathscr{F}_x) \tag{15.5.2}$$

We remark in passing that $\mathscr{F}_x$ itself is a negative number for the situation depicted in Fig. 15.5.1, so that the last term on the right-hand side of Eq. (15.5.2) is positive. We

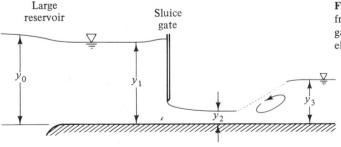

Large reservoir

Sluice gate

Figure 15.5.1 Channel flow emanating from a reservoir, proceeding under a sluice gate, and subsequently increasing its surface elevation quite abruptly.

now rewrite the terms on the left-hand side of Eq. (15.5.2) as

$$\gamma\left(\frac{y^2}{2} + \frac{\tilde{Q}^2}{gy}\right)$$

and define the term in parentheses as the *momentum function* MF. Thus, Eq. (15.5.2) is written as

$$\mathrm{MF}_1 = \mathrm{MF}_2 + \frac{\mathcal{R}_x}{\gamma} \qquad (15.5.3)$$

in which

$$\mathrm{MF} = \frac{\tilde{Q}^2}{gy} + \frac{y^2}{2} \qquad (15.5.4)$$

and $\mathcal{R}_x$ is the reaction of the *fluid on the gate,* which is a positive quantity in this case.

Just as in the discussion of specific energy, we shall determine how MF varies with y, for fixed $\tilde{Q}$. The result is shown in Fig. 15.5.2a. Point C, the nose of the curve, can be specified by differentiating Eq. (15.5.4) with respect to y and setting the derivative to zero. The result is $y_c = (\tilde{Q}^2/2g)^{1/3}$. *This means that the depth for minimum specific energy is also the depth for a minimum of the momentum function.* Figure 15.5.2a can be used in conjunction with Fig. 15.5.2b to analyze certain situations. Although we were concerned with a sluice gate in writing Eq. (15.5.1), the result would have been similar had we been concerned with the flow past a submerged hump or a hydraulic jump (see Chap. 7). In those cases, $\mathcal{R}_x$ would be the force of the fluid on the surroundings, a "drag" force; or the result of fluid friction on the bottom of the channel, a "bottom shear" force.

Point I' in Fig. 15.5.2b corresponds to location 1 in Fig. 15.5.1. Point II" has the same specific energy as point I' in Fig. 15.5.2b, because it was assumed that there are no losses in going past the gate. This determines the fluid depth, which is the same as in Fig. 15.5.2a. Figure 15.5.2a shows a difference between the momentum functions upstream and downstream of the gate that is determined by the reaction on the gate and γ. Point III' in Fig. 15.5.2a results from assuming no bottom shear in the case of a hydraulic jump, so that $\mathrm{MF}_{\text{before}} = \mathrm{MF}_{\text{after}}$, or $\mathrm{MF}'' = \mathrm{MF}'$ for the conditions designated

Figure 15.5.2 (a) Free-surface elevation versus momentum function for flow depicted in Fig. 15.5.1. $\mathcal{R}_x$ = reaction force on gate. $\mathcal{S}_x$ = bottom shear. $\mathrm{MF}_1 - \mathcal{R}_x/\gamma = \mathrm{MF}_2$. (b) Free-surface elevation versus specific energy of flow depicted in Fig. 15.5.1.

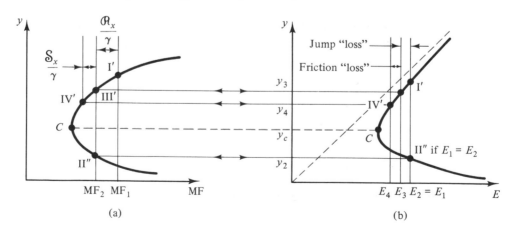

II″ and III′. The two depths associated with the same value of the momentum function are called *conjugate depths.* Had there been some shear or drag due to large stones, point IV′ would have been achieved because $MF_{after} = MF_{before} - drag_x/\gamma$. The value of y for point IV′ would be translated to increased losses on the Ey diagram (Fig. 15.5.2b), where there is a point IV′ corresponding to that on the (MF)y diagram (Fig. 15.5.2a).

15.6 UPSTREAM AND DOWNSTREAM EFFECTS

In treating the situation depicted in Fig. 15.5.2, we assumed that the open channel flow was "triggered" from supercritical at location 2 to subcritical at location 3 by means of a hydraulic jump. (The operating condition went from point II″ to III′ in this figure.) The reasons why this might have occurred are now discussed. Suppose, as in Fig. 15.6.1a, there is a hump at a large distance downstream of the sluice gate. The flow is *first assumed* to approach the hump in a supercritical state. As long as the height of the hump, h, is sufficiently low, the flow over the hump remains supercritical. If the hump's height were to increase until the flow was critical at this section, nothing would occur to alter the oncoming supercritical condition. Ultimately, the height of the hump could be increased so that the specific energy ahead of the hump would be insufficient to maintain the prescribed conditions. (In Fig. 15.1.3 or 15.5.2b, one would have proceeded to the left of point C.)

In order to maintain the flow, the specific energy upstream of the hump, for the prescribed flow, would have to increase. (Since the flow is supercritical, with Fr > 1, just downstream of the sluice gate, it is assumed that the disturbances that occur at the hump are not communicated upstream to the gate, so that $\tilde{Q}$ remains unaffected.) From an Ey diagram such as Fig. 15.5.2b, we can see that an increase in E, for constant $\tilde{Q}$, can occur for the case of supercritical flow, if the depth decreases. This would

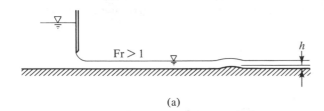

Fr > 1

h

Figure 15.6.1 (a) Supercritical flow over a low hump: h too low to produce y_c at crest. (b) Supercritical flow over a high hump: h higher than needed to give y_c at crest.

(a)

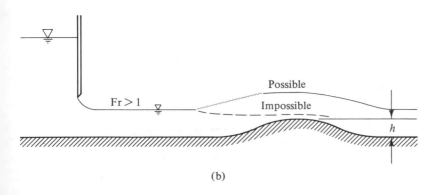

Fr > 1

Possible

Impossible

h

(b)

require that, between the outlet of the sluice gate and just ahead of the hump, the depth decrease, yielding the needed increase of specific energy of the liquid. This is impossible, since the total energy would increase in such a drop in the level of the free surface and there is no agent to cause this increase in total energy. Thus, some other way must be found to resolve the situation so that the specific energy immediately ahead of the hump can increase.

If *instead* the flow ahead of the hump were to be subcritical, instead of supercritical—how that could be accomplished will be described next—the necessary increase in the specific energy could be achieved by increasing the depth. This subcritical condition ahead of the hump would mean that the fluid would have to increase its depth between the sluice gate and the hump. If such a positive increase in liquid surface elevation occurred, it would be accompanied by a positive energy loss somewhere between the gate and the crest of the hump. This is a possible resolution of the situation. Figure 15.6.1b depicts the situation.

If a hydraulic jump were to occur downstream of the sluice gate, it is natural to inquire about the location. The place must be such that conditions of continuity, energy, and momentum are satisfied. The "solution points" would have to be on an appropriate set of Ey and $(MF)y$ diagrams. Finally, this requires that the level of the free surface change in a realistic way. We shall illustrate these considerations by referring to the situation that is illustrated in Fig. 15.6.2a.

Assume for the sake of the following discussion that after the liquid flows under a sluice gate, it proceeds along a channel that is almost horizontal until it moves over an edge and plummets downwards. We shall analyze the case even though the conditions at the end of the channel do not completely fulfill those that are implicit in our previous development (e.g., hydrostatic conditions cannot prevail at the end of the channel, because of the impending downward acceleration). At the end of the channel, we shall

Figure 15.6.2 (a) Supercritical flow after a sluice gate in a channel terminating in an overflow. (b) Overflow for channel conditions that require an approaching supercritical flow to become subcritical in a hydraulic jump. y_4 is an arbitrary position between stations 3 and 5.

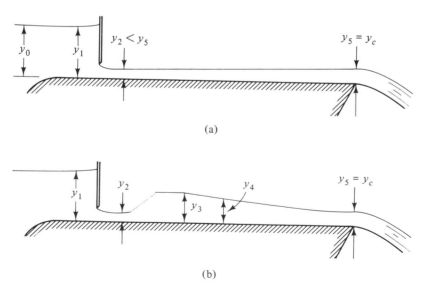

(a)

(b)

assume $y = y_c$, because at the beginning of the overflow the shear stress on the fluid will vanish.* [Here $\mathcal{R}_x/\gamma$ is zero, and so $d(\text{MF})/dx$ will vanish, as one can infer from Eq. (15.5.3). It follows that we are at—or at least near—the critical point of the $(\text{MF})y$ curve or the Ey curve.] We must now determine whether the integrated result of the bottom shear along the channel floor for the assumed (or calculated) local velocity would decrease the momentum function sufficiently. (Station 2 is supercritical, and $\text{MF}_2 > \text{MF}_5$.) It is necessary that the depth increase to y_c from the sluice gate to the end of the channel. Figure 15.6.2a shows this very special case.

If the estimated velocities indicate that y_c would be reached before the end of the channel, then a hydraulic jump could occur, so that the change of the momentum function per unit of channel length would decrease. The hydraulic jump would be accompanied by an energy loss, and point C would be approached from the subcritical portion of the Ey and $(\text{MF})y$ curves. Clearly, the depth downstream of the hydraulic jump would not necessarily be equal to y_c. The liquid's depth directly after the gate would be the value which would be assumed upstream of the jump. Now we know from Sec. 7.3 that for a hydraulic jump, using the notation of Fig. 15.6.2b,

$$2\left(\frac{y_3}{y_2}\right) = -1 + \sqrt{1 + 8\,\text{Fr}_2^2} \tag{15.6.1}$$

This fact can be included in the energy equation, and the loss term can then be evaluated for the hydraulic jump:

$$\text{Loss}_{\text{jump}} = \frac{(y_3 - y_2)^3}{4y_3 y_2} \qquad \text{ft (or m)} \tag{15.6.2}$$

This estimate of the losses in the hydraulic jump and a knowledge of the losses per unit length due to fluid friction in supercritical flow ahead of the jump, as well as the subcritical flow losses downstream, serve to determine the position of the "jump." It must be located so that the total decrease in specific energy will result in $\text{Fr} = 1$ at the end of the channel that was chosen in this example. We shall defer a discussion of the viscous effects in a channel and the "bottom shear," which would aid us in determining the value of $\mathcal{R}_x$, or in specifying the energy dissipated per unit length. At this time we shall mention briefly some events that could occur in the vicinity of a sluice gate when it acts to regulate a flow.

Until now, we have considered cases in which the events downstream of such a gate did not influence the flow through the gate. For the gate shown in Fig. 15.6.2a, values of the discharge $\tilde{Q}$ and the height y_1 are assumed to be known. Because $E_1 = E_2$, we had, from the Ey curve, a set of values for y_2. We chose the smaller of the values. The larger value, because $E_1 = E_2$, was equal to y_1. Of course, the higher height can be approached in practice if the hydraulic jump moves upstream to the gate. The "drowned" outflow would then appear as a submerged jet. A great deal of mixing could then occur, just as in a hydraulic jump, so that the assumption of constant energy would be unrealistic.

In reaching our solution for station 2 in Fig. 15.6.2a, we were given our conditions at station 1. These could have been determined by using the energy equation

---

* Measurements indicate that y_c occurs slightly upstream of the overflow location, at which place the depth is about 72 percent of y_c.

between the reservoir and our place of interest, station 1:

$$E_2 = y_2 + \frac{V_2^2}{2g} = E_1 = y_1 + \frac{V_1^2}{2g}$$

We shall now expand our discussion through Fig. 15.6.3a to c. The Ey curve in Fig. 15.6.4 summarizes the situation. The asymptotes $y = 0$ and $y = E$ are in effect the Ey curves for $\tilde{Q} = 0$. When the gate is closed, as it is in Fig. 15.6.3a, the possible fluid depths at station 1 are $y_{A'}$ and $y_{A''}$, and the former is selected from Fig. 15.6.4. It is the height in the reservoir. Since $E_1 = E_2$, the solution $y_{A''}$ is the appropriate value for y_2, the depth downstream of the gate. As the gate opens, water starts to flow underneath it. Then $\tilde{Q}$ is no longer zero. Nevertheless, $E_0 = E_1 = E_2$ if no losses are assumed, so that acceptable values of y result from the intersection of a suitable $\tilde{Q}$, say, $\tilde{Q}_B$, with the $E_1 = E_2$ line. We let $y_1 = y_{B'}$ and $y_2 = y_{B''}$. This means that, as the water begins to flow, the upstream level begins to drop as $\tilde{Q}$ increases while the downstream level rises. (See Fig. 15.6.3b.) This changing of the depth will continue to occur while the gate is opened. As the gate opening is increased until $\tilde{Q}_c$ flows (see Figs. 15.6.3c and 15.6.4), the level will become the associated critical value, y_c.

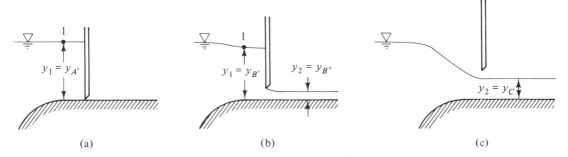

(a)   (b)   (c)

Figure 15.6.3 Stages in initiation of flow from a reservoir: (a) a closed sluice gate with no flow; (b) partially opened sluice gate; (c) sluice gate that has been raised so that critical depth occurs below it.

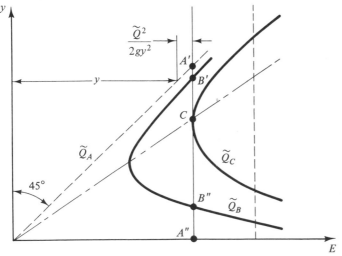

Figure 15.6.4 Channel depths and specific energy for flow depicted in Fig. 15.6.3. $E = \tilde{Q}/2gy^2 + y$. For the conditions of Fig. 15.6.3, $E_0 = E_1 = E_2$.

If the gate is now lowered, the volume flow rate will not necessarily change, since critical conditions will have been reached between the reservoir and the gate. (Thus, any disturbances created at the gate will not propagate to the reservoir.) This value $\tilde{Q}_c$ must pass below the gate, which has now been partially closed again. This will require the E value just upstream of the gate, E_1, to be larger than before so that $y_1 > y_2$. The increase could perhaps be that associated with the vertical dashed line in Fig. 15.6.4. How such a thing can occur—without creating energy—requires that we consider how the specific energy in a channel decreases as a result of fluid friction at the boundaries (i.e., the riverbed) and how E increases as a result of the slope of the channel. (Note that an increase in E need not imply an increase in the total energy of the system.) Figure 15.6.5 shows the structures that are used to induce hydraulic jumps. A laboratory test with such elements is shown in Fig. 15.6.6.

Figure 15.6.5 Hydraulic structures in a channel to initiate a hydraulic jump. *(Courtesy of the Swedish Power Association)*

Figure 15.6.6 Model test of structures used to initiate a hydraulic jump in a basin downstream of a dam. *(Courtesy of the Swedish Power Association)*

15.7 FLOW RESISTANCE IN OPEN CHANNELS

Previously, it was mentioned that the momentum function, $\tilde{Q}^2/gy + y^2/2$, will change from section to section if there is friction along the bottom or a component of gravity in the flow direction. (There is usually a slope to the channel's bottom.) Figure 15.7.1a shows a short section of a channel along with lines that represent the original energy level, the energy grade line, and the hydraulic grade line. The last of these lines coincides with the free surface. The cross-sectional area of the channel (Fig. 15.7.1b) is designated A, and the wetted perimeter of the channel is termed P. For wide, shallow channels, P will be approximately equal to the width of the channel. We shall divide the channel's cross section into segments with an area of $\hat{a}$ (see Fig. 15.7.1b) and lateral width $\hat{P}$. The momentum equation for the x direction can be written for a typical section with cross-sectional area $\hat{a}$, length dx, and width $\hat{P}$ (with a flow in this section of $\hat{Q}$) as

$$\gamma \frac{y}{2}(\hat{a}) - \gamma \frac{y+dy}{2}\left(\hat{a} + \frac{d\hat{a}}{dx}dx\right) - \tau_0(dx)(\hat{P}) + \gamma \hat{a}\,dx\,(\sin\theta)$$

$$= \rho\hat{Q}\left(V + \frac{dV}{dx}dx - V\right) \tag{15.7.1}$$

if one recalls that $\gamma y/2$ is the pressure at the centroid of the upstream section. The need to include $\hat{a} + (d\hat{a}/dx)\,dx$ in the second term results from the possibility that the channel area can be changing. This will certainly be the case if the width stays constant but the depth changes. We can observe that some of the expressions that result from the second term of this equation are products of derivatives. We shall retain only the differential terms that are linear, on the assumption that the differentials are infinitesimal, in accordance with the spirit of the calculus. The reader can verify that the following terms result, where $\sin\theta$ is replaced by $-dh/dx$ and $\hat{Q}$ by $V\hat{a}$:

$$-\frac{\gamma}{2}y\frac{d\hat{a}}{dx}dx - \frac{\gamma}{2}\hat{a}\,dy - \tau_0\,dx\,(\hat{P}) - \gamma\hat{a}\frac{dh}{dx}dx = \rho V\hat{a}\left(\frac{dV}{dx}dx\right)$$

Figure 15.7.1 (a) Control volume for flow in a channel with a sloping bottom. (b) Arbitrary channel cross section of area A with wetted perimeter P.

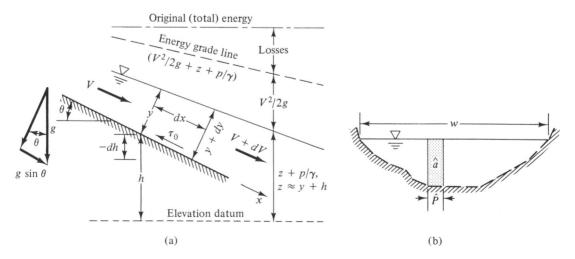

(a)

(b)

We obtained the sections with area $\hat{a}$ by dividing the channel at the section of interest into n parts of equal width with area $\hat{a}_i$. If the channel changes width, the $\hat{a}_i$ will change for a fixed value of n. Also, *for a fixed width and value of* n, the $\hat{a}$ could all be different due to a variation in liquid depth, y. Since $y\hat{P} = \hat{a}$, we have *for fixed* $\hat{P}$ the result that $dy = d\hat{a}/\hat{P} = y\,d\hat{a}/\hat{a}$. Then $d\hat{a}/dx = (dy/dx)(\hat{a}/y)$. When this expression is substituted for the first term of the last equation, we obtain, for any section,

$$-\gamma\hat{a}\,dy - \tau_0\,dx\,\hat{P} - \gamma\hat{a}\frac{dh}{dx} = \rho V\hat{a}\left(\frac{dV}{dx}\,dx\right)$$

or
$$-\tau_0 = \frac{\hat{a}}{\hat{P}}\left[\gamma\frac{d}{dx}\left(y + h + \frac{V^2}{2g}\right)\right] \tag{15.7.2}$$

after we have divided through by dx. (The ratio of the area of an element to its wetted perimeter is $\hat{a}/\hat{P}$.) We now add the results for all n elements. We *assume* that τ_0 and the sum in the parentheses are constant for each section. (If there is no lateral variation in y, h, and V, then $d(y + h + V^2/2g)/dx$ will be independent of width.) The result is, for n segments,

$$-n\tau_0 = \frac{1}{\hat{P}}\frac{d}{dx}\left(y + h + \frac{V^2}{2g}\right)\sum_1^n \hat{a}_i$$

because $\hat{P}$ is constant for each of the n parts with the value P/n, in which P is the wetted perimeter of the entire channel. The summation is simply the total area of the channel, A. With this equipartitioning, we can now revise Eq. (15.7.2) to read

$$\tau_0 = -\frac{A}{P}\gamma\left[\frac{d}{dx}\left(y + h + \frac{V^2}{2g}\right)\right] \tag{15.7.3}$$

It is useful to examine the term in brackets. A moment's reflection upon it and Fig. 15.7.1 is all that is needed to convince oneself that the bracketed term is the slope of the energy grade line, as one moves in the direction of flow. *We shall call this slope S.* One can easily perceive that $d(y + h + V^2/2g)/dx$ is a negative quantity. (The energy grade line and the hydraulic grade line are parallel if the fluid speed V remains constant.)

It is useful to define a situation in which the channel velocity, the mean flow depth, the channel cross section, and the channel slope dh/dx are *constant* in the direction of flow. This is a state of *uniform flow*. It follows from this definition that, for uniform flow and with the definition $dh/dx = -K_0$,

$$\tau_0 = \frac{A}{P}\gamma[-(0 - K_0 + 0)]$$

The term in the brackets is the positive quantity which represents the magnitude of the slope of the hydraulic grade line, or the slope of the free surface, if the velocity does not change along the channel. It is, of course, also the slope of the channel, because y is assumed to remain constant. We shall call this *particular* "slope" S_u. Then, for uniform flow, we have the following *definition*

$$\tau_0 = \frac{A}{P}\gamma S_u = \frac{A}{P}\gamma\left(-\frac{dh}{dx}\right) \tag{15.7.4}$$

for uniform flow. [Equation (15.7.4) can also be interpreted by recalling that $\tau_0 P(1)$ is the force resisting the flow, per unit length, and $A \gamma K_0$ is the component of gravitational attraction that causes the flow. These forces are evidently in balance if no changes in y and V are present.] Uniform flow will seldom occur in nature, because of the requirements that are inherent in its definition. Nevertheless, we shall often employ it as a criterion for comparing various channel designs.

The usual situation is to have, among other things, $V^2/2g$ varying with x. Thus, we write, instead of Eq. (15.7.4),

$$\tau_0 = \frac{A}{P} \gamma S \tag{15.7.5}$$

in which S *is the magnitude of the slope of the energy grade line.* [S_u is the slope of the energy grade line when V and y do not vary. Equation (15.7.5) is, in effect, Eq. (15.7.3).] We have encountered a wall shear stress previously and can profit from this experience. In the case of the flow over a flat plate, we wrote [see the definition of C_f that follows Eq. (8.6.9)]

$$\tau_0 = C_f \left(\frac{\rho V^2}{2} \right) = \frac{\text{drag}}{\text{area}}$$

in which τ_0 was an average wall stress. For pipe flow, we write [see Eq. (8.3.12b)]

$$4\tau_0 = f \left(\frac{\rho V^2}{2} \right) \tag{15.7.6}$$

in which f is the friction factor. Thus, it is natural to consider the bottom shear stress τ_0 in Eq. (15.7.5) as

$$\frac{\tau_0}{\rho V^2} = C' = \frac{g}{C^2} \tag{15.7.7}$$

in which C' is a nondimensional constant but C^2 will depend on the system of units in which the magnitude of the gravitational constant g is expressed. Upon comparing Eqs. (15.7.6) and (15.7.7), we can conclude that

$$\tau_0 = f \left(\frac{\rho V^2}{8} \right) = C'(\rho V^2) = \frac{g}{C^2} (\rho V^2)$$

so that $C' = f/8$ and

$$C^2 = \frac{8g}{f} \tag{15.7.8}$$

The constant C is important to us because Eq. (15.7.7) tells us that

$$\tau_0 = \frac{\gamma V^2}{C^2}$$

which, when using Eq. (15.7.5) to eliminate τ_0, gives

$$\gamma \frac{A}{P} S = \gamma \frac{V^2}{C^2}$$

or
$$V = C \sqrt{\frac{AS}{P}} \tag{15.7.9}$$

which is called the *Chézy equation.* A knowledge of C—called the *Chézy coefficient*—will give us information about V, the flow velocity in the channel. We might be able to tell something about C if we use data that have been collected from other conduits. In this case, we shall use pipe data because they are extensive, partially due to some pioneering work in Germany and Great Britain of Prandtl, von Kármán, Colebrook, and Stanton between 1900 and 1940. For smooth pipe, Blasius established that

$$f = \frac{0.316}{\text{Re}^{1/4}} \qquad \text{Re} = \text{Reynolds number}, \tag{15.7.10}$$

after reviewing a large amount of the available experimental data that were available for turbulent flows in smooth pipes. This fact could give a value of C in Eq. (15.7.8) for smooth channels, provided that the Reynolds number is properly interpreted. For pipe flow, the diameter is the length parameter in the Reynolds number. In open channel flow, the length parameter is A/P, which is called the *hydraulic radius,* which is one half the radius for a circular conduit. Hence, in correlating pipe flow data, the value 4 times the hydraulic radius, or $4A/P$, should be used in all places where D occurs. (See also Sec. 10.5.)

But the smoothness of a glass or brass pipe is almost never realized in an open channel outside of the laboratory. More often, such flows are over rough and complex surfaces. We can correlate C for flows at high Reynolds numbers in pipes by noting from the Moody diagram (Fig. 9.4.2) that, in this regime, f depends not on Re but solely on ϵ/D. This portion of the Moody diagram is called the *fully rough* region. If one plots the values of f against ϵ/D for, say, Re $= 10^7$, one would conclude that

$$f = (\text{constant})\left(\frac{\epsilon}{D}\right)^p = (\text{constant})\left(\frac{\epsilon P}{4A}\right)^p \tag{15.7.11}$$

fitted the data reasonably well and that $p = \frac{1}{3}$. (Try it on log-log paper!) Through this empirical result, we have found an expression for f for "fully rough" conditions from which, according to Eq. (15.7.8), we can determine C. After referring to Eq. (15.7.9), we write C as

$$f = (\text{constant})\left(\frac{\epsilon P}{4A}\right)^{1/3} = \left(\frac{8g}{C}\right)^{1/2}$$

and upon combining terms,

$$C = \frac{C_1(A/P)^{1/6}}{n} \tag{15.7.12}$$

in which n accounts for the differences in surface roughness in a manner proportional to $\epsilon^{1/6}$ and C_1 is another constant. On the basis of our conclusions that led to Eq. (15.7.12), we can express Eq. (15.7.9) as

$$V = \frac{C_1(A/P)^{1/6}}{n}\left(\frac{A}{P}\right)^{1/2} S^{1/2}$$

$$= \frac{C_1(A/P)^{2/3} S^{1/2}}{n} \tag{15.7.13}$$

Now Eq. (15.7.13), while useful in its present form, is not nondimensional. If one wishes to use it in "English units" (slug, foot, second), one would need a conversion factor of (Number of feet/meter)$^{1/3}$. For this reason, we write, when using the slug-foot-second system,

$$V = \frac{1.49(A/P)^{2/3} S^{1/2}}{n} \tag{15.7.14}$$

This result is called the *Manning equation* and is widely used in the turbulent regime of flow that is called "fully rough" (i.e., the region where f is independent of Re for pipe flow). Table 15.7.1 lists values of n to use with the Manning equation.

TABLE 15.7.1 MANNING ROUGHNESS FACTORS

Channel Material	n
1. Concrete, finished	0.012
2. Concrete, unfinished	0.014
3. Cast iron	0.015
4. Brick	0.016
5. Riveted metal	0.018
6. Corrugated metal	0.022
7. Gravel	0.029
8. Earth	0.025
9. Earth with stone and/or weeds	0.035

Example 15.7.1

A rectangular channel with a width of 4 ft and a depth of 2 ft is lined with brick ($n = 0.016$). It has a drop in elevation of 36.25 ft in 1 mile. What velocity can be expected if uniform flow conditions were to prevail (i.e., $S = S_u$)?

Solution   Equation (15.7.14) can be used:

$$V = \frac{1.49\left(\frac{A}{P}\right)^{2/3}\left(\frac{36.25}{5280}\right)^{1/2}}{0.016}$$

$$\frac{A}{P} = \frac{4(2)}{2(2) + 4} = 1$$

$$V = 1.49(1)(0.082)/0.016 = 7.64 \text{ ft/s}$$

If V or Q is given and one asks about the depth that corresponds to uniform flow, it will be necessary to proceed somewhat in the manner of the following example.

Example 15.7.2

A rectangular channel with a width of 12 ft must carry a flow of 600 cfs. It has a slope of 1 ft in 150 and will be lined with masonry ($n = 0.014$). What will be the depth of the water in the channel? (Assume $S = S_u$.)

Solution   We can use the Manning equation in the form

$$600 = AV = A\frac{(1.49(A/P)^{2/3}(1/150)^{1/2})}{n}$$

With $A = 12y$ and $P = 12 + 2y$, we can write

$$\frac{600(0.014)}{1.49(0.082)} = \frac{y^{5/3}(12)^{5/3}}{(y+6)^{2/3}(2)^{2/3}}$$

and solve iteratively for y to obtain a value of 3.4 ft for the depth.

The value of y calculated in this example is the depth for uniform flow only. Consequently, we shall *designate such values as* y_u. Section 15.8 will make extensive use of this defined quality.

One may wish to optimize the flow situation by having, for a given slope, the maximum mean velocity or maximum Q. The latter situation is the more likely. If A is kept constant, one can conclude that a semicircle would be the best, in the sense of providing a maximum flow. It seems reasonable that the polygon shapes that closely approximate a semicircle should be selected if flat sections of channel wall are preferred. Other constraints are present during the planning of a channel's design. The cost of excavation will be influenced by the area A, and the cost of facing the channel, by the perimeter P. Some compromise must be reached.

Example 15.7.2 illustrated the manner in which a uniform flow velocity could be calculated. However, as mentioned previously, uniform flow rarely occurs in nature. Nevertheless, the concept provides a useful criterion when comparing alternative designs for a channel.

In developing the Manning equation—Eq. (15.7.14)—we utilized a parameter C and some results for circular pipes to relate C with the hydraulic radius and the roughness of the surface. Most of such pipe data are less than 100 years old—Darcy's values are older—and open channels have been designed for centuries before that. The experience that was obtained from earlier efforts led to the accumulation of a great quantity of data that were correlated near the beginning of the twentieth century into what was called, subsequently, the Manning equation. Hence, the Manning equation is empirical in its origin. It is to the credit of engineers and scientists who have been able to assemble a consistent theory to explain the laboratory and field data. Because Chézy tried to correlate the resistance force in the channel with the gravitational "driving" force in a manner similar to that which led to Eq. (15.7.9), it is natural that C is called the *Chézy coefficient.*

For the flow rate and slope that were prescribed in Example 15.7.2, the Froude number associated with the value of $\tilde{Q}^2/gy_u^3$ is found to be less than 1. The flow is subcritical. If we maintained the value of $\tilde{Q}$ at 50 ft²/s and increased the slope of the channel, the value of y_u would decrease and the Froude number would increase. For a slope that is in excess of 1 ft in 20, y_u will be less than 1.98 ft. The associated Froude numbers would be greater than 1. The slope would be steep enough for $y_u < y_c$. We shall now explore the consequences of the channel slope in greater detail.

15.8 FREE-SURFACE PROFILES

Our development of the concepts leading to the Chézy equation, $V = C\sqrt{AS/P}$, with the associated concept of uniform flow, provides us with the experience to deal with more complex situations. For uniform flow, one can calculate a depth y_u. This uniform depth would remain constant along the channel for the conditions that produce y_u. The free surface would have the same slope as the channel's bottom. The Chézy equation, or the Manning equation, which was used in Example 15.7.2, leads us to conclude that for a given value of $\tilde{Q}$, the depth y_u will vary with the channel's slope so that the greater the value of S, the smaller the value of y_u. This would indicate that if a flow were proceeding along a gentle slope, the uniform depth might be great enough that the Froude number would be less than 1 (i.e., a case for which y_u would be greater than y_c). If the continuation of the channel bottom were a much steeper slope, a uniform-flow calculation could indicate that $y_u < y_c$, so that Fr would be greater than 1. Somewhere in the transition region between the two slopes, critical conditions would have been reached. In drawing the profile of the elevation of the free surface, one would be inclined to draw a curve that blended smoothly from the greater depth to the lesser. This is one example of a varying free-surface profile that is already within the realm of our experience, even if we are combining two uniform, nonvarying situations that are assumed to occur on either side of a transition in slope.

Another situation where depth can change along a watercourse has already been mentioned. For the flow of liquid in a horizontal channel, the friction on the bottom can be the cause for the fluid depth to vary in the flow direction. (We saw this when we utilized the (MF)y diagram and observed that the bottom drag would decrease the value of MF along a horizontal channel so that the depth would have to change.) In the remainder of this section, we shall investigate the manner in which bottom shear and the slope of the channel's bottom can influence the profile of the free surface.

We can start our detailed discussion with the aid of Fig. 15.7.1 and Eq. (15.7.3), which relates the streamwise momentum changes with the forces applied in the direction of flow. This equation gives the definition of S in the form

$$\frac{d}{dx}\left(h + y + \frac{V^2}{2g}\right) = -S \tag{15.8.1}$$

(The term in parentheses is the ordinate of the energy grade line.) This result can be made additionally useful by writing it as

$$\frac{d}{dx}\left(y + \frac{V^2}{2g}\right) = \left(-\frac{dh}{dx}\right) - S$$

We have come to know $y + V^2/2g$ as the specific energy E, and $-dh/dx$ as the value of S for uniform flow, which we designated S_u [see Eq. (15.7.4)]. Consequently, we write Eq. (15.8.1) as

$$\frac{dE}{dx} = S_u - S$$

This gives the dependence of E on x; however, we know from Eq. (15.2.1) that

$$\frac{dE}{dy} = 1 - \text{Fr}^2$$

If we eliminate dE from these last two equations, we obtain

$$\frac{dy}{dx}(1 - \text{Fr}^2) = S_u - S \tag{15.8.2}$$

This result is similar to that given by Eq. (15.3.2), in which, because of the limited conditions that were considered, S was zero.

Equation (15.8.2) can be used for a variety of channel shapes, but we shall *restrict* ourselves to wide rectangular channels. In this case, the side walls contribute relatively little to the shearing of the fluid. Per unit width of such a channel, the area is $y(1)$ and the wetted perimeter—of the bottom—is unity. Consequently, S in Eq. (15.8.2) can be conveniently rewritten. Equation (15.7.9), the Chézy equation, will have $A/P = y/1$ and can be written as

$$S = \frac{V^2}{C^2 A/P} = \frac{V^2}{C^2 y} = \frac{\tilde{Q}^2}{C^2 y^3} \tag{15.8.3}$$

The square of the Froude number in Eq. (15.8.2) is, as before, V^2/gy, or $\tilde{Q}^2/gy^3$. Thus, for the restricted case of *a wide channel that is essentially rectangular*, we have another form of Eq. (15.8.2):

$$\frac{dy}{dx} = \frac{S_u - \tilde{Q}^2/C^2 y^3}{1 - \tilde{Q}^2/gy^3} = \frac{S_u - \tilde{Q}^2/C^2 y^3}{1 - \mathrm{Fr}^2} \tag{15.8.4}$$

(Recall that S_u is the slope of the channel's bottom.) A large value of S_u would make dy/dx in Eq. (15.8.4) greater than 1, while a small value of S_u would require a small value of the derivative. Values of dy/dx could even be negative, depending on the signs of the numerator and denominator. For the case $dy/dx = 0$, we require that $S_u = \tilde{Q}^2/C^2 y^3$, which is in accord with Eq. (15.7.9), so that $S = S_u$ and $y = y_u$.

We can be somewhat more specific when discussing this Eq. (15.8.4). In this way, the profiles of the free surface could be computed and, if one wished, compared with the drawings in Fig. 15.8.1. If we write Eq. (15.8.4) as

$$\frac{dy}{dx} = S_u \frac{1 - \tilde{Q}^2/S_u C^2 y^3}{1 - \tilde{Q}^2/gy^3} \tag{15.8.5}$$

we can substitute $\tilde{Q} = C_u^2 S_u A_u^3/P_u$ from Eq. (15.8.3) (even though uniform flow—note the inclusion of u as a subscript—may not be occurring) in the numerator. We shall also employ Eq. (15.2.2), $\tilde{Q}^2/gy_c^3 = 1$, in the second term of the denominator. The result of these substitutions is

$$\frac{dy}{dx} = S_u \frac{1 - (C_u/C)^2 (A_u^3/P_u)/y^3}{1 - (y_c/y)^3} \tag{15.8.6}$$

The terms C_u, A_u, and P_u will depend upon y_u, whereas C will depend on y. For a very wide channel, for which the analysis on a unit width basis can proceed without difficulty, $A/P = y(1)/(1) = y$. Obviously, $A_u/P_u = y_u$ and $A_u^3/P_u^3 = y_u^3$, since $P_u = 1$ because of the unit width. The ratio C_u/C in Eq. (15.8.6) can be deduced to equal $(y_u/y)^{1/6}$ by referring to the definition of C in Eq. (15.7.12). These results allow us to write Eq. (15.8.6) as

$$\frac{dy}{dx} = S_u \frac{1 - (y_u/y)^{10/3}}{1 - (y_c/y)^3} \tag{15.8.7}$$

It is apparent that the sign of dy/dx will depend on the algebraic signs of the numerator and denominator in Eq. (15.8.7). These will be determined by whether the ratios y_u/y and y_c/y are greater or less than unity for a given situation. For each case, there will be an ordering of y, y_u, and y_c, and their interrelationship will determine the sign of dy/dx. In an effort to classify the slopes in a convenient way, the case of $y_u > y_c$

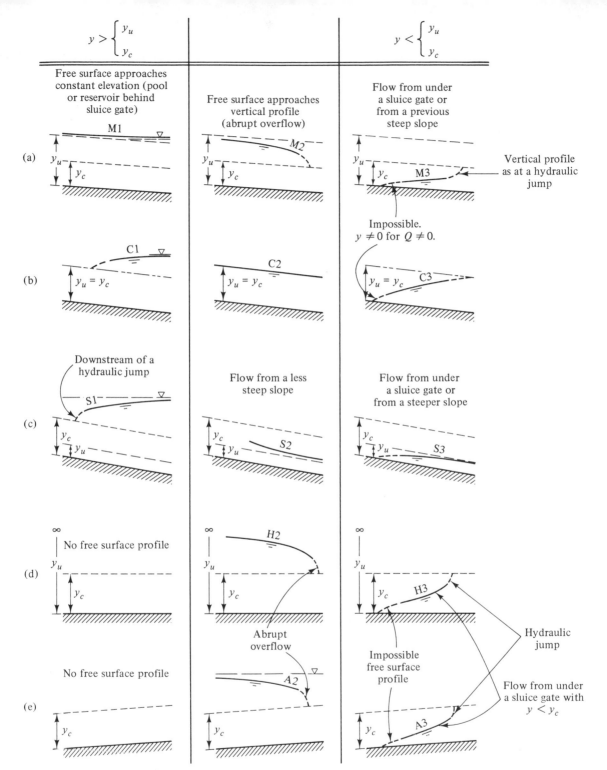

Figure 15.8.1 Classification of flow profiles for gradually varied flow: (a) mild slope, $y_u > y_c$; (b) critical slope, $y_u = y_c$; (c) steep slope, $y_u < y_c$; (d) horizontal slope, $y_u > y_c$; (e) adverse slope.

is termed a *mild slope,* $y_u < y_c$ is called a *steep slope,* and $y_u = y_c$ is a *critical slope.* These definitions would give a Froude number for a uniform flow that would be less than 1 on a mild slope. Figure 15.1.3 indicates that a uniform flow would be supercritical on a steep slope.

We begin by considering a mild slope, which requires $y_u > y_c$, and write

M1. $y > y_u > y_c$ requires that $dy/dx > 0$.
M2. $y_u > y > y_c$ requires that $dy/dx < 0$.
M3. $y_u > y_c > y$ requires that $dy/dx > 0$.

For a steep slope, $y_c > y_u$, and we can write

S1. $y > y_c > y_u$ requires that $dy/dx > 0$.
S2. $y_c > y > y_u$ requires that $dy/dx < 0$.
S3. $y_c > y_u > y$ requires that $dy/dx > 0$.

The use of the letters M and S to designate the cases follows from the words *mild* and *steep.* The number 1 denotes that y is greater than both y_u and y_c, while the number 3 signifies that y is less than both y_u and y_c.

Between the mild and the steep slope is the critical slope, for which $y_u = y_c$. The possibilities inherent in Eq. (15.8.7) reduce to

C1. $y > y_u = y_c$ requires that $dy/dx > 0$.
C2. $y = y_u = y_c$ requires uniform flow with $dy/dx = 0$.
C3. $y < y_u = y_c$ requires that $dy/dx > 0$.

It was noted at the beginning of this section that if the channel bottom is horizontal, the depth must change in the flow direction and a uniform flow is not possible. The (MF)y diagram in Fig. 15.5.2a tells us that for $S_u = 0$, the following situations can occur:

H1. Does not exist.
H2. $y > y_c$, subcritical flow, requires that $dy/dx < 0$.*
H3. $y < y_c$, supercritical flow, requires that $dy/dx > 0$.

Figure 15.8.1 catalogs all of these possibilities, including the shape of these curves as y approaches y_u and y_c. When $y = y_u$, the numerator in Eq. (15.8.7) tends to vanish. Hence, it is not surprising that the depth approaches y_u asymptotically. For values of y just above and below y_c, the profiles have slopes of opposite signs. This suggests that as $y \to y_c$, the slope approaches either zero or infinity. In this case, an infinite slope is the choice, because the numerator of Eq. (15.8.7) does not vanish as $y \to y_c$. This situation would be approached by an abrupt change in surface elevation due to a hydraulic jump or a very steep overflow.

The case of an adverse slope—$dy/dx > 0$—is also given in Fig. 15.8.1. For this situation, S_u would be negative. Then the sign of dy/dx in Eq. (15.8.1), the source of Eq. (15.8.7), would be determined solely by the sign of $1 - \mathrm{Fr}^2$. For $y > y_c$, $\mathrm{Fr} < 1$, and the denominator would be negative. Because the right-hand side of Eq. (15.8.5) is negative for an adverse slope, we can write

---

* Note that y_u would have to approach infinity for a nonzero velocity, since the slope is zero in the Chézy equation. Hence, in H2, y is not greater than y_u; it is also less than y_u for case H3.

A1. Does not exist, because uniform flow does not exist.*
A2. $y > y_c$ produces $dy/dx < 0$.
A3. $y < y_c$ produces $dy/dx > 0$.

We can now use the results given in Fig. 15.8.1 to determine qualitatively the elevation in a channel in which weirs or gates are inserted to prescribe the depth or to make it locally critical. Such devices have been called *controls*. Figure 15.5.1 showed a control in the form of a sluice gate. When we observed that the flow would increase as the gate was opened, we concluded that ultimately the depth below the gate would reach y_c. Previously, the consequences of subsequently lowering the gate were discussed qualitatively and speculatively. Now, with Fig. 15.8.1 available to us, we can examine the situation anew with greater assurance. Figure 15.8.2a could represent an intermediate gate position during the opening process. (Compare this with Fig. 15.6.3a to c.) If, after a sufficient gate opening y_c was reached, the gate were then closed (Fig. 15.8.2b), the critical flow rate would continue to occur. The gate would cease to regulate the flow and serve as a control.* (Despite the lowered gate position, there will be some position upstream where the depth will remain at y_c and the flow will remain at $\tilde{Q}_c$. The disturbance of the gate would not propagate far upstream for $y = y_c$, because then Fr = 1.) In order that this critical flow might flow beneath the gate, the water depth immediately ahead of the gate must rise. A hydraulic jump would provide the means of raising the upstream fluid level to the required height. Such a jump is shown in Fig. 15.8.2b. This phenomenon is possible for either a mild or a steep slope. Only the position along the channel where the upstream jump would occur would be affected by the slope.

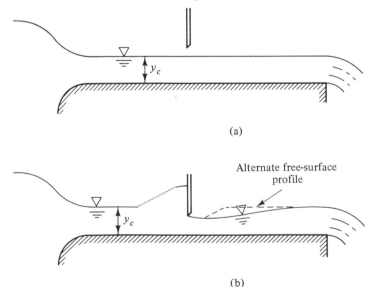

(a)

(b)

Alternate free-surface profile

Figure 15.8.2 (a) Flow under a sluice gate that has been raised sufficiently to allow critical depth to occur beneath it. (b) Channel flow configurations after the sluice gate is lowered into the moving stream.

* For an adverse slope, the Chézy equation would give a value of y_u that was negative. Certainly, y would be different from this unrealistic value. Some authors list only H1, H2, A1, and A2 curves, which is in contrast to the system for ordering the curves that has been adopted here. This should cause no difficulty to the reader, but one should be careful when comparing the notation from different books.

† This use of the word *control* implies the "accounting" connotation that one associates with the term *control volume*.

Downstream of the gate in both parts of Fig. 15.8.2, there is an abrupt overflow. At this point, the water depth will approach y_c. From this point, working backward toward the gate, one must determine the profile so that the slope and the supercritical conditions below the gate are accounted for. Of course, the possibility exists that the length of the mild slope is such that the required increase in the surface elevation shown in Fig. 15.8.2b occurs at the downstream side of the gate (this is the alternative profile shown). Then the gate would be *drowned*.

In the remaining part of this section, we shall use Eq. (15.8.7) to calculate the profile of the free surface. The channel characteristics—its slope, the discharge, and the initial depth—are prescribed. This will be done by determining the slope at the initial depth location. Then for a channel that has been subdivided into Δx steps, the change in depth, Δy, can be given for each step, where $\Delta y \approx (dy/dx)\,\Delta x$. For this new value of depth, another derivative can be found, and with it another change Δy for the next channel increment Δx. The procedure can be continued in this way.

Example 15.8.1

A wide channel with a discharge of 50 cfs per foot of width has a slope of 1/800 and a Manning friction value of $n = 0.014$, a value appropriate to roughly finished concrete. This channel has a sluice gate, and the water flows into an 8-ft-deep reservoir behind the gate. Determine the profile of the free surface that exists upstream of gate.

Solution (a) First, y_u and y_c must be found. The Manning equation, Eq. (15.7.14), is, in effect,

$$Q = \frac{1.49A\,(A/P)^{2/3}\,S^{1/2}}{n}$$

with $A = y_u(1)$ and $P = 1$ for $Q = \tilde{Q}$. By rearranging this equation to solve for y_u, one obtains

$$y_u = \left[\frac{(0.014)(50)}{(1/800)^{0.5}(1.49)}\right]^{0.6} = 4.72 \text{ ft}$$

The critical depth is, from Eq. (15.2.2),

$$y_c = \left(\frac{50^2}{32.2}\right)^{1/3} = 4.266 \text{ ft}$$

This means that $y_u > y_c$, so that a mild slope exists. The final elevation is greater than y_u. This is the condition shown at the left in Fig. 15.8.1a.

(b) After determining the classification of the flow, M1, in this case, one chooses the number of steps to be used to proceed from the depth at the gate to the uniform-flow depth. Assume that five steps are selected. Then the uniform changes of elevation are $(8.00 - 4.72)/5 = 0.656$. This means that the depth at the position corresponding to the first station will have a depth of 7.34 ft ($= 8 - 0.656$).

(c) Equation (15.8.7) can now be used to find the slope of the profile at this depth. S_u in this equation is the slope of the channel, 1/800. Hence,

$$\left(\frac{dy}{dx}\right)_{\text{station 1}} = \left(\frac{1}{800}\right)\left[\frac{1 - (4.72/7.34)^{10/3}}{1 - (4.266/7.34)^3}\right]$$

This value could be used as an estimate of the horizontal distance required to realize the 0.656-ft change of elevation. If this value of the slope at the first station is used as representative of the entire region between it and the gate, then the distance between the gate and the first station is

$$\frac{0.656}{dy/dx} = 547 \text{ ft}$$

This procedure is wanting, and a more representative value of dy/dx would be the average of the slopes at the beginning and end of the interval. This would be

$$\frac{0 + 1.1984(10^{-3})}{2}$$

and would produce a distance to the first station, where the elevation had decreased 0.656 ft, of 1094 ft. The following table summarizes the results that are based upon this (improved) approximation for the average slope:

Station	0	1	2	3	4	5
Depth, ft	8	7.34	6.69	6.03	5.38	4.72
(Slope)(10^3)	0	1.1984	1.1597	1.0798	0.8783	0
Distance between stations, ft		1094	555	586	670	1494
Total distance of transition, ft						4399

If the same calculation scheme is employed but 10 steps are used, the total distance is 4089 ft. For 20 steps, this distance is 4040 ft. The procedure apparently converges. The fact that the slope is zero at $y = y_u$, the last station, influences the horizontal distance significantly. (Indeed, if one did not use the average value, the final slope would have a disproportionate influence on the total transition length.) It would seem reasonable to minimize the effect of this zero slope in the last calculation by setting its depth somewhat higher, in this case, than y_u. A value of $1.002\, y_u$ would appear to be reasonable, given the uncertainties in the measurements. (Similar considerations apply in problem to the initial position, at the gate, where the slope is also zero.)

The beneficial effect on the answer due to increasing the number of steps seems localized in the stations for which the slope is nearly zero. (Some situations have stations at which the slope changes very much; here, too, the calculations would be improved with a large number of steps.) One may wish to have a variable step size, with the smallest steps at the ends of the flow transition region.

15.9 SUMMARY

The flow in open channels obeys the basic principles that have been developed in previous chapters. The energy equation has yielded the concept of *specific energy*, which proves to be useful in a variety of ways. The momentum principles gives the (MF)y diagram for open channel flow. Because both gravity and shear forces can alter the value of the *momentum function*, it was necessary to examine these forces in detail. The Chézy equation, which results from such an analysis, is useful because it provides a basis for justifying the utility of the Manning equation. This development leads naturally to the concept of a uniform flow. While a uniform flow seldom occurs in nature, and the channel depth should approach it only asymptotically, we are able to generalize our analysis using the *uniform depth* y_u to model the actual situation. The Chézy equation was reexamined, with the result that an equation for the slope of the free surface, Eq. (15.8.7), could be derived. The results were cataloged in Fig. 15.8.1.

We can use this information to predict the surface elevations qualitatively for a variety of situations.

A quantitative determination of the surface elevations requires an integration of Eq. (15.8.7). This can be done by a variety of numerical methods that are enhanced by the capabilities of the digital computer. The interested reader can examine books that specialize in open channel flow for additional procedures to determine free-surface profiles.

EXERCISES

15.1.1(a) What would be the Froude number at location 1 and the flow rate if $z_2/z_1 = 1.7$?

(b) What would be the Froude number at location 1 and the flow rate if $z_2/z_1 = 0.8$?

(c) What would be Fr_2 for $z_2/z_1 = 1.7$ and 0.8?

15.2.1(a) What is the value of y_c for $\tilde{Q} = 10$ m²/s? What is the corresponding E_c?

(b) What is the Froude number corresponding to the original conditions? What is the Froude number corresponding to $\tilde{Q}_{max}$ in the narrowed channel?

(c) If the channel were narrowed so that its width was 45 percent of that at section 1, instead of the 49 percent for $\tilde{Q}_{max}$, conditions at section 1 might have to change. Discuss the possible result.

15.2.2(a) For the case of Exercise 15.1.1a, would one expect the flow at location 2 to be tranquil or shooting?

(b) For the data in Exercise 15.1.1b, is the flow at location 2 subcritical or supercritical?

(c) For the contracted channel in Exercise 15.2.1c, which of the alternate depths would occur?

15.3.1(a) Had the free surface risen 0.115 m as the bottom rose 0.1 m over a 2-m length, what would be the expected local Froude number?

(b) Had the free surface lowered 0.115 m as the bottom lowered due to a gradual depression 0.1 m over the distance of 2 m, what would be the expected local Froude number?

(c) The Froude number at a particular location in a horizontal channel is 0.5. At this point the bottom begins to lower with a slope whose magnitude is $|-0.5/2|$. What should be the slope of the free surface?

15.7.1(a) If the channel has a width of 8 ft and all other dimensions are as in Example 15.7.1, what velocity can be expected for uniform flow conditions?

(b) If the channel in Example 15.7.1 has a drop in elevation of 72.5 ft in 1 mile, what should be the velocity for uniform flow?

(c) If instead of bricks, the channel in Example 15.7.1 is lined with earth and stones, what should be the velocity for uniform flow?

15.8.1(a) If all other conditions are the same, what are y_u and y_c if $\tilde{Q} = 40$ cfs per foot of width?

(b) If all other conditions are the same, what are y_u and y_c if the slope of the decline is 1/400?

(c) Determine the profile of the free surface for the case in Exercise 15.8.1b.

PROBLEMS

15.1 A hump at the bottom of a horizontal channel has a height of 1 ft. The depth of the water upstream of it is 4 ft. What are the expected changes in surface elevation if the upstream Froude number is 0.5 in one case and 1.2 in another? What is the volume rate of flow (per unit of channel width) in each case?

15.2 If a hump in a horizontal channel is 0.4 m high and the upstream flow is 0.8 m deep, what is the upstream velocity when the fluid depth at the crest is 0.24 m?

15.3 The flow in a horizontal water channel is 1.8 m deep and passes over a hump that is 0.9 m above the channel floor. The depth of water at the crest of the hump is 0.36 m.

What is the volume rate of flow (per unit width of channel)? What is the alternate depth for this flow?

15.4 In Fig. 15.1.2, determine the value of z_2/z_1 for a minimum value of Fr_1 (but greater than 1) with $h/z_1 = 0.5$.

15.5 If a horizontal channel of width w_1 narrows to w_2, elevation changes of the free surface could occur just like those described by Fig. 15.1.2. An analysis similar to that which led to Eq. (15.1.1a) yields

$$\frac{1}{2}\frac{V_1^2}{gz_1}\left[1 - \left(\frac{z_1}{z_2}\right)^2\left(\frac{w_1}{w_2}\right)^2\right] + \left(1 - \frac{z_2}{z_1}\right) = \frac{\text{losses}}{z_1}$$

Derive this equation, and solve it for the values of z_2/z_2 when $0 < Fr_1^2 < 2$ and $w_2/w_1 = 1$, 0.75, and 0.5, under the assumption that the losses are negligibly small. What conditions determine $z_2/z_1 < 1$, if any?

15.6 A hump has water flowing over it at a depth of 0.24 m, and the depth upstream is 0.8 m. For channel currents of 0.6 and 3.6 m/s, determine the specific energy both upstream and at the hump's apex. What is the hump's height in each case? Is the flow supercritical at either of the stations for any of the stated conditions?

15.7 A horizontal channel of constant width has a flow per unit width of 18 ft²/s when its depth is 6 ft. Is this flow tranquil? What is the alternate depth?

15.8 For the situation described in Prob. 15.7, a variety of depths are possible. Plot Eq. (15.1.4) as E versus y for $\tilde{Q} = 18$ ft²/s to obtain a curve that is similar to those in Fig. 15.1.3. What does the plot give for the minimum specific energy for this flow? If the channel is originally 6 ft deep, can a hump be used to obtain the minimum specific energy in the channel?

15.9 What is the critical depth for a horizontal channel that has a water depth of 5 m and $\tilde{Q} = 10$ m²/s? What is the specific energy at this critical depth? What is the alternate depth for this flow? (*Hint:* Refer to Example 5.2.3.)

15.10 A channel 38 ft wide has a flow of 480 cfs, and the water depth is 4.5 ft. What is the critical depth? Is the flow in the channel sub-

critical or supercritical? What is the Froude number for the flow?

15.11 Determine the alternate depth for the flow described in Prob. 15.10. What is its Froude number?

15.12 A 38-ft-wide channel has a uniform flow of 480 cfs at a depth of 4.5 ft, and it is proposed to bridge it using a structure that would narrow the channel by 6 ft. How will this change the depth of water in the channel at the location of the bridge?

15.13 Use the data in Prob. 15.12 and Table 15.3.2 to describe the way the free-surface elevation will change as the flow in the channel approaches the bridge and its supports.

15.14 The area of a symmetrical trapezoidal channel is $y(w + cy)$, in which y is the depth, w is the width of the bottom, and c depends on the slope of the sides. Find dA/dy. Is its value equal to the width of the free surface? What is y_c for such a channel?

15.15 A trapezoidal channel has walls that slope at 30° with respect to the horizontal. The depth of the water is 8 ft, the free surface is 87 ft wide, and the flow is 9000 cfs. What is the critical depth for this flow, and is the flow supercritical?

15.16 The area of a triangular channel for depth y is $cy^2/2$, in which c is a constant associated with the included angle. What is dA/dy for the channel? Is $y_c = (8\tilde{Q}^2/g)^{1/3}$ for such a channel?

15.17 The flow in a channel whose cross section is a circular segment is an application of Eq. (15.4.1). It is quite easy to show that the area of a circular segment of radius a is $a^2(2\alpha - \sin 2\alpha)/2$ and $da/dy = b = B/2$ as defined in Fig. P.15.17. Show that it follows directly that $dA/dy = (dA/d\alpha)(d\alpha/dy) = B$.

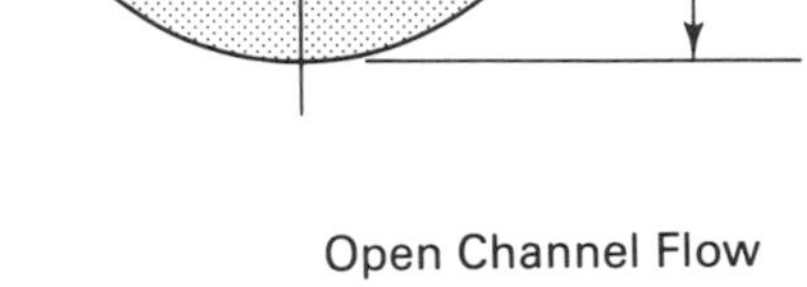

Figure P.15.17

Determine y_c for $y = 15$ ft and $Q = 1000$ ft³/s in a channel with a radius of 20 ft.

15.18 Derive the equation that corresponds to Eq. (15.3.2) for a channel with a triangular cross section. (*Suggestion:* Use the y_c given in Prob. 15.16 as appropriate. dh/dx will now be the slope of the channel bottom.)

15.19 A triangular channel whose included angle is 150° has a maximum depth of 10 ft and transports 1000 cfs of water. Use the formula in Prob. 15.18 to determine the critical depth for this flow. Would you expect that the 10-ft depth corresponds to a supercritical flow?

15.20 A channel carrying a flow of 30 cfs/ft is backed up by a sluice gate to a depth of 10 ft. Assume constant energy at the stations just ahead of and behind the gate. What is the downstream water depth?

15.21 For the flow described in Prob. 15.20, what is the momentum function upstream and downstream of the gate? What is the significance of the difference in the two values of the momentum function?

15.22 The speed of water downstream of a sluice gate is 12 ft/s, and the depth is 2 ft. What is the value of the momentum function at this location? Plot the y(MF) curve for the $\tilde{Q}$ of this problem. What is y at the minimum value of the MF? What is the conjugate depth to the 2-ft elevation that is given? If a hydraulic jump were to occur at this location, what would be the elevation of the free surface after the jump?

15.23 A hydraulic jump occurs such that the upstream and downstream depths are 3 and 7 ft, respectively. What is the value of $\tilde{Q}$ if the momentum function is the same at these two locations? (This assumes negligible shear forces.) What is the specific energy at these two depths? What is the significance of the two different values for specific energy?

15.24 For a supercritical channel flow, with Fr = 1.8 at a depth of 15 ft, what hump elevation would cause critical conditions at the apex? If a hump that is 10 percent higher were there instead, what would be the Froude number just upstream of this hump? What would be the Froude number at its apex?

15.25 A wide, rectangular channel is nearly horizontal. Water flows in it at 30 cfs/ft with a depth of 2 ft shortly after it passes under a sluice gate. If the flow has y_c approximately at an abrupt overflow (see Fig. 15.6.2), what should be the maximum distance between the gate and the overflow if there is no intermediate hydraulic jump? (Assume values of the wall shear stresses from the Moody diagram of Fig. 9.4.2 based upon a rough concrete channel lining. $D_{\text{hydraulic}} = 8$, because for most of the channel width the wetted perimeter is only the unit width along the bottom. Use 10-ft steps along the channel, so that $\mathcal{R}_x = 10\tau_w$. Then, $\text{MF}_x - \text{MF}_{x+10}$ will equal $\mathcal{R}_x/\gamma$, so that MF_{x+10} can be found, which in turn specifies y_{x+10}. When this value equals y_c, the required length will have been found.)

15.26 If the flow described in Prob. 15.25 underwent a hydraulic jump shortly after passing the sluice gate, estimate the distance between the jump and the overflow. [This problem describes the flow situation if the distance between the sluice gate and the overflow is too long to permit a supercritical flow to the overflow. If this distance is very short, the supercritical flow will be one for which significant acceleration and other (nonhydrostatic) effects could occur. These are beyond the scope of the discussion in this book.]

15.27 The depth changes from 1 m to 2.8 m in a hydraulic jump. What are the initial and final Froude numbers? What is the energy loss per newton of fluid flowing? If this jump could be made to occur in six stages of equal elevation change, what would be the total energy loss per newton of fluid flowing?

15.28 Show that a hydraulic jump can occur in a channel with a triangular cross section with a 90° included angle. What is the ratio of the (maximum) depths across such a jump?

15.29 A comparison of Eqs. (15.7.12), (15.7.13), and (15.7.14) reveals that the Chézy coefficient $C = 1.49(A/P)^{1/6}/n$. What value of C is appropriate for a channel 12 ft wide and 9 ft deep that is lined with rough concrete?

15.30 Show that the Manning formula, Eq. (15.7.14), will maximize Q for a given A if

P is at a minimum. What geometric shape has the smallest perimeter for a given area? (This shape produces the "most efficient" channel, but it is not necessarily the shape that is employed, because construction and maintenance considerations are simplified with trapezoidal sections.)

15.31 A triangular flume (included angle = 90°) has a uniform flow of water when the depth is 3 ft. What is the average shear stress along the wetted perimeter for a slope of 1 in 250?

15.32 A rectangular channel 6 ft deep and 10 ft wide changes elevation 3 ft in 500 ft of run. If water is flowing uniformly at a depth of 5 ft, what is the average shear stress on the wetted surfaces?

15.33 What is the hydraulic radius of a round duct 8 ft in diameter that is running full? What is the hydraulic radius if it is running half full?

15.34 Corrugated metal tubes 3 m in diameter are used with a slope of 1 in 500. What is the uniform flow rate when the tubes are flowing half full and completely full?

15.35 A canal with rectangular cross section 5 m wide and 4 m deep is lined with unfinished cement. What slope is necessary if the uniform flow is to be less than 1 m/s at a water depth of 3 m? (This speed is a navigational requirement, in this case.)

15.36 A channel with a trapezoidal cross section is lined with earth. The water surface breadth is 26 ft for a depth of 6 ft. The bottom is 8 ft wide. Sufficient water flows uniformly at a slope of 1 ft in 800 in a debris-free channel, but the flow becomes too meager if the channel becomes cluttered with weeds and stones. What is the flow reduction (in percent) in this unmaintained condition?

15.37 The canal described in Prob. 15.36 no longer fulfills the flow requirements, even in the "clean" condition. It is proposed to widen the canal to double the flow rate. The slope of the side walls would remain the same as before. What would be the breadth of the water surface in this new condition? How many cubic yards of dirt would have to be removed for each 3 ft of canal length?

15.38 A trapezoidal channel made from corrugated metal has side walls that slope at 20° to the horizontal. What are the "best" dimen-

sions for it if its slope is 8 ft in 1000 and it should deliver 1500 cfs?

15.39 A concrete pipe 1000 ft long is on the ground where the slope is 5 ft in 100. The pipe takes water from a reservoir that is 70 ft above the center of the pipe's terminus. The water that discharges from the pipe enters a concrete triangular flume (90° included angle) that has the same slope as the pipe. What is the uniform depth of water in the flume for steady-state conditions?

15.40 Is the 90° included angle for the triangular flume in Prob. 15.39 the "most efficient" for a triangular cross section?

15.41 A trapezoidal channel with side walls at 30° to the horizontal is lined with rough concrete. It is 30 ft wide at the bottom and has a slope of 2 ft in 1000. What uniform depth can be expected if the flow in it is to have a maximum velocity of 2.5 ft/s?

15.42 A hydraulic jump is caused to occur in a channel that is 3 m wide. The water is 80 cm deep ahead of the jump and 300 cm downstream. What is the rate of energy dissipation in the jump? What is the expected temperature rise of the water as it passes through the jump?

15.43 A rectangular spillway chute with a slope of 0.05 is 120 ft wide and is made of smooth concrete. The chute is very long and ends in a stilling basin with a horizontal bottom. What is the depth of the water in the chute just prior to entering the basin? Is this flow subcritical or supercritical?

15.44 What must be the depth in the stilling basin of Prob. 15.43 for a hydraulic jump to occur after the chute?

15.45 A rectangular channel with a slope of 0.0015 is 12 ft wide. Water flows uniformly at a depth of 7 ft upstream of a smooth hump that causes the flow to assume a critical depth. What should be the height of the hump?

15.46 What is the diameter of a concrete pipe flowing full that would transport the same quantity of water per unit time as a rectangular concrete channel that is 3 ft deep and 20 ft wide? Assume a slope of 0.006. Is the value of the slope important in the analysis?

15.47 A concrete irrigation ditch whose sloping

sides (30° with the vertical) are 2 ft long and whose bottom is 4 ft wide should carry a flow of 4 ft³/s during storm conditions. What should be its slope? How many feet of elevation difference is this per mile?

15.48 A channel is attached to an opening in the wall of a reservoir by means of a bottom that slopes upward 1 m vertically for 4 m of horizontal distance. At the point that is 4 m away from the reservoir, the bottom is horizontal and the water gradually assumes a depth of 1.6 m. The free surface of the reservoir is 2.5 m above the floor of the horizontal channel. What is the depth of the water halfway along (at 2 m) the sloping transition section? What is the critical depth in the channel?

15.49 A channel that is lined with rough concrete is 6 ft wide and has a slope of 2 ft in 1/4 mile. An obstruction raises the depth of the water from 2.4 ft (for a uniform flow) to 3.2 ft. What is the critical depth for this situation? How far upstream of the obstruction has the water depth increased 0.4 ft? Draw the backwater curve.

15.50 A flow of 480 cfs occurs in a 12-ft-wide channel lined with finished concrete. Draw the profile of the water's surface after the slope changes abruptly from 0.0015 to 0.015.

15.51 The spillway from a large reservoir is, in effect, a broad-crested weir that slopes upward from a point at the reservoir to a point on the weir which begins a horizontal section that leads to the weir's abrupt end and the overflow. The horizontal length of the incline is 20 ft, and the vertical rise is 6 ft. The elevation of the reservoir is 5 ft above the horizontal portions of the weir. Determine the profile of the water's surface over the weir. What is the length of the horizontal section of the weir if the depth at the end is $0.72y_c$?

15.52 A rectangular chute of smooth concrete has a slope of 0.003. The water's depth is measured as 1 m just before the end of the chute. (The water falls to a river below.) What is the discharge per foot of width? What are y_c and y_u for this case? What is the expected depth 200 ft upstream of the end of the chute?

15.53 The channel shown in Fig. P.15.53 is very wide, has a slope of 0.0002, and has $n = 0.025$. The water depth downstream of the gate is adjusted to be 2 ft. What is the depth of water behind the gate? Draw the backwater curve upstream of the gate. What should be the distance between the gate and the brink to ensure that a hydraulic jump will occur? What is the depth just downstream of such a jump?

15.54 A trapezoidal channel has a base 10 ft wide and side walls that slope 30° with the horizontal. It is lined with unfinished concrete and has a discharge 540 cfs. Depths of 5 and 6 ft are reported. Are these depths possible in a long channel with a slope of 0.0002? If so, is the greater depth to be expected upstream or downstream of the lower depth?

15.55 A length of concrete pipe is used to drain a pond. It is 500 ft long and 4 ft in diameter. The center of the end of the pipe is 6 ft below the surface of the pond. If the water cascades out of this pipe into a canyon below, does the pipe run full? At what rate does the pipe drain the pond?

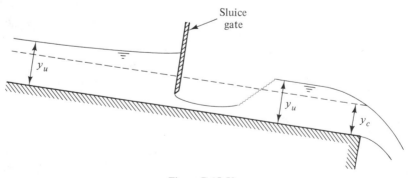

Figure P.15.53

16

The Navier-Stokes Equations

16.1 INTRODUCTION

In Chap. 8, we successfully solved a number of problems by utilizing (1) a free-body diagram, (2) associated shear stresses and normal stresses with the appropriate signs, and (3) Newton's laws of motion. This process resulted in a differential equation that related the pressure gradient, the shear stresses, and the product of the mass and acceleration. Finally, we introduced a special form of Newton's law of viscosity into the work. Thereby, we were able to derive a final equation that could be solved to yield the velocity profile for the case at hand. Although this procedure was productive, it has some drawbacks.

There are many problems for which a free body of the fluid element cannot be chosen at the outset of the problem with the necessary insight to enable a straight-forward analysis to ensue. (The flow between two circular cylinders rotating about two different axes is a case in point.) However, the possibility exists that a simple, but general, free body might be envisioned and the appropriate equations of motion developed for it. The result would be a set of differential equations that applies to a general set of situations concerning the flow of fluids. The development of such equations was the work of many scientists, and two whose lives spanned an important period of research have had their names associated with the final form of these equations: Louis M. H. Navier (1785–1836) and George G. Stokes (1819–1905).

16.2 EQUATIONS OF MOTION

We begin the process of formulating the equations of motion for a fluid by following the work of Stokes. A small element of volume, $dx\,dy\,dz$, is used. This material volume is pictured in Fig. 16.2.1. Point P is located at the center of this element, and it is

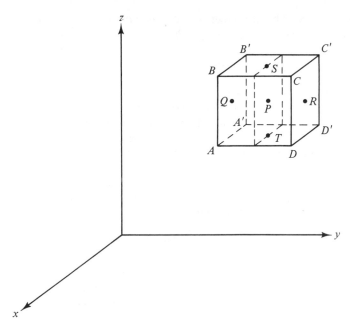

Figure 16.2.1 Infinitesimal control volume with some designated points for orientation.

assumed that the stresses in the fluid at that point are completely specified. This means that the normal and shear stresses are known by some means. Later we shall find that particular values of these stresses are not critical; only their derivatives will play a role in the equations that we shall develop. However, boundary conditions can require specific values of normal and/or shear stresses.

In general, the stresses between point P and points on surfaces $ABCD$ and $CC'D'D$ are different. The stress component is different not only because of the different displacement between the two planes and point P (in one case it is $dx/2$, and in the other, $dy/2$), but also because of the different orientations of these two planes. It is different yet again for plane $BB'C'C$, which is parallel to the xy plane, whereas the other two planes are parallel to the yz plane and the xz plane, respectively. The fact that the stresses vary not only with location but also with orientation is discussed in books dealing with solid mechanics or strength of materials.* The reader may be reminded of this variation of stress with direction if the concepts behind Mohr's circle and principal stresses are recalled.

This variability of the shear and normal stresses at a point, depending on the orientation of the plane on which they act, is the reason for specifying the orientation of the plane. In Fig. 16.2.1, a plane through point P has been indicated with dotted lines. This plane is parallel to that designated $CC'D'D$. It is also parallel to the xz plane. Hence it is perpendicular to the y axis. We use the direction of the normal to a plane to specify its orientation. There are also planes through P that have normals parallel to the x and z directions. The stresses on these three, mutually perpendicular planes through P can all be different. We designate all the stresses at P by the symbol τ, with two appropriate subscripts to make its meaning unequivocal. *The first subscript on*

---

*See, for example, E. P. Popov, *Mechanics of Materials,* 2nd ed. (Prentice-Hall, Englewood Cliffs, N.J., 1976).

τ *designates the plane on which it acts.* The direction of the *normal* to this plane identifies this subscript. *The second subscript designates the direction parallel to that in which the stress acts.*

Example 16.2.1

The symbol τ_{yy} has two subscripts, the first of which indicates that the stress component acts on a plane whose normal is parallel to y. The second subscript tells us that the stress acts in a direction that is parallel to y. The point at which this stress is to be evaluated must be specified. In Fig. 16.2.1, the point could be P, Q, or R, depending on which point is prescribed. Because τ_{yy}, a force per unit area, acts parallel to the y direction, the sign of τ_{yy} will determine whether it is in the plus or minus y direction. One must remember that τ_{yy} has magnitude and direction; it is one component of the stress vector acting on a plane whose normal is parallel to the y axis. (A stress vector has components that are stresses. These act parallel to the corresponding component direction.)

We have encountered in this example an instance of a normal stress, because the stress acts perpendicular to the surface. Any stresses acting parallel to the surface can be resolved into two components that are mutually perpendicular but still in the plane. (This decomposition is a property of vectors.) We choose to specify these tangential, or shear, stresses by two components that are parallel to the coordinate axes. Thus, for the dotted plane of Fig. 16.2.1, the other two components of the stress vector acting on that plane are τ_{yx} and τ_{yz}. The former acts on a plane whose normal is aligned with the y direction, and acts parallel to the x direction. (Again, the sign of τ_{yx} determines whether it is in the plus or minus x direction.) The third component of the stress vector is τ_{yz}, which is the component of the shear stress that acts parallel to the z axis. The complete stress vector can be written as

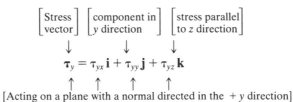

[Acting on a plane with a normal directed in the $+y$ direction]

Of course, there are at any point, for example, P, a myriad of stress vectors, depending upon the orientation of the plane through P that is considered. For the other coordinate planes through P, we have

$$\boldsymbol{\tau}_x = \tau_{xx}\,\mathbf{i} + \tau_{xy}\,\mathbf{j} + \tau_{xz}\,\mathbf{k}$$

and

$$\boldsymbol{\tau}_z = \tau_{zx}\,\mathbf{i} + \tau_{zy}\,\mathbf{j} + \tau_{zz}\,\mathbf{k}$$

Example 16.2.2

Draw $\boldsymbol{\tau}_z$ at point S in Fig. 16.2.1 if $\tau_{zx} = 1$ kPa, $\tau_{zy} = 3$ kPa, and $\tau_{zz} = -4$ kPa.

Solution The stress components and the stress vector are shown in Fig. 16.2.2.

The stresses that are given in this way act on planes that have two sides. For example, on $BB'C'C$ in Fig. 16.2.3, there is a side with a normal that points away from the plane in the $+z$ direction—*away* from the cube. There is also a normal for the surface "inside" the cube. This "inside" normal points toward the cube in the $-z$

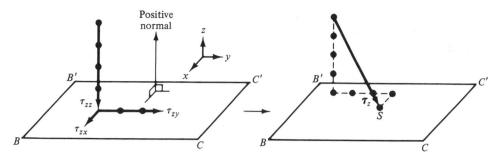

Figure 16.2.2 Components of a stress vector acting across an infinitesimal plane.

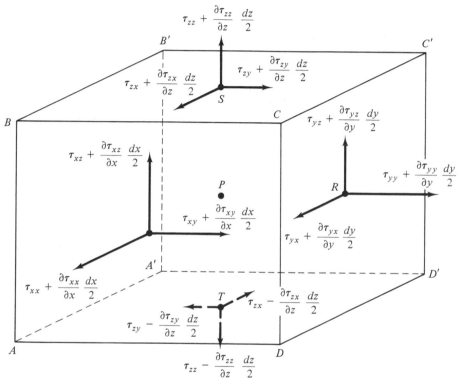

Figure 16.2.3 Stress components on three faces of the complete control surface for an infinitesimal control volume.

direction for the plane $BB'C'C$. There are also two normals on surface $AA'D'D$ at point T. However, the normal in the $+z$ direction acts on the surface "inside" the cube in this case. Each plane with normals parallel to the coordinate axes will be designated with a positive and a negative side. The sign will be determined by the positive or negative coordinate direction in which the normal points. Subsequently, relationships will be specified that relate the stresses in the fluid with the associated velocity distribution.

When the value of a stress component is given, it is taken to be the stress induced *by* the material located on the *positive* side of the plane *upon* the material located on the *negative* side of the plane. This will be our *sign convention*. In Example 16.2.2, τ_{zz}, with its negative value at point S, acts toward the fluid within the material volume, and accordingly, such a normal stress is similar to the stress that we have called *pressure*. (Had this value of τ_{zz} been specified at point T, with the positive side of the plane within the cube, we would have a stress component acting upon the fluid outside the parallelepiped that comprises the material volume.)

With the stresses given at P, we inquire about the magnitude of the various stresses on the faces of the material volume. These are shown in Fig. 16.2.3, where the magnitudes of some of the stresses are shown. On surface $CC'D'C$, the normal stress is given by $\tau_{yy} + (\partial \tau_{yy}/\partial y)(dy/2)$, where τ_{yy} is the value of this component at P and $\partial \tau_{yy}/\partial y$ is the rate of variation of this stress due to the change of the y coordinate in the interval from P to R. This coordinate change is $+dy/2$. Similarly, the shear stress that acts on this plane in the z direction is $\tau_{yz} + (\partial \tau_{yz}/\partial y)(dy/2)$. In expressions such as these, neither the values of the stresses, τ_{yz}, for example, nor the values of the rate of change, such as $\partial \tau_{yz}/\partial y$, are known as yet. The stress and/or its derivative could be positive, zero, or negative.

The stress components on three of the faces of the parallelepiped in Figure 16.2.3 have been drawn as solid lines. A fourth stress vector, the one on $AA'D'D$, has had its components drawn with dashed lines. The directions of the arrows for these components are opposite to those for the other three vectors shown. This is because the stress vector acting at S on $BB'C'C$ is presumed to be in the opposite direction to the stress vector on the opposite surface, $AA'D'D$. Observe that if the rates of change of the stresses in the z direction, $\partial \tau_{zy}/\partial z$, $\partial \tau_{zy}/\partial z$, and $\partial \tau_{zz}/\partial z$, were to be zero, then the stress vector on the bottom face of the material volume of Fig. 16.2.3 would have the same magnitude as but opposite direction to that in the upper face. This would be in agreement with our ideas about the relationship between action and reaction. Note also that on surface $AA'D'D$ the positive side of the surface is within the region of fluid particles whose dynamics are being modeled. This means that in writing $\tau_{zx} - (\partial \tau_{zx}/\partial z)\, dz/2$, we have the stress acting upon the material exterior to the body. We need the negative of this quantity. In Fig. 16.2.3, this was accomplished with the reversed direction of the arrows.

The fact that the stresses on the two surfaces may be different (i.e., $\partial \tau_{zx}/\partial z \neq 0$, and so on) means that a net unbalanced surface force on the material within the parallelepiped can exist in the z direction. Such an unbalanced force requires the existence of some other forces or acceleration so that $\mathbf{F} = m\mathbf{a}$ is satisfied. One such force is the force due to gravitational acceleration. We shall *assume* gravity to act in the $-z$ direction.

We now sum all of the forces acting on the parallelepiped in Fig. 16.2.3 in the positive z direction. (Subsequently, we shall equate the result to the product of the mass of the fluid in the material volume and the fluid's acceleration.) This summation will incorporate the body forces and the surface stresses. In the latter group, we include the normal stresses and shear stresses. Gravitational attraction will be the only body force considered. On surface $BB'C'C$, the normal stress $\tau_{zz} + (\partial \tau_{zz}/\partial z)(dz/2)$ contributes to a force in the $+z$ direction. The magnitude of this force will depend on the area on which the stress acts. In this case, the area is $dx\, dy$. Although this same area applies to surface $AA'D'D$, the magnitude of the stress there is $\tau_{zz} - (\partial \tau_{zz}/\partial z)(dz/2)$. The associated force acts in the minus z direction. Hence, the sum of the z-directed forces on these two surfaces is

$$\left(\tau_{zz} + \frac{\partial \tau_{zz}}{\partial z}\frac{dz}{2}\right)(dx\,dy) - \left(\tau_{zz} - \frac{\partial \tau_{zz}}{\partial z}\frac{dz}{2}\right)(dx\,dy)$$

For surfaces $ABCD$ and $A'B\,C'D'$, the contribution to the z force component is

$$\left(\tau_{xz} + \frac{\partial \tau_{xz}}{\partial x}\frac{dx}{2}\right)(dy\,dz) - \left(\tau_{xz} - \frac{\partial \tau_{xz}}{\partial x}\frac{dx}{2}\right)(dy,\,dz)$$

The last two of the six surfaces, $DCC'D'$ and $ABB'A'$, provide the following additional terms to the summation:

$$\left(\tau_{yz} + \frac{\partial \tau_{yz}}{\partial y}\frac{dy}{2}\right)(dx\,dz) - \left(\tau_{yz} - \frac{\partial \tau_{yz}}{\partial y}\frac{dy}{2}\right)(dx\,dz)$$

The final force to consider is that due to gravitational attraction. The weight per unit volume is ρg, so that for our parallelepiped the force in the $+z$ direction due to the weight is $-\rho g\,dx\,dy\,dz$. These steps facilitate writing the summation of all the forces in the z direction as

$$\Sigma F_z = \left[\left(\frac{\partial \tau_{xz}}{\partial x} + \frac{\partial \tau_{yz}}{\partial y} + \frac{\partial \tau_{zz}}{\partial z}\right) - \rho g\right](dx\,dy\,dz)$$

These forces are acting on the fluid that is instantaneously within the minute region around point P in Fig. 16.2.3. Such forces and the acceleration of this fluid mass are related by

$$\text{(mass of fluid)}(z\text{ component of acceleration}) = \Sigma F_z$$

or

$$[\rho(dx\,dy\,dz)]a_z = \left(\frac{\partial \tau_{xz}}{\partial x} + \frac{\partial \tau_{yz}}{\partial y} + \frac{\partial \tau_{zz}}{\partial z} - \rho g\right)(dx\,dy\,dz)$$

Although the elemental volume is very small, it is not zero and can, therefore, be canceled from this equation. The result is

$$\rho a_z = \frac{\partial \tau_{xz}}{\partial x} + \frac{\partial \tau_{yz}}{\partial y} + \frac{\partial \tau_{zz}}{\partial z} - \rho g \qquad (16.2.1)$$

The same procedure can be used for the x and y directions. Such steps would yield

$$\rho a_x = \frac{\partial \tau_{xx}}{\partial x} + \frac{\partial \tau_{yx}}{\partial y} + \frac{\partial \tau_{zx}}{\partial z} \qquad (16.2.2)$$

and

$$\rho a_y = \frac{\partial \tau_{xy}}{\partial x} + \frac{\partial \tau_{yy}}{\partial y} + \frac{\partial \tau_{zy}}{\partial z} \qquad (16.2.3)$$

These last two results could also be inferred from the first, Eq. (16.2.1). In that equation, for the z direction, the subscript z was on the acceleration. This is the "direction" appropriate to the equation. The second subscript on the stress components is also z for the direction of the equation. The first subscript is the same as the coordinate appearing in the derivative of the term. This correspondence between

Eqs. (16.2.1), (16.2.2), and (16.2.3) is not accidental. It occurs because of the basic physics of the problem and a system of notation that enhances one's insight.

If the Eulerian statements for acceleration, Eqs. (4.6.6b) to (4.6.8), are substituted for a_x, a_y, and a_z, one has

$$\rho\left(\frac{\partial u}{\partial t} + u\frac{\partial u}{\partial x} + v\frac{\partial u}{\partial y} + w\frac{\partial u}{\partial z}\right) = \frac{\partial \tau_{xx}}{\partial x} + \frac{\partial \tau_{yx}}{\partial y} + \frac{\partial \tau_{zx}}{\partial z} \qquad (16.2.4)$$

$$\rho\left(\frac{\partial v}{\partial t} + u\frac{\partial v}{\partial x} + v\frac{\partial v}{\partial y} + w\frac{\partial v}{\partial z}\right) = \frac{\partial \tau_{xy}}{\partial x} + \frac{\partial \tau_{yy}}{\partial y} + \frac{\partial \tau_{zy}}{\partial z} \qquad (16.2.5)$$

$$\rho\left(\frac{\partial w}{\partial t} + u\frac{\partial w}{\partial x} + v\frac{\partial w}{\partial y} + w\frac{\partial w}{\partial z}\right) = \frac{\partial \tau_{xz}}{\partial x} + \frac{\partial \tau_{yz}}{\partial y} + \frac{\partial \tau_{zz}}{\partial z} - \rho g \qquad (16.2.6)$$

These three equations contain as unknowns the three velocity components and the nine stress components. Clearly, there are too many unknowns for only three equations. If, however, one could relate stress to the velocity, it might be possible to add sufficient equations that the number of unknowns would equal the number of equations. Thus our goal is to seek a set of relations between the stress components and, on the basis of our earlier experience, the rates of shear and volume deformation (see Sec. 4.4). The simplest such relationship would be a linear one. In proposing such a convenient form, however, we are not sure that measurements will show a direct proportion between the stresses and the rates of deformation. If our model does not produce reasonable agreement with experimental results, a different—possibly more complicated—relationship would have to be proposed. Fortunately, the linear relationship has proved useful in a great many engineering situations that involve commonly used fluids.

The situation is also somewhat brighter than 3 equations and 12 unknowns. An analysis, similar to one that has been encountered before, allows one to show that $\tau_{xy} = \tau_{yx}$, $\tau_{xz} = \tau_{zx}$, and $\tau_{yz} = \tau_{zy}$. To prove this statement, one takes moments, successively, about axes through point P in Fig. 16.2.3 that are parallel to the coordinate axes. These moments are due to the shear stresses acting on the surface elements of the material volume. Such a moment equation must include the distances dx, dy, and dz to specify the areas associated with the stresses and the moment arms. These moments are equal to the angular acceleration of the fluid particles within the element and their mass moment of inertia. In a procedure that is similar to that of Sec. 2.2, where the pressure was shown to be independent of direction in a static fluid (or a fluid without shear stresses), the size of the element is allowed to shrink to zero. The result is that the ordering of the subscripts on the shear stress is unimportant (i.e., $\tau_{ij} = \tau_{ji}$).

No attempt will be made to justify completely here the linear relationships between stress and the rate-of-strain in the *constitutive equations* that will be used. These equations do, however, satisfy the requirement that the fluid is *isotropic*. There are no directions for which the physical properties are not the same. (This would not be the case in liquids that contain long-chain molecules. The lack of isotropy, called *anisotropy*, is characterized by materials such as wood. The tensile strength in samples of wood cut parallel to the grain is greater than the value for samples cut across the grain.) We *assume* a form for the equations of the stresses that is a generalization of Newton's law of viscosity. The resulting expressions are the following:

$$\tau_{xx} = -p + 2\mu\frac{\partial u}{\partial x} \qquad \tau_{xy} = \mu\left(\frac{\partial v}{\partial x} + \frac{\partial u}{\partial y}\right)$$

$$\tau_{yy} = -p + 2\mu\frac{\partial v}{\partial y} \qquad \tau_{xz} = \mu\left(\frac{\partial w}{\partial x} + \frac{\partial u}{\partial z}\right) \qquad (16.2.7)$$

$$\tau_{zz} = -p + 2\mu\frac{\partial w}{\partial z} \qquad \tau_{yz} = \mu\left(\frac{\partial w}{\partial y} + \frac{\partial v}{\partial z}\right)$$

(The fluid has also been *assumed* to be incompressible for the sake of simplifying the development. Thus a bulk viscosity—similar to a bulk modulus of elasticity—has been eliminated from further consideration. Such a bulk viscosity would cause the damping of compressive motions that occur in a model for sound waves.) In these equations, p is the *thermodynamic pressure* at the point where the normal stress is to be determined. The coefficient μ is the constant that determines the relationship between the relative motion and the shearing stress, just as before.

We now add the *condition* that μ does not depend on position. (The fluid has a homogeneous viscosity.) This allows μ to be factored out of the derivative terms in Eqs. (16.2.4 through (16.2.6). The coefficient of μ in the equations that result from using Eq. (16.2.7) in these equations contains spatial derivatives of $(\partial u/\partial x + \partial v/\partial y + \partial w/\partial z)$. This term is zero for an incompressible fluid and so are derivatives of it. The final results are

$$\rho\left(\frac{\partial u}{\partial t} + u\frac{\partial u}{\partial x} + v\frac{\partial u}{\partial y} + w\frac{\partial u}{\partial z}\right) = -\frac{\partial p}{\partial x} + \mu\left(\frac{\partial^2 u}{\partial x^2} + \frac{\partial^2 u}{\partial y^2} + \frac{\partial^2 u}{\partial z^2}\right) \qquad (16.2.8)$$

$$\rho\left(\frac{\partial v}{\partial t} + u\frac{\partial v}{\partial x} + v\frac{\partial v}{\partial y} + w\frac{\partial v}{\partial z}\right) = -\frac{\partial p}{\partial y} + \mu\left(\frac{\partial^2 v}{\partial x^2} + \frac{\partial^2 v}{\partial y^2} + \frac{\partial^2 v}{\partial z^2}\right) \qquad (16.2.9)$$

and $\quad \rho\left(\frac{\partial w}{\partial t} + u\frac{\partial w}{\partial x} + v\frac{\partial w}{\partial y} + w\frac{\partial w}{\partial z}\right) = -\frac{\partial p}{\partial z} + \mu\left(\frac{\partial^2 w}{\partial x^2} + \frac{\partial^2 w}{\partial y^2} + \frac{\partial^2 w}{\partial z^2}\right) - \rho g \quad (16.2.10)$

These are the *Navier-Stokes equations*. They can be solved for u, v, w, and p when they are coupled with the equation of continuity for an incompressible fluid. (In this case, ρ is a constant in the equations given above.)

The solution of these equations for prescribed flow situations requires a specification of appropriate boundary conditions and initial conditions if the flow is unsteady. In our work, we *assume* that there is no fluid motion relative to a solid boundary, because of viscosity, and that the fluid does not penetrate an impermeable boundary. This will require that $u = 0 = v = w$ on such a stationary boundary.

Example 16.2.3

What are the boundary conditions for velocity for case A, shown in Fig. 16.2.4, and case B, shown in Fig. 16.2.5?

Solution   Case A:

$$u(y = 0) = 0, \qquad u(y = k) = -\frac{U}{2}$$

Case B:

$$u(y = 0) = U \cos \omega t \qquad u(y \to \infty) \to 0$$

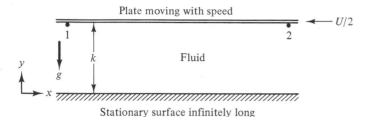

Plate moving with speed

$\longleftarrow U/2$

Fluid

Stationary surface infinitely long

Figure 16.2.4 Flow induced by a moving plate (case A for Example 16.2.3). Fluid viscosity μ and density ρ are to be considered.

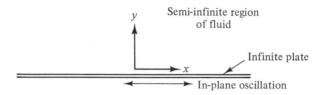

Semi-infinite region of fluid

Infinite plate

In-plane oscillation

Figure 16.2.5 Flow induced by a plate that oscillates in its own plane (case B for Example 16.2.3). Plate oscillates with speed $U_0 = U \cos \omega t$. As $y \to \infty$, the fluid is at rest. Consider viscosity μ and density ρ.

In Fig. 16.2.4, for a time-independent solution (perhaps a long time after the initial motion of the plate), $a_x = 0$ for each particle. An examination of Eq. (16.2.8) yields the same conclusion. The temporal derivative, $\partial u/\partial t$, is zero because steady state is *assumed*. From the continuity equation for a two-dimensional incompressible flow, one has $\partial u/\partial x = -\partial v/\partial y$. Since v is assumed everywhere zero, $\partial v/\partial y$ is zero. Thus $\partial u/\partial x$ must be zero. This means that Eq. (16.2.8) becomes, under the conditions outlined in Fig. 16.2.4,

$$\rho(0) = 0 + \mu\left(0 + \frac{\partial^2 u}{\partial y^2} + 0\right)$$

The three values of zero that are included on the right-hand side of this equation are due to (1) the *assumed* absence of pressure changes in the x direction; (2) continuity, which requires that $\partial u/\partial x$ be zero everywhere because v is identically zero, so that $\partial^2 u/\partial x^2$ is zero; and (3) the fact that all derivatives with respect to z are zero, since no variations are *assumed* for the z direction. The consequence of these considerations is a reduction of the complete Navier-Stokes equation for the x direction to

$$0 = \frac{d^2 u}{dy^2}$$

so that
$$u = Ay + B$$

The constants A and B are determined from the boundary conditions that have already been listed.

For Fig. 16.2.5, motion parallel to the plate is *assumed* once again. By examining the terms comprising a_x, one concludes that a simplification can be made because $\partial u/\partial x$ and v are zero. Because this flow is unsteady, $\partial u/\partial t$ is a nonzero quantity. It is the only contribution to a_x in this case.

A zero pressure variation in the x direction is *assumed* for this case. The *assumed* absence of a y component of velocity and any z-direction variations permits Eq. (16.2.8) to be written as

$$\frac{\partial u}{\partial t} = \frac{\mu}{\rho}\frac{\partial^2 u}{\partial y^2}$$

The Navier-Stokes Equations   Chap. 16

This was the differential equation that was solved in Sec. 8.5 for the boundary conditions that are associated with Fig. 16.2.5.

Example 16.2.4

The boundary condition at $y = 0$ for Fig. 16.2.5 is pictorially represented in Fig. 16.2.6 along with the boundary condition for a plate whose motion is initiated impulsively. In this case, the plate was at rest prior to $t = 0$. The plate is suddenly accelerated to a speed of U_0 that is subsequently kept constant. Write the applicable equations of motion.

Solution A consideration of the problem of determining the velocity distribution at times greater than $t = 0$ could be accompanied by the *assumptions* that $\partial p / \partial x = 0$ and $v = 0$, as well as no variations in the z direction. Under these conditions, the governing equation is achieved by reducing the x component of the Navier-Stokes equations to the form

$$\frac{\partial u}{\partial t} = v \frac{\partial^2 u}{\partial y^2} \tag{16.2.11}$$

This is the same equation that was used to solve the problem given in Fig. 16.2.5. However, as indicated in Fig. 16.2.6, the boundary condition at $y = 0$ is new to us. The solution to the governing differential equation is achieved by introducing a new independent variable

$$\eta = \frac{y}{2\sqrt{vt}}$$

A check on the dimensions of this relationship shows that η is nondimensional. [Dimensional reasoning indicates that $u = \phi(U_0, v, y, t)$, so that there are five variables, with the two dimensions of length and time. But u should be directly proportional to U_0, so that $u/U_0 = \hat{\phi}(v, y, t)$. This leads directly to $u/U_0 = \tilde{\phi}(y/\sqrt{vt})$ and the two dimensionless parameters of the problem. The factor of 2 in η is introduced for convenience.] The solution to the differential equation and the boundary conditions is shown in Fig. 16.2.7. To solve the problem, one assumes that $u = U_0 f(\eta)$. Then the equation of motion demands that the function $f(\eta)$ satisfy the equation $f''/f' = -2\eta$, which leads to $\ln f' = -\eta^2 - \ln C_1$, or $C_1 f' = \exp(-\eta^2)$. The solution for f through another integration gives $f = 1 - \mathrm{erf}\,\eta$, in which $\mathrm{erf}\,\eta$ is the *error function* that is often encountered in statistics and probability theory. One denotes $1 - \mathrm{erf}\,\eta$ as the *complimentary error function* and this function is plotted in Fig. 16.2.7.

Naturally, in the study of fluid mechanics, when one attempts to solve the differential equation that is associated with a particular problem, an analytical solution may not be possible. Recently, the digital computer and the techniques of numerical analy-

Figure 16.2.6 (a) Velocity history of a plate that oscillates sinusoidally in its own plane with simple harmonic motion and (b) velocity history of a plate that impulsively moves with constant speed in its own plane. U_0 = velocity of each plate at $y = 0$.

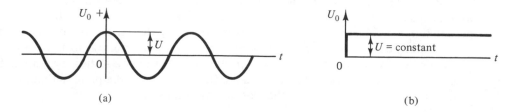

(a) (b)

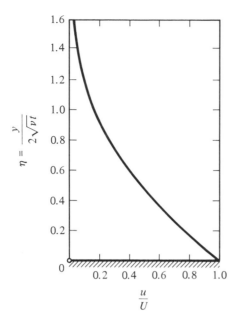

Figure 16.2.7 Velocity distribution in the upper half plane due to an impulsively started plate. (Adapted from H. Schlichting's "Boundary-Layer Theory," McGraw-Hill Book Company, by permission)

sis have been successfully combined to obtain solutions to some otherwise "unsolvable" problems. This capability for solving extremely complex problems is rapidly increasing.

Example 16.2.5

Determine the boundary-layer thickness for the case of a suddenly accelerated plate that was treated in Example 16.2.4.

Solution   The boundary layer thickness δ is often defined as the position that is normal to the boundary, at which the local velocity is a small fraction—usually 1 percent—of the velocity of the moving plate. Figure 16.2.7 is too limited to give information in such fine detail. Nevertheless, one can estimate from this figure that, at $\eta = 1.6$, u/U_0 is about 0.02. From the definition of η, and letting δ for this example be the location where $u/U_0 = 0.02$, we have

$$\eta = 1.6 = \left(\frac{y}{2\sqrt{vt}}\right)_{y\,=\,\delta}$$

The result for this specially defined δ is

$$\delta = 3.2\sqrt{vt}$$

Thus the boundary layer thickens with time. The longer the fluid is exposed to the movement of the plate, the thicker the boundary layer becomes. The fluid motions that are a consequence of the moving plate diffuse into the fluid through the viscosity as time passes. The process has been described as one of diffusion because the y component of velocity was assumed to be zero, so that there is no change of x momentum due to vertical mass transport. Such motions normal to the plate were present when we considered the similar problem of flow over a stationary plate in Sec. 8.6. Because of the motions that carried the fluid away from the plate in that case, the boundary layer was thicker than that for the present example.

Example 16.2.6

The flow between two parallel plates can be extended in an enlightening way by assuming the bounding surfaces to be porous. In addition to the velocity parallel to the walls that is induced by the pressure gradient $-\partial p/\partial x = \rho C$, there is a constant vertical component V that is perpendicular to the porous boundaries. Solve for the velocity distribution in the channel. What does this solution predict as $\mu \to 0$? What does the differential equation predict when $\mu = 0$?

Solution   Let the x axis be embedded in the lower, stationary plate. The upper plate, also stationary, is at $y = k$. The flow is assumed to be steady, and the fluid is incompressible. The latter requirement gives, from the continuity equation, for a two-dimensional flow,

$$\frac{\partial u}{\partial x} + \frac{\partial v}{\partial y} = 0$$

Since $v = V$ (a constant), $\partial v/\partial y = 0$, so that $\partial u/\partial x$ must also be zero. An examination of the terms contained in Eq. (16.2.8) shows that only $\rho v(\partial u/\partial y)$, $-\partial p/\partial x$, and $\mu\,\partial^2 u/\partial y^2$ are nonzero. Hence, one can write for this momentum equation

$$\rho V \frac{du}{dy} = \rho C + \mu \frac{d^2 u}{dy^2}$$

or

$$v \frac{d^2 u}{dy^2} - V \frac{du}{dy} = -C$$

This linear equation can be solved easily. The general solution is

$$u = C_1 e^{Vy/v} + \frac{C}{V} y + C_2$$

The corresponding boundary conditions are $u(0) = 0$ and $u(k) = 0$. The former condition requires

$$0 = C_1 + C_2$$

while the latter one requires

$$0 = C_1 e^{Vk/v} + \frac{C}{V} k + C_2$$

A subtraction of these two equations yields

$$\frac{C}{V} k = C_1 (1 - e^{Vk/v})$$

so that

$$C_1 = \frac{-Ck/V}{e^{Vk/v} - 1} = -C_2$$

These constants result in

$$u = \frac{Ck}{V} \left[\frac{y}{k} - \left(\frac{e^{Vy/v} - 1}{e^{Vk/v} - 1} \right) \right]$$

If $\mu \ll 1$, the exponential in the denominator will be large and dominate the other term. At $y = 0$, the numerator in the term in parenthesis vanishes, but its value slowly increases with y. For moderate values of y, the large denominator in the term in parentheses—due to the value of Vk/v—makes the entire fraction small with respect to y/k. One would expect that u to vary nearly linearly with y in this range of y values. Nevertheless, in the remaining portion of the channel—near the upper plate—the velocity distribution changes rapidly. The velocity component must go to zero at the upper plate. Figure 16.2.8 displays these characteristics of the velocity profile.

However, if $\mu = 0$ is used in the differential equation, one has

$$\frac{\partial u}{\partial y} = \frac{C}{V} \quad \text{or} \quad u = \frac{C}{V}y + C_3$$

Depending on which boundary condition is satisfied, C_3 is either 0 or $-(C/V)k$. However, both boundary conditions cannot be concurrently satisfied. This is the result of setting μ to be exactly zero in the differential equation with the consequence that the second derivative of u is dropped from the governing equation.

The following three equations are the Navier-Stokes equations for cylindrical coordinates without the inclusion of a body force term if $(u, v, w) = (u_r, u_\theta, u_z)$:

$$\frac{\partial u}{\partial t} + u\frac{\partial u}{\partial r} + \frac{v}{r}\frac{\partial u}{\partial \theta} + w\frac{\partial u}{\partial z} - \frac{v^2}{r} = -\frac{1}{\rho}\frac{\partial p}{\partial r} + v\left(\nabla^2 u - \frac{u}{r^2} - \frac{2}{r^2}\frac{\partial v}{\partial \theta}\right) \quad (16.2.12)$$

Figure 16.2.8 u velocity distributions in a channel of width k for different values of the transverse velocity V. Magnification factor of 50 for $Vk/v = 0.1$ curve.

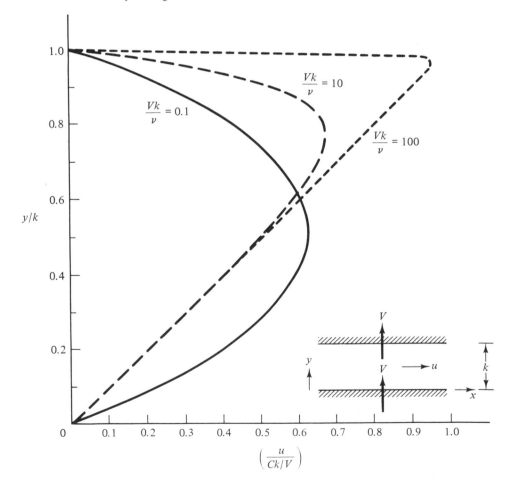

The Navier-Stokes Equations   Chap. 16

$$\frac{\partial v}{\partial t} + u\frac{\partial v}{\partial r} + \frac{v}{r}\frac{\partial v}{\partial \theta} + w\frac{\partial v}{\partial z} + \frac{uv}{r} = -\frac{1}{\rho}\frac{1}{r}\frac{\partial p}{\partial \theta} + \nu\left(\nabla^2 v + \frac{2}{r^2}\frac{\partial u}{\partial \theta} - \frac{v}{r^2}\right) \qquad (16.2.13)$$

and
$$\frac{\partial w}{\partial t} + u\frac{\partial w}{\partial r} + \frac{v}{r}\frac{\partial w}{\partial \theta} + w\frac{\partial w}{\partial z} = -\frac{1}{\rho}\frac{\partial p}{\partial z} + \nu\nabla^2 w \qquad (16.2.14)$$

where
$$\nabla^2 = \frac{\partial^2}{\partial r^2} + \frac{1}{r}\frac{\partial}{\partial r} + \frac{1}{r^2}\frac{\partial^2}{\partial \theta^2} + \frac{\partial^2}{\partial z^2} \qquad (16.2.15)$$

Note that the first two terms on the right-hand side of Eq. (16.2.15) can be expressed as follows:
$$\frac{\partial^2}{\partial r^2} + \frac{1}{r}\frac{\partial}{\partial r} = \frac{1}{r}\frac{\partial(r\,\partial/\partial r)}{\partial r}$$

The equation of continuity,
$$\operatorname{div} \mathbf{V} = \frac{1}{r}\frac{\partial}{\partial r}(ru) + \frac{1}{r}\frac{\partial v}{\partial \theta} + \frac{\partial w}{\partial z} = 0 \qquad (16.2.16)$$

completes this system of equations to determine the velocity components and pressure from the prescribed boundary and initial conditions. These equations readily give the velocity distribution for a laminar flow in a round conduit that were found in Sec. 8.3. The results of Sec. 8.4 also follow without difficulty.

The above form of the Navier-Stokes equations follows from a complete statement of the acceleration vector in polar coordinates. This includes the centripetal and Coriolis acceleration terms. This coordinate system and the Cartesian system are connected by equations that relate r and θ with x and y. Consequently, derivative terms in one coordinate system are obtained by chain-rule differentiation. The following example should give the reader an appreciation of what is involved.

Example 16.2.7

Cylindrical polar coordinates were used in Sec. 8.4, and the flow pattern was assumed to be circular. An expression for the shear stresses in polar coordinates was used. Justify using this expression.

Solution The shear stress τ_{yx} is $\mu(\partial v/\partial x + \partial u/\partial y)$, with $u = -V_\theta \sin\theta$ and $v = V_\theta \cos\theta$ for $V_r = 0$. This follows from a coordinate transformation similar to that indicated in Fig. 4.6.1. Using these terms in the expression for τ_{yx}, and remembering that $\sin\theta = y/r$, $r^2 = x^2 + y^2$, and $V_\theta = Ar + B/r$, one obtains

$$\tau_{yx} = \mu\left[\frac{\partial}{\partial x}\left(V_\theta\frac{x}{r}\right) - \frac{\partial}{\partial y}\left(V_\theta\frac{y}{r}\right)\right]$$

$$= \mu\left\{\frac{x}{r}\frac{\partial V_\theta}{\partial x} + V_\theta\left[\frac{(x^2+y^2)^{1/2}}{x^2+y^2} - \frac{x^2}{(x^2+y^2)^{3/2}}\right]\right.$$

$$\left. - \frac{y}{r}\frac{\partial V_\theta}{\partial y} - V_\theta\left[\frac{(x^2+y^2)^{1/2}}{x^2+y^2} - \frac{y^2}{(x^2+y^2)^{3/2}}\right]\right\}$$

This expression for circular flow can be simplified in general; and *for the point $x = 0$ and $y = r$,*

$$\tau_{yx} = \mu\left(-\frac{\partial V_\theta}{\partial y} + \frac{V_\theta r^2}{r^3}\right)$$

At that point, $\partial/\partial y = \partial/\partial r$, so that

$$\tau_{yx} = \mu \left(\frac{V_\theta}{r} - \frac{\partial V_\theta}{\partial r} \right)$$

which can be written in the form used in Sec. 8.4 as

$$\tau_{yx} = -\mu r \left[\frac{\partial}{\partial r} \left(\frac{V_\theta}{r} \right) \right]$$

Finally, one must note at $(x = 0, y = r)$ that $\tau_{yx} = -\tau_{r\theta}$, because the positive θ direction is counterclockwise while x is positive in the opposite direction at this point. This results in Eq. (8.4.4), for circular flow, which was written without subscripts:

$$\tau_{r\theta} = \mu r \left[\frac{\partial}{\partial r} \left(\frac{V_\theta}{r} \right) \right]$$

The work above constitutes a plausible argument for the form of the shear stress in polar coordinates. A more exact derivation would be more lengthy and would include the resolution of the stresses, *and* the areas on which they act, in the directions associated with the new coordinates.

16.3 BOUNDARY-LAYER EQUATIONS

The purposes for presenting the examples of the previous section were threefold. The first goal was to show the application of the Navier-Stokes equations and the importance of the boundary conditions in obtaining solutions. The other two purposes are to show characteristics of solutions and to motivate a simplification of the Navier-Stokes equations for problems where the existence of a boundary layer is expected.

The variable η was defined as $y/2\sqrt{\nu t}$ in Example 16.2.4. One can see that the same value of η results in a particular problem for the same values of $y/\sqrt{t}$. It is an example of a similarity variable, a concept that was introduced in Sec. 9.7. Thus, for $y = 0.5$ and $t = 4$, and $y = 0.75$ and $t = 9$, the same value of η results. Fig. 16.2.7 plotted u/U_0 against the other variable, η, which is in fact the combination of two obvious variables of the problem, y and t.

Because inertia effects are one factor that is considered to be important in cases where boundary layers are thought to occur, the acceleration terms in the Navier-Stokes equations are important. If one seeks to simplify these equations in order to facilitate the solution process for a problem, it is necessary to determine whether any terms can be considered negligible. Often the pressure gradient is the agent that causes the flow to exist. This means that it must be included in the problem in some way. The terms that are associated with μ in Eq. (16.2.8) might be thought to be unimportant because of the smallness of the coefficient μ. This was the mode of thought until 1904, when Ludwig Prandtl demonstrated the rationale for neglecting all the terms that multiply μ in Eq. (16.2.8) except $\partial^2 u/\partial y^2$. This term, when multiplied by μ, could be of the same order of magnitude as $\partial p/\partial x$ and a_x and should be retained when considering a viscous flow near a boundary. (Example 16.3.1 will provide the basis for the argument to neglect $\partial^2 u/\partial x^2$ with respect to $\partial^2 u/\partial y^2$.) This important step led to the *boundary-layer equation* in two dimensions,

$$\rho\left(\frac{\partial u}{\partial t} + u\frac{\partial u}{\partial x} + v\frac{\partial u}{\partial y}\right) = -\frac{1}{\rho}\frac{\partial p}{\partial x} + v\frac{\partial^2 u}{\partial y^2} \tag{16.3.1}$$

The justification of the form of this equation follows.

Example 16.3.1

Show that Eq. (16.3.1), in which the second derivative term is multiplied by v, usually a very small number, can be transformed by proper scaling of the variables into an equation in which the second derivative term has a coefficient of order unity. (As a result, the second derivative term—because the viscosity is included—will affect the solution.) Use the scaling for the flow over a flat plate of length L as follows.

$$x = Lx^* \qquad y = (\alpha L)y^*$$

in which the vertical scale factor α must be determined. The starred terms in these definitions are nondimensional and represent the fraction of the plate's length or the fraction of the vertical extent over which the solution will depend significantly on y. Also let

$$u = U_\infty u^* \qquad v = (\beta U_\infty)v^* \qquad p = (\rho U_\infty^2)p^*$$

in which the factor β is unrestricted but one expects $\beta \ll 1$ because v should be nearly zero for small and large y.

Solution The term $u\,(\partial u/\partial x)$ can be written as

$$(U_\infty u^*)\left[\frac{\partial(U_\infty u^*)}{\partial x}\right]$$

or

$$U_\infty^2 u^*\left(\frac{\partial u^*}{\partial x^*}\right)\left(\frac{\partial x^*}{\partial x}\right) = \frac{U_\infty^2 u^*(\partial u^*/\partial x^*)}{L}$$

because $\partial x^*/\partial x = 1/L$, a fact that can be verified from the definition of x^*. In a similar fashion, one can express $v\,\partial u/\partial y$ as

$$U_\infty^2 \beta v^*\left(\frac{\partial u^*}{\partial y^*}\right)\left(\frac{\partial y^*}{\partial y}\right) = \frac{U_\infty^2 \beta}{\alpha L}v^*\left(\frac{\partial u^*}{\partial y^*}\right)$$

The pressure gradient term, $(-1/\rho)(\partial p/\partial x)$, is rewritten as

$$-\left(\frac{\rho U_\infty^2}{\rho L}\right)\left(\frac{\partial p^*}{\partial x^*}\right)$$

and the viscous term, $v(\partial^2 u/\partial y^2)$, becomes

$$\frac{v U_\infty}{(\alpha L)^2}\left(\frac{\partial^2 u^*}{\partial y^{*2}}\right)$$

If the new expressions are substituted in Eq. (16.3.1), one obtains, after dividing by U_∞^2/L,

$$u^*\frac{\partial u^*}{\partial x^*} + \frac{\beta}{\alpha}v^*\frac{\partial u^*}{\partial y^*} = -\frac{\partial p^*}{\partial x^*} + \left(\frac{v}{U_\infty L}\right)\left(\frac{1}{\alpha^2}\right)\frac{\partial^2 u^*}{\partial y^{*2}} \tag{16.3.2}$$

The coefficient of the second term in this equation will be set equal to 1 by letting $\alpha = \beta$. (It is convenient to have a unity coefficient, but if some good purpose would be served, it could be any other constant.) The term in the first parenthesis is the reciprocal of the Reynolds number Re_L based upon the plate's length.

If the entire coefficient of the second derivative is required to be approximately unity, then

$$\alpha^2 = \frac{1}{U_\infty L/\nu} \quad \text{or} \quad \alpha = \text{Re}_L^{-1/2} = \beta \qquad (16.3.3)$$

This is the condition that was sought at the outset in order to focus attention on the highest derivative in the governing equation.

The fact that α will be small for large values of Re_L implies that, if y^* influences the solution when it is less than some constant C, the distance y that corresponds to it will be very much less, $C/\text{Re}_L^{1/2}$. From these observations, one can infer the extent of the distance from the plate where the no-slip condition influences the velocity distribution through the viscosity. We can conclude that the distance decreases with increasing Re_L. (Figure 16.2.8 illustrates this point.)

The boundary-layer equation, Eq. (16.3.1), differs from the corresponding Navier-Stokes equation by the term $\nu \, \partial^2 u/\partial x^2$. It is easy to show why this is justified for a flow with a high Reynolds number. The dimensionless variables used in Example 16.3.1 transform $\nu \, \partial^2 u/\partial x^2$ initially into $\nu \, U_\infty/L^2 (\partial^2 u^*/\partial x^{*2})$ and ultimately into $(\nu/U_\infty L)(\partial^2 u^*/\partial x^{*2})$ after a division by U_∞^2/L. Whereas in Eq. (16.3.2) one could make the coefficient of the second derivative with respect to y^*, a viscous term, equal to unity by setting $\alpha = \text{Re}_L^{-0.5}$, no such step is possible for $\partial^2 u^*/\partial x^{*2}$, because there is no disposable constant in its coefficient. Thus, this derivative's coefficient, $1/\text{Re}_L$, will vanish as $\text{Re}_L \to \infty$. The possibility of preserving the second derivative of u^* with respect to y^* at high Re_L, even though the second derivative with respect to x^* is lost, is a *salient feature* of the boundary-layer equations.

In a manner similar to that of Example 16.3.1, one can use the dimensionless variables presented there in the steady-state Navier-Stokes equation for the y direction to see what can be learned. The following forms can be verified easily by using the results of chain-rule differentiation used in Example 16.3.1. Starting with

$$u\frac{\partial v}{\partial x} + v\frac{\partial v}{\partial y} = -\frac{1}{\rho}\frac{\partial p}{\partial y} + \nu\left(\frac{\partial^2 v}{\partial x^2} + \frac{\partial^2 v}{\partial y^2}\right)$$

one can nondimensionalize it to obtain

$$\beta\frac{U_\infty^2}{L}\left(u^*\frac{\partial v^*}{\partial x^*} + \frac{\beta}{\alpha}v^*\frac{\partial v^*}{\partial y^*}\right) = -\frac{U_\infty^2}{\alpha L}\frac{\partial p^*}{\partial y^*} + \frac{\nu U_\infty}{L^2}\beta\left(\frac{\partial^2 v^*}{\partial x^{*2}} + \frac{1}{\alpha^2}\frac{\partial^2 v^*}{\partial y^{*2}}\right)$$

Therefore, with $\beta = \alpha$,

$$\frac{1}{\text{Re}_L}\left(u^*\frac{\partial v^*}{\partial x^*} + v^*\frac{\partial v^*}{\partial y^*}\right) = -\frac{\partial p^*}{\partial y^*} + \frac{1}{\text{Re}_L^2}\left(\frac{\partial^2 v^*}{\partial x^{*2}} + \text{Re}_L\frac{\partial^2 u^*}{\partial y^{*2}}\right) \qquad (16.3.4)$$

This means that $\partial p^*/\partial y^*$, the change of pressure across the boundary layer, begins to vanish as $\text{Re}_L \to \infty$. This is the *important result* that is contained in the Navier-Stokes equation in the y direction for very high Reynolds numbers. Equation (16.3.1), the boundary-layer equation, was the consequence of letting Re_L approach infinity in the equation for the velocity component that is parallel to the surface at the outer edge of the boundary layer. In going to this limit, the second derivative with respect to y was retained so that two boundary conditions could be satisfied. These transformations of the complete equations of motion to the special case of very large Re_L facilitated the process of obtaining useful theoretical results. Because $\partial p^*/\partial y^* \to 0$ as $\mathrm{Re}_L \to \infty$, p will be a function only of x in the boundary layer. The value of dp/dx will be taken from the expression in the streamline at the outer edge of the boundary layer. On this line, the effects of viscosity are assumed to be zero.

The selected values of α and β produced unity coefficients in Eq. (16.3.2). However, if the plate had been semi-infinite in length ($0 \le x \le \infty$), then a suitable characteristic length, L, would not exist. Thus, we seek a similarity variable η that combines x^* and y^* ($= x/L$ and $y\,\mathrm{Re}_L^{1/2}/L$) in such a way that L does not appear. This makes u^* a function of x^* and y^* that is independent of any arbitrarily chosen L. The ratio $y^*/(x^*)^{1/2}$ is such a similarity variable. To show this, we write

$$\eta = y^*/(x^*)^{1/2} = y\sqrt{\frac{U_\infty}{L\nu}}/\sqrt{x/L} = y\sqrt{\frac{U_\infty}{x\nu}}$$

In the case of the flow over a flat plate, one expects that $u/U_\infty \to 1$ at a value of y that is sufficiently far from the plate. This is the *boundary-layer thickness*, and it has been denoted by δ.

The continuity equation for a two-dimensional incompressible flow is $\partial u/\partial x + \partial v/\partial y = 0$. The previous steps to determine α and β yield $\partial u^*/\partial x^* + \partial v^*/\partial y^* = 0$. This equation indicates that our scaling of an equation of motion has preserved the ratio of changes of u to x and v to y. From this result, one concludes that, if $\partial u^*/\partial x^* \approx 1$, then changes in v^* will be approximately equal to changes in y^*. Since $v^*(y^* = 0) = 0$, one would have a (nearly) linear change of v^* with y^*. Additional information about v and, in effect, v^* can be gained from an examination of other consequences of the continuity equation. It is convenient to do this by using the stream function, which, when differentiated with respect to y, produces u, according to Eq. (4.7.4). This differential relationship leads to the integral statement

$$\psi = \int_0^y u\,dy$$

A change of variables to η ($= y/\delta$, with $\delta = \sqrt{x\nu/U_\infty}$) and u^* ($= u/U_\infty$) for the case of a similarity solution leads to

$$\psi = U_\infty \delta \int_0^\eta u^*\,d\eta$$

We see that δ can be factored out of the integral because it is independent of y, the original variable of integration. It is now assumed that u^* depends on the combination of x and y that is given by η. We could call this relationship $G(\eta)$; however, in order that our discussion might correspond closely to the existing literature, we write instead

$$\frac{u}{U_\infty} = u^* = F'(\eta) \tag{16.3.5}$$

Thus, u^* will be known only after $F(\eta)$ is differentiated. As a consequence of our *assumption* that a similarity solution exists,

$$\psi = U_\infty \delta \int_0^\eta F'(\eta)\,d\eta$$

This expression can be integrated—that was the reason for introducing the derivative of F—and the constant $F(0)$ is taken as zero. The result is

$$\psi = U_\infty \delta F(\eta)$$

The y component of velocity follows from Eq. (4.7.5), $v = -\partial\psi/\partial x$. This requires that $v = -\partial(U_\infty \delta F)/\partial x$, in which δ and F both depend on x. The differentiation is carried out with the chain rule:

$$v = -U_\infty \frac{\partial(\delta F)}{\partial x} = -U_\infty \left(\delta \frac{\partial F}{\partial \eta} \frac{\partial \eta}{\partial x} + F \frac{\partial \delta}{\partial x} \right)$$

This expression can be simplified, because

$$\frac{\partial F}{\partial \eta} = \frac{dF}{d\eta} = F'$$

and

$$\frac{\partial \eta}{\partial x} = \frac{\partial(y/\delta)}{\partial x} = -\frac{y \delta'}{\delta^2}$$

$$= -\frac{y}{\delta} \frac{\delta'}{\delta} = -\frac{\eta \delta'}{\delta}$$

if

$$\frac{\partial \delta}{\partial x} = \delta'$$

(Our choice of $\delta = \sqrt{x\nu/U_\infty}$ yields $\delta' = \sqrt{\nu/U_\infty x}/2$.) Therefore,

$$v = -U_\infty \left[\delta F' \left(\frac{-\eta}{\delta} \right) \delta' + F\delta' \right] = U_\infty \delta'(F'\eta - F) \qquad (16.3.6)$$

The use of $u = U_\infty F'$ in Eq. (16.3.1) produces

$$\nu F''' \left(\frac{d\eta}{dy} \right)^2 = \frac{\nu F'''}{\delta^2} \quad \text{for the term} \quad \frac{\nu \partial^2 u}{\partial y^2}$$

Similar terms are found for those on the left-hand side of Eq. (16.3.1) with the use of u from Eq. (16.3.5) and the form of v that is given by Eq. (16.3.6). (Some of these terms cancel). We have been considering a flat plate without a pressure gradient, so that $dp/dx = 0$. When the previous steps are completed for such a case, the resulting equation is

$$FF'' + 2F''' = 0 \qquad (16.3.7)$$

with $F(0) = 0$, $F'(0) = 0$, and $F'(\infty) = 1$. [Equations (16.3.5) and (16.3.6) can be used to justify the boundary conditions.] The numerical solution of this equation is the source of Fig. 16.3.1. This solution was first found by Prandtl's student H. Blasius in 1908. His solution was approximated by $u/U = \sin(\pi y/2\delta)$ in Example 8.6.1. Experiments to verify this solution are difficult to perform, because of the disturbance to the flow that is caused by the leading edge of the plate. Nevertheless, the velocity profile has been measured and the data confirm Blasius' results.

From the time of Prandtl's derivation of the boundary-layer equations to the present, boundary-layer research has played an important role in many engineering problems related to meteorology, naval architecture, and aeronautical, civil, chemical, and mechanical engineering. It is clear that Prandtl's thoughts and observations that led to Eq. (16.3.1) are as important to our present knowledge as Bernoulli's work, which was done about 200 years earlier.

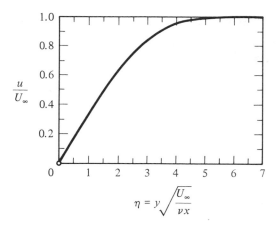

Figure 16.3.1 Distribution of the velocity component parallel to a flat plate, after Blasius. (From "Fluid Mechanics," by Chia-Shun Yih, West River Press, 1977)

$$\eta = y\sqrt{\dfrac{U_\infty}{\nu x}}$$

Outside the boundary layer, the effects of $\mu\,\partial^2 u/\partial y^2$ are taken as negligible. When this term is neglected, the remaining part of Eq. (16.3.1) is identical with the Euler equation for the x direction. One of the extensive uses of the Euler equation is to study the dynamics of irrotational flows. (These are also called *potential flows,* because irrotationality implies the existence of the velocity potential function Φ.) Thus, if one can find a suitable function Φ which gives a velocity component that is zero normal to a prescribed solid boundary, one can use it with the Euler equation, or the Bernoulli equation in this case, to specify $\partial p/\partial x$. This is the pressure variation near the solid boundary, along which the potential flow has a nonzero tangential velocity. (The Euler equation does not have second derivatives to accommodate the no-slip boundary condition.) This $\partial p/\partial x$ from the potential flow is used in Eq. (16.3.1). (The *assumption* that the pressure does not vary across the thickness of the boundary layer is an important feature of the boundary-layer theory.) Then $\partial p/\partial x$ is no longer an unknown. The corresponding equation for the x direction and the continuity equation permit progress toward a solution for the remaining unknowns, u and v.

For the two-dimensional flow into a (line) sink shown in Fig. 16.3.2, the streamlines are radial lines given by $-Q\theta/2\pi$, with

$$u_r = -\frac{Q}{2\pi r} \tag{16.3.8}$$

in which $Q/2$ is the volume rate of flow (per unit of length normal to the page) *into* the sink from the region $y > 0$. The Bernoulli equation can be used from any point with radius r to a point at $r \to \infty$. There, u_r approaches 0. This means that

$$\frac{u_r^2}{2} + \frac{p}{\rho} = \frac{0^2}{2} + \frac{p_\infty}{\rho}$$

and

$$-\frac{\partial p}{\partial r} = \frac{\partial}{\partial r}\left(\frac{u_r^2}{2}\right) = \rho u_r \frac{\partial u_r}{\partial r}$$

In this case, with $u_r = -Q/2\pi r$,

$$-\frac{1}{\rho}\left(\frac{\partial p}{\partial r}\right) = \left(\frac{Q}{2\pi}\right)^2 \frac{1}{r}\left(-\frac{1}{r^2}\right) = -\left(\frac{Q}{2\pi}\right)^2 \frac{1}{r^3}$$

Near the boundary at the right in Fig. 16.3.2 ($x > 0, y \approx 0$), one would have $r \approx x$, so that

$$-\frac{1}{\rho}\left(\frac{\partial p}{\partial x}\right) = -\left(\frac{Q}{2\pi}\right)^2 \frac{1}{x^3}$$

This would be the value of the pressure gradient to use in Eq. (16.3.1) if one wished to solve for the velocity near the solid boundary for a viscous two-dimensional sink flow.

Example 16.3.2*

Use the pressure gradient near the right-hand boundary of Fig. 16.3.2 that was just given to determine the flow in the boundary layer.

Solution   The entire flow into the sink will be assumed to be completely radial. The continuity equation, in plane polar coordinates, prescribes that

$$u_r = -\frac{F(\theta)}{r} \qquad (16.3.9)$$

in which F is a function that depends only on the angle θ. [This result has the general form of Eq. (16.3.8).] Thus,

$$u = u_r \cos\theta = -\frac{F(r \cos\theta)}{r^2} = -\frac{Fx}{x^2 + y^2}$$

$$v = u_r \sin\theta = -\frac{Fy}{x^2 + y^2}$$

Near the boundary in question, y will be very small, so that $y^2 \ll x^2$ if one stays sufficiently far to the right of the sink. This is the reason why we shall employ the approximations

$$u = -\frac{F}{x} \quad \text{and} \quad v = -F\frac{y/x}{x}$$

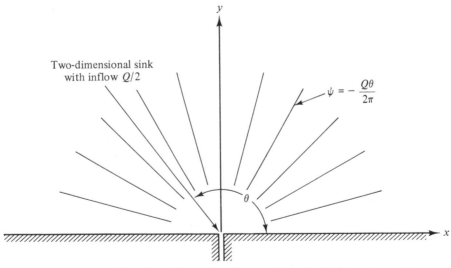

Figure 16.3.2   Two-dimensional viscous flow due to a sink embedded in a plane surface.

*This example follows broadly the presentation of D. Jankowski in "A non-linear example in elementary boundary layer theory," *Int. J. Mech. Eng. Ed.,* Vol. 2, pp. 19–22, 1974.

In general, $\tan \theta = y/x$, but for sufficiently small y/x one can use $\theta = y/x$. This gives θ the role of a similarity parameter. For the sake of clarity, and consistency with the previous presentation, we shall use η instead of θ. Then

$$u = -\frac{F(\eta)}{x} \quad \text{and} \quad v = \eta \left[-\frac{F(\eta)}{x} \right]$$

with

$$\eta = \frac{y}{x} \tag{16.3.10}$$

Furthermore, chain-rule differentiation provides us with

$$\frac{\partial u}{\partial x} = \frac{F(\eta)}{x^2} - \frac{1}{x} \frac{\partial}{\partial x} (F(\eta))$$

$$= \frac{F(\eta)}{x^2} - \frac{1}{x} \frac{dF}{d\eta} \frac{\partial \eta}{\partial x}$$

$$= \frac{F(\eta)}{x^2} - \frac{1}{x} F' \left(-\frac{y}{x^2} \right) = \frac{1}{x^2} (F + \eta F')$$

and

$$\frac{\partial u}{\partial y} = -\frac{1}{x} F' \frac{\partial \eta}{\partial y} = -\frac{1}{x^2} F' \qquad \frac{\partial^2 u}{\partial y^2} = -\frac{F''}{x^3}$$

Now one can substitute into Eq. (16.3.1), with the result that

$$\left(-\frac{F}{x} \right) \left(\frac{F + \eta F'}{x^2} \right) + \eta \left(-\frac{F}{x} \right) \left(-\frac{F'}{x^2} \right) = -\left(\frac{Q}{2\pi} \right)^2 \left(\frac{1}{x^3} \right) + v \left(-\frac{F''}{x^3} \right)$$

Note that $1/x^3$ is a common factor, and the resulting equation,

$$vF'' - F^2 = -\left(\frac{Q}{2\pi} \right)^2 \tag{16.3.11}$$

does *not* depend on *x and* η, the latter being the same as x and y. It depends only on η and the parameters v and $Q/2\pi$. (This is the goal in choosing a similarity parameter, as we have seen before.) The reader should verify by appropriate differentiation that the solution to Eq. (16.3.11) is

$$F(\eta) = 3 \left(\frac{Q}{2\pi} \right) \tanh^2 \left[\eta \left(\frac{Q}{4\pi v} \right)^{1/2} + B \right] - \frac{Q}{\pi} \tag{16.3.12}$$

in which $\tanh^2 B = \frac{2}{3}$, so that $B \approx 1.1462$. Figure 16.3.3 is a display of $F(\eta)$. The boundary conditions for $F(\eta)$ are at $y = 0$ and $y = $ large. At $y = 0$, $\eta = 0$ and $u = 0$, so that $F(0) = 0$. For $y = $ large, one lets $\eta \to \infty$, so that $F(\infty) = Q/2\pi$.

Although Eq. (16.3.1) is an approximate form of the Navier-Stokes equations, it contains all of the terms that were needed for the solution of the problems of Figs. 8.5.1 and 16.2.6, which led naturally to the occurrence of boundary layers. However, prior to closing this brief introduction to the Navier-Stokes equations, it is necessary to emphasize that not all problems with inertia terms contain the possibility of reducing Eq. (16.2.8) to Eq. (16.3.1). Consider for a moment the flow between parallel walls when the angle θ between the walls is not small. (This situation is similar to the problem depicted in Fig. 16.3.2.) The same complete equations and boundary conditions apply for the case of radial inflow as for radial outflow. However, because the equations of motion are nonlinear (i.e., because products of the unknowns like $u \, \partial u/\partial x$

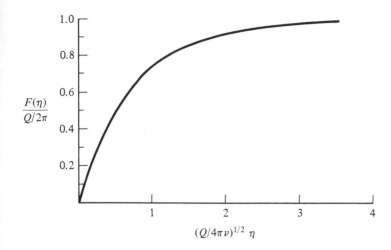

The y-axis is labeled $\dfrac{F(\eta)}{Q/2\pi}$ with values 1.0, 0.8, 0.6, 0.4, 0.2. The x-axis is labeled $(Q/4\pi\nu)^{1/2}\,\eta$ with values 1, 2, 3, 4.

Figure 16.3.3 Solution to Eq. (16.3.11), velocity distribution associated with flow described in Fig. 16.3.2.

appear), these two situations are vastly different. The result is that for the converging flow, for which $\partial p/\partial x < 0$, the boundary-layer approximation appears to be valid. This conclusion is based upon the exact solution for the case. (The boundary-layer approximation for this case was discussed in Example 16.3.2.) However, for diverging flow, with an associated positive $\partial p/\partial x$, the reduced Navier-Stokes equations (i.e., the boundary-layer equations) do not give a solution that is in agreement with that obtained by using the complete formulation of the equations of motion. Experience has shown the classes of problems to which the boundary-layer equations can be suitably applied. Generally, conditions for which the pressure decreases in the direction of flow along the body, and for which the Reynolds number is high, are sought.

16.4 SOME SIMILARITY SOLUTIONS

Equation (16.3.7) requires that $F'''(0) = 0$ for the boundary conditions prescribed at the origin. If this equation were differentiated with respect to η, one would generate terms containing the first and fourth derivatives of F, as well as others. At the origin, the boundary conditions would demand that the fourth derivative also be zero. Higher derivatives could also be found by successive differentiation of the differential equation and utilizing the boundary conditions and their implications. However, one would soon find that the value of $F''(0)$ was needed to specify the value of these derivatives at the origin. But this boundary condition was not prescribed and cannot be inferred. Instead of this boundary condition, one has a constraint at a second point, $\eta = \infty$. This apparent difficulty is overcome by the numerical technique called *shooting*. It is used for most numerical solutions to *two-point boundary-value problems*.

 This procedure requires that one *assume* a trial value for each of the missing derivatives at one point of the independent variable's interval. We call this point η_1 for our convenience. If a suitable integration algorithm is available—and one usually is—one can then obtain a numerical solution to the differential equation. If that solution is consistent with the boundary condition that is required by the problem statement at the end of interval η_2, the problem is solved. But if the assumed derivative values do not produce a solution that complies with the boundary condition at η_2, one

The Navier-Stokes Equations   Chap. 16

must use new trial values and begin again. It is helpful to observe the difference between the computed boundary condition at η_2 and the value of the prescribed boundary condition. This difference, or *error,* can be correlated with the values that are assumed at η_1 to aid in the selection of subsequent trial values. In this way, one improves one's "aim" in trying to hit the "target."

After values have been assumed at η_1 for any derivatives that are unprescribed, a Taylor series expansion of the function could be made near that point to specify values of the function at neighboring points. These points should be within the radius of convergence of the series. It would be possible in this way to integrate the differential equation for increasingly large values of $\eta - \eta_1$ and proceed toward values of the function at η_2. This sequence of steps assumes that the point at the end of the interval of integration is within the radius of convergence of the series. Moreover, the series should converge sufficiently rapidly to make the computations practical.

If the differential equation is nonlinear, which is the case for Eq. (16.3.7), the successively higher derivatives of the function will involve more and more and more combinations of the lower derivatives. This process can become cumbersome and time-consuming. In an attempt to speed up the solution process, other numerical methods have been developed to solve linear and nonlinear differential equations. A widely used algorithm is the *Runge-Kutta method,* which was named after its original developers. The book by Carnahan, Luther, and Wilkes, *Applied Numerical Methods* (John Wiley & Sons, Inc., New York, 1969), among others, discusses the use of this method for solving nonlinear differential equations that have an order greater than 1. The small amount of programming necessary to solve Eq. (16.3.7), with its boundary conditions, and the relatively short running time are inducements for the reader to become knowledgeable of the Runge-Kutta algorithm and to attempt a solution to Eq. (16.3.7). A BASIC program using this method is contained in Appendix H. This appendix also gives a program that evaluates the first 10 derivatives associated with Eq. (16.3.7) and uses a Taylor series to evaluate the function F and its first two derivatives at the next increment of the independent variable, η. A small digital computer permits this program to be implemented quite easily. It is interesting to see the effect that the assumed value of $F''(0)$ has on the solution.

This brief excursion into the subject of numerical methods for the solution of differential equations was undertaken to emphasize the utility of the similarity variable associated with Eq. (16.3.7). This variable allowed the partial differential equation to be reduced to an ordinary differential equation. Thus a solution that is valid in the entire xy plane can be sought by using only one variable. This observation provides the incentive to find other cases for which the boundary-layer equations reduce to a similarity form.

Such cases also arise if dp/dx is not zero, as it was for the flow past a flat plate that yielded Eq. (16.3.7). A nonzero pressure gradient would occur if the value of the velocity component parallel to the plate, U, were not a constant along the streamline that forms the outer limit of the boundary layer. This was the case in Example 16.3.2. Because the limiting streamline of the boundary layer is the first streamline of the exterior region in which the viscous effects in the flow are ostensibly absent, we use the Bernoulli equation along it and write

$$\frac{U^2}{2} + \frac{p}{\rho} = \text{constant} \quad \text{or} \quad UU' + \frac{dp/dx}{\rho} = 0 \tag{16.4.1}$$

in which U' is the x derivative of $U(x)$. We begin our search for a family of similarity solutions by writing Eq. (16.3.1) and employing the expressions for the dependent variables that were developed in Eqs. (16.3.5) and (16.3.6). The results follow.

The *steady-state* boundary-layer equation is

$$u\frac{\partial u}{\partial x} + v\frac{\partial u}{\partial y} = -\frac{1}{\rho}\frac{\partial p}{\partial x} + \nu\frac{\partial^2 u}{\partial y^2}$$

and we have seen that

$$\psi = \int_0^y u\, dy = U\delta\int_0^y \frac{u}{U}\, d\left(\frac{y}{\delta}\right)$$

With $y/\delta = \eta$ and $u = UF'(\eta)$, this integral becomes

$$\psi = (U\delta)\int_0^\eta F'(\eta)\, d\eta = ((U\delta)F$$

We also know that

$$v = -\frac{\partial\psi}{\partial x} = -\left[(U\delta)'F + (U\delta)\frac{dF}{d\eta}\frac{\partial\eta}{\partial x}\right]$$

$$= -[(U\delta)'F - (U\delta')\eta F']$$

These results convert the boundary-layer equation to

$$UF'\left[U'F' - (U)\frac{\delta'}{\delta}\eta F''\right] - [(U\delta)'F - (U\delta')\eta F']\frac{UF''}{\delta} = -\frac{1}{\rho}\frac{\partial p}{\partial x} + \frac{\nu}{\delta^2}UF'''$$

which, upon simplifications due to cancellations and the use of Eq. (16.4.1), becomes

$$UU'(F')^2 - \frac{U}{\delta}(U\delta)'FF'' = UU' + \frac{\nu U}{\delta^2}F'''$$

This last equation can be simplified to read

$$F''' + \frac{\delta}{\nu}(U\delta)'FF'' + \frac{\delta^2 U'}{\nu}[1 - (F')^2] = 0 \qquad (16.4.2)$$

We have required that the viscous term have a unity coefficient. This will keep that term a significant part of the solution near a stationary boundary. In order to obtain a differential equation that can have a similarity solution, all of the remaining coefficients must be independent of x. We denote these coefficients, which are then constants, as

$$a = \frac{\delta(U\delta)'}{\nu} = \text{constant}_1 \quad\text{and}\quad b = \frac{\delta^2 U'}{\nu} = \text{constant}_2 \qquad (16.4.3)$$

Then $$F''' + aFF'' + b[1 - (F')^2] = 0 \qquad (16.4.4)$$

The set of equations defining a and b, Eq. (16.4.3), can be solved easily if we write

$$a = \frac{\delta^2 U'}{\nu} + \frac{U\delta\delta'}{\nu} = b + c$$

with
$$c = \text{constant}_3 = \frac{U\delta\delta'}{\nu} = \frac{U(\delta^2)'}{2\nu}$$

Then
$$\frac{2U'}{U} = \frac{b}{c}\frac{(\delta^2)'}{\delta^2}$$

$$\delta^2 = K^2 U^{2c/b} \quad \text{or} \quad \delta = KU^{c/b} \quad \text{for } b \neq 0$$

$$b = \frac{K^2}{\nu}(U^{2c/b})U' = \frac{d}{dx}\left[\frac{K^2}{\nu}\frac{(U^{2c/b+1})}{2c/b+1}\right]$$

$$U = \left[\frac{\nu b}{K^2}\left(\frac{2c}{b}+1\right)x\right]^{b/(2c+b)}$$

$$= \left[\frac{\nu}{K^2}(2a-b)x\right]^{b/(2a-b)} = \left[\frac{\nu b}{K^2}\left(\frac{2a}{b}-1\right)x\right]^{1/(2a/b-1)} \tag{16.4.5}$$

Let
$$1/(2a/b - 1) = m \quad \text{and} \quad b/a = \beta \quad \text{if } b \neq 0$$

Then, Eq. (16.4.4) can be written as

$$U = \left[\frac{\nu}{K^2}\frac{b}{m}x\right]^m \quad \text{with} \quad b \neq 0 \tag{16.4.6a}$$

or
$$U = Cx^m \quad \text{with} \quad C = \left(\frac{\nu}{K^2}\frac{b}{m}\right)^m \tag{16.4.6b}$$

Because $c = a - b$,

$$\delta = KU^{(a-b)/b} = K\left[\left(\frac{\nu b}{K^2 m}\right)^m x^m\right]^{(1-m)/2m} = K^m\left(\frac{\nu b}{m}\right)^{(1-m)/2}x^{(1-m)/2} \tag{16.4.7a}$$

or
$$\delta = \left(\frac{\nu b}{Cm}\right)^{1/2}x^{(1-m)/2} \quad \text{with} \quad b \neq 0^* \tag{16.4.7b}$$

The solutions to Eq. (16.4.3) for various values of m are given in Fig. 16.4.1. They are usually referred to as the *Falkner-Skan* solutions, after V. M. Falkner and Sylvia Skan, who developed these solutions to the boundary-layer equations. When $m = 0$, $\beta = 0$, and one has the case of the flat plate that originated this discussion, as well as boundary-layer theory itself. For $m > 0$, $dp/dx = -UU' = -mU^m U^{m-1}$, from which one can infer that the pressure decreases in the flow direction. However, for $m < 0$ the pressure would increase. The lowest value of m for which a velocity profile is portrayed in Fig. 16.4.1 is $m = -0.091$. Lower values of m correspond to flows for which there is *separation* along the boundary. At the inception of separation, $\partial u/\partial y$ is zero. If this derivative becomes negative, there would be a region near the boundary where u, the velocity component parallel to the boundary, would be in the opposite direction of the upstream flow. This is the definition of separation that has been used previously in discussing the flow past bodies.

* If $b = 0$, then $U = $ constant would be the result. Also, then $\delta = \sqrt{2\nu x/U}$.

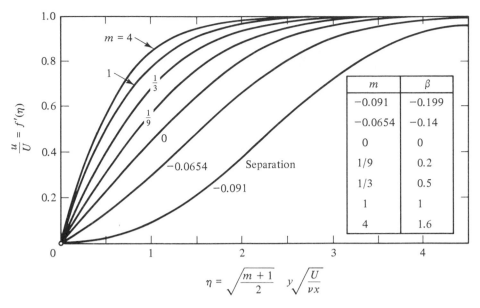

Figure 16.4.1 Velocity distributions in a laminar boundary layer in a flow whose asymptotic velocity is prescribed by $U(x) = u_1(x^m)$. (From H. Schlichting's "Boundary-Layer Theory," McGraw-Hill Book Company, by permission)

The shape of the boundary that corresponds to the various values of m, which determines the potential flow that is parallel to the boundary, follows from a study of such irrotational flows. The streamlines given by $\psi = Axy$ correspond to a flow with $u = Ax$. This stream function can be used to model the flow very near the stagnation point on a cylinder that is in a uniform fluid stream. ($\theta = 0$ in Fig. 5.4.3.) If the fluid is viscous, one would use this potential flow to prescribe the flow away from the surface with $U = Ax$. This suggests that m equal to 1 would be used in the equation for the solution of the velocity in the boundary layer near the stagnation point. The solution would also apply to the flow against a flat plate that was positioned normal to the upstream flow, if one stayed away from its edges, where separation would surely occur. This problem was first solved by K. Hiemenz in 1911, the year after Blasius solved the flat-plate problem of Eq. (16.3.7). The flow in the Hiemenz problem moves along the boundary and away from the stagnation point in two directions, so that the angle between the boundaries is π. This requires β in Fig. 16.4.2 to be 1. It is obvious from

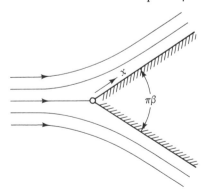

Figure 16.4.2 Viscous flow past a wedge for which the streamline that is δ away from the plate has a velocity that varies as $U(x) = u_1(x^m)$. Wedge angle $\pi\beta$ and m in Fig. 16.4.1 are connected through Eq. (16.4.8). (From H. Schlichting's "Boundary-Layer Theory," McGraw-Hill Book Company, by permission)

this figure that β would be zero for the Blasius problem. Thus, β depends on m, and the relationship can be shown to be

$$\beta = \frac{2m}{m+1} \quad \text{or} \quad m = \frac{\beta}{2-\beta} \tag{16.4.8}$$

Negative values of m would require negative β's. This means that the flow would be one in which the region is expanding in size, in contrast to the situation explicitly portrayed in Fig. 16.4.2, with an associated decrease of U in the direction of flow.

There are a few other similarity solutions for the boundary-layer equations. They are different from the Falkner-Skan solutions, which required only the constants a and b. In the case of the two-dimensional jet emerging perpendicularly from a plane wall, there is a constraint on the form of the solution. This restriction comes from a dynamic condition that includes the fact that $dp/dx = 0$ for an ambient fluid that is at rest. Because the external flow is assumed to be without shear stresses, there are no external forces on the jet to change its momentum if gravity is perpendicular to the flow direction. Without such forces, the momentum of the jet is constant. The x momentum at any position x along the axis of Fig. 16.4.3 is, if U is the speed on the centerline,

$$\int_{-\infty}^{+\infty} \rho u^2 \, dy = \rho U^2 \delta \int_{-\infty}^{+\infty} (F')^2 \, d\eta$$

and it must be independent of x. The constraint that must be incorporated into the solution is $[\delta(x)][U(x)]^2 = $ constant. This restriction interrelates the constants a and b. It is easy to verify that $a = -b$, and the equation that must be solved is

$$F''' + FF'' + (F')^2 = 0 \tag{16.4.9}$$

with the boundary conditions

$$F(0) = 0 \qquad F''(0) = 0 \qquad F'(\infty) = 0$$

The reader should be convinced that the first two conditions require that $v = 0$ along the jet's axis and that the axial velocity profile be symmetric there.

Figure 16.4.4 is the flow pattern for a jet at a Reynolds number for which the boundary-layer approximation is not applicable. The complete Navier-Stokes equations were needed to simulate the flow.

The flow at the interface between two parallel streams having different speeds can also be modeled with a similarity solution. A boundary layer develops because of the shear stresses that are generated by the differential speed. This boundary-layer velocity is the transitional velocity between the two streams. Once again, we assume

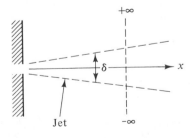

Figure 16.4.3 Two-dimensional jet issuing out of a plane wall.

Figure 16.4.4 Flow pattern for a jet showing entrainment streamlines for a Reynolds number at which the boundary-layer equations are inadequate. *(By permission of Dr. E. Zauner, Technical University of Vienna, Austria,* J. Fluid Mechanics, *(1985), figure 2(a).)*

that the v component of velocity is small, but we still include it in the convection terms to affect the boundary-layer growth. If it is assumed that $dp/dx = 0$ in a horizontal flow, the differential equation that must be solved is the one given by Eq. (16.3.7). However, now the boundary conditions are different from those used previously. While F' could approach 1 at $+\infty$, it should approach U_2/U_1 at $-\infty$. The last needed boundary condition prescribes that the line $y = 0$, the original interface, is a streamline. If the value of the streamline constant is taken so that $\psi = 0$, then $F(0) = 0$. [Along this horizontal streamline, $v = 0$, and Eq. (16.3.6) demands then that $F(0) = 0$ for $\eta = 0$.]

The *wake* behind a flat plate aligned with a flow also leads to a similarity solution. In this case, the *velocity defect,* $U - u$, that characterizes the wake is taken to depend on a similarity variable. (Figure 9.7.5 relates to this case.) The velocity defect is nondimensionalized with respect to W, the maximum defect at the local x position along the wake. This occurs along the axis, $\eta = 0$. The differential equation that must be solved for F', which is $(U - u)/W$, is

$$F''' + \eta F'' + F' = 0 \qquad (16.4.10)$$

if no pressure gradient is assumed. One boundary condition is that $U - u = 0$ far from $\eta = 0$. The velocity profile should also be symmetric at $\eta = 0$. These observations require that

$$F'(0) = \text{constant} \qquad F''(0) = 0 \qquad F'(\infty) = 0$$

Equation (16.4.10) can be immediately reduced in order by setting $G = F'$ and adjusting the boundary conditions appropriately. The constraint condition requires that the net momentum efflux be a constant for a control volume surrounding the plate and extending above and below the plate to infinity, where $v = 0$. The constant is, of course, the drag that the plate exerts on the fluid. This constraint, as well as Eq. (16.4.10), results from a linearization procedure in which terms such as W^2 and WW' are neglected with respect to terms having UW or UW' as coefficients.

These last two similarity equations were not developed in detail, because that would greatly extend this introduction to boundary-layer theory and its similarity solutions. However, the equations relate to important physical problems, and it is important to remember once again that similarity variables are valuable in related experimental research. This last observation is made obvious by an examination of Fig. 9.7.5.

16.5 SUMMARY

The Navier-Stokes equations are an expression of Newton's laws of motion and a *constitutive law* that relates the stresses in the fluid with the volume and shape distortions that a material element can undergo in a Newtonian fluid. This relationship has been restricted in this chapter to incompressible fluids, but no appreciable complexity would be introduced if gases had been included in the treatment. A second viscosity coefficient would then have been necessary to account for the dissipation associated with normal stresses, such as the attenuation of sound waves in gases or ultrasonic waves in liquids. One would also need an equation of state in order to make the system of equations complete.

The unknown quantities that must be found in a problem associated with an incompressible fluid are the three components of velocity and the pressure. For an incompressible fluid, the volume of a material element must be constant; this requires that there be a specific relationship between certain derivatives of the velocity components, the continuity equation. Hence, there are as many equations available as there are unknowns when one studies incompressible flows. The boundary conditions that must be applied are the usual ones of no slip and no penetration at solid boundaries. At fluid-fluid interfaces, one requires continuity of the velocity and the stresses across the interface, but surface tension can modify the latter condition through Eq. 1.6.1.

The form of the Navier-Stokes equations and the associated boundary conditions for a particular problem can be solved analytically in only a few cases. Numerical integration of these equations has become increasingly easier with the progress that has been made in computing machines and the interest in numerical procedures.

Despite the availability of such problem-solving tools, one should be careful in formulating a problem in order to realize the maximum benefit from them. Even before the advent of electronic computers, it was realized by Ludwig Prandtl that in a high–Reynolds number flow there would be only a small region near a solid surface in which the presence of the boundary would significantly affect the flow velocity. His concept of the *boundary layer* permitted one to simplify the Navier-Stokes equations into the *boundary-layer equations*. These latter equations permit the no-slip condition to be imposed, even though the fluid's viscosity is very small, and provide the means to solve many important problems.

The differential equations in this chapter were solved numerically in most cases. The concept of a *similarity solution* was applied to the boundary-layer equations by V. Falkner and S. Skan. While one must still solve the differential equations for a complete solution, one can determine quite readily the manner in which the characteristic velocity and boundary layer have changed with distance. This is a great deal of information for relatively little effort.

The most satisfying aspect of the theory outlined in this chapter is the agreement in the results that one can observe in experiments and the corresponding mathematical models that can be analyzed. This reinforcement of theory with experiment is not simply a case of duplicating numerical values. More important, it is the ability of one approach to uncover an aspect of a problem that can be verified by another means. It is this mutual "unlocking of doors" and removal of obstacles that has made theory and experiment such valuable partners.

EXERCISES

16.2.1(a) If $\tau_{yy} = 17$ kPa at point P, draw a picture showing the manner with which this stress would act on a fluid element.

(b) If $\tau_{yy} = -24$ kPa at point P, draw a picture showing the manner with which this stress would act on a fluid element.

16.2.2(a) Draw τ_z if $\tau_{zx} = 1$ kPa, $\tau_{zy} = 3$ kPa, and $\tau_{zz} = 4$ kPa.

(b) Draw τ_z if $\tau_{zx} = 1$ kPa, $\tau_{zy} = -3$ kPa, and $\tau_{zz} = 4$ kPa.

16.2.3(a) For case A, should one include $v(y = 0) = 0$ and include $v(y = h) = 0$? Explain.

(b) For case B, should one include $v(y = 0) = 0$ and include $v(y \rightarrow \infty) \rightarrow 0$? Explain.

(c) If, in case B, the plate is porous so that a velocity in the y direction of -6 ft/s is induced at the plate by suction devices, what are the boundary conditions for the problem?

16.2.4(a) Write the applicable equations of motion if $v = V = $ constant, something that could occur if the plate were porous.

(b) Consider a two-dimensional channel infinitely long consisting of parallel plates spaced a distance k apart. One plate is stationary, and the second is suddenly accelerated. Write the equations of motion and the boundary conditions for this case.

16.2.5(a) Compare the boundary-layer thicknesses for water and air at 20°C and 101.3 kPa.

(b) Figure 16.2.6b portrays the velocity of the plate remaining constant. What would be the solution at $t = 10$ s if the velocity were to change abruptly at 4 s from U_0 to $2U_0$? Use $U_0 = 5$ ft/s and $v = 10^{-4}$ ft^2/s. [*Hint:* Could the solution have the form $u = u_1 + u_2 = U_0 f(\eta_1) + U_0 f(\eta_2)$ for $t \geq 4$, in which $\eta_1 = y/2\sqrt{vt}$ and $\eta_2 = y/2\sqrt{v(t - 4)}$?]

(c) Approximate the solution for the velocity u for the case of a plate whose speed increases U_0 every 0.1 s for 2 s. Your solution need be applicable only at $t = 1.6$ s. (*Suggestion:* First plot the speed of the plate as a function of time and then refer to Exercise 16.2.5b.)

16.2.6(a) The solution is not valid for $Vk/v = 0$, something that could occur for $V = 0$. Use $V = 0$ initially in the equations of motion to determine $u(y)$.

(b) Plot the solution for $Vk/v = 0.05$.

(c) An estimate of the solution at $Vk/v = 0$ could be gotten by comparing the solution at $Vk/v = +0.1$ and -0.1. Do so. How could the solution for $Vk/v = -0.1$ be realized in practice?

16.2.7(a) What is $\tau_{r\theta}$ for the case of the circular flow between two concentric circular cylindrical surfaces for which $\Omega_1 = 0$ at $r_1 = 2$ in. and $\Omega_2 = 2\pi$ rad/s at $r_2 = 3$ in.? Plot the result versus the radial distance for $r_1 \leq r \leq r_2$. Is $\tau_{r\theta}$ a constant in the interval? If not, is it higher or lower than what you expect it to be? Explain.

(b) Plot $\tau_{r\theta}$ for the data in Exercise 16.2.7a, except that $r_1 = 2.95$ in. Is $\tau_{r\theta}$ a constant in the interval $r_1 \le r \le r_2$? How does this velocity profile compare with the solution between two parallel plates, one of which is stationary and the other moving at the speed U? (No pressure gradient is allowed in the flow direction.)

(c) $\tau_{xy} = \mu\left(\dfrac{\partial u}{\partial y} + \dfrac{\partial v}{\partial x}\right) = \mu\left[\dfrac{\partial}{\partial y}\left(-V_\theta\dfrac{y}{r}\right) + \dfrac{\partial}{\partial x}\left(V_\theta\dfrac{x}{r}\right)\right]$

Evaluate this expression, as was done for the expression in Example 16.2.7, at $x = r$ and $y = 0$. Is $\tau_{xy} = \tau_{r\theta}$ at this point? Compare your result with that in Example 16.2.7.

16.3.1(a) Let $\mathrm{Re}_L = 10^5$ and assume that at $y^* = 5$, $u \approx U_\infty$. What is the boundary-layer thickness for $L = 10$ ft and $L = 10$ m?

(b) For a plate with a length of 10 ft with $U_\infty = 12$ ft/s in air at 70°F, what is the boundary-layer thickness halfway along the plate, and at the end of the plate? What would be the boundary-layer thickness if the plate were in water at 70°F?

(c) Write the continuity equation in starred variables. Start with the unstarred variables and substitute appropriately.

16.3.2(a) Show that differentiation of Eq. (16.3.12) produces Eq. (16.3.11).

(b) Why must $\tanh^2 B = 2/3$?

(c) What is the radial inflow within the boundary layer associated with Fig. 16.3.3? (*Hint:* Start with $Q = \int_0^\delta u\,dy$ at some station x and use $u = -F/x$ and $dy = x\,d\eta$ for fixed x. The value of δ follows from $(Q/4\pi\nu)^{1/2}\,\delta/x = 3$.)

PROBLEMS

16.1 For a point symmetric flow, $|\mathbf{V}| = V_R = k/R^2$ in which k is a constant. Hence, $u = xk/(x^2 + y^2 + z^2)^{3/2}$, $v = yk/(x^2 + y^2 + z^2)^{3/2}$, and $w = zk/(x^2 + y^2 + z^2)^{3/2}$. Determine the nine components of stress for this flow.

16.2 Verify Eq. (16.2.8) by differentiating the stress terms of Eq. (16.2.7) as required by Eq. 16.2.4.

16.3 Equations (16.2.8) to (16.2.10) can be re-written in vector form as

$$\rho\left[\frac{\partial \mathbf{V}}{\partial t} - \mathbf{V} \times (\mathrm{curl}\,\mathbf{V})\right]$$

$$= -\mathrm{grad}\,p - \rho\left[\mathrm{grad}\left(gz + \frac{\mathbf{V}\cdot\mathbf{V}}{2}\right)\right]$$

$$- \mu[\mathrm{curl}\,(\mathrm{curl}\,\mathbf{V})]$$

Recall that $\mathrm{curl}\,\mathbf{V} = $ vorticity. Verify this vector equation by demonstrating its correspondence with Eqs. (16.2.8) to (16.2.10).

16.4 Because $\mathrm{curl}\,(\mathrm{grad}\,p) = 0$, the vector equation in Prob. 16.3 can be written in the terms of the vorticity, in which the pressure term is absent. For a two-dimensional flow (x, y) with constant ρ, the result is

$$\rho\frac{D\zeta}{Dt} = \mu\nabla^2\zeta$$

in which ζ is the z component of vorticity. Verify this statement.

16.5 Show that Eqs. (16.2.12) to (16.2.14) follow from the vector equation given in Prob. 16.3. (Hint: use only (u_r, u_θ, u_z).)

16.6 Verify Eq. (16.2.16) by using the general definition of the divergence.

16.7 Use Eqs. (16.2.8) to (16.2.10) to derive the one-dimensional flow between two parallel plates that are separated by the distance k. The lower plate is stationary, and the upper plate moves horizontally with the speed U.

16.8 Use Eqs. (16.2.8) to (16.2.10) to derive the axial velocity in a round tube of radius a. Assume at the outset that $u = 0 = v$ and $w = K(x^2 + y^2 - a^2)$. Are these reasonable assumptions if one considers boundary conditions? Does this assumption satisfy the equations of motion?

16.9 Use Eqs. (16.2.8) to (16.2.10) to derive the axial velocity in an elliptical tube $(x^2/a^2 + y^2/b^2 = 1)$. Assume initially that $w = K(x^2/a^2 + y^2/b^2 - 1)$ to see whether such a distribution is suitable.

16.10 Use Eqs. (16.2.8) to (16.2.10) to determine the volume rate of flow in a duct whose cross section is an equilateral triangle. For convenience, let the side of the triangle consist of the lines $y = \pm\sqrt{3}x/3$ and $x = b\sqrt{3}/2$.

16.11 For the solution displayed in Fig. 16.2.7 for an impulsive start, what is the location where $u/U = 0.2$ in a fluid with a viscosity of 3 centistokes (cSt)? Use $t = 0.1$, 1.0, and 10. Estimate the speed at which the locus of the points corresponding to $u/U = 0.2$ advances in the y direction.

16.12 For the velocity profile developed in Example 16.2.6, how is the volume rate of flow (in the direction parallel to the surface of the plates) affected by the value of V? Does the shear stress on the upper plate vary linearly with V? Explain.

16.13 Water flows down a gentle slope in a thin layer of depth k. The seepage of water into the ground, the lower surface, is exactly the same as the falling upon the free surface of a fine mist (of rain). (Assume that the mist passes perpendicularly into the water layer.) What is the velocity distribution in the water?

16.14 Give the equation that must be solved for the motion of a fluid above an oscillating porous plate that has the in-plane motion $U_0 \sin \omega t$, for which the normal velocity at the plate is $-V$.

16.15 Solve for the velocity distribution in Example 16.2.6 if the upper plate is also translating at speed U.

16.16 Consider the flow between two plane walls that are inclined at an angle of 2α to one another. Only u, the radial velocity, is nonzero. What does Eq. (16.2.16) require? Use Eqs. (16.2.12) to (16.2.14) to derive a differential equation for the θ variation of u. (The pressure terms can be eliminated by cross differentiation.) The result should be

$$f'' + 4f + \frac{f^2}{\nu} - c_1 = 0$$

with $f(\pm\alpha) = 0$. What is the velocity distribution if the nonlinear term can be neglected because of low velocities and large ν? (Note: $u = f/r$ for radial outflow.)

16.17 Use the α and β scales of Example 16.3.1 to show that $\partial p/\partial y$ will be small if α is small.

16.18 Prepare a drawing showing the boundary-layer thickness as a function of x and

$(Q/4\pi\nu)$ by using the information contained in Fig. 16.3.3.

16.19 Test the Runge-Kutta program given in Appendix H., or one of your own design, to find the velocity within the boundary layer on a flat plate in a flow without a pressure gradient. At what value of η does the velocity become constant, in effect? (Call this value η_∞.) Is this velocity profile well approximated by $u/U_\infty = \sin(\pi\eta/2)$? Experiment with the step size for the Runge-Kutta method to determine the largest one that still gives a satisfactory solution to the Blasius problem.

16.20 What is the boundary-layer value η_∞ at which $u \to U$ for $m = -0.091$? (Use Runge-Kutta.)

16.21 What is the velocity distribution for stagnation flow? This is also called the *Hiemenz problem.* (Use Runge-Kutta.)

16.22 Use the Taylor series program in Appendix H. to find the velocity within the boundary layer on a flat plate in a flow without a pressure gradient. Vary the number of derivatives used and the step size to see their effect on the solution.

16.23 Carry out the steps specified in Prob. 16.22 for $\beta = 1$, corresponding to stagnation, or Hiemenz flow.

16.24 Write a computer program that uses the Runge-Kutta technique to solve Eq. (16.4.9) and the associated boundary conditions. How does the jet broaden with distance?

Write a computer program to solve Eq. (16.4.10) by the method of Runge-Kutta. How does the wake broaden with distance?

16.25 Equation (16.4.9) can be differentiated to give the fourth and higher derivatives of F. Find the first eight derivatives, and construct a Taylor series solution in the manner of the one in Appendix H. (*Suggestion:* Compare the solutions obtained with four, six, and eight derivatives.)

16.26 Use a Taylor series to solve Eq. (16.4.10) after differentiating it to find the higher derivatives in terms of the lower ones.

The Navier-Stokes Equations   Chap. 16

17

Flow Measurement

17.1 INTRODUCTION

The measurement of fluid velocities, temperatures, pressures, volume rates of flow, and forces is necessary in order to provide information on the performance of equipment such as pumps or turbines. The safe and efficient operation of canals and other commercial waterways depends upon the gathering and assessment of such information, as well as the determination of free-surface elevations. Weather prediction and the control of production in the chemical industry are directly tied to the reliability of such measurements. The control characteristics and the economical performance of automobiles demand that sufficient sensitivity be inherent in these measurements, so that decisions concerning design changes can be made with confidence. The demand for precision in measurement may be very high in laboratory tests which seek to calibrate a flow meter or provide new phenomenological information such as the viscosity of liquid metals at high temperature. Such laboratory tests can also include model studies of the wave forces on breakwaters, or a critical test so that one can decide upon the applicability of one or more theories.

As suggested above, data are gathered for a variety of reasons and under greatly different conditions. Laboratory tests on the oscillations of the free surface in an inlet structure that should provide cooling water to a power plant yield elevations in the order of centimeters to meters. In the actual installation, the variations can be between 1 and 10 meters. The forces, using Froude scaling, will be correspondingly greater in the prototype than in the model. Accordingly, the type of instrumentation used in model and field tests may be different. Even if the type of equipment is the same, the robustness and maintenance requirements will vary.

The kind of equipment that one chooses to obtain the desired information will be based upon purpose, cost, environment, amount and frequency of the data to be retrieved, precision, and availability of trained personnel to service the equipment, as well as convenience. The ultimate choice of a method or technique will result from trade-offs among most of the items above, and will include others such as speed and reliability. In the sections which follow, no attempt has been made to order the methods or techniques according to specific criteria. Experience and the user's need will provide these constraints. The references that are provided at the end of the book should give the reader the answer to very specific questions.

17.2 FLOW VISUALIZATION

One has a natural desire to "see what is going on." In a particular situation, the engineer or scientist benefits greatly by viewing regions in a flowing fluid where separation or stagnation occur. The position and extent of vortices in a flow are important to the design of the system. It has been said that one of da Vinci's most valuable contributions to science was his penchant for recording realistically the phenomena that he observed (e.g., vortices similar to the von Kármán vortex street were drawn by him). Today we record such observations upon photographic film instead of using pencil drawings. Nevertheless, the problem is the same. One needs to find an effective way to make the flow visible to the eye of the artist or to the photographic emulsion.

Smoke from a chimney that indicates the direction of a blowing wind is a common experience to many of us. The particles of soot or condensed moisture are objects that can be followed by the eye (or a television camera). It is only natural when dealing with the flow of air that smoke (hot oil of wintergreen is useful, and heated kerosene is also effective) will be introduced to show the streamlines for steady flow. Separation zones will also become evident due to obvious backflow. "Smoke" that is somewhat corrosive can be produced by the chemical titanium tetrachloride, which reacts with the moisture in the air to produce a chemical cloud that can be photographed. Figure 17.2.1 is an example of this technique.

It is effective to add plastic or aluminum particles to a water or oil flow to help visualize the flow. Such particles are about 10 microns (μm) in length, or width, and considerably less in thickness. While they are not neutrally buoyant, their mass is sufficiently small that they tend to follow the path of the fluid particles that are being controlled by the general stress environment. As they move, they reflect light differently, so that bright and dark regions will appear on photographic film. These regions can be interpreted to indicate flow directions. Refer to Fig. 1.6.4.

Other uses can be made of such flow tracers. If the camera that records the scene has a proper time setting on the shutter, a streak will appear on the film corresponding to the displacement of one of the particles in the flow that reflect light into the camera's lens. The length of such streaks will be an indication of the displacement during the interval of the exposure. Consequently, one can get an indication of the fluid's speed at the location in question. The streak's direction is approximately that of the local velocity. This method is illustrated in Fig. 1.8.10a and b.

When particles are not readily wetted by the liquid, those that are on a free surface will tend to remain there through surface tension. This was the case with the

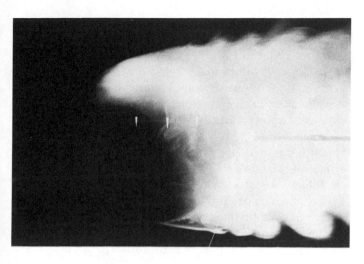

Figure 17.2.1 Flow past a sphere at Re = 2.3(10⁴). Titanium tetrachloride was brushed on the surface of the sphere. *(Courtesy of Prof. S. Taneda, Kyushu University, Japan)*

mica flakes that Prandtl used in his first experiments with a flow table that was powered by a hand-cranked paddle wheel. Aluminum (or bronze) flakes are used to color paints and are available at an artist's supply shop. They tend to be slightly oily, so that a homogeneous layer of the powder is difficult to obtain on a water surface without a cleansing operation. If the aluminum flakes are thoroughly mixed with a liquid and one wishes to observe the motion within the fluid, a well-designed lighting system is important. See Figs. 11.3.5 and 12.2.5. Through the use of parabolic lenses and slits, one can create a "curtain" of light. If a laser beam is reflected from a mirror with a varying angle, such a "slice" would also be produced. This can be accomplished by rotating the mirror about an axis that is parallel to its reflection surface. When the light passes through the liquid only those particles that are within the plane of the light will reflect it to the viewer. In this way one obtains a sectioned view of the flow—a "slice."

But solid particles are not the only things that give useful information. If the flow of water sweeps past a platinum wire (which is noncorrodible and, at 0.05 mm or less in diameter, reasonable in price) that is connected to a direct-current power supply, and another conductor—perhaps a graphite rod—downstream, one will generate hydrogen bubbles; or oxygen, depending on the polarity; but hydrogen, with two atoms per molecule of water, is the better. These bubbles are swept downstream, and with proper illumination (the light source should be about 90° from the viewer), the path and motion of the bubbles can be observed. Because the bubbles are buoyant, they tend to rise, so that the method is not useful in the case of very slow flows. Figure 10.3.1 provides an example of this technique.

For slow cases, one can use the method that was first published by Baker, in which a solution of thymol blue has had its pH value adjusted so that it is maize in color and any slight increase in the acidity will make the solution turn blue. The necessary ions to achieve this are provided by platinum wire, a direct-current power supply, and a ground connection downstream to complete the circuit. A blue streak that conforms to the wire geometry forms on the wire, and like the hydrogen bubbles that were mentioned previously, it is swept downstream by the flow past the wire. If the flow is slow enough, the concentration of blue liquid (with the same specific gravity as the yellow medium, so that no buoyancy effects are introduced) is sufficient to produce good photographs of the scene. F. X. Wortmann had great success in water with thin tellurium wires that emit a fine cloud after a chemical reaction is initiated by a brief electrical impulse. The result of such an experiment is shown in Fig. 1.8.7.

In liquids, one is inclined to add dye. Refer to Fig. 17.2.2. When using water, a solution of potassium permanganate or nigrosin will produce dense lines when introduced into the system by means of a syringe with a small-diameter needle. This produces streak lines that simulate streamlines for steady flow. Fluorescein dyes are useful because of their tendency to fluoresce upon proper illumination. (A laser can be the light source.)

Recently, liquids which contain photosensitive dyes have been used to study jets and the flow past bodies. The experiment is illuminated with a series of pulses from a very strong light source which has a slitlike geometry. When a light pulse occurs, the segment of fluid that receives this light changes color. Subsequent photographs record the position and character of this volume of dyed fluid. See Fig. 1.1.5. For the method to be practical, it is necessary to focus the light upon a small region. The brief duration of the light pulse and the large number of photons that must be transmitted in this time require that one have a high-intensity light source.

A small amount of phenolphthalein in a slightly acidic solution will turn red when mixed with a basic solution. Sodium carbonate and sodium hydroxide were used to produce the effect shown in Fig. 17.2.3.

Figure 17.2.2 Surge wave moving in a liquid layer with density intermediate to that of layers above and below it. Photograph shows a dyed layer that has been reflected from the vertical wall at the left. Just previously it had been flowing entirely from right to left. Slope of surge wave is affected by surface tension between layers, which affects entrainment rate into wave.

Figure 17.2.3 Liquids from two different sources separated by splitter plate (end visible at left). One reservoir had dilute solution of NaOH; the other, some phenolphthalein. Because chemical turns red when mixed with a base, dark region downstream of splitter plate indicates extent of mixing zone. Experiment shown here simulated a section of a power plant's chimney where flue gases from several boiler units were to be mixed.

In a specific case, it may be necessary to make the flow visible only in the neighborhood of a surface. In other cases, such a location may be the only place where visualization is practical. For example, dye can be allowed to seep into a liquid flow through one of the boundaries. As it is swept along by the flow, the dye will tend to congregate in stagnation areas or in separation zones where it is trapped in closed vortices. See Fig. 17.2.4. While photographs can be taken of such streak lines, it is also possible to coat the surface with a latex-based white paint that will be stained by certain chemicals (e.g., potassium permanganate). In some applications, it may be more convenient to bond crystals of the dye to the surface prior to immersing it into water. When the object is towed through the water, the crystal dissolves, leaving an indelible streak upon the white paint. Such paints are also stained by hydrogen sulfide gas, which, of course, has an obnoxious odor.

Partially dried painted surfaces have been exposed to high-speed water flows to test the efficiency of turning vanes in the elbow corners of water tunnels. The manner in which the paint was eroded was used as an indicator for the type of flow conditions that prevailed. Fluorescent paints that are applied to one side of a model ship's propeller can show the locations where the flow has passed from one side of the blade to the other. Dots of ink applied to the surface of an automobile model in a wind tunnel will flow as a result of the wind shear to reveal the surface flow pattern. Figure 17.2.5 shows this.

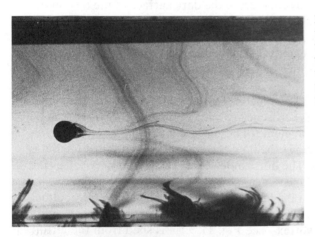

Figure 17.2.4 Circular cylinder moving horizontally in liquid with density varying in vertical direction. Small crystals of potassium permanganate were placed on topmost part of cylinder. As particles dissolved, a portion of the solution remained trapped in attached vortices on downstream side of the cylinder.

Figure 17.2.5 Ink streaks produced in wind tunnel on clay model of proposed car design. Spots of ink can be a non-Newtonian liquid that flows considerably more when wind shear is high. *(Photograph courtesy of the Design Staff, General Motors, Corp.)*

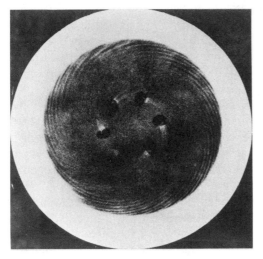

Figure 17.2.6 Central portion of this rotating disk had a laminar boundary layer, while outermost region had a turbulent boundary layer that dried the china clay coating. Spiral streaks indicate presence of vortices whose axes are spirals. This laminar, secondary flow is unstable, as experiment technique demonstrates. *(N. Gregory, T. Stuart, and S. Walker, Phil, Trans, Roy. Soc. London Ser/A, 248, 1955, by permission)*

A suspension of fine clay, called *china clay,* can be sprayed upon a dark surface. A white coating results that can subsequently be sprayed with oil of wintergreen. The oil is absorbed into the clay but does not remove it from the solid surface below. Light will pass through the clay via the oil and reflect from the dark surface of the test object back to the viewer. As the oil evaporates, the clay once again occludes the dark surface from view. Hence, if such a prepared surface is placed in an airstream, the appearance of white portions will indicate those regions that have been preferentially dried. This usually results from locally high velocities or turbulence levels. A result of using this method is shown in Fig. 17.2.6.

The flow of air over surfaces can be made visible by a coating of a pigment that has been mixed with an oil, sometimes kerosene. A trace of olive or linseed oil can be added to aid in the dispersion of the pigment. (These oils contain oleic acid, which is the useful ingredient.) Figure 17.2.7 shows the flow at the base of a cone frustum that is attached to the floor of a wind tunnel. The horseshoe vortex can be seen, as well as the effect of the splitter plate along the plane of symmetry.

The wing whose tip can be seen at the far left of Fig. 17.2.8 had its angle of attack changed suddenly just before the photograph was taken. The effect is a change of lift, so that something akin to a starting vortex—see Fig. 11.3.5—is shed from the airfoil. The effect of this on the flow pattern is made visible with a *smoke wire* that was placed

Figure 17.2.7 Flow at base of cone frustum to which a splitter plate is affixed that has been made visible with particles of pigment in form of oleic acid applied to floor of this special wind tunnel. *(Courtesy of Prof. H. H. Fernholz, Technical University Berlin)*

Figure 17.2.8 Streak lines formed by smoke wire located near trailing edge of airfoil in a wind tunnel. When angle of attack was suddenly altered, flow field accommodated to this change with formation of vortex that was swept downstream *(Courtesy of J. Świrydczuk, Institute of Fluid-Flow Machinery, Polish Academy of Sciences, Gdańsk, Poland, from "Optical Methods in Dynamics of Fluids and Solids, IUTAM Symposium Lubice, Czechoslovakia, 1984," p. 246, Springer-Verlag 1985, by permission.)*

vertically just aft of the trailing edge of the wing. Small beads of a grease (petrolatum) were placed at intervals along the wire before each test run. These beads were heated and emitted smoke due to the 25 volts passing through the wire, which had a resistance of 10 ohms.

Tufts of yarn or strands of thread can be attached to a surface, and the direction of these streamers will correspond to the direction of flow in the same way that a windsock or weather vane functions. When yarn is attached to the surface of an airfoil, it is possible to detect the critical angle of attack which, when exceeded, results in a spanwise flow, instead of a chordwise flow, with a resulting loss in lift. On models of automobiles, such tufts can indicate areas where the airflow is unsatisfactory or where openings for ventilation systems should be placed. Refer to Fig. 17.2.9. Threads can be attached to the surfaces of building models, which can be tested in a large water channel to show areas where the wakes from neighboring buildings interact to form gusty regions. Once such locations are identified, additional, and perhaps more detailed, measurements can be undertaken. Figure 17.2.10 shows how the extent of the wake behind a sphere can be detected by this method.

Figure 17.2.9 Another means to indicate flow at surface of a body. Unsteady character of flow adjacent to driver's window of automobile, due to a separation cavity, is evident. Phenomenon can also be recognized in the film of water on such a window when driving on a rainy day. *(Courtesy of the Volkswagen Corporation, Wolfsburg, W. Germany)*

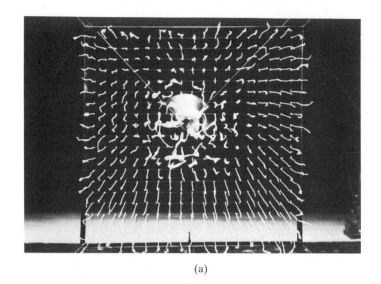

(a)

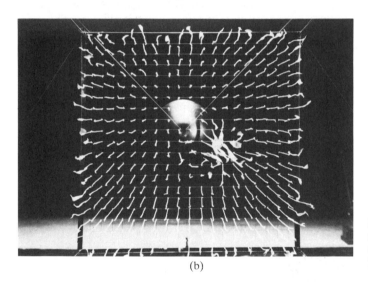

(b)

Figure 17.2.10 (a) Extent of wake from a sphere with Re = 2.3(10⁵). (b) Sphere's wake at Re = 4.7(10⁵). *(Courtesy of Prof. S. Taneda, Kyushu University, Japan)*

Whenever one makes a flow measurement, care must be taken so that the measuring technique itself does not influence the flow field. (For this reason, one uses nearly neutrally buoyant particles, including droplets from solutions of oils and other organic compounds, to obtain the needed specific gravity. In this way, the tracer elements have about the same inertia as the medium that they replace. Hopefully, they will follow the same paths as the fluid itself would.) In fluids that have density differences, one can use the variable index of refraction to give an indication of the fluid motions. It is not uncommon on a hot summer day to be able to look across an asphalt street and see the background as though it were slightly moving and a bit out of focus. (The same is true when looking across the top of a chimney on a winter day.) The heated air that rises from the pavement is nonhomogeneous in temperature and, therefore, of variable refractive index. Particles of air with different temperatures rise in a nonsteady way, so that the light that is transmitted from the background is passing through a constantly changing medium. In this case, the light transmission is sufficiently altered to be noticeable. In the laboratory, one employs a parallel light source that can be produced, depending on the desired diameter and available funds, by a lens or parabolic mirror. (Telescope mirrors for the hobbyist are suitable to use and suitably priced. In diameters below 15 cm, they are often ground spherical to make them inexpensive, but this is not a serious objection.) The light passes through the test section where the density changes could be induced by a dissolving substance, a heat source, or the like, and is made visible on a translucent screen that could be made of Mylar sheet. This technique is called *shadowgraph* and delineates regions where there is a change in the density gradient (i.e., the second derivative of the density profile). Figure 14.7.8 employed this method, in effect.

More information can be had from an optical technique if one invests in more care and equipment. If, after passing through a homogeneous test section, a parallel beam of light is focused by a lens or a parabolic mirror onto a knife edge, a portion will pass over it and expand subsequently. Again, the light will be projected upon a screen for viewing. Afterward, the test section can be subjected to an experiment that alters the index of refraction within it. Then there will be rays of light that do not pass through it in a parallel fashion. These light rays will be reflected from different portions of the mirror, and when they arrive at the plane of the knife edge, they will be at positions that are different from those for a homogeneous test section. Thus, some light rays that originally passed over the knife edge will not intersect this opaque surface. Other rays that formerly were cut off by the knife edge from reaching the screen beyond will pass over it and onto the screen. The result is a pattern of light and dark areas on the screen that correspond to regions where there is a change of density (i.e., the first derivative of the density distribution is made visible). A schematic drawing of the apparatus is given in Fig. 17.2.11. This method was used to test the

Figure 17.2.11 Schematic diagram of a schlieren system

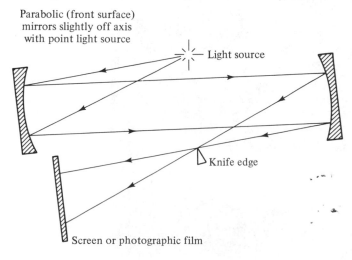

Parabolic (front surface) mirrors slightly off axis with point light source

Light source

Knife edge

Screen or photographic film

quality of glass that had striations in it that were not obvious to the naked eye. The technique is called a *schlieren* system after the German word for such striations. Figure 17.2.12 was made in this way.

The shadowgraph technique gave us the second derivative of the density distribution, and by paying the "price" of additional complexity, one can obtain the first derivative with a schlieren technique. What must be done to get the density distribution itself? One must use an *interferometer.* Such a device was perfected in order to study the nature of light, including its speed. It is highly sensitive to adjustments, and hence vibration, but in the hands of a competent investigator, excellent results can be obtained. (Refer to Fig. 17.2.13 during the following description.) In this method, part of the light is split away from the original beam by a partially silvered mirror (appropriately called a *beam splitter*). Thus, part of the light passes through the test section and part of it passes around it. This second light path includes other mirrors and a second partially silvered mirror. The light rays that pass through the test section recombine with the other rays from the other path at the second partially silvered mirror. If the lengths of the two optical paths are painstakingly adjusted, the two beams will recombine so that there is no phase shift between them. When the combined beam reaches a screen, the screen will be brightly illuminated. (Should the optical paths be so adjusted that the beams recombine with a 180° phase shift, the screen would be

Figure 17.2.12 Schlieren photograph of flow past a two-dimensional body with upper and lower surfaces of different roughness. *(Courtesy of Dr. A. Keller, Oskar Von Miller-Institute of the Technical University of Munich)*

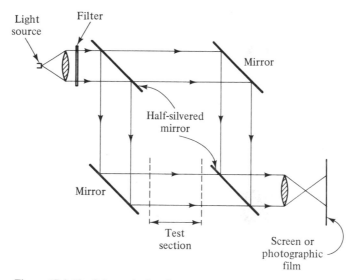

Figure 17.2.13 Schematic drawing of a Mach-Zehnder interferometer system.

dark.) After obtaining a bright screen, an experiment in the test section can commence. If density differences occur in the test section, the resulting changes of refractive index are equivalent to a local change in optical path length. Thus, light rays that pass through these regions will arrive at the second half-silvered mirror at a different phase angle than before. They will not recombine with the portion of the second beam that has gone around the test section with a zero phase shift. The spot where they meet the reviewing screen will not be as bright as before. (The probability that the phase shift will be 360° is small.) It can be shown that the white and black lines that appear on the screen correspond to surfaces of constant optical path length. If the experiment in the test section is two-dimensional (i.e., there is no variation in refractive index in the direction of the light beam), it is possible to correlate these lines with the fluid's density. Such density differences in the test section can occur due to heat or mass diffusion.

Recently, the laser has been used as a light source for this technique. With it, one can make optical filters on photographic plates that are called *holograms,* which are optical records of the characteristics of the light beam that passed through the photographic emulsion. Thus it is possible to substitute holograms for the reference beam of classical interferometry. This is called *holographic interferometry.* It is not the change of name that is important. Rather, the latter technique is much less sensitive to the quality of the transparent components that constitute the test section. This is an important advantage to the user in many applications. Figure 12.2.7 was produced in this manner. Fig. 17.2.14 is a schematic diagram that shows the components of a holographic interferometer, and Fig. 17.2.15 shows another configuration of the optical elements. Another type of interferometer, a shearing interferometer, can make visible the gradients of density. Figure 17.2.16 shows the density waves that are radiated from a circular cylinder that is rising in a stratified liquid.

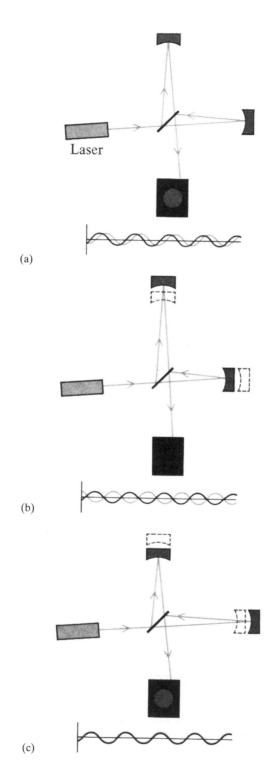

(a)

(b)

(c)

Figure 17.2.14 Simple interferometer consisting of laser, beam splitter, two mirrors defining two arms at right angles, and photodetector. Graph of electric fields of two returning light beams as function of time is shown next to photodetector. (a) Two arms differ in length by $\frac{1}{8}$ wavelength. Electric fields recombine $\frac{1}{4}$ cycle out of phase, and only half of light appears at output port. (b) Gravitational wave shifts path length in one sense so that light interferes more destructively at output port and output is less. (c) Shift in opposite sense causes light to interfere constructively, so that output is brighter. *(Adapted from illustration on p. 55, by George Retseck in "Gravitational Wave Observatories" by Andrew Jeffries et al. Copyright © 1987 by Scientific American, Inc. All rights reserved.)*

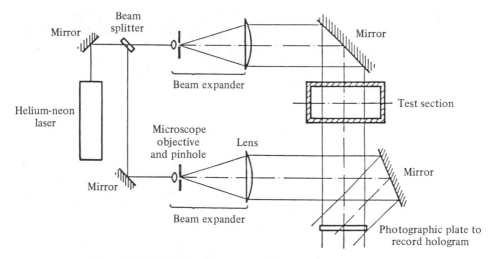

Figure 17.2.15 One arrangement of a holographic interferometer.

Figure 17.2.16 Internal waves generated by cylinder moving vertically through a fluid with a vertical density gradient as recorded with a shearing interferometer. Note also the wake pattern. *(Courtesy of Prof. W. Merzkirch, Essen University, F.R.G)*

17.3 PRESSURE MEASUREMENT

Pressure is measured by allowing the medium in which the pressure is unknown to communicate with some device from which one can make an observation and deduce the pressure therefrom. Such a device is generally called a *transducer,* a *pressure transducer* in this case. A manometer can perform this function when the level of the manometer liquid changes with fluid pressure. For a given pressure difference, the sensitivity of the manometer can be enhanced by using an inclined tube and by using a liquid whose density approximates that of the medium in which the pressure is desired. (Recall that the pressure difference is proportional to the product of the reading and the difference in density between the two fluids. As this difference decreases, the reading must increase.)

The manner in which one observes the differences in the liquid level in a manometer can determine the sensitivity of the system. A very sensitive instrument called the *Chattock gauge* observes the change in form of the interface, which is always brought to a reference position. The interface geometry is controlled by the size of the associated glassware and the surface tension of the two liquids, both of which are fixed. The sensitivity can be further enhanced by having the interface reflect a beam of light, which then appears on a scale some distance away. A small change in liquid level will give a large reading on this scale. Devices called *micromanometers* are usually single-tube manometers in which the manometer well can be moved on a micrometer screw. In this way, the changes of position of the well can be determined with great accuracy. This is important because for a given manometer it is only the difference in liquid elevations (between the well and the liquid in the vertical or inclined tube) that is important. Hence, if one connects the micromanometer to a system in which the pressure difference is zero, one could raise (or lower) the manometer well until the liquid level in the manometer tube was exactly at a reference mark. (This is made easily visible with a lens and suitable lighting.) If, subsequent to this "zeroing," a positive pressure difference occurs in the system to which the manometer is connected, the liquid in the manometer tube will rise. Then, one does not read the amount that the liquid has risen, but one records the height that it was necessary to lower the reservoir so that the liquid in the manometer tube once again is at the reference mark. This lowering of the reservoir for a fixed location of the liquid in the manometer tube is the same as the rise in elevation of the liquid for a fixed position of the reservoir. By operating in this way, one has provided a constant viewing position in the tube that is uninfluenced by differences in bore size and capillarity variations due to surface contamination of the liquid. This is called a *null* instrument.

Often, the pressure in the fluid is communicated to a pressure gauge. This is an instrument with a dial and a needle that is connected by a linkage to a curved tube. As the pressure increases, this tube begins to become straighter. In so doing, it moves to change the position of the needle on the dial. Sometimes the element which deforms is not a curved tube (called a *Bourdon* tube) but a metallic bellows. This is the case in an aneroid barometer.

The deformation of elastic members is utilized in other pressure transducers. The extent of the deformation, and consequently the extent of the pressure changes, is sensed by some change in electrical properties of associated elements. If a pressure differential exists across a circular diaphragm, it will deform, with the result that one side of the diaphragm will have tensile stresses induced and the other, compressive stresses. If electric strain gauges are mounted on these surfaces, the wires of these gauges will be extended or contracted. This changes their electrical resistivity. A calibration would reveal the correlation between pressure difference and the value of the electrical resistance of the gauges that are mounted on the diaphragm. This kind of pressure transducer is used in measuring the resistance changes with a circuit called a

Wheatstone bridge. The electrical signals can be subsequently amplified to achieve satisfactory sensitivities. The amplifiers can be made so that they are stable and do not give spurious results due to "drift." The entire system, including the diaphragms, can be constructed so that an acceptable frequency response results. The placement of the strain gauges in a Wheatstone bridge facilitates the construction of a transducer that is insensitive to the temperature, or temperature changes in the fluid.

The elastic deflection of a membrane is exploited in another kind of pressure transducer. As the membrane moves, it changes the distance between it and another surface. This surface and the membrane form a kind of plate capacitor whose capacitance is a function of the distance between them. Hence, the capacitance depends upon the pressure difference across the membrane. Such transducers are manufactured to meet a variety of needs, and extremely high sensitivities can be achieved. The pressure fluctuations due to the presence of turbulence in air can be detected with a capacitance microphone. (Indeed, the rumbling due to turbulent airflow near a surface can be detected with a stethoscope; however, the electrical recording of this signal will require a transducer similar to the last two that have been described.)

The manner in which the pressure is transmitted to the transducer from the location of interest is a matter of great importance. It can affect the magnitude and frequency response of the final result. If the connection is a long one, the fluid in the tubing will have an inertia that can cause high-frequency signals to be completely filtered out before they reach the transducer. If the holes through a surface and into the region of fluid that is of interest are not properly manufactured, erroneous readings can result. If the hole has a burr at its edge, the reading may be too high (if the burr is on the downstream side of the hole) or too low (if the burr is on the upstream side, so that the pressure tap falls in the wake behind the burr). If the hole is not bored normal to the surface, it is possible that the pressures that are sensed will be influenced by the direction of the flow near the wall. If the hole is too large in diameter, the flow that sweeps past the hole will induce a circulating flow within the hole. A spurious pressure reading will result.

17.4 VELOCITY MEASUREMENT

In Example 5.2.4, we saw a way in which the stagnation pressure that appears at the tip of a small, open tube that has been inserted into a flow can be measured. The manometer to which the tube is connected produces a reading from which the fluid's speed can be inferred. Such an *impact* tube (also called a *Pitot* tube) is widely used in fluid measurement to determine the velocity of a flow. (See Fig. 17.4.1.) The tube can be small in diameter and traversed across a flow section to establish the velocity profile in the region. In addition to the stagnation pressure that the tube senses, it is necessary that one know the ambient fluid pressure. This pressure must be detected through some properly constructed and located pressure tap. The "Pitot-static" tube provides a surface on the instrument itself on which the static pressure of the flowing fluid can be sensed. The static pressure is measured by some holes, or a slit, aft of the tip. This pressure is transmitted to the transducer through concentric, or otherwise separate, tubes to the two sides of a manometer or other transducer. The static pressure holes are placed along the instrument at a place where the flow is not accelerating, as it is

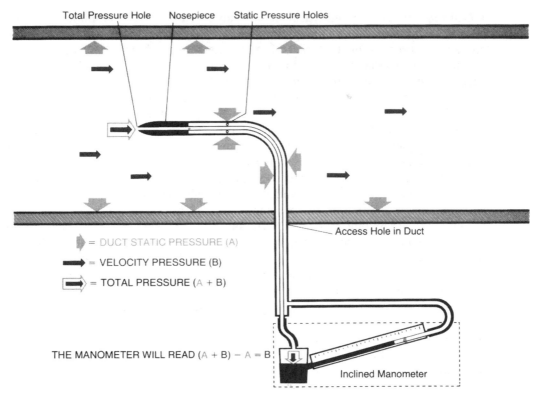

Total Pressure Hole   Nosepiece   Static Pressure Holes

Access Hole in Duct

= DUCT STATIC PRESSURE (A)

= VELOCITY PRESSURE (B)

= TOTAL PRESSURE (A + B)

THE MANOMETER WILL READ (A + B) − A = B

Inclined Manometer

Figure 17.4.1 Principle of operation of a Pitot-static tube. *(Courtesy Airflow Technical Products Incorporated, Landing, New Jersey)*

near the tip. They are also placed so that the effects of the strut that is perpendicular to the flow do not extend upstream to them. The static pressure readings are affected by the alignment of the probe with the direction of flow. The impact pressure readings are relatively insensitive to the flow direction, provided that the tube's axis is within $\pm 5°$ of that direction.

Impact tubes can be made quite small when one is working with air. In liquids, however, one must be certain that the holes are large enough that an air bubble, or dirt, does not fill the hole and render the instrument useless or, worse, give misleading readings. It is possible to squeeze a hypodermic needle so that only a slit that is the thickness of this page remains. The flattened surfaces of the tube can be honed to the thickness of a hair. With such a probe, measurements in an air boundary layer can be made.

If the struts of two impact tubes are placed together so that the tubes themselves form a vee, a device for sensing the direction of an oncoming flow is the result. Although this is one of several geometries that are in use, the principle of operation is always the same. If the flow is aligned with the bisector of the angle between the two impact tubes, each tube should sense the same impact pressure. This is because each tube is exposed to the same magnitude of a component of the velocity. If these two impact tubes are connected to the opposite sides of a U-tube manometer, a zero reading should occur. As these two impact tubes are rotated about an axis that is perpendicular to the velocity, the reading of the manometer will consequently change. In this situation, each tube will sense a different proportion of the dynamic pressure associated with the speed of the flow. Sometimes a hollow circular cylinder is used to sense the flow direction. A small radial hole is bored through the wall, and on the inside of the tube a connection is made to a manometer that senses the stagnation pressure. At angles of $\pm 30°$ around the circumference of this first hole, two additional holes are drilled and connected to each side of a second manometer. When the cylinder is placed in a flow and the manometer that is connected with the $\pm 30°$ holes reads zero, the middle hole should be aligned with the flow and sense the stagnation pressure. The $30°$ angle was chosen because experience, and theory, indicate that at this point the pressure should be the ambient pressure for the flow that is aligned with the central hole. Thus, if one also connects this pressure from the "direction-seeking manometer" to the second leg of the first manometer, a reading corresponding to the difference of the stagnation and ambient pressures should result when the "direction-reading manometer" reads zero. Such a reading is, of course, proportional to the square of the local velocity.

Instead of boring three holes in a cylinder, five holes can be drilled into a small, hollow sphere. The necessary internal connections to tubes that lead to manometers must then be provided. There are one central hole, two symmetrically placed lateral holes on the "equator," and two symmetrically placed holes on a longitudinal plane through the central hole. It appears that such a device could be capable of specifying the flow direction with respect to two directions. This instrument can be used successfully to determine the velocity distribution in such complicated three-dimensional flows as occur in the wake of a ship's propeller.

The Pitot-static tube is a rugged instrument, and where it is suitable, it is a valuable probe. The pressure readings can be amplified to give great sensitivity, but the frequency response of the instrument is low. In performing turbulence measurements, one is interested in a frequency response for the measuring system that is in excess of 10 kHz. One is also interested in the characteristics of the u, v, and w components of the velocity and not just the magnitude. The development of electronic circuits, beginning first with the vacuum tube and leading to solid-state devices, permitted such measurements to be made. For these reasons the period between 1920 and 1970 saw the development of the hot-wire anemometer.

A *hot-wire anemometer* operates by sensing the cooling effect that a gas has on a heated wire when it flows past the wire. The wire is heated electrically, and one can infer its temperature from its resistance. (The resistance of most materials, except carbon, germanium, and a few others, increases with temperature. The decrease in resistance with temperature of some semiconductors containing germanium is so great that one employs this characteristic to measure the temperature.)

If a platinum wire about one-tenth the diameter of the reader's hair and about 2 mm in length is inserted as one element in a Wheatstone bridge, it will have a resistance of about 4 ohms. One can supply a current to this bridge with a power supply that is connected in series with a resistance that is relatively large with respect to the equivalent series resistance of the bridge network. If a sufficiently high current is provided to the bridge—about 100 mA—all of the components in the bridge, including the platinum wire, will change their resistance due to Joule heating, which is equal to $I^2 R$,

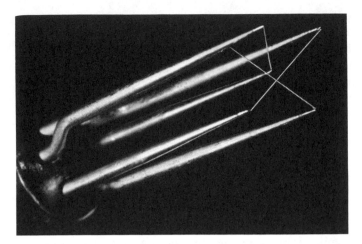

Figure 17.4.2 Magnified view of triple-sensor hot-film probe to determine simultaneously the three components of velocity. Distance between prongs of probe can be as small as 2 mm; length of minute cylinder sensitive to flow is even less. This permits one to minimize spatial averaging and approach a "point" measurement. *(Photograph courtesy of TSI, Incorporated)*

the power dissipated in the wire. However, only the resistance of the wire sensor will appreciably change. The resistance of the platinum wire will be designated R_{oh}, with the subscript denoting *overheat*. This wire is attached to the tips of two parallel supports like fine sewing needles by spot welding. (See Fig. 17.4.2.) This probe is connected to a Wheatstone bridge and inserted into an airstream. The wire will then be cooled by the forced convection. The resistance of the wire will no longer be R_{oh}, and this will be obvious from the "unbalance" of the Wheatstone bridge. One can now raise the resistance of the platinum wire in the probe to R_{oh} by increasing the electric current to the bridge through a decrease in the resistance in the large variable resistor that is in series with it and the power supply. Once the resistance of the wire returns to R_{oh} and there is no "unbalance" in the bridge network, the value of the air speed and the current to the bridge are recorded. This is a part of the calibration procedure. The air speed is next increased, and a new current is supplied to the bridge so that the resistance of the probe returns to R_{oh}. The final result is a calibration curve that relates the current and wind speed relationship that is necessary in order to keep the probe at the desired "overheat." Subsequently, one could put the device into an airstream and record the amount of electric current needed to maintain R_{oh}. From the calibration curve, one could infer the velocity of the air.

In another application, where the air speed was momentarily constant, on the average, so that a fixed value of bridge current could be provided, there could be turbulent fluctuations in the air. This would imply that there were high-frequency velocity fluctuations about the mean value of the velocity. (This is the air speed associated with the value of the electric current needed to keep the probe at an average value of R_{oh}.) Accordingly, one would expect high-frequency fluctuations of resistance about the selected value of R_{oh} for the prescribed value of input current to the bridge at a given air speed. It is for this reason that this type of system is called a *constant-current hot-wire anemometer.* The platinum wire, depending upon its diameter, has a small thermal inertia, so that there is a relatively small phase lag between the velocity and resistance fluctuations. However, above a certain frequency that is determined by the wire's characteristics, electronic compensation networks are necessary to follow the transients well.

The electronic circuits for the constant-current system were not too complex and were achievable by the state of electronic technology prior to 1950. The advent of the solid-state era that followed this date opened up, in a practical way, the path to the *constant-temperature hot-wire anemometer.* In this system, one continuously senses the resistance of the wire, and when it deviates from the selected value of R_{oh}, the power to the bridge circuit is changed so that the wire's resistance does not noticeably vary. Because the wire's resistance is coupled with its temperature, it is understandable why this system is so named. In order to provide this virtually instantaneous control of the power to the Wheatstone bridge that contains the probe as one element, stable electronic feedback circuits had to be achieved. The design of such circuits is conceptually easy; however, the availability of components to produce the effect is another matter. Vacuum tube circuits tended to "drift" from their prescribed voltage and current levels with time and temperature. This seriously handicapped the control system. The appearance of solid-state amplifier components made it possible to construct the stable direct-current amplifiers that were needed. The constant-temperature system has gained great favor with those who need to measure airflows and is, at this time, the most widely used system.

This usage has increased since 1965 because of the ever-improved reliability of operational amplifiers and related systems, as well as the availability of commercially produced probes. Formerly, the platinum wires were attached to the prongs of the probe by a few scientists and technicians in the laboratories that were concerned with research in turbulent flows. These probes were, and still are, very sensitive to mechanical shock and contamination by dust. When a small dust particle, perhaps 5 μm, collides with a 10-μm wire and remains there, the local heat transfer is appreciably changed and the wire can heat up instantaneously and melt at this point. In the atmosphere, dust particles of much greater size are common. Hence, only a few meteorologists used hot-wire probes to explore wind currents during the years that preceded 1960. The effort to produce reliable probes was very great. The calibration curves of probes that are used in flows where the contaminant level is not extremely low will tend to vary with time. As more and more soot, or an oil film, is deposited on the platinum wire (tungsten wire can also be used), the heat transfer characteristics of the system are affected and the calibration changes. Today, one can change the commercially available probes relatively often, clean them in a solution that is agitated ultrasonically, and replace them during the next maintenance period without the necessity of having an "in-house" production facility.

The hot wire is ideal for studying gas flows because of the low inertia loads due to the fluid. Also, gases are not usually electrically conductive; the electric currents do not stray from the wire. This would not be true for liquids, in general. Technology and ingenuity were able to overcome the difficulties with using a thermal probe in liquids, however, and the *hot-film probe* was produced. A constant-current system is used, and the probe is a metallic film that is deposited upon an insulating substratum. This thin film has the needed high electrical resistance so that small changes of resistance per ohm of probe resistance can be measured. This film is insulated from the fluid, usually electrically conductive, by an extremely thin film of silica. The film is so thin that it does not appreciably affect the heat transfer characteristics of the probe. (Naturally, some penalty in frequency response must be expected.) These probes are made in a variety of shapes to meet specific needs, including those associated with measurements in liquid metals. Refer again to Fig. 17.4.2.

There are a variety of other probes that employ heat flow as part of their operating principle. Two wires that are displaced a definite distance could be used as an anemometer if heat were transferred from one wire in pulses to the fluid, and the time needed to sense the associated temperature changes by the second wire were recorded. This time interval and the separation distance between the heat emitter and sensor can be calibrated to form an anemometer. A variation of such a device can be used to measure fluctuations in very turbulent flows where linearity requirements that are inherent in other systems are no longer realizable.

Another measurement technique involves introducing other substances at known locations in the fluid and observing them as they appear at specified downstream positions, in order to determine the "time of flight." Salt is one such substance, and conductivity probes can sense when it arrives at a downstream location in a river or canal. Fluorine can also be readily detected. Other physical principles that are associated with specific probes, such as the thermocouple, can be used in specific applications. The thermocouple is a temperature-sensitive device. The greater the temperature difference between it and a reference thermocouple, the greater the electromotive force (emf) produced. It is obvious, then, that one should try to heat a system of thermocouples in such a way that an airstream will cool them. The change of probe temperature would give a change in emf that can be calibrated to give a relationship with the speed of the air.

But one need not employ specialized physical principles or elaborate electronics to observe the speed of a fluid. In a sense, the speed of rotation of a windmill does this. A device employing this principle is shown in Fig. 17.4.3. It is natural, then, that

Figure 17.4.3 A portable air-velocity measuring system. *(Courtesy of Omega Engineering, an Omega Technologies Company)*

probes are constructed with small dimensions and very small bearing friction to record the speed of flows in air and liquids. This could involve the velocity distribution near a hydraulic structure, or the velocity distribution near a ship model. Vane anemometers that are made out of lightweight plastics (to minimize inertia effects and decrease the velocity level to activate them) are used to measure wind speeds at meteorological stations. What is needed in these situations is a "frictionless" method for determining the speed of rotation of the propeller. This can be done with an optical method that records the number of times a light beam is reflected from a portion of the rotating propeller. Often, the position where this is detected is on the hub of the propeller or its shaft.

If aluminum flakes or plastic particles have been added to the flow for flow visualization, and suitable lighting is available, an indication of fluid velocity can be gained. If the shutter of a camera is opened for a specific interval, say, $\frac{1}{10}$ s, a streak will appear on the film corresponding to the path of a particle that reflects light during this interval. The length of the streaks generated by particles, and the selected time interval, will give a measure of the velocity at various points when the photograph is taken. Thus, a velocity field could be determined from a photograph such as Fig. 1.8.10a.

The motion of a ship model as a function of time, and consequently its velocity, have been recorded in a variety of ways. Three tracking telescopes can keep a light source on the model constantly in view. By means of trigonometry, its position with time can be established. (This is similar to the way in which balloons are tracked and the winds aloft are observed.) Another scheme would be to photograph the ship in its basin or canal with a television camera and display the result on a television screen. Visually, or electronically, the coordinates of a reference light on the model could be recorded from the screen of the cathode-ray tube (or electronically by similar means). The result would be a system to record the velocity or maneuvering characteristics of a ship model.

It is evident in the foregoing that the development of the field of electronics has greatly expanded the scope and capabilities of those interested in measuring velocity. Currently, the Doppler phenomenon has seen widespread application. Aircraft are located and controlled in their flight by the technique called *radar* (*ra*dio *d*etection *a*nd *r*anging). In this system, radio waves of high frequency are reflected from a moving object. If this object is not moving, the frequency of the reflected signal will be equal to that of the transmitted signal. If the object is moving, the frequency of the reflected signal will be dependent upon the speed of movement as well as its direction. Appropriate encoding of the characteristics of the reflected signal provides information about the object's location and velocity. A similar scheme is available for detecting the speed of minute particles that have been introduced into a flowing fluid. Plastic or latex particles reflect light from a laser beam into a detector called a *photomultiplier*. The frequency of these reflections is compared with a reference signal from a laser light beam that arrives at the measurement location via a different path. The result, when properly processed and interpreted, is the time history, including the fluctuations of the velocity at the point where the two laser beams intersect. Such a device is called a *laser-Doppler anemometer* (LDA) or a *laser-Doppler velocimeter* (LDV). This device has evolved from a "laboratory curiosity" to an instrument of considerable industrial and commercial importance. Its ability to measure in air and liquids, including those with suspensions, along with the fact that the scheme does not introduce an intrusive probe into the measurement region, makes it attractive in a variety of applications.

17.5 VOLUME FLOW RATE MEASUREMENT

A Pitot-static tube can be positioned at various radii in a circular duct. If the velocity is assumed to be axisymmetric about the duct's axis, one would be able to determine the volume rate of flow because it would be $\Sigma V_i(\Delta A_i)$, in which ΔA_i would undoubtedly be $2\pi r_i(\Delta r_i)$. While such a system is possible, and sometimes necessary in ventilation systems, it is cumbersome. In this section we will discuss some other methods for determining the volume rate of flow of a fluid.

Volume flow rate meters are an everyday experience at gasoline stations. A volume-measuring device records the volume of gasoline pumped. If one divided the amount listed on the pump by the time interval, one would have the volume per unit time. Similarly, the water meter in a house records the volume of water that is consumed. This meter is read monthly, in many cases. The differences in the recorded readings should relate to the volume consumed per unit time (the month, perhaps). The ancients recognized the ability of a hole in a vessel to serve as a metering device for liquids. In the water clock, they constructed a device which would allow a definite amount of water to flow out of (or into) a vessel in a prescribed amount of time. The exact volume of fluid and the height of it in the vessel were subject to calibration. An important feature of such devices, and an important requirement in fluid metering devices today, is that they gave the same readings (i.e., volume rate of flow) regardless of which water supply was employed. They were reliable, calibrated devices. Today, we use orifice meters in another way. From the pressure difference between the two sides of the orifice plate, one can calculate—and calibrate—the associated volume rate of flow that can be expected. Once an orifice has been fabricated, it ought to be installable in a variety of locations with the assurance that the pressure differences that occur can be directly correlated with the volume rates of flow. More often than not, it is not convenient or possible to calibrate the orifice after installation and one must rely on the calibration curve and hope that it is applicable to the installation at hand. For this reason, design standards for orifices have been established by the various engineering societies around the world. Definite requirements for the geometry and installations are prescribed. If these are adhered to, one can expect that the relationship between volume flow rate and the measured pressure differentials will be within about 5 percent of the correct values.

Orifice plates, while relatively simple in appearance, require considerable care and expense in their manufacture. A flow nozzle will be more costly to produce. Nozzles have also been carefully studied and designed so that they, too, will pass a prescribed volume rate of flow for a given pressure difference. One would choose a nozzle instead of an orifice if the higher energy losses associated with the flow through an orifice, which would usually be manifested as a pressure drop in the flowing system, would be too high to be practical or efficient. (For a given value of $A_{opening}/A_{pipe}$, the losses with an orifice are 3 to 5 times those with a nozzle.) Both nozzles and orifices can be used with liquids and gases. In the latter case, compressibility effects must be taken into account if the speed of the flow is high. (Anything greater than $Ma = 0.3$ would require that the equations for compressible flow be used, instead of the simpler incompressible energy or Bernoulli equation. The more accurate the results desired, the more these compressibility effects would need to be accounted for, even at lower Mach numbers.) Nozzles can also be useful to measure liquid flows with entrained solids. Figures 17.5.1a, c, and d show these devices, and Fig. 17.5.1e shows an elbow meter that can be readily installed in most piping systems.

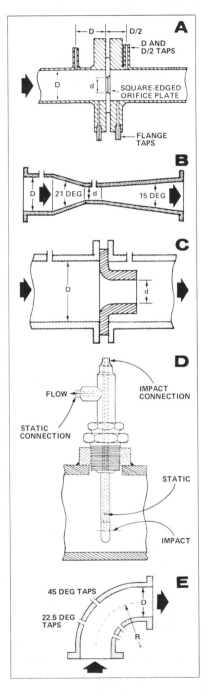

Figure 17.5.1 Common differential-pressure flow meters: (a) orifice, (b) Venturi tube, (c) flow nozzle, (d) Pitot tube, and (e) elbow-tap meter. All require secondary elements for measuring the differential pressure and for converting the data to flow values. *(Courtesy of Omega Engineering, an Omega Technologies Company)*

One can employ the appropriate formulas for determining the theoretical velocity through a nozzle or orifice. Usually, the measured velocity will be less than this. It is quite common to quote a *velocity coefficient* for such devices. This is designated as C_v and is equal to V_{actual}/V_{theory}. Other coefficients are also used. The discharge coefficient C_d is the ratio between the actual flow rate and the theoretical rate, Q_{actual}/Q_{theory}. In this case, Q_{theory} equals V_{theory} times the area of the theoretical stream tube. This area is not the area of the opening. The flow usually contracts somewhat when passing through the orifice (a bit less so when going through a nozzle), so that the flow area for which V_{theory} might apply is a fraction of the opening in the nozzle. We define a *contraction coefficient* C_c equal to the effective area of the flow divided by the actual area of the opening. The velocity, discharge, and contraction coefficients are interrelated, as the following definition clearly shows:

$$Q_{actual} = (V_{avg} A_{actual}) = (C_v V_{theory})(C_c A_{opening})$$

in which V_{theory} is obtained from the Bernoulli equation for incompressible flows. For compressible flows, one could assume an isentropic process.

We define the product $C_v C_c$ as the *discharge coefficient, C_d*. All three coefficients will vary with the conditions of the installation (i.e., the ratio of the pipe diameter to the orifice diameter, and the surface roughness of the pipe). It is desirable that for a given installation the various coefficients be nearly independent of Reynolds number, so that one constant can be applied over a large range of Q. For standard orifice and nozzle designs, the three coefficients are well tabulated.

The discharge coefficient for an orifice or a nozzle is also sensitive to the physical condition (e.g., dents or burrs) on the piece. Of great importance is the velocity profile just upstream of the metering device. A sufficient run of straight pipe should precede an orifice or nozzle so that any asymmetries in the velocity profile that result from bends and elbows can be damped out. It may be necessary to install "honeycomb" sections upstream of the metering device to align the flow. Figures 10.7.1 and 10.7.2 give discharge coefficients for orifices and nozzles.

When the energy losses from an orifice or a nozzle are too much to be tolerated, one can use the more costly Venturi tube (Fig. 17.5.1b). (The losses are about one-fourth those in a nozzle.) This device is characterized by a relatively sudden decrease in diameter, followed by a short section of the reduced diameter—the throat—where the pressure of the high-speed flow is sensed. Downstream of the throat, the diameter gradually returns to that of the conduit upstream of the Venturi tube. The gradual increase in the flow's cross-sectional area minimizes the production of energy-consuming eddies. The tapered section often has an included angle between 5° and 15°, so that the expanding flow does not appreciably separate from the walls. This angle is somewhat of a compromise, because a more gradual expanding section would certainly restrict the separation losses; it would also result in a longer section in which the flow was moving very rapidly and which would be subject to high wall shear stresses. There is no need to have a long, tapering inlet section, because separation is not a critical factor here. The contraction section is designed so that there is no separation in the throat. The discharge coefficient for a Venturi tube is somewhat dependent upon the Reynolds number, but it is generally constant above Re = 10^5 at a value of about 0.98. (See Fig. 10.7.3.) Sometimes more than one pressure tap is inserted around the periphery of the throat and there is a tap in the section upstream of the contraction as well, where the higher pressure is sensed. These upstream or throat taps are interconnected (among themselves) so that an average pressure at the appropriate section results. This is considered advisable when upstream conditions do not ensure a symmetrical velocity profile. The absence of anything to obstruct the flow in a Venturi meter makes it suitable for measuring liquids with suspended solids.

Another kind of flow meter consists of an axial turbine mounted in a cage that has low-friction bearings and flow straighteners at each end. See Fig. 17.5.2. This assembly is placed within the conduit. The speed of rotation of the turbine is directly proportional to the volume rate of flow through it over the designed range of the instrument. This speed can be sensed with an electromagnetic device. The resulting electrical pulses are given as input signals to a suitable counter that displays the pulse rate, from which the flow rate can be deduced.

Electromagnetic effects can be utilized to measure the flow rate itself. A large magnetic coil is contained in a housing which surrounds a segment of the conduit. The flowing fluid is slightly electrically conductive (or it can be so made by the addition of ions), so that one has, in effect, a moving conductor in an electromagnetic field. An electromotive force is generated in the flow meter in this way and can be measured and correlated with the flow rate through a calibration. See Fig. 17.5.3. Such flow meters can be used to observe the conditions in industrial pipelines; they also have medical uses where the flow of blood to artificial kidneys must be accurately monitored. Ultrasonic devices and the Doppler effect are currently being used to measure blood flow.

A device called a *rotometer,* shown in Fig. 17.5.4, is widely used to measure the flow rates of liquids or gases. It consists of a tapered bore in a glass tube that is mounted vertically. A "float" moves in this bore so that a sufficient area is available to pass the flow. A balance is reached between the drag on this float and its weight. The position of the float in the tube corresponds to a calibrated flow rate.

Figures 17.5.5 to 17.5.8 show other flow-rate transducers. All produce electrical signals that can be easily monitored. The methods discussed previously in this section

Figure 17.5.2 Turbine flow meter consists of multiple-bladed free-spinning permeable metal rotor housed in nonmagnetic stainless steel body. In operation, rotating blades generate a frequency of signal proportional to the liquid flow rate, which is sensed by the magnetic pickup and transferred to a readout indicator. *(Courtesy of Omega Engineering, an Omega Technologies Company)*

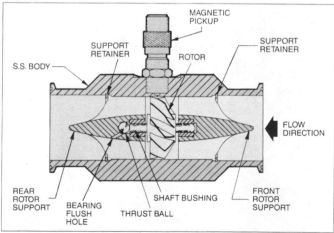

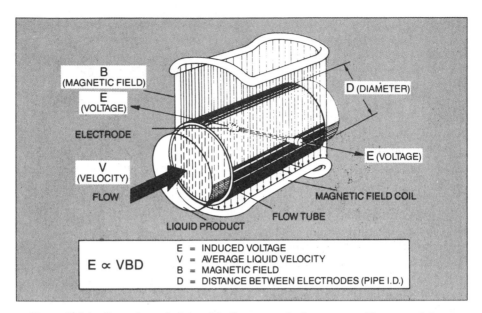

$$E \propto VBD$$

E = INDUCED VOLTAGE
V = AVERAGE LIQUID VELOCITY
B = MAGNETIC FIELD
D = DISTANCE BETWEEN ELECTRODES (PIPE I.D.)

Figure 17.5.3 Operating principle of in-line magnetic flow meter. *(Courtesy of Omega Engineering, an Omega Technologies Company)*

Figure 17.5.4 Variable-area flow meter. Glass tube has tapered bore, and float rises to a location within it at which hydrodynamic drag on float is equal to its net weight. *(Courtesy of Omega Engineering, an Omega Technologies Company)*

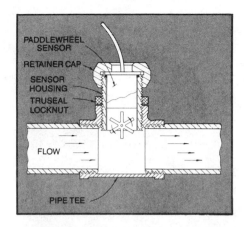

Figure 17.5.5 Paddle wheel flow sensor installation and operation. *(Courtesy of Omega Engineering, an Omega Technologies Company)*

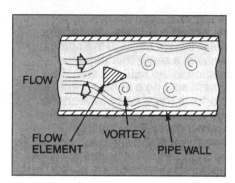

Figure 17.5.6 Vortex meters operate on principle that when a nonstreamlined object is placed in the middle of a flow stream, a series of alternating vortices are shed downstream of the object. Frequency of vortex shedding is directly proportional to velocity of liquid flowing in pipeline. *(Courtesy of Omega Engineering. an Omega Technologies Company)*

Figure 17.5.7 Doppler meters use sound-pulse reflection principle to measure liquid flow rates. Solids or bubbles in suspension in liquid reflect sound back to receiving transducer element. *(Courtesy of Omega Engineering, an Omega Technologies Company)*

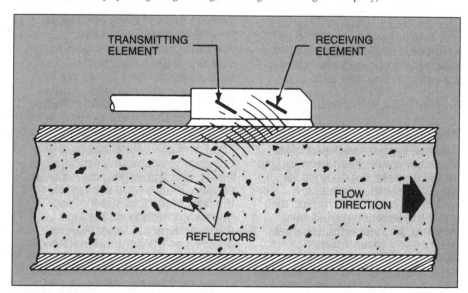

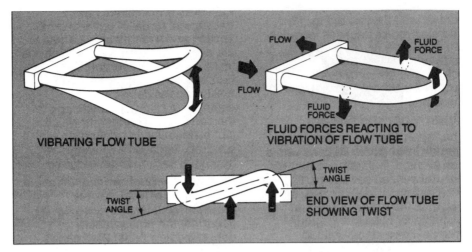

VIBRATING FLOW TUBE

FLUID FORCES REACTING TO
VIBRATION OF FLOW TUBE

END VIEW OF FLOW TUBE
SHOWING TWIST

Figure 17.5.8 Vibrating U-shaped flow tube is heart of popular Coriolis mass flow meter. Tube vibration, coupled with fluid's force, causes tube deflection directly proportional to mass flow rate. *(Courtesy of Omega Engineering, an Omega Technologies Company)*

have emphasized measuring flow rates in closed conduits. In open channels, one does not use pressure as the parameter to be sensed; one usually measures the elevation of the free surface to determine the volume rate of flow past a particular metering device. A weir performs this function in many applications. The liquid, usually water, is required to have a free-surface elevation upstream of the weir that is higher than the weir's lowest edge. The actual upstream excess of elevation will determine the velocity of the liquid that cascades over the edge of the weir, as well as the cross-sectional area of the flow. Once again, a calibration of the weir is recommended.

In Sec. 15.1, it was shown that as a liquid flows over a hump, the free surface changes elevation. The measurement of the elevation change can be used to meter the volume rate of flow. It is often the case that the hump has an extended horizontal summit so that the region where the maximum depth change occurs is not limited. This facilitates the measurement of liquid elevation on such broad-crested weirs. A special section called a *Parshall flume* can also be built into a channel. Its geometry permits a determination of the flow rate in accordance with the principles in Chap. 15.

17.6 SURFACE ELEVATION

The position of the water's surface at a weir is often recorded on a vertical scale attached to the weir itself. This assumes that the surface disturbances and undulations are sufficiently small that they do not introduce appreciable errors. In special cases, one uses a *point gauge* to locate the level of the free surface. This is simply a pointed rod which, when lowered, ultimately comes into contact with the free surface. The instant that it does so can be readily observed, because as the point pierces the surface, surface tension causes a small "dimple" to form. This position of strong surface curvature reflects light in such a fashion that a bright spot appears when the angle of viewing is optimum. A vernier scale is an aid to the observer in reading the small elevation changes that can be detected in this way.

Contact with the surface of water can be sensed electrically also. When a probe with a small electric potential comes into contact with water in which another electrode has been inserted, a complete circuit can be realized. The associated flow of current can be observed in a variety of ways, including the illumination of an electric light bulb. This system is sensitive to the location of the free surface, and its electrical nature allows it to be incorporated into automatic systems for measuring the elevation in hydraulic models or prototypes.

The position of the free surface is also important in the measurement of wave heights. Here the problem may center upon one's ability to continually follow the position of the free surface as it moves up and down with time. One way to do this in a somewhat discrete fashion is to have many probe tips, say, 1 mm apart, attached to a vertical strut. The time when each tip begins to conduct electricity can be recorded, as well as the instant when it ceases to conduct. From this information, and a knowledge of the position of the probe tips, one can construct the time history of the elevation at the measurement location. (The fact that a thin film may remain on the probe assembly for a brief interval after the water has receded is a problem that is common to all of the probes that are being discussed in this section.)

Another system employs two parallel vertical conductors. An electric potential exists between them, but the air gap that separates the wires prevents any flow of current. If the tips of these two conductors are immersed in water, electric current will flow. The amount of this current will depend upon the depth of submergence of the probes, because this will determine the volume of water that can transfer ions. As the water level changes while the probe remains stationary, so will the electric current. This can be calibrated to make a wave gauge.

The calibration of such devices will depend upon the electrical properties of the water in which the wave gauge is used and the chemical condition at the surface of the probe's sensing elements. If the surface resistance changes with time, an unacceptable deviation from previous calibration values can result. Temperature compensation for such devices can also be employed to advantage.

The two parallel wires can be used to sense the height in another way. In this type of gauge, they form the surfaces of a capacitor, the capacitance of which changes with the depth of immersion. The electronic circuits necessary for a capacitance system can also be miniaturized so that the sensor and its amplifier can be installed together. This minimizes the tendency to have spurious signals interfere with the measurement. Some workers believe that this system can be made to operate for long periods of time without the instrument's calibration values drifting because of surface contamination on the probes.

Ultrasonic reflections are now being sensed and calibrated to determine the surface elevations of liquids in storage tanks.

17.7 TEMPERATURE MEASUREMENT

Often, a simple mercury thermometer is all that is needed to specify the fluid's temperature. A variation of 0.1°C is easy to detect on a suitable thermometer. The ease of use, and relatively low cost, of such an instrument is sometimes offset by other considerations. One may wish to sense the temperature in a remote location. This would require some means of transmitting the temperature signal to a recorder or an observer. There are several methods to do this; usually, the systems are rugged and reliable.

A thermocouple is a loop which is formed when two metallic pieces, usually wires, are joined at their ends to form a closed loop. If the two junctions are kept at different temperatures, an emf will exist between the two junctions to an extent that is dependent upon the temperature difference and upon the two different materials that form the junctions. Previously, one sensed these small potentials (measurable in microvolts) with a null method that employed a precision potentiometer. This made the readings inconvenient to take, because a balancing procedure was necessary for each reading. Today it is possible to use a digital voltmeter with a high input impedance to sense the small voltages that arise. It is usual to keep one of the junctions at the temperature of melting ice, so that a calibration of a thermocouple will give the temperature of the second junction with respect to 0°C or 32°F. More than one thermocouple can be connected to a single readout device, but extreme care must be taken with the attendant switches, because they involve the breaking and making of junctions at which thermoelectric effects can occur. If this does happen, an unwanted potential can be generated which would give an erroneous reading. (As mentioned previously, the thermocouple could be heated by an external power source and the effect of a wind detected by the decrease in temperature of the tip when it was placed in the flow.) If the wire diameters are kept small so that the junction (often it is welded) is small, the thermocouple can have a frequency response up to about 1 Hz.

The change of resistance of metals with temperature has been used for many years as a means of detecting temperature changes. (Bimetallic strips which bend due to differential thermal expansivities are also used in temperature controllers.) Platinum is an excellent material to use, because it has a high resistivity and the changes in resistance with temperature are also quite substantial and therefore easy to detect. For a given voltage difference across a platinum resistor, the electric currents can be in the milliampere range, so that there is virtually no self-induced heating (at a rate equal to $I^2 R$) to affect the measurement process. The changes in resistance are usually sensed by having the thermometer as one of the resistances in a Wheatstone bridge. (This would be similar to a constant-current hot-wire anemometer, but with currents so low that an "overheat" does not occur.) Complete commercial units are available, including shields to prevent mechanical and chemical damage to the platinum sensing element. High-frequency response can be gained by having the resistance element made of very fine wire (about 1 μm in diameter). One can purchase resistance elements that are made of etched foil, just as in the case of electric strain gauges. These resistors can be mounted on surfaces conveniently. Their small size permits the measurement of a local temperature with a high rate of response.

Thermistors, solid-state devices, show exceedingly high resistance changes with temperature changes. In contrast with resistance thermometers, for which the variation is nearly linear, the temperature-resistance characteristics for thermistors is quite nonlinear. Also, the resistance decreases with increasing temperature. Linearizer circuits are now available to obviate this difficulty. One can use thermistors to measure small temperature differences to detect liquid levels, because their resistance changes greatly when they are cooled by the liquid, and to measure fluid velocity when they are heated by an applied current.

Certain "liquid crystals" now can be used to measure temperatures because their color depends on their temperature. When minute liquid crystals are "seeded" in a water flow, they can give a visual display of the temperature field.

17.8 VISCOSITY MEASUREMENT

The ability to predict the volume rate of flow in a round tube under laminar conditions was one of the great successes of the mathematical theory of fluid mechanics. This result means that if one knew the flow rate and geometric dimensions of the tube, one could calculate the viscosity, provided the pressure gradient was also known. The tube can be installed in a vertical orientation and be open at each end. The pressure difference between the beginning and end of the tube would be nearly zero. This would imply that the pressure gradient was zero. Still, there is a "driving force" for the flow, and this is the weight per unit volume of the liquid. (See Chap. 8.) Various devices that relate the time for a prescribed volume of liquid to pass through a glass capillary tube are used to determine the viscosity of liquids. One type, called the *Cannon-Fenske* viscometer, is shown in Fig. 17.8.1. Standards for its use are prescribed by the American Society for Testing Materials.

Another kind of viscometer which employs a flowing liquid is the *Saybolt* viscometer. A prescribed amount of liquid is poured into a vessel that has a calibrated orifice at its bottom. The liquid can be released through this orifice and the time recorded for it to completely drain from the container. This time, in seconds, is a measure of the viscosity in SSU units. (See Fig. A.2. caption for a conversion factor.) In all viscosity measurements, the temperature of the fluid must be known and held constant during the experiment. Temperature-controlled water baths serve this purpose.

The torque that is required to rotate one cylinder with respect to a concentric one can be used to measure the viscosity of the liquid in the annulus. The inner cylinder, or cup, is held stationary, and the outer cylinder can be rotated so that the speed is increased each time a torque measurement is made. This torque can be sensed by noting the deflection of the vertical wire that supports the cup. (The slope of the

Figure 17.8.1 Cannon-Fenske routine viscometer which utilizes the flow in a capillary tube to measure kinematic viscosity. Capillary tube is in leg at right. *(Courtesy of Cannon Instrument Co.)*

Figure 17.8.2 Brookfield viscometer. Moment induced on rotating cylinder by the viscous liquid is indicated on an easily read dial. *(Courtesy of Brookfield Engineering Laboratories, Inc.)*

torque-speed curve is proportional to the viscosity.) When the unit is operated in this fashion, it is possible to have a velocity distribution in the annulus which is primarily in the azimuthal direction (i.e., direction of rotation) for a wide range of speeds. Ultimately, the flow will cease to be laminar and become turbulent. However, if the inner cylinder were to be rotated and the outer one restrained, one would find that after a critical speed of rotation, determined by the geometry and viscosity, the fluid motion would abruptly become more complex. The result would be a break in the slope of the speed-torque curve. While these new motions may be laminar, the manner in which one can deduce the viscosity from the torque is much more complex after these new, or secondary, motions occur. It is for this reason that the less ambiguous system, a rotating outer cylinder, is preferred. Portable commercially made instruments that operate on the rotating cylinder principle are available; one is shown in Fig. 17.8.2.

Very completely instrumented units with rotating plates and cones as accessories are called *rheogoniometers*. They are used to tell the relationships between shear stress and rates of deformation for liquids that are more complex than Newtonian fluids. Such materials could include latex paints or elastic liquids that have some of the characteristics of silicone putty, a substance which when rolled into a ball can bounce like one, but when allowed to rest on a table will become flattened over a period of 24 hours. (It has been pointed out that whipped egg yolks and mayonnaise appear to move toward the center of a bowl for an electric mixer, in contrast to milk, which moves outward. Hence, egg yolks give evidence of non-Newtonian behavior.)

In very slow motion (Re < 0.1), a sphere experiences a drag force of $3\pi\mu UD$, in which U is the speed of the fluid at points far from the sphere and D is the diameter. If the sphere is falling in a liquid in a large stationary container, U becomes the terminal velocity. Under these conditions, the force causing the motion is the effective weight of the sphere, $\pi D^3 g (\rho_{ball} - \rho_{liquid})/6$. This means that with a knowledge of the sphere's diameter and density, as well as the density of the liquid (from a hydrometer), one should be able to determine the viscosity from a measurement of U (from distance and time). This is true, in principle, but one must account for the influence that the walls of the container will have on the motion. Consequently, appropriate corrections must be made. This method of determining viscosity is termed, appropriately, the *falling-ball viscometer*.

18

Hydraulic Machinery

18.1 INTRODUCTION

The scope of this chapter's topic is broad enough that a complete book could be written on it without examining all of its facets. This chapter will be limited to the major factors that affect the performance of pumps and turbines. Most of the work will be an extension of the principles discussed in Sec. 7.6, in which the concept of angular momentum was applied to *centrifugal* and *axial machines*. These are the two extreme categories of devices through which a fluid flows while concurrently changing its angular momentum. In the former, the fluid moves in a direction that is radial to the axis of rotation, while in the latter the fluid moves parallel to the axis. (In most machines, both radial and axial flow occur to some extent, but if a combination of these is present throughout most of the machine, it is called a *mixed-flow* pump or turbine.) Much of the material that is discussed at the beginning of this chapter is quite general and applies to both liquids and gases; later, however, the presentation is restricted to pumps and turbines for liquids. The only exception to this bias is in the examples and problems involving fans that operate at low speed. In Chap. 14, we saw that in such situations the air can be considered to be incompressible.

   If the angular momentum of the fluid is altered by the machine as it rotates, energy is either transferred to or extracted from the fluid. In the case of a pump, the transfer of energy usually takes place in order to increase the fluid's pressure. A hydraulic turbine converts the potential energy of water into energy that is transferred to a generator, where it is transformed into electrical energy. The changes of pressure that occur between the inlet and the outlet of machines that alter the angular momentum of the fluid are associated with the laws of fluid dynamics that describe their motions. Such machines could be called *hydrodynamic* machines, but the term *turbo-*

machine is usually used to identify them. A second category of fluid machinery is called *hydrostatic*. Machines in this category depend largely upon the actions of an incompressible fluid on pistons and other devices without taking into account the dynamic processes within the fluid while it is in motion. (Sometimes these are impossible to ignore, however, because if a piston accelerates too rapidly away from the liquid with which it is in contact, the liquid may vaporize. The rigid-body accelerations mentioned in Sec. 2.7 bear upon this point, and Sec. 8.8 presents the way in which a hydrodynamic effect can be used to support loads in a bearing.) Gear and piston pumps will be described at the close of this chapter to round out a discussion that will be primarily concerned with turbomachines.

18.2 SIMILARITY (OR AFFINITY) LAWS FOR HYDRODYNAMIC MACHINES

In Sec. 7.6 we established the relationship between the torque applied to a fluid passing in steady flow through a control volume and the rate of angular momentum change as

$$T_z = T = \rho Q[(rV_\theta)_{\text{out}} - (rV_\theta)_{\text{in}}] \tag{18.2.1}$$

The rate of energy transfer to the fluid must be $T\Omega$, or

$$\dot{W} = T\Omega = \rho Q \Omega[(rV_\theta)_{\text{out}} - (rV_\theta)_{\text{in}}] \tag{18.2.2}$$

which can be rewritten as

$$\frac{\dot{W}}{\dot{m}} = \Omega[(rV_\theta)_2 - (rV_\theta)_1] \tag{18.2.3}$$

In many pumps, the fluid enters without a V_θ component, and one speaks of the absence of *preswirl*. This condition leads to writing the right-hand side of Eq. (18.2.3), with the expressions for V_θ and V_r that can be inferred from Fig. 18.2.1, as

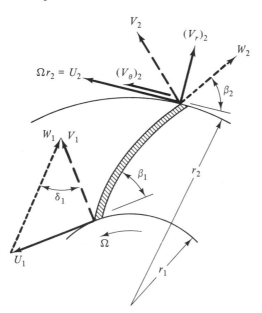

Figure 18.2.1 Impeller geometry for inlet flow without preswirl.

$$\left(\frac{\dot{W}}{g\dot{m}}\right)g = \Omega\{r_2[\Omega r_2 - (V_r)_2 \cot \beta_2] - r_1(0)\} \tag{18.2.4}$$

The radial velocity of the fluid leaving the impeller is related to the volume flow rate by

$$Q = (2\pi r_2 b_2)(V_r)_2 \tag{18.2.5}$$

With this fact, Eq. (18.2.4) can now be rewritten as

$$\frac{(\dot{W}/g\dot{m})g}{(\Omega r_2)^2} = 1 - \frac{Q}{\Omega r_2^3} \frac{\cot \beta_2}{2\pi b_2/r_2} \tag{18.2.6}$$

The term $\dot{W}/g\dot{m}$ has the dimensions of length and is referred to as the *Euler pump head*, H_{Euler}. Only in an ideal, 100 percent efficient system is this head transferred to the fluid in a completely usable form. Usually, there are viscous (frictional) effects and secondary flows that will limit the conversion of the energy added by the impeller into a change in the kinetic energy and pressure head of the fluid. The sum of the change in the kinetic and pressure head of the fluid, as measured from the horizontal inlet to the horizontal outlet of the pump, is simply called the *pump head*. This pump head is a rate of energy transfer per unit weight of fluid flowing, and the product of it and the weight rate of flow is called the *water horsepower* if water is the medium that is being pumped. (One horsepower is equivalent to 550 ft-lb/s, while 737 ft-lb/s equals one kilowatt.)

The term on the left-hand side of Eq. (18.2.6), $gH_{\text{Euler}}/(\Omega r_2)^2$, is dimensionless. It is often denoted by Ψ_{Euler}. The subscript is an indication that the Euler head from Eq. (18.2.6) is to be used. Ψ_{Euler} is commonly called the *Euler head coefficient*. (Sometimes the factor g is not included. The coefficient is then, of course, not dimensionless, but still useful.) The parameter $Q/\Omega r^3$ is called a *flow coefficient* and is designated by the symbol Φ. (A subscript is sometimes added to identify the location of Q and r. The former could vary from the pump's inlet to the outlet if there is leakage.) Figure 18.2.2

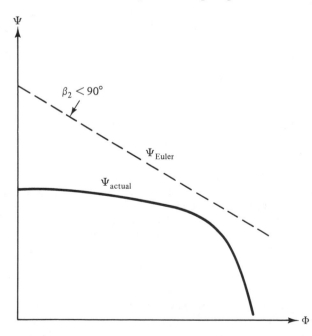

Figure 18.2.2 Relationships between head and flow coefficients.

shows how these coefficients interrelate from the theoretical point of view and in practice. The latter case requires a head coefficient that is based upon the (measured) pump head.

The two curves in Fig. 18.2.2 differ, and the ratio of the ordinates is a measure of the theoretical efficiency of a unit. (A discussion of the factors that affect the efficiency of hydrodynamic machines will be the main theme of Sec. 18.3.) The theoretical curve is for a machine with *backward-turning* vanes, so that β_2 in Fig. 18.2.1 and Eq. (18.2.6) is less than 90°. That equation is an explicit relationship between the flow coefficient and the Euler head coefficient. For each value of Φ, a value of Ψ_{Euler} ensues. A similar result holds for Ψ_{actual} in Fig. 18.2.2, even though an explicit formula that relates the coefficients would have to be obtained from a curve that is fitted to the experimental data. If this curve is almost independent of other operating parameters, such as the Reynolds number, one can use the ideas of dimensional analysis that were developed in Chap. 9 to say that for two different operation conditions or experiments, called A and B, if $(H/\Omega^2 r_2^2)_A = (H/\Omega^2 r_2^2)_B$, then $(Q/\Omega r_2^3)_A = (Q/\Omega r_2^3)_B$. These similarity relationships are sometimes called the *pump affinity laws*. When using these relationships, it is usual to base the rotational speed on the number of revolutions per minute, N, and to use the diameter D instead of the radius. The flow rate is commonly based on gallons per minute (i.e., gpm) in the United States, so that the relationships

$$\left(\frac{H}{N^2 D^2}\right)_A = \left(\frac{H}{N^2 D^2}\right)_B \tag{18.2.7}$$

and

$$\left(\frac{Q}{N D^3}\right)_A = \left(\frac{Q}{N D^3}\right)_B \tag{18.2.8}$$

are not dimensionless with such units, but that does not diminish their utility. A third quantity that is dependent upon the head and flow rate is the power, P. The affinity law for this variable follows from first principles, or by multiplying the previous two equations together in the proper fashion. The result is

$$\left(\frac{P}{N^3 D^5}\right)_A = \left(\frac{P}{N^3 D^5}\right)_B \tag{18.2.9}$$

Example 18.2.1

A centrifugal pump produces 310 ft of head while delivering 350 gpm of water. Its operating speed is 3500 rpm, and the impeller diameter is 10 in. These values lead to related values for the head and flow coefficients. What are the associated values of H and Q for the same coefficients if the speed of the pump is halved?

Solution Because the same pump diameter will be used in both operating conditions, Eqs. (18.2.7) and (18.2.8) tell us that the head would be decreased by a factor of 4 and the flow rate would be halved. This result is shown in Fig. 18.2.3 along with the transformation of the other operating points at speed N_A to comparable ones at N_B. The dotted curves indicate the locus of points that give the same head and flow coefficients as the original operating point.

Example 18.2.2

The pump impeller described in Example 7.6.1 has inlet and outlet radii of 20 and 170 mm, respectively, with a blade breadth of 15 mm. The (backward) outlet blade angle was

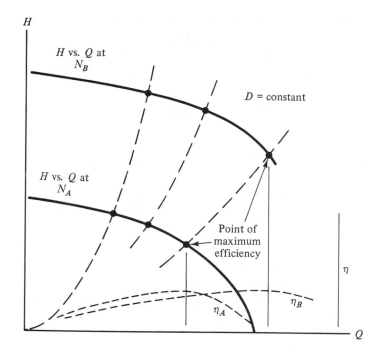

H

H vs. Q at N_B

D = constant

H vs. Q at N_A

Point of maximum efficiency

η

η_B

η_A

Q

Figure 18.2.3 Application of affinity laws for a change of pump speed.

45°. The pump delivered 0.01 m³/s at 1750 rpm. The power consumption and pressure rise across the impeller are 968 N-m/kg and 454 kPa. Determine the comparable performance figures for outlet blade angles of 90 and 135°.

Solution   For the 90° outlet angle:

$$Q = VA = 0.01 = 2\pi(0.17)(0.015)V_{r_2}$$

$$V_{r_2} = 0.624 \text{ m/s}$$

$$U = \Omega r_2 = 31.2 \qquad W = V_{r_2}V_2 \qquad V = U + W$$

$$|V|_2 = \sqrt{31.2^2 + 0.624^2} = 31.2 \text{ m/s}$$

$$T = \rho Q(rV_\theta) = 1000(0.01)(0.17)(31.2) = 53.04 \text{ N-m}$$

$$\dot{W} = T\Omega = 53.04(1750)\frac{(2\pi)}{60} = 9720 \text{ watts}$$

$$p_2 - p_1 = 1000\left(\frac{9720}{10} + \frac{5.24^2}{2} - \frac{31.2^2}{2}\right) = 499 \text{ kPa}$$

For the 135° outlet angle:

$$W = \frac{V_{r_2}}{\cos 45°} = 0.883$$

$$|V|_2 = \sqrt{(31.2 + 0.624)^2 + (0.624)^2} = 31.83 \text{ m/s}$$

$$T = \rho Q(rV_\theta) = 1000(0.01)(0.17)(31.2 + 0.624) \text{ N-m}$$

$$\dot{W} = 985 \text{ W/kg/s}$$

$$p_2 - p_1 = 1000\left(985 + \frac{5.24^2}{2} - \frac{31.83^2}{2}\right) = 492 \text{ kPa}$$

Results	Outlet Blade Angle		
	45°	0°	135°
Outlet V, m/s	30.6	31.2	31.8
Power input, kW	9.68	9.72	9.85
$p_2 - p_1$, kPa	454	499	492
Δp per unit power, kPa/kW	46.9	51.3	50

The high outlet kinetic energy of the forward-turning vanes is also inefficiently converted to pressure within the pump, so that the Δp's may be unrealistically high. (One does see such vanes on centrifugal fans where the high speed is used in cooling and not pressurization functions.) The radial vane is used where the strength of the material in the impeller dictates a minimization of the effects of the stresses due to centrifugal forces.

Example 18.2.3

The Kaplan turbine described in Example 7.6.3 operated at 300 rpm at a water flow of 12.5 m³/s. For the "off-design" condition of 12 m³/s, what should be the inlet blade angle? Use an impeller hub diameter of 600 mm and a blade tip diameter of 1500 mm. Assume that the inlet swirl velocity remains at 2.55/r. (This relationship—the reciprocal of r—gives each particle entering the turbine the same angular momentum.)

Solution $U_{tip} = 23.56$ m/s, as in Example 7.6.3.

$$V_{axial} = \frac{8.42(12)}{12.5} = 8.0832 \text{ m/s}$$

$$\text{At the tip, } V_\theta = \frac{2.55}{0.75} = 3.39 \text{ m/s}$$

Then $$\tan \beta_i = \frac{8.083}{23.56 - 3.39}$$

$$\beta_i = 21.8°$$

At the design conditions of Example 7.6.3, the inlet blade angle is 22.6° at the blade tip. Consequently, to avoid inlet blade "shock," the pitch of the blades should be adjusted 0.8°. "Postswirl" should result from this change of "pitch," with a corresponding effect on power. But this is the small price for avoiding inlet "shock."

Figure 18.2.3 is a common representation of pump performance and is called a *pump characteristic curve* for the speeds N_A, N_B, and so on. Figure 18.2.4 shows pump characteristic curves for a series of similar pumps. The numbers in the regions of Fig. 18.2.4 refer to the pump's outlet diameter, inlet diameter, and nominal impeller diameter. All are expressed in inches. Because each impeller's diameter can be reduced somewhat by machining it, the curves in Fig. 18.2.4 for a nominal impeller diameter contain an implicit group of characteristic curves that are associated with the range of diameters that are reasonable to fabricate. (The collection of curves is indi-

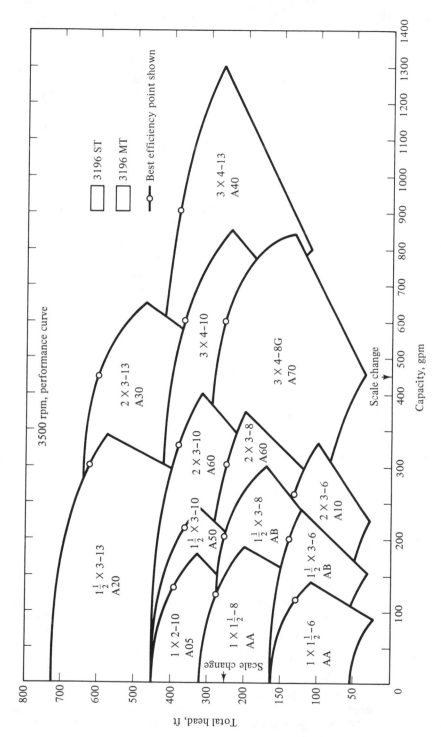

Figure 18.2.4 Range of pump characteristic curves for a particular series of pumps. Numbers in the different regions denote the discharge, inlet, and impeller diameter, respectively. *(From "Pump Selection for the Chemical Process Industries," R. Neerken, Chemical Engineering, Feb. 18, '74, Vol. 81, p. 104)*

cated by the band surrounding each specified value of discharge, inlet, and rotor diameter.) The ability of the manufacturer to adjust the diameter of an impeller permits a better matching of the requirements of the application and the operating characteristics of a pump.*

The relationship given by Eq. (18.2.9) is not the only useful one that can be formed from the previous two. In particular, we can eliminate the speed between Eqs. (18.2.7) and (18.2.8). Thus, if the fourth root of the first equation is divided by the square root of the second equation, a dimensionless term results, provided that the volume flow rate, head, and diameter are in the proper units. This ratio leads directly to the definition of the *specific diameter* D_s of a pump of turbine:

$$D_s = \frac{D(Hg)^{0.25}}{Q^{0.5}} = \frac{\Psi^{0.25}(2)}{\Phi^{0.5}} \tag{18.2.10}$$

In practice, the gravitational constant is omitted and Q is usually specified in gpm. The value of the diameter is sometimes specified in inches and sometimes in feet. Care must be taken to establish the units of the parameters, because they will affect the magnitude of D_s. (In countries using the metric system, the SI units might not be used when giving a pump parameter, which would make D_s not dimensionless.)

One can also eliminate the diameter between Eqs. (18.2.7) and (18.2.8). One form of such an effort is given by the expression for the *specific speed*, N_s:

$$N_s = \frac{NQ^{0.5}}{H^{0.75}} \tag{18.2.11}$$

While a hydrodynamic machine can have many values of N_s corresponding to the speed, flow rate, and head at which it operates, one often specifies this parameter under conditions that produce a maximum operating efficiency of the unit. This fact will be brought out in the next section.

18.3 HYDRAULIC LOSSES AND EFFICIENCY

The ratio of the useful pump head to the Euler head is a measure of the efficiency of the machine, using the theoretical result as the reference value. It has already been pointed out that this efficiency is the ratio of the ordinates in Fig. 18.2.2. We shall call this the theoretical efficiency of a pump, η_{Euler}, such that

$$\eta_{Euler} = \frac{\Psi_{actual}}{\Psi_{Euler}} \times 100 \tag{18.3.1}$$

If one were able to measure the actual rate of energy transferred to the rotor with a dynamometer or watt-hour meter, one could establish the useful measure for pumps

$$\eta_{overall} = \frac{[(V_2^2 - V_1^2)/2g + (p_2 - p_1)/\gamma]Q\gamma(100)}{\text{power input to pump}} \tag{18.3.2}$$

---

*If *only* the diameter of the impeller is reduced, the pump's head will closely approximate Eq. (18.2.7), even though implicitly the geometric scaling has been changed. However, the discharge Q will not follow Eq. (18.2.8). Under these circumstances, Q will vary almost directly with D as the diameter is reduced.

in which the term in square brackets is the actual head developed. Comparable expressions can be written for turbines.

The fact that Ψ_{actual} is not near the value of 1 at $\Phi = 0$, but more usually between 0.5 and 0.7 should make one suspect that in diagrams such as Fig. 18.2.2 parameters other than Ψ and Φ are of some importance. (Diagrams that represent Ψ_{actual} often refer to water as the pumping medium and are, therefore, a broad but special case.) A general explanation of the difference between the theoretical and actual heads or flow coefficients at low flow can begin by considering the effects of fluid viscosity and turbulence. These effects can manifest themselves in the frictional torque necessary to spin a disk, which is a simple model of a rotor, at constant speed. This means that we must expect *disk friction* due to the rotation of the impeller relative to the housing to be an important aspect in the operation of a pump. If the space surrrounding the disk is large, the required power is less than that when the region is small. The magnitude of the torque that is a consequence of such friction will depend significantly upon the presence of secondary flows in the clearance spaces between the impeller and the housing.

In the material that follows we shall see that a given geometry for the impeller has been based upon a definite rotational speed and the volume rate of flow that the machine is designed to supply. If either this speed or this volume flow rate is not possible in the preferred combination, then the geometry of the pump that has been based upon such combinations will result in increased losses at the *off-design points* of operation. Because zero flow is such an operating point, it is not surprising that the actual head coefficient is drastically different from the Euler coefficient.

Nevertheless, experience has shown that the flow coefficient is the most important parameter to consider when discussing Ψ_{actual} for a particular type of turbomachine. This is the reason why the affinity laws, Eqs. (18.2.7) and (18.2.8), are so useful for making initial estimates of performance. Some of the principal parameters that also influence the operation of turbomachines are the Reynolds number, the roughness of the flow passages, and the Mach number. The last of these variables is important in compressible flow, but we have limited ourselves to incompressible flows in this chapter. (Our results could still apply to the flow of air that would be induced by a fan if the air speed were below 300 ft/s, because, as we saw in Chap. 14, the air would then have the flow characteristics of an incompressible fluid.) At the outset, we assume that the flow surfaces of the machines that we shall study are smooth, but later we have commented upon this assumption. This leaves the Reynolds number as the only parameter that will be examined in detail. Figure 18.3.1 shows the effect of viscosity on the characteristic curves, and this should implicitly give the significance of the Reynolds number for pump performance. This figure suggests that Fig. 18.3.2 is a logical generalization of a $\Psi\Phi$ relationship to include the influence of the Reynolds number.

Fig. 18.3.2 makes it obvious that the ratio of the ordinates, η_{Euler} from Eq. (18.3.1), will depend on Φ and Re, or, equally well,

$$\eta_{\text{Euler}} = \text{function}(\Phi, \Psi_{\text{actual}}) \tag{18.3.3}$$

Figure 18.3.3 shows measured overall pump efficiencies as a function of H and Q. (This information could be used to infer the form of the Ψ_{actual} surface in Fig. 18.3.2, if bearing and seal friction could be properly estimated and subtracted.) The contours of constant η in Fig. 18.3.3 result from considering the viscous effects of some standard

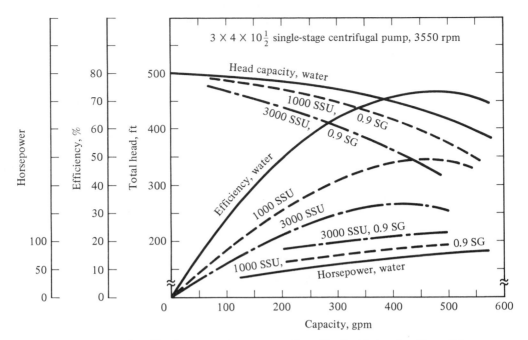

Figure 18.3.1 Centrifugal pump performance as affected by liquid viscosity. *(From "Pump Selection for the Chemical Process Industries," R. Neerken, Chemical Engineering, Feb. 18, '74, Vol. 81, p. 104)*

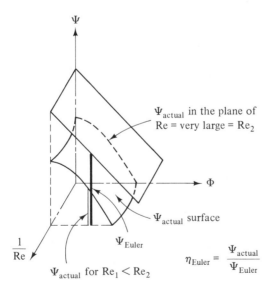

Ψ_{actual} in the plane of
Re = very large = Re_2

Ψ_{actual} surface

$\dfrac{1}{Re}$

Ψ_{Euler}

Ψ_{actual} for $Re_1 < Re_2$

$\eta_{Euler} = \dfrac{\Psi_{actual}}{\Psi_{Euler}}$

Figure 18.3.2 Effect of Reynolds number on head and flow coefficients.

liquid, usually water, as well as the additional viscous effects that result from altering the rotor's diameter. When this is done, most of the other dimensions within the pump housing are not changed, so that one is not comparing exactly geometrically similar pumps in Fig. 18.3.3. (This means that similarity conditions are not strictly maintained. The result is that the affinity laws give better predictions for speed changes

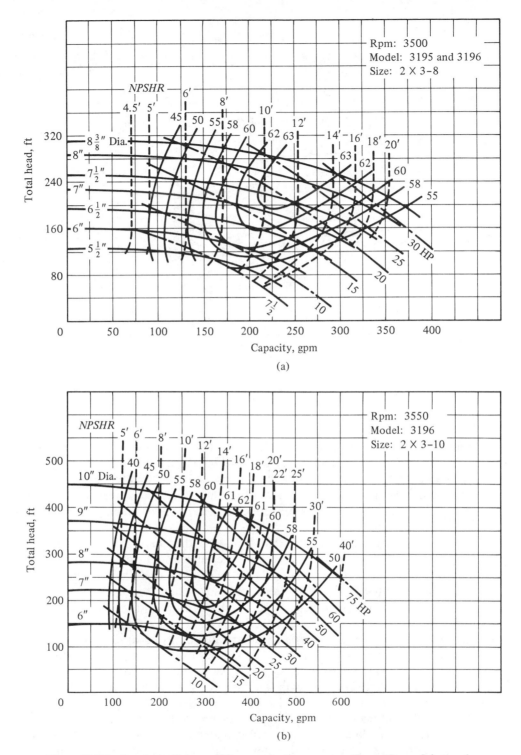

Figure 18.3.3 Constant-efficiency *HQ* curves for three pumps. *(From "Pump Selection for the Chemical Process Industries," R. Neerken, Chemical Engineering, Feb. 18, '74, Vol. 81, p. 104)* (continued on next page)

Sec. 18.3   Hydraulic Losses and Efficiency

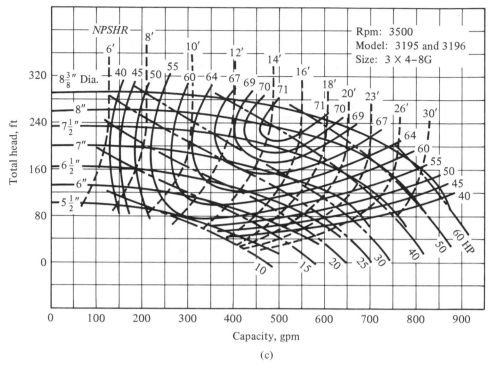

Figure 18.3.3 (Continued)

than for changes of diameter only.) The effect of increasing the clearance between the rotor and the housing can be an increase in the amount of secondary flow within the pump itself. This is inefficient, because such flows ultimately decrease the pressure and velocity head changes across the rotor. Obviously, the clearances can be too small, so that very high shear stresses are induced in the liquid between stationary and rapidly moving surfaces.

If viscosity were absent, what would be the shape of the constant-efficiency curves? Equation (18.3.3) would have to be interpreted to mean that η_{Euler} would be constant for a constant value of Φ. Thus the dashed curves in Fig. 18.2.3, which were constructed with Φ equal to different constants, would be curves of constant η, albeit different constants. The efficiency curves that are drawn at the bottom of Fig. 18.2.3 reflect this fact. It should be mentioned that we have been dealing with $\Phi_2 = \text{constant}$, as measured at r_2, which gives a constant value of Φ_1 because of the relationship between Q and r_1 that is implicit in Eq. (18.2.5) for no leakage. If we include an expression for the radial velocity in the Q of this parameter, it becomes evident that Φ_1 is the ratio of the radial velocity of the fluid and the peripheral velocity of the impeller at the inlet. This is shown in Fig. 18.2.1 for the case of no preswirl. The angle δ_1, which is the angle that the relative velocity W_1 makes with the radial direction, is proportional to cot Φ_1. If the blading design is well matched to the flow rate and angular velocity of the pump, the relative velocity moves smoothly (i.e., tangentially) onto the vane, which makes an angle β_1 with respect to the $-U_1$ vector according to Fig. 18.2.1. The result is $\delta_1 + \beta_1 = 90°$. Thus, if β_1 is fixed, there is only one value of Φ_1 for which there are very favorable inlet conditions. Any other value of Φ_1 should lead to a lower efficiency. Similar inefficiencies could result at the end of the blading, where a value of Φ_2 that was not for the optimum operating conditions could result in an inefficient conversion of the kinetic energy of the liquid at the rotor's outer edge into pressure. The mismatch of the direction of the relative velocity at the inlet and the blade angle is called *inlet shock*. The lowering of the efficiency of the pump, with the associated difference between the actual and Euler heads in Fig. 18.2.2, can be partially explained by such inlet and exit shocks. Of course, values of Q greater than that at the *best efficiency point* (BEP) will not only account for inlet shock, but will result in high frictional losses as the liquid moves relative to the rotor, as well as in the stationary passages. At Q's between *shut-off* and the BEP, a pump recirculates some of the liquid within the housing, and the noticeable temperature rise that occurs attests to the poor operating condition.

Characteristic curves similar to those in Fig. 18.3.3, with the associated efficiency values, are useful once a decision has been made about the general type of pump that should be considered further. But how should one make the preliminary choice? Graphs similar to those shown in Figs. 18.3.4 and 18.3.5 for turbines and pumps are helpful in this initial selection process. Figure 18.3.4 presents the maximum efficiency

Figure 18.3.4 Specific speed as a function of reservoir elevation for various turbines. *(Courtesy of J. M. Voith GmbH, Heidenheim, F.R.G.)*

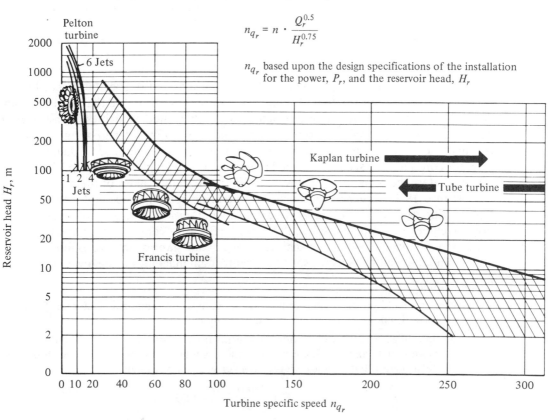

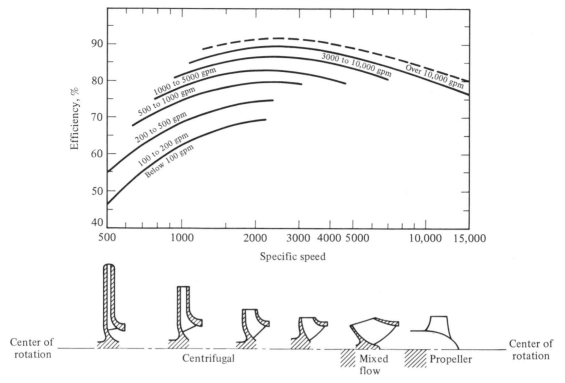

Figure 18.3.5 Variation of efficiency with specific speed for various sizes of pumps. *(Courtesy of the Dresser Pump Division, Dresser Ind., Inc.)*

that can be expected for a turbine installation as a function of specific speed. Figure 18.3.5 gives similar information for pumps. One notes in this figure that the efficiency depends upon the capacity of the machine. (The higher rated capacities are associated with larger machines that lead to reduced viscous losses.) In order to bring the size of the machine more clearly into the picture, one can introduce the specific diameter and produce contour plots such as the one in Fig. 18.3.6. (In all plots having specific speed and diameter, check the units that were used to obtain the values of the parameters!)

Figure 18.3.6 can be used once the operating speed, head, and flow rate are known. The value of N_s will be an indicator of whether a radial, mixed-flow, or axial-flow machine should be considered seriously. (Because centrifugal pumps are high-head and relatively low-flow machines, while axial pumps are the opposite, it is understandable that the specific speed for centrifugal machines is the lower.)

Example 18.3.1

A pump should produce 240 ft of head at 500 gpm while running at 3500 rpm. Estimate the best efficiency and the diameter of the impeller.

Solution   Figure 18.3.6 will provide the starting point for the study. The specific speed is determined first, and the figure will provide a guide to the best efficiency. From the specific diameter that is indicated, the impeller size can be found:

$$N_s = \frac{3500\sqrt{500}}{240^{3/4}} = 1283$$

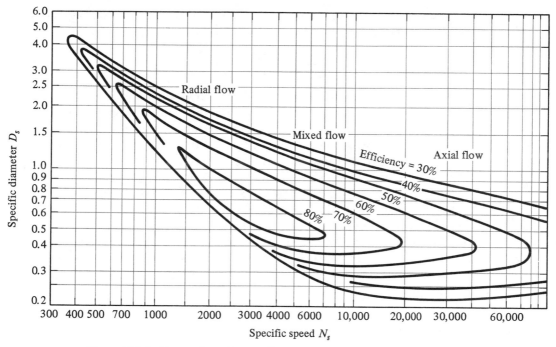

Figure 18.3.6 Expected maximum efficiency contours as a function of specific speed and specific diameter. $N_s = N\sqrt{Q}/H^{3/4}$, $D_s = DH^{1/4}/\sqrt{Q}$. N in rpm; Q, U.S. gpm; H, ft; D, in. *(From "Select the Right Pump," R. Neerken, Chemical Engineering, Apr. 3, '78, Vol. 85, p. 87)*

Figure 18.3.6 indicates $\eta = 75$ percent and $D_s = 1.5$. Hence,

$$D = \frac{1.5\sqrt{500}}{240^{1/4}} = 8.52 \text{ in.}$$

The 3×4–8G pump whose characteristics are given in Fig. 18.3.3 has a peak efficiency of about 72 percent for an 8-in. impeller.

The head and flow rate should be available from an analysis of the flow network in which the pump is to operate. (Chapter 10 and Appendix D treat this topic.) Figure 18.3.6 will give the maximum efficiency that can be expected, on the basis of extensive correlations of pump data, for each specific diameter that should be considered. With a general idea of the size and type of pump that will operate efficiently, the designer can consult with suppliers to obtain detailed characteristic curves, such as Fig. 18.3.3, that should be useful to specify the pump for a design proposal.

Until now, the relative roughness of the pump's surfaces has been assumed to be zero. While most pumps are made of castings, some of these can come extremely smooth from the mold. The large pumps will have the impressions of the sand in the cores that formed their passageways. When pumps with not perfectly smooth surfaces are scaled up to larger sizes, the roughness is generally unaffected. It is determined by the grain size of the foundry sand. The result is that large pumps have a lower relative roughness, and one would expect a higher efficiency as a consequence of this and the smaller influence of the walls, because they are more remote, on the entire fluid. One estimate of the effect of size on the efficiency of a turbomachine is given by Davis, Kottas, and Moody as

$$\frac{1 - \eta_A}{1 - \eta_B} = \left[\frac{(\text{Re})_B}{(\text{Re})_A} \right]^n \qquad (18.3.4)$$

in which the Reynolds numbers—Re_A and Re_B—are based upon the rotor tip speed, U_2. The exponent n is determined from experimental data and is usually in the range from 0.1 to 0.25. The value 0.2 is often used. Similar relationships have been developed by other investigators but give a lower efficiency than Eq. (18.3.4) for a prototype when the efficiency of a smaller model is used.

Figure 18.3.7 shows some model efficiencies for Kaplan turbines. Because on this type of turbine one can adjust the inlet blade angles for different operating conditions, it shows good efficiency over a wide range of these conditions.

Besides knowing information about bearing types (and maintenance), seals, and corrosion resistance, the designer must be sure that the vapor phase of the liquid will not be present within the pump. If this happens in very small amounts, the pump's efficiency and its HQ characteristic could be adversely affected. Cavitation damage as shown in Fig. 1.8.3 could also occur. The result would be poor performance and an

Figure 18.3.7 Improvements of performance for Kaplan turbines. *(Adaptation courtesy of the Swedish Power Association)*

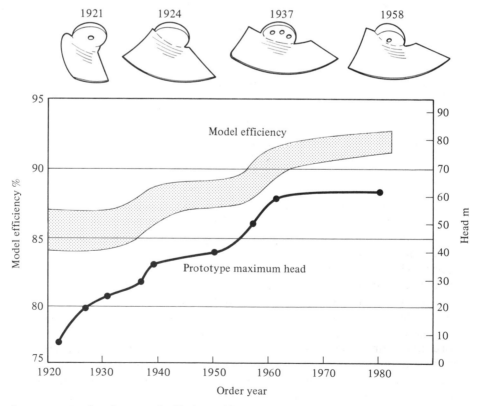

Improvements of performance for Kaplan turbines.

eventual structural failure of the unit, if excessive vibration did not become objectionable first. If large amounts of vapor occur, the pump would cease to transport the liquid along the pipe. This condition is referred to as a loss of *prime*. The presence of a liquid within most of the pump's passages is a necessary condition to achieve the hydrodynamic effects necessary to increase the liquid's total head at the discharge port of the pump.

18.4 NET POSITIVE SUCTION HEAD (NPSH)

The design of a pumping circuit will determine the sum of the pressure and velocity head at the inlet of the pump. This sum represents the height to which a column of the liquid that is in the pipe could rise if an impact tube were inserted into the flow. Such a view of the locally available head implicitly recognizes the possibility that the pressure could decrease significantly within the pump if the velocity increased greatly at some point within it. Should the pressure decrease to the vapor pressure of the liquid that is flowing, vapor bubbles would normally appear and the pump's performance would deteriorate. This possibility will be reduced if the vapor pressure head, p_v/γ, is sufficiently less than the sum of the velocity and pressure heads at the pump inlet. The difference between the sum of the velocity and pressure heads and the vapor pressure head is called the *available net positive suction head* and abbreviated NPSHA. This definition is expressed in the notation of Fig. 18.4.1 as

$$\text{NPSHA} = \frac{V_1^2}{2g} + \frac{p_1}{\gamma} - \frac{p_v}{\gamma} \tag{18.4.1}$$

In using this definition, one must specify the liquid's pressure in absolute units, which is the usual scale to measure the vapor pressure. This establishes the pressure datum of the NPSHA; the elevation datum is taken at the inlet of the pump. For the simple configuration given in Fig. 18.4.1, one calculates the NPSHA by first finding the total head at the inlet. This is

$$\frac{V_1^2}{2g} + \frac{p_1}{\gamma} = \frac{V_0^2}{2g} + \frac{p_0}{\gamma} + z_0 - \text{losses}_{0\text{-}1} \tag{18.4.2}$$

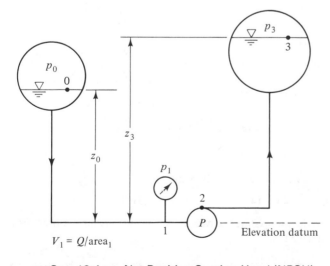

Figure 18.4.1 Typical pump-piping arrangement. Reading of p_1 corrected for distance from gauge to centerline of pump.

Remember that p_0 is absolute pressure and z_0 is positive if it specifies a location that is above the pump. Once the value of the sum on the left-hand side of Eq. (18.4.2) is known, the NPSHA is the difference between this value and p_v/γ. Because the vapor pressure of a liquid depends upon its temperature, one must specify this value too.

However, the NPSHA, even though it is a positive value, may not be sufficiently large to prevent the liquid's pressure in some of the passages from being less than its vapor pressure. [Recall that in Reynolds' experiment, shown in Fig. 1.1.1, the pressure ahead of the contraction must have been close to atmospheric pressure. This means that the NPSHA would have been about (14.7)(144)/62.4 ft if the velocity head were neglected. Nevertheless, the high velocity at the minimum section of the contraction reduced the pressure sufficiently that the visual and audible effects of cavitation were observed.] Each design of a hydrodynamic pump or turbine must be tested to establish the margin of head required at the inlet in excess of the vapor pressure. This value is the *required net positive suction head,* the NPSHR. The manufacturer of a pump can provide the customer with data of NPSHR versus Q, just as in Fig. 18.3.3. At the design point for the pump-piping combination, the condition NPSHR < NPSHA should prevail. (The inequality has been given to provide a margin of safety.)

Example 18.4.1

Consider an installation similar to that in Fig. 18.4.1 for which 100 ft of 4-in. galvanized pipe connects a reservoir to a pump that has an NPSHR of 15 ft at 500 gpm. The free surface of the water (160°F) in the reservoir is 14 ft below the pump, and the reservoir pressure is 1.3 in. Hg vacuum. What is the NPSHA for this installation at 500 gpm? Do you expect the pump to cavitate at the inlet?

Solution Using point 1 as the elevation datum, the energy equation can be written as

$$\frac{V_1^2}{2g} + \frac{p_1}{\gamma} = \frac{V_0^2}{2g} + \frac{p_0}{\gamma} + z_0 - \text{losses}$$

or

$$\frac{V_1^2}{2g} + \frac{p_1}{\gamma} = 0 + \frac{(29.92 - 1.3)(13.6)(62.4) - 14}{12\gamma}$$

with

$$\text{Losses} = f\frac{L}{D}\frac{V^2}{2g}$$

$$V = \frac{Q}{A} = \frac{500/448}{\frac{1}{9}\pi/4} = 12.8 \text{ ft/s}$$

since 1 cfs = 448 gpm. The NPSHA can now be found easily.

$$\left(\frac{V_1^2}{2g} + \frac{p_1}{\gamma}\right) - \frac{p_v}{\gamma} = \text{NPSHA} = (32.44 - 14) - f\frac{(100)(12.8)^2}{\frac{1}{3}}$$

or

$$\text{NPSHA} = \left[(32.44 - 14) - f\frac{100}{\frac{1}{3}}\frac{12.8^2}{2g}\right] - 10.95$$

$$f \approx 0.021$$

from the Moody diagram (Fig. 9.4.2) for Re = $4.3(10^5)$ and $\varepsilon/D = 0.0015$.

$$\text{NPSHA} = (32.44 - 14 - 16) - 10.95 = -8.51 \text{ ft}$$

Hence NPSHA < NPSHR, and the pump should cavitate. At the given reservoir pressure, the pump should be higher than 9.51 ft.

Two types of tests have been used to determine the NSPHR of a unit. The older test employs an inlet and exit configuration that is similar to that shown in Fig. 18.4.1. One begins with a pump that is operating satisfactorily for the existing elevations of the inlet and exit reservoirs and induces cavitation by a procedure that is now outlined.

While the pump is running normally, the elevation of the inlet reservoir is raised a small amount. This decreases the head that the pump must produce. If the centrifugal impeller has backward-turning vanes, the flow rate should slowly increase, with the result that the losses in the inlet and outlet pipes increase. These changes in the inlet reservoir elevation and the losses in the inlet pipe will counteract one another in the calculation of the NPSHA for the new condition. This new value of the NPSHA can be made the same as the original value by now adjusting the elevation of the second reservoir to increase or decrease the flow, and consequently changing the losses in the inlet pipe, appropriately. This procedure keeps the NPSHA constant while increasing the flow through the pump. [The value of the NPSHA would be found from the measured values of the terms on the left-hand side of Eq. (18.4.2) and not from estimates of the terms on the right-hand side of this equation.] The process can be continued until the head produced by the pump begins to decrease drastically for a small change in flow.

When the head with cavitation occurring is 97 percent of the head produced without cavitation at the Q in question, the inlet pressure and velocity are recorded. These values and the vapor pressure head specify the NPSHR for the machine at the value of Q where the pump's performance is diminished. Figure 18.4.2 shows how

Figure 18.4.2 Cavitation in a small centrifugal pump. *(Courtesy of Matley, J. and Chemical Engineering Staff, in Fluid Movers—Pumps, Compressors, Fans and Blowers, McGraw-Hill Publications Co., 1979)*

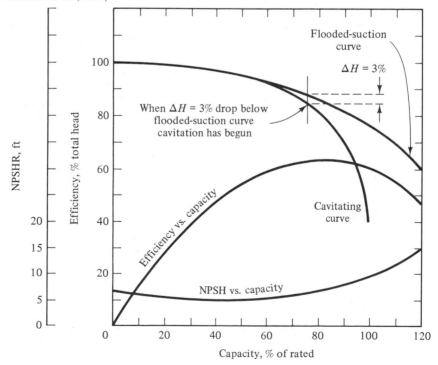

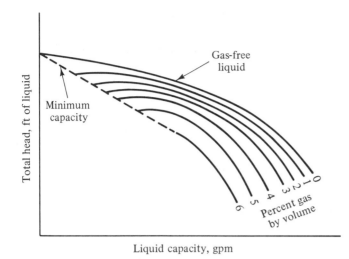

Figure 18.4.3 Effect of entrained gas on the performance of centrifugal pumps. *(From "Inert Gas in Liquid Mars Pump Performance," R. Penny, Chemical Engineering, July 3, 1978, V. 85, p. 63)*

the characteristic curve of a pump is affected by cavitation. The effect of gases in the liquid is shown in Fig. 18.4.3.

A more recent type of test to determine the NPSHR for a pump places the pump, in effect, at intermediate positions between the inlet and outlet reservoirs. These are kept at constant conditions. Hence, the head required by the pump is constant but the pump's inlet conditions are varied—its elevation, for example—so that the NPSHA changes. By adjusting the inlet conditions to the pump, at a fixed pump head and Q,

Figure 18.4.4 Cavitation characteristics of a high-performance centrifugal pump. *(After Pearsall and Scobic, "Superactivating Process Pumps for the Chemical and Petroleum Industries," 1973, reprinted by permission of the Council of the Institution of Mechanical Engineers)*

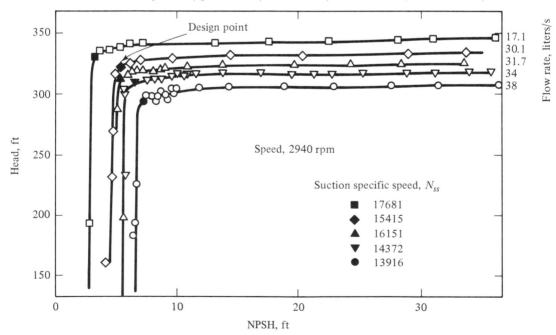

cavitation can be induced within the pump so that the flow cannot be maintained. When this occurs, the measured NPSHA of the tests is taken as the NPSHR of the pump for the particular flow rate. The results of such tests at different (constant) flows are shown in Fig. 18.4.4.

The NPSHR for a pump leads naturally to the definition of a *suction specific speed* that incorporates this parameter. It is

$$N_{ss} = \frac{NQ^{0.5}}{(\text{NPSHR})^{0.75}} \tag{18.4.3}$$

in which the second subscript s should remind one that suction conditions are the criteria. (One also encounters the symbol S for suction specific speed.) The correlation of N_{ss} with the operating specific speed given by Eq. (18.4.3) is helpful to the designer of a pumping system. Such an empirical relationship is given in Fig. 18.4.5.

Example 18.4.2

The NPSHR at the BEP for the 3×4–8G pump in Fig. 18.3.3 is 15 ft. What is the associated suction specific speed? Compare this result with Fig. 18.4.5.

Figure 18.4.5 Cavitation limits of pump operation with $H_{sv} = $ NPSHR and $S = N_{ss}$. *(Courtesy of Dover Publications, Inc.)*

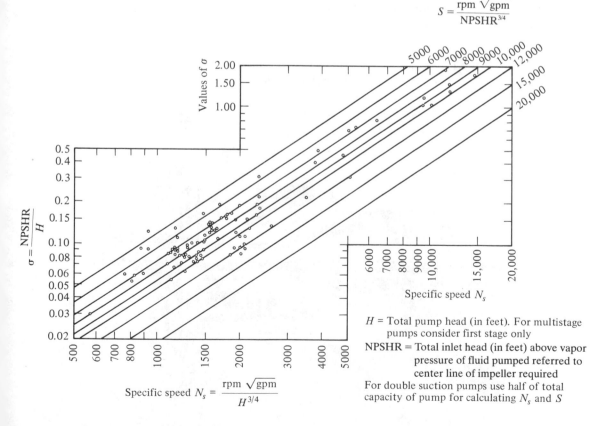

$$S = \frac{\text{rpm } \sqrt{\text{gpm}}}{\text{NPSHR}^{3/4}}$$

Values of σ

$\sigma = \dfrac{\text{NPSHR}}{H}$

Specific speed N_s

Specific speed $N_s = \dfrac{\text{rpm } \sqrt{\text{gpm}}}{H^{3/4}}$

H = Total pump head (in feet). For multistage pumps consider first stage only

NPSHR = Total inlet head (in feet) above vapor pressure of fluid pumped referred to center line of impeller required

For double suction pumps use half of total capacity of pump for calculating N_s and S

Solution

$$N_{ss} = \frac{3500(500)^{1/2}}{(15)^{3/4}} = 10,267$$

for comparison with Fig. 18.4.5,

$$N_s = \frac{3500(500)^{1/2}}{240^{3/4}} = 1283$$

$$\sigma = \frac{\text{NPSHR}}{H} = \frac{15}{240} = 0.0625$$

For these values, $S = N_{ss} = 10,000$; hence, Fig. 18.4.5 can be used to give reasonable estimates of NPSHR when they are not available from other sources.

18.5 PERFORMANCE OF PUMPS CONNECTED IN SERIES AND PARALLEL

The pumps that are installed at intervals along a petroleum pipeline operate in *series,* because the outlet of one is, in effect, directly connected to the inlet of the next. The same volume rate of flow passes through each pump, but the energy addition to the fluid is cumulative. It is usual to use a series arrangement to pump water in large quantities from reservoirs that are located below the main pumping unit. A vertical pump would be lowered into the well pit without submerging the electric motor. (Totally submersible units are available for special applications.) Its discharge pipe would be connected to a high-pressure pump that would normally cavitate because of the low value of NPSHA associated with the deep reservoir. In a sense, all pumps in series act as "boosters." An *inducer,* shown in Fig. 18.5.1, is a small pump that raises

Figure 18.5.1 A centrifugal pump with an inducer in the inlet to improve low-NPSHA performance. *(Courtesy of the Dresser Pump Division, Dresser Ind., Inc.)*

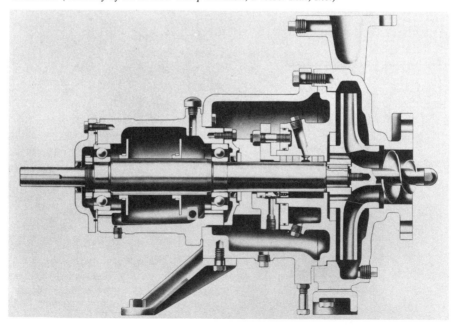

the pressure slightly so that the larger impeller, to which it attached, will not cavitate for the existing NPSHA.

Sometimes two nearly identical pump impellers are connected in series within one housing. Passages lead from the outlet of the first pump to the second impeller's inlet, which may be slightly different from that of the first because of the different inlet conditions. One speaks of a *two-stage* pump, and Fig. 18.5.2 shows such a unit that has been disassembled to reveal the internal passageways. The case can be opened as shown because it is *horizontally split*. This allows the impeller, shaft, seals, and bearings to be inspected relatively easily. Pumps with six or more stages are constructed for high-pressure applications.

The characteristic curve for a single-stage pump relates the energy added to the liquid and its flow rate. If two single-stage pumps are installed in series, the net energy addition at a specified value of Q will be the sum of the individual contributions, barring any appreciable losses due to the interconnection. A combined charcteristic curve for two such pumps is shown in Fig. 18.5.3. The head required for the pumps in a

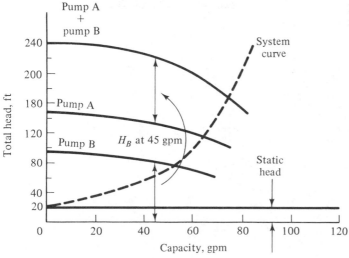

Figure 18.5.2 A disassembled double-inlet two-stage centrifugal pump used to supply drinking water at the rate of $\frac{3}{4}$ m³/s at a head of 305 m. *(Courtesy of J. M. Voith GmbH, Heidenheim, F.R.G.)*

Figure 18.5.3 Performance of two pumps in series for a prescribed system curve.

pipeline between two elevations separated by Δz is also shown in this figure. This is the *system curve*. The liquid must receive enough energy to increase its potential energy and to compensate for the frictional losses along the pipeline. These latter losses are approximately proportional to the square of the fluid speed, and for a pipe of constant size this would make the losses proportional to Q^2. This accounts for the general shape of the system curve in Fig. 18.5.3. The intersection of the system and characteristic curve determines the operating point of the pipeline-pump combination. Appendix D discusses in detail the way one can construct system curves for pipes in series and parallel.

The pump in Fig. 18.5.2 not only was multistaged, but also had a *double inlet*. Figure 18.5.4 shows a drawing of a single-stage double-inlet pump. The impeller for such a pipe is designed so that the liquid can enter on both sides of the plane of the impeller. This can be accomplished by having an annulus around the drive shaft through which the liquid can flow onto the impeller from its "front" and "back" sides. Figure 18.5.5 shows such an impeller. The passages within the casing of the pump shown in Fig. 18.5.2 provide the means for the liquid to reach the impeller from both its sides.

When multiple-stage dual-inlet pumps are used, one must be careful in calculating the specific speed and the suction specific speed. In the case of N_s, one uses the head increase of a single stage and the Q passing through the impeller of that stage.

Figure 18.5.4   Double-inlet single-stage centrifugal pump showing the shrouded impeller and other design details. *(Photograph courtesy of the Dresser Pump Division, Dresser Ind., Inc.)*

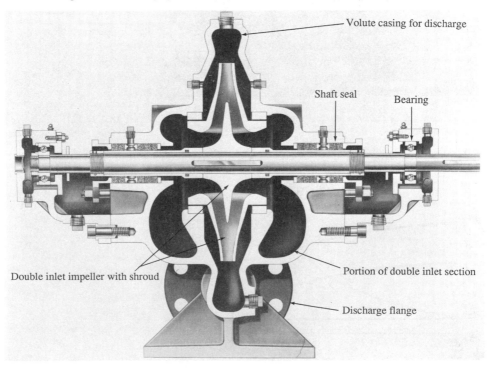

Figure 18.5.5 Impeller for a double-inlet centrifugal pump. *(Courtesy of COLENCO Ltd. (formerly Motor-Columbus, Consulting Engrs Inc., Baden/Switzerland)*

However, when considering N_{ss}, one must take the Q through the pump and divide this value by the number of inlets to specify properly the cavitation conditions at each inlet.

If two pumps with separate inlets are directly coupled to a common source and pump a liquid to a common point without appreciable losses—their discharge pipes could be large and short, for example—the pumps are said to be operating *in parallel.* Each of the pumps is operating against the same head difference. It is the difference in head from their common discharge location to the location given by their common inlet conditions. This is equivalent to saying that the head supplied by each pump is the same relative to the amount of Q passing through each pump. Because the pumps need not be identical, the flows through them may be different. We shall call the two flows Q_1 and Q_2 for convenience and observe that the values of Q_1 and Q_2 will come from the characteristic curve of the particular pump at the head that is specified. At this required head, the volume rate of flow through both pumps would be $Q_1 + Q_2$. Thus, to estimate the overall performance of the combination of the two pumps—i.e., to relate the total flow delivered and the head required—one would use the procedure shown in Fig. 18.5.6 for two pumps in parallel operation.

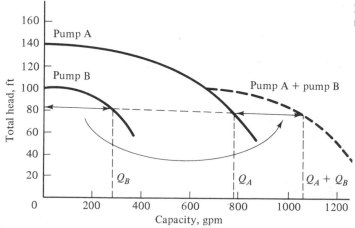

Figure 18.5.6 Performance of two pumps in parallel operation.

Example 18.5.1

In Fig. 18.3.3, a model 3196 pump (2×3–10) produces a head of 380 ft at 375 gpm at its BEP. What is the corresponding performance if two of these pumps are connected in series? What should be the result if the pumps are in parallel?

Solution　In series operation, the total head is the sum of the individual pump heads for the prescribed flow rate. For Q at BEP,

$$H_{\text{series}} = 380 + 380 = 760 \text{ ft}$$

$$Q_{\text{series}} = 375 \text{ gpm}$$

For parallel operation, the head for each pump is the same and the total flow rate is the sum of the individual flows. For BEP conditions,

$$H_{\text{parallel}} = 380 \text{ ft}$$

$$Q_{\text{parallel}} = 375 + 375 = 750 \text{ gpm}$$

18.6 MISCELLANEOUS CONSIDERATIONS FOR TURBOMACHINES

The casing around a centrifugal impeller includes a section in which the discharge is ducted to the outlet flange of the pump. The function of this portion of the pump is more than one of "corralling" the flow. It also acts as a diffuser to convert a portion of the kinetic energy of the fluid as it leaves the impeller into pressure head. This redistribution of the total energy should be accomplished with a minimum of unavailable energy (i.e., losses). The volute form of the discharge passage minimizes the flow separation losses by gradually expanding toward the discharge flange. Figure 18.6.1 shows the exterior of this part of a pump, called the *scroll case*, from which the internal volute section can be inferred. The narrowest section of this passage is immediately adjacent to its widest section, and separated from it by the segment of the housing called the *cutwater*. The shape of this piece and the clearance between it and the impeller can affect the noise produced by the pump.

Figure 18.6.1　A welded pump housing that shows its volute design and internal guide vanes. (Weight = 85 metric tons.) *(Courtesy of J. M. Voith GmbH, Heidenheim, F.R.G.)*

Not only will the shape of the volute section determine the losses that are incurred therein, but it will influence the pressure distribution around the periphery of the impeller. An unbalanced pressure distribution can result in excessive bearing loads and associated repairs. Vanes are often included in large pumps to guide the flow and reduce the losses. They also produce more uniform pressure distributions around the impeller. Such vanes can be seen in Fig. 18.6.1. The pressure distributions that are generated between the impeller's vanes produce an axial thrust on the drive shaft that must be supported by an appropriate thrust bearing.

In multistaged pumps, the discharge from one stage does not pass through a volute section that surrounds the impeller. Instead of this, a constant-area toruslike chamber is used. Vaned passages that act as diffusers connect it to the inlet of the next stage. The vanes in the expanding passages not only aid in the effective production of pressure head, but they also serve to align the flow for passage into the inlet of the subsequent stage of the pump.

The shape of the characteristic curve will be determined by the geometry of the impeller's vanes. If the head varies little over a broad range of flow, the design provides a pressure that is nearly independent of the flow rate. If such a pump supplied the oil for a hydraulic lift, its velocity would be almost independent of the load carried by it. Of course, good efficiency should also be possible for the range of conditions that are expected. A steep (negative) slope of the HQ curve would be useful if nearly constant flow were desired over a broad range of pump heads. A pump that supplies water to a filter, whose pressure drop changes with the amount of dirt absorbed, might have such a characteristic curve.

Axial-flow pumps usually have steep slopes for their characteristic curves, and this can result in high torques at low flow rates. This is especially important when starting such pumps, because of the requirements that it places on the motor. In order to allow the motor to come up to its operating speed without demanding excessively high starting torque capability, axial-flow pumps often include devices that permit them to begin turning without pumping the liquid. This could be accomplished by opening a bypass valve in a passage that connects the pump's discharge port with its inlet. Once the motor reaches its designed speed, the valve is closed and the pumping begins without overloading the motor. The inlet of an axial pump could also be vented to atmospheric pressure. When the pump reached its operating speed, the atmospheric vent would be closed and the liquid forced by auxiliary means—often pneumatically—from the reservoir to the pump's inlet. Pumping could then commence.

The need to provide liquid to the inlet of a pump at the outset to permit the unit to function is shared by axial and centrifugal pumps alike. Units that draw their liquid from unpressurized reservoirs that are below the pump's elevation will have their inlet pipes drained dry when the pump is shut off. Without liquid at the inlet (i.e., with *loss of prime*), the unit will not begin pumping once its motor is started. Some centrifugal pumps are designed to obviate this problem. One design has a housing near the inlet, like a U tube, that traps a small amount of liquid so that the impeller inlet is partially submerged when stopped. When the pump is started, this liquid moves over the impeller and decreases the pressure at the inlet of the U tube sufficiently to draw up a column of water from the reservoir. If this occurs before the impeller has completely expelled the liquid trapped at its inlet, the pump will ingest the water from the reservoir and commence normal operation.

The shape of the characteristic curve can also affect the stability of the pumping process. If, as can happen, the characteristic curve has a slow rise in head as Q increases at very low flow rates and this is followed at higher flows by the normal negative slope, unstable flow can occur. The reason that this situation is unsatisfactory can be easily understood. Let us assume that the pump is operating at a flow rate that gives the maximum head of the characteristic curve. If some disturbance in the discharge line decreases the head that the pump needs to supply, it will accommodate the disturbance by changing the volume flow rate according to the shape of the characteristic curve. But will that change be an increase or a decrease in flow as one moves away from the maximum of the HQ curve? Either situation is possible, and usually the flow jumps back and forth between the lower and the higher Q. This surging is unsatisfactory in a single pump but can be worse when a pump with such a rising head curve is combined in parallel with another pump. If surges are initiated by the unstable pump, they will affect the operation of even a stable pump in parallel with it. This unsteadiness will be communicated back to the surging pump by the parallel passages and will aggravate the already unsatisfactory performance.

The number of vanes on a centrifugal impeller varies from very many (i.e., *high solidity*) to very few (*low solidity*). Sewage is often pumped by impellers having only one or two vanes to minimize the possibility of clogging the pump. The efficiency of low-solidity impellers is lower than that of comparable impellers with more vanes; however, their values are still quite satisfactory. The impeller of the centrifugal compressor shown in Figure 18.6.2 is *shrouded*. The air that enters the vane passages is isolated from the stationary compressor housing while it moves toward the outlet. This improves the guiding of the flow and the efficiency, even though a small price must be paid for the additional disk friction. Shrouded impellers are also used in pumps when the chances of clogging the confined passageways are remote in the application.

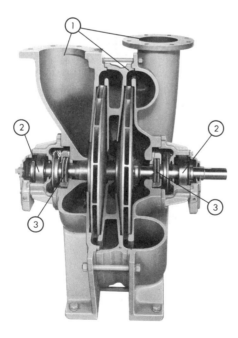

Figure 18.6.2 Centrifugal air compressor with shrouded impeller. *(Photograph courtesy of Hoffman Air and Filtration Systems)*

Food stuffs such as sardines and potatoes have been pumped satisfactorily, along with large quantities of water, with low-solidity open impellers that allow the solids to move freely through them. (The size of the *eye* of the impeller, r_1 in Fig. 18.2.1, must then be adequate for such applications.) The impellers may be rubber-coated to protect the produce. Pumps for some consumable liquids may have impellers of stainless steel so that they do not corrode and spoil the pumped medium. Bronze impellers are used in other, less critical applications for the same reason. The materials in *slurry pumps* and pumps used for dredging must be resistant to abrasion. If the solids in the slurry are long fibers, one must be concerned that they do not entangle themselves on the leading edges of vanes and struts. This could clog the pump. Sometimes such fibrous materials are "spun" together by a pump to form long filaments that may entangle themselves in the flow passages or produce an inferior product. Stainless steel is able to withstand the erosion process that accompanies cavitation better than many other metals. For this reason it is used in pumps and turbines that must operate occasionally under cavitating conditions.

The intake manifold, or joint intake reservoir, of parallel operating pumps should be carefully designed to prevent one pump from starving another. This can occur if the pump inlets are too close together, so that the eddies produced by the flow of one pump influence the flow at the inlet of another one. Baffles may be necessary to isolate the inlets, and there are recommended designs that specify the spacing of inlets in common reservoirs. If the liquid level in such reservoirs is not sufficiently high above the submerged inlet duct of a pump, a (bathtub) vortex can be generated that extends from the free surface into the pump's inlet. This will bring air into the pump, and its performance can be appreciably affected. Parallel inlets that are too close to one another can magnify this problem.

Figure 18.6.3 shows a small gas turbine with a single-stage axial compressor that is in series with a centrifugal compressor. Radial vanes are used in the latter.

Figure 18.6.3 Gas turbine with 2-stage compressor (axial and centrifugal). *(Courtesy of Williams International)*

18.7 POSITIVE-DISPLACEMENT PUMPS AND MOTORS

When a piston moves in a cylinder and forces a liquid ahead of it and out of the cylinder's open end, one can speak of a *positive-displacement* device. Calling it a positive-displacement pump would require the addition of a valve to close off the discharge area when the piston was adjacent to it and had come to rest. A second valve would be necessary to admit new liquid to the cylinder after the piston had reached the discharge end of it. This valve would be connected to a reservoir of liquid and would remain open until the piston had been moved to the other end of the cylinder. Then this *inlet valve* would have to be closed and the *discharge valve* would be opened to allow the liquid to leave the cylinder as the piston was made to approach it. This sequence of operations contains a feature that is common to all *positive-displacement pumps*: an object moves in a confined passage and forces the liquid ahead of it toward a location called the *discharge port*. This object could be a piston, as in the case just discussed, or it could be a gear tooth that conveys a liquid in the clearance between it and a housing.

A gear pump is shown in Fig. 18.7.1. Gear pumps are constant-discharge pumps for a specific rotational speed. They are available for pressures as high as 2000 psi, but in most cases gear pumps operate below 1500 psi. In the pump shown in the figure, oil is trapped in the pockets (*A,* in Fig. 18.7.1c) between the case and the teeth of the revolving gears and is carried around the periphery of both gears from the suction side of the pump to the discharge side. Return of oil to the suction side is prevented by the meshing of the teeth. Gear pumps such as that shown use spur, helical, or herringbone gears. Special gears, however, are employed in certain makes of pumps, especially internal-gear types.

Figure 18.7.1 Gear pump: (a) external housing, (b) helical glass used in such a pump, and (c) cross-sectional view of gears illustrating pumping action. *(Reprinted by permission. Copyright 1959, 1967, Mobil Oil Corporation.)*

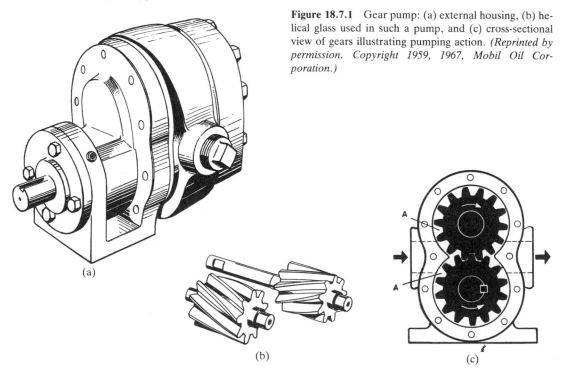

(a)

(b)

(c)

The pressure of the liquid in the cylinder that was used previously as an example is not determined by the motion of a piston pump that operates at its rated speed. The pressure is determined by the resistance to flow that the fluid encounters beyond the piston. The discharge port acts like an orifice that requires a pressure drop that will depend upon the velocity of the liquid through it. This velocity will depend upon the piston speed and the port area. The fluid viscosity will also be a factor in the loss of available energy across the discharge port. The passages and restrictions that occur in the hydraulic circuit downstream of the discharge port are the primary ingredients to be considered in estimating the pressure that is induced within the positive-displacement pump. If such a pump is running at its normal operating speed and a restriction in the flow conduit is further constricted, the pressure in the pump will increase accordingly.

If the prime mover of the pump does not change speed with the associated increase in load, the discharge of the pump will remain almost unaffected by this increase in pressure. This is a major difference between the operation of a positive-displacement pump and a hydrodynamic pump, for which a characteristic curve relates the flow and the head produced by the pump. The slight deterioration in the delivery rate of the liquid from a positive-displacement pump as the pressure increases is called *slip*. This is due to the increased leakage past seals within the device as the pressure rises within it.

In order to regulate the pressure within the flow circuit, a *pressure relief valve* must be installed. (A pressure-regulating device may be necessary in any hydraulic circuit, including, of course, one supplied by a turbopump, in order to protect certain components.) Such a valve consists of a spring-loaded plunger that is exposed to the fluid's pressure. A rising fluid pressure increases the spring's reaction on the opposite side of the plunger. This can occur because the spring changes length and the plunger moves accordingly. When the movement is sufficient, the plunger passes by a *relief port* that is connected to the reservoir for the pump itself. (The reservoir is usually at atmospheric pressure.)

Positive-displacement pumps have some uses that cannot be met satisfactorily by turbopumps. When the centrifugal pump was described, it was characterized as a relatively high-head machine, if multistaged. However, higher heads can be achieved by positive-displacement pumps. This feature makes them useful in many high-pressure applications, including diesel fuel injectors. (Pressures of 2500 psig are not uncommon in such pumps.) Centrifugal pumps are also relatively low-flow turbomachines; however, they usually have higher flow rates than positive-displacement pumps of comparable size. Thus, when volume and not pressure is the primary factor, the turbomachine may be preferred.

Because the discharge of positive-displacement pumps is relatively insensitive to the discharge pressure, they are often used as metering pumps in the chemical industry to supply carefully controlled quantities of materials to a chemical process. (In fact, the diesel fuel injector is primarily a fuel-metering device.)

Positive-displacement pumps do not depend upon high operating speeds to deliver their liquid. This is a distinct advantage in some applications. (While the volume rate of flow will depend upon the pump's speed, it does not depend upon a hydrodynamic effect associated with speed, and associated accelerations, in order to function.) This is a factor in the pumping of high-viscosity liquids as well as slurries that contain a considerable volume of suspended solid material. The efficiency of a centrifugal pump can be reduced 50 percent if the volume of solids in the liquid is 5 percent; but the positive-displacement pump would not suffer a comparable decrease. Moreover, the high operating speed of the turbomachine makes abrasion from the solids in the slurry an important consideration. The relatively slow speed of large piston pumps permits coal slurries to be conveyed over long distances in many parts of the world. (Such pumps have hardened components to make the process feasible as well as economical.)

The application of positive-displacement pumps is broad but not universal or obvious. Sometimes one could select one for a particular task or equally well use a low–specific speed centrifugal pump. First-cost and maintenance considerations could be the deciding factors determining which unit is selected. Often, the choice turns on the special features of positive-displacement pumps. They are used to transport blood, which could be damaged in a centrifugal pump. Other extremely sensitive liquids are conveyed in *peristaltic* pumps, which keep the liquid within a flexible tube and never allow it to be contaminated by the chemicals in shaft seals or their lubricants. (See Fig. 18.7.2.) The characteristic of being self-priming is an important, and often critical, advantage of positive-displacement pumps. This is the case for bilge and lubricating pumps.

Nevertheless, positive-displacement pumps have some of the same problems as turbopumps, and some unique problems besides. Cavitation must be considered at their inlets. Poorly designed fuel systems can result in vaporized fuel at the inlet under conditions of elevated temperature. The result is *vapor lock,* a condition whereby the engine receives the vapor but not enough liquid to operate satisfactorily. The many teeth of a gear pump produce a flow rate with a small, high-frequency variation in time. In the case of a piston device, this frequency is lower and the amplitude higher. If the device is an air compressor, it is not surprising that its discharge goes to a large tank that serves to dampen the surges in the output line.

Figure 18.7.2 Disassembled peristaltic pump showing the rollers that deform the flexible tubing and force the fluid through the pump. *(Courtesy of Masterflex®, Palmer Instrument Co.)*

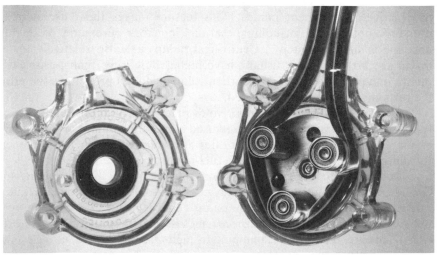

This section concludes with a brief mention of positive-displacement motors, a topic that was included in the section's heading. They are, in effect, the counterparts of positive-displacement pumps. If a pressurized liquid is admitted to the lower end of a vertical hydraulic cylinder, the piston will tend to move upward and raise a platform, for example. The piston-cylinder combination acts as a motor to perform the work of raising a weight. Many forms of hydraulic motors exist and find many uses in industry today. For many years, they were the agents for raising and lowering the wheels on aircraft, as well as moving rudders and wing flaps. The machine-tool industry currently employs many positive-displacement motors.

18.8 SUMMARY

Devices that transfer energy to or from a liquid are called *pumps* and *turbines*. *Compressors,* or *fans,* and *turbines* are the terms used for these machines when gases are involved. The manner in which the energy is transferred determines the general classification of the machine. In a *turbomachine,* the angular momentum of the fluid is changed. The associated torque that accompanies this action is applied to a rotating member that is the agent for the transfer of energy. *Positive-displacement machines* induce normal stresses within the fluid in the unit as some surface is displaced that alters the space occupied by the fluid. The normal stresses that are applied to the moving surfaces result in a transfer of energy to or from the fluid.

The process of energy transfer is not completely efficient. Viscous effects directly reduce the amount of energy transferred. Moreover, viscosity is usually involved in the generation of unwanted secondary flows whose energy is largely useless. Separation and misdirected flow within the machine lead to eddies whose eventual dissipation is counterproductive.

Diagrams with plots of specific speed versus specific diameter are useful in the initial planning phase for a new pump or turbine installation.

The flow through the machine is accompanied by different total heads at different points within it. Equally well, the fluid pressure within the machine changes; it can reach very low values, and in liquids it can become equal to the vapor pressure. When this occurs, the liquid locally changes into its vapor, and the performance of the device is adversely affected. The delivered head is reduced in pumps, but in all machines there will be erosion of the internal surfaces due to the physical processes associated with *cavitation* damage. The required NPSH of a particular machine determines the constraints on its installation for a definite application.

Hydraulic machinery is connected to other components through pipes and ducts to convey the fluid. Such a system of pipes will affect the net energy transferred to the fluid. In order to determine the operating condition of a pump and the associated piping system, one must match the pump's *characteristic curve* and the system's *demand curve*. This can be done graphically, or numerically. Appendices D and E give the details of this.

There are a variety of applications for which a positive-displacement machine is superior to a turbomachine. The very high pressures and other advantages that one can realize with it can be accompanied by a combination of relative simplicity, good efficiency, reliability, and reasonable cost. The internal-combustion engine for automobiles is a case in point.

18.2.1(a) For the same flow coefficients (similarity), what are H and Q if the speed is 1150 rpm? How does the pump power compare between 3500 and 1150 rpm?

(b) The pump in the example was developed from a geometrically similar unit that was one-fourth the size. Water was used in the model and the same Reynolds achieved as in the prototype. What was the speed of the model? What values of H, Q, and power should have been measured for the model?

(c) If the 10-in. impeller is reduced in diameter to 9 in. and run in the same housing, geometric similarity with the original pump configuration will not be maintained. Experience shows, however, that $H/N^2D^2 = $ constant will still occur, but that $Q/ND = $ constant will be the rule. Find H and Q for the 9-in. impeller at 3500 rpm.

18.2.2(a) What should be the inlet blade angle for the 90° and 135° outlet blade angles?

(b) What would be the performance figures for the pump at an outlet blade angle of 90°, assuming the correct inlet blade angle, if Q were doubled?

(c) Determine the ratio of the kinetic energy associated with a backward-facing vane with a 25° blade angle and the vane of Exercise 18.2.2b. If half of each of these kinetic energies is converted into pressure in the diffuser, which pump impeller will produce the higher pressure at the outlet flange for the same Q?

18.2.3(a) The rotation of the blade by 0.8° $(= 22.6 - 21.8)$ yields some postswirl. How much is it? What is the power produced by the turbine at a flow of 12 m³/s with the new blade position?

(b) What would be the power produced at 300 rpm if the flow were to increase to 13 m³/s? Again, assume that the inlet swirl velocity remains at $2.55/r$.

(c) What would be the means to keep a constant inlet swirl velocity for different flow rates?

18.3.1(a) What would be the best efficiency of a pump that produced the same head and flow rate but ran at half the speed? What would be its diameter?

(b) What would be the expected efficiency and size of a pump that produces 240 ft of head at 1000 gpm when running at 3500 rpm?

(c) What would be your suggestion if, given the speed, head, and flow rate, the resulting specific speed and diameter were at the very far left-hand side of Fig. 18.3.6?

18.4.1(a) If the pump were connected to the reservoir by 25 ft of pipe, would the pump cavitate?

(b) Would the pump cavitate with water at 60°F?

(c) If the 100 ft between the pump and the reservoir must be maintained, what diameter of pipe should be specified to avoid the possibility of pump cavitation?

18.4.2(a) What is the suction specific speed for the 2×3–8 and 2×3–10 pumps in Fig. 18.3.3 at BEP?

(b) Compare the results of Fig. 18.4.5 with the NPSHR at 260 ft, 170 gpm for 45 percent efficiency and at 260 ft, 275 gpm for 60 percent efficiency with an 8-in. impeller in the 3×4–8G pump in Fig. 18.3.3.

(c) What NPSHR does Fig. 18.4.5 give for the 3×4–8G pump in Fig. 18.3.3 when it is operating at 750 gpm, 140 ft?

18.5.1(a) Draw the (effective) characteristic curve for two model 3196 $(2 \times 3$–10) pumps connected in series. Repeat for parallel operation.

(b) What length of 6-in. wrought iron pipe would permit a single model 3196 pump to operate at its BEP? What length would permit two pumps in series or two pumps in parallel to run at their BEP? (Assume a horizontal pipe.)

(c) A model 3196 $(2 \times 3$–10) pump and a model 3195 & 3196 $(3 \times 4$–86) pump could be connected in series and in parallel. Draw the (effective) pump characteristic curve for each operation. (Assume a 10-in. impeller for one pump and an 8-in. impeller for the other.)

PROBLEMS

18.1 A centrifugal pump of the type shown in Figure (P.18.1) has $r_1 = 20$ mm, $r_2 = 170$ mm, and $\beta_1 = 55°$. There is no preswirl, and the pump turns at 1750 rpm. The width of the impeller is nearly constant at 15 mm. Determine the volume rate of flow. What is

V_2 for $\beta_2 = 25°$? What is the minimum power supplied to the rotor? What pressure rise can be expected across the impeller?

18.2 Water enters a centrifugal pump with $V_\theta = 8.7$ ft/s and $V_r = 30$ ft/s. At that point, the radius is 4 in. and the value of $r\Omega$ is 42 ft/s. Refer to Fig. P.18.1.

(a) What angle should the blade make with the tangent to the circle inscribing the leading edges of the blades? Assume shockless entry onto the blades.

(b) The blades are of constant depth (perpendicular to the page), and at the outer radius of 8 in., the blade makes an angle of 135° as shown. What is the magnitude of the absolute velocity leaving the blade?

(c) The mass rate of flow through the pump is ρQ. What torque is being applied to the fluid within the pump?

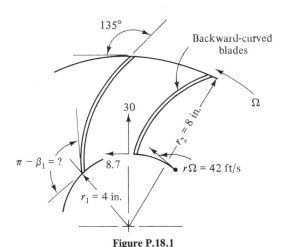

Figure P.18.1

18.3 An axial-flow turbine has a diameter of 7800 mm. It operates with a head of 17.4m—the elevation difference between the upstream and downstream reservoirs—with a flow of 270 m³/s. The turbine produces 41,350 kW at 626 rpm. What is the overall efficiency of the turbine system? Estimate the inlet swirl velocity. What should the trailing-edge blade angle be at the tip if $r_{tip}/r_{root} = 3.9$?

18.4 The model 3196 pump $(2 \times 3-10)$ whose characteristics are shown in Fig. 18.3.3 is available to meet an emergency requirement for 600 gpm at 845 ft of head. Can this flow be produced by this pump by some means?

If it can be done, what should be the power rating of the motor?

18.5 If a pump of model 3196 $(2 \times 3-10)$ were operating at 1775 rpm with a 10-in. impeller, what would be the H, Q, and η corresponding to $H = 375$ ft, $Q = 370$ gpm at 3550 rpm?

18.6 It is desired to pump 400 gpm of water from an open reservoir to an open tank that is 200 ft higher in elevation. A 2-in. pipe should be used. Can a model 3196 $(2 \times 3-10)$ pump—see Fig. 18.3.3—be used with a suitable motor?

At what elevation must the lower reservoir be located with respect to the pump to ensure satisfactory operation with the 3-in.-diameter suction line.

18.7 What is the NPSHA for the pump under the conditions shown in Fig. P.18.7?

18.8 Figure P.18.8 shows the result of a test on a hydrodynamic pump. What is the specific speed of this machine? On the basis of this

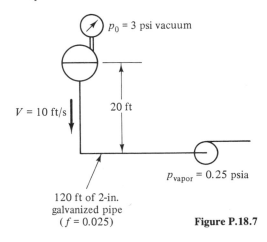

Figure P.18.7

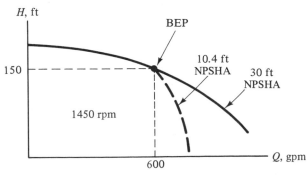

Figure P.18.8

result, state whether the machine is a centrifugal, mixed-flow, or axial-flow machine. What is the suction specific speed? Is this value an expected one for a machine of this type? What should be the head and delivery of the pump at the best efficiency point if the pump is run at 2900 rpm?

18.9 Determine the NPSHA at the pump's inlet for water at 40°F and 200°F for the conditions given in Fig. P.18.9. Three cases of elevations should be used and the results summarized: $H_1 = 10$ ft, $H_2 = 60$ ft; $H_1 = 0$, $H_2 = 50$ ft; and $H_1 = -10$ ft, $H_2 = 40$ ft. What is the required pump head in each case?

18.10 A pump (3500 rpm) is directly connected to a reservoir as shown in Fig. P.18.10. It can produce 230 ft of head while delivering 1.11 cfs of water in a 3-in. pipe. (Vapor pressure = 2.3 psia, and kinematic vis-

cosity = 0.00001 ft²/s.) At point 4, a second pump will be installed that has a NPSHR of 16 ft at 1.11 cfs.

(a) For the pipe described in the figure, what is the maximum length of horizontal pipe between points 2 and 4?

(b) What should be the efficiency of the generic pump at location 2?

18.11 The pump in Prob. 18.10 is connected to a reservoir (at atmospheric pressure) that has a free surface 10 ft below it. At 500 gpm, the losses between the reservoir and the pump and the kinetic energy head in the fluid total to 5 ft. (The vapor pressure is 0.8γ, with $\gamma = 62.4$ lb/ft³.) Will the pump operate at the design point? If the answer to this question is no, what should be the level in the reservoir? If the answer to this question is yes, how far can the level drop before difficulties are encountered?

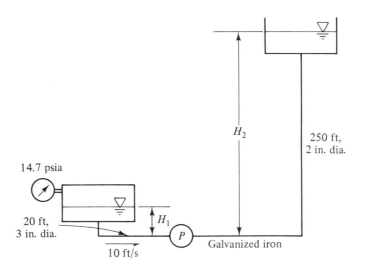

Figure P.18.9

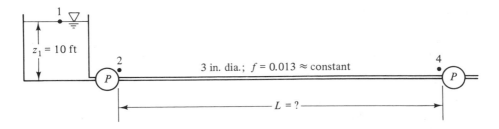

Figure P.18.10

18.12 What impeller diameter should you expect for the pump at location 2 in Prob. 18.10? What power is being supplied to the pump itself? What torque is being applied to the impeller? Estimate the radial velocity of the fluid leaving the impeller if the vanes are 0.5 in. deep. Estimate the θ component of the velocity leaving the impeller. Estimate the blade angle at the periphery of the impeller.

18.13 A wind tunnel in the form of a closed loop is to be designed in which automobiles can be placed to ascertain their aerodynamic characteristics. The test section is to be 7.5 m wide and 5 m high. A maximum wind speed of 50 m/s should be achieved in the test section. This test section is downstream from a contraction section—a nozzle with a 4 to 1 area ratio. The length of the loop is proposed to be 150 m. The largest section would be 13.8 m in diameter just upstream of the contraction section, and the portion just downstream of the test section with its diameter of 8 m would be the smallest part. It is estimated that an electric motor rated at 2600 kW driving an axial-flow fan at constant speed (175 rpm) through a power train that is 85 percent efficient will be sufficient to achieve the desired maximum operating speed. This fan would have 10 blades and be 9 m in diameter at the tip. The hub would have a diameter of 4.5 m.

Determine the entrance and exit blade angles necessary at maximum speed in the test section. (Assume that all of the power is being absorbed uniformly by all the particles of air under this condition. The air should enter the fan axially.) Prescribe both root and tip angles.

Determine the shape of the guide vanes that must be installed downstream of the fan to achieve axial flow without swirl further downstream.

What must be the blade angles when 25 m/s is desired in the test section? The blades on the fan can be rotated to change its performance. Through what angle must they be rotated to achieve optimum *entrance* conditions at maximum and at half speed? For the "optimum" angle, what will be the angle of the flow relative to the guide vanes at half speed, if they are fixed and have been

set in place at the position prescribed by full-speed operation? (Assume that at half speed the absorbed power is one-fourth that at full speed.) Remember that the geometric relationship between the blade's leading and trailing edges was established by the conditions of maximum flow.

18.14 Assume that the model 3196 (2×3–10) pump in Fig. 18.3.3 operates without preswirl and that $V_{r_1} = 16.8$ ft/s at the radius $r_1 = 1.5$ in. At what angle with respect to the radius is the relative velocity oriented? What is the ideal entrance angle β_1? If the effective blade depth is 0.75 in., what is the volume rate of flow? What should be the outlet angle for a 10-in. impeller? What head is produced, according to Fig. 18.3.3? What is the Euler head? Why do these two values for head not agree?

18.15 The model 3196 (2×3–10) pump in Fig. 18.3.3 has its inlet and outlet blade angles, β_1 and β_2, set for the flow at BEP with no preswirl (about 370 gpm at 385 ft). Assume that the operation point goes to 300 gpm. Compare the directions of the relative velocity at the inlet for these two cases. Does the difference between these angles serve as a cause for the decreased efficiency at the lower flow? (Assume a radial inlet velocity of 16.8 ft/s at BEP.)

18.16 If one did not have Fig. 18.3.3, with its constant-efficiency curves, one could estimate the efficiency from Fig. 18.3.6. What efficiency should the three pumps described in Fig. 18.3.3 have at BEP according to Fig. 18.3.6? Do these values agree with those given by the curves in Fig. 18.3.3 at BEP?

18.17 Estimate the efficiency at the BEP point for the 3×4–13, 1.5×3–13, and 1×1.5–6 pumps in Fig. 18.2.4. (The discharge and inlet pipe diameters, along with the impeller diameters, are given in inches.)

18.18 Compare the efficiency predicted by Fig. 18.3.6 for the 3×4–13 pump in Fig. 18.2.4 with the efficiency given by Fig. 18.3.5.

18.19 A vertical mixed-flow pump is used to supply water for cooling purposes in an industrial complex. (The impeller is 36.8 in. in diameter and rotates at 490 rpm.) It delivers 60,000 gpm at a head of 69 ft while absorbing 889 kW. How does its operating efficiency agree with the estimates of Figs. 18.3.5 and 18.3.6?

18.20 A propeller pump in a steam power plant delivers 5.0 m³/s at a head of 10.1 m while consuming 700 kW. The impeller has a diameter of 1100 mm and turns at 490 rpm. Do Figs. 18.3.5 and 18.3.6 reasonably estimate its efficiency?

18.21 Do the data in Fig. 18.4.4 reasonably agree with those presented in Fig. 18.4.5?

18.22 Use Fig. 18.3.1 to estimate the BEP for the model 3196 (2 × 3–10) pump in Fig. 18.3.3 when pumping 3000 SSU oil with a specific gravity of 0.9. What would be the corresponding H and Q?

18.23 Figure 18.3.7 indicates that a model efficiency of 92 percent can be obtained for a Kaplan turbine. What should be the efficiency for a prototype that is 5 times the size of the model? Assume for the present that the model and the prototype will operate at the model speed that and water is the liquid common to both. Should cavitation requirements dictate the speed of the model tests? How would such considerations affect the estimation of the prototype efficiency?

18.24 Compare the (effective) characteristic curve of a model 3196 (2 × 3–8) pump, with the standard 8-in. impeller, operating in series or in parallel with a 2 × 3–10 pump with a 10-in. impeller. If 300 gpm were to be pumped through both configurations, what would be the ratio of the heads and power consumption?

18.25 If two model 3196 (2 × 3–8) pumps with 8-in. impellers are connected in parallel, how would the characteristic curve for the combination compare with that of the 3 × 4–8G pump? How would the heads and power consumptions compare at 750 gpm?

18.26 For a pump, the speed, flow, and head intended for a particular application make ideal variables for specifying the specific speed. For a turbine, it is common to calculate the specific speed as $nP^{1/2}/H^{5/4}$, in which n is the speed in rpm, P is the (design) power and H is the nominal available head in meters. Verify this form for the specific speed and indicate why it is preferred for turbines. (Note that in turbines as well as in pumps, various units are used, so that the specific speed is neither dimensionless nor unique in numerical value.)

18.27 The following information applies to turbines that have been installed at various places around the world:

Turbine Type	Head (m) at Design Point	Flow (m³/s)	rpm	Power (MW) at Design Point
Pelton	398	45.27	180	158.2
Pelton	559	12.48	400	61.2
Francis	107.8	10.95	500	10.3
Francis	265	51.3	300	120
Kaplan	7.95	256	60	18.6
Kaplan	74.5	16.5	610	11.05
Bulb	13.57	334.8	103.4	41.22
Bulb	5.2	99.5	85	4.65

What are the efficiency and specific speed for these units? How do these values agree with the information in Fig. 18.3.4?

Appendix A

Physical Properties

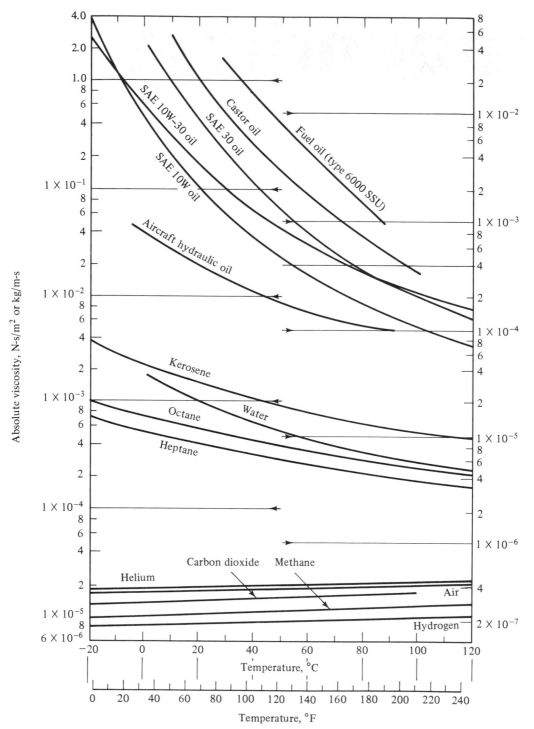

Figure A.1 Absolute viscosity versus temperature for several fluids.

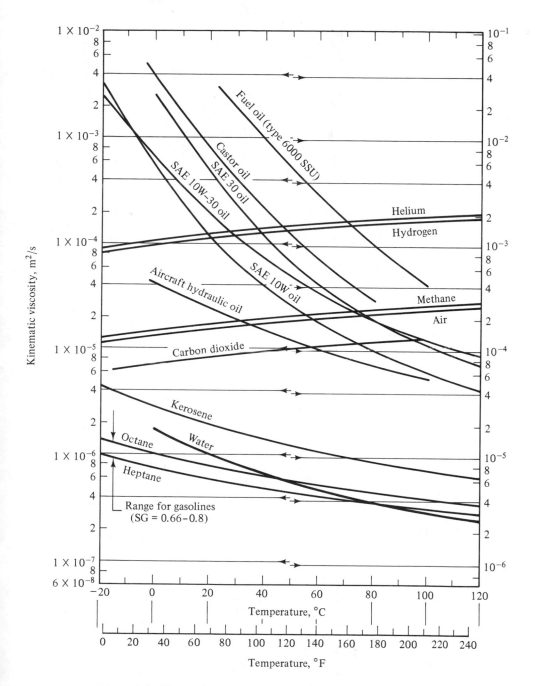

Figure A.2 Kinematic viscosity versus temperature at 1 atmosphere. An approximate relationship between Saybolt seconds (universal) and centistokes is SSU $- 29.74 = 1.96 \, cSt^{1.185}$.

TABLE A.1 PROPERTIES OF WATER AT 1 ATM

T, °C	T, °F	ρ, kg/m³	ρ, slugs/ft³	$\mu\,(10^3)$,† kg/m-s	$\mu\,(10^5)$, slugs/ft-s	$\sigma\,(10^3)$, N/m	$\sigma\,(10^3)$, lb/ft	p_v, kPa	p_v, lb/ft²
0		1000		1.788		75.6*		0.6108	
	32		1.94		3.73		5.18		12.75
10		1000		1.307		74.22		1.2276	
	50		1.94		2.73		5.08		25.6
20		998		1.003		72.75		2.339	
	68		1.937		2.09		4.98		49.8
30		996		0.799		71.18		4.246	
	86		1.932		1.67		4.88		88.6
40		992		0.657		69.56		7.384	
	104		1.925		1.37		4.76		154
50		988		0.548		67.91		12.349	
	122		1.917		1.14		4.65		258
60		983		0.407		66.18		19.940	
	140		1.908		0.975		4.53		416
70		978		0.405		64.4		31.19	
	158		1.897		0.846		4.41		651
80		972		0.355		62.6		47.39	
	176		1.886		0.741		4.29		989
90		965		0.316		61.2		90.714	
	194		1.873		0.660		4.19		1894
100		958		0.283		58.9		101.35	
	212		1.859		0.591		4.03		2116

* Dynes/cm.   † If $\mu(10^3) = 1.788$ kg/m-s, then $\mu = 1.788(10^{-3})$ kg/m-s.

TABLE A.2 PROPERTIES OF AIR AT 1 ATM

T, °C	T, °F	ρ, kg/m³	$\rho\,(10^3)$, slugs/ft³	$\mu\,(10^5)$, kg/m-s	$\mu\,(10^7)$, slugs/ft-s
−40		1.52		1.51	
	−40		2.94		3.16
0		1.29		1.71	
	32		2.51		3.58
50		1.09		1.95	
	122		2.12		4.08
100		0.946		2.17	
	212		1.94		4.54
150		0.835		2.38	
	302		1.62		4.97
200		0.746		2.57	
	392		1.45		5.37
250		0.675		2.75	
	482		1.31		5.75
300		0.616		2.93	
	572		1.20		6.11
400		0.525		3.25	
	752		1.02		6.79
500		0.457		3.55	
	932		0.89		7.41

TABLE A.3 STANDARD ATMOSPHERE

Measured Altitude, (m)	Gradient, K/km	Temperature, K	p/p_0*	ρ/ρ_0†	Measured Altitude, (ft)	Temperature, °R
−500		291.4	1.061	1.049	−1,640	524.5
0		288.2 (15°C)	1.000	1.000	0	518.8
500		284.9	0.9421	0.9529	1.640	512.8
1,000		281.7	0.8870	0.9075	3,280	507.1
1,500		278.4	0.8345	0.8638	4,921	501.1
2,000		275.2	0.7846	0.8217	6,560	495.36
2,500		271.9	0.7372	0.7812	8,202	489.4
3,000	−6.5	268.7	0.6920	0.7423	9,843	483.7
3,500		265.4	0.6492	0.7048	11,483	477.7
4,000		262.2	0.6085	0.6689	13,123	472.0
4,500		258.9	0.5700	0.6343	14,764	466.0
5,000		255.7	0.5334	0.6012	16,404	460.3
6,000		249.2	0.4660	0.5389	19,685	448.6
7,000		242.7	0.4057	0.4817	22,966	436.9
8,000		236.2	0.3519	0.4292	26,247	425.2
9,000		229.7	0.3040	0.3813	29,528	413.5
10,000		223.3	0.2615	0.3376	32,808	401.9
11,000		216.8	0.2240	0.2978	36,089	390.2
12,000		216.7	0.1915	0.2546	39,370	390.1
13,000		216.7	0.1636	0.2176	42,651	390.1
14,000		216.7	0.1399	0.1860	45,932	390.1
15,000	0.0	216.7	0.1195	0.1590	49,213	390.1
16,000		216.7	0.1022	0.1359	52,493	390.1
17,000		216.7	0.08734	0.1162	55,774	390.1
18,000		216.7	0.07466	0.09930	59,055	390.1
19,000		216.7	0.06383	0.08489	62,336	390.1
20,000		216.7	0.05457	0.07258	65,617	390.1
22,000		218.6	0.03995	0.05266	72,178	393.5
24,000		220.6	0.02933	0.03832	78,740	397.1
26,000	+1.0	222.5	0.02160	0.02797	85,302	400.5
28,000		224.5	0.01595	0.02047	91,864	404.1
30,000		226.5	0.01181	0.01503	98,425	407.7
32,000						
40,000	+2.8	250.4	0.002834	0.003262	131,234	450.7
47,000						
50,000	0.0	270.7	0.0007874	0.0008383	164,042	487.3
52,000						
60,000	−2.0	255.8	0.0002217	0.0002497	196,850	460.4
61,000						
70,000	−4.0	219.7	0.00005448	0.00007146	229,659	395.5
79,000						
80,000	0.0	180.7	0.00001023	0.00001632	262,467	325.3
88,743						
90,000		180.7	0.000001622	0.000002588	295,276	325.3

* $p_0 = 1.01325(10^5)$ N/m² abs (= 14.696 psia).

† $\rho_0 = 1.2250$ kg/m³ (= 0.002377 slug/ft³).

Source: U.S. Committee on Extension to the Standard Atmosphere (COESA).

TABLE A.4 PROPERTIES OF LIQUID METALS AT ATMOSPHERIC OR PUMPING PRESSURES

Metal, (Melting Point, °F)	Temperature °F	Temperature °C	Specific Gravity	Thermal Conductivity Btu/h-ft-°F	Thermal Conductivity cal/s-cm-°C*	Absolute Viscosity lb_m/ft-s	Absolute Viscosity cP
Aluminum	1250	677	2.38				
(1220)	1300	704	2.37	60.2	0.249	1.88×10^{-3}	2.8
	1350	732	2.36	63.4	0.262	1.61×10^{-3}	2.4
	1400	760	2.35	64.3	0.266	1.34×10^{-3}	2.0
	1450	788	2.34	69.9	0.289	1.08×10^{-3}	1.6
Lead	700	371	10.5	9.3	0.038	1.61×10^{-3}	2.39
(621)	850	454	10.4	9.0	0.037	1.38×10^{-3}	2.05
	1000	538	10.4	8.9	0.036	1.17×10^{-3}	1.74
	1150	621	10.2	8.7	0.036	1.02×10^{-3}	1.52
	1300	704	10.1	8.6	0.035	9.20×10^{-4}	1.37
Mercury	50	10	13.6	4.8	0.020	1.07×10^{-3}	1.59
(−38)	200	93	13.4	6.0	0.025	8.4×10^{-4}	1.25
	300	149	13.2	6.7	0.028	7.4×10^{-4}	1.10
	400	204	13.1	7.2	0.030	6.7×10^{-4}	0.997
	600	316	12.8	8.1	0.033	5.8×10^{-4}	0.863
Zinc	600	316	6.97	35.4	0.146		
(787)	850	454	6.90	33.7	0.139	2.10×10^{-3}	3.12
	1000	538	6.86	33.2	0.137	1.72×10^{-3}	2.56
	1200	649	6.76	32.8	0.136	1.39×10^{-3}	2.07
	1500	816	6.74	32.6	0.135	9.83×10^{-4}	1.46

* For W/cm-°C, multiply by 418.4.

TABLE A.5 PROPERTIES OF COMMON LIQUIDS AT 1 ATM AND 20°C = 68°F

Liquid	Specific Gravity	μ/μ_{H_2O}	σ/σ_{H_2O}	$p_v/(p_v$ for $H_2O)$	K/K_{H_2O}
Ammonia	0.576	0.22	0.29†	380	
Benzene	0.881	0.65	0.40*†	4.2	0.478
Carbon tetrachloride	1.595	0.97	0.37†	5	0.438
Ethanol	0.789	1.2	0.31†	1.4	0.406
Gasoline	0.680	0.29		1950–1368‡	
Glycerin	1.260	1500	0.87*	$6(10^{-6})$	1.97
Kerosene	0.804	1.91	0.38*	1.3	
Mercury	13.6	1.56	7.04*	$4.6(10^{-7})$	11.88
Methanol	0.81	0.6	0.31*	5.6	0.375
Motor oils	0.92	81–440	0.5*		0.608
Water	1	1	1*	1	1
Sea water	1.025	1.07	1*	1	1.03

* In contact with air.
† In contact with vapor.
‡ Vapor pressure is altered for the season.

Properties of water at 1 atm and 68°F (20°C):

$\rho = 1.94$ slugs/ft^3 = 1000 kg/m^3

$\mu = 2.09(10^{-5})$ slug/ft-s = $1(10^{-3})$ kg/m-s = $1(10^{-3})$ N-s/m^2

$\sigma = 0.0050$ lb/ft = 0.073 N/m (73 dynes/cm is sometimes quoted)

$p_v = 48$ lb/ft^2 = 2.34 kPa

$K = 320,000$ lb/in^2 = $4.61(10^7)$ lb/ft^2 = $2.21(10^9)$ Pa = 2.21 GPa

TABLE A.6A PROPERTIES OF COMMON GASES AT 1 ATM AND 20°C IN SI UNITS

Gas	Molecular Weight	R, J/kg − K or m²/s² − K	c_p, J/kg − K or m²/s² − K	c_v, J/kg − K or m²/s² − K	Specific Heat Ratio k	ρ, kg/m³
He	4.003	2,078	5,225	3,174	1.66	0.1664
H₂	2.016	4,120	14,180	10.060	1.40	0.0839
H₂O (steam)*	18.02	460	2,000	1,540	1.33	0.7518
Dry air†	28.96	287	1,004	717	1.40	1.2050
CO₂	44.01	189	840	651	1.30	1.8298
CO	28.01	297	1,039	742	1.40	1.1644
N₂	28.02	297	1,039	742	1.40	1.1644
O₂	32.06	259	909	650	1.40	1.5353
CH₄	16.04	518	2,190	1,672	1.32	0.6676

* Approximate values
† A standard atmosphere has p = 101.325 kPa and T = 288.2 K yielding ρ = 1.2250 kg/m³.

TABLE A.6B PROPERTIES OF COMMON GASES AT 1 ATM AND 68°F IN ENGINEERING (ENGLISH) UNITS

Gas	Molecular Weight	R^*, ft-lb/slug-°R [Btu/slug-°R]	c_p^*, ft-lb/slug-°R [Btu/slug-°R]	c_v^*, ft-lb/slug-°R [Btu/slug-°R]	Specific Heat Ratio k	$(\rho)(1000)$, slugs/ft³
He	4.003	12,420 [15.96]	31,276 [40.19]	18,856 [24.20]	1.66	0.323
H₂	2.016	24,650 [31.68]	84,902 [109.1]	60,252 [77.42]	1.40	0.163
H₂O (steam)†	18.02	2,759 [3.56]	11,985 [15.4]	8,225 [11.7]	1.33	1.453
Dry air	28.96	1,717 [2.21]	5,992 [7.70]	4,275 [5.49]	1.40	2.335
CO₂	44.01	1,129 [1.57]	5,027 [6.46]	3,898 [5.01]	1.30	3.551
CO	28.01	1,774 [2.28]	6,218 [7.99]	4,444 [5.71]	1.40	2.260
N₂	28.02	1,774 [2.28]	6,218 [7.99]	4,444 [5.71]	1.40	2.260
O₂	32.06	1,553 [2.00]	5,440 [6.99]	3,887 [4.99]	1.40	2.582
CH₄	16.04	3,099 [3.98]	17,961 [16.84]	17,651 [12.86]	1.32	1.294

* 778.2 ft-lb = 1 Btu
† Approximate values

TABLE A.7 ISOTHERMAL BULK MODULUS OF LIQUIDS

Liquid	Bulk modulus*, lb/in²			
	1 atm†	250 atm	1000 atm	5000 atm
Acetic acid	163,000		—	
Benzene	155,000		294,000	
Carbon tetrachloride	140,000		283,000	
Kerosene	—		327,000	
Mercury	3,680,000		3,680,000	
Methanol	123,000		294,000	
Oils, vegetable	294,000		—	
Toluene	163,000		—	
Water	320,000	368,000	420,000	920,000

* The bulk modulus varies nearly linearly with pressure.
† 1 atm = 14.696 psi = 1.013 bars = 0.1013 MN/m².

Physical Properties

Appendix B
Total Differentials

Since we are interested in the differences in pressure, temperature, density, and velocity between two points in a flow, we shall now focus our discussion on the determination and meanings of such differences. If we wished to know the x component of velocity, u, at the neighboring points $(0,0,0)$ and $(0.5, 0.5, 0)$ from the expression $u = -2y + 7x$, we could simply substitute the value of the coordinates for these two points in this expression for u. The difference between these two values—$u(0.5, 0.5, 0) - u(0,0,0)$—has, in general, no particular significance. If, however, we calculate $u(0.5, 0, 0) - u(0,0,0)$, we will have found the difference in u corresponding to a change in x only. This numerical value is $3.5 - 0$ if one uses the appropriate values of u specified by Eq. (3.2.3), $u = -2y + 7x$. This could be converted into a difference of u per unit length in the x direction by dividing it by the distance Δx, which is $0.5 - 0$. The ratio $(3.5 - 0)/(0.5 - 0)$ is a finite difference approximation of the value of the partial derivative $\partial u/\partial x$ in the neighborhood of $(0.25, 0.0, 0.0)$.

Such partial derivatives are used in the Taylor series expansion of the dependent variable—u, in this case—about a particular point (x_0, y_0, z_0) at t_0:

$$du = \left(\frac{\partial u}{\partial x}\right) dx + \left(\frac{\partial u}{\partial y}\right) dy + \left(\frac{\partial u}{\partial z}\right) dz + \left(\frac{\partial u}{\partial t}\right) dt \tag{B.1}$$

The derivatives in this expression are to be evaluated at the specified coordinates $x_0, y_0, z_0, t_0)$. If the x velocity component is steady, $\partial u/\partial t = 0$. Whenever $u = u(x, y)$, then $\partial u/\partial z = 0$. This is the case for $u = 7x - 2y$, for which

$$du = 7dx - 2dy \tag{B.2}$$

The magnitude and sign of the displacement dx and dy in this expression will determine the difference between the values of u at $(x_0 + dx, y_0 + dy)$ and (x_0, y_0); this will be $u - u_0 = du$.

Example B.1

What are the expressions for the three partial derivatives associated with $p = $ constant $-3x^2 + 5y^3$?

Solution If one differentiates this expression, one has

$$\begin{aligned} dp &= d(\text{constant}) - 3d(x^2) + 5d(y^3) \\ &= 0 - 3(2x\,dx) + 5(3y^2\,dy) \\ &= -6x\,dx + 15y^2\,dy \end{aligned} \qquad (B.3)$$

Consequently, $\partial p / \partial x = -6x$; $\partial p / \partial y = 15y^2$. These derivatives are interpreted in Fig. B.1. The values of the partial derivatives, given above, will determine the slopes at any point (x, y).

One important observation must be made about the results given in Eqs. (B.2) and (B.3). They state that one would obtain the same value for du if one proceeded to $(x_0 + dx, y_0 + dy, z_0 + dz)$ from (x_0, y_0, z_0) by any of the four paths shown in Fig. B.2. This is because even though the order of taking dx, dy, and dz is different, the magnitude and sign of Δx, and so forth, are the same on each path.

The fact that the value of u at a point depends only on the value of the point's coordinates leads us to define it as a *point function*. The pressure distribution within a medium will be written as $p(x, y, z, t)$ and is a point function because we shall consider t to be simply another coordinate. The same characteristic of a point function is true of the density ρ. But not all the quantities that are of interest to us are point functions.

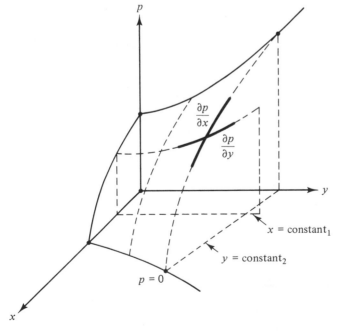

Figure B.1 Plot of $p = \text{constant} - 3x^2 + 5y^3$ showing intersections of the generic surface $f(p, x, y) = 0$ with planes $x = \text{constant}_1$ and $y = \text{constant}_2$, and associated partial derivatives.

Total Differentials

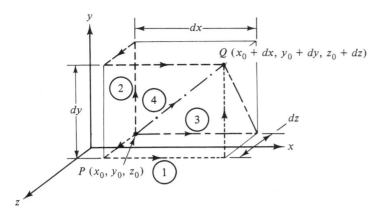

Figure B.2 Four possible paths connecting neighboring points.

Example B.2

If $M = x^2 + 4y + \sin 3z - 9t^3$, what are the associated partial derivatives?

Solution

$$dM = 2x\,dx + 4dy + (3\cos 3z)\,dz - (27t^2)\,dt$$

Then
$$\frac{\partial M}{\partial x} = 2x \qquad \frac{\partial M}{\partial y} = 4 \qquad \frac{\partial M}{\partial z} = 3\cos 3z \qquad \frac{\partial M}{\partial t} = -27t^2$$

Example B.3

The pressure at point R in Fig. B.3 is known. Determine the pressure at points P, Q, S, and T. In this example, we shall restrict ourselves to the case in which there is no variation in the direction perpendicular to the page (i.e., $\partial p/\partial z = 0$).

Solution Using the form of Eq. (B.1) as a guide, we write in this case

$$dp = \frac{\partial p}{\partial t}\,dt + \frac{\partial p}{\partial x}\,dx + \frac{\partial p}{\partial y}\,dy + \frac{\partial p}{\partial z}\,dz$$

For a steady-state condition, the temporal derivative is zero. Along lines of constant x and z, dx and dz are zero. Writing $p_Q - p_R$ for dp, one has

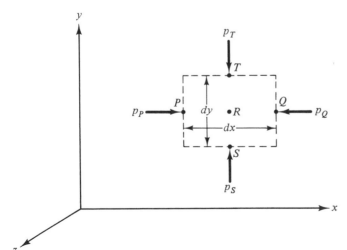

Figure B.3 Pressure at four points on the surface of an infinitesimal control surface, and the pressure within the control volume.

$$p_Q = p_R + \left(\frac{\partial p}{\partial x}\right)\left(\frac{dx}{2}\right) + \left(\frac{\partial p}{\partial y}\right)(0)$$

because the excursion from point R to point Q is $dx/2$ and not dx. Similarly,

$$p_P = p_R + \frac{\partial p}{\partial x}\left(\frac{-dx}{2}\right)$$

since there is no variation of y between point R and P or Q. Also, for point P, the displacement in the x direction is negative. The corresponding values for the pressure at points T and S are

$$p_T = p_R + \frac{\partial p}{\partial y}\frac{dy}{2}$$

and

$$p_S = p_R + \frac{\partial p}{\partial y}\left(\frac{-dy}{2}\right)$$

(Note carefully the minus sign in the last expression and assure yourself that you know the reason for employing it.)

The work that is done in a thermodynamic cycle does not depend solely on the extremes of pressure and density that the gas, the working medium, experiences. The work per cycle, the integral $\int p[d(1/\rho)]$, will depend upon the interrelationship between p and $1/\rho$ during the various portions of the thermodynamic cycle. We need to know all of the values of p and $1/\rho$ intermediate to their end-point values as determined by the path of integration. Different paths will give different intermediate values, and consequently different amounts of work. Thus, we must specify both the end points and the path between them. Hence, work is not a point function. It is called a *path function*.

Example B.4

If a gas is expanding against a piston in a cylinder, the expansion could take place in such a way that the process is closely modeled by the equation $pV^n = $ constant. What work is produced in going from V_1 to V_2?

Solution The work performed by the gas would be $\int p\, dV$, in which dV is the product of the gas's mass and its specific volume. The result is

$$\frac{p_2 V_2 - p_1 V_1}{n - 1}$$

for n not equal to 1. Each value of n would be associated with a different thermodynamic path. Hence, work is a path function.

If the (frictionless) piston were restrained by a linear spring, one would have $p = p_0 + kV$, in which k incorporates the spring constant and the piston's area. This pressure relationship would give a different result for the work from the (polytropic) result given above.

Appendix C
Moving Control Volumes

In this book, we usually use a reference frame for the coordinate system that has a fixed location. The variables associated with a flow, such as ρ, u, and p, were functions of the locations specified by this coordinate system and time. In some cases it is convenient to adopt another point of view. In Chap. 7, it was observed that, if a vane translated at constant speed, the flow would be unsteady. However, if one moved with the vane, a liquid jet could appear to be deflected by the vane in a steady manner. The following was stated at that time:

> If the vane moves at constant velocity, the equations associated with the momentum principle will be unaltered, if one uses velocities that are measured with respect to the moving frame of reference instead of absolute velocities.

In the following paragraphs, we shall explore the truth of this statement, as well as the consequences when the reference frame is not moving at constant velocity. (Some of the steps in the development will entail a sequence of rearrangements and substitutions that may seem tedious. Nevertheless, the reader is encouraged to persevere to the end. Only then can one have confidence when using the final result.)

In writing the conservation laws using variables associated with a moving frame, we shall use the following convention:

$$x = \mathbb{O}_x(t) + \hat{x}$$
$$y = \mathbb{O}_y(t) + \hat{y}$$
$$z = \mathbb{O}_z(t) + \hat{z}$$

in which $\mathbb{O}_x$, $\mathbb{O}_y$, and $\mathbb{O}_z$ are the instantaneous locations of the origin of the moving reference frame. In writing these expressions, we are asserting that the position vector for a fixed coordinate system is the sum of the vector determining the location of the moving reference frame and the position vector taken with respect to this moving frame. This is stated as

$$\mathbf{x} = \underset{\rightarrow}{\mathbb{O}}(t) + \hat{\mathbf{x}} \tag{C.1}$$

in which $\mathbf{x}$ is the (absolute) position vector and $\hat{\mathbf{x}}$ is the position vector relative to the coordinate system with location $\underset{\rightarrow}{\mathbb{O}}(t)$. All vectors appropriate to our subject could be decomposed in a similar way. For example,

$$\mathbf{V} = \underset{\rightarrow}{\dot{\mathbb{O}}}(t) + \hat{\mathbf{V}} \tag{C.2}$$

in which $\underset{\rightarrow}{\dot{\mathbb{O}}}(t)\,(= \dot{\mathbb{O}}_x\mathbf{i} + \dot{\mathbb{O}}_y\mathbf{j} + \dot{\mathbb{O}}_z\mathbf{k})$ is the velocity of the reference frame and $\hat{\mathbf{V}}$ is a velocity relative to this frame. (We used this idea, as obvious, when angular momentum was discussed in Chap. 7.)

When one considers a fixed mass of fluid in a fashion similar to that in Chap. 3 and wishes to account for its parts during the time interval $t_1 - t_0$, one could use a control volume similar to that in Fig. 3.6.4 if that geometry were appropriate. We shall do the same for a moving control volume with an origin located at a point within it, say, $\mathbb{O}$. Moreover, after a moment's reflection, it becomes apparent that the flux of mass out of this moving control volume is $\int \rho \hat{\mathbf{V}} \cdot \mathbf{n} \, dA$, in which ρ is specified at points within the moving control volume by the coordinate vector $\hat{\mathbf{x}}$, whose origin is moving with point $\mathbb{O}$ at an absolute velocity $\underset{\rightarrow}{\dot{\mathbb{O}}}.$* Once again, $\mathbf{n}$ is the outer normal to the control volume, a vector determined solely by the surface geometry. (If the origin of the coordinate system is fixed, then $\mathbf{x} = \hat{\mathbf{x}}$ and $\mathbf{V} = \hat{\mathbf{V}}$.) With ρ as $\rho(\hat{\mathbf{x}}, t)$, the equation of mass conservation for a system whose origin is moving as $\underset{\rightarrow}{\dot{\mathbb{O}}}(t)$ is

$$\frac{\partial}{\partial t}\left(\int_{CV} \rho \, d\mathbf{V}\right) + \int_{CS} \rho \hat{\mathbf{V}} \cdot \mathbf{n} \, dA = 0 \tag{C.3}$$

[This statement will be employed often in the remainder of this chapter, and the reader is urged to review it, and the development of Eq. (3.6.6), to be certain that it is correct.] Thus, we see that the continuity equation appears to be unchanged whether the relative velocity $\hat{\mathbf{V}}$ or the absolute velocity $\mathbf{V}$ is used. (In each case, the instantaneous control volume and surface geometries would be the same.)

The constancy of form when going from a fixed coordinate frame to a moving one is not always preserved. The momentum and energy equations will appear to be different in the two coordinate systems. We shall investigate these differences by beginning with the momentum equation.

The vector form of the momentum equation will be inferred by generalizing the result for the z direction. We begin by writing

$$\Sigma F_z = \frac{\partial}{\partial t} \int w \rho \, d\mathbf{V} + \int w \rho \, \hat{\mathbf{V}} \cdot \mathbf{n} \, dA$$

in which
$$\mathbf{V} = \text{absolute velocity} = u\mathbf{i} + v\mathbf{j} + w\mathbf{k}$$
$$\hat{\mathbf{V}} = \text{relative velocity} = \hat{u}\mathbf{i} + \hat{v}\mathbf{j} + \hat{w}\mathbf{k}$$
$$\mathbf{V} = \underset{\rightarrow}{\dot{\mathbb{O}}} + \hat{\mathbf{V}}$$

---

* The use of $\hat{\mathbf{V}}$ to account for mass efflux would be natural when, for example, one was watching a smoke plume while aboard a ship. From the deck, one could gauge the expansion of the plume by considering the velocity of the particles relative to the smokestack. [Note that $\int (\mathbf{V} \cdot \mathbf{n} \, dA) \, dt$ or $\int (\hat{\mathbf{V}} \cdot \mathbf{n} \, dA) \, dt$ gives the volume efflux from the control volume in which ρ is specified because $\int \underset{\rightarrow}{\dot{\mathbb{O}}} \cdot \mathbf{n} \, dA = 0$.]

Moving Control Volumes

in accordance with Eq. (C.2). If one does not use w but uses its equivalent, the momentum equation is

$$\Sigma F_z = \frac{\partial}{\partial t} \int \rho (\hat{w} + \dot{\mathcal{O}}_z)\, d\forall + \int (\hat{w} + \dot{\mathcal{O}}_z)\rho \hat{\mathbf{V}} \cdot \mathbf{n}\, dA$$

The integrals can be expanded to read

$$\Sigma F_z = \frac{\partial}{\partial t} \int \rho \hat{w}\, d\forall + \frac{\partial}{\partial t} \int \dot{\mathcal{O}}_z \rho\, d\forall + \int \hat{w} \rho \hat{\mathbf{V}} \cdot \mathbf{n}\, dA + \int \dot{\mathcal{O}}_z \rho \hat{\mathbf{V}} \cdot \mathbf{n}\, dA$$

The second integral can be expressed as

$$\int \ddot{\mathcal{O}}_z \rho\, d\forall + \int \dot{\mathcal{O}}_z \frac{\partial \rho}{\partial t}\, d\forall = \ddot{\mathcal{O}}_z \int \rho\, d\forall + \dot{\mathcal{O}}_z \frac{\partial}{\partial t} \int \rho\, d\forall$$

Note that $\ddot{\mathcal{O}}_z$ and $\dot{\mathcal{O}}_z$ do not depend upon the locations within the control volume or the control surface. Thus, they are constant in the appropriate integrals and can serve as a coefficient for them. This observation permits us to write

$$\Sigma F_z = \frac{\partial}{\partial t} \int \rho \hat{w}\, d\forall + \ddot{\mathcal{O}}_z \int \rho\, d\forall + \int \hat{w} \rho \hat{\mathbf{V}} \cdot \mathbf{n}\, dA + \dot{\mathcal{O}}_z \left(\frac{\partial}{\partial t} \int \rho\, d\forall + \int \rho \hat{\mathbf{V}} \cdot \mathbf{n}\, dA \right)$$

The term in parentheses vanishes because of Eq. (C.3). The second integral is simply the mass of fluid within the control volume. Its coefficient, $\ddot{\mathcal{O}}_z$, is the vertical acceleration of the relative coordinate frame. The equation now has the form

$$\Sigma F_z = \frac{\partial}{\partial t} \int \hat{w} \rho\, d\forall + \int \hat{w} \rho \hat{\mathbf{V}} \cdot \mathbf{n}\, dA + (\text{fluid mass in CV}) \ddot{\mathcal{O}}_z \tag{C.4}$$

If the control volume in which the relative coordinate frame is located is not accelerating in the z direction, the equation governing the dynamics of the fluid is the same in absolute and in relative coordinates.[*]

This was assumed to be the case when Example 7.3.5 was presented, and Eq. (C.4) supports the assertion in that example.

We can almost exclusively use Eq. (C.4) for problems involving moving vanes and rockets. The pressure on, and in, the jet as it enters and leaves the control volume in such problems will be atmospheric pressure. Accordingly, it will be *assumed* that the only surface forces acting on the fluid will be due to the vane or the walls of the rocket. We can call this resultant force $\vec{\mathcal{F}}$ ($= \mathcal{F}_x \mathbf{i} + \mathcal{F}_y \mathbf{j} + \mathcal{F}_z \mathbf{k}$). There will be a body force acting on the fluid. This is $-(g \int \rho\, d\forall)\mathbf{k}$. With these *restrictions* in mind, we write

$$\mathcal{F}_z = g \int \rho\, d\forall + \frac{\partial}{\partial t} \int \hat{w} \rho\, d\forall + \int \hat{w} \rho \hat{\mathbf{V}} \cdot \mathbf{n}\, dA + \ddot{\mathcal{O}}_z \int \rho\, d\forall$$

The generalization of this equation is

$$\vec{\mathcal{F}} = \mathbf{k} \int \rho g\, d\forall + \frac{\partial}{\partial t} \int \hat{\mathbf{V}} \rho\, d\forall + \int \hat{\mathbf{V}} \rho \hat{\mathbf{V}} \cdot \mathbf{n}\, dA + \ddot{\vec{\mathcal{O}}} \int \rho\, d\forall \tag{C.5}$$

The first term incorporates the assumption that $\hat{z}$ is positive upward.

---

[*] Only the last term in Eq. (C.4) is different for the two coordinate systems; it represents the product of the fluid mass within the control volume and its acceleration. (It remains on the right-hand side of this equation because its origin is in the momentum terms and not the force terms.)

The power supplied by the surroundings, $\dot{W}$, that is expended upon the fluid, which is now *assumed* to be incompressible, is $\overrightarrow{\mathscr{F}} \cdot \overrightarrow{\dot{\mathbb{O}}}$. We shall now expand this scalar (dot) product by using Eq. (C.5), with the result that

$$\dot{W} = \dot{\mathbb{O}}_z \textstyle\int \rho g\, d\forall + \overrightarrow{\dot{\mathbb{O}}} \cdot \overrightarrow{\dot{\mathbb{O}}} \int \rho\, d\forall + \overrightarrow{\dot{\mathbb{O}}} \cdot \frac{\partial}{\partial t} \int \hat{\mathbf{V}} \rho\, d\forall + \overrightarrow{\dot{\mathbb{O}}} \cdot \int \hat{\mathbf{V}}(\rho\hat{\mathbf{V}}\cdot\mathbf{n})\, dA \qquad \text{(C.6)}$$
$$\underset{\text{①}}{} \qquad \underset{\text{②}}{} \qquad \underset{\text{③}}{} \qquad \underset{\text{④}}{}$$

The reason for identifying the integrals with numbers will be apparent shortly.

The energy equation for the fluid in the control volume is

$$\dot{W} + \dot{\mathbb{2}} = \frac{\partial}{\partial t} \int \left(\frac{V^2}{2} + gz + e \right) \rho\, d\forall + \int \left(\frac{V^2}{2} + gz + e + \frac{p}{\rho} \right) \rho \hat{\mathbf{V}} \cdot \mathbf{n}\, dA$$

(Observe that the kinetic energy is based upon V, the absolute velocity.) This expression is in mixed notation (i.e., $\mathbf{V}$ and $\hat{\mathbf{V}}$). It would seem desirable to use $\overrightarrow{\dot{\mathbb{O}}} + \hat{\mathbf{V}}$ for $\mathbf{V}$. Then

$$\frac{V^2}{2} = \frac{1}{2}(\mathbf{V}\cdot\mathbf{V}) = \frac{\dot{\mathbb{O}}^2}{2} + \frac{\hat{V}^2}{2} + \overrightarrow{\dot{\mathbb{O}}}\cdot\hat{\mathbf{V}}$$

The energy equation would then have an apparently lengthy form,

$$\dot{W} + \dot{\mathbb{2}} = \frac{\partial}{\partial t} \int \frac{\dot{\mathbb{O}}^2}{2} \rho\, d\forall + \frac{\partial}{\partial t} \int \overrightarrow{\dot{\mathbb{O}}}\cdot\hat{\mathbf{V}} \rho\, d\forall + \frac{\partial}{\partial t} \int \left(\frac{\hat{V}^2}{2} + gz + e \right) \rho\, d\forall$$
$$\underset{\text{⑤}}{} \qquad\qquad \underset{\text{⑥}}{} \qquad\qquad \underset{\text{⑦}}{}$$

$$+ \frac{\dot{\mathbb{O}}^2}{2} \int \rho\hat{\mathbf{V}}\cdot\mathbf{n}\, dA + \int \overrightarrow{\dot{\mathbb{O}}}\cdot\hat{\mathbf{V}}(\rho\hat{\mathbf{V}}\cdot\mathbf{n})\, dA + \int \left(\frac{\hat{V}^2}{2} + gz + e + \frac{p}{\rho} \right) \rho\hat{\mathbf{V}}\cdot\mathbf{n}\, dA \qquad \text{(C.7)}$$
$$\underset{\text{⑧}}{} \qquad\qquad \underset{\text{⑨}}{} \qquad\qquad \underset{\text{⑩}}{}$$

The variable z appears in integral 7, where it is to be differentiated with respect to t, and also in integral 10. This coordinate is with respect to some fixed frame of reference, and $z = \mathbb{O}_z + \hat{z}$, in accordance with the previous definition. It follows directly that if one substitutes this definition of z into the two integrals, terms like

$$\dot{\mathbb{O}}_z \int \rho g\, d\forall + g\mathbb{O}_z \left(\frac{\partial}{\partial t} \int \rho\, d\forall + \int \rho\hat{\mathbf{V}}\cdot\mathbf{n}\, dA \right)$$
$$\underset{\text{⑦'}}{} \qquad\qquad \underset{\text{⑦''}}{} \qquad \underset{\text{⑩'}}{}$$

will be part of the result, as well as terms in $\hat{z}$ not shown. (The integral labeled 7 previously would contribute to the integrals labeled 7' and 7''. The part of integral 10 that contains z is the source of 10'.) The sum of the integrals labeled 7'' and 10' is zero, from the principle of mass conservation. Integral 7' will subsequently be combined with another integral to yield a simplification.

The first two integrals in Eq. (C.7) can be fruitfully examined after performing the indicated differentiation. This will be done inside the integral, at first:

$$\frac{\partial}{\partial t} \int \frac{\dot{\mathbb{O}}^2}{2} \rho\, d\forall = \int \dot{\mathbb{O}}\cdot\ddot{\mathbb{O}} \rho\, d\forall + \frac{\dot{\mathbb{O}}^2}{2} \frac{\partial}{\partial t} \int \rho\, d\forall$$
$$\underset{\text{⑤}}{} \qquad\qquad \underset{\text{⑤'}}{} \qquad\qquad \underset{\text{⑤''}}{}$$

and

$$\frac{\partial}{\partial t}\int \underset{\rightarrow}{\dot{\mathbb{O}}}\cdot\hat{\mathbf{V}}\rho\,d\mathbb{V} = \int \underset{\rightarrow}{\dot{\mathbb{O}}}\cdot\hat{\mathbf{V}}\rho\,d\mathbb{V} + \underset{\rightarrow}{\dot{\mathbb{O}}}\cdot\frac{\partial}{\partial t}\int\hat{\mathbf{V}}\rho\,d\mathbb{V}$$

$$\underset{\text{⑥}}{} \qquad\qquad \underset{\text{⑥'}}{} \qquad\qquad \underset{\text{⑥''}}{}$$

Now it is time to reduce our long expression by suitable cancellations. First, we shall *assume* that the flow is adiabatic—something that may not be strictly true in a rocket problem. Then it is observed that $\dot{W}$ in Eqs. (C.6) and (C.7) is the same if we apply the energy equation to the case of a vane-jet system. Consequently,

1. The terms labeled 1 and 7' cancel.
2. Similarly, terms 2 and 5' cancel.
3. Terms 3 and 6'' annul one another.
4. Integral 4 cancels with integral 9, because $\dot{\mathbb{O}}$ is a constant in the latter form.
5. The sum of integrals 5'' and 8 is zero, because of mass conservation.

The final result is

$$0 = \frac{\partial}{\partial t}\int_{\text{CV}}\left(\frac{\hat{V}^2}{2} + g\hat{z} + e\right)\rho\,d\mathbb{V} + \int_{\text{CS}}\left(\frac{\hat{V}^2}{2} + g\hat{z} + e + \cancelto{0}{\frac{p}{\rho}}\right)\rho\hat{\mathbf{V}}\cdot\mathbf{n}\,dA + \underset{\rightarrow}{\ddot{\mathbb{O}}}\cdot\int\hat{\mathbf{V}}\rho\,d\mathbb{V} \qquad \text{(C.8)}$$

Note that p/ρ has been set to zero on the flow boundaries of the incompressible jet. Recall that the pressure is atmospheric pressure—zero gauge—there. If e is associated with the product of a specific heat and a temperature rise, one might expect the role of e in this equation to be minor in some cases. (This is not true in some torque converters for which provisions must be made to cool the fluid.) If the energy losses in the jet as it moves onto the vane are small, one speaks of a "shockless" entry, just as in Sec. 7.6. It is customary to consider changes in $\hat{z}$ to be minor in jet and nozzle problems, so that one has, in effect,

$$0 = \frac{\partial}{\partial t}\int\frac{\hat{V}^2}{2}\rho\,d\mathbb{V} + \int\frac{\hat{V}^2}{2}\rho\hat{\mathbf{V}}\cdot\mathbf{n}\,dA + \underset{\rightarrow}{\ddot{\mathbb{O}}}\cdot\int\hat{\mathbf{V}}\rho\,d\mathbb{V} \qquad \text{(C.9)}$$

The first two integrals can be interpreted as the total change of "relative kinetic energy" for the fluid that is instantaneously on the control volume. If the third term is zero, the rate of change of this "relative kinetic energy" would be zero.

When a vane is moving at constant velocity, the coordinate system attached to it will have $\underset{\rightarrow}{\ddot{\mathbb{O}}} = 0$. For such a case, one could expect that the kinetic energy within the control volume would be time-independent, since $\hat{V}$ would necessarily be constant on the vane for a constant (absolute) jet velocity. This would make the first integral in Eq. (C.9) equal to zero. The second integral in Eq. (C.9) would give

$$\left[\frac{\hat{V}^2}{2}(\rho\hat{V}A)\right]_{\text{entering}} = \left[\frac{\hat{V}^2}{2}(\rho\hat{V}A)\right]_{\text{leaving}}$$

so that $\hat{V}$ is constant at these two portions of the moving control surface. This constancy of the relative speed while the fluid is on the vane is consistent with the deduction that was just made about the first integral in Eq. (C.9), and it was assumed, and employed, in Example 7.3.5.

Equations (C.3) to (C.9) will now be applied to examples that examine the special aspects of the terms in the equations.

Example C.1

A vane on a cart is currently being propelled horizontally at 30 ft/s by a water jet that issues out of a stationary nozzle at 75 ft/s and strikes the vane. See Fig. C.1. Can $\hat{V}$ be considered constant on the vane? What is the velocity of the vane as a function of time?

Solution (a) The continuity equation, Eq. (C.3), will be used initially. In doing so we shall *assume* that the amount of fluid on the vane remains constant, or nearly so. Then, using the notation of Fig. C.1, $\hat{V}_Q A_Q = \hat{V}_P A_P$ for $\rho = $ constant. It follows that, if $\hat{V}_Q \neq \hat{V}_P$, then $A_Q \neq A_P$. Since $\hat{V}_P$ would be changing with time, all of the other terms could also change.

(b) Next we shall consider the "relative energy" statement, Eq. (C.9). In this instance, $\overset{\circ}{\mathcal{O}}$ is not zero. Because it does not depend upon the location within the control volume, the last of the integrals of Eq. (C.9) can be written as $\int_{cv} \mathcal{O}_x \hat{u} \rho \, dV$ for the situation depicted in Fig. C.1. From P to R, $\hat{u}$ will be greater than or equal to zero; from R to Q, it will be less than or equal to zero. Thus, the totality of the product $\mathcal{O}_x \hat{u}$ within the control volume should be a very small quantity, in view of the apparent cancellations. This would imply that the sum of the first two integrals of Eq. (C.9) vanishes (or nearly so). Hence, the total "relative kinetic energy" of a set of fluid particles does not change (appreciably, if at all) with time.

Because the relative fluid speed $\hat{V}$ will decrease with time, the first of the remaining nontrivial integrals in Eq. (C.9) will be negative. This means that the second will be positive, or

$$\frac{\hat{V}_Q^2}{2} (\rho \hat{V}_Q A_Q) > \frac{\hat{V}_P^2}{2} (\rho \hat{V}_P A_P)$$

In view of what the continuity equation has told us about the terms in parentheses, we can conclude that

$$|\hat{V}_Q| > |\hat{V}_P|$$

Thus the speed is not constant along the vane if $\hat{V}$ is changing with time (i.e., the vane is accelerating).

Now we shall attempt to estimate the magnitude of this difference in $\hat{V}_Q$ and $\hat{V}_P$. While we believe that the first integral in Eq. (C.9), $\partial[\int (\hat{V} \rho/2) \, dV]/\partial t$, is negative, we shall expand it to assess its magnitude conveniently. Thus

$$\frac{\partial}{\partial t} \int \frac{\hat{V}^2}{2} \rho \, dV = \int \frac{\hat{V}^2}{2} \frac{d(\text{mass})}{dt} + \int \hat{V} \frac{d\hat{V}}{dt} \rho \, dV$$

Figure C.1 Control volume attached to a moving vane.

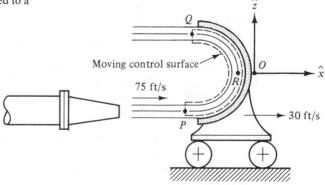

In concert with the initial assumption associated with mass conservation, we employ $d(\text{mass})/dt = 0$. This means that $\int \hat{V}(d\hat{V}/dt)\rho\,dV$ will give us the magnitude we seek. We proceed by assuming that V_R and dV_R/dt will be representative values for the integrand. Hence, $\int \hat{V}(d\hat{V}/dt)\rho\,dV$ will be taken to be approximated by $\hat{V}_R(d\hat{V}_R/dt)m$, in which m is the mass of liquid on the vane. Note that $\hat{V}_R$, the relative speed at R, is always positive, and $d\hat{V}_R/dt$ will be negative. We can estimate its value by noting at P that $\hat{V}_P = V_P - U$, in which U is the cart's speed. V_P is a constant, and so

$$\frac{d\hat{V}_P}{dt} = -\frac{dU}{dt} = -\ddot{\mathcal{O}}_x$$

This agrees with our intuition that $d\hat{V}_P/dt$ should be negative for an accelerating vane. If $\hat{V}_Q$ is 90 percent of $\hat{V}_P$, then $d\hat{V}_Q/dt = -0.9\ddot{\mathcal{O}}_x$.

Similarly, for point R, $d\hat{V}_R/dt = -K\ddot{\mathcal{O}}_x$, with $K \le 1$. Our work up to this point allows us to write Eq. (C.9) as

$$0 = \hat{V}_R \frac{d\hat{V}_R}{dt} m + \frac{V_Q^2}{2}(\hat{\dot{m}}) - \frac{V_P^2}{2}(\hat{\dot{m}})$$

in which $\hat{\dot{m}}$ is $\rho\hat{V}_p A_p$. By factoring out $\hat{\dot{m}}$ and setting $\hat{V}_R \approx (\hat{V}_Q + \hat{V}_p)/2$, we obtain

$$\hat{V}_Q - \hat{V}_p = K\ddot{\mathcal{O}}_x \frac{m}{\hat{\dot{m}}}$$

The factor $m/\hat{\dot{m}}$ should always be small to have $\hat{V}_Q \approx \hat{V}_p$ regardless of the choice of K and $\ddot{\mathcal{O}}_x$. *This makes our conclusion about $\hat{V}$ when $\ddot{\mathcal{O}} = 0$ more general.* We employ this conclusion in the "relative momentum equation" as the last step in this analysis.

(c) Equation (C.5) gives

$$\mathcal{F}_x = \frac{\partial}{\partial t} \int \hat{u}\rho\,dV + \int \hat{u}\rho\hat{V}\boldsymbol{\cdot}\mathbf{n}\,dA + \ddot{\mathcal{O}}_x \rho\, V_{\text{H}_2\text{O}}$$

The integral $\int \hat{u}\rho\,dV$ should be nearly zero, since $\hat{u}$ will change sign in various parts of the control volume. (The positive values should find almost an equal number of negative counterparts in the control volume.) This zero, or nearly zero, value should change little with time, so that the term with the time derivative in the last equation should vanish.

The x component of the force on the vane due to resistance (i.e., air resistance and/or friction against a rolling or sliding surface) is designated F_{Rx}. Equilibrium of the vane requires that

$$-\mathcal{F}_x + F_{Rx} = m_{\text{vane}}\,\ddot{\mathcal{O}}_x$$

This result for the vane is now used in the momentum equation for the fluid that was written a few lines previously. This gives

$$F_{Rx} - \ddot{\mathcal{O}}_x(m_{\text{vane}} + m_{\text{H}_2\text{O}}) = \int \hat{u}\rho\hat{V}\boldsymbol{\cdot}\mathbf{n}\,dA = -(\hat{u}\hat{\dot{m}})_{\text{out}} - (\hat{u}\hat{\dot{m}})_{\text{in}}$$

The last and next-to-last terms in this result have almost the same magnitude, but each has a negative sign for a different reason. (What are the reasons?) For a retarding force on the vane, F_{Rx} would be negative, and the last equation would then require that $\ddot{\mathcal{O}}_x$ be less positive if a retarding force were present.

A special case that could give a limiting result can be considered. Let F_{Rx} be zero, as it might be in a vacuum and for a frictionless sliding surface. Then, if $|\hat{u}|_{\text{in}} = |\hat{u}|_{\text{out}} = \hat{u}$,

$$\ddot{\mathcal{O}}_x = \frac{(2\rho\hat{u}A_{\text{jet}})(\hat{u})}{m_{\text{vane} + \text{H}_2\text{O}}}$$

or
$$\frac{d\dot{\mathbb{O}}_x}{dt} = \frac{(V_{jet} - \dot{\mathbb{O}}_x)^2 (2\rho A_{jet})}{m_{vane + H_2O}}$$

This equation can be integrated to give $\dot{\mathbb{O}}_x(t)$ and $\mathbb{O}_x(t)$ after another integration. These expressions can be used to find $\hat{V}(t)$ entering the moving control volume, because it is $V_{jet} - \dot{\mathbb{O}}_x$.

Example C.2

High-speed boats are built so that they ride out of the water on submerged foils. This can be seen in Fig. C.2a. A lift force is exerted on the foil as a result of its shape and the speed of the water past the foil. In this respect, the foil acts very much like an airfoil, or wing, of an airplane. The lift keeps the hull of a hydrofoil boat out of water so that the drag on the vessel is significantly reduced. This lower drag permits the power plant of the vessel to propel it at a much higher speed. At these higher speeds, a propeller is inefficient, and a water jet is an effective propulsion system. Water is taken into the boat at the forward foil and pumped into the boat and ejected rearward at high speed. The momentum of the fluid stream is changed, and the reaction on the hull propels the vessel forward. A method of analyzing the propulsion system is to adopt a coordinate system that translates at the constant speed of the boat. Such a coordinate system does not change the dynamics of the system, as has already been demonstrated in this section.

When one is moving with the boat, the stationary water ahead of the boat appears to be approaching the boat at speed U. This is shown in Fig. C.2b. The change of horizontal momentum per unit mass of the fluid is $(V_{jet} - U)$ in Fig. C.2c. For a mass rate of flow of ρQ, the $+x$ force on the fluid is $(\mathscr{F}_x + \bar{p} A_{inlet} - p_{atm} A_{jet}$, in which $\bar{p}$ is the average *absolute* pressure at the inlet. $\mathscr{F}_x$ is the $+x$ component of the force on the water that goes through the pump-jet system. Equation (C.5) prescribes that for the $+x$ direction,

$$\mathscr{F}_x = \rho Q (V_{jet} - U) - \bar{p} A_{inlet} + p_{atm} A_{jet}$$

because the coordinate frame is not accelerating. An effect of the fluid within the duct, on the duct itself, is $\mathscr{R}_x = -\mathscr{F}_x$. The forces on the vessel will include $\mathscr{R}_x$, as well as the air drag on the parts above water and the water drag on the submerged parts.

The drag on the submerged section must be related to the momentum change of the water that flows past, but not into, the foil assemblies. In addition, it is necessary that one account for the forces on the portion of the control surface labeled $ADEFA$, *excluding* the portion A_{inlet} along AF. (This section was appropriate to the fluid that passed through the pump-jet system and must be treated accordingly.) The $+x$ forces on these surfaces are $\mathscr{F}_x' - \bar{p} A_{inlet}$, in which $\mathscr{F}_x'$ is the force of the foil assemblies on the water. This quantity is equivalent to the rate of momentum change of the water, $\Delta \dot{M}_{H_2O}$, flowing past the foils and struts. The principle of action and reaction relates $\mathscr{F}_x'$ to the drag on the water and hydrofoil surfaces:

$$Drag_{H_2O} = -\Delta \dot{M}_{H_2O} - \bar{p} A_{inlet}$$

This result suggests that in the final summation of forces on the system, the value of $\bar{p}$ will not play a role.

Similarly, for the effects due to the air, which experiences $\mathscr{F}_x''$ from the boat, one has

$$\mathscr{F}_x'' + p_{atm} A_{jet} = \Delta \dot{M}_{air}$$

The second term in this equation accounts for the exclusion of area JK for the air's control surface along surface CD but inclusion along AB. Since $drag_{air} = -\mathscr{F}_x''$, we can write

$$Drag_{air} = -\Delta \dot{M}_{air} + p_{atm} A_{jet} \,^*$$

---
* There is assumed to be no interaction between the air and the water in the jet.

(a)

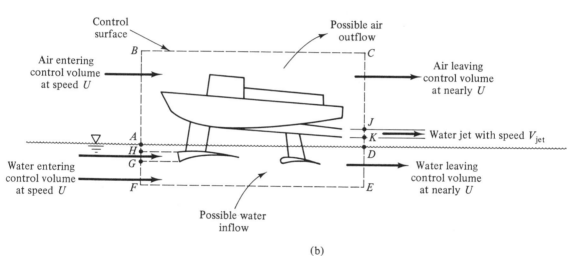

(b)

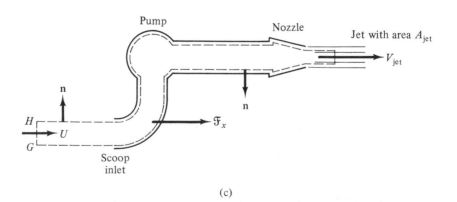

(c)

Figure C.2 (a) Hydrofoil boat operating with its hull out of the water. (b) Moving control volume around hydrofoil boat. G, H: points along control surface FA. (c) Schematic diagram of control volume encompassing inlet, pump and discharge sections of the propulsion system. *(Photograph courtesy of the Boeing Co.)*

The total x-directed force on the vessel system is called ΣR_x and would be

$$\text{Drag}_{H_2O} + \text{drag}_{air} + \text{propulsive force}$$

The last of these three terms is $\mathcal{R}_x$. Hence,

$$\Sigma R_x = \rho Q(U - V_{jet}) - [\Delta \dot{M}_{H_2O} + \Delta \dot{M}_{air}] = 0$$

for a constant speed of the hydrofoil, U. (The effect of p_{atm} does not appear in this result as one would expect.) The terms in square brackets could be found from towing and wind tunnel tests to determine the drag on hull and superstructure. These would have to be conducted so that the flows past the model structures, including the inlet to the pump-jet system, are similar to the prototype. Under this condition in which $p_{ambient}$ surrounds the body, the momentum changes will be directly related to the drag measurements obtained from force transducers.

Example C.3

When a stream (with speed V_1 and surface elevation h_1) is suddenly dammed, an *undular bore* (i.e., a moving hydraulic jump) propagates upstream with constant speed c. Use a control volume moving with the constant speed of the bore to determine the ratio of downstream to upstream depths, h_2/h_1. (Observe that the absolute velocity far behind the bore is zero.)

Solution   By using the control volume shown in Fig. C.3a, the continuity equation is written as

$$(V_1 + c)h_1 = ch_2$$

or

$$c = \frac{V_1 h_1}{h_2 - h_1}$$

The momentum equation for this flow (made "steady" by a coordinate system translating at a constant speed of c) is

$$\frac{\gamma}{2}(h_1^2 - h_2^2) = \rho h_2 c[c - (V_1 + c)]$$

The speed of propagation, c, can be eliminated by using the expression that was previously obtained from the continuity equation, so that

$$V_1^2 = \frac{g}{2}\left[\frac{(h_2 - h_1)^2(h_2 + h_1)}{h_1 h_2}\right]$$

This equation can be nondimensionalized to read

$$\frac{V_1^2}{gh_1} = \text{Fr}_1^2 = \frac{(h_2/h_1 - 1)^2(h_2/h_1 + 1)}{2h_2/h_1}$$

or, in a rewritten form using the continuity equation,

$$c^2 = \frac{gh_1}{2}\left(\frac{h_2}{h_1} + 1\right)\left(\frac{h_2}{h_1}\right)$$

The former equation gives h_2/h_1 for given values of Fr_1, while the latter will yield c for the appropriate depth ratio. These relationships are shown in Fig. C.3b.

To show the role of the energy equation in this example, one can now deduce the unavailable energy, $e_2 - e_1$, associated with this moving wave, sometimes called a surge wave. A stationary control volume will be used. This will make it relatively easy to compute the different energy terms.

Moving Control Volumes

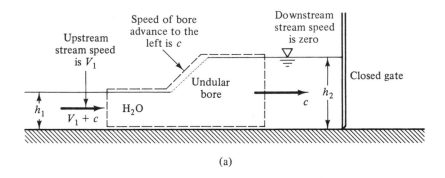

(a)

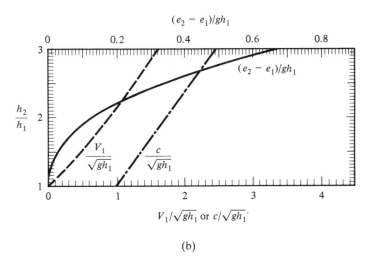

(b)

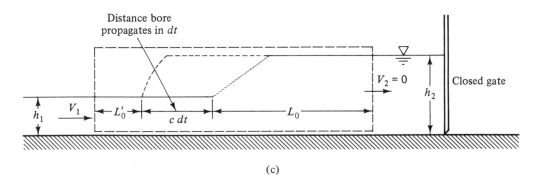

(c)

Figure C.3 (a) Control volume translating to the left with constant speed c, with associated fluid velocities relative to it. (b) Undular bore depth ratios as a function of Froude number and energy loss. (c) Stationary control volume for undular bore moving with speed c to the left in a channel with upstream speed V_1.

For the situations under consideration,

$$\dot{W} + \dot{\mathfrak{Q}} = 0 = \frac{\partial}{\partial t} \int \left(\frac{V^2}{2} + gz + e \right) dV + \int \left(\frac{V^2}{2} + gz + e + \frac{p}{\rho} \right) \rho \mathbf{V} \cdot \mathbf{n} \, dA$$

This becomes for a unit width

$$0 = \frac{\partial}{\partial t} \int \left(\frac{V^2}{2} + gz + e \right) dV - \rho V_1 \left[\frac{V_1^2}{2} h_1 + g\frac{h_1}{2} h_1 + e_1 h_1 + \frac{\gamma h_1^2}{2\rho} \right]$$

(Note that the second term in the square brackets is $\int z \, dA = \bar{z} A$.) Consequently,

$$\frac{\partial}{\partial t} \int \left(\frac{V^2}{2} + gz + e \right) dV = \rho V_1 h_1 \left(\frac{V_1^2}{2} + gh_1 + e_1 \right)$$

so that the rate of change of energy within the control volume is positive (i.e., energy increases with increasing time).

Figure C.3c shows the position of the surge wave at two different times, t_0 and $t_0 + dt$. At t_0, the value of the integral is

$$\rho \left\{ (1)h_1[L_0' + c(dt)] \left(\frac{V_1^2}{2} + \frac{gh_1}{2} + e_1 \right) + (1)h_2 L_0 \left(\frac{0^2}{2} + \frac{gh_2}{2} + e_2 \right) \right\}$$

In this expression, (1) $h_1[L_0' + c(dt)]$ is the volume of fluid whose internal energy is e_1. At $t_0 + dt$, the integral is

$$\rho \left\{ (1)h_1 L_0' \left(\frac{V_1^2}{2} + \frac{gh_1}{2} + e_1 \right) + (1)h_2 [L_0 + c(dt)] \left(\frac{0^2}{2} + \frac{gh_2}{2} + e_2 \right) \right\}$$

The difference in the value of the integral during the time interval dt is

$$\rho \left\{ h_1[-c(dt)] \left(\frac{V_1^2}{2} + \frac{gh_1}{2} + e_1 \right) + h_2[c(dt)] \left(\frac{gh_2}{2} + e_2 \right) \right\}$$

Consequently, the change per unit time is

$$c\rho \left[h_2 \left(\frac{gh_2}{2} + e_2 \right) - h_1 \left(\frac{V_1^2}{2} + \frac{gh_1}{2} + e_1 \right) \right]$$

Because mass conservation for this unsteady case requires that $c(h_2 - h_1) = V_1 h_1$, one can write for the energy equation

$$h_2 \left(\frac{gh_2}{2} + e_2 \right) - h_1 \left(\frac{V_1^2}{2} + \frac{gh_1}{2} + e_1 \right) = (h_2 - h_1) \left(\frac{V_1^2}{2} + gh_1 + e_1 \right)$$

This expression can be solved for $(e_2 - e_1)/g$, a measure of the "loss," in the form

$$\frac{e_2 - e_1}{g} = \left(\frac{V_1^2}{2g} + h_1 \right) - \frac{1}{2} h_1 \left(\frac{h_1}{h_2} + \frac{h_2}{h_1} \right) = h_1 \left[\left(\frac{\mathrm{Fr}_1^2}{2} + 1 \right) - \frac{1}{2} \left(\frac{h_1}{h_2} + \frac{h_2}{h_1} \right) \right]$$

For a small surge wave, let $h_2 = h_1 + \Delta h$ with $\Delta h / h_1 \ll 1$. Then

$$\frac{e_2 - e_1}{g} = h_1 \left[\left(\frac{\mathrm{Fr}_1^2}{2} + 1 \right) - \frac{1}{2}(2) \right] = h_1 \left(\frac{\mathrm{Fr}_1^2}{2} \right)^2$$

Appendix D

Pipe Networks

SYSTEM CURVES

A network consists of pipes, fittings, reservoirs, and pumps that are interconnected to move fluid from one place to another. Some of the elements in the network can be installed so that flow circuits are formed that are similar to the pipes in parallel that were mentioned in Sec. 10.2. A relatively simple example can introduce the method of analysis that can be applicable to more complicated networks.

Example D.1

Two reservoirs are connected by 1500 ft of cast iron pipe that is 2 ft in diameter and branches into two pipes, each 1000 ft long, prior to the lower reservoir. This reservoir is 80 ft below the upper reservoir. One of these two parallel pipes is made of cast iron, and the other, of concrete. The former is 2 ft in diameter, and the latter is 3 ft. Determine the volume rate of water flow. Figure D.1 shows the system.

Solution   If we use the energy equation between the free surfaces of the reservoirs, we can realize that the loss per pound of fluid flowing per second, h_L, is simply Δz, or 80 ft. This means that $80 = \Sigma(\text{all losses})$, or

$$80 \text{ ft} = (\text{entrance loss}) + (h_f \text{ for 1500 ft of 2-ft cast iron pipe}) + (\text{branching loss})$$
$$+ [\text{loss in parallel circuit of (1000 ft of 2-ft cast iron pipe and exit loss)}$$
$$\text{and (1000 ft of 3-ft concrete pipe and exit loss)}]$$

It is recognized that the sum of the volume rates of flow in the parallel pipes must equal the flow rate of the 1500-ft pipe that supplies them.

We shall define Q_{ij} as the flow from point i to point j. Thus,

$$Q_{23} = Q_{34} + Q_{35} \tag{D.1}$$

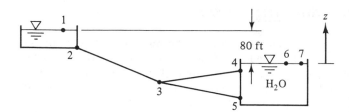

Figure D.1 Series and parallel pipes connecting two reservoirs.

$$\text{and} \qquad 80 \text{ ft} = (h_L)_{13} + (h_L)_{34} + [(h_L)_{46}; \text{ neglect sometimes as being small}]$$
$$= (h_L)_{13} + (h_L)_{35} + [(h_L)_{56}; \text{ neglect sometimes as being small}] \qquad \text{(D.2)}$$

$$(h_L)_{13} = (0.5)\frac{V_{23}^2}{2g} + f_{23}\left(\frac{1500}{2}\right)\frac{V_{23}^2}{2g} \qquad \text{(D.3)}$$

$$(h_L)_{36} = (K_{\text{branch}})\frac{V_{34}^2}{2g} + f_{34}\left(\frac{1000}{2}\right)\frac{V_{34}^2}{2g} + (1)\frac{V_{34}^2}{2g} \qquad \text{(D.4)}$$

in which the last term is due to the sudden expansion into the reservoir; and

$$(h_L)_{37} = (K_{\text{branch}})\frac{V_{35}^2}{2g} + f_{35}\left(\frac{1000}{3}\right)\frac{V_{35}^2}{2g} + (1)\frac{V_{35}^2}{2g} \qquad \text{(D.5)}$$

We can solve this system of equations by assigning an arbitrary value to either $(h_L)_{13}$ or V_{23} in Eq. (D.3) and solving for the term that is not assumed. Initially, f_{23} could be taken as the asymptotic value for infinite Reynolds number. Subsequently, this value could be refined with Eq. (10.2.3). Once $(h_L)_{13}$ is assumed, or obtained from Eq. (D.3), the value of $(h_L)_{36}$ is prescribed through Eq. (D.2). V_{34} and V_{35} can be found from Eqs. (D.4) and (D.5). But these V's and their associated Q's must obey Eq. (D.1). If this equality is not satisfied, one would have to return to Eq. (D.3) and begin the procedure anew with another assumed value for $(h_L)_{13}$ or V_{23}.

For the cast iron pipe, $\epsilon = 0.00085$, so that $\epsilon/D = 0.0004$, approximately. The roughness for concrete pipe can range from 0.01 to 0.001. We shall choose an intermediate value of 0.003, so that $\epsilon/D = 0.001$. These relative roughnesses give values of f of 0.016 for the cast iron conduit and nearly 0.02 for the concrete pipe from the Moody diagram (Fig. 9.4.2). If we write $Q/(\pi D^2/4)$ for V, in Eqs. (D.2) through (D.5) we must solve

$$(h_L)_{13} = \frac{Q_{23}^2}{2g}\left\{\frac{1}{[\pi(2)^2/4]^2}\right\}\left[0.5 + 0.016\left(\frac{1500}{2}\right)\right]$$

$$(h_L)_{36} = \frac{Q_{34}^2}{2g}\left\{\frac{1}{[\pi(2)^2/4]^2}\right\}\left[30 + 0.016\left(\frac{1000}{2}\right) + 1\right]$$

$$(h_L)_{37} = \frac{Q_{35}^2}{2g}\left\{\frac{1}{[\pi(3)^2/4]^2}\right\}\left[30 + 0.02\left(\frac{1000}{3}\right) + 1\right]$$

$$\text{and} \qquad Q_{23} = Q_{34} + Q_{35}$$

In writing these equations, the loss coefficient for the branch was taken as that for a 90° elbow. These last written equations can be rewritten as

$$(h_L)_{13} = Q_{23}^2(0.0008 + \underline{0.019}) \qquad \text{(D.6)}$$

$$(h_L)_{36} = Q_{34}^2(0.047 + \underline{0.013} + 0.0016) \qquad \text{(D.7)}$$

Pipe Networks

$$(h_L)_{37} = Q_{35}^2(0.0093 + \underline{0.0021} + 0.0003) \tag{D.8}$$

in which the underlined terms are often referred to as the *resistances* for the pipe in question. Thus, $r_{23} = 0.019$, $r_{34} = 0.013$, and $r_{35} = 0.0021$. The solution could start by assuming $(h_L)_{13} = 50$ ft, since the one pipe should have more than half the total loss of 80 ft. This would require $(h_L)_{36} = 80 - 50$. The use of a personal computer (PC) allows one to increment this initial estimate easily and to find that Eq. (D.1) is satisfied when $(h_L)_{13} = 62.1$ ft. The solution is $Q_{23} = 56.19$ cfs, with $Q_{34} = 17.08$ cfs and $Q_{35} = 39.11$ cfs.

This example would have been very tedious to solve by the method given above prior to the age of electronic computation. One method that was employed, however, closely parallels what has just been done. In presenting that method, we do not intend to provide an alternative solution scheme, but to indicate the manner by which instructive diagrams can be drawn. Equation (D.6) could be plotted as shown in Fig. D.2a. In the same way, Eqs. (D.7) and (D.8) can be represented as in Fig. D.2b and c. These last two plots show that for any value of $(h_L)_{36} = (h_L)_{37}$, there is a prescribed value of Q_{34}, abbreviated as Q_2, and $Q_{35} (= Q_3)$. Thus, a value of the total head at point 3, H_3, determines a value of $Q_{34} + Q_{35} (= Q_2 + Q_3)$. This is the volume rate of flow that issues from—or enters—the two pipes in parallel. We can draw a curve for H_3 for this total flow by "adding" the lower curves given in Fig. D.2b and c as in Fig. D.3. In Fig. D.3, for a value of Q_1, one would have the h_L in the parallel network for the same *flow* as through the single pipe. The sum of $(h_L)_{13}$ and $(h_L)_{36}$, which is equal to $(h_L)_{37}$, would be the total head loss from the upper reservoir to the lower reservoir. The total head at station 1 has been plotted in Fig. D.4.

With Fig. D.4 prepared, the solution to the problem of finding the flow for the case shown in Fig. D.1 is at hand. Equation (D.2) states, in effect, that $H_1 = 80$. The series equivalent curve for the *entire* system from Fig. D.4 has been replotted in Fig. D.5 solely for clarity. The solution for $Q_{23} (= Q_1)$ is associated with the location on the

Figure D.2 (a) Head loss as a function of flow in a pipe and the associated head available at point 3. (b) Head loss as a function of flow in a pipe and the associated head required at point 3. (c) Head loss in a pipe and the head required at its inlet as a function of flow through the pipe.

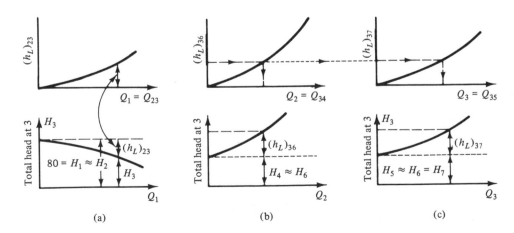

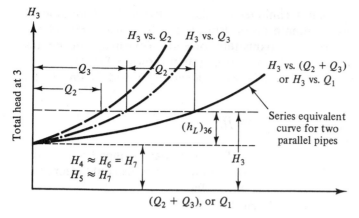

Figure D.3 Construction of series equivalent curve for two pipes in parallel operation.

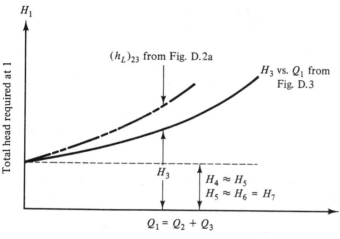

Figure D.4 Equivalent system curve for a pipe in series with two others that are in parallel. Head required at reservoir 1 is shown as a function of total flow between it and a reservoir whose head is maintained at H_6.

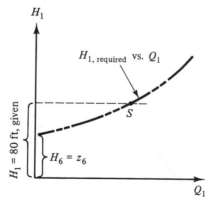

Figure D.5 Solution to problem posed in Fig. D.1.

curve where the ordinate of the curve is 80 ft. This point is marked as point S in Fig. D.5.

From a knowledge of Q_1, which is $Q_2 + Q_3$, one could go to Fig. D.3 and determine Q_2 and Q_3. From the curve, one would find the value of H_3 for the abscissa value Q_1 that corresponds to point S in Fig. D.5. This ordinate value for H_3 would also be related to Q_2 and Q_3 by the appropriate curves in Fig. D.3.

Pipe Networks

The advantage of this graphic technique was the relatively direct method of solution that avoided a tedious iteration process. But the result shown in Fig. D.5 is useful even if the figure is drawn only approximately correctly. We can appreciate this utility by momentarily examining the pump-reservoir system shown in Fig. D.6.

If the energy equation were applied between locations 1 and 5, one could conclude that

$$\frac{\dot{W}}{Q\gamma} + (\text{total head at location } 1 = H_1) = H_5 + \Sigma \, \text{losses}_{1\text{-}5}$$

The term $\dot{W}/Q\gamma$ is called the *pump head*. We shall assume that the pipe friction loss, $(h_f)_{34}$, is the dominant loss term. Our previous work in this appendix could prompt us to graph this energy equation in the manner shown in Fig. D.7.

For most centrifugal pumps, the head that is produced,

$$\left(\frac{V_3^2}{2g} + z_3 + \frac{p_3}{\gamma}\right) - \left(\frac{V_2^2}{2g} + z_2 + \frac{p_2}{\gamma}\right)$$

will vary with the flow rate approximately as shown by the appropriate solid line in Fig. D.8. This performance curve is called the *pump characteristic curve*, and it will be supplied by the pump's manufacturer. The dotted lines represent curves of constant pump efficiency.

If one merges the information given in Figs. D.7 and D.8, one should be able to predict the performance of a pump-pipe system. A superposition of the figures is shown in Fig. D.9. The predicted flow rate is at point P. The head developed by the pump and its expected efficiency can be read from the graph. One could equally well examine the consequences of installing a pump at point 2 in Fig. D.1 to increase the discharge from reservoir 1. Perhaps it would be necessary to install a pump at point 3

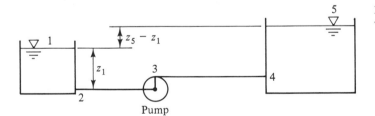

Figure D.6 Pump connecting two reservoirs at different elevations.

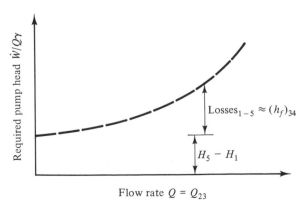

Figure D.7 Pump head required to produce a flow through a pipe network.

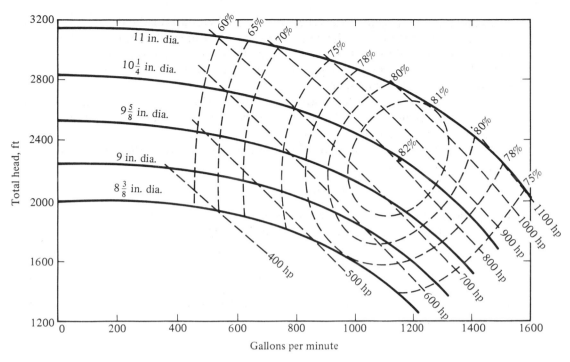

Figure D.8 Head versus flow for a pump with various impeller diameters with curves of constant efficiency superimposed.

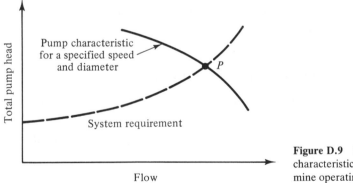

Figure D.9 Superposition of pump characteristic and system curves to determine operating point of the installation.

and deliver the water into reservoir 1. Only a minor rearrangement of the curves that led to Fig. D.5 would be necessary to learn what would be required to do this. Moreover, one could ask about the effect of an alteration of the pipe network on the operating point P in Fig. D.9. This would affect the degree of operating efficiency of the pump.

NUMERICAL SOLUTION

The system of pipes and reservoirs depicted in Fig. D.1 has served two purposes. First, it has provided the basis for discussing and solving a problem concerned with pipes in series and parallel. Second, it has supplied the background for the development of a "system curve" that predicts the head necessary for a prescribed flow rate. We shall now extend the analysis of this case in a new direction. The result will be a general method of analyzing complex piping systems that can be effectively executed on a relatively small digital computer.

To this end we have redrawn the piping system with a few minor modifications in Fig. D.10. Previously, point 7 was unnecessary because point 6 was at the same elevation. In the present case, this is not so. Once again, H_i is the total head, $V_i^2/2g + p_i/\gamma + z_i$, at point i. With this notation, we can write the energy equation between locations 1 and 3 as

$$H_3 = H_1 - \text{loss}_{13} = z_1 - r_{23}Q_{23}^2 \tag{D.9}$$

if only the pipe losses are considered and r_{23} is the resistance in the pipe connecting locations 2 and 3. The equations for the other two pipes are

$$H_3 - \text{loss}_{34} = H_6 = z_6 \tag{D.10}$$

and

$$H_3 - \text{loss}_{35} = H_7 = z_7 \tag{D.11}$$

if, as before, only pipe friction is considered. Continuity must be satisfied at the junction, so that

$$Q_{34} + Q_{35} - Q_{23} = 0 \tag{D.12}$$

The system of Eqs. (D.9) to (D.11) is similar to Eqs. (D.6) to (D.8) and must be solved for the unknown Q's in the loss terms by assuming a value for H_3. The value of H_3 is altered until the Q's that are found satisfy Eq. (D.12). The term H_3, however, is not essential to the solution of the problem. It can be eliminated by subtracting Eq. (D.10) from Eq. (D.9) and writing the result as

$$H_1 - H_6 - r_{34}Q_{34}^2 - r_{23}Q_{23}^2 = 0 \tag{D.13}$$

or subtracting Eq. (D.11) from Eq. (D.10) to produce

$$H_6 - H_7 - r_{35}Q_{35}^2 + r_{34}Q_{34}^2 = 0 \tag{D.14}$$

Figure D.10 Schematic drawing of three reservoirs and connecting pipes in circuits ("loops") employed to facilitate the analysis.

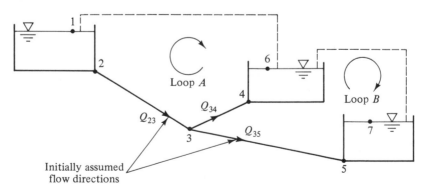

Appendix D

Although Eqs. (D.13) and (D.14) have been written as direct applications of the energy equation, they can be "generated automatically" by the following procedure:

1. Connect the reservoirs in the schematic drawing by a dotted line to form a "pseudoconnection." (The end points of this dotted line connect locations of specified pressure. These pressures do not depend on the flow rates.) Such "pseudoconnections" are necessary only for reservoirs.

2. Assume a direction for the flow in each pipe. (Mark it on the drawing!) Sometimes the choice will be obvious, while at other times, it will be almost impossible to choose the direction of the final flow with confidence. In the latter case, make your best estimate of the situation. A good solution method should be relatively insensitive to an error in the sign of the assumed Q's.

3. Examine the schematic drawing of the complete piping system and subdivide it into a collection of closed pipe circuits called *loops*. Each loop is usually a simple closed polygon on whose perimeter—composed of lengths of pipe that join at the vertices of the polygon—one can circumnavigate that subsection of the pipe network. The loops should be sufficient in number to include all of the pieces of pipe, but redundant loops must be avoided. Thus, in Fig. D.10, the circuit 1-6-4-3-2-1 is one loop and 6-7-5-3-4-6 is another, but the loop 1-6-7-5-3-2-1 is redundant. It is the "sum" of the previous two, a fact that will be soon apparent.

4. Trace a clockwise path around a loop and sum the losses that occur in the segments. Each loss is expressed as the product of the resistance in a pipe segment, r_{mn}, and Q_{mn}^2. The algebraic sign associated with the loss is *positive* if the assumed direction of flow is in the *same direction* as the portion of the *clockwise* circuit in the particular pipe segment. Otherwise, the sign is negative. The sum of the losses, with the proper algebraic signs as coefficients, is zero, as we have seen from Eqs. (D.13) and (D.14). This process is repeated for all of the loops. The number of such loop equations that can be written will be less than the number of segments of pipe, and, thus, less than the number of unknown Q's.

When writing an equation for a "pseudoloop," *begin* with the total head at the reservoir that will permit you to proceed clockwise, via the "pseudoconnection," directly to the second reservoir.

Equations that express mass conservation in the form of volume conservation can be written at each junction. The total number of such continuity equations that can be written will usually be more than that which is necessary to produce the J independent equations for the J unknown Q's. (The redundant equations occur because continuity must be satisfied for the entire pipe network also. This will happen if the continuity equations at the various junctions are added together in the proper manner. The presence of an extra continuity equation can readily arise, but this possibility is easily checked and accommodated.) In this presentation, the continuity equations are written using the convention that the Q's are *positive* if their assumed direction is *away* from the junction. This accounts for the form of Eq. (D.12).

Equations (D.12) through (D.14) are now written in symbolic form as

$$F_1(Q_{23}, Q_{34}, Q_{35}) = 0$$

$$F_2(Q_{23}, Q_{34}, Q_{35}) = 0$$

$$F_3(Q_{23}, Q_{34}, Q_{35}) = 0$$

to facilitate the discussion. If these equations were linear in the Q's, one could obtain the solution vector by standard methods. But two of the equations are quadratic in the Q's. This was the reason the iterative and graphic methods were used earlier. Consequently, a more efficient solution method for the nonlinear equations will be used shortly. However, another difficulty with the equations must also be mentioned.

If the assumed flow directions are not correct—for example, if the liquid flows from reservoir 6 to reservoir 7—it would not simply be a matter of changing the sign of the Q's in the energy equations. The formulated equations would be incorrect because the sign of the affected loss term would be wrong, and this would not be remedied by changing the sign of Q when that quantity is quadratic. (The reader can verify this statement by changing the assumed direction of flow in the pipe connecting locations 3 and 4. Then write the corresponding loop equation and note the signs of the result.) The difficulty associated with the sign of the loss term changing with the direction of flow can be circumvented by writing Q^2 as $Q|Q|$. Then, if Q changes its sign during the computations, $Q|Q|$ does also, whereas Q^2 would not. This is the reason we shall write our loop equations as

$$(z_1 - z_6) - r_{34} Q_{34}|Q_{34}| - r_{23} Q_{23}|Q_{23}| = 0 = F_1 \tag{D.13}$$

$$(z_6 - z_7) - r_{35} Q_{35}|Q_{35}| + r_{34} Q_{34}|Q_{34}| = 0 = F_2 \tag{D.14}$$

and
$$Q_{34} + Q_{35} - Q_{23} = 0 = F_3 \tag{D.12}$$

This system of equations will now be solved by the Newton-Raphson method. (An explanation of the logic behind this procedure is given in Appendix E.)

The first step is to write the equivalent equations in matrix notation as

$$\begin{pmatrix} a_{11} & a_{12} & a_{13} \\ a_{21} & a_{22} & a_{23} \\ a_{31} & a_{32} & a_{33} \end{pmatrix} \begin{pmatrix} \tilde{Q}_1 \\ \tilde{Q}_2 \\ \tilde{Q}_3 \end{pmatrix} = - \begin{pmatrix} B_1 \\ B_2 \\ B_3 \end{pmatrix} \tag{D.15}$$

The solution vector for Eq. (D.15) is *not* the desired flow rates, but rather the correction terms needed to form the solution. The meaning of the $\tilde{Q}_1$, and so forth, will be explained shortly. The elements of the coefficient matrix have subscripts according to the row and column in which they occur. If these a_{ij} were constants, the methods of linear algebra would lead directly to a solution. However, the a_{ij} are not constants; they are functions of the Q's. This requires that we use an iterative method to obtain a solution.

It simplifies the description of the steps in the procedure if we change the notation and write the variables as Q_j, as implied in Eq. (D.15), instead of using a double subscript as in Q_{23}. The single subscript on the Q's lets us write a_{12} as $\partial F_1/\partial Q_2$, in which F_1 is an abbreviation for Eq. (D.13). In general, we write $a_{ij} = \partial F_i/\partial Q_j$. In the problem that serves as the basis for our discussion, we let $Q_1 = Q_{23}$, $Q_2 = Q_{34}$, and $Q_3 = Q_{35}$. Thus,

$$a_{12} = \frac{\partial[(z_1 - z_6) - r_{34} Q_2|Q_2| - r_{23} Q_1|Q_1|]}{\partial Q_2}$$

The derivative of $Q_2|Q_2|$ follows from the product rule,

$$\frac{\partial(Q_2|Q_2|)}{\partial Q_2} = \frac{Q_2\,\partial(|Q_2|)}{\partial Q_2} + |Q_2|\,(1)$$

If Q_2 is positive, the derivative is $+1$; -1 is the derivative's value if $Q_2 < 0$. This result is apparent if one plots the value of $|Q|$ as a function of Q. Figure D.11 is the result, and the slope of the curve is ± 1, depending on whether Q is greater than or less than 0.

We now can write with confidence that

$$\frac{\partial(Q_j|Q_j|)}{\partial Q_j} = |Q_j| + \left\{\begin{array}{l} Q_j(+1)\text{ for }Q_j > 0 \\ Q_j(-1)\text{ for }Q_j < 0 \end{array}\right\} = 2|Q_j|$$

Thus, for the problem at hand,

$$\begin{pmatrix} -2r_{23}|\overline{Q}_1| & -2r_{34}|\overline{Q}_2| & 0 \\ 0 & +2r_{34}|\overline{Q}_2| & -2r_{35}|\overline{Q}_3| \\ -1 & +1 & +1 \end{pmatrix}\begin{pmatrix} \tilde{Q}_1 \\ \tilde{Q}_2 \\ \tilde{Q}_3 \end{pmatrix} = -\begin{pmatrix} B_1 \\ B_2 \\ B_3 \end{pmatrix}$$

One begins the solution process by assuming a set of values for the unknown Q_j's. (This could be any reasonable flow rate, if the value of the solution is impossible to estimate. However, the values should not all be zero. This often makes the determinant singular, and then the solution process cannot proceed.) The assumption of the Q_j's, whatever it is, should satisfy the continuity equation, Eq. (D.12). This set will be designated by an overbar. $\overline{Q}_j^n$ is the assumed solution for Q_j at the nth stage of the iteration process. The values of B_i in the constant matrix are the values of the functions F_i [see Eqs. (D.12) through (D.14)] when the values $\overline{Q}_j^n$ are used. These F_i will be zero only when the solution has been obtained. Hence, the B_i for $\overline{Q}_j^n$ will usually not be zero. The assumed $\overline{Q}_j^n$ are used in the coefficient matrix to complete the evaluation of the a_{ij}. This step permits an evaluation of the $\tilde{Q}_j^n$ terms, whose meaning will now be discussed.

These $\tilde{Q}$'s are the correction terms to the $\overline{Q}$'s that are present at the nth stage of the iteration. For the next stage in the procedure, one sets

$$\overline{Q}_j^{n+1} = \overline{Q}_j^n + \tilde{Q}_j^n \tag{D.16}$$

The procedure has started with the various $\overline{Q}_j^1$ equal to their assumed values. Thereafter, the process proceeds automatically, with Eq. (D.16) changing the values for the next iteration of Eq. (D.15). The method continues until the correction terms $\tilde{Q}_j$ decrease to a value near zero that is deemed satisfactory by the problem solver.

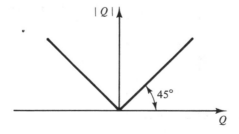

Figure D.11 Plot of absolute value of a quantity as a function of that quanitty.

Even if one does not have any special insight into the final answer, the procedure just outlined should converge regardless of how one chooses all the $\overline{Q}_j^1$ at the outset. Our particular problem produces a 3×3 matrix with a solution for the three unknown Q's. The Newton-Raphson method is capable of giving solutions to systems with a very large number of unknowns. The method is simple to execute with a desk-top digital computer. This method is explained in Appendix E.

The need for solutions to problems associated with pipe networks predated the digital computer, as we have seen in part I of this appendix. Thus the solution of complex pipe networks was simplified to meet the available computational capability. One simplification involved the assumption that the system equations represented by Eq. (D.15) could be solved sequentially with all the $\tilde{Q}_j$ equal in magnitude at each *stage*. (In the present case there are three stages, corresponding to the number of equations, for each iteration step n.) Because of this simplification, one obtains values for $\tilde{Q}_j^n$ from each equation. These $\tilde{Q}_j$ are added to the $\overline{Q}_j$ that are appropriate for the equation that is being solved. If the $\overline{Q}_j^1$ in a loop is in the clockwise direction, the associated $\tilde{Q}_j$ at a particular stage in the nth iteration is added to $\overline{Q}_j$ without any change of sign. However, if the $\overline{Q}_j^1$ for the initial value were counterclockwise in a loop, the $\tilde{Q}_j$ for the stage would be multiplied by -1 prior to its addition to $\overline{Q}_j$. The solution at the end of the nth iteration is the sum of all the $\tilde{Q}_j^n$ from each stage of the particular iteration—with the ± 1 multipliers just mentioned. This solution technique is called the *Hardy-Cross method,* to honor its developer. The method, while important prior to the era of the small, relatively inexpensive digital computers, will not be discussed further here. The Newton-Raphson method and the availability of the means to implement it have diminished the importance of this simplification to the complete equation system.

Figure D.12 shows another piping network. The flows at junctions 2 and 3 and the reservoir elevation are prescribed values. (Positive-displacement pumps could control the flows, regardless of the pressure at points 2 and 3, by adjusting their speeds.) For the three unknowns, there is an equation for loop A,

$$r_{13} Q_{13}|Q_{13}| - r_{23} Q_{23}|Q_{23}| - r_{12} Q_{12}|Q_{12}| = 0$$

and two continuity equations,

$$2.0 + Q_{23} - Q_{13} = 0$$

and

$$4.2 - Q_{13} - Q_{23} = 0$$

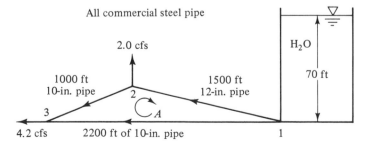

All commercial steel pipe

2.0 cfs

1000 ft
10-in. pipe

1500 ft
12-in. pipe

H_2O

70 ft

3

2

A

4.2 cfs   2200 ft of 10-in. pipe

1

Figure D.12 Pipe network attached to a reservoir.

A continuity equation for junction 1 would be

$$-6.2 + Q_{12} + Q_{13} = 0$$

but it would encumber the equation system, since it is the sum of the two previous equations.

The network shown in Fig. D.13 can be analyzed with equations for loops A to D. A moment's reflection will reveal that loop E is redundant, as are a few others. Continuity equations at junctions 1 to 4 will provide the remaining equations to solve for the eight unknown Q's. These continuity equations can be added to yield

$$Q_{51} - Q_{26} + Q_{73} = 0$$

which is understandable but superfluous.

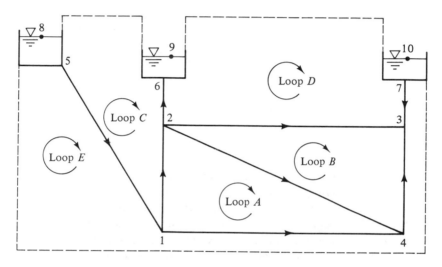

Figure D.13 Multiple pipes connecting three reservoirs.

Appendix E

Newton-Raphson Method of Solution for Pipe Networks

Currently, the Newton-Raphson method is one of the most widely used numerical methods for solving pipe network problems. Appendix D contained examples that illustrated its use. A brief discussion follows that should indicate why the technique can yield the desired results. We begin with a one-dimensional case and see that the Newton-Raphson method is merely an extension of this simple development. Only the number of the variables is increased in more complex situations.

A function of x called $f(x)$ is portrayed in Fig. E.1. If one desires the roots of this function, one would seek the values of x at the intersections of $f(x)$ with the x axis. If an estimate to the root x_R is made with the value x_1, one could evaluate $f(x_1)$ and $f'(x_1)$. Thereby, one could determine the distance Δx to locate the position where the tangent, $f'(x_1)$, intersects the x axis. The geometry shown in Fig. E.1 allows us to write

$$\tan \theta_1 = f'(x_1) = -\frac{f(x_1)}{\Delta x} \tag{E.1}$$

so that $\Delta x = -f(x_1)/f'(x_1)$. The minus sign takes care of the fact that if $\Delta x > 0$ and $f(x_1) > 0$, then $f'(x_1) < 0$ in Fig. E.1. A next choice for x to obtain the root x_R could have the value of the position where the tangent line intersects the x axis. This is x_2, and the location has the value $x_1 + \Delta x$. Out of these steps, we recognize that the $(i + 1)$st trial will be $x_{i+1} = x_i - f(x_i)/f'(x_i)$. This procedure, called *Newton's method,* continues until $\Delta x \rightarrow 0$.

If the function $f(x)$ oscillates a great deal (i.e., if there are many roots in a region), there will be difficulties. Thus, if the root x_R is desired but the first estimate were made as x_P, the final result would undoubtedly be $x_{R'}$. Moreover, if x_S had been chosen, it is highly probable that the root $x_{R'}$ would be obtained also. Another unwanted root would occur if x_Q had been the initial guess at a solution. In addition, the

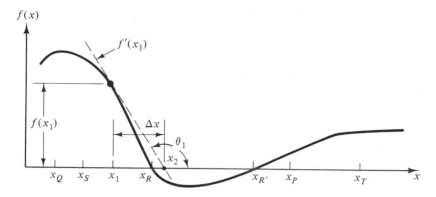

Figure E.1 Notation for finding the roots of $f(x)$ by Newton's method.

procedure need not converge in some cases. This could be true for the function depicted in Fig. E.1 if a poor choice of the initial estimate, x_T, for example, were made. The reader should consult a reference on numerical analysis to review the associated convergence criteria.

It is instructive now to restate the method by which Δx was found, because it will be analogous to the approach that will be used subsequently for higher-dimensioned spaces such as surfaces, volumes, and higher dimensional surfaces. The slope of the curve at x_1 is $f'(x_1)$. Hence, the tangent line has the equation $y = f(x_1) + f'(x_1)\,\Delta x$, in which Δx is a coordinate that is measured *from* x_1. For $y = 0$,

$$\Delta x \left[\frac{df(x_1)}{dx} \right] = -f(x_1) \tag{E.2}$$

in accordance with what was found previously.

We shall now turn our attention to our principal task. This is related to finding the locations of the intersections of $f(x, y) = 0$ and $g(x, y) = 0$. The Newton-Raphson method is a way of doing this. Such intersections are shown in Fig. E.2. We shall start our discussion by considering not $f(x, y)$ but $\hat{F}(x, y)$ and a constant K such that $\hat{F}(x, y) = K - f(x, y) = 0$. For purposes of illustration, we shall let $f(x, y) = xy - 1$. The constant K can be 1, 3, 8, -9, or any number. Each point (x, y) will give a value for K from the requirement that $K - f(x, y) = 0$. For $f(x, y) = xy - 1$, the point $(1, 2)$ yields $K = 1$, and $(2, 5)$ requires that $K = 9$. One can conclude that the value of K depends on x and y. Such a dependency will be written as $F(x, y, K) = 0$ instead of the previous notation, $K - f(x, y) = 0$. For a particular value of K, there is a set of points $\{x, y\}$ that

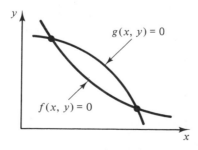

Figure E.2 Intersections between two functions $f(x, y) = 0$ and $g(x, y) = 0$.

Newton-Raphson Method of Solution for Pipe Networks

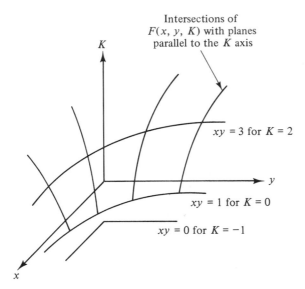

Intersections of $F(x, y, K)$ with planes parallel to the K axis

K

$xy = 3$ for $K = 2$

y

$xy = 1$ for $K = 0$

$xy = 0$ for $K = -1$

x

Figure E.3 Intersections of $F(x, y, K) = 0$ with planes of constant K.

satisfies $F(x, y, K) = 0$. We shall plot this variation of K on an axis that is perpendicular to the xy plane, as shown in Fig. E.3.

Note that for each value of K = constant, there is a corresponding relationship between x and y. The curve $K = 0$ is $f(x, y) = xy - 1 = 0$. Similar things could be said for $g(x, y)$ and the associated functions $G(x, y, K)$ that satisfy $G(x, y, K) = 0$. This function could be $K - (x^2 + y^2 - 9)$. Keeping in mind that we seek the intersection of $F(x, y, 0)$ and $G(x, y, 0)$—refer to Fig. E.2, in which $F(x, y, 0)$ and $G(x, y, 0)$ are $f(x, y)$ and $g(x, y)$ when $K = 0$—we make an initial trial for the solution. Let us designate this estimate (x_1, y_1). This value and its consequences are shown in Fig. E.4.

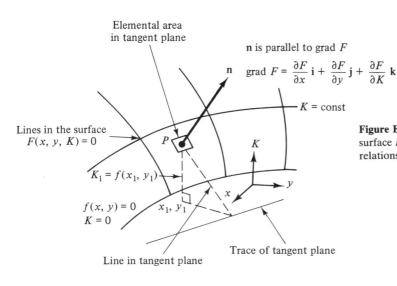

Elemental area in tangent plane

$\mathbf{n}$ is parallel to grad F

$\mathbf{n}$ grad $F = \dfrac{\partial F}{\partial x}\mathbf{i} + \dfrac{\partial F}{\partial y}\mathbf{j} + \dfrac{\partial F}{\partial K}\mathbf{k}$

K = const

Lines in the surface $F(x, y, K) = 0$

P

K

y

$K_1 = f(x_1, y_1)$

x

$f(x, y) = 0$ x_1, y_1

$K = 0$

Line in tangent plane

Trace of tangent plane

Figure E.4 Geometric features of the surface $F(x, y, K) = 0$ and their interrelationships near the plane $K = 0$.

The choice of (x_1, y_1) can be used to find $K = f(x_1, y_1)$ and to determine the normal to the surface $F(x, y, K) = 0 = K - f(x, y)$ at the point $(x_1, y_1, f(x_1, y_1))$. This normal has the same direction as the gradient vector at the point in question.

The expression for the gradient in *our current* Cartesian coordinates, (x, y, K), is

$$\left(\frac{\partial F}{\partial x}\right)\mathbf{i} + \left(\frac{\partial F}{\partial y}\right)\mathbf{j} + \left(\frac{\partial F}{\partial K}\right)\mathbf{k} \tag{E.3}$$

which, in view of our choice of $F(x, y, K) = K - f(x, y)$, is

$$-\left(\frac{\partial f}{\partial x}\right)\mathbf{i} - \left(\frac{\partial f}{\partial y}\right)\mathbf{j} + (1)\mathbf{k} \tag{E.4}$$

The tangent plane to the surface at point P will have the same normal as the surface at that point. Hence, the equation of the plane with such a normal, the tangent plane must be

$$K - f(x_1, y_1) = \left(\frac{\partial f}{\partial x}\right)_1 \Delta x + \left(\frac{\partial f}{\partial y}\right)_1 \Delta y \tag{E.5}$$

In this expression, Δx and Δy are the coordinate excursions from x_1 and y_1. This tangent plane will intersect the xy plane when $K = 0$ and will make the trace

$$-f(x_1, y_1) = \left(\frac{\partial f}{\partial x}\right)_1 \Delta x + \left(\frac{\partial f}{\partial y}\right)_1 \Delta y \tag{E.6}$$

This is a straight line.* In the same way, the tangent plane to the surface $G(x, y, K) = 0$ will make a trace in the xy plane that has the form

$$-g(x_1, y_1) = \left(\frac{\partial g}{\partial x}\right)_1 \Delta x + \left(\frac{\partial g}{\partial y}\right)_1 \Delta y \tag{E.7}$$

The solution of Eqs. (E.6) and (E.7) is the intersection of these two traces. This intersection, with the associated Δx and Δy, yields new values $x_2 = x_1 + \Delta x$ and $y_2 = y_1 + \Delta y$. Again, $f(x_2, y_2)$, $g(x_2, y_2)$, $(\partial f/\partial x)_2$, $(\partial g/\partial x)_2$, and so on, are evaluated to obtain a new Δx and Δy which would provide x_3 and y_3. The process ceases when Δx and $\Delta y \to 0$.

For $f(x, y) = xy - 1 = 0$, $\partial f/\partial x = y$ and $\partial f/\partial y = x$. Using the function $g(x, y) = x^2 + y^2 - 9 = 0$, one writes $\partial g/\partial x = 2x$ and $\partial g/\partial y = 2y$. The equations can be written in matrix notation as

$$\begin{vmatrix} \dfrac{\partial f}{\partial x} & \dfrac{\partial f}{\partial y} \\ \dfrac{\partial g}{\partial x} & \dfrac{\partial g}{\partial y} \end{vmatrix}\begin{pmatrix} \Delta x \\ \Delta y \end{pmatrix} = \begin{pmatrix} -f \\ -g \end{pmatrix} \quad \text{or} \quad \begin{pmatrix} y & x \\ 2x & 2y \end{pmatrix}\begin{pmatrix} \Delta x \\ \Delta y \end{pmatrix} = \begin{pmatrix} -(xy - 1) \\ -(x^2 + y^2 - 9) \end{pmatrix} \tag{E.8}$$

If, for (x_1, y_1), one chooses $(3, 1)$, this general matrix equation would read

$$\begin{pmatrix} 1 & 3 \\ 6 & 2 \end{pmatrix}\begin{pmatrix} \Delta x \\ \Delta y \end{pmatrix} = \begin{pmatrix} -[3(1) - 1] \\ -(3^2 + 1^2 - 9) \end{pmatrix}$$

---

* Notice the similarity with Eq. (E.2) for Newton's method.

Newton-Raphson Method of Solution for Pipe Networks

which could be solved for Δx and Δy. The next iteration would employ for x the value $3 + \Delta x$, using the value of Δx just found. Similarly, the next value of y would be $1 + \Delta y$. The Δx and Δy become very small after three iterations, giving the solution $(2.981, 0.335)$. The initial choice of $(5, 4)$ takes five iterations to obtain the solution.

While the manner of solution for the 2×2 system of equations, Eqs. (E.6) and (E.7), is inconsequential, it becomes a matter of considerable importance when solving larger systems of equations. Then the method of determinants can be costly in terms of computer storage requirements, and significant difficulties can occur due to "round-off" errors. Consequently, it would behoove anyone interested in solving systems of linear equations to become acquainted with the techniques, such as Gauss elimination, that can be effectively used. In the present case, for example, a transformation of the general coefficient matrix and constant vector given as Eq. (E.8) is

$$\begin{pmatrix} 2y^2 - 2x^2 & 0 \\ 0 & 2x^2 - 2y^2 \end{pmatrix} \begin{pmatrix} \Delta x \\ \Delta y \end{pmatrix} = -\begin{pmatrix} 2y(xy - 1) - x(x^2 + y^2 - 9) \\ 2x(xy - 1) - y(x^2 + y^2 - 9) \end{pmatrix} \quad \text{(E.9)}$$

This expression occurs if each of the equations of the system is multiplied by suitable factors so that one of the two variables is eliminated when the equations are subsequently added to or subtracted from one another. Such a form makes it quite easy to find Δx and Δy at each step of the iteration process. Using $(3, 1)$ as an initial trial for (x, y), the matrix equation becomes

$$\begin{pmatrix} -16 & 0 \\ 0 & 16 \end{pmatrix} \begin{pmatrix} \Delta x \\ \Delta y \end{pmatrix} = \begin{pmatrix} -1 \\ -11 \end{pmatrix}$$

One can use the logic of this discussion, and the similar reasoning used in the development of Newton's method for the one-variable case, to find solutions not only for $f(x, y) = 0$ and $g(x, y) = 0$, but also $u(x, y, z) = 0$, $v(x, y, z) = 0$, and $w(x, y, z) = 0$. One must then solve a 3×3 system of equations for Δx, Δy, and Δz. The procedure for J functions in J variables is now just an extension of the 2×2 system of equations given by Eqs. (E.6) and (E.7). Such systems of equations occur in pipe networks, where the variables are Q_1, Q_2, Q_3, and so forth. The corresponding correction terms, such as ΔQ_1, ΔQ_2, and ΔQ_3, were given the symbol $\tilde{Q}_j$ in Appendix D for a more compact notation.

Example E.1

The piping system shown in Fig. D.10 was used to develop the equations that would have to be solved to determine the flow in each branch. The BASIC computer code listed below shows most of the elements that would be needed to solve this problem on a PC. (Note that the subroutine for the solution to the system of linear equations has not been included. Such programs are generally available to PC users.) The results of several initial trial values are shown directly following the computer code.

```
10 ' prgm is called trespipe.3
12 ' prgm solves a 3 reservoir problem (Fig. A.D.10) using Newton-Raphson
13 ' prgm needs a linear equation solver to function successfully.
15 DIM A(20,20),B(20,20),D(20,20)
16 DIM M(20),VARIABLE(20,20),N(20,20)
17 DIM X(20),DIA(20),R(20),F(20),L(20),DX(20),FF(20),DX#(20)
18 '
19 DEFINT E-M : 'useful in equation solver section
20 DEFDBL A-D,S,X
34 '
```

Appendix E

```
11    35 TOLER = .05: 'this is the allowable error limit
12    36 '
13    38 NU = 1/10^5 : 'this is kinematic viscosity for water in ft^2/s
14    40 PI= 3.1415962#
15    42 TEE = 32.2: ' Gravitational constant
16    45 CLS
17    60 INPUT "How many pipes in network"; PMAX: 'PMAX=3 in this special case
18    61 ' Above statement would be more useful in a more general program
19    62 PRINT "": PRINT ""
20    63 PMAX = 3
21    65 Q = PMAX:' needed for information transer to set up equation matrix
22    66 QUE = Q
23    70 '    FOR E = 1 TO PMAX
24    80 '      PRINT "pipe = "E:PRINT "Use inches for diameter and feet for length"
25    82 '      PRINT ""
26    85 '      INPUT "Give pipe diameter, length, & friction factor";DIA(E),L(E),nF(E)
27    90 '    NEXT E
28    92 DIA(1)=4: L(1)=500
29    93 NF(1) = .023: 'An assumed value for f.  Use Eq. 10.2.3 for generality.
30    94 DIA(2)=3: L(2)=300
31    95 NF(2) = .026: 'Assumed value for the friction factor.
32    96 DIA(3)=4: L(3)=400
33    97 NF(3) = .023: 'Assumed value.
34    100 INPUT "What is the elevation (in feet) of reservoir 1"; Z(1)
35    110 INPUT "What is the elevation of reservoir 6"; Z(6)
36    120 INPUT "What is the elevation of reservoir 7"; Z(7)
37    125 PRINT "The elevation of reservoir 1 is "; Z(1)
38    126 LPRINT "The elevation of reservoir 1 is "; Z(1)
39    130 PRINT "The elevation of reservoir 6 is "; Z(6)
40    132 LPRINT "The elevation of reservoir 6 is "; Z(6)
41    135 PRINT "The elevation of reservoir 7 is "; Z(7)
42    137 LPRINT "The elevation of reservoir 7 is "; Z(7)
43    145 '
44    150 A1= PI*(DIA(1)^2)/576: A2= PI*(DIA(2)^2)/576: A3= PI*(DIA(3)^2)/576
45    160 R1 = NF(1)*L(1)/(DIA(1)/12)/(2*TEE)/A1^2: 'Pipe "resistance"
46    170 R2 = NF(2)*L(2)/(DIA(2)/12)/(2*TEE)/A2^2: ' "         "
47    180 R3 = NF(3)*L(3)/(DIA(3)/12)/(2*TEE)/A3^2: ' "         "
48    190 '
49    200 CLS: PRINT "Give flow rates in c.f.s.": PRINT ""
50    205 PRINT "Do not give 0's": PRINT ""
51    210 INPUT "what is the initial estimate of Q2 and Q3";X(2),X(3)
52    212 X(1) = X(2) +X(3)
53    214 ' x(1) is Q1 = Q23  in Eqs. A.D.12 - A.D.14, and Q2 = Q34; Q3 = Q35.
54    220 FLAG = 0: 'This is an indicator to show if the tolerance limit is achieved.
55    230 GOSUB 5300
56    240 GOSUB 1100
57    260 '
58    270    FOR E = 1 TO PMAX
59    275    IF ABS( DX#(E) ) > TOLER THEN FLAG = 1
60    280    X(E) = X(E) +DX#(E)
61    290    NEXT E
62    300 IF FLAG > 0 THEN GOTO 220
63    400 GOSUB 4432
64    495 '
65    500 END
66    502 '
67    510 '.................... SUBROUTINES .......................
68    560 '
69    1095 ' Input of coefficients of unknowns, x(j).  These are called a(i,j).
70    1096 'These a(i,j) and the constant matrix are processed in a "solver".
71    1097 'Such a linear equation solver often utilizes Gauss elimination.
72    1100    FOR I= 1 TO Q
73    1900 '
74    2000 ' Statements 1100-4470 are the elements of a linear equation solver
75    3000 ' that must be provided to solve the system of equations that result
76    4000 ' from the Newton-Raphson method of solution.
77    4100 '
78    4465 LPRINT ""
79    4470 RETURN
80    5295 ' subroutine to calculate matrix coefficients
81    5300 DF1DX1= -2*R1*ABS( X(1) )
82    5302 A(1,1) = DF1DX1
83    5305 DF1DX2 = -2*R2*ABS( X(2) )
84    5307 A(1,2) = DF1DX2
```

Newton-Raphson Method of Solution for Pipe Networks   723

```
85      5310 DF1DX3 = O
86      5312 A(1,3) = DF1DX3
87      5315 DF2DX1 = O
88      5317 A(2,1) = DF2DX1
89      5320 DF2DX2 = +2*R2*ABS( X(2) )
90      5322 A(2,2) = DF2DX2
91      5325 DF2DX3 = -2*R3*ABS( X(3) )
92      5327 A(2,3) = DF2DX3
93      5330 DF3DX1 = -1
94      5332 A(3,1) = DF3DX1
95      5335 DF3DX2 = +1
96      5337 A(3,2) = DF3DX2
97      5340 DF3DX3 = +1
98      5342 A(3,3) = DF3DX3
99      5350 F(1)= ( Z(1) -Z(6) ) -R2*X(2)*ABS( X(2) ) -R1*X(1)*ABS( X(1) )
100     5360 F(2)= ( Z(6) -Z(7) ) -R3*X(3)*ABS( X(3) ) +R2*X(2)*ABS( X(2) )
101     5370 F(3) = +X(2) +X(3) -X(1)
102     5400 RETURN
```

The elevation of reservoir 1 is 100
The elevation of reservoir 6 is 110
The elevation of reservoir 7 is 10
Initial trials: Q1 = 2   Q2 =   .5   Q3 =   1.5

THE FOLLOWING IS THE SOLUTION SET TO THE SYSTEM: Q1 = x(1). etc.
(1) = +6.231523D−01
(2) = −4.315538D−01
(3) = +1.054706D+00
number of steps =   5
****

The elevation of reservoir 1 is 100
The elevation of reservoir 6 is 110
The elevation of reservoir 7 is 10
Initial trials: Q1 = 1   Q2 =   −.5   Q3 =   1.5

THE FOLLOWING IS THE SOLUTION SET TO THE SYSTEM: Q1 = x(1). etc.
(1) = +6.251189D−01
(2) = −4.306529D−01
(3) = +1.055772D+00
number of steps =   3
****

The elevation of reservoir 1 is 100
The elevation of reservoir 6 is 90
The elevation of reservoir 7 is 10
Initial trials: Q1 = 2   Q2 =   .5   Q3 =   1.5

THE FOLLOWING IS THE SOLUTION SET TO THE SYSTEM: Q1 = x(1). etc.
(1) = +6.780437D−01
(2) = −3.323021D−01
(3) = +1.010346D+00
number of steps =   4
****

The elevation of reservoir 1 is 100
The elevation of reservoir 6 is 90
The elevation of reservoir 7 is 10
Initial trials: Q1 = 1   Q2 =   −.5   Q3 =   1.5

THE FOLLOWING IS THE SOLUTION SET TO THE SYSTEM: Q1 = x(1). etc.
(1) = +6.792302D−01
(2) = −3.317348D−01
(3) = +1.010965D+00
number of steps =   3
****

The elevation of reservoir 1 is 100
The elevation of reservoir 6 is 70
The elevation of reservoir 7 is 10
Initial trials: Q1 = 2   Q2 =   .5   Q3 =   1.5

THE FOLLOWING IS THE SOLUTION SET TO THE SYSTEM: Q1 = x(1). etc.
(1) = +7.454904D−01
(2) = −2.092671D−01
(3) = +9.547575D−01
number of steps =   4
****

The elevation of reservoir 1 is 100
The elevation of reservoir 6 is 70
The elevation of reservoir 7 is 10
Initial trials: Q1 = 1   Q2 =   −.5   Q3 =   1.5

THE FOLLOWING IS THE SOLUTION SET TO THE SYSTEM: Q1 = x(1). etc.
(1) = +7.438656D−01
(2) = −2.116092D−01
(3) = +9.554748D−01
number of steps =   3
****

The elevation of reservoir 1 is 100
The elevation of reservoir 6 is 50
The elevation of reservoir 7 is 10
Initial trials: Q1 = .9   Q2 =   .1   Q3 =   .8

THE FOLLOWING IS THE SOLUTION SET TO THE SYSTEM: Q1 = x(1). etc.
(1) = +8.465827D−01
(2) = +5.661174D−03
(3) = +8.409215D−01
number of steps =   2
****

The elevation of reservoir 1 is 100
The elevation of reservoir 6 is 50
The elevation of reservoir 7 is 10
Initial trials: Q1 = 4   Q2 =   1   Q3 =   3

Newton-Raphson Method of Solution for Pipe Networks   **725**

THE FOLLOWING IS THE SOLUTION SET TO THE SYSTEM: Q1 = x(1). etc.
(1) = +8.431602D−01
(2) = −3.039115D−03
(3) = +8.461993D−01
number of steps =   5
****

The elevation of reservoir 1 is 100
The elevation of reservoir 6 is 30
The elevation of reservoir 7 is 10
Initial trials: Q1 = 2.5   Q2 =   .5   Q3 =   2

THE FOLLOWING IS THE SOLUTION SET TO THE SYSTEM: Q1 = x(1). etc.
(1) = +9.307145D−01
(2) = +2.142588D−01
(3) = +7.164558D−01
number of steps =   4
****

The elevation of reservoir 1 is 100
The elevation of reservoir 6 is 30
The elevation of reservoir 7 is 10
Initial trials: Q1 = 2.5   Q2 =   −.5   Q3 =   3

THE FOLLOWING IS THE SOLUTION SET TO THE SYSTEM: Q1 = x(1). etc.
(1) = +9.314310D−01
(2) = +2.138422D−01
(3) = +7.175888D−01
number of steps =   4
****

The elevation of reservoir 1 is 100
The elevation of reservoir 6 is 10
The elevation of reservoir 7 is 10
Initial trials: Q1 = 1   Q2 =   .9   Q3 =   .1

THE FOLLOWING IS THE SOLUTION SET TO THE SYSTEM: Q1 = x(1). etc.
(1) = +9.751522D−01
(2) = +3.369260D−01
(3) = +6.382262D−01
number of steps =   3
****

The elevation of reservoir 1 is 100
The elevation of reservoir 6 is 10
The elevation of reservoir 7 is 10
Initial trials: Q1 = −3   Q2 =   2   Q3 =   −5

THE FOLLOWING IS THE SOLUTION SET TO THE SYSTEM: Q1 = x(1). etc.
(1) = +9.756081D−01
(2) = +3.388964D−01
(3) = +6.367117D−01
number of steps =   7
****

Appendix F

Specific Heats, Enthalpy, and Entropy

We employ the energy equation, in the context of both a closed and an open system, in Chap. 6. It is useful to consider a closed system (Fig. F.1) which consists of a rigid container and the enclosed gas into which a quantity of energy δq^* has been transferred as heat. The first law of thermodynamics is, for this closed system,

$$\delta q = de + p(d\tilde{v}) \tag{F.1}$$

in which e is the internal energy per unit mass and $d\tilde{v}$ is the change of specific volume $\tilde{v} = 1/\rho$ of the gas. Because the walls of the container are assumed to be rigid, $d\tilde{v} = 0$. This means that $\delta q = de$. Because this is true exactly at all instants of time during the energy transfer, we write $\delta q/\delta t = \dot{q} = de/dt$. If we had recorded the temperatures inside the vessel before and after δq was transferred, we would expect that, for an addition of energy in the form of heat, the temperature would increase. A

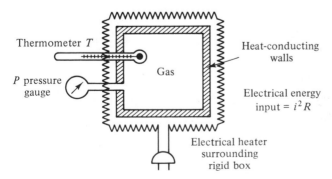

Thermometer T

P pressure gauge

Gas

Heat-conducting walls

Electrical energy input $= i^2 R$

Electrical heater surrounding rigid box

Figure F.1 Thermodynamic experiment in a rigid box surrounded by an electric heater.

---

* The symbol δ has been used to indicate that the associated variable is not a total differential.

graph of $\dot{q}$ (e.g., from an electric current and a resistor) versus temperature might look like Fig. F.2.

The slope of the curve in Fig. F.2 is called the *specific heat at constant volume, c_v*. Often, there is a temperature range over which c_v is independent of the temperature. In addition, c_v may be independent of p. This temperature and pressure independence of c_v will be assumed for the remainder of this discussion. In view of the constant-volume process which is under discussion,

$$\frac{\delta q}{\delta T} = c_v = \text{constant} = \frac{de}{dT} \tag{F.2}$$

which gives, for constant c_v,

$$e = c_v T + e_0 \tag{F.3}$$

The term e_0 is a constant of integration and is the internal energy at some reference temperature.

Naturally, all changes of conditions for a gas are not associated with a constant-volume process. Figure F.3 shows a situation in which the pressure is kept constant while an amount of energy as heat, δq, is transferred to the gas. (Why is the pressure constant in Fig. F.3?) In this case, Eq. (F.1), although applicable, is not convenient for

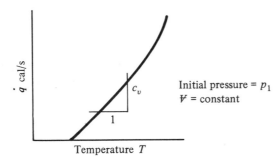

Figure F.2 Plot of heat input versus temperature for the gas in the box shown in Fig. F.1.

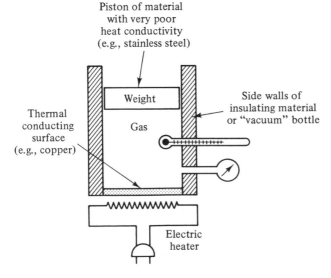

Figure F.3 Thermodynamic experiment in a container that is maintained at constant pressure.

a preliminary discussion. Instead, we choose to rearrange the equation by introducing the product of the pressure and specific volume, $p\tilde{v}$. In the case of an ideal gas, $p\tilde{v} = RT$, in which R is a constant, so that the product $p\tilde{v}$ is proportional to the temperature of the gas.

The ideal gas law need *not* be assumed in what follows immediately below; nevertheless, the differential of $p\tilde{v}$ will be assumed to exist, so that $d(p\tilde{v}) = \tilde{v}\,dp + p\,d\tilde{v}$. This result can be combined with Eq. (F.1) to give

$$\delta q = de + d(p\tilde{v}) - \tilde{v}(dp) \tag{F.4}$$

For the situation illustrated in Fig. F.3, where $dp = 0$,

$$\delta q = de + d(p\tilde{v}) = d(e + p\tilde{v})$$

The term $e + p\tilde{v}$ is called the *enthalpy* and is designated by the letter h. Once again, one could measure the temperature before and after the addition of δq. We could make a graph similar to Fig. F.2, but in this case the slope would be c_p, the specific heat at constant pressure. Because δq for a constant-pressure process can be associated with the enthalpy, we write*

$$\frac{\delta q}{\delta T} = c_p = \frac{d(e + p\tilde{v})}{dT} = \frac{dh}{dT} \tag{F.5}$$

Again, if c_p is temperature-independent,

$$h = e + p\tilde{v} = c_p T + h_0 \tag{F.6}$$

in which h_0 is the enthalpy at some reference temperature. This last result for the enthalpy of an ideal gas permits Eq. (F.4) to be written as

$$\delta q = d(e + p\tilde{v}) - \tilde{v}\,dp = c_p\,dT - \tilde{v}\,dp \tag{F.7}$$

Finally, it is possible to interrelate the two specific heats and the gas constant R. This can be done by considering a closed vessel that is filled with an ideal gas. There is a unit mass of gas, so that all the volumes will be specific volumes. A given quantity of energy in the form of heat, $_1\delta q_2$,† is added to the gas while it is confined at constant volume. The pressure increases from p_1 to p_2. This process is shown in Fig. F.4. After state point 2 is reached, the gas is allowed to expand at constant temperature to a

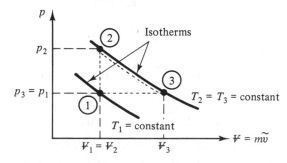

Figure F.4 Thermodynamic cycle with paths of constant volume, temperature, and pressure.

---

*The total derivative has been used because e and pv are assumed to be functions of temperature only.

†The subscripts denote the initial and final state points.

Specific Heats, Enthalpy, and Entropy

specific volume $\tilde{v}_3$ at which the pressure p_3 is equal to the original pressure p_1. Last, the gas is returned to its original state at constant pressure. The dotted line in Fig. F.4 describes this thermodynamic cycle. For this *closed* system, we can write the first law of thermodynamics, $\delta q + \delta w = de$, which states that the internal energy will be changed by energy transfers as heat and work to the fluid. This equation is true between any two of the states of the cycle shown in Fig. F.4. Thus, $_1\delta q_2 + {_1\delta w_2} = {_1de_2}$, or $_1\delta q_3 + {_1\delta w_3} = {_1de_3}$. The energy additions to the gas are additive, so that $_1\delta q_3 = {_1\delta q_2} + {_2\delta q_3}$. In going around the complete path, we would return to state point 1, so that $_1\delta q_1 = {_1\delta q_2} + {_2\delta q_3} + {_3\delta q_1}$. A similar subdivision could be made for $_1\delta w_1$ and $_1de_1$. The latter term is the difference in the internal energy of the gas at two different times but at the same thermodynamic state. Since the value of the internal energy is dependent only on the thermodynamic state and not the path by which that state was reached, $_1de_1$ is zero. The same is not true for $_1\delta w_1$, since the amount of energy transfer as work to the gas will depend on the path that constitutes the thermodynamic cycle. Consequently, we shall examine $_1\delta w_2$, $_2\delta w_3$, and $_3\delta w_1$ and sum them together to obtain $_1\delta w_1$.

From the definition of work, $\delta W = \mathbf{F} \cdot \delta \mathbf{l}$, we move to $\delta W = -\int p \, (dA)(\text{distance}) = -\int p \, dV$, in which dA is the area on which p acts and $dV \, (= d\tilde{v})$ is the volume change. The distance is the displacement of the dA surface in the direction of the surface's normal. This results in a displaced volume $dV \, (= (\text{unit mass})(d\tilde{v}))$. The work from point 1 to point 2 follows by integration:

$$_1\delta w_2 = -\int_1^2 p \, d\tilde{v}$$

Since there was no change in volume in this part of the cycle, $_1\delta w_2 = 0$. It is convenient to evaluate $_3\delta w_1$ now, and so we write

$$_3\delta w_1 = -\int_3^1 p \, d\tilde{v} = -p_1(\tilde{v}_1 - \tilde{v}_3) = +p_1(\tilde{v}_3 - \tilde{v}_1)$$

We will not explicitly evaluate $_2\delta w_3$. However, we know from the first law of thermodynamics that it will equal $-{_2\delta q_3}$, because $_2de_3$ is zero. This conclusion is a result of the fact that, for an ideal gas, the internal energy is a function of the temperature only. The path from state 2 to state 3 is given to be isothermal, and so the constancy of the internal energy follows. We now have all the quantities we need, since

$$_1\delta q_2 = c_v(T_2 - T_1) \quad \text{and} \quad {_3\delta q_1} = c_p(T_1 - T_3)$$

Thus, for the complete cycle, the expression

$$({_1\delta q_2} + {_2\delta q_3} + {_3\delta q_1}) + ({_1\delta w_2} + {_2\delta w_3} + {_3\delta w_1}) = 0$$

can be written on a unit mass basis as

$$c_v(T_2 - T_1) + c_p(T_1 - T_3) + 0 + p_1(\tilde{v}_3 - \tilde{v}_1) = 0$$

In the last term, we have $p_1 \tilde{v}_1$, which, from the equation of state, is RT_1. Since $p_1 = p_3$, the product $p_1 \tilde{v}_3$ is $p_3 \tilde{v}_3$, or RT_3. Last, in the first parenthesis we can substitute $T_3 = T_2$, since an isothermal process is being considered. A reworking of the last equation then gives

$$R(T_3 - T_1) = c_p(T_3 - T_1) - c_v(T_3 - T_1)$$

It follows directly that for any temperature difference,

$$R = c_p - c_v \qquad \text{(F.8a)}$$

By introducing the ratio of the specific heats $c_p/c_v = k$, one can rewrite the equation as

$$\frac{c_p}{R} = \frac{k}{k-1} \qquad \text{(F.8b)}$$

Until now, we have not made use of the second law of thermodynamics with its related concept of entropy. We shall now summarize some elementary consequences of this law. The formal definition of the change of entropy that a substance undergoes during a reversible process,

$$s_2 - s_1 = \int_1^2 \frac{dq}{T} \qquad \text{(F.9)}$$

is the result of a development that starts with the Clausius statement of the second law of thermodynamics:

> No process is possible whose sole result is the removal of heat from a reservoir at one temperature and the absorption of an equal amount of heat by a reservoir at a higher temperature.

From this concept, one proceeds to the Carnot cycle, for which the Clausius inequality is a relation between the temperature of an arbitrary number of heat reservoirs that transfer heat to the substance while it undergoes changes during an arbitrary heat-engine process. These transfers of energy in the form of heat occur at $T_1, T_2, \ldots, T_N$ for each of the N reservoirs that are selected to transfer the appropriate amount of heat, Δq_n, at each step. We can designate the amount of energy added to a material volume at the higher temperature as Δq_h—a negative number if energy were withdrawn from the system at this temperature. Then, for each Carnot cycle that operates between T_h and T_l, $\Delta q_h/T_h + \Delta q_l/T_l = 0$. From such a model follows the statement

$$\sum_{n=1}^{N} \left(\frac{\Delta q_n}{T_n} \right) \leq 0 \qquad \text{(F.10)}$$

with the equality holding only if the process is reversible. *If* an infinite number of steps are used during which dq is transferred *reversibly* at temperature T, we write, instead of Eq. (F.10),

$$\oint \frac{dq}{T} = 0 \qquad \text{(F.11)}$$

The circle has been included on the integral sign to convey the meaning that the integration is for a complete cycle. The substance, which is assumed to be a gas in this discussion, starts at one state point and, after proceeding through a range of other state points, ultimately returns to the original point. A possible cycle is shown in Fig. F.5. Because the integral for the complete cycle is zero, one can reach the conclusion that, according to Fig. F.5,

$$\int_1^2 \left(\frac{dq}{T} \right)_{\text{along I}} = -\int_2^1 \left(\frac{dq}{T} \right)_{\text{along II}}$$

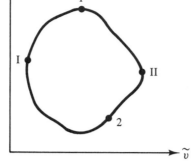

Figure F.5 Diagram of a cycle with pressure and specific volume as coordinates.

Now, because the line integral in one direction should be the negative of that in the opposite direction, we write

$$\int_1^2 \left(\frac{dq}{T}\right)_{\text{along II}} = -\int_2^1 \left(\frac{dq}{T}\right)_{\text{along II}}$$

The result of these two steps is

$$\int_1^2 \left(\frac{dq}{T}\right)_{\text{along I}} = \int_1^2 \left(\frac{dq}{T}\right)_{\text{along II}}$$

or that the integral $\int_1^2 (dq/T)$ is independent of the path—either path I or path II—between points 1 and 2. This means that the *entropy is a function only of the end points* 1 and 2.

We can calculate the difference in entropy between two thermodynamic state points by choosing a reversible process that connects them and performing the integration of Eq. (F.9). In fact, the reversible process that is chosen *need not* be the same as the one that originally took the gas from point 1 to point 2. Also, the process that actually accompanied the change of state might not have been reversible. However, the entropy at the initial and final points is uniquely determined by these state points, so that the choice of any convenient *reversible* process that connects states 1 and 2 can be employed for finding $s_2 - s_1$. The use of Eq. (F.9) will determine the change of entropy that is associated with the change of state.

An examination of Eq. (F.9) makes it apparent that if the reversible process is associated with an absence of heat transfer—i.e., if it is an adiabatic process—then the change of entropy would be zero. This situation is called, accordingly, an *isentropic process*. Other possibilities exist, and the associated entropy changes can be calculated with the aid of the first law of thermodynamics,

$$dq = de + \frac{p\,(d\tilde{v})}{J}$$

in which $p\,(d\tilde{v})$ is the energy supplied to the surroundings in the form of work done *by* the system [refer to Eq. (6.3.2)] and J is the Joule constant. J represents the equivalent mechanical work that can be performed by the transfer of the same amount of energy that would cause the temperature of 1 lb of water ($g = 32.2$ ft/s^2) to rise 1°F. The water would be insulated, and standard pressure is assumed. Joule deduced this equivalence by measuring the amount of chemical energy that was transformed to electrical energy in a battery and subsequently to the forms of heat and work. ($J = 778$ ft-lb/Btu, or 1.0 in SI units.) We also know that $de = c_v(dT)$. This quantity can now be substituted in Eq. (F.9) to give

$$s_2 - s_1 = c_v \int_1^2 \frac{dT}{T} + \frac{1}{J} \int_1^2 \frac{p}{T} d\tilde{v}$$

for a constant value of c_v. Then

$$s_2 - s_1 = c_v \int_1^2 \frac{dT}{T} + \frac{R}{J} \int_1^2 \frac{d\tilde{v}}{\tilde{v}}$$

if the equation of state for an ideal gas is used in the second integral. This last expression can be readily integrated to give

$$s_2 - s_1 = c_v \ln \frac{T_2}{T_1} + \frac{R}{J} \ln \frac{\tilde{v}_2}{\tilde{v}_1} \tag{F.12}$$

The coefficients on the logarithms can be converted into exponents of the arguments of the logarithms. Hence

$$s_2 - s_1 = \ln \left[\left(\frac{T_2}{T_1} \right)^{c_v} \left(\frac{\tilde{v}_2}{\tilde{v}_1} \right)^R \right]$$

for $J = 1$. The relationship between R and the specific heats that is given by Eq. (F.8a) permits the rewriting of this last result as

$$s_2 - s_1 = \ln \left[\left(\frac{T_2}{T_1} \right)^{c_v} \left(\frac{\tilde{v}_2}{\tilde{v}_1} \right)^{c_v(c_p - c_v)/c_v} \right]$$

$$= \ln \left[\left(\frac{T_2}{T_1} \right) \left(\frac{\tilde{v}_2}{\tilde{v}_1} \right)^{k-1} \right]^{c_v}$$

$$= c_v \ln \left[\left(\frac{T_2}{T_1} \right) \left(\frac{\tilde{v}_2}{\tilde{v}_1} \right)^{k-1} \right] \tag{F.13}$$

Now, if $s_1 = s_2$, as it would for an isentropic process,

$$\frac{T_2}{T_1} \left(\frac{\tilde{v}_2}{\tilde{v}_1} \right)^{k-1} = \frac{T_2}{T_1} \left(\frac{\rho_1}{\rho_2} \right)^{k-1} = 1 \tag{F.14}$$

A use of $p_1/\rho_1 = RT_1$ and $p_2/\rho_2 = RT_2$ allows Eq. (F.14) to be rewritten as

$$\frac{p_1}{p_2} = \left(\frac{\rho_1}{\rho_2} \right)^k \tag{F.15}$$

and

$$\frac{T_1}{T_2} = \left(\frac{p_1}{p_2} \right)^{(k-1)/k} \tag{F.16}$$

for isentropic processes.

Specific Heats, Enthalpy, and Entropy

Example F.1

Some reversible processes (with heat transfer) can be characterized by the equation $p/\rho^n = $ constant, for which $n \neq k$. (See the section on polytropic processes in a standard thermodynamics textbook.) The change of entropy should then be nonzero. This can easily be demonstrated as follows. The equation of state can be combined with the stated process equation to yield

$$\frac{p_1}{p_2} = \left(\frac{\rho_1}{\rho_2}\right)^n = \frac{R\rho_1 T_1}{R\rho_2 T_2}$$

with the consequence that

$$\left(\frac{\rho_1}{\rho_2}\right)^{n-1} = \frac{T_1}{T_2} = \left(\frac{\tilde{v}_2}{\tilde{v}_1}\right)^{n-1}$$

Equation (F.12) can then be written in terms of only T_2/T_1 as

$$s_2 - s_1 = c_v \ln\left(\frac{T_2}{T_1}\right) + \frac{R}{J} \ln\left(\frac{T_2}{T_1}\right)^{1/(1-n)}$$

$$= c_v \ln\left(\frac{T_2}{T_1}\right) + \frac{c_p - c_v}{1-n} \ln\left(\frac{T_2}{T_1}\right)$$

for $J = 1$. These two logarithms can be combined to give

$$s_2 - s_1 = c_v\left[\ln\left(\frac{T_2}{T_1}\right)\right]\left(1 + \frac{k-1}{1-n}\right) = c_v\left(\frac{n-k}{n-1}\right)\ln\left(\frac{T_2}{T_1}\right)$$

EXERCISES

F.1(a) What is the change of entropy if $n = k$? According to the relationship developed in the example, if $n = k$, what kind of process is being considered? What should be the Δs for such a process on theoretical grounds?

(b) Air ($k = 1.4$) flowing between a reservoir at 77°C and a location in a pipe where the temperature is 20°C experiences a transfer of heat amounting to 0.17776 Btu/slug–°R. Is the flow isentropic? What is the appropriate value of n for this process?

(c) If in doubling the length between the reservoir of Exercise F.1b and the location in the pipe under examination the total entropy change also doubles, what is the expected temperature at this location? Assume that $n = 1.1$.

Appendix G

Integration Methods in Computer Programs

Three computer programs are listed below. The first two implement a "simple" Euler integration method and an "improved" Euler method for comparison. Figure 8.7.2 might be useful to review when studying these programs.

The third program, titled KP&RK.1, uses Runge-Kutta integration to apply Eq. (8.7.18) for the flow past a circular cylinder. While Runge-Kutta methods are very widely used because of their efficacy and accuracy, the integration of Eq. (8.7.18) can be effectively done with one of the Euler methods that are given in the first two progams.

```
12  'prgm is called Euler1.1
13  'This is the simplest form of integration
14  '
20  ' Program integrates dy/dx = f(x,y) = x*x + y.
21  ' initial point is (0,1)
22  '
25  'Program requires a printer to produce the output.
26  '
30  INPUT "What is the increment on x"; DX
33  INPUT "What is the maximimum value if x"; XMAX
34  '
40  POINTS = INT(XMAX/DX):   'Determines number of steps between O and Xmax.
90  X=O: Y=1
100      FOR I= 1 TO POINTS
120      GOSUB 500
130      DY1 = DX*FXY
140      Y = Y + DY1
150      X = X + DX
160      LPRINT "x = "X,"y= "Y,"dy1= "DY1,"i= "I
200      NEXT I
300  '
400  END
425  '  ************ SUBROUTINES *************
500  FXY = X*X + Y: 'FXY is the derivative dy/dx.
530  RETURN
```

```
12 'Program is called Euler2.1 .
14 'Program is a modified Euler method that improves the derivative within dx.
20 'Program integrates dy/dx = f(x,y) = x*x + y.
21 '
22 'Program requires a printer to produce output.
23 '
25 ' Initial point is (0,1).
26 '
30 INPUT "What is the increment on x"; DX
33 INPUT "What is the maximimum value if x"; XMAX
34 '
40 POINTS = INT(XMAX/DX): 'This determines number of steps between 0 and Xmax.
90 X=0: YO=1
100     FOR I= 1 TO POINTS
110     Y = YO
120     GOSUB 500
130     DY1 = DX*FXY
135     FXY1 = FXY
140     Y1 = YO + DY1
150     X = X + DX
160     LPRINT "x = "X,"y1= "Y1,"dy1= "DY1,"i= "I
162     J=0
165     DWHY = DY1: 'Saves y increment estimate from slope at initial point
166 '                  of interval.
168     Y = YO +DY1: 'Prepares to solve for slope at end of interval.
170     GOSUB 500
180     DY2 = DX*FXY: 'Finds y increment based upon slope at end of interval.
200     DYAVG = (DY1 +DY2)/2: 'Finds average dy based upon initial slope and
201 '                  and an estimate of slope at end of interval.
204 '
205     LPRINT "x= "X,"dy2= "DY2,"dyavg= "DYAVG,"j= "J
206 '
208 '   Will compare "improved" y increment with value from previous step.
210     IF ABS(DYAVG -DWHY) < .00005 THEN GOTO 300
220     Y = YO +DYAVG: 'Prepares to evaluate slope at an "improved" y point
221 '                  at the end of the interval, (x +dx).
230     J=J+1
232     IF J>10 THEN GOTO 395
235     DWHY = DYAVG: 'Saves previous estimate of dy for future comparison.
240     GOTO 170
300 '
310     LPRINT "","x= "X,"y= "Y,"j= "J
315     YO= Y
330     NEXT I
390 GOTO 400
395 LPRINT "avg slope in interval not found"
399 '
400 END
425 '      ************* SUBROUTINES ****************
500 FXY = X*X + Y: 'FXY is dy/dx.
505 LPRINT "","","",FXY
530 RETURN
```

```
5 'Program is called KP&RK.1
6 ' Program solves for separation point on a circular cylinder
8 ' a modified Karman-Pohlhausen method with Runge-Kutta integration.
11 '
12 INPUT "how many steps should be taken along body's surface"; IMAX
16 '
17 DIM L(IMAX,4),G2(IMAX,4),EWED2(IMAX),UD2(IMAX,4),DUD2(IMAX),UPRIMOU(4)
18 DIM DELTA1(IMAX),DELTA2(IMAX),DELTA3(IMAX),DELTA4(IMAX),LA(IMAX)
```

```
19  '
25  ' ******** Explanation of the notation used **************
27  ' The Runge Kutta method employs a weighted average of the increments
28  ' of the dependent variable, say y, in an interval of the independent
29  ' variable, e.g., x.  These increments are obtained by determining the slope,
30  ' called f(x,y), at various points in the interval.  The 4 increments are
31  ' labeled DELTA1(N), DELTA2(N), DELTA3(N), & DELTA4(N).  N denotes the
32  ' interval of the independent variable in question.
33  '
34  ' In the following x0 and y0 are the coordinates at the beginning of the
35  ' interval whose length is dx.
36  ' DELTA1 = [ f(x0,y0)  ]*dx
37  ' DELTA2 = [ f(x0 +dx/2, y0 +DELTA1/2) ]*dx
38  ' DELTA3 = [ f(x0 +dx/2, y0 +DELTA2/2) ]*dx
39  ' DELTA4 = [ f(x0 +dx, y0 +DELTA3) ]*dx
40  ' dy = [DELTA1 +2*DELTA2 +2*DELTA3 +DELTA4]/6
42  '
44  '                    ******
46  ' The solution method is associated with Equation 8.7.18
47  '
50  LO = 7.052: 'This is lambda at stagnation point, x =0.
52  '
54  PI = 3.141593
55  LENGTH = PI
56  DX = LENGTH/IMAX
57  XZERO = 0
58  '
59  ' ******* Solution for lambda for first step from x = 0 *********
60  '
61  X = DX
62  L(1,1) = LO : 'this is the lambda for finding delta1(1), i.e., at point 1
63  LAMB = LO: 'Subroutine requires LAMB as a parameter.
64  GOSUB 620: G2(1,1) = G2: DELTA1(1) = G2*DX
65  '
66  EX = DX/2: GOSUB 1500 : 'Evaluation at the MIDDLE of first interval
67  'UD2 is Y in Equation 8.7.17
68  UD2(1,2) = 0 +DELTA1(1)/2: 'This is an argument in DELTA1(2).
69  UPRIMOU(2) = UPRIME/U
70  L(1,2) = UD2(1,2)*UPRIMOU(2)*2/D2SQR: 'Lambda from Equation 8.7.17.
71  LAMB = L(1,2)
72  GOSUB 620: G2(1,2) = G2: DELTA2(1) = G2*DX
73  '
75  EX = DX/2: GOSUB 1500: 'Evaluation is at the MIDDLE of first interval
76  UD2(1,3) = 0 +DELTA2(1)/2: 'Argument for Y in DELTA3(1).
77  UPRIMOU(3) = UPRIME/U
78  L(1,3) = UD2(1,3)*UPRIMOU(3)*2/D2SQR
79  LAMB = L(1,3)
80  GOSUB 620: G2(1,3) = G2: DELTA3(1) = G2*DX
81  '
82  EX = DX: GOSUB 1500
83  UD2(1,4) = 0 +DELTA3(1): 'Argument for Y in DELTA4(1).
84  UPRIMOU(4) = UPRIME/U
86  L(1,4) = UD2(1,4)*UPRIMOU(4)*2/D2SQR
87  LAMB = L(1,4)
88  GOSUB 620: G2(1,4) = G2: DELTA4(1) = G2*DX
89  '
90  'DUD2 is the increment on Y in Equations 8.7.17 and 18.
91  DUD2(1) = (DELTA1(1) +2*DELTA2(1) +2*DELTA3(1) +DELTA4(1) )/6
92  EWED2(1) = 0 +DUD2(1): 'This is Y at the end of the first interval.
93  LA(1) = EWED2(1)*UPRIMOU(4)*2/D2SQR: 'From Equation 8.7.17
94  '
95  PRINT "lambda(1) ="LA(1)
96  LPRINT "lambda(1) ="LA(1)
97  '        ***** End of first interval evaluation of Y *****
98  '
99  'Statements #60-96 could have been incorporated into the next FOR-NEXT loop.
100 '
103     FOR I= 2 TO IMAX: 'xmax will be pi which corresponds to 180 degrees
```

Integration Methods In Computer Programs

```
105      LPRINT ""
106      X = XZERO +DX*I : 'This is the place to which one wants to go
107  '
109  '   This trial is at INITIAL point of current interval.
110      UD2(I,O) = EWED2(I-1)
112      L(I,1) = LA(I-1)
114      LAMB = L(I,1) : PRINT "","","lamb= "LAMB
116      GOSUB 620
118      G2(I,1) = G2
120      DELTA1(I) = G2*DX
121  '
122      EX = (X -DX) +DX/2 : 'This evaluation is at MIDDLE of current interval
124      GOSUB 1500
126      UPRIMOU(2) = UPRIME/U
128      UD2(I,2) = UD2(I,O) + DELTA1(I)/2
130      L(I,2) = UD2(I,2)*UPRIMOU(2)*2/D2SQR
132      LAMB = L(I,2) : PRINT "","","lamb= "LAMB
134      GOSUB 620
136      G2(I,2) = G2
138      DELTA2(I) = G2*DX
139  '
142      UPRIMOU(3) = UPRIMOU(2): 'Also evaluated at MIDDLE of current interval
144      UD2(I,3) = UD2(I,O) +DELTA2(I)/2
146      L(I,3) = UD2(I,3)*UPRIMOU(3)*2/D2SQR
148      LAMB = L(I,3) : PRINT "","","lamb= "LAMB
150      GOSUB 620
152      G2(I,3) = G2
154      DELTA3(I) = G2*DX
155  '
158      EX = (X -DX) +DX  : 'This point is at the END of current interval
160      GOSUB 1500
162      UPRIMOU(4) = UPRIME/U
164      UD2(I,4) = UD2(I,O) +DELTA3(I)
166      L(I,4) = UD2(I,4)*UPRIMOU(4)*2/D2SQR
168      LAMB = L(I,4) : PRINT "","","lamb= "LAMB
170      GOSUB 620
172      G2(I,4) = G2
174      DELTA4(I) = G2*DX
175  '
180      DUD2(I) = (DELTA1(I) +2*DELTA2(I) +2*DELTA3(I) +DELTA4(I) )/6
184      EWED2(I) = EWED2(I-1) +DUD2(I)
185      ' LPRINT "ewed2(i)= "EWED2(I)
186      LA(I) = EWED2(I)*UPRIMOU(4)*2/D2SQR
187      LPRINT "x= "X,"lambda= "LA(I)
188      PRINT "lambda(i)= "LA(I)
190  '   STOP: 'Originally used to assess progress of program.
196      IF LA(I) < -18 THEN GOTO 600
200      NEXT I
255  '
600 END
604  '
605 '*************** SUBROUTINES ******************
606  '
615  ' subroutine to calculate Karman-Pohlhausen parameters
617 'D1 is the displacement thickness and D2 is the momentum thickness.
620 D1= (.3 -LAMB/120): D2= (37/5 -LAMB/15 -LAMB*LAMB/144)/63
621 D2SQR = D2*D2
625 H = D1/D2: 'Shape factor
632 G2 = D2*(  (2 +LAMB/6) -LAMB*(3/2 +H)*D2  ): 'Right hand side of Eqn 8.7.18
633 '                                           This is dY/dx.
635 'LPRINT "g2 = "G2,"lambda = "LAMB
695 RETURN
698  '
1500 'subroutine to calculate U(ex) and U'(ex)
1520 U = 2*SIN(EX)
1530 UPRIME = 2*COS(EX)
1540 'LPRINT "x= "X,"ex= "EX,"U = "U,"U' = "UPRIME
1550 RETURN
```

Appendix H

Solutions to Falkner-Skan Equation by Runge-Kutta Method and Taylor Series Expansion

The Blasius equation, Eq. (16.3.7), can be integrated readily with one of the Runge-Kutta algorithms (of which many variations exist). One must assume a value for the second derivative of the function at the origin and see whether a solution converges to the desired value of the function, or one of its derivatives, at another point in the domain of interest. If the specified boundary condition is not achieved, a new value of the second derivative must be used to determine this number's influence on the solution at the second boundary point. The desired boundary value at the second point of the *two-point boundary-value problem* is sought in this way. The method is described as *shooting,* in the sense that the first assumed values are used to assess the "range" and subsequent ones should improve one's "aim."

The first computer program given below solves the Blasius equation using Runge-Kutta integration.

```
1     10 'prgm is called Rungek1.2
2     11 '
3     12 'Runge-Kutta solutions to Falkner-Skan equations are sought.
4     13 'Refer to development leading to Eqn. 16.4.4, F"' + aFF" + b{1 -F'F'} = O.
5     15 'G1 is F(y).   G2 = dF/dy, or G2 = dG1/dy.   G3 = dG2/dy
6     16 ' F-S gives dG3/dy = -a*G1*G3 -b*(1 -G2*G2)
7     17 ' Thus, the second order O.D.E. has been converted into 3 first order ones
8     18 ' to solve for G1, G2, and G3.
9     19 '
10    20 N = 500
11    30 DIM G1(N),G2(N),G3(N),K1(4),K2(4),K3(4)
12    68 '
13    69 CLS: PRINT "For Blasius problem, a = 0.5, b = O.": PRINT ""
14    70 INPUT "What are the values of a and b"; A,B
15    74 INPUT "What is the assumed value of the second derivative"; G3(O)
16    76 G1(O) = O: G2(O) = O
17    79 PRINT "the eta increment should be greater than or equal to 0.05"
18    80 PRINT "": INPUT "What is the increment on eta"; H
19    90 LPRINT "":LPRINT "****************":LPRINT ""
```

739

```
20      92 LPRINT "constant a= "A,"constant b= "B,"stepsize ="H
21      93 LPRINT "assumed second derivative= "G3(0)
22      130 '
23      145 IMAX = INT(10/H)
24      146 Y =0
25      150     FOR I = 1 TO IMAX
26      160     Y = Y +H
27      170     GOSUB 1000
28      178 '   A type of "Simpson's Rule" follows.
29      179 '   The K's that follow are similar to the DELTAS in program KP&RK.1.
30      180     G1(I) = G1(I-1) +( K1(1) +2*K1(2) +2*K1(3) +K1(4) )/6
31      190     G2(I) = G2(I-1) +( K2(1) +2*K2(2) +2*K2(3) +K2(4) )/6
32      200     G3(I) = G3(I-1) +( K3(1) +2*K3(2) +2*K3(3) +K3(4) )/6
33      250     PRINT "i= "I,"y= "Y
34      255     PRINT "g1(i)= "G1(I),"g2(i)= "G2(I),"g3(i)= "G3(I)
35      260     IF I/10 = INT(I/10) THEN GOSUB 2500
36      270     NEXT I
37      438 '
38      450 INPUT "Do you wish to try another value of the second derivative"; SI$
39      452 IF SI$ = "y" THEN GOTO 74
40      454 INPUT "Do you wish to alter the value of the variable increment"; SI$
41      456 IF SI$ = "y" THEN GOTO 80
42      495 '
43      500 END
44      505 '
45      1000 'subroutine to calculate k's for G1, G2, G3
46      1010 J = I -1: DY = H
47      1015 '   The K's that follow are similar to the DELTAS in program KP&RK.1.
48      1020 K3(1) = (-A*G1(J)*G3(J))*DY
49      1025 K3(1) = K3(1) +(-B*( 1 -G2(J)*G2(J) ))*DY
50      1030 K2(1) = (G3(J))*DY
51      1040 K1(1) = (G2(J))*DY
52      1050 K3(2) = (-A*(G1(J) +K1(1)/2)*(G3(J) +K3(1)/2))*DY
53      1055 K3(2) = K3(2) +(-B*( 1 -(G2(J) +K2(1)/2)*(G2(J) +K2(1)/2) ))*DY
54      1060 K2(2) = (G3(J) +K3(1)/2)*DY
55      1070 K1(2) = (G2(J) +K2(1)/2)*DY
56      1080 K3(3) = (-A*(G1(J) +K1(2)/2)*(G3(J) +K3(2)/2))*DY
57      1085 K3(3) = K3(3) +(-B*( 1 -(G2(J) +K2(2)/2)*(G2(J) +K2(2)/2) ))*DY
58      1090 K2(3) = (G3(J) +K3(2)/2)*DY
59      1100 K1(3) = (G2(J) +K2(2)/2)*DY
60      1110 K3(4) = (-A*(G1(J) +K1(3))*(G3(J) +K3(3)))*DY
61      1115 K3(4) = K3(4) +(-B*( 1 -(G2(J) +K2(3))*(G2(J) +K2(3)) ))*DY
62      1120 K2(4) = (G3(J) +K3(3))*DY
63      1130 K1(4) = (G2(J) +K2(3))*DY
64      1190 RETURN
65      2500 LPRINT "i= "I,"y= "Y
66      2505 LPRINT "g1(i)= "G1(I),"g2(i)= "G2(I),"g3(i)= "G3(I)
67      2550 RETURN
```

The Blasius equation is a specific case of the more general Falkner-Skan equation, Eq. (16.4.2). Values of the constants in Eq. (16.4.4) determine the case that is being treated. The second program listed below solves this equation by using a Taylor series expansion. The number of terms in the series can be varied to determine the utility and stability of the method. (Numerical integration algorithms usually compare their results with the number of terms that must be taken in a Taylor series, 4 or 5, in order to gain comparable accuracy.) The higher-order derivatives that are needed in the Taylor series are obtained by differentiating the differential equation itself. This produces higher derivatives in terms of lower ones. Again, *shooting* is used.

```
10 'prgm is called flkskan1.3
15 'prgm solves Falkner-Skan problems by Taylor series
16 'equation: F3 +A*F*F2 +B*(1- F1*F1) = 0
17 'F is the unknown function and F3 is its third derivative.
18 ' A and B are constants. (For Blasius case A= 1/2, B=0; c.f. Eqn 16.3.7.)
19 '
29 N = 500: 'This is the maximum number of increments that can be stored.
30 DIM F(N),F1(N),F2(N)
```

```
31  '
33  F(0)=0:F1(0)=0: 'These are boundary conditions.
35  '
36  CLS: PRINT "You can carry from 4 to 10 derivatives in the solution.":PRINT ""
37  INPUT "Please indicate the number of derivatives that you wish."; LIMIT
38  '
41  PRINT"": INPUT "What are the constants A and B"; A,B: PRINT ""
43  '
44  CLS: PRINT "choose d(eta) >  or = 0.05"
45  INPUT "what is the increment on eta"; DN
46  PRINT "": PRINT ""
49  '
50  INPUT "What is the assumed value of the F2(0)"; F2(0)
51  GOSUB 550: 'Is a printer activated?
53  '
80  IMAX = INT(10/DN): 'This sets the number of increments in program.
98  '
135 F5=0:F6=0:F7=0:F8=0:F9=0:F10=0: 'Initialization of 5th-10th derivatives.
136 '
139 'Statements #140-320 comprise a single FOR-NEXT loop.
140     FOR I=0 TO IMAX
150 F = F(I): F1 = F1(I): F2 = F2(I)
155 F3 = -A*(F*F2)
156 F3 = F3 -B*(1 -F1*F1)
160 F4 = -A*(F*F3 +F1*F2)
161 F4 = F4 +B*2*(F1*F2)
162 IF LIMIT = 4 THEN GOTO 230
165 F5 = -A*(F*F4 +2*F1*F3 +F2*F2)
166 F5 = F5 +B*2*(F1*F3 +F2*F2)
167 IF LIMIT = 5 THEN GOTO 230
170 F6 = -A*(F*F5 +3*F1*F4 +4*F2*F3)
171 F6 = F6 +B*2*(F1*F4 +3*F2*F3)
172 IF LIMIT = 6 THEN GOTO 230
175 F7 = -A*(F*F6 +4*F1*F5 +7*F2*F4 +4*F3*F3)
176 F7 = F7 +B*2*(F1*F5 +4*F2*F4 +3*F3*F3)
177 IF LIMIT = 7 THEN GOTO 230
180 F8 = -A*(F*F7 +5*F1*F6 +11*F2*F5 +15*F3*F4)
181 F8 = F8 +B*2*(F1*F6 +5*F2*F5 +10*F3*F4)
182 IF LIMIT = 8 THEN GOTO 230
185 F9 = -A*(F*F8 +6*F1*F7 +16*F2*F6 +26*F3*F5 +15*F4*F4)
186 F9 = F9 +B*2*(F1*F7 +6*F2*F6 +15*F3*F5 +10*F4*F4)
187 IF LIMIT = 9 THEN GOTO 230
190 F10= -A*(F*F9 +7*F1*F8 +22*F2*F7 +42*F3*F6 +56*F4*F5)
191 F10= F10 +B*2*(F1*F8 +7*F2*F7 +21*F3*F6 +35*F4*F5)
228 'Taylor series for F
230 F = F +F1*DN +F2*DN^2/2 +F3*DN^3/6 +F4*DN^4/24 +F5*DN^5/120 +F6*DN^6/720
240 F = F + F7*DN^7/5040 +F8*DN^8/40320! +F9*DN^9/362880! +F10*DN^10/3628800!
248 'Taylor series for F1
250 F1 = F1 +F2*DN +F3*DN^2/2 +F4*DN^3/6 +F5*DN^4/24 +F6*DN^5/120 +F7*DN^6/720
260 F1 = F1 + F8*DN^7/5040 +F9*DN^8/40320! +F10*DN^9/362880!
268 'Taylor series for F2
270 F2 = F2 +F3*DN +F4*DN^2/2 +F5*DN^3/6 +F6*DN^4/24 +F7*DN^5/120 +F8*DN^6/720
280 F2 = F2 + F9*DN^7/5040 +F10*DN^8/40320!
289 '
290 F(I+1)= F: F1(I+1)= F1: F2(I+1)=F2: 'Sets values for F, F1, F2 at (I+1)th
291 '                                    increment from Taylor series.
295 PRINT "i= "I,"f= "F,"f1= "F1,"f2= "F2
296 IF (I+1)/10 = INT( (I+1)/10 ) THEN GOSUB 600
320     NEXT I
322 '
350 GOSUB 1000
352 '
400 INPUT "do you wish to try again with another F2(0)"; SI$
405 IF SI$ ="y" THEN GOTO 50
410 INPUT "Do you wish to try another eta increment"; SI$
415 IF SI$ ="y" THEN GOTO 45
420 '
500 END
525 ' *********** SUBROUTINES **************
526 '
550 LPRINT "*************": 'Program requires a printer to execute this step.
```

```
551 LPRINT "constant A is "A,"constant b is "B
552 LPRINT ""
553 LPRINT "assumed f2(0)= "F2(0),"eta increment= "DN,"last derivative= "LIMIT
554 LPRINT ""
555 '
600 ETA = (I+1)*DN
610 LPRINT "eta= "ETA,"f= "F,"f1= "F1,"f2= "F2
640 RETURN
641 '
1000 'integration subroutine
1002 'The use of some1 and some2 result from integration of original D.E.
1003 'to obtain f2(infinity)-f2(0).
1005 'Some3 integrates f2 (i.e., f") to obtain f'(infinity)-f'(0).
1007 'See note at end of program for justification of calculations here.
1010 H = DN
1012 KMAX = IMAX -5
1018 SUM1 = 0: SUM2 = 0: SUM3 = 0
1020    FOR K = 0 TO KMAX STEP 5
1025 '   IF ABS(F1(K))> 1.02 THEN GOTO 1050
1030    SOME1 = 19*(F1(K  ))*(1 -F1(K  ))
1035     SOME2 = 19*(1 +F1(K  ))*(1 -F1(K  ))
1037      SOME3 = 19*F2(K  )
1040    SOME1 = 75*(F1(K+1))*(1 -F1(K+1)) +SOME1
1045     SOME2 = 75*(1 +F1(K+1))*(1 -F1(K+1)) +SOME2
1047      SOME3 = 75*F2(K+1)  +SOME3
1050    SOME1 = 50*(F1(K+2))*(1 -F1(K+2)) +SOME1
1055     SOME2 = 50*(1 +F1(K+2))*(1 -F1(K+2)) +SOME2
1057      SOME3 = 50*F2(K+2)  +SOME3
1060    SOME1 = 50*(F1(K+3))*(1 -F1(K+3)) +SOME1
1070     SOME2 = 50*(1 +F1(K+3))*(1 -F1(K+3)) +SOME2
1072      SOME3 = 50*F2(K+3)  +SOME3
1075    SOME1 = 75*(F1(K+4))*(1 -F1(K+4)) +SOME1
1080     SOME2 = 75*(1 +F1(K+4))*(1 -F1(K+4)) +SOME2
1082      SOME3 = 75*F2(K+4)  +SOME3
1085    SOME1 = 19*(F1(K+5))*(1 -F1(K+5)) +SOME1
1090     SOME2 = 19*(1 +F1(K+5))*(1 -F1(K+5)) +SOME2
1092      SOME3 = 19*F2(K+5)  +SOME3
1095    SUM1 = SUM1 +5*H*SOME1/288
1100     SUM2 = SUM2 +5*H*SOME2/288
1105      SUM3 = SUM3 +5*H*SOME3/288
1150    NEXT K
1200 '
1353 F1INFIN =  SUM3: ' This is the integration of F" and gives F' at infinity.
1355 F2ZERO = A*SUM1 +B*SUM2: 'For SOLUTION, this is F" at 0.
1356 '                        See explanation after program.
1357 PRINT ""
1358 PRINT "Assumed value of F2(0) was "; F2(0)
1359 PRINT "F1(infinity) from calculations of assumed F2(0) is "; F1INFIN
1360 PRINT "integral condition for 2nd derivative at origin yields "; F2ZERO
1363 LPRINT ""
1365 LPRINT "integral condition for 2nd derivative at origin yields "; F2ZERO
1366 LPRINT "Assumed value of F2(0) was "; F2(0)
1368 LPRINT "F1(infinity) from calculations of assumed F2(0) is "; F1INFIN
1370 LPRINT "integral condition for 2nd derivative at origin yields "; F2ZERO
1375 RETURN
```

The equation from line 16 can be integrated (by parts):

$$F''(\infty) - F''(0) + A\left(FF' \Big|_0^\infty \right) - A\int_0^\infty (F')^2 \, d\eta + B\int_0^\infty \{1 - (F')^2\} d\eta = 0.$$

But $F'(\infty) = 1$ and since $F(0) = 0$, $F(\infty) = \int_0^\infty F' \, d\eta$. Hence,

$$F''(0) = A\int_0^\infty F'(1 - F')d\eta + B \int_0^\infty (1 + F')(1 - F') \, d\eta.$$

Bibliography

GENERAL INTEREST

ARIS, R., H.T. DAVIS, and R. STUEWER. *Springs of Scientific Creativity*. Minneapolis: University of Minnesota Press, 1983.

BOYS, C.V. *Soap Bubbles, Their Color and Forces Which Mold Them*. New York: Dover Publications, 1959.

CLANCY, E.P. *The Tides, Pulse of the Earth*. Garden City: Anchor Books, 1969.

FEATHER, N. *Mass, Length and Time*. Baltimore: Penguin Books, 1959.

————. *Matter and Motion*. Baltimore: Penguin Books, 1970.

GAMOW, G. *Gravity*. Garden City: Anchor Books, 1962.

ROUSE, H. AND S. INCE. *History of Hydraulics*. Iowa City: Iowa Institute of Hydraulic Research, 1957.

SUTTON, O.G. *The Science of Flight*. Harmondsworth, Middlesex: Penguin Books, 1949.

TABOR, D. *Gases, Liquids and Solids*. Baltimore: Penguin Books, 1969.

VALLENTINE, H.R. *Water in the Service of Man*. Baltimore: Penguin Books, 1967.

VAN DYKE, M. *An Album of Fluid Motion*. Stanford: Parabolic Press, 1982.

INTRODUCTORY FLUID MECHANICS

BERTIN, J.J. *Engineering Fluid Mechanics*. Englewood Cliffs: Prentice-Hall, Inc., 1987.

DAILY, J.W. and D.R.F. HARLEMAN. *Fluid Dynamics*. Reading: Addison-Wesley Publishing Company, 1973.

Fox, R.W. and A.T. McDonald. *Introduction to Fluid Mechanics*. New York: John Wiley & Sons, 1985.

Long, R.R. *Mechanics of Solids and Fluids*. Englewood Cliffs: Prentice-Hall, Inc., 1961.

Olson, R.M. *Essentials of Engineering Fluid Mechanics*. New York: Harper & Row Publishers, 1980.

Prandtl, L. *Essentials of Fluid Dynamics*. New York: Hafner Publishing Company, 1952.

Sabersky, R. H., A.J. Acosta, and E.G. Hauptmann. *Fluid Flow*. (3rd ed.) New York: Macmillan Publishers Company, 1989.

Shames, I. *Mechanics of Fluids*. New York: McGraw-Hill Book Company, 1982.

Streeter, V.L. and E.B. Wylie. *Fluid Mechanics*. New York: McGraw-Hill Book Company, 1985.

Tritton, D.J. *Physical Fluid Dynamics*. Oxford: Oxford University Press, 1988.

Vallentine, H.R. *Applied Hydrodynamics*. London: Butterworths Scientific Publications, 1959.

Vennard, J.K. and R.L. Street. *Elementary Fluid Mechanics*. New York: John Wiley & Sons, 1975.

White, F.M. *Fluid Mechanics*. New York: McGraw-Hill Book Company, 1986.

FLUID PROPERTIES

Engineering Sciences Data
    Item No. 65009, "Viscosity of Water," 1965
    Item No. 66024, "Approximate Data on the Viscosity of Some Common Liquids," 1966
    Item No. 67015: "A Guide to the Viscosity of Liquid Petroleum Products," 1967
The Institution of Mechanical Engineers, London, U.K.

Marks, L.S. *Mechanical Engineers' Handbook*. New York: McGraw-Hill Book Company, 1951.

Perry, R.H. and C.H. Chilton. *Chemical Engineers' Handbook*. Tokyo: McGraw-Hill Book Company Japan, 1974.

Reid, R.C., J.M. Prausnitz, and B.E. Poling. *The Properties of Gases and Liquids*. New York: McGraw-Hill Book Company, 1987.

Rex, D.F. *Climate of the Free Atmosphere—World Survey of Climatology*. Vol. 4. Amsterdam: Elsevier Publishing Company, 1969.

The U.S. Standard Atmosphere. Washington, D.C.: U.S. Government Printing Office, 1962.

Weast, R.C. *Handbook of Chemistry and Physics*. Cleveland: CRC Press, 1973.

DIMENSIONAL ANALYSIS

Bridgman, P.W. *Dimensional Analysis*. New Haven: Yale University Press, 1931.

Sedov, L.I. *Similarity and Dimensional Methods in Mechanics*. New York: Academic Press, 1959.

Zierep, J. *Similarity Laws and Modeling*. New York: Marcel Dekker, Inc., 1971.

LIFT AND DRAG

Abbott, I.H. and A.E. von Doenhoff. *Theory of Wing Sections*. New York: Dover Publications, Inc., 1959.

HOERNER, S.F. *Fluid-Dynamic Drag*. Midland Park: Published by the Author, 1965.

SACHS, P. *Wind Forces in Engineering*. Oxford: Pergamon Press, 1978.

SHAPIRO, A.H. *Shape and Flow: The Fluid Dynamics of Drag*. New York: Doubleday and Company, Inc., 1961.

VON MISES, R. *Theory of Flight*. New York: Dover Publications, Inc., 1959.

COMPRESSIBLE FLOW

ANDERSON, J.D. *Modern Compressible Flow*. New York: McGraw-Hill Book Company, 1982.

THOMPSON, P.A. *Compressible-fluid Dynamics*. New York: McGraw-Hill Book Company, 1972.

LIEPMANN, H.W. and A.E. PUCKETT. *Aerodynamics of a Compressible Fluid*. New York: John Wiley & Sons, 1947.

LIEPMANN, H.W. and A. ROSHKO. *Elements of Gasdynamics*. New York: John Wiley & Sons, 1958.

SHAPIRO, A.H. *The Dynamics and Thermodynamics of Compressible Fluid Flow*. New York: Ronald Press, 1953.

OPEN CHANNEL FLOW

BAKHMETEFF, B.A. *Hydraulics of Open Channels*. New York: McGraw-Hill Book Company, 1932.

CHOW, V.T. *Open-Channel Hydraulics*. New York: McGraw-Hill Book Company, 1959.

FLOW MEASUREMENT

ASHRAE Standard 41.8.78. *Standard Methods of Measurement of Flow of Fluids—Liquids*. New York: Am. Soc. Heat., Refrig. & Air-Cond. Engrs., Inc., 1978.

BECKWITH, T.G. and N.L. BUCK. *Mechanical Measurements*. Reading: Addison-Wesley Publishing Company, 1973.

BRADSHAW, P. *Experimental Fluid Mechanics*. Oxford: Pergamon Press, 1964.

GOLDSTEIN, RICHARD J., ed. *Fluid Mechanics Measurements*. Washington: Hemisphere Publishing Corporation, 1983.

MERZKIRCH, W. *Flow Visualization*. New York: Academic Press, 1974.

OWER, E. and R.C. PANKHURST. *The Measurement of Air Flow*. Oxford: Pergamon Press, 1966.

TUVE, G.L. and L.C. DOMHOLDT. *Engineering Experimentation*. New York: McGraw-Hill Book Company, 1966.

PUMPS

HENSHAW, T.L. *Reciprocating Pumps*. New York: Van Nostrand Reinhold Company, 1987.

KARASSIK, I.J., W.C. KRUTZSH, W.H. FRASER, and J.P. MESSINA. *Pump Handbook*. New York: McGraw-Hill Book Company, 1986.

MATLEY, J. and *Chemical Engineering* staff. *Fluid Movers—Pumps, Compressors, Fans and Blowers*. New York: McGraw-Hill Publications Co., 1979.

SHEPHERD, D.G. *Principles of Turbomachinery*. New York: The Macmillan Company, 1956.

STEPANOFF, A.J. *Centrifugal and Axial Flow Pumps*. New York: John Wiley & Sons, 1957.

WISLICENUS, G.F. *Fluid Mechanics of Turbomachinery*. Vols. I & II, New York: Dover Publications, Inc., 1965.

MATHEMATICS AND MATHEMATICAL METHODS

CARNAHAN, B., H.A. LUTHER, and J.O. WILKES. *Applied Numerical Methods*. New York: John Wiley & Sons, 1969.

FADDEEVA, V.N. *Computational Methods of Linear Algebra*. New York: Dover Publications, Inc., 1959.

WYLIE, C.R. and L.C. BARRETT. *Advanced Engineering Mathematics*. New York: McGraw-Hill Book Company, 1982.

COMPUTERS

OLFE, D.B. *Fluid Mechanics Programs for the IBM PC*. New York: McGraw-Hill Book Company, 1987.

FILMS FOR FLUID MECHANICS

Aerodynamic Generation of Sound (44 min, principals: M.J. Lighthill, J.E. Ffowcs-Williams)

Boundary Layer Control (25 min, principal: D.C. Hazen)

Cavitation (31 min, principal: P. Eisenberg)

Channel Flow of a Compressible Fluid (29 min, principal: D.E. Coles)

Deformation of Continuous Media (38 min, principal: J.L. Lumley)

Eulerian and Lagrangian Descriptions in Fluid Mechanics (27 min, principal: J.L. Lumley)

Flow Instabilities (27 min, principal: E.L. Mollo-Christensen)

Flow Visualization (31 min, principal: S.J. Kline)

The Fluid Dynamics of Drag (4 parts, 120 min, principal: A.H. Shapiro)

Fundamentals of Boundary Layers (24 min, principal: F.H. Abernathy)

Low-Reynolds-Number Flows (33 min, principal: Sir. G.I. Taylor)

Magnetohydrodynamics (27 min, principal: J.A. Shercliff)

Pressure Fields and Fluid Acceleration (30 min, principal: A.H. Shapiro)

Rarefied Gas Dynamics (33 min, principals: F.C. Hurlbut, F.S. Sherman)

Rheological Behavior of Fluids (22 min, principal: H. Markovitz)

Rotating Flows (29 min, principal: D. Fultz)

Secondary Flow (30 min, principal: E.S. Taylor)

Stratified Flow (26 min, principal: R.R. Long)

Surface Tension in Fluid Mechanics (29 min, principal: L.M. Trefethen)

Turbulence (29 min, principal: R.W. Stewart)

Vorticity (2 parts, 44 min, principal: A.H. Shapiro)

Waves in Fluids (33 min, principal: A.E. Bryson)

Source:

Encyclopaedia Britannica Educational Corporation
780 South Lapeer Road
Lake Orion, Michigan 48035

Other Sources:

Engineering Societies Library
United Engineering Center
345 East 47th Street
New York, N.Y. 10017

University of Iowa
The Audiovisual Center
Iowa City, Iowa 52240

University of Minnesota
Saint Anthony Falls Hydraulics Laboratory
Mississippi River and 3rd Avenue, SE
Minneapolis, Minn. 55414

Film catalogues available upon request.

Index

A

Absolute temperature, 6
Absolute viscosity (coefficient of viscosity), 23, 139
Acceleration:
   introduction to, 144–46
   linear, rigid-body, 74–77
   pressure and, in rotating fluids, 77–82
   substantial derivative and, 146–49
Adiabatic flow in pipes, 533–38
Adiabatic process, 17
   reversible, 504
Advection, definition of, 413
Adverse (positive) pressure gradients, 418
Affinity (similarity) laws, for turbomachines, 646–50, 652
Alcyone, 443
Alternate depth, 549
American Society for Testing Materials, 643
Anemometer:
   hot-wire, 467, 474, 629–31, 642
   Laser-Doppler (LDA), 633–34
   makeshift, 632
   vane, 633
Angle of attack, 363, 365, 429, 444
Angular deformation, 139–40, 142
Angular momentum, definition of, 259–61
Angular momentum conservation:
   detailed explanation of, 259–62
   pump and turbine applications of, 262–73
Anisotropy, 586
Applied Numerical Methods (Carnahan, Luther, and Wilkes), 338, 603
Archimedes' principle, 440
Atmospheres, 6
Atmospheric pressure, 6
Attack, angle of, 363, 365, 429, 444
Available net positive suction head (NPSHA), 661–67
Average value, of velocity, 101, 467
Avogadro's number, 8
Axial-flow machines, 262–63, 287–88, 645, 658, 671
Axisymmetric flow, 315

B

Backflow, 336
Baker, D. James, 615
Bar, definition of, 7
Beam splitter, 622–23
Bernoulli, Daniel, 158–59, 598
Bernoulli constant, 193
   applications of, 186–88
   definition of, 163

Bernoulli equation, 158, 193, 217, 274, 333, 366, 392, 545, 603–4, 635, 636
  compressible flow and, 511
  energy conservation and, 203–4, 220
  irrotational flow and, 171–76, 182, 599
  pipe flow and, 410, 411, 416
  steady flow and, 159, 161–71, 203
Best efficiency point (BEP), 657
Bingham fluid, 19
Blasius, H., 472, 565, 598, 606–7
Bluff bodies, 429
Bodies, flow around. *See* External flow
Bottom shear stress, 556–57, 564, 568–69
Boundary layer(s), 347–48
  definitions of, 29, 124, 323
  laminar, 323–39, 390, 394, 487, 496
  momentum-integral equation for a, 333–35
  Navier-Stokes equations and, 581, 587, 589, 590, 592, 594–610
  shear stress and, 581
  thickness of, 323, 390, 487, 496, 590, 597, 599
  transition length of, 486–87
  turbulent, 487, 489–91, 494–95
Boundary-layer control (BLC), separation and, 441–44
Boundary layer theory, Prandtl's, 426, 429, 431–32, 594, 598, 609
Boundary Layer Theory (Schlichting), 338–39
Bourdon tube pressure gauge, 7, 626
Boussinesq, Joseph, 472
Boyle, Robert, 500
Boyle's law, 9, 499
Bremen, thrust and drag measurements of, 346
Broad-crested weir, 553, 640
Buckingham, E., 360
Buffer (production, generation) zone, 481
Bulk modulus, 17
Buoyant force/buoyancy, 71

C

California Institute of Technology, 473
Cambridge University, 473
Cannon-Fenske viscometer, 643
Carnahan, Brice, 338, 603
Cavitation, 499
  damage caused by, 25–26, 29, 660–61, 677
  definitions of, 25, 222
Cavitation coefficient, definition of, 383
Celsius scale, 6

Centrifugal-flow machines, 262–63, 288, 645, 658, 675
Channel flow. *See* Open channel flow
Characteristic curve of pump, 652, 672, 677
  definition of, 650
Charles' law, 9, 499
Chattock gauge, 626
Chézy, Antoine, 567
Chézy coefficient, 565, 567
Chézy equation, 565, 567–69, 574
Choking, 509, 554
Chord, definition of, 429
Churchill, S. W., 408
Circulation, 437
Closure, 470
Coefficient:
  cavitation, 383
  Chézy, 565, 567
  contraction, 636
  discharge, 636
  drag, 258, 385–89, 429, 430, 433, 435, 441, 533
  Euler head, 647–48
  flow, 647–48
  frictional, 388–89
  lift, 430
  moment, 318
  pressure, 182
  skin friction, 258, 326–27, 329–30
  turbulent exchange, 472
  velocity, 636
  of viscosity (absolute viscosity), 23, 139
  wave drag, 388–89
Coherent structures, 494
Colebrook, C. F., 314, 485, 565
Colebrook equation, 407–8
Complimentary error function, 589
Compressibility, measurement of, 17–18
Compressible flow:
  energy equation for gases, 505–11
  introduction to, 499–500
  Mach number and, 511–13, 516–18, 533, 536–38
  shock waves in nozzles, 522–31, 533, 538, 539
  sound waves, 500–504
  summary of, 538–39
  supersonic flow, 27, 513–22, 538
  viscosity and, 533–39
Compression shock wave, 520, 525
Compressor, definition of, 677
Computer programs, integration methods in, 735–38
Conduit flow. *See* Pipe flow
Conformal mapping, 437
Conjugate depth, 557

Conservation of energy. *See* Energy
   conservation
Conservation of mass. *See* Mass
   conservation
Conservation of momentum. *See*
   Momentum conservation
Constant-current hot-wire anemometer,
   630–31, 642
Constant-stress region, 481, 496
Constant-temperature hot-wire
   anemometer, 631
Constitutive equations, 19, 586, 609
Continuity, equation of mass, 154
Continuity equation for steady flow,
   117–24, 140, 154
Continuum, definition of, 12
Contraction coefficient, 636
Controls:
   broad-crested weir, 553, 640
   definitions of, 553, 572
   Parshall flume, 640
   sluice gate, 149, 557–61, 572–74
   wicket gate, 269
Control surfaces, 105–7, 118
Control volume(s), 116
   definition of, 118
   moving, 694–701, 703, 705
Convection, definition of, 413
Core-flow region, 477
Correction factor:
   kinetic energy, 206
   momentum, 244
Creeping flow, 347, 348, 427
Critical depth, 550
Critical temperature, 10
Cross section, channel, 552-54
Curved surfaces, introduction to forces on,
   51–53
Curvilinear squares, 183
Cutwater, 670
Cylinder:
   drag on a, 362–63
   pressure distribution on a, 361–62
   wake shedding and a, 365–67

D

Darcy, Henri, 474, 567
Deformation:
   angular, 139–40, 142
   rotation and, 140–44
   volume, 139–40
Demand curve, pump, 677
Density, measurement of, 8–9

Depth:
   alternate, 549
   conjugate, 557
   critical, 550
   uniform, 574
Diameter:
   hydraulic, 414
   specific, 652
Die-swell, 420
Differential form of mass conservation,
   153–55
Differentials, total, 690–93
Diffusion, 453–54
   double, 453
   thermal, 458–59, 461
Dimensionless parameters, use of:
   choosing the experimental parameters,
     360–63, 365–67
   drag on a cylinder, 362–63
   interpretation of dimensionless
     parameters, 381, 383, 385
   introduction to, 354–57
   lift and, 363, 365
   in modeling, 385–91
   pi parameters, calculation of, 367–69,
     371–74, 376, 378–79, 381
   pipe flow and, 360–61, 383
   pi theorem, 357–60
   pressure distribution on a cylinder,
     361–62
   Reynolds number and, 354, 374, 381,
     383, 385, 387–89
   self-similar fluid motions, 392–95
   shear stress and, 354, 356, 362
   summary of, 396
   wake shedding and, 365–67
Discharge coefficient, 636
Discharge port, 674
Discharge valve, 674
Disk friction, 653
Displacement length, 334
Doppler, C. J., 501
Doppler effect, 501-2, 633, 637
Double diffusion, 453
Downstream and upstream effects, in open
   channel flow, 557–61
Downwash, 431, 444
Draft tube, 270
Drag:
   on bodies, 431–36
   comparisons between theory and
     experiment, 487, 489–90
   on a cylinder, 362–63
   definitions of, 324, 427
   dimensionless parameters and, 354–55,
     385
   form, 428, 444

Drag (*cont.*)
  induced, 431
  lift-drag polar diagram, 434
  measurement of, 29–32, 258–59, 329–30,
      345, 346
  pressure, 428
  roughness paradox and, 3–5
  wave, 28–29, 330
  See also Friction factor and Shear stress
Drag coefficient, 429, 430, 433, 435
  definition of, 258
  dimensionless parameters and, 385–89
  spoiler's effect on, 441
  wall, 533
  wave, 388–89
Drowned gate, 573
Duct flow. *See* Pipe flow
Dynamic pressure, 310

E

Eddy viscosity, 472
Efficiency and losses, of hydraulic
      machinery, 652–54, 656–61
Einstein, Albert, 98, 496
Elevation head, 404
Energy conservation, 97–98, 356, 499, 538
  Bernoulli equation and, 203–4, 220
  flux of kinetic energy, 204–7
  general energy equation, detailed
      explanation of, 207–24
  interrelationship of energy equation and
      momentum equation, 274–83, 545
  introduction to, 203–4
  Newton's laws of motion and, 203–4
  Reynolds transport theorem and, 208,
      230
  summary of, 230–31
  unsteady-flow applications of general
      energy equation, 225–30
Energy equation, general, 204
  compressible flow and, 500, 514, 522–25,
      530, 534, 538
  detailed explanation of, 207–24
  interrelationship of momentum equation
      and, 274–83, 545
  open channel flow and, 545–46, 548,
      553–54, 559–60, 574
  unsteady-flow applications of, 225–30
Energy equation for gases, 505–11
Energy grade line, 404–5, 564
Enthalpy, specific heats, entropy and,
      727–34
Entropy:
  conservation of, 522, 538

specific heats, enthalpy, and, 727–34
Equation of state:
  compressible flow and, 499–500, 514,
      524, 535, 538
  introduction to, 9–10
Equivalent force, 59-60
Equivalent loading, statically, 63
Euler, Leonhard, 259, 262, 287, 496
Euler equations, 193
  applications of, 187–92, 512, 586
  compressible flow and, 512
  introduction to, 186–87
  irrotational flow and, 189, 599
  steady flow and, 188
Euler head coefficient, 647–48
Eulerian (spatial) flow, 99–100, 144–46,
      155
Euler pump head, 647
Euler's improved method, 338
Euler's method, 338
Expansion shock wave, 509, 520
Extensive properties of fluids, 11, 128
External flow (flow around bodies):
  drag on bodies, 431–36
  introduction to, 426–31
  laminar boundary layers over two-
      dimensional bodies, 331–39, 394
  lift on bodies, 436–40
  Reynolds number and, 426, 428, 429,
      435
  separation and boundary-layer control,
      441–44
  shear stress and, 426–27
  summary of, 444

F

Fahrenheit scale, 6
Falkner, V. M., 605, 610
Falkner-Skan solutions, 605–7, 610
  by Runge-Kutta method, 739–40
  by Taylor series expansion, 740–42
Falling-ball viscometer, 644
Fan, definition of, 677
Fanno line, 534
Flat-plate flow:
  applications of momentum equation to,
      256–59
  turbulent, 486–87, 489–91, 494–95
Flow:
  adiabatic, in pipes, 533–38
  axial, 262–63, 287–88, 645, 658, 671
  axisymmetric, 315
  centrifugal, 262–63, 288, 645, 658, 675
  compressible, 499–531, 533–39

creeping, 347, 348, 427
external, 426–44
flat-plate, 256–59, 486–87, 489–91, 494–95
Hagen-Poiseuille, 309, 464
irrotational (potential), 171–83, 189, 599
laminar, 3, 24, 258, 309, 313, 360–61, 413, 461, 464, 465, 473, 487, 489–91, 496
Mach number and, 511–13, 516–18, 533, 536–38
material (Lagrangian), 99, 144
mixed, 645, 658
open channel, 545–69, 571–75
pipe, 2–3, 23–25, 306–15, 360–61, 383, 404–20, 470–74, 476–77, 479, 481–85, 533–38
radial, 262–63, 287–88, 658
secondary, 264, 451–55, 457–59, 461–62
shockless, 266
sinuous, 24
spatial (Eulerian), 99–100, 144–46, 155
steady, 100, 117–24, 159, 161–71, 188, 203, 244–56, 500
Stokes, 347, 348, 435
subcritical (tranquil), 547, 549–52, 557–58
subsonic, 27
supercritical (rapid, shooting), 547, 549–51, 557–58, 573
supersonic, 27, 513–22, 538
swirling, 315–20
turbulent, 3, 24, 360–61, 413, 461–62, 464–65, 467–74, 476–77, 479, 481–87, 489–91, 494–96
uniform, 100, 563–64, 566–68, 571, 574
unsteady, 124–27, 225–30, 283–87, 320–23
volume rate of, 100–105
See also Velocity
Flow, kinematics of:
acceleration, introduction to, 144–46
acceleration and the substantial derivative, 146–49
angular and volume deformation with associated consequences, 139–40
deformation and rotation, 140–44
differential form of mass conservation, 153–55
introduction to, 136
rotation, introduction to, 137–38
stream functions, 149–52
summary of, 155
Flow, measurement of:
introduction to, 613–14
pressure measurement, 6–8, 625–27
surface elevation and, 640–41
temperature measurement, 6, 641–42
velocity measurement, 627–34
viscosity measurement, 18–23, 643–44
visualization of flow, 614–23
volume flow rate measurement, 100–105, 634–37, 640
Flow, theoretical effects of viscosity on fluid. See Viscosity, theoretical effects of, on fluid flow
Flow coefficient, 647–48
Flow nets, 183–86
Flow of inviscid fluid. See Inviscid fluid, flow of
Flow phenomena. See Fluid properties and flow phenomena
Flow-rate transducers, 637, 640
Flow resistance, in open channels, 562–67
Flow stability:
basic ideas of, 451–55, 457–59, 461–62
introduction to, 451
summary of, 462
Flow work, definition of, 209
Fluctuating (turbulent) component, of velocity, 467
Fluid:
Bingham, 19
definitions of, 2, 18, 19
incompressible, 17, 118
Newtonian, 20, 82, 609
strain-thickening, 20
strain-thinning, 20
Fluid flow, theoretical effects of viscosity on. See Viscosity, theoretical effects of, on fluid flow
Fluid motions, self-similar, 392–95
Fluid properties and flow phenomena:
density, 8–9
equation of state, 9–10, 499–500, 514, 524
examples of common flows, 23–33
extensive properties, 11, 128
general physical properties, 13–18
ideal gas law, 9–10, 499, 522, 535
intensive properties, 10–12, 128
introduction to, 1–6
pressure, 6–8
temperaturre, 6
viscosity, 18–23
Fluid velocity. See Velocity
Flume, Parshall, 640
Flux of kinetic energy, 204–7
Force(s):
buoyant, 71
equivalent, 59–60
London, 10
resultant, 63
Van der Waals, 10

Force, shear. *See* Shear stress
Forces, distributions of. *See* Pressure
    distributions, in fluids without
    relative motions
Form drag, definitions of, 428, 444
Francis turbines, 269–70, 288
Free jets, 418
Free-surface profiles, 567–69, 571–74
Friction:
  disk, 653
  frictional coefficient, 388–89
  friction factor, 310, 374, 405, 472, 533,
    536–37
  friction velocity, 474
  pipe, 405
  skin friction coefficient, 258, 326–27,
    329–30
  See also Drag and Shear stress
Froude, William, 330, 388, 487
Froude number, 330
  definition of, 383
  dimensionless parameters and, 383, 385
  effects of increases and decreases in,
    546–48, 567, 568
  open channel flow and, 546–48, 552–53,
    555, 567, 568, 571
Froude scaling, 388–90, 613

G

Gas, definition of, 2
Gate:
  drowned, 573
  sluice, 149, 557–61, 572–74
  wicket, 269
Gauge pressure, 7
Gay-Lussac's law, 499
General energy equation. *See* Energy
  equation, general
General physical properties of fluids,
  13–18
Generation (buffer, production) zone, 481
Goethe, Johann Wolfgang von, 539
Grade line:
  energy, 404–5, 564
  hydraulic, 404
Gradient operator, 176

H

Hagen, G., 24, 309, 464
Hagen-Poiseuille flow, 309, 464
Hairpin vortex, 491

Head:
  elevation, 404
  kinetic energy, 404
  pressure, 404
  pump, 647
  total, 220, 404
Head loss:
  definitions of, 374, 404
  entire, 405
Heat added, definition of, 208–9
Heat transferred, definition of, 208–9
Hiemenz, K., 339, 606
Holograms, 623
Holographic interferometer/interferometry,
  623
Homogeneous turbulent flow, 473
Horsepower, water, 647
Horseshoe vortex, 418, 451
Hot-wire anemometer, 467, 474, 629–30
  constant-current, 630–31, 642
  constant-temperaturre, 631
Hydraulic diameter, definition of, 414
Hydraulic grade line, 404
Hydraulic machinery:
  introduction to, 645–46
  losses and efficiency of, 652–54, 656–61
  Mach number and, 653
  miscellaneous considerations for
    turbomachines, 670–73
  net positive suction head (NPSH),
    661–66, 677
  performance of pumps connected in
    series and parallel, 666–70
  positive-displacement pumps and motors,
    674–77
  Reynolds number and, 653, 660
  similarity (affinity) laws for
    turbomachines, 646–50, 652
  summary of, 677
Hydraulic radius, 565
Hydrodynamic machines. *See* Hydraulic
  machinery
Hydrodynamic stability, 496
Hydrostatic forces and moments on planar
  surfaces, 53–63
Hydrostatic machines, 646
Hydrostatic pressure variation, 39–47

I

Ideal gas law, 9–10, 499, 522, 535
Impact tube (Pitot tube), 168, 627–29
Impulse machines, description of, 282
Inception of separation, definition of, 336
Incompressible fluid, definitions of, 17, 118

Indirect results, 181
Induced drag, 431
Inducer, 666–67
Inflow, 107
Inlet shock, 657
Integration methods in computer
    programs, 735–38
Intensive properties of fluids, 10–12, 128
Interferometer/interferometry, 159, 622
    holographic, 623
    shearing, 623
Internal flow. *See* Pipe flow
Inviscid fluid, flow of:
    Bernoulli equation and irrotational flow,
        171–76, 182
    Bernoulli equation and steady flow, 159,
        161–71
    Euler equations, applications of, 187–92
    Euler equations, introduction to, 186–87
    flow nets, 183–86
    introduction to, 158–59
    irrotational flow, examples of, 176–83
    summary of, 193
Irrotational (potential) flow:
    Bernoulli equation and, 171–76, 182,
        599
    Euler equations and, 189, 599
    examples of, 176–83
Irrotationality, definition of, 138
Isentropic process, 504
Isothermal bulk modulus, 17
Isothermal process, 503–4
Isotropy, 586

J

Jets:
    free, 418
    submerged, 418
Joule, J., 208

K

Kaplan turbines, 270–71, 288, 650, 660
Kármán, T. von, 337, 472–74, 483, 484,
        565, 614
Kármán-Pohlhausen method, 337, 339
Kelvin scale, 6
Kilopascals (kPa), 6
Kinematics of flow. *See* Flow, kinematics
        of
Kinematic viscosity, 23

Kinetic energy, flux of, 204–7
Kinetic energy correction factor, 206
Kinetic energy head, 404
Kingsbury, A., 345
Kutta condition, 438–39

L

Lagrangian (material) flow, 99, 144
Laminar boundary layers, 347–48
    detailed explanation of, 323–31
    thickness of, 390, 487, 496
    over two-dimensional bodies, 331–39,
        394
Laminar flow, 258, 473
    definitions of, 3, 24, 309
    dimensionless parameters and, 360–61
    Reynolds number and, 313, 464, 465,
        467, 496
    transition in, 413, 461, 465, 467, 487,
        489–91
Laminar sublayer, 479
Lanchester, F. W., 438–39
Lanchester-Prandtl theory, 439
Laplace equation, 193
    direct method and, 182–83
    indirect method and, 182
    statement of, 178–79
Laser-Doppler anemometer (LDA),
        633–34
Laser-Doppler velocimeter (LDV), 633–34
Laufer, J., 474, 482
Lavoisier, Antoine, 114
Leonardo da Vinci, 2, 614
L'Hôpital's rule, 513
Lift:
    on bodies, 436–40
    definitions of, 363, 427, 444
    dimensionless parameters and, 363, 365
Lift coefficient, 430
Lift-drag polar diagram, 434
Line source, 179
Line (two-dimensional) sink, 179
Liquid, definition of, 1
London forces, 10
Loss:
    of available energy, 215
    head, 374, 404, 405
    minor, 408–12
    of prime, 661, 671
Losses and efficiency, of hydraulic
        machinery, 652–54, 656–61
Lubrication equations, 348
Luther, H. A., 338, 603

M

Mach number, 552
  definitions of, 26, 383, 386
  dimensionless parameters and, 383, 385,
    386
  flow and, 511–13, 516–18, 533, 536–38
  hydraulic machinery and, 653
Manning equation, 566–68, 573, 574
Manometers, 47–50, 625–29
  micro-, 626
  single-tube, 50
Mapping, conformal, 437
Mass conservation, 275, 356, 499, 538
  continuity equation for steady flow,
    117–24, 140, 154, 500
  control surfaces, 105–7, 118
  general ideas regarding, 114–17
  introduction to, 97–98
  Reynolds transport theorem and,
    127–28, 153, 155
  streamlines, streaklines, and pathlines,
    108–14
  summary of, 128
  velocity and, 98–100
  volume rate of flow, 100–105
Mass conservation equation, compressible
    flow and, 512, 514, 523–25
Mass continuity, equation of, 154
Material (Lagrangian) flow, 99, 144
Material point, definition of, 12
Material volume, 114
Maxwell, James C., 472
Mechanical energy, 204
Megapascals (MPa), 6
Micromanometers, 626
Millikan, C., 484
Minor loss, 408–12
Mitchell, A. G. M., 345
Mixed-flow machines, 645, 658
Mixing-length theory, Prandtl's, 472–74
Modeling, dimensionless parameters in,
    385–91
Mohr's circle, 581
Moment coefficient, definition of, 318
Moment of momentum, 262
Momentum, definitions of, 12, 240
Momentum conservation, 97–98, 356, 499,
    538
  angular, detailed explanation of, 259–62
  flat-plate flow applications of momentum
    equation, 256–59
  interrelationship of energy equation and
    momentum equation, 274–83, 545
  introduction to, 238–41

momentum equation, detailed
    explanation of, 241–44
Newton's laws of motion and, 246,
    249–53, 258, 287
pump and turbine applications of
    angular momentum conservation,
    262–73
Reynolds transport theorem and, 242,
    287
steady-flow applications of momentum
    equation, 244–56, 500
summary of, 287–88
unsteady-flow applications of momentum
    equation, 283–87
Momentum correction factor, definition of,
    244
Momentum equation, 240
  compressible flow and, 514, 521–22, 524,
    525, 533–34
  detailed explanation of, 241–44
  dimensionless parameters and, 356
  flat-plate flow applications of, 256–59
  interrelationship of energy equation and,
    274–83, 545
  steady-flow applications of, 244–56, 500
  unsteady-flow applications of, 283–87
Momentum function, open channel flow
    and, 555–57, 562, 574
Momentum-integral equation for a
    boundary layer, 333–35
Momentum thickness, 334
Moody, L. F., 314, 485, 660
Moody diagram, 314, 374, 388, 537, 565
  pipe flow and, 405–8, 411, 414, 415
Motion, equations of, 580–94
Motion, Newton's laws of, 76, 98, 238,
    239, 580, 609
  energy conservation and, 203–4
  general applications of, 13, 40, 41, 78,
    136, 145, 158, 159, 161–63, 171,
    186–88, 193, 321, 346, 454, 512
  momentum conservation and, 246,
    249–53, 258, 287
  statements of, 20–21, 240–42

N

National Bureau of Standards, 490
Navier, Louis M. H., 580
Navier-Stokes equations, 331, 470
  boundary layers and, 581, 587, 589, 590,
    592, 594–610
  introduction to, 580
  motion, equations of, 580–94
  no-slip condition and, 596, 609–10

Reynolds number and, 595–97, 602, 609
  shear stress and, 581–87
  similarity solutions using, 602–10
  summary of, 609–10
Net positive suction head (NPSH), 677
  available (NPSHA), 661–67
  definition of, 661
  required (NPSHR), 662–66
Nets, flow, 183–86
Networks, pipe. *See* Pipe networks
Newton, Isaac, 20, 158, 496
Newtonian fluid, 20–22, 82, 609
Newton-Raphson method, 718–26
Newton's laws. *See* Motion, Newton's laws
  of *and* Viscosity, Newton's law of
Newtons (*N*) per square meter (pascals
  [Pa]), 6
Nikuradse, J., 485, 487
Noncircular conduits, 413–15
Nonrectangular channel section, 554–55
Normal component, of velocity, 104
Normal stress, 40, 581–87
No-slip condition:
  definitions of, 22, 102
  Navier-Stokes equations and, 596,
    609–10
Null instrument/method, 626, 642
Nusselt number, definition of, 461

O

"On the Conditions at the Boundary of a
  Fluid in Turbulent Motion" (Stanton,
  Marshall, and Bryant), 470–71
Open channel flow:
  critical conditions of, 550–52
  effect of channel cross section on
    constant flow, 552–54
  free-surface profiles, 567–69, 571–74
  Froude number and, 546–48, 552–53,
    555, 567, 568, 571
  introduction to, 545–50
  momentum function and, 555–57, 562, 574
  nonrectangular channel selection, 554–55
  resistance in, 562–67
  Reynolds number and, 565
  shear stress and, 556–57, 559, 564,
    568–69, 574
  summary of, 574–75
  upstream and downstream effects,
    557–61
Open system, 118
Orifices, pipe flow through, 418–20
Outflow, 107

P

Parshall flume, 640
Pascals (Pa) (newtons [N] per square
  meter), 6
Pathlines, 108–14
Pelton turbines, 282
Peristaltic pumps, 676
Photomultiplier, 633
Pi parameters:
  calculation of, 367–69, 371–74, 376,
    378–79, 381
  definition of, 359
Pipe flow:
  adiabatic, 533–38
  Bernoulli equation and, 410, 411, 416
  dimensionless parameters and, 360–61,
    383
  introduction to, 2–3, 404-5
  measurement of, 23–25
  minor losses in, 408–12
  Moody diagram and, 405–8, 411, 414,
    415
  Newton's law of viscosity and, 299, 302,
    307, 311
  in noncircular conduits, 413–15
  through orifices, 418–20
  Reynolds number and, 409–11, 413, 420
  separation in passages, 415–18
  in series and parallel, 405–8
  shear stress and, 306–7, 310–13, 405,
    412, 415–16
  theoretical effects of viscosity on, 306–15
  transition to fully developed, 412–13
  turbulent flow in pipes, 470–74, 476–77,
    479, 481–85
  viscosity and, 299, 302, 307, 311, 415–18
Pipe friction, 405
Pipe networks:
  construction of system curves for,
    706–11
  definition of, 706
  Newton-Raphson method of solution for,
    718–26
  numerical solution for, 711–17
Pi theorem:
  dimensionless parameters and, 357–60
  statement of, 360
Pitot-static tube, 168, 474, 627, 629, 634
Pitot tube (impact tube), 168, 627–29
Planar surfaces:
  hydrostatic forces and moments on,
    53–63
  introduction to forces on, 51–53
Plane, definition of, 44

Planes, theoretical effects of viscosity on
flow between:
    almost parallel, 340–47
    parallel, 300–306
Plane shock wave, 525–27, 538, 539
Plates, flow over flat. *See* Flat-plate flow
Pohlhausen, K., 337, 339
Point:
    definition of, 12
    material, definition of, 12
    pressure at a, definition of, 39
Point gauge, 640
Point (three-dimensional) source, 180, 393
Poiseuille, J., 24, 158, 309, 464
    Hagen-Poiseuille flow, 309, 464
Positive-displacement pumps and motors,
    674–77
Positive (adverse) pressure gradients, 418
Potential flow. *See* Irrotational flow
Potentiometer, 642
Pounds per square inch (psi), 6
Power, definitions of, 209, 210
Power laws, Prandtl's, 479
Prandtl, Ludwig, 159, 323, 347, 418, 429,
    484, 490, 565, 615
    boundary layer theory of, 426, 429,
        431–32, 594, 598, 609
    Lanchester-Prandtl theory, 439
    mixing-length theory of, 472–74
    power laws of, 479
Pressure:
    atmospheric, 6
    dynamic, 310
    gauge, 7
    measurement of, 6–8, 625–27
    at a point, definition of, 39
    stagnation, 31
    vacuum, 7
Pressure coefficient, 182
Pressure distribution, on a cylinder,
    361–62
Pressure distributions, in fluids without
    relative motions:
    acceleration and pressure in rotating
        fluids, 77–82
    forces and moments due to pressure
        distributions on surfaces, in general,
        63–74
    forces on planar and curved surfaces,
        introduction to, 51–53
    hydrostatic forces and moments on
        planar surfaces, 53–63
    hydrostatic pressure variation, 39–47
    introduction to, 38–39
    linear, rigid-body acceleration, 74-77
    manometers, 47–50, 625–29
    summary of, 82–83

Pressure drag, definition of, 428
Pressure drop, 360
Pressure gradients, positive (adverse), 418
Pressure head, 404
Pressurre-relief valve, 675
Pressure transducers, 625–27
Preswirl, 646
Prime, loss of, 661, 671
Production (buffer, generation) zone, 481
Properties of fluids. *See* Fluid properties
    and flow phenomena
Pump, definition of, 677
Pump affinity (similarity) laws, 646–50,
    652
Pump characteristic curve. *See*
    Characteristic curve of pump
Pump demand curve, 677
Pump head, 647
    Euler, 647
Pump-storage plants, 217

Q

Quantity per unit time, 100

R

Radar (radio detection and ranging), 633
Radial-flow machines, 262–63, 287–88, 658
Rankine-Hugoniot equations, 526
Rankine temperature, 6
Rapid (supercritical, shooting) flow, 547,
    549–51, 557–58, 573
Rayleigh, J. W. S., 485
Rayleigh number, 461
Reaction, definition of, 282
Relief port, 675
Required net positive suction head
    (NPSHR), 662–66
Resistance, flow, 562–67
Resultant force, 63
Reynolds, Osborne, 2, 3, 24, 314, 390,
    462, 465, 467–68, 474, 495, 662
Reynolds number, 28, 32, 318, 326, 342,
    367, 369, 648
    compressible flow and, 537
    definitions of, 24, 310
    derivation of, 314, 465, 467–68
    dimensionless parameters and, 354, 374,
        381, 383, 385, 387–89
    effects of increases and decreases in,
        311–14, 323, 327–29, 331, 347, 348,
        383, 420, 426, 428, 474, 476, 485,
        489, 596–97, 602, 609, 636

external flow and, 426, 428, 429, 435
hydraulic machinery and, 653, 660
laminar flow and, 313, 464, 465, 467,
    496
Navier-Stokes equations and, 595–97,
    602, 609
open channel flow and, 565
pipe flow and, 409–11, 413, 420
turbulence, 483
turbulent flow and, 465, 467–68, 474,
    476–77, 483, 485, 489, 496
velocity and, 476
Reynolds stress, 470, 472, 476, 481, 483,
    496
Reynolds transport theorem:
    energy conservation and, 208, 230
    mass conservation and, 127–28, 153, 155
    momentum conservation and, 242, 287
Rheogoniometer, 644
Rigid-body acceleration, linear, 74–77
Rigid-body rotation, 77
Root-mean-square (rms), 476
Rotating fluids, acceleration and pressure
    in, 77–82
Rotation:
    deformation and, 140–44
    introduction to, 137–38
Rotometer, 637
Roughness paradox, 3–5
Runge-Kutta method, 603
    Falkner-Skan solution by, 739–40
Runner, 269

S

Saybolt viscometer, 643–44
Schlichting, Herman, 338–39, 490–91, 496
    Tollmien-Schlichting waves, 328, 454,
        491
Schlieren system, 622
Schubauer, G., 490–91
Scirocco, 441
Scroll case, 670
Secondary flow:
    basic ideas of, 451–55, 457–59, 461–62
    definition of, 264
    introduction to, 451
    summary of, 462
Self-similar fluid motions, 392–95
Separation:
    boundary-layer control and, 441–44
    inception of, 336
    in passages, 415–18
Separation point/line/position, definitions
    of, 31, 329, 365

Shadowgraph, 621–22
Shape factor, 336
Shearing interferometer/interferometry,
    623
Shear layer, 494
Shear stress, 20–21, 256–59, 277–78, 347,
    348, 580, 593–94
    almost parallel planes and, 342, 345
    bottom, 556–57, 559, 564, 568–69
    compressible flow and, 533
    dimensionless parameters and, 354, 356,
        362
    external flow and, 426–27
    laminar boundary layers and, 323–24,
        326–28, 336
    Navier-Stokes equations and, 581–87
    negligible, 276
    open channel flow and, 556–57, 559,
        564, 568–69, 574
    parallel planes and, 300–306
    pipe flow and, 306–7, 310–13, 405, 412,
        415–16
    swirling flow and, 315, 316, 318
    turbulent flow and, 470, 473, 474, 481
    velocity and, 360, 362, 471–72, 474
    See also Drag and Friction factor
Shock:
    definition of, 264
    inlet, 657
Shockless flow, 266
Shock wave(s):
    compression, 520, 525
    definitions of, 27, 509
    expansion, 509, 520
    in nozzles, 522–31, 533, 538, 539
    plane, 525–27, 538, 539
    two-dimensional, 525, 539
Shooting, numerical technique of, 602–3
Shooting (supercritical, rapid) flow, 547,
    549–51, 557–58, 573
Similarity (affinity) laws, for
    turbomachines, 646–50, 652
Similarity solution(s), 392, 395
    Navier-Stokes equations and, 602–10
Single-tube manometer, 50
Sink, two-dimensional (line), 179
Sinuous flow, definition of, 24
Skan, Sylvia, 605, 610
    Falkner-Skan solutions, 605–7, 610,
        739–42
Skin friction coefficient, 258, 326–27,
    329–30
Skramstad, H., 490–91
Slip:
    definition of, 675
    no-slip condition, 22, 102, 596, 609–10
Sluice gate, 149, 557–61, 572–74

Slurry pumps, 673
Sommerfeld, Arnold, 467
Sound waves:
  introduction to, 500–502
  speed of, 502–4
Source:
  line, 179
  strength of the, 179, 180
  three-dimensional (point), 180, 393
  two-dimensional, 179
Spatial correlation, 494
Spatial (Eulerian) flow, 99–100, 144–46,
  155
Specific diameter, 652
Specific gravity (SG), 8
Specific heats, enthalpy, entropy and,
  727–34
Specific speed, 652
Specific volume, 10
Speed, specific, 652
Spoiler, 441
Squares, curvilinear, 183
Stagnation points, definition of, 31
Stagnation pressure, definition of, 31
Stalling, 430, 439, 441, 444
Stanton, T. E., 470–73, 483, 485, 565
Starting vortex, 439
State, equation of. *See* Equation of state
Statically equivalent loading, 63
Steady flow:
  Bernoulli equation and, 159, 161–71,
    203
  continuity equation for, 117–24, 140,
    154, 500
  definition of, 100
  Euler equations and, 188
  momentum equation applications,
    244–56, 500
Stokes, George G., 24, 158, 309, 347, 427,
  464–65, 580
  Stokes flow, 347, 348, 435
  See also Navier-Stokes equations
Stopping vortex, 439
Strain, volume, 142
Strain-thickening fluid, 20
Strain-thinning fluid, 20
Streaklines, 108–14
Stream functions, 149–52
Streamlines, 108–14, 149, 151
Streamlining, 441, 444
Strength, of the source, 179, 180
Stress:
  constant, 481, 496
  normal, 40, 581–87
  Reynolds, 470, 472, 476, 481, 483, 496
Stress, shear. *See* Shear stress

Subcritical (tranquil) flow, 547, 549–52,
  557–58
Submerged jets, 418
Subsonic flow, 27
Substantial derivative, acceleration and,
  146–49
Supercritical (rapid, shooting) flow, 547,
  549–51, 557–58, 573
Supersonic flow, 27, 513–22, 538
Surface elevation, flow measurement and,
  640–41
Surface tension:
  definition of, 13
  measurement of, 13–16
Surge waves, 274–75
Swirling flow, theoretical effects of
  viscosity on, 315–20
System:
  definition of, 98
  open, 118
System curve(s):
  construction of, 706–11
  definition of, 668

T

Tailrace, 218, 270
Tailwater, 218
Tangential component, of velocity, 104
Taylor, Geoffrey I., 318, 458, 472–74, 484
Taylor number, 458
  definition of, 318
  derivation of, 318
Taylor series expansion, 139, 186, 603
  Falkner-Skan, solution by, 740–42
Temperature:
  absolute, 6
  critical, 10
  measurement of, 6, 641–42
  Rankine, 6
Temperature field, 100
Temporal mean component, of velocity,
  467
Thermal diffusion, 458–59, 461
Thermistor, 642
Thermocouple, 632, 642
Thermodynamics:
  first law of, 204, 208, 230
  second law of, 504, 538
Thermometer, 641–42
Three-dimensional (point) source, 180, 393
Tidal bores, 275
Tiplets, 444
Tollmien, W., 496
Tollmien-Schlichting waves, 328, 454, 491

Total differentials, 690–93
Total head, 220, 404
Tranquil (subcritical) flow, 547, 549–50, 557–58
Transducer(s):
  definition of, 625
  flow-rate, 637, 640
  pressure, 625–27
Transition:
  to fully developed pipe flow, 412–13
  in laminar flow, 413, 461, 465, 467, 487, 489–91
  in turbulent flow, 413, 461–62, 487, 489–91
Transition length, boundary layer, 486–87
Transport theorem, Reynolds. *See* Reynolds transport theorem
Turbine, definition of, 677
Turbomachines. *See* Hydraulic machinery
Turbulence Reynolds number, 483
Turbulent boundary layer, 487, 489–91, 494–95, 496
Turbulent (fluctuating) component, of velocity, 467
Turbulent exchange coefficient, 472
Turbulent flow:
  definitions of, 3, 24
  dimensionless parameters and, 360–61
  over flat plates, 486–87, 489–91, 494–95
  homogeneous, 473
  introduction to, 464–65, 467–70
  Newton's law of viscosity and, 470, 473
  in pipes, 470–74, 476–77, 479, 481–85
  Reynolds number and, 465, 467–68, 474, 476–77, 483, 485, 489, 496
  shear stress and, 470, 473, 474, 481
  summary of, 495–96
  transition in, 413, 461–62, 487, 489–91
  velocity fluctuations and, 494
Turbulent spots, 491
Two-dimensional shock wave, 525, 539
Two-dimensional (line) sink, 179
Two-dimensional source, 179
Two-dimensional velocity field, 100
Two-dimensional vortex, 180

U

Uniform depth, 574
Uniform flow:
  definitions of, 100, 563–64
  of open channel flow, 566–68, 571, 574
Unsteady flow:
  energy equation applications, 225–30

general equations for, 124–27
momentum equation applications, 283–87
theoretical effects of viscosity on, 320–23
Unsteady velocity, definition of, 100
Upstream and downstream effects, in open channel flow, 557–61

V

Vacuum pressure, 7
Van der Waals forces, 10
Vane anemometer, 633
Velocity:
  average value of, 101, 467
  derivatives of, 473
  fluctuating (turbulent) component of, 467
  friction, 474
  introduction to, 98–100
  measurement of, 627–34
  normal component of, 104
  Reynolds number and, 476
  sheer forces and, 360, 362, 471–72, 474
  tangential component of, 104
  temporal mean component of, 467
  turbulence and fluctuations in, 494
  two-dimensional, 100
  unsteady, 100
  See also Flow
Velocity coefficient, 636
Velocity defect law, 483–85
Velocity field, 100
Velocity signals, 491
Venturi tube, 165, 636–37
Vernier scale, 640
Viscometer:
  Cannon-Fenske, 643
  falling-ball, 644
  Saybolt, 643–44
Viscosity:
  absolute (coefficient of), 23, 139
  compressible flow and, 533–39
  eddy, 472
  kinematic, 23
  measurement of, 18–23, 643–44
  pipe flow and, 299, 302, 307, 311, 415–18
Viscosity, Newton's law of, 38, 139, 158, 347, 580
  pipe flow and, 299, 302, 307, 311
  statement of, 20–21
  swirling flow and, 316, 317, 320
  turbulent flow and, 470, 473

Viscosity, theoretical effects of, on fluid
    flow
  on flow between almost parallel planes,
    340–47
  on flow between parallel planes, 300–306
  introduction to, 298–300
  laminar boundary layers, detailed
    explanation of, 323–31
  laminar boundary layers, over two-
    dimensional bodies, 331–39, 394
  on pipe flow, 306–15
  summary of, 347–48
  on unsteady flow, 320–23
Viscous sublayer, 479
Visualization of flow, 614–23
Voltmeter, 642
Volume(s):
  control, 116, 118, 694–701, 703, 705
  material, 114
  rate of volume expansion, 139
  specific, 10
Volume deformation, 139–40
Volume flow rate, measurement of,
    100–105, 634–37, 640
Volume strain, 142
Von Mises, Richard, 413–14
Vortex:
  hairpin, 491
  horseshoe, 418, 451
  starting, 439
  stopping, 439

  two-dimensional, 180
  wing-tip, 430, 444

W

Wake:
  definitions of, 282, 365
  similarity solution for, 608–9
Wake shedding, 365–67
Wall, law of the, 484, 485, 495
Wall drag coefficient, 533
Wall layer:
  buffer (generation, production) zone of,
    481
  constant-stress region of, 481, 496
  definition of, 477
  viscous sublayer of, 479
Water horsepower, 647
Wave drag, 28–29, 330
Wave drag coefficient, 388–89
Weber number, definition of, 385
Weir, broad-crested, 553, 640
Wheatstone bridge, 627, 629–31, 642
Wicket gate, 269
Wilkes, James O., 338, 603
Wing-tip vortex, 430, 444
Work done on the fluid, definition of,
    208–9
Work-energy equation, 163
Wortmann, F. X., 615–16

CONVERSION TABLES

To convert a dimension in inches to feet, for example, one divides the number of inches by the factor 12, an operation that is suggested by ÷ ÷ ÷ . Similarly, to convert a dimension in feet to inches, one multiplies the number of feet by the factor 12, in the manner indicated by × × × . The same pattern should be used to convert the other entries in the tables below.

ENGLISH UNITS:

	÷ ÷ ÷ Divide by ÷ ÷ ÷	
inches (in)	**12**	feet (ft)
	× × × Multiply by × × ×	

	÷ ÷ ÷ Divide by ÷ ÷ ÷	
feet	**5280**	miles (statute)
	× × × Multiply by × × ×	

	÷ ÷ ÷ Divide by ÷ ÷ ÷	
feet	**6076**	miles (nautical, Int.)*
	× × × Multiply by × × ×	

	÷ ÷ ÷ Divide by ÷ ÷ ÷	
gallons (gal)	**7.48**	cubic feet
	× × × Multiply by × × ×	

	÷ ÷ ÷ Divide by ÷ ÷ ÷	
pound** (lb)	**1**	slug-foot/s^2
	× × × Multiply by × × ×	

	÷ ÷ ÷ Divide by ÷ ÷ ÷	
pounds	**2000**...... **2240**......	tons (short) tons (long)***
	× × × Multiply by × × ×	

	÷ ÷ ÷ Divide by ÷ ÷ ÷	
gallons per minute (g.p.m. or gpm)	**448.8**	cubic feet/s (c.f.s or cfs)
	× × × Multiply by × × ×	

	÷ ÷ ÷ Divide by ÷ ÷ ÷	
ft-lb	**778.2**	British thermal unit (btu)
	× × × Multiply by × × ×	

÷ ÷ ÷ Divide by ÷ ÷ ÷

	÷ ÷ ÷ Divide by ÷ ÷ ÷	
ft-lb/s	**550**	horsepower (U.S.A)
	× × × Multiply by × × ×	

	÷ ÷ ÷ Divide by ÷ ÷ ÷	
miles per hour (m.p.h. or mph)	**0.682**	ft/s
	× × × Multiply by × × ×	

	÷ ÷ ÷ Divide by ÷ ÷ ÷	
knots (Int.)	**0.5925**	ft/s
	× × × Multiply by × × ×	

ENGLISH UNITS   **S.I. UNITS**

	÷ ÷ ÷ Divide by ÷ ÷ ÷	
feet	**3.28**	meters
	× × × Multiply by × × ×	

	÷ ÷ ÷ Divide by ÷ ÷ ÷	
slugs	**0.0685**	kilograms
	× × × Multiply by × × ×	

	÷ ÷ ÷ Divide by ÷ ÷ ÷	
pounds	**0.225**	newtons
	× × × Multiply by × × ×	

	÷ ÷ ÷ Divide by ÷ ÷ ÷	
pounds/ft^2	**0.0209**	pascals (Pa)
	× × × Multiply by × × ×	

	÷ ÷ ÷ Divide by ÷ ÷ ÷	
slugs/ft s	**0.00209**	poise
	× × × Multiply by × × ×	

	÷ ÷ ÷ Divide by ÷ ÷ ÷	
ft^2/s	**0.00108**	stokes
	× × × Multiply by × × ×	

Fahrenheit	°F = 32 + 9/5 (°C)	Celsius
Rankine	°R = 9/5 (K)	Kelvin

* The nautical mile is one minute of arc along a great circle (meridian) of the earth.

** This is a force unit. The use of the word pound to connote mass is discouraged in this book. Accordingly, whenever the reader encounters the term "pound mass," it is implicit that a unit of mass equivalent to 1/32.2 of a slug is, in fact, being referred to.

*** Used in shipping some commodities in the United States of America, but more usually used in Great Britain where a ton is 20 hundredweight (cwt) and the cwt is 112 pounds there.